International Association for Wind Engineering

WIND ENGINEERING

RETROSPECT AND PROSPECT

VOLUME I

PUBLISHING FOR ONE WORLD

NEW AGE INTERNATIONAL (P) LIMITED, PUBLISHERS

New Delhi • Bangalore • Chennai • Cochin • Guwahati • Hyderabad
Jalandhar • Kolkata • Lucknow • Mumbai • Ranchi
Visit us at **www.newagepublishers.com**

Published by New Age International (P) Ltd., Publishers
First Edition: 1994
Reprint: 2011

Branches:

- 37/10, 8th Cross (Near Hanuman Temple), Azad Nagar, Chamrajpet, **Bangalore**-560 018. Tel.: (080) 26756823, Telefax: 26756820, E-mail: bangalore@newagepublishers.com
- 26, Damodaran Street, T. Nagar, **Chennai**-600 017. Tel.: (044) 24353401, Telefax: 24351463 E-mail: chennai@newagepublishers.com
- CC-39/1016, Carrier Station Road, Ernakulam South, **Cochin**-682 016. Tel.: (0484) 2377004, Telefax: 4051303. E-mail: cochin@newagepublishers.com
- Hemsen Complex, Mohd. Shah Road, Paltan Bazar, Near Starline Hotel, **Guwahati**-781 008. Tel.: (0361) 2513881. Telefax: 2543669, E-mail: guwahati@newagepublishers.com
- 105, 1st Floor, Madhiray Kaveri Tower, 3-2-19, Azam Jahi Road, Nimboliadda, **Hyderabad**-500 027. Tel.: (040) 24652456, Telefax: 24652457, E-mail:hyderabad@newagepublishers.com
- RDB Chambers (Formerly Lotus Cinema)106A, 1st Floor, S.N. Banerjee Road, **Kolkata**-700 014. Tel.: (033) 22273773, Telefax: 22275247, E-mail:kolkata@newagepublishers.com
- 16-A, Jopling Road, **Lucknow**-226 001. Tel.: (0522) 2209578, 4045297, Telefax: 2204098 E-mail: lucknow@newagepublishers.com
- 142C, Victor House, Ground Floor, N.M. Joshi Marg, Lower Parel, **Mumbai**-400 013. Tel.: (022) 24927869. Telefax: 24915415, E-mail: mumbai@newagepublishers.com
- 22, Golden House, Daryaganj, **New Delhi**-110 002. Tel.: (011) 23262370, 23262368, Telefax: 43551305. E-mail: sales@newagepublishers.com

ISBN: 978-81-224-0714-3

₹ 695.00

C-10-06-4758

Printed in India at Glorious Printers, Delhi.

PUBLISHING FOR ONE WORLD

NEW AGE INTERNATIONAL (P) LIMITED, PUBLISHERS
4835/24, Ansari Road, Daryaganj, New Delhi-110002
Visit us at **www.newagepublishers.com**

PREFACE

The Ninth International Conference on Wind Engineering (9ICWE) will be held at New Delhi, India from January 9 to 13, 1995 to discuss subjects related to the research and practice of Wind Engineering. This 4-yearly World Conference rotates amongst the Asia-Pacific region, Americas, and, Europe and Africa, and, is being held for the first time in South East Asia, a region frequently lashed by strong cyclones and wind storms, which has witnessed some of the worst wind disasters of the century. It is significant that the event is taking place at a crucial time when the International community is most conscious of the need for urgent efforts towards natural disaster reduction and the U.N. has declared the current decade as the International Decade for Natural Disaster Reduction. The purpose of most of the research in Wind Engineering is indeed to improve the quality of engineering effort and thus reduce failure; the need is to direct this effort towards application in disaster reduction. The conference is being organised by the Indian Society for Wind Engineering under the auspices of the International Association for Wind Engineering. The event is sponsored by the University of Roorkee, Roorkee and the Department of Science and Technology, New Delhi, and co-sponsored by, the IDNDR programme of the United Nations, Geneva, the Third World Academy of Sciences, Trieste, The Association of Geoscientists for International Development, Sao Paulo, and, a number of teaching, research and industrial organisations in India.

Summary papers of about 1000 words or equivalent, running into two A3 size sheets were invited for the conference. The intention was to get a reasonably good idea of the contents expected to be contained in the full papers. As a result of this invitation, over 260 papers were received from about 450 authors representing 25 countries. These covered a wide spectrum of themes related to wind engineering, which include Wind Characteristics, Extreme Winds, Bluff Body Aerodynamics, Wind Tunnel Techniques & Studies, Field/Proto-type Measurements, Computational Fluid Dynamics & Fluid-Structure Interaction, Bridges and Cable Structures, Roofs and Low Rise Buildings, Tall Buildings, Towers, Chimneys and Transmission Lines, Vibrations and Dampers, Atmospheric Dispersion and Particle Drifting, Wind Breaks, Wind Power and Waves, Wind Codes and Design Problems, and, Disaster Assessment and Reduction.

The papers received were reviewed by a panel of experts worldwide. Those summary papers which offered the promise of valuable technical content in the full papers, were accepted and printed in a separate volume. The authors of accepted Summary Papers were then requested to prepare full length papers. Those full length papers which have been accepted (till the time of going to press), after further review, for presentation at the conference, are complied in these volumes.

As a post-conference exercise, out of the papers presented at the conference, some selected ones will be published by the Elsevier Science Publishers, Netherlands, in a special volume of their Journal of Wind Engineering and Industrial Aerodynamics, and Wiley Eastern Ltd., New Delhi plan to bring out thematic volumes.

PREM KRISHNA
Chairman, 9ICWE
and
Chairman, International Association of Wind Engineering

NATIONAL ADVISORY COUNCIL

A. Trehan, *Director (Technical), National Thermal Power Corporation, New Delhi.*
H.C. Visvesvaraya, *Vice Chancellor, University of Roorkee, Roorkee*
K.K. Madan, *Director General, Central Public Works Department, New Delhi*
N.C. Nigam, *Professor & Former Director, Indian Institute of Technology, New Delhi*
P. Rama Rao, *Secretary, Department of Science and Technology, New Delhi.*
R. Narasimha, *Hony. Professor, Indian Institute of Science, Bangalore.*
Raj Kumar, *Member (Engg.), Railway Board, New Delhi.*
S.K. Joshi, *Director General, Council of Scientific and Industrial Research, New Delhi.*
S.K. Khanna, *Chairman, All India Council of Technical Education, New Delhi.*
S. Varadarajan, *President, National Academy of Engineering, New Delhi.*
T.N. Gupta, *Executive Director, Building Materials Technology Promotion Council, New Delhi.*
U.R. Rao, *Former Chairman, Space Commission, Bangalore.*

SCIENTIFIC COMMITTEE

A.G. Davenport, *Canada*
A.N.L. Chiu, *USA*
Carl Kramer, *Germany*
G.N.V. Rao, *India*
H.von Koten, *Netherlands*
J.D. Holmes, *Australia*
K.C. Mehta, *USA*
N. Shiraishi, *Japan*
Prem Krishna, *India*
S. Murakami, *Japan*
T.F. Sun, *China*
Y.K. Cheung, *Hong Kong*
A.J. Bowen, *New Zealand*
B.J. Vickery, *Canada*
G. Augusti, *Italy*
H.J. Niemann, *Germany*
H. Ruscheweyh, *Germany*
J.E. Cermak, *USA*
Manabu Ito, *Japan*
N.J. Cook, *UK*
R.H. Scanlan, *USA*
T. Miyata, *Japan*
T. Stathopoulos, *Canada*
W.H. Melbourne, *Australia*

REVIEWERS

A.R. Santhakumar, *Anna Univ, Madras*
G.N.V. Rao, *IISc, Bangalore*
Krishen Kumar, *Univ., Roorkee*
N.L. Arora, *IIT, Kanpur*
P.K. Pande, *Univ., Roorkee*
P.S. Rao, *IIT, Madras*
B. Venkateswarlu, *SERC, Madras*
K.K. Chaudhry, *IIT, New Delhi*
N.C. Nigam, *IIT, New Delhi*
P.N. Godbole, *Univ., Roorkee*
Prem Krishna, *Univ., Roorkee*
S. Durvasula, *IISc, Bangalore*
T.V.S.R. Appa Rao, *SERC, Madras*

ORGANISING COMMITTEE

Chairman
Prem Krishna, *University, Roorkee*

Vice-Chairman
Gopal Ranjan, *University, Roorkee*

Organising Secretary
P.N. Godbole, *University, Roorkee*

Jt. Organising Secretaries
A.K. Ahuja, *Univ., Roorkee* **M.N. Viladkar**, *Univ., Roorkee*

Members

A.K. Gupta, *Univ., Roorkee*
H. Sinvhal, *Univ., Roorkee*
K.K. Chaudhry, *IIT, New Delhi*
P.K. Pande, *Univ., Roorkee*
R.K. Midha, *DST, New Delhi*
Babu Jacob, *Min. of Agriculture, New Delhi*
Krishen Kumar, *Univ., Roorkee*
M.P. Jain, *Univ., Roorkee*
R.P. Mathur, *Univ., Roorkee*
R.S. Bajaj, *NTPC, New Delhi*
S.K. Agarwal, *SERC, Ghaziabad*

COSPONSORS

- Association of Geoscientists for International Development, (AGID), Sao Paulo, Brazil.
- United Nations' IDNDR Programme, Geneva.
- Third World Academy of Sciences (TWAS), Trieste, Italy.
- Building Materials Technology Promotion Council (BMTPC), New Delhi.
- Central Building Research Institute (CBRI), Roorkee.
- Central Public Works Department (CPWD), New Delhi.
- Housing & Urban Development Corporation (HUDCO), New Delhi.
- Indian Institute of Technology (IIT), New Delhi.
- Oil and Natural Gas Corporation (ONGC), Dehradun.
- Railway Board, Ministry of Railways, New Delhi.
- Structural Engineering Research Centre (SERC), Ghaziabad.
- Structural Engineering Research Centre (SERC), Madras.
- National Thermal Power Corporation (NTPC), New Delhi.
- Nuclear Power Corporation (NPC), Bombay.
- Water and Power Consultancy Services (WAPCOS), New Delhi.

CONTENTS

Volume — I

Theme I — Wind Characteristics

Theme II — Extreme Winds

Theme III — Aerodynamics of Bluff Bodies and Cylinders

Theme IV — Wind Tunnel Techniques, Studies

Theme V — Field/Prototype Measurements

Volume — II

Theme VI — Computational Fluid Dynamics & Fluid Structure Interaction

Volume — III

Theme VIII — Roofs and Low Rise Buildings

Theme IX — Tall Buildings

Theme X — Towers, Chimneys and Transmission Lines

Volume — IV

Theme XI — Vibrations and Dampers

Theme XII — Atmospheric Dispersion and Particle Drifting

Theme XIII — Wind Breaks, Wind Power and Waves

Theme XIV — Wind Codes and Design Problems

Theme XV — Disaster Assesment and Reduction

Theme XVI — Miscellaneous

Papers Received Late

THEME I
Wind Characteristics

9ICWE, 1995, New Delhi,

VERTICAL DISTRIBUTION OF GUST IN THE LOWER ATMOSPHERE OVER A COASTAL REGION

NAITO, G. and KOBAYASHI, F.

Department of Geoscience, National Defense Academy, Hashirimizu, Yokosuka 239, Japan.

Abstract

The spatial distribution and behavior of strong gusts were experimentally evaluated over a coastal region during the passage of a developed extratropical cyclone. The wind field in the lower atmosphere was measured using a Doppler sodar. Strong gusts were generated by severe meso- and micro-scale wind disturbances located far from the cyclone center. The vertical profiles of the mean wind direction and speed do not exhibit constant dependence on height in a severe storm.

1. INTRODUCTION

Very high winds are almost always caused by severe storms. These are strong meso-scale or micro-scale atmospheric disturbances which are neither homogeneous nor stationary, and which affect the structure of the boundary layer near the ground. Therefore, the wind profile in the lower atmosphere (up to several hundred meters) does not depend on height alone. Naito [1] has reported various characteristics of wind velocity in the sea when a strong disturbance lasting many hours passes over. For example, a severe gust storm exhibits much different turbulence features in high winds in the boundary layer, and the fluctuating intensity is very large.

The vertical profile of mean wind can be given by the logarithmic law for height of 30 m or less above a grass field, or by the power law for heights of 200 m or less, when wind blows for a long time over a homogeneous plain. The aerodynamic roughness or the index of the power predominantes in controlling the wind field. Wieringa [2] reviewed the surface roughness for various kinds of terrain. However, the surface roughness probably plays a less important role at heights up to 10 m during very high winds induced by a severe storm.

Generally, air flow is affected by topography, and local wind exhibits characteristic features. Harada [3] discovered a nocturnal cyclonic vortex in a plain, and Kimura and Arakawa [4] conducted a numerical simulation on a low level jet to investigate the topographical effect.

However, there have been few long-term observations of the vertical structure of a severe storm in the layer below 1,000 m. In the present study, the spatial structure of gusts induced by disturbances is experimentally defined during the passage of a strong extratropical cyclone.

2. EXPERIMENTAL OBSERVATION

2.1 Topography of a coastal region

Gust distribution in the lower atmosphere was estimated over a coastal region in the southern Kanto Plain, Japan. Figure 1 shows the topography of the southern area of the

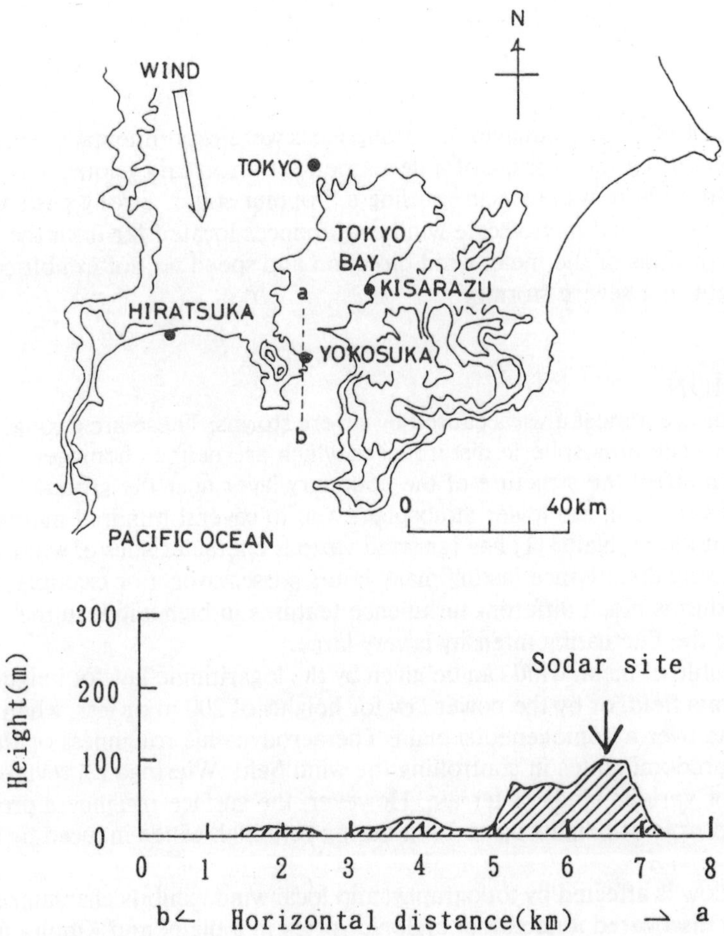

Figure 1. Topography of the coastal region near Doppler sodar site and surface observation stations. Contors represent 100 m and 200 m above mean sea level. Lower portion is vertical cross-section from north to south around the sodar site.

Kanto Plain. Mountains are located on the west side of the plain, and a peninsula in the south-east has some hills about 200 m high. Another peninsula in the center has a few hills about 100 m high. The present Doppler sodar site is located at Yokosuka on this center peninsula.

The lower portion of Figure 1 shows a detailed vertical cross-section from north to south around the site. The sodar sensor was installed on the top of the laboratory building on an 80-m hill. The sodar was 100 m above the mean sea surface. Observation stations of Hiratsuka, Yokosuka and Kisarazu are approximately lined up from west to east. The Hiratsuka observation station is a marine tower located about 1,000 m off the shoreline [1], and the wind sensor is set on the top of the tower 23 m above the sea surface. The Kisarazu station is in the airport facing Tokyo Bay, and the wind sensor is installed on the top of the control tower. There are no small scale topographic changes which influence the surface wind around either of the observation stations.

2.2 Measurement using a Doppler sodar

Vertical profiles of the wind vector were measured using a Doppler sodar (Kaijo AR-410). This sodar is a mono-static acoustic radar with three beams. It has a transmitting frequency of 2,400 Hz, and a maximum transmitting power of 900 W (300 W during this observation). Therefore, acoustic echoes can be received from a maximum height of about 1,000 m, if the air flow and the atmospheric stratification are optimum for Doppler sodar sensing. However, it is difficult to measure strong wind beyond 15 m/s in layers above 500 m even in clear sky conditions. The 10-min mean wind profile was obtained every 10 min at 25 m intervals.

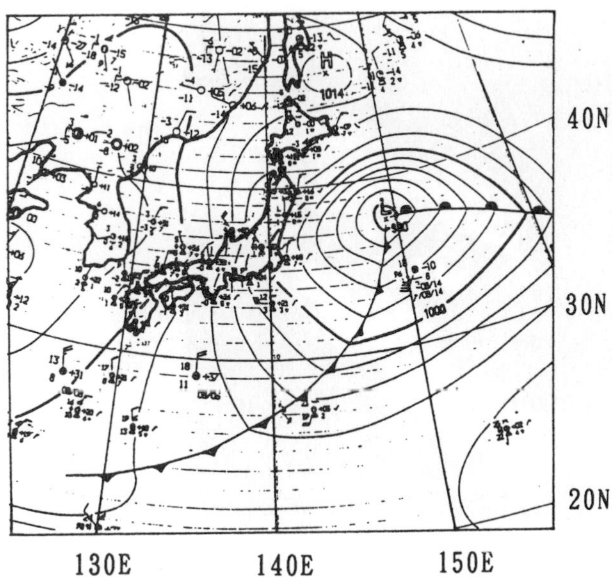

Figure 2. Surface weather map at 0900 JST, March 29, 1993.

3. RESULTS AND DISCUSSIONS

3.1 Synoptic feature during a severe storm

Strong gusts occurred in the coastal region around Tokyo on March 29, 1993. The 10-min mean wind speed remained above 10 m/s for over 10 hours at Hiratsuka and Kisarazu, about 30 km from the sodar site in Yokosuka. Two extratropical cyclones passed the north and south sides of the Japanese Islands on the previous day. These cyclones merged and became a single large cyclone which developed violently in the east sea off the Kanto Plain and left from the coast region on that day.

Figure 2 shows the surface weather map at 0900 JST (00Z), March 29. The center of the cyclone (pressure 980 hPa) was located about 800 km east of Yokosuka. The Kanto Plain was covered by the back side, that is, the cold region of the cyclone, for a long time. This extratropical cyclone is the maximum grade of a synoptic disturbance in middle latitudes.

A satellite image of the area near the Japanese Islands is shown in Figure 3. This visible photograph was taken at 1200 JST on the 29th. The corresponding weather map is approximately the same as shown in Figure 3, even though the satellite image was produced 3 hours before the weather map.

We can find a large cloud area which corresponds to the developed cyclone in the east sea, and many line clouds from northwest to southeast which correspond to the severe northern monsoon cold outbreak from the continent. Additionally, we can see that

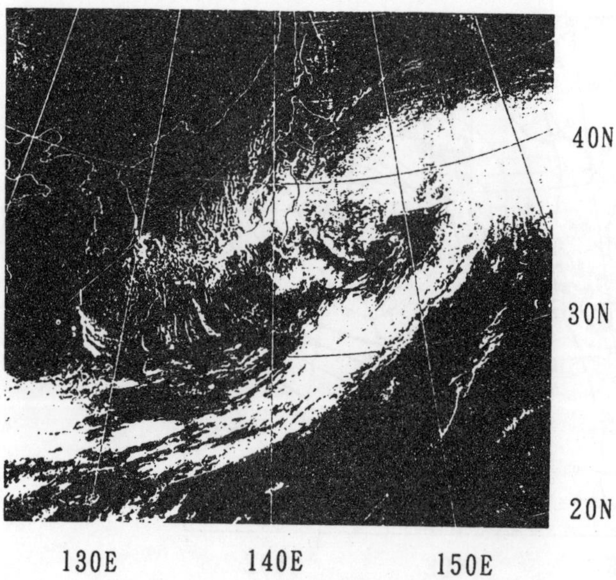

Figure 3. Satellite image around Japanese Islands at 1200 JST, March 29.

the strong northern wind field dominates over the Kanto Plain. This wind field probably consists of strong meso- and micro-scale disturbances. Thus, strong gusts attack the coastal region.

3.2 Surface wind over a coastal region

Local circulation in the Kanto Plain has been investigated by Kawamura [5]; Fujibe [6]; Yoshikado, Mizuno and Shimogoya [7] and others, in connection with the extensive air polution and the land- and sea-breeze circulation. These reports mostly show the atmospheric structure of the relatively low wind field, and do not estimate behaviors of strong gusts.

Figure 4 shows time variations of the 10-min mean surface wind during passage of the cyclone over the Hiratsuka off-shore tower and the Kisarazu airport tower on March 29. The wind direction is between north and northwest at both locations. The wind speed at Kisarazu is much higher than that at Hiratsuka throughout most of the day. This is because that the air flow is transformed for large scale topographical features (50 km wide) such as mountain ranges. In the figure, we find that several peaks appear at both locations, but those do not always appear at the same time at both locations. Therefore, we can conclude that the meso- and micro-scale disturbances behave violently in a small area. As seen in the whole distribution, the gusts peak at 0900 JST at Kisarazu then at

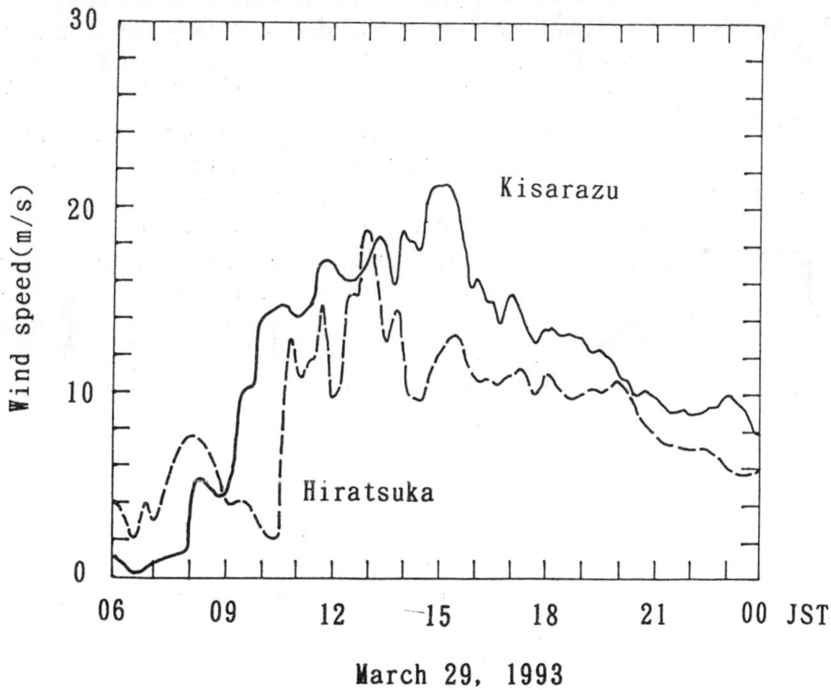

Figure 4. Surface winds at Hiratsuka off-shore tower and Kisarazu airport tower.

1030 JST at Hiratsuka. However, the highest wind occurs at 1500 JST at Kisarazu and at 1300 JST at Hiratsuka. The latter feature suggests that the most severe disturbance moved westward with the cyclone center.

The distance between Kisarazu and the cyclone center was 1,140 km at the highest wind. We notice that the gust structure of an extratropical cyclone differs from that of a tropical cyclone in which the highest winds arise several kilometers to 100 km from the center.

Figue 5 shows time variations of the geostrophic wind at Hiratsuka and Kisarazu. The geostrophic wind U_G can be considered as the wind in the upper atmosphere in this case, and is described as tollows:

$$U_G = \frac{\partial p / \partial n_H}{\rho \cdot 2\Omega \sin \phi} \qquad (1)$$

where Ω is the angular speed of rotation of the earth, ρ the air density, ϕ the geographic latitude and P the atmospheric pressure. $\partial p / \partial n_H$ represents the horizontal pressure gradient perpendicular to the isobaric line. To obtain the pressure gradient, a detailed local weather map was drawn every three hours. The geostrophic wind direction along the isobar is north-northeast, and is shifted about 30 degrees clockwise to the surface wind direction. This phase lag is reasonable for the structure of the boundary.

The maximum U_G is 30 m/s at 1500 JST at Kisarazu and 34 m/s at 1200 JST at Hiratsuka. These speeds agree well with each other, considering the accuracy of the isobar. The surface wind velocity is about 30 % less than the maximum U_G at Kisarazu. The time of the highest wind velocity corresponds well to the highest surface wind velocity at both locations. Table 1 shows statistics of the peak gusts.

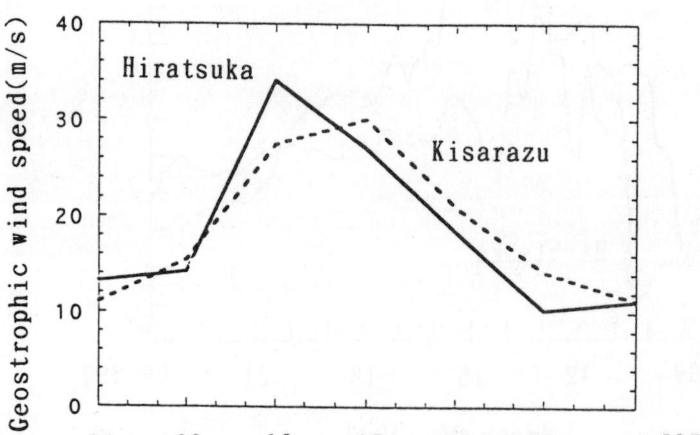

Figure 5. Geostrophic wind at Hiratsuka and Kisarazu.

Table 1. Wind speed statistics for peak gusts.

Location	Development time (JST)	Peak time (JST)	Mean wind (m/s)	Instantaneous wind (m/s)	Geostrophic wind (m/s)
Hiratsuka	10:30	13:10	18.0	25.5	34
Kisarazu	09:00	15:00	21.2	27.0	30

3.3 Wind velocity field obtained using a Doppler sodar

The spatial distribution of wind velocity determined using the Doppler sodar at Yokosuka is shown in Figures 6 and 7. The shaded part indicates values above 13 m/s. The upper figure expresses behaviors of the disturbance on a time scale of about an hour. The lower figure expresses behaviors on a scale of about 10 min to analyze the passage of the strongest micro-scale disturbance.

We find that the strong disturbance reaches a height of about 30 m above the ground regardless of the surface roughness. Strong gusts suddenly occur over the whole lower boundary after 0900 JST. In the early hours, small regions of strong wind do not necessarily appear at higher altitudes, and the vertical profile of the horizontal wind does not depend on height, as if the disturbance does not sense the surface roughness. The highest wind region passed around 1500 JST.

As seen in the pattern of the height-time cross-section, the maximum wind axis inclines slightly as the high wind appears late near the ground surface. Therefore, the surface wind is relatively strong yet, although the upper wind decreases after the passage of the severe disturbance.

The hourly change of the horizontal wind profile in Yokosuka agrees very well with that of the surface wind in Kisarazu shown in Figure 4, and the highest wind occurs simultaneously in both locations. In section 3.2, we find that the isobar runs along the line from Kisarazu to Yokosuka. As a result, we can conclude that the severe disturbance expands along the isobar, and that the strong surface wind appears on the line along the upper wind direction. We can see in the lower portion of Figure 6 that the strong meso-scale disturbance consists of some micro-scale disturbances, which approach the ground and cause the very strong instantaneous wind velocity.

Vertical wind velocity in Figure 7 is upward all day in this case. We consider that this observational result is probably due to the influence of the topography around the sodar site. The distribution pattern corresponds well to that of the horizontal wind. Accordingly, the updraft increases as the horizontal wind becomes high, and reaches about 2 m/s for a horizontal wind of 13 m/s at 100 m height on the peak of the disturbance. The detailed distribution in the lower portion also corresponds well to that of the horizontal wind, and the large updraft appears due to the strong micro-scale disturbance.

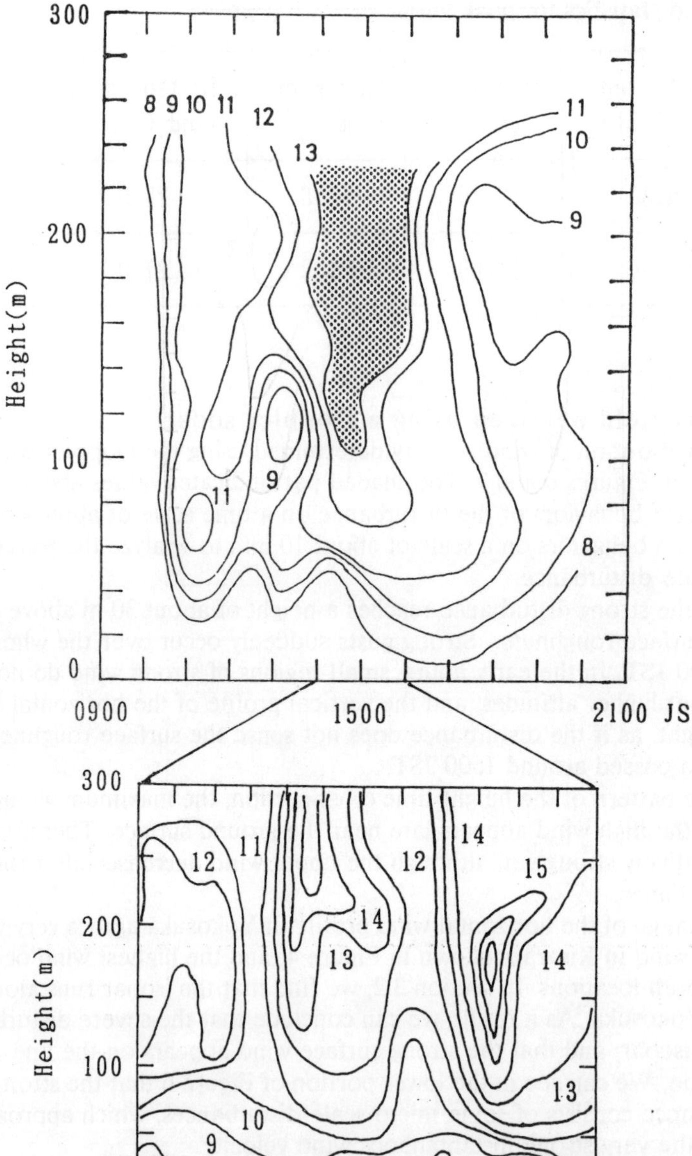

Figure 6. Height-time cross-section of horizontal wind speed (m/s) obtained using a Doppler sodar. Lower portion expresses the detailed process. Ground surface is 100 m above mean sea level.

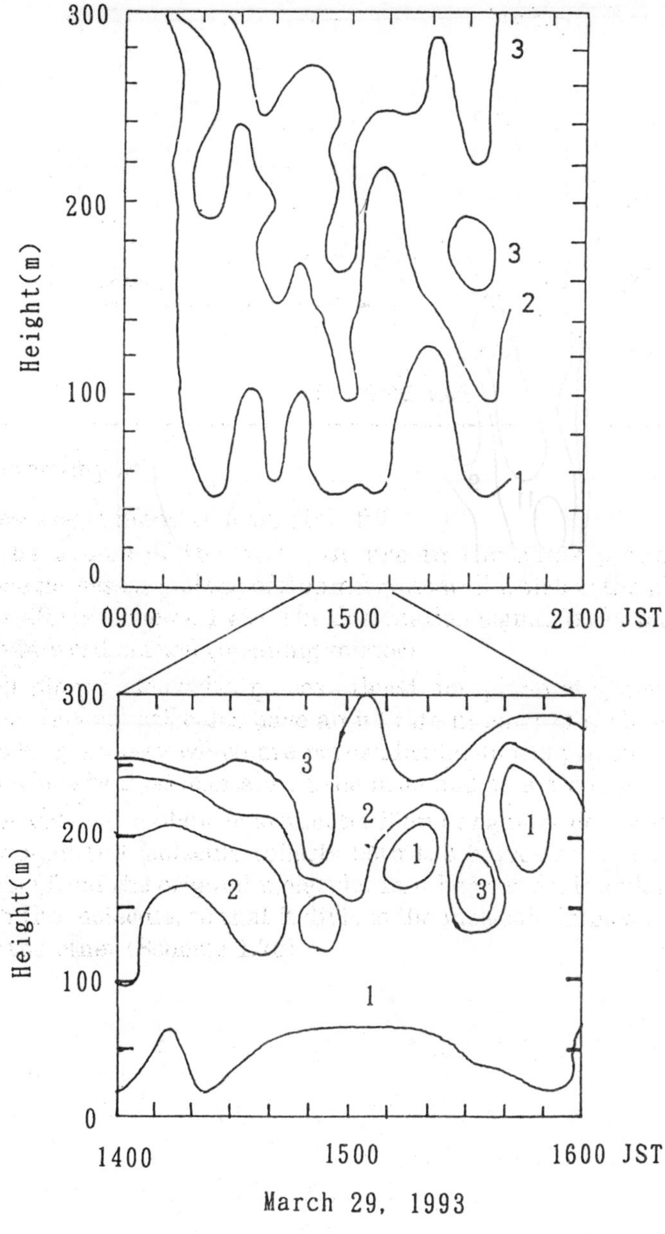

Figure 7. Same as Figure 6 except for vertical wind speed.

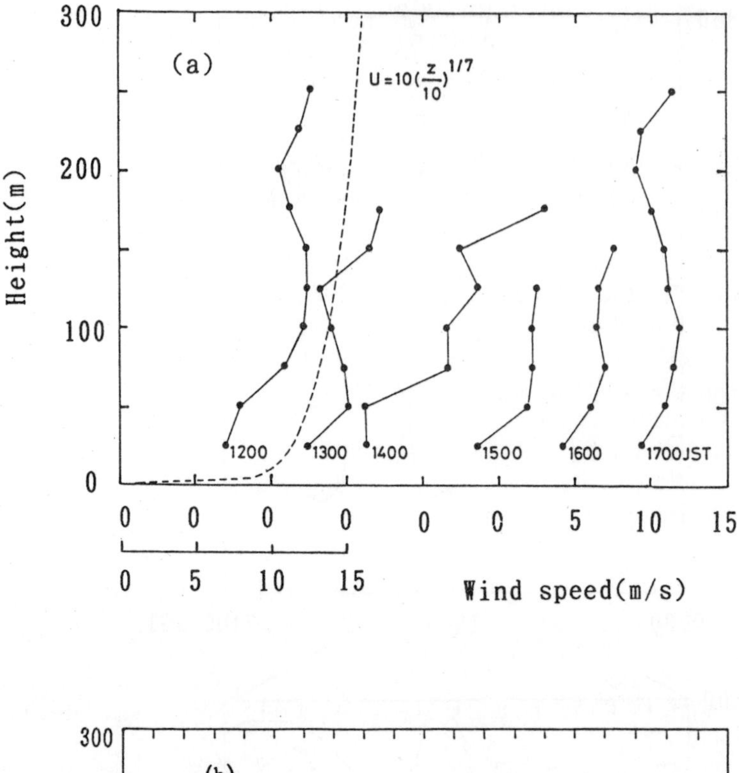

Figure 8. Hourly change of vertical profile of wind speed (a) and direction (b).

3.4 Hourly change of vertical wind velocity profile

We can describe the vertical mean wind speed profile using the following power law:

$$U = U_1(z/z_1)^{\alpha} \qquad (2)$$

where U is the mean wind speed at height z, U_1 the mean wind speed at reference height z_1, and α the power index controlled by the surface roughness. Generally, $\alpha = 1/8$ for the sea, and 1/7 for a grass field.

Figure 8 shows the hourly change of vertical wind speed and direction profiles during high winds as measured by the sodar. The reference profile in the figure is for $\alpha = 1/7$ and $U_1 = 10$ m/s at $z_1 = 10$ m. Wind speed profiles do not show the constant dependence on height as described in (2). Thus, we cannot apply the power law to the vertical profile of gusts in the lower atmosphere during severe storms, although the power law is widely used for design of large structures.

Profiles of the direction change considerably around 1500 JST, and do not depend on height. However, the directions at 1200 and 1700 JST shift slightly clockwise with hight, and approach the geostrophic wind direction.

4. CONCLUSION

The wind field in the lower atmosphere was experimentally estimated during the passage of severe storm. Remarkable conclusions are as follows:

(1)The vertical distribution of wind velocity in the lower atmosphere can be successfully evaluated using a Doppler sodar.

(2)Strong gusts occur in coastal regions during the passage of a developed extratropical cyclone due to severe meso- and micro-scale disturbances located about 1,000 km away from the cyclone center.

(3)Severe disturbances expand along the isobar of the cyclone, and strong gusts occur simultaneously along the isobar, that is, the upper wind direction.

(4)During the passage of a severe disturbance, the vertical wind direction profile changes considerably, and the wind speed profile does not exhibit constant dependence on height. Therefore, the power law cannot be applied to profiles in the lower atmosphere.

ACKNOWLEDGMENTS

The authors would like to thank Mr. N.Tsuzuki for his assistance in assembling the surface wind data and in analyzing the weather map. A part of surface wind data in this paper was provided by the Kisarazu Station, GSDF, and the Hiratsuka Labolatory, NRIES.

REFERENCES

1 G. Naio, Turbulent properties and spectral behaviours of ocean winds observed at an off-shore tower. J. Wind Eng. and Ind. Aerodynamics, 28(1988) 51-56.

2 J. Wieringa, Updating the Davenport roughness classification. J. Wind Eng. and Ind. Aerodynamics, 41-44(1992) 357-368.
3 A. Harada, An analysis of nocturnal cyclonic vortex in the Kanto Plain. J. Meteorol. Soc. Jpn, 59(1981) 602-610.
4 F. Kimura and S. Arakawa, A numerical experiment on the nocturnal low level jet over the Kanto Plain. J. Meteorol. Soc. Jpn, 61(1983) 848-861.
5 T. Kawamura, Climatology of the land and sea breezes. Report of the Atmospheric Environmental Experiment in the South Kanto area I, JMA, 241pp (in Japanese), 46-52.
6 F. Fujibe, An effect of pressure gradient on the diurnal variation of wind in the atmospheric boundary layer. J. Meteorol. Soc. Jpn, 63(1985) 52-59.
7 H. Yoshikado, T. Mizuno and S. Shimogoya, Terrain-induced air stagnation over the southern Kanto Plain. Boundary-Layer Meteorol., 68(1994) 159-172.

TERRAIN RELATED PARAMETERS FOR THE EVALUATION OF TURBULENCE AND THEIR APPLICATION TO THE RESPONSE OF TRANSMISSION STRUCTURES

VILAR, J.I. and LÓPEZ, A.

Researchers, Instituto de Investigaciones Eléctricas (IIE), Cuernavaca, México

Abstract

The dynamic response of transmission structures and cables is discussed. Expressions and parameters describing terrain roughness and turbulence are proposed and calibrated. The results so obtained are compared with other studies and field tests. An adjacent span reduction factor is proposed for the combination of the peak fluctuating responses of adjacent cables.

1. INTRODUCTION

The current transmission structure design in Mexico and other countries is based on a pseudo-static procedure. Design wind speeds are obtained from regional values having a return period which is related to the line importance. These values are corrected to account for terrain roughness and height.

Although an infinite number of terrain roughnesses can be found, four terrain types are usually designated in codes for the evaluation of wind forces over the structures. Several parameters such as terrain roughness height and surface roughness coefficient have been proposed as a measure of roughness for each terrain type. Other terrain related parameters are used in practice such as the power law variation coefficient, gradient height, turbulence scales and turbulence intensity. Several expressions have also been separately proposed to relate them. However, little agreement can be found when trying to interrelate them. The expressions and values adopted and proposed for such parameters for the four terrain types are included and discussed.

Near the surface design winds are linked to high turbulence levels, which bring about dynamic effects on the response of structures and cables. Davenport (1979) faced the problem with a sensible probability basis. The force values obtained on cables using the adopted parameters were compared with the field measurements obtained by Kempner (1981) and Krishnasamy (1981), showing acceptable agreement. However, the design pressures obtained applying the dynamic response factors appear very high, according to the previous design experience of transmission structures in Mexico. A proposal for an adjacent span reduction factor is discussed in this paper.

2. TERRAIN DEFINITIONS

The simplest way to define a terrain is according to the characteristics of the individual obstacles and the surface roughness found. This is the most practical way to clasify the structure's upwind terrain characteristics, specially for design purposes, since no argument can usually be made as whether a certain terrain in a particular location corresponds to one definition or another.

Four terrain types have traditionally been used to describe terrain roughness in design codes, going from very flat (snow surface) to city centres. These definitions correspond to certain values of terrain related parameters which have been used and calibrated, such as the hourly power law variation coefficient and gradient height. Intermediate terrain types found in practice are ignored and usually referenced as the next lower category. Hence, the pressures used for the design of structures will be higher than the values corresponding to the true roughness of the terrain, since the wind speeds are obtained from regional values and corrected to match the terrain definition.

Terrain definitions were adjusted to be consistent with the common terrain types in Mexico, and to obtain a good agreement between terrain parameters, as explained below.

3. TERRAIN RELATED PARAMETERS

While terrain definitions are the most practical way to clasify upwind terrain roughness and obstacle characteristics, they have to be related to some measure of the effect they have on the wind flow near the surface.

Surface roughness and obstacles are the main causes of turbulence and wind speed variations with height. The rougher the surface and the higher the obstacles, the higher the turbulence level and wind speed variation with height above ground.

There is an infinite number of combinations of terrain roughnesses and obstacles. Several parameters are commonly used as a measure of terrain roughness, and reference values have been set for each terrain type. Expressions have been proposed in different studies to relate them. However, these expressions and values do not always match, so we end up with the question of : Which one can we accept? Extensive testing is required for this and the results will not be ready right away. Therefore, we have to select the most reliable parameters.

Commonly used measures of terrain roughness are the roughness height (or length), z_0 and the surface drag (or roughness) coefficient, K. They are related as $K=[k/\ln(10/z_0)]^2$, in which $k = 0.4$, refered by Venkateswarlu (1989) as proposed by Davenport. The Standards Association of Australia (SAA, 1989) and Venkateswarlu (1989) relate the terrain types to roughness heights which differ one another by an order of magnitude (from the flattest 0.002 m up to 2.0 m in city centres).

The wind velocity fluctuations passing a particular point in space can be seen as a set of eddies carried away by the mean wind. The transverse integral scale of turbulence, L_S, is a measure of the average size of the eddies. The value of L_S mainly depends on terrain roughness and at lesser extend on height. Davenport (1979) proposed a reference value of 65 m (increasing slightly with height and decreasing with roughness). ASCE (1984) provides values of L_S for all four terrain types and Simiu (1986) includes expressions for all integral scales of turbulence, but the values obtained using Simiu's expressions only approximately agree with the values in ASCE. A fairly linear relationship between L_S and $\ln(z_0)$ can be established.

One of the main effects of terrain on the wind is the delay of the flow near the surface. Design speeds are obtained from regional values, which are usually given for a flat (crop) terrain, normalized to a 10 meter height above ground. Variation of wind speed with height is evaluated using either the logarithmic or the power law. For 3 second gust wind speeds, we express the latter as:

$$V_Z = 1.56\,(Z/\delta)^{\beta}$$

It is in the power law variation coefficient, ß, where the best agreement is found, mainly for larger averaging times (10 minutes). Besides, there is more experience about it and more confidence can be expected.

ß values are highly sensitive to the averaging time interval. Lower values of ß are found for shorter average times. 3 second (β_s) and 10 minute (β_m) values against hourly ß are shown in Figure 1. Values for other averaging times can be obtained via the gust factor described below. Linear relationships between the gradient height and L_S against the power law coefficient are found for higher averaging intervals.

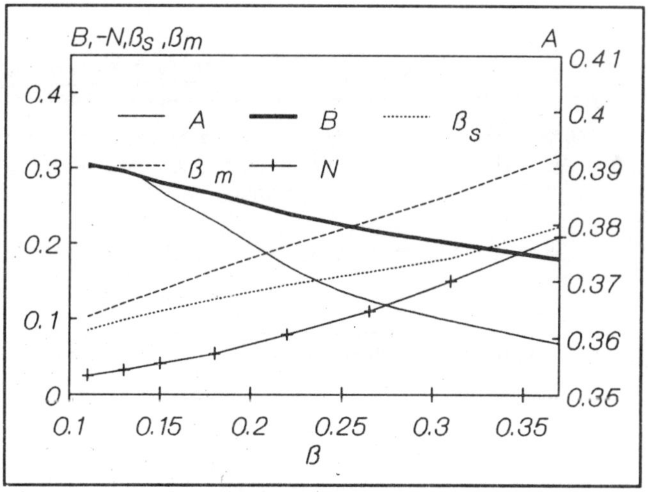

Figure 1 Terrain related parameters

Turbulence intensity values can better be obtained from measurement. If we record wind speed over a sample time interval, the intensity of turbulence is the ratio of the standard deviation of the wind velocity divided by the mean value of the sample. Typical values can be obtained from sampling in different terrain types. Turbulence levels are higher near the ground surface and decrease with height. Very good agreement is found when using the formulae given by Simiu (1986), SAA (1984) and Venkateswarlu (1988), for given values of z_0. The following expression is proposed for its evaluation:

$$I_v = A\ (Z/\delta)^{-N} - B$$

A,N and B are terrain related non-dimensional parameters, Z is the height above ground and δ is the gradient height, both in metres. Figure 2 shows a comparison between turbulence intensity values using this expression an the one proposed by Venkateswarlu (1989) (and hence, Simiu's proposal). This is done for the z_0 value that corresponds to the terrain definitions. The values of A, B and N were selected to match turbulence intensity values at 10 metres above ground and at the gradient height. As can be seen, an excellent match can be found on terrain types 1 to 3. Turbulence intensity at 10 metres above ground was limited to 0.45, for terrain type 4, to fit the values of intermediate to small cities. Figure 1 shows the variation of the values of A, B, N against hourly ß.

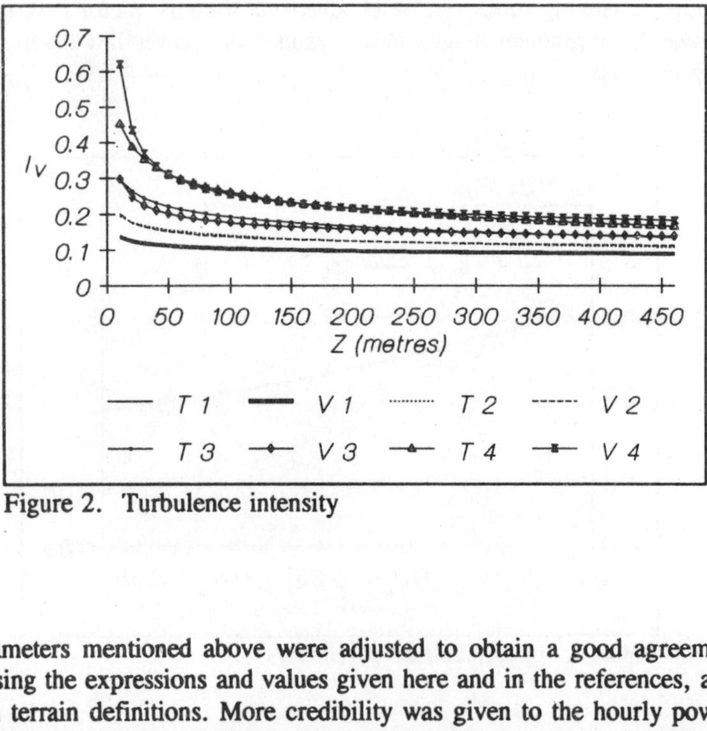

Figure 2. Turbulence intensity

The parameters mentioned above were adjusted to obtain a good agreement among their definitions by using the expressions and values given here and in the references, as shown in table 1, to agree with terrain definitions. More credibility was given to the hourly power law variation coefficient and the intensity of turbulence, since this parameters can be obtained more accurately from field measurement and more experience has been gained for them as well as the very good agreement found in the references. These factors are also recommended in Lopez et al (1993).

4. GUST FACTOR

The gust factor is defined as the ratio of the maximum wind speeds between two different averaging times, t and T. The expression proposed by Mackey (1970) relates to the turbulence intensity at any height as:

$$g = 1 - 0.6226\ I_v^{1.2716}\ \ln(t/T)$$

T is the storm duration. It is usually taken as one hour, so the wind speed ratio between two averaging times t_1 and t_2, is the ratio of the gust factors g_1/g_2. We should mention that t must be lower than T. This definition was used to derive the power law variation coefficients in table 1.

For t and T values of 3 seconds and 10 minutes, the following simplified expression can be applied:

$$g = b\ (Z/\delta)^n$$

The values of terrain related constants b and n are also given in table 1.

Table 1. Terrain related parameters

Param.	Terrain type			
	I	II	III	IV
$\beta_{3\ sec}$	0.099	0.128	0.156	0.17
$\beta_{10\ min}$	0.121	0.164	0.216	0.262
$\beta_{1\ hour}$	0.13	0.18	0.25	0.29
K	0.003	0.0065	0.0142	0.0318
L_S (m)	72.1	63.7	53.5	44.5
A	0.391	0.382	0.369	0.363
N	0.032	0.054	0.096	0.151
B	0.295	0.265	0.227	0.195
δ (m)	245	315	390	455
b	1.224	1.288	1.369	1.457
n	-0.032	-0.054	-0.096	-0.151

5. DYNAMIC RESPONSE

Wind speeds are not steady with time. Wind speed variations can be considered random. Hence, since the pressure (or drag) that the wind produces over the obstacles is proportional to the square of the velocity, the presence of gusts does cause dynamic effects on towers and cables of

transmission lines. The possibility of resonance and dynamic amplification can be estimated by using gust response factors, as proposed by Davenport (1979). Then, the wind pressure would be computed as:

$$p_Z = F_g\, C\, q_Z$$

This is based on a statistical prediction of the structure's response, where the peak response can be expressed as:

$$R_p = R_m + g_p\sigma_R = R_m[1+g_p(\sigma_R/R_m)]$$

where the term in the square parenthesis is the gust response factor, R_m is the mean response, σ_R^2 is the mean square fluctuating response and g_p is a statistical factor. Davenport (1979) developed response factors for both towers and cables. These were expressed in terms of their dynamic properties and characteristic dimensions.

5.1 Structure Response

To predict the response of towers, the response factor for 3 second gust wind speeds is:

$$F_{gt} = \frac{1}{g^2}[1+12.86K^{½}(\frac{10}{h_o})^{\beta_{10\,min}}(\frac{1}{1+0.563h_o/L_S} + 0.0194\frac{f_t h_o g}{V_o})^{½}]$$

The term $1/g^2$ was included because Davenport's expression was developed for 10 minute averaging time and we are proposing to use 3 second gust wind speeds (in kilometres per hour). Units for L_S and reference height, h_o are metres, and the tower fundamental frequencies are in Hz.

The fluctuating response term includes two parts: the first considers the mean square (quasi-static) response at frequencies well below the structure's fundamental frequency, f_t. The second involves the resonant response at frequencies around f_t.

Design wind speeds usually exceed 15 m/s and transmission tower frequencies range between 1 and 4 Hz. Lopez et al (1994) found that the fundamental frequencies of five typical transmission towers used in Mexico for 115 kV, 230 kV and 400 kV obtained from modal analysis using a commercial computer code ranged between 2.0 Hz and 3.1 Hz. For these values, the resonant term has little influence on the response factor for transmission towers. This agrees with Kempner's (1981) argument that the response is quasi-static, and with Mozer's (1989) recommendation to consider exclusively the background response.

Figure 3 shows the envelope of the tower response factor for wind speeds of 100 km/h and 200 km/h, for a natural frequency of 2 Hz, which would be conservative for design.

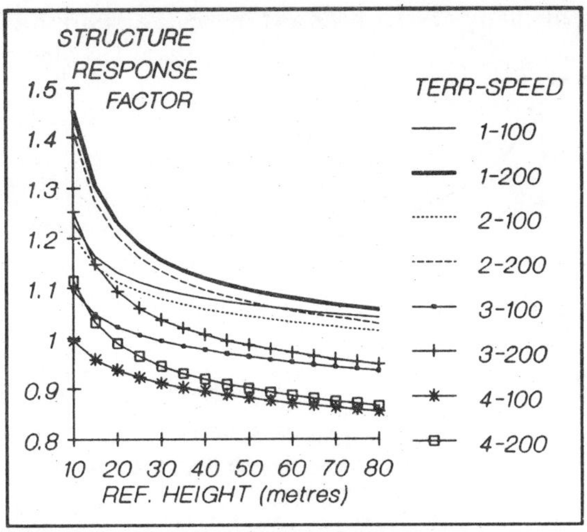

Figure 3. Tower response factor

5.2 Cable response

The cable parameters that influence the response to gusty wind are the diameter, the swing frequency (which depends on the sag) and the span. The wind parameters are the velocity and the transverse lenght scale of turbulence. Davenport's expression for the response factor, expressed in terms of these parameters, in the units described for the towers is:

$$F_{gc} = \frac{1}{g^2}\left[1+12.86K^{1/2}\left(\frac{10}{Z}\right)^{\beta_{10\,min}}\left(\frac{1}{1+0.8L/L_s} + 1.393\frac{d}{L}\left\{\frac{S^{1/2}V_Z}{L\,g}\right\}^{2/3}\right)^{1/2}\right]$$

The cable diameter is in centimetres and the sag is in metres.

In cables the resonant term is more important than in structures, since the swing frequencies are much lower than the structures'. Figures 4 and 5 show envelopes of the cable gust response factors for shield wires and conductors, for wind speeds between 22 and 56 metres per second and sag over span ratio between 3 and 5 percent. They are common ranges for design of transmission towers. Cable diameters were taken between 0.5 and 2.0 centimetres for shield wires and between 3.0 and 5.0 centimetres for conductors.

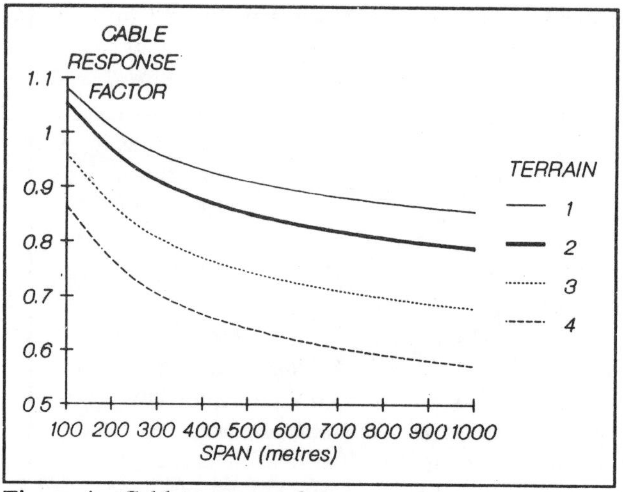

Figure 4. Cable response factor

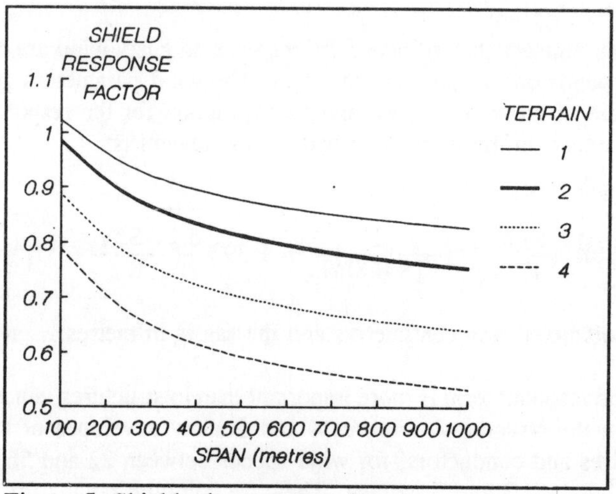

Figure 5. Shield wire response factor

A comparison with field tests might not be a good measure of the accuracy of the gust response factors. This is due to the fact that the response factors were developed to take into account the randomness of the wind pressure fluctuations. Therefore, it should be expected that the theoretically obtained forces exceed the measured values. If the field measured forces were higher than the computed values using response factors, it would either mean that the worst possible load fluctuations were registered or that the peak factor was selected for a low reliability level. It could also be that the terrain type suppossed did not agree with the real roughness of the site.

To evaluate the response obtained theoretically, a comparison is made between the estimated and measured forces on cables. Theoretical values are obtained as the wind pressure, computed as described above, times the exposed area of the cables (span times diameter). Measured values are accepted from Krishnasamy (1981) and from Kempner et al (1981).

To compare with Krishnasamy's tests, the referenced one minute wind speeds were converted to 3 second gust speeds via the gust factor described above. The theoretical values were evaluated using these speeds and the design drag coefficient proposed in ASCE (1984). The measured values were obtained from the swing angle of the insulator chains, Θ. The wind force was calculated on the assumption that tan Θ = Wind force/total weight. For this, the total weight includes half the weight of the insulators. The comparison is made for measured forces corresponding to the upper 95 % confidence limit. The results of measured over theoretical forces are shown on Figure 6, as the curve labeled KrD. The measured over estimated force ratio goes from a minimum 0.44 to a 0.8 maximum. As can be seen in Davenport (1979), conductor drag coefficients drop around a Reynolds number of 20000, and start increasing at about 40000. If we evaluate the forces using an average drag coefficient of the values shown by Davenport, the range of the measured over calculated responses raises to 0.71-0.8, with an average of 0.76. This results are shown on Figure 6, on the curve labeled as KrA.

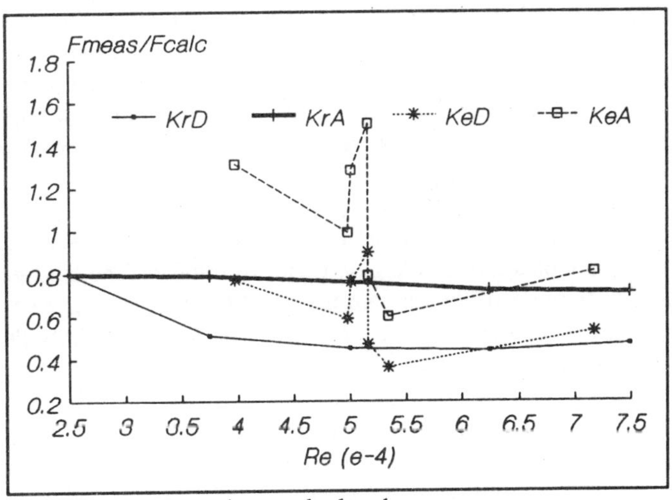

Figure 6. Measured vs. calculated response

Kempner et al (1981) obtained surface drag coefficients, which show a significant difference between maximum and minimum, for a relatively small difference in wind incidence angle. They also evaluated turbulence intensity and gust factors for several recordings. To establish a comparison between measured versus theoretical responses, terrain types were selected to agree with these parameters. As done with the previous comparison, the measured wind forces were evaluated from the transverse swing of the conductors. For the design drag coefficients, the force ratio ranges between 0.36 and 0.9, with an average of 0.63, these results are labeled as KeD on Figure 6. When using an average drag coefficent the measured over computed response varies between a 0.6 minimum and a maximum value of 1.5. These values are presented on Figure 6 as KeA.

As mentioned before, the ratio of measured over computed forces generally lies under unity. This results should be expected, since the response factor must take into acount the random nature of the loading. Besides, other uncertainties exist, such as the actual terrain roughness, which influences the exposure term on the response factors and affects the wind speed corrections for the calculations, since velocity measurements were not made at the average height of conductors. In both cases, conductor heights were approximate. For the comparison with Kempner, the recording time is only six minutes and surface drag coefficients cannot be related to specific wind incidence angles, but the turbulence intensity and the gust factor are specified for some of the recordings. In this case, the comparison values were taken from the maximum oscillation measured. It must be noted that there is a notable difference between the maximum and minimum values. Two of them even have a negative angle in the minimum value. This shows the importance of the dynamic response of the cables.

6. ADJACENT SPAN FACTOR

There has always been the argument that gusts do not cover the whole span and that a span reduction factor should be used to reduce the wind forces over cables. However, this has already been considered in the response factor developed by Davenport. Nevertheless, span reduction factors have been used, and values as low as 0.3 can be seen (Kelly, 1987). Note that load factors are high, but the product still falls below unity. Conversely, the pressures obtained applying cable response factors appear very high, according to the previous experience on the design of transmission structures in Mexico.

As mentioned on the response factor, the peak response of a system can be viewed as the sum of a mean (static) force plus a fluctuating part with zero mean. The treatment of the fluctuating part is done statistically, because of the random nature of gust formation. As a result, the response of adjacent cables will be independent. Moreover, it will be highly unlikely that the peak response values of both cables be at the same time.

Hence, it will be overly conservative to consider peak values for adjacent cables for the design of the insulators and supporting structures. A combination of the fluctuating part of the responses could be acceptable, such as is done with modal responses of structures in earthquake engineering. Following this procedure we evaluated the total force on adjacent spans and obtained the factor that should be applied to the shorter span when considering the peak value on the larger one. The envelope values for all terrain types, for adjacent span lenghts of half the larger span and for equal spans are shown on Figure 7.

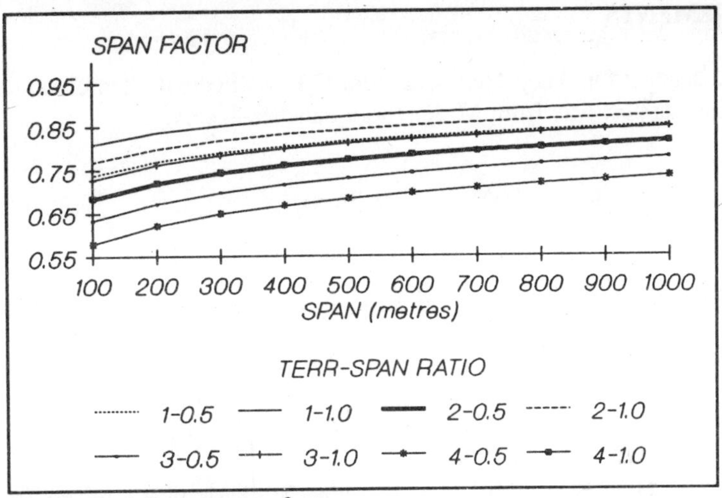

Figure 7. Adjacent span factor

CONCLUSIONS

Turbulence intensity and power law variation coefficients for large averaging times are the most reliable parameters to define terrain roughness. All other parameters can be obtained from these by using the expressions proposed here and in the references.

The values of the terrain related parameters were calibrated to satisfy the interrelations between terrain definitions, wind velocity, turbulence intensity and gust factor variations.

For design purposes, the dynamic effects of gusty winds on transmission structures can be estimated by the use of response factors. The differences between computed values and measured responses can be explained by the fact that terrain definitions do not always agree and that the response factors take into account the random nature of wind pressure fluctuations.

In order to avoid overestimating peak responses an adjacent span reduction factor could be applied for the design of supports.

ACKNOWLEDGEMENTS

Financial support for this project was from C.F.E. (Federal Commission of Electricity of Mexico). Consulting support was from M.Sc. Jorge Sanchez-Sesma.

REFERENCES

ASCE (1984), "Guidelines for transmission structures", Commitee on Electrical Transmission Structures

Davenport, A. (1979), "Gust Response Factors for Transmission Line Loading", Proceed., Fifth Internat. Conf. on Wind Engineering, Colorado State Univ., Forth Collins, CO., p899-909

Kelly, A.G. (1987), "Wind Gust Phenomena in the Design of Transmission Lines", Power Engineering Journal, V4, N3, pp 129-132

Kempner, L., Laursen, H. (1981), "Measured Dynamic Response of a Latticed Transmission Line Tower and Conductors to Wind Loading", fourth U.S. National Conference on Wind Engineering Research, Seattle Washington, p348-355

Krishnasamy, S. (1981), "Dynamic Response of Overhead Transmission Lines to Wind and Ice Loading", CEA International Symposium on Overhead Conductor Dynamics, Toronto Ontario, Canada, p37-44

Lopez, A., Vilar, J.I., Muñoz, C., Alanís, A., De Buen, P., (1993), "Manual de Diseño por Viento", Comisión Federal de Electricidad, IIE, México

Lopez, A., Neri, E., Vilar, J.I., Bicuña, A., (1994), "Unificacion de criterios en el cálculo de presiones, flechas y tensiones para el diseño mecánico de torres de transmisión", Instituto de Investigaciones Eléctricas, Research Report IIE/42/18/I 01/94 ALL/ENB/IVR/ABM, México

Mackey, S. Cheong, S., Lam, R., (1970), "Gust Factors", Proceedings of the seminar: Wind Loads on Structures, National Science Foundation, Japan Society for Promotion of Science, University of Hawaii, pp 191-202

Mozer, J. (1989), "Transmission Line Loading", Proceed. of the sessions related to Steel Struct. at Struct. Congress, San Francisco, CA, New York, New York, ASCE, p51-60

Simiu, E., Scanlan, R., (1986), Wind Effects on Structures: an introduction to wind engineering, John Wiley & Sons, NY

Standards Association of Australia (1989), "Minimum design loading loads", part 2: Wind loads, Std. House, N. Sydney, Australia

Venkateswarlu, B., Arunachalam, S., Shanmugasundaram, J., Annamalai, G., (1989), "Variation of Wind Speed with Terrain Roughness and Height", IE(I) Journal-CI, V69,p228-234

Vilar R., J.I., Muñoz B., C.J. (1992), "Procedimiento para la obtención de las velocidades regionales de viento a utilizar en el diseño de estructuras", Instituto de Investigaciones Eléctricas, Research Report IIE/42/2919/I/05/P, México

FULL-SCALE MEASUREMENTS OF TOPOGRAPHIC MULTIPLIERS FOR A STEEP ESCARPMENT

BANKS, R.W.[1], HOLMES, J.D.[2], and PAEVERE, P.[3]

Experimental Scientist — *Project Area Manager* — *Technical Officer*

CSIRO Division of Building, Construction and Engineering, P.O. Box 56, Highett, Victoria, 3190, Australia

Abstract

The instrumentation and measurements of topographic multipliers and turbulence characteristics for a steep escarpment, with a maximum average slope of 0.48, are described. The values are compared with corresponding values from current Australian Standards. The measurements indicate significant errors in some of the values specified by the Standards, believed to be caused by the neglect of flow separation at the leading edge of the escarpment. For one wind direction, significant mean wind direction changes were produced by the escarpment.

1. INTRODUCTION

The effect of hills, ridges and escarpments on wind speeds for the design of structures has been well recognised in recent years with the inclusion of 'topographic multipliers' or similar factors in a number of design codes and standards for wind loading (e.g. [1],[2],[3]). Most of these factors have been based on full-scale, wind-tunnel, theoretical and numerical studies for ***shallow*** hills with extrapolations to the steeper slopes. Very little study of wind flow and topographic multipliers for *steep* (slope greater than 0.3) ridges and escarpments has been carried out, even though such sites are common for structures such as communication and broadcasting towers.

In this paper, we describe the instrumentation for, and some results from a full-scale study of topographic multipliers and turbulence characteristics for an escarpment with a maximum average slope over the top half of 0.48.

2. TOPOGRAPHIC MULTIPLIERS

The definition of Topographic Multiplier used in the Australian Standards [1],[2], and used in this paper is as follows :

$$\text{Topographic Multiplier} = \frac{\text{(Wind speed at height z, above the escarpment)}}{\text{(Wind speed at height z, above the flat ground upwind)}}$$

This definition applies to mean, gust and r.m.s. fluctuating wind speeds, and these will be denoted by $\bar{M}_t$, $\hat{M}_t$ and M_t', respectively, in this paper.

3. CURRENT DESIGN VALUES

Up to the nineteen-seventies some codes and standards for wind loading included an 'escarpment rule', in which the wind speed above the escarpment was calculated from the same velocity profile as that assumed upwind, but the height was measured, not above the actual ground level, but from an artificial height datum, lying below the actual terrain. Apart from being a concept with little physical basis, the rule, as formulated in the British Code of Practice of 1972 [4], was found to give much lower mean multipliers than those measured in wind-tunnel tests by de Bray [5].

The speed-up of wind flow over topographic features was the subject of extensive theoretical and experimental work in the nineteen-seventies and nineteen-eighties, with the incentive provided by energy crises and the growth in the exploitation of wind energy. Jackson and Hunt [6] developed a theory for the flow of a turbulent boundary layer over a shallow hill. The main result of this theory is that the maximum 'fractional speed-up' of the mean wind speed, equal to $(\bar{M}_t-1)$, is directly proportional to the average upwind slope approaching the crest of the escarpment. The constant of proportionality depends upon the shape of the topography. This relationship has been adopted for the specification of topographic multipliers, or topography factors, in several codes and standards for wind loading.

For practical purposes, the upwind slope may be defined as $H/2L_u$, where H is the height of the crest above the level terrain upwind, and L_u is the horizontal distance from the crest to where the elevation drops to H/2. In the Australian Standard AS 1170.2 [1], the constant of proportionality is assumed to be 1.4 for escarpments at ground level, with a value of 3.2 for ridges and hills. In other design guides and codes, (e.g. [3],[7]), a value of 2.0, independent of the shape of the topography, is assumed.

For upwind slopes greater than 0.3, arbitrarily assumed to be the slope for which separation effects are important, the value of the multiplier for the slope equal to 0.3 is assumed. The basis for this is that the separation 'bubble' upwind of the crest is regarded as presenting an effective slope with a value independent of the actual slope. However, the effects of separations *downwind* of the leading edge of the escarpment are ignored.

To determine the topographic multiplier for gust wind speeds, it is necessary to include the variation of the r.m.s. turbulence over the escarpment, i.e. to know the multiplier, M_t', as defined previously. For the values specified in AS 1170.2, it was assumed that the r.m.s. wind speed remains constant over topographic features [8]. This results in topographic multipliers for gust speeds that are lower than those for the mean speeds. For escarpments with upwind slopes less than 0.3 the multiplier at ground level is given by the upwind slope, defined as described previously, multiplied by a constant of 0.8. For steeper slopes, the value for the 0.3 slope applies.

4. INSTRUMENTATION

The instrumentation used for the measurements described in this paper, has been developed by CSIRO to conduct long term environmental monitoring at remote sites. The modular system is designed for easy installation and for severe environments with sub zero temperatures, and high radiation fields that often occur on transmission and broadcasting towers. The system consists of four modular parts: transducers, instrumentation module, base station and power supply. Vaisala cup anemometers, with a distance constant of 2.0 metres, and Vaisala wind vanes were used in the present study. The instrumentation module is microprocessor-controlled and converts signals to a single digital data stream, these modules are connected to the base station by a single cable carrying both the digital data at 9600 baud and power for the module. This module also provides isolation and lightning protection.

The base computer consists of an IBM compatible 80486 computer with a 250 megabyte hard disk. Data is received as a serial stream at 9600 baud through an optically isolated data line. This data stream is decoded and wind speed and direction are displayed in real time and monitored for trigger levels. When these levels are exceeded, all data is logged to the hard disk for a preset time.

A modem connects the computer by telephone line to a control site from where parameters can be set, and data can be transferred from the remote site. Interaction from the control site is only required on an infrequent basis, i.e. to transfer any data files, or to change trigger parameters.

Power is supplied to the instrumentation module at 70 Volts and 20 kHertz; this allows transformer isolation of 4 kVolt while maintaining a small size.

A block diagram of the instrumentation system is shown in Figure 1.

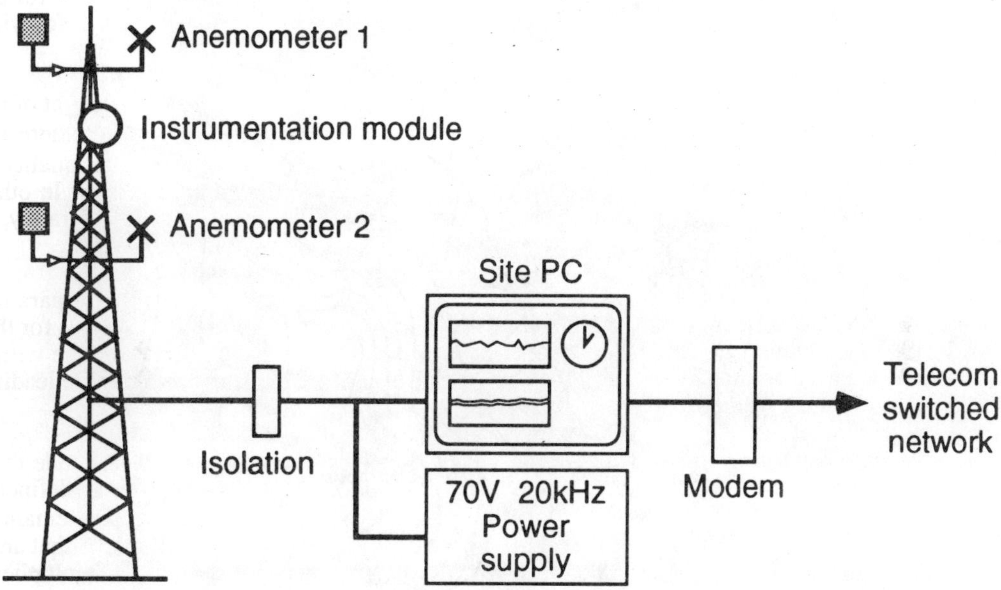

Figure 1. Instrumentation block diagram.

5. SITE DESCRIPTION

The top of the escarpment (Mount Dandenong), at which several television towers are located, is 600 metres above sea level and about 480 metres above the plain where the base tower is located (Kilsyth).

Figure 2 shows a simplified contour map of the topography of Mount Dandenong, and lines indicating the directions of the winds for which data was collected.

A terrain profile on a line through the two measurement positions is shown in Figure 3. This line (on a bearing of 308°) is also approximately the line of maximum slope for the escarpment. Strong winds from this general direction, i.e. approximately north-west, are quite common in the Melbourne area, generated both by synoptic winds (gales), and by local thunderstorms usually associated with cold fronts.

At the Mount Dandenong site, cup anemometers and direction vanes were mounted at two levels on a lattice-type freestanding television tower – 32 and 69 metres above the local ground datum. This tower is about 80 metres from the 'leading edge' of the escarpment. The land elevation falls by 100 metres in 2.4 kilometres downwind of the leading edge so the topography would qualify as an escarpment according to the Australian Standard for Wind Loads [1] and similar codes.

At the Kilsyth site, which is approximately 4.5 kilometres from the Mount Dandenong site, a single cup anemometer and direction vane was installed at a level of 34 metres above local ground level, on top of a mobile telephone tower of circular cross-section.

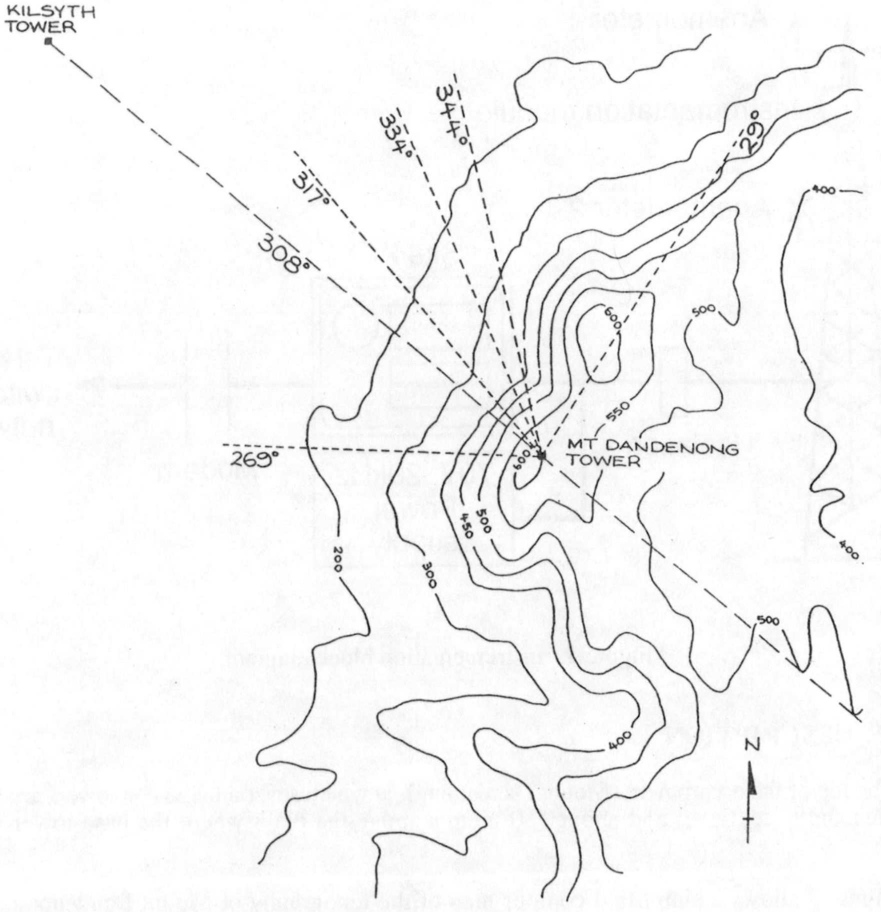

Figure 2. Contour plan of the Mount Dandenong area, and mean wind directions.

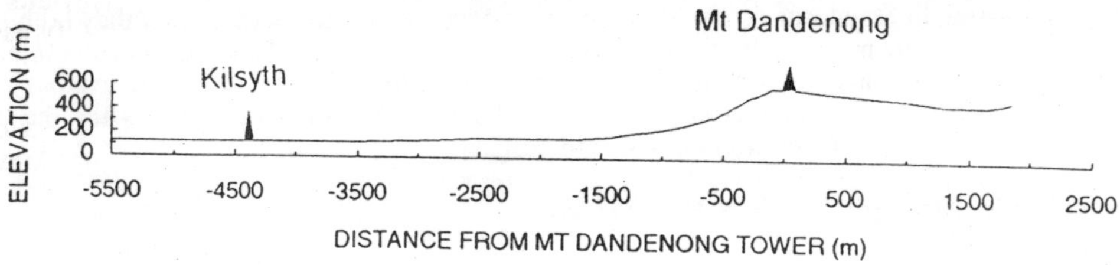

Figure 3. Terrain profile on a bearing of 308 degrees.

6. RESULTS

The results described here were derived from measurements taken between April and July 1994. The main purpose of the measurements was to determine topographic multipliers, and these are presented in Sections 6.1 to 6.4 following.

6.1 Topographic Multipliers – N.W. winds

Measured values of the topographic multipliers for the mean and r.m.s. fluctuating wind speeds (averaged over 5–10 minutes), and for the maximum, or peak gust, wind speeds are given in Table 1.

For the results in Table 1, the average mean wind direction at Kilsyth is 317° for the 8 selected runs, i.e. the wind direction is close to parallel to the line between the towers. Although values of wind speed at a height of 69 metres at Kilsyth could not be directly measured, they were inferred from the 34 metres values, by multiplying by 1.15 for the mean and 1.10 for the gust. These factors are based on an appropriate value of value of roughness length, for the approach terrain, of 200 mm [1]. No correction was made to the r.m.s. velocities from 34 to 69 m.

No adjustment for the difference in heights between the Kilsyth anemometer (34 metres) and the lower one at Mount Dandenong (32 metres) was made. Values derived from Australian Standards AS 1170.2 (Wind Loads), [1], and AS 3995 (Design of Lattice Towers and Masts), [2], are also shown in Table 1.

Table 1 shows that a considerably higher value was obtained for the mean multiplier at 69 metres compared with that at 34 metres. This is believed to be caused by the separation 'bubble' from the front of the escarpment; the lower anemometer (34 metres) is probably beneath the average position of the separating shear layer, whereas the upper anemometer (69 metres) may be above the separation bubble. The high value of the multipliers for the r.m.s. velocities at 34 metres are strong evidence for this. Further evidence is in the spectra of the wind speeds which are discussed in Section 6.6.

Table 1. Topographic Multipliers for N.W. Winds (317°)

	Measured values 25–26/5/94*	AS 1170.2	AS 3995
Mean 34m	1.01	1.39	1.53
R.m.s. 34 m	1.19	(1.0)	(1.0)
Gust 34 m	1.19	1.22	1.53
Mean 69m	1.51	1.37	1.46
R.m.s. 69 m	0.86	(1.0)	(1.0)
Gust 69 m	1.28	1.21	1.46

* Averaged over 8 runs.

The two Australian Standards do not allow for the effect discussed above; in fact they both give slightly higher values for the mean multipliers for the *lower* height, whereas the measurements indicate the opposite. Both Standards are slightly unconservative for the mean multiplier at 69 metres, but quite conservative at 34 metres. The Standards give values closer to the measured values for the gust multipliers, although AS 3995 is conservative for both heights.

6.2 Topographic Multipliers (N.N.W. Winds)

Table 2 shows the average topographic multipliers for two N.N.W. directions (average direction 334° and 344°). The values in this Table are generally higher than those in Table 1 for the mean and gust multipliers. This is a little surprising because the approach slope is slightly lower for these wind directions. However, there may be some effect of the second crest to the north of Mount Dandenong tower (Figure 2). Both Standards are unconservative for the mean multiplier at 69 metres, and conservative at 34 metres. AS 1170.2 is unconservative at both heights for the gust multiplier.

Table 2. Topographic Multipliers for N.N.W. Winds

	Measured values		AS 1170.2	AS 3995
	27–28/4/94* (344°)	25/5/94† (334°)		
Mean 34m	1.23	1.17	1.39	1.53
R.m.s. 34 m	1.51	1.42	(1.0)	(1.0)
Gust 34 m	1.37	1.37	1.22	1.53
Mean 69m	1.78	1.70	1.37	1.46
R.m.s. 69 m	0.82	0.92	(1.0)	(1.0)
Gust 69 m	1.38	1.39	1.21	1.46

* Averaged over 13 runs

† Averaged over 6 runs

6.3 Topographic Multipliers (W. Winds)

Table 3 shows the average values of the topographic multipliers for a series of 9 recordings when the wind direction at Kilsyth was westerly. For this direction, the Kilsyth tower does not directly measure the approach flow, but it can be assumed that wind characteristics are spatially uniform on the flat terrain, so that the Kilsyth data was taken for the reference values to calculate the topographic multipliers. The same correction factors of 1.15 for the mean and 1.10 for the gust as used for the N.W. winds were used to correct from a height of 34 metres to 69 metres at Kilsyth.

Again much lower values of mean multiplier, and a higher r.m.s. multiplier were obtained at 34 metres compared with the corresponding values at 69 metres height, indicating a separation region affecting the lower elevation on Mount Dandenong. The gust multiplier was quite similar at the two heights. The values were higher than those in Table 1, but lower than those in Table 2.

Table 3. Topographic Multipliers for W. Winds

	Measured values (28/4/94)* (269°)
Mean 34m	1.22
R.m.s. 34 m	1.35
Gust 34 m	1.31
Mean 69m	1.60
R.m.s. 69 m	0.98
Gust 69 m	1.31

* Averaged over 9 runs.

6.4 Topographic Multipliers (N.E. Winds)

Using the same techniques described previously for the N.W. and W. directions, topographic multipliers were obtained for 10 runs from an average direction close to north-easterly at Kilsyth (bearing 29°). The average value of the multipliers are shown in Table 4.

Table 4 shows that lower values of both mean and gust multipliers were obtained for this direction, however similar values of M'_t, were recorded. Some shielding by the second peak is provided in this direction (see Figure 2).

For this wind direction, there was also a significant *change* in mean wind direction at Mount Dandenong compared with that at Kilsyth, of about –27°, i.e. the wind direction was nearly northerly at Mount Dandenong.

Table 4. Topographic Multipliers for N.E. Winds

	Measured values (29/7/94)* (29°)
Mean 34m	0.99
R.m.s. 34 m	1.35
Gust 34 m	1.17
Mean 69m	1.18
R.m.s. 69 m	1.09
Gust 69 m	1.13

* Averaged over 10 runs.

6.5 Spectral Densities

Figures 4 and 5 shows the non-dimensional spectral density for representative records from all three anemometers for a north-north-westerly wind direction. Figure 4 shows the spectrum for the level site at Kilsyth; a von Karman spectral form as in Equation (1), with an integral length scale, L_u, of 70 metres has been included for reference purposes.

$$n.S_u(n)/\sigma^2 = (4n\ L_u/\bar{u}) / [1 + 70.78(n\ L_u/\bar{u})^2]^{5/6} \qquad (1)$$

where n denotes frequency

σ is the r.m.s. wind speed

$\bar{u}$ is the mean wind speed.

In Figure 5, the spectra for the two different heights at Mount Dandenong have been shown; the turbulence at the lower level contains significantly higher frequencies than that at the 69 metres level, as would be expected for turbulence generated by separated flow.

7. CONCLUSIONS

Measurements of topographic multipliers and turbulence characteristics for a steep escarpment have been presented. Corresponding values obtained from current Australian Standards indicate significant errors caused by neglect of the flow separation downwind of the leading edge of the escarpment.

At a height of 69 metres, the Australian Standards underestimate the mean topographic multipliers. Significant mean direction shifts were produced by the escarpment for one wind direction.

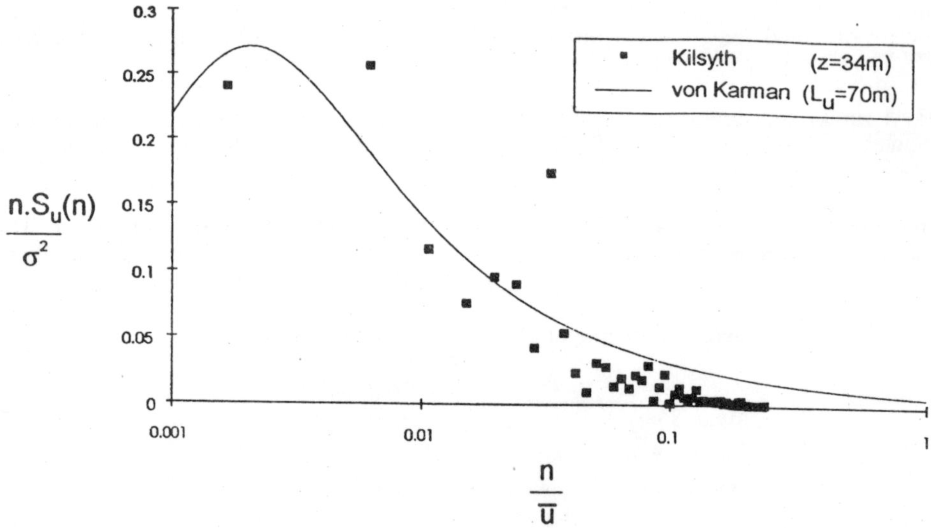

Figure 4. Sample spectral densities for wind speed measurements at Kilsyth.

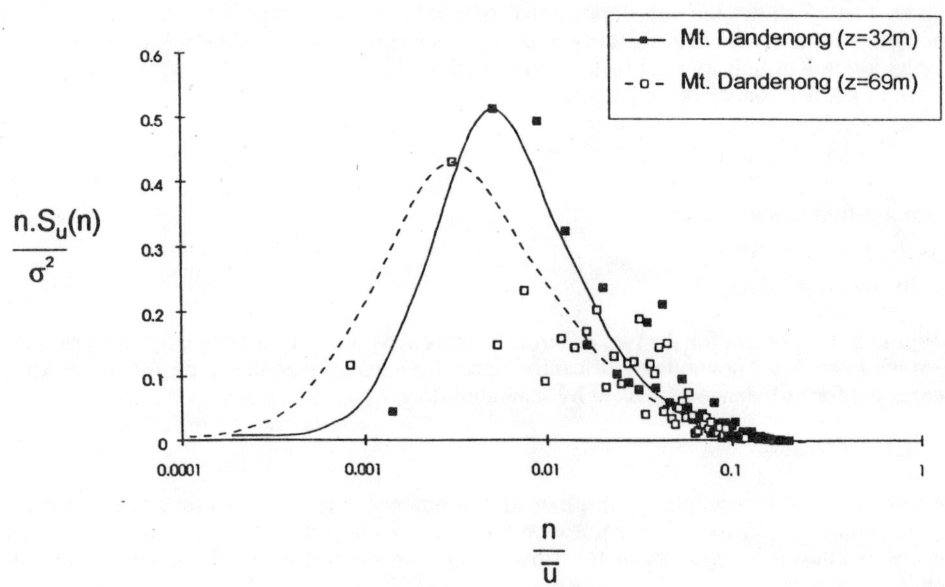

Figure 5. Sample spectral densities for wind speed measurements at Mount Dandenong.

8. ACKNOWLEDGMENTS

The participation and assistance of Mr. Tom Glass (Telecom Australia) is gratefully acknowledged. Associate Professor K.C.S. Kwok and Matthew Glanville (University of Sydney) assisted in the planning of these experiments and will carry out further analysis of the data and conduct wind-tunnel tests.

9. REFERENCES

1. Standards Association of Australia (1989)."SAA Loading Code Part 2: Wind loads", Australian Standard AS 1170.2–1989.
2. Standards Association of Australia (1994). "Design of Steel Lattice Towers and Masts", Australian Standard AS 3995–1994.
3. Bureau of Indian Standards (1989). "Indian Standard Code of Practice for Design Loads for Buildings and Structures, Part 3: Wind loads", IS:875(Part 3)–1989.
4. British Standards Institution (1972). "Basic Data for the Design of Buildings. Chapter V. Part 2 Wind loads". CP3: Chapter V: Part 2:1972.
5. de Bray, B.G. (1973). "Atmospheric Shear Flow over Ramps and Escarpments", Industrial Aerodynamics Abstracts, Vol. **5**, pp. 1–4.
6. Jackson, P.S. and Hunt, J.C.R. (1975). "Turbulent Wind Flow over a Low Hill", Quarterly Journal of the Royal Meteorological Society, Vol. **101**, pp.929–955.
7. Building Research Establishment (1989). "The Assessment of Wind Loads. Part 5: Assessment of wind speed over topography", BRE Digest 346.
8. Holmes, J.D., Melbourne, W.H. and Walker, G.R. (1990). "A Commentary on the Australian Standard for Wind Loads", Australian Wind Engineering Society.

THE ANALYSIS OF WIND SPECTRUM CHARACTERISTIC IN SURFACE LAYER IN BADALING AREA OF BEIJING

LUO, J.Y.[1] and ZHU, R.Z.[2]

[1] *Human Province Meteorological Service, Changsha 410007, P.R. China.*
[2] *Chinese Academy of Meteorological Science, Beijing 100081, P.R. China.*

ABSTRACT

In this paper, we calculate 21 examples of horizontal wind power spectrum and cross-spectrum in different conditions at 8 heights. The mean figures of power spectrum and cross-spectrum at 8 levels are presented in the frequency range from $2.25*10^{-4}$ to 0.5 cycles per second. There appears to be obvious peaks in the period range from several seconds to more than ten minutes in the spectrum,which differ from the former traditional results. It is found that there is evident relationship in horizontal wind power spectrums at different heights. The relationship is more evident in low frequencies than in high frequencies, and the shorter the distance between two levels,the better the relationship. In addition, in order to compare with the classical results in engineering mechanics we also calculate the reduced power spectrum and simulate its equation.

1. INTRODUCTION

In the recent decades, many scholors studied the turbulent fluctuation spectrum based on the similarity theory. Van der Hoven (1957) measured the horizontal wind power spectrum with the wide range from 0.0007 to 900 cycle per hour at about 100-m height and found that the length of the obvious period is 4 days,12 hours and about 1 minute, respectively. Davenport (1963) summarized the spectrum of horizontal wind speed in strong wind from more than 90 examples which were measured at different height and at different location all around the world and presented the empirical formula of wind spectrum.

In our nation,there are many research works on the spectrum of trubulence. Lu et al (1979) presented that there are some principal periods with the range from several minutes to more than ten minutes in the mesoscale range,whether or not the stratification is stable,data gotten by acoustic radar. This result is different from the traditional knowledge. In the past, it was considered that there are no obvious peaks in this frequency range. In recent year, with using of the wind energy,it was requested to know of the wind spectrum systematically. But there are not special observation in the region with plentiful wind energy and there are few detailed research work on this field.

Considering previous research works,even though many works dealt with the turbulent structure of wind speed and characteristic of turbulent spectrum,the systematic research of wind spectrum at different levels in different regions is default because of the default of data and the limitation of the methods of the detecting. In our nation,the data served in previous works, which were obtained in middle or big urban regions,were affected by kinds of urban efects and lack of representativeness. Bowne et al (1970) found that characteristic of the wind spectrum and the turbulent structure between the rural and urban region diveres remarkablely.

This paper presents recent systematic analyses of horizontal wind spectrum,specially in strong wind, and cross-spectrum with the wide frequency range and compares them with the former typical results. The data were obtained at 8 levels (5,10,15,20,30,50,70,100 m) at the 104-m meteorological tower suituated at Badaling area of Beijing.

2. DATA COLLECTING

2.1 Site and instrument

Located 5 Km north of Badaling Great Wall,the tower is situated in a well-exposed area free of near-by structures which could interfere with the air flow.

The tower is an open-lattice structure of steel with 3 legs spaced same distance at all heights. It is instrumented at 8 levels with sensors to measure the wind speed and direction. Sensors are on the movable arms aparted from the tower body 2 m and the affection of the tower is reduced greatly. The instrument collecting data is controlled by microcomputer TP805 with memory size 16K byte and may sample at 1 Hz or 1/6 Hz at 8 levels at same time. The data were automatically recorded on digital magnetic tape while memory is full. The wind speed and direction sensors are remade from EL cup-contact anemometer.

2.2 Data analysis

The data were obtained from October of 1989 to July of 1990.Normally,data only in last ten minutes of each hour were recorded,sampled at 1 Hz. While strong wind occurred data were collected at sampling rate 1 or 1/6 Hz. Because of the limitation of the memory size of computer the data series might be broken while strong wind last long enough. Using autoregression method we make the complete data series.

3. CHARACTERISTICS OF POWER SPECTRUM

3.1 Computation

Suppose a stationary time series Xt (t=1,2,.....,n), its lag correlation coefficient is $r(\tau)$. Using the standard correlation Fourier transform with Hanning smoothing method we may obtain power spectrum S_l:

$$S_l = \frac{B_l}{m}\left[r(0) + \sum_{\tau=1}^{m-1} r(\tau)\cdot\left(1+\cos\frac{\pi\tau}{m}\right)\cdot\cos\frac{l\pi\tau}{m}\right] \qquad (l = 0, 1, 2, \cdots\cdots, m)$$

where m is maxium time lag and $B_l = \begin{cases} 1 & (l \neq 0, m) \\ 1/2 & (l = 0, m) \end{cases}$

In order to obtain spectrum in wide frequency range with detailed descripitation in both high and low frequency range,we estimated them for each different portions separately and then linked them.The result of using this method is the same as using a filter over the original data series to filt parts of high frequency peaks.

3.2 The characteristic of power spectrum at 8 levels

All 21 runs in strong wind were analyzed including strong wind of cold front and thunderstorm in summer. Detailed information about all runs is in Tab 1.

Tab 1. Introduction of 21 Runs

No.	1	2	3	4	5	6	7
Time	Nov. 17 13 - 18	Nov. 13 00 - 10	Dec. 18 12 - 15	May 10 10-12	May 10 13-15	May 19 09-12	May 19 14-16
Rate *	A	A	A	B	B	B	B
Type **	C	C	C	C	C	C	C
No.	**8**	**9**	**10**	**11**	**12**	**13**	**14**
Time	May 20 10 - 13	May 23 19 - 21	June 8 16	June 9 09 - 11	June 22 16 - 17	June 23 16 - 17	June 27 02
Rate *	B	B	B	B	B	B	B
Type **	C	D	D	D	D	D	D
No.	**15**	**16**	**17**	**18**	**19**	**20**	**21**
Time	June 27 22 - 23	June 28 15	June 29 16 - 17	July 4 19	July 6 19	July 7 17	July 9 19 - 20
Rate *	B	B	B	B	B	B	B
Type **	D	D	D	D	D	D	D

* : A represents 1/6 Hz, B is 1 Hz.
**: C represents strong wind of cold front, D thunderstorm wind.

According to selected runs the frequency range analyzed is from 2.25*10 to 0.5 cycle per second, corresponding to period range from 2 to 4447 seconds. Traditionally considered, this range is across the fields from mic- to mesoscale meteorological power spectrum. For the convenience of discussion we called the part of spectrum with period length more than 1 minute as low frequency spectrum and period length about or shorter than one minute as high frequency spectrum. Calculating the autocorrelation function of horizontal wind at different levels we found that the autocorrelation coefficient decreased on exponential law with the increasing of lag time length and smoothed well in short time. So the red noise spectrum was used to be a standard spectrum for testing.

The curse of power spectrum at three levels were in Fig. 1 (other levels absent). In this system the area under the curve corresponding to frequency range n - n+ n equals the energy in this frequency range.

It is found in Fig. 1 that, in low frequency range, a peak appears in period 9 to 12

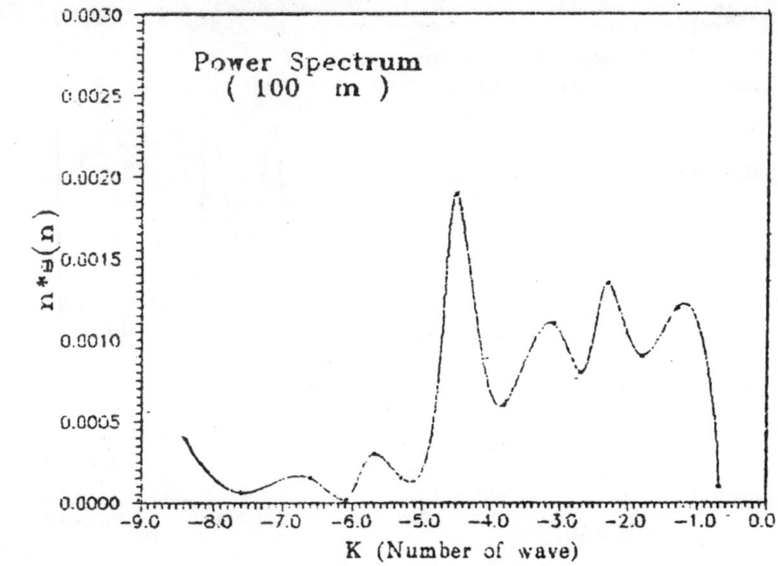

Fig. 1 (a) Power Spectrum

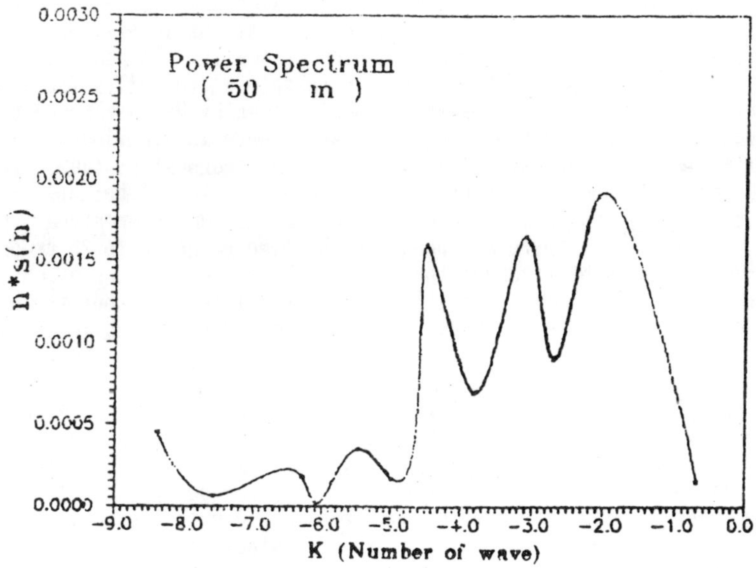

Fig. 1 (b) Power Spectrum

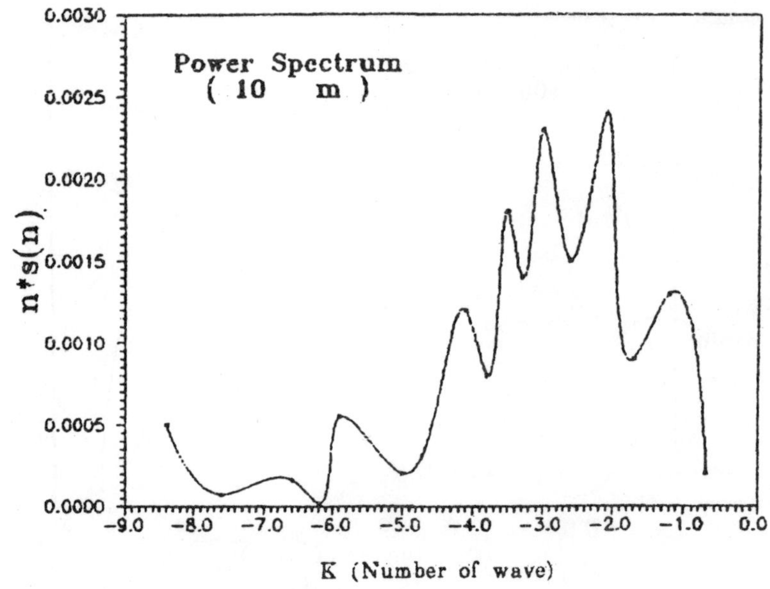

Fig. 1 (c) Power Spectrum

minutes and 4 to 6 minutes. The period at about 10 minutes is more obvious which may pass the red noise spectrum test with 95% and gradually decreases while decreasing of the height with the shortening of the length of the period. At low levels, while period of 10 minutes decreasing the period of 6 minutes enhances gradually. In addition, there is a center vally in the period length of 30 minutes corresponding to valley field between the peaks of micrometeorological scale and synoptic scale given by Van der Hoven but drifting to the high frequency range. In high frequency range there is a obvious period at 60 to 100 seconds, i.e. 1 to 2 minutes. This period is the strongest at 100m, and then the length of this period is shorten gradually with the decreasing of height. And only 60 seconds under 10 m level, which corresponds to the Van der Hoven spectrum's peak in high frequency range. In addition, there are peaks in the time range 10 to 25 seconds, 8 to 10 seconds,respectively, and the stronger the amptitude the lower of the height. At the end part of the high frequency, there are still a peaks at all levels, but the peaks is the weakest at middle level 50 m. At low level, 5 m and 10 m, there are obvious peaks at about 33 seconds.

Generally speaking, the peaks in low frequency range with period length of several minutes are more obvious at high levels than at low levels, at the same level the energy in low frequency is much less than it in high frequency, and at the high levels the peak with period of about 1 minute is the strongest in all frequency and with the height's decreasing, the energy with the period shorter than 1 minute gradually enhances and eventually exceeds the peak with the period of about 1 minute,especially, the peaks with the period of 16 to 25 seconds and about 10 seconds are the strongest. The spetrums at 8 levels is shown in Fig. 2.

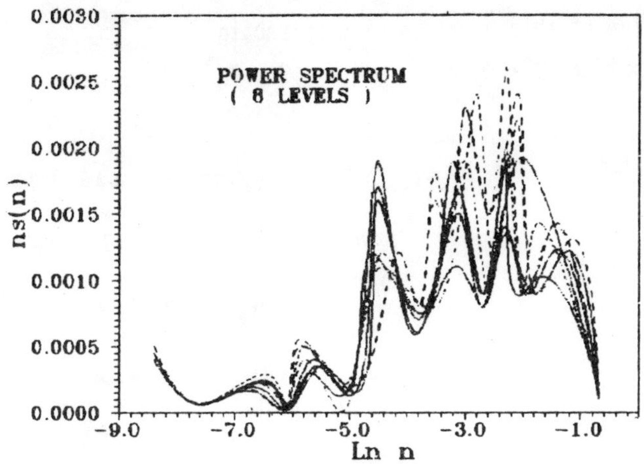

Fig. 2 Solid line corresponding to high levels, dashed line low levels

Constrasting the horizontal spectrums of different kinds of strong wind, we found that place of peaks in all frequency range have almost not obvious difference, but undulation is much more and the amptitude is stronger in high frequency range in thunderstorm strong wind.

3.3 The characterastic of reduced power spectrum

In order to compare with Davenport(1963)'s empirical spectrum which is widely used in engineering mechanics, we calculate reduced power spectrum, and obtain curve spectrum with wide range of wave-length. this spectrum is double-peak, two peaks correspond to wave length of 833m ane 105m respectively, which differs from the classic Davenport's empirical spectrum,which is single peak corresponding to the wave length of about 555m, The reduced power spectrums at 8 levels are shown in Fig. 3.

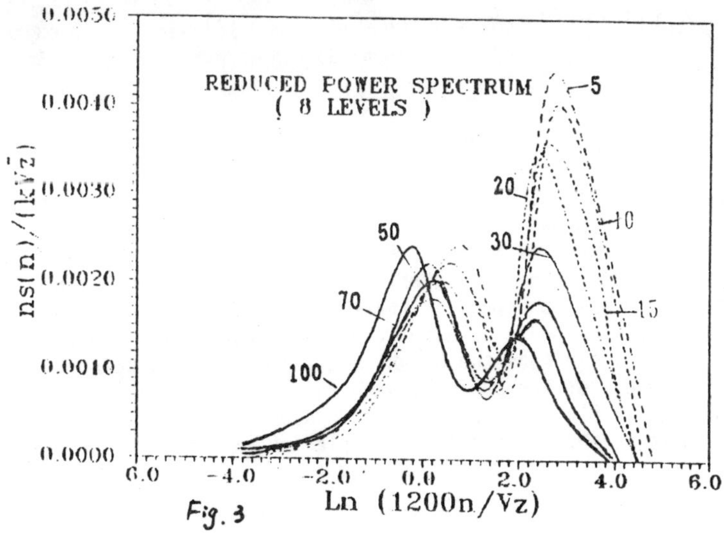

Fig. 3

we simulate the equation of this spectrum with polynomia

$$y = a_0 + a_1 x + a_2 x^2 + a_3 x^3 + \cdots\cdots + a_{10} x^{10}$$

where

$a_0 = 1.996 \times 10^{-3}$

$a_1 = -6.888 \times 10^{-5}$

$a_2 = -1.061 \times 10^{-3}$

$a_3 = 1.957 \times 10^{-4}$

$a_4 = 3.580 \times 10^{-4}$

$a_5 = -1.804 \times 10^{-5}$

$a_6 = -4.436 \times 10^{-5}$

$a_7 = 1.267 \times 10^{-7}$

$a_8 = 2.272 \times 10^{-6}$

$a_9 = 1.618 \times 10^{-8}$

$a_{10} = -4.157 \times 10^{-8}$

4. THE CHARACTERISTIC OF CROSS-SPECTRUM

Suppose two stationary time series X_{1t}, X_{2t} (t=1,2,...,n), their mean values are $\bar{X}_1, \bar{X}_2$, their variances are S1, S2, and lag cross-correlation function are $r_{12}(\tau)$, $r_{21}(\tau)$, using binomial smoothing, the cospectrum $P_{12}(l)$, coherence spectrum $R_{12}(l)$ and lag phase spectrum $L(l)$ are, respectively:

$$P_{12}(l) = \frac{B_1}{m}\gamma_{12}(0) + \frac{1}{2}\sum_{\tau=1}^{m-1}\left[\gamma_{12}(\tau) + \gamma_{21}(\tau)\right]\left(1 + \cos\frac{\pi\tau}{m}\right)\cos\frac{l\pi\tau}{m}$$

$$Q_{12}(l) = \frac{B_1}{m}\sum_{\tau=1}^{m-1}\left[\frac{1}{2}\left(1 + \cos\frac{\pi\tau}{m}\right)\cdot\sin\frac{\pi l\tau}{m}\right]\left[\gamma_{12}(\tau) - \gamma_{21}(\pi)\right]$$

$$(l = 0, 1, 2, \cdots\cdots, m)$$

$$R^2_{12}(l) = \left[P^2_{12}(l) + Q^2_{12}(l)\right] / \left[P_{11}(l)\cdot P_{22}(l)\right]$$

$$(l = 1, 2, \cdots\cdots, m-1)$$

$$Q(l) = tg^{-1}\left[Q_{12}(l)/P_{12}(l)\right]$$

$$L(l) = mQ(l)\cdot\Delta t/\pi l$$

where Δt is sampling interval. Coherence spectrums and lag phase spectrums between two typical different levels are shown in Fig. 4. For significance test 95%, the order of critical value of coherence spectrum is 10^{-2}. From Fig. 4 it is found that the coherence spectrum greatly surpass the critical value, especially, in low frequency range, the shorter the distance between two levels, the greater the coherence. From the lag phase spectrum, in high frequency range, the time phase in high levels lags behind the one in low levels for about 10 to 40 seconds, and the more the height's difference the greater the lag scope of time phase. The difference of the phase between two levels gradually smallen from high frequency to low frequency. Fig.5 gives the superimposed cospectrum between level 100, 10 m and other levels. There is obvious positive correlation in phase in low frequency.

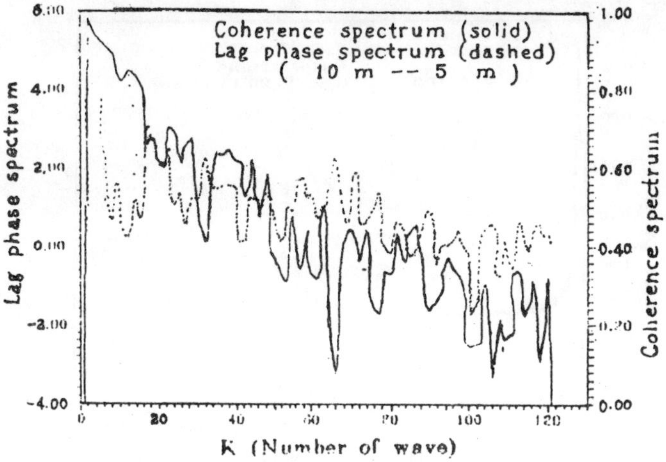

Coherence spectrum (solid)
Lag phase spectrum (dashed)
(10 m -- 5 m)
Lag phase spectrum
Coherence spectrum
K (Number of wave)

Fig. 4 (a)

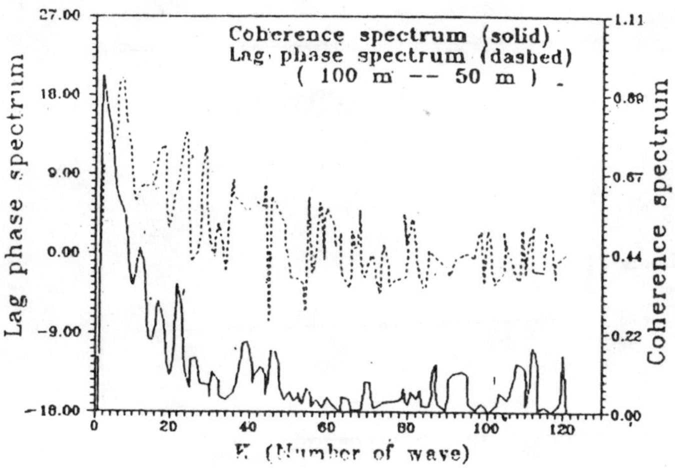

Coherence spectrum (solid)
Lag phase spectrum (dashed)
(100 m -- 50 m)
Lag phase spectrum
Coherence spectrum
K (Number of wave)

Fig. 4 (b)

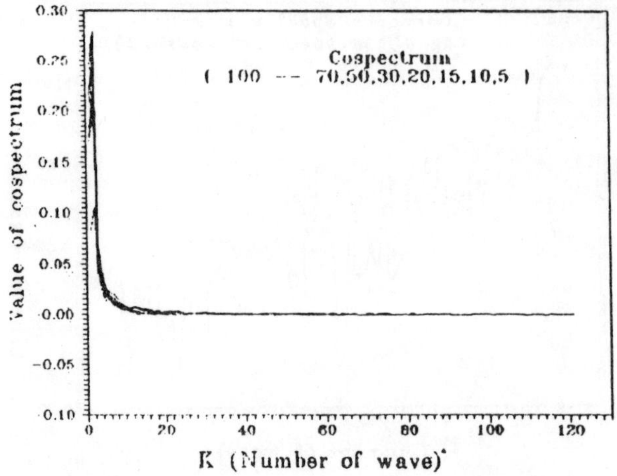

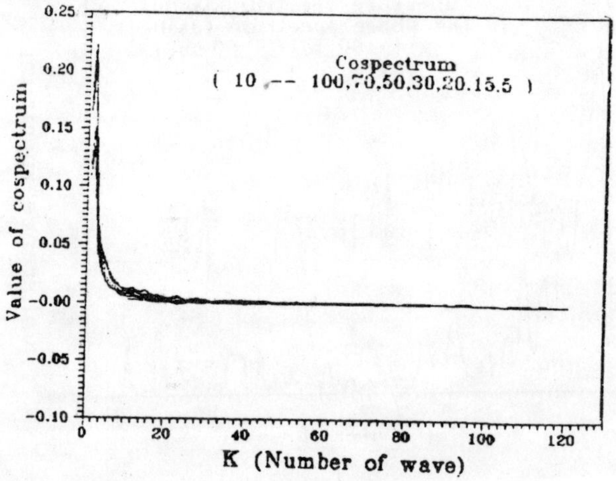

5. CONCLUSION AND DISCUSSION

5.1 The horizontal wind power spectrums of 8 levels (100,70,50,30,20,15,10,5 m) at Badaling area of Beijing are given and the principal periods are seen in Tab 2.

Tab 2 Principal periods of power spectrum at all levels (Second)

100 m	70 m	50 m	30 m	20 m	15 m	10 m	5 m
735	665	544	544	544	735	735	601
306	270	244	284	330	330	365	365
90	90	90	100	81	81	60	60
22	22	22	25	20	16	33	33
10	10	8	10	10	8	20	20
4	5		6	6	5	8	10
						3	5

The characteristic of horizontal power spectrum: The turbulent energy in low frequency range with period of several minutes is more obvious at high levels than at low levels. At the same height the energy in low range is much less than it in high range. And at high levels the peaks with period of about 1 minute is the strongest in all frequency range and with the height's decreasing, the energy with the period length shorter than 1 minute gradually enhances and eventually exceeds the peak of about 1 minute,especially, the peaks with the period of 16 to 25 seconds and about 10 seconds are the strongest.

5.2 The mean figure of reduced power spectrum is given.This spectrum is double-peak and two peaks correspond to wavelength of 833 m and 105 m, respectively, which differs from the classic Davenport's empirical spectrum with the single-peak corresponding to the wavelength of about 555 m. We still simulate the equation of this spectrum curse with polynomial.

5.3 There is evident relationship of wind spectrum between different heights. It appears to be obvious positive correlation in phase in low frequency range. The time phase at high levels lags behind the one at low levels for about 10 to 40 seconds, and the more the height's difference, the greater the lag scope of time phase. From the coherence spectrum, the longer the distance between two levels, the smaller the coherence, otherwise,the shorter the distance, the greater the coherence. In brief, the relationship of the spectrums in low frequency range among all levels is more obvious than it in the other range.

5.4 The frequency range ($0.5 - 2.25*10^{-4}$ $second^{-1}$)analyzed is in the meso- and microscale of turbulence spectrum. Generally considing that the turbulence energy is low in this range, it is still active in this range from our answer. There are obvious periods from several minutes to more than 10 minutes at different heights in different typical weather condition. The similiar results were obtained on analyzing other elements by Lu(1979). In addition to the energy source created by horizontal nonuniformity on synoptical scale and perpendicular nonuniformity by surface boundary layer effect, the mesoscale motion with periods about 10 minutes is also a important factor and the source of the energy resulting in air turbulent motion.

Because of the quite wide range of horizontal wind spectrum, it is a difficult problem how to rationally determine obtaining average collecting time of stable horizontal wind in the conventional observation. Generally, when average time obviously exceeds the period of peak with meso- and micro-scale, we can obtain stable value. The period of microscale peak usually is more than ten seconds, so it is thought that 10 minutes average collecting time is enough.But from our results, it is known that the horizontal wind usually have the peak with period from several minutes to more than ten minutes in surface layer, so that, in this condition, It is possible that it is not enough to use 10 minutes as average collecting time.

REFERENCE

Bowne, N. E. et al (1970), "Observational comparison of rural and urban boundary layer turbulence", J.A. Met, 1970,9

Davenport, A. G.(1963),"The relationship of wind structure to wind loading", W.E.B.E.,1963

Lu,Naiping et al (1979), "Remote sensing of the turbulent characteristics in the atmospheric boundary layer by acoustic radar", ACTA METEOROLOGICA SINICA, 1979, 37 (3)

Van der Hoven(1957), "Power spectrum of horizontal wind speed in the frequency range from 0.0007 to 900 cycles per hour", J. Met., 1957, 14

NUMERICAL SIMULATION OF STATIONARY AND NONSTATIONARY STOCHASTIC PROCESSES: A COMPARATIVE ANALYSIS FOR TURBULENT WIND FIELDS

BORRI, C., CROCCHINI, F., FACCHINI, L., and SPINELLI, P.
Professor *Ph D Students* *Professor*

Department of Civil Engg. University of Florence, Florence, Italy, I-50139

Abstract

Multi-correlated and multi-dimensional stochastic processes, which simulate wind velocity or pressure on a given structure, are an important tool to predict the dynamic response of the structure itself, particularly if pronounced non-linear behavior and/or interaction phenomena arise.

Attention has thus been mainly focused on the generation of stationary processes which best fit the characteristics of natural wind under the assumption of a stationary or non stationary turbulent component.

This paper is aimed to a comparison of the performances and efficiencies of multi-regressive and parametric generation procedures.

1. MULTI-REGRESSIVE MODELS

A first way to approximately reproduce a realization of a stochastic process is to approximate it by means of a linear combination of the previous values of each component process plus a random shock which displays an appropriate correlation structure:

$$X_i(t) = \sum_{k=1}^{p}\sum_{j=1}^{M}\Phi_{ij}(k)X_j(t-kDt) + N_i(t) \qquad 1.1$$

where:

p	:	order of the filter
M	:	number of processes
X(t)	:	$[X_1(t), ... X_M(t)]^T$ M values at each time
Φ	:	$[\Phi_{ij}(1), ... \Phi_{ij}(p)]$ p matrices MxM
N(t)	:	$[N_1(t), ... N_M(t)]$ M values at each time

Equation 1.1 can be ascribed to Akaike; Iwatani (1982) later proposed a very efficient way to compute all unknown parameters arising in the expression. They are the p matrices (of order MxM) and the matrix $R^N_{ij}(0)$.

The solution algorithm takes the following steps:

1) determination of the auto regressive parameters $\Phi_{ij}(k)$. Post multiplication of eq. 1.1 by the vector $X_m(t\text{-}lDt)$ leads to:

$$X_i(t)X_m(t-lDt) = \sum_{k=1}^{p}\sum_{j=1}^{M}\Phi_{ij}(k)X_j(t-kDt)X_m(t-lDt) + N_i(t)X_m(t-lDt)$$

and integrating and averaging both sides over the time T yields:

$$\frac{1}{T}\int_0^T X_i(t)X_m(t-lDt)dt =$$

$$= \frac{1}{T}\int_0^T \sum_{k=1}^{p}\sum_{j=1}^{M}\Phi_{ij}(k)X_j(t-kDt)X_m(t-lDt)dt + \frac{1}{T}\int_0^T N_i(t)X_m(t-lDt)dt$$

moreover we have: $\int_0^T N_i(t)X_m(t-lDt)dt = 0$; when l≠0, and so:

-) for i=1,...p

$$R_{im}^X(-lDt) = \sum_{k=1}^{p}\sum_{j=1}^{M}\Phi_{ij}(k)R_{jm}^X((k-l)Dt)$$

this equation contains only the unknowns Φij(k), then this system can be firstly solved

-) and for l=0 we have:

$$R_{im}^X(0) = \sum_{k=1}^{p}\sum_{j=1}^{M}\Phi_{ij}(k)R_{jm}^X(kDt) + R_{im}^N(0)$$

and so, introducing the computed auto regressive coefficients into the previous system this can be easily solved, founding the $R^N(0)$.

2) After computing the desired correlation for the random shock processes, these can now be generated in the following manner:

$$N_i(t) = \sum_{j=1}^{i} L_{ij}\varepsilon_j(t)$$

where:

$\varepsilon_j(t)$: unrelated random series with zero mean and unit variance

L_{ij} : Cholesky factor of the matrix $R^N(0) = \frac{R^N(0)}{\sigma_\varepsilon^2} = LL^T$

2. STATIONARY PARAMETRIC MODELS

Another way to approximately reproduce a realization of a stochastic process is to approximate it by means of a projection in a proper subspace; in other words, a scalar, or vector, function of time can be expanded in a convergent series of functions, like harmonic functions, or orthogonal polynomials, and so on.

We are currently studying the various properties of each class of functions which can be chosen, and throughout this work elementary white noises were employed.

A fundamental approach to the generation of stationary processes based on the results of a sampling theorem were made by Grigoriu (19..); in his work, a simple, accurate and efficient method was developed, which had the advantages of both the discrete spectral representation and time series techniques.

Suppose that the power spectral density of the stationary stochastic process $X(t)$, $S_{XX}(f)$, vanishes outside a bounded frequency range $(-\bar{f}:\bar{f})$, $0<\bar{f}<\infty$; then, the sampling theorem the auto covariance function of $X(t)$ can be described by

$$R_{XX}(\tau)=\sum R_{XX}(t_0+kT)\frac{sin(\pi(t-t_0-kT)/T)}{\pi(t-t_0-kT)/T} \tag{2.1}$$

where $T=1/2\bar{f}$ and t_0 is any real number.

Thus, an approximation of the process was generated in the form

$$X_N(t)=\sum_{k=-N}^{N}A_k\alpha_k(t) \tag{2.2}$$

where $N=1,2,\ldots$, $\alpha_k(t)=\dfrac{sin(\pi(t-kT)/T)}{\pi(t-kT)/T}$ and A_k are properly correlated random variables.

The main results obtained with such a process are that $X_N(t)$ has the same first two moments as $X(t)$ as $N\to\infty$; then it is not stationary in the wide sense, but approaches stationarity for increasing N.

It can also be seen that

$$E[A_k A_l]=R_{XX}((k-l)T) \tag{2.3}$$

The main problem when a simulation is carried out, is that $\bar{f}$ can be large, so that T is a very short time lag, and if a long simulation is desired, it could be difficult to properly correlate the coefficients A_k altogether, and then generate as many approximations as desired; in this way the process $X_N(t)$ can be replaced by

$$Y_N(t)=\sum_{k=n_t-N}^{n_t+N+1}A_k\alpha_k(t) \tag{2.4}$$

where $n_t=[t/T]$; that is, $Y_N(t)$ is a *local* approximation of $X(t)$ and requires a smaller amount of properly correlated coefficients.

3. THE SIMULATION OF TURBULENT WIND FIELDS

If an orthonormal Cartesian reference system is chosen such that the x axis is parallel to the mean wind velocity, then power spectral density functions can be estimated in the following way:

$$\frac{fS_{UU}(f)}{u_*^2}=\frac{1+a_1f_r}{(1+a_2f_r)^{a_3}} \tag{3.1}$$

where:

f is the frequency;

u is the fluctuating velocity in one of the three reference directions;

$f_r=\dfrac{fz}{\bar{u}(z)}$ is a reduced frequency;

$\bar{u}(z)$ is the mean velocity at height z above ground;

u_* is the so-called shear velocity.

For the along-wind component of velocity, u_x, we usually take

$a_1=105$ $a_2=33$ $a_3=5/3$

When a velocity vector field has to be simulated, power cross-spectral densities are needed too; cross-correlation functions are no longer symmetric, so that a quadrature component in power spectral densities arises:

$$S_{U_1U_2}(f)=S^c_{U_1U_2}(f)+iS^q_{U_1U_2}(f) \tag{3.2}$$

where $i=\sqrt{-1}$; coherence function and phase angle are defined as

$$Coh(f)=\sqrt{\frac{\left[S^c_{U_1U_2}(f)\right]^2+\left[S^q_{U_1U_2}(f)\right]^2}{S_{U_1U_1}(f)S_{U_2U_2}(f)}}\quad ;\quad \varphi(f)=tan^{-1}\left(-\frac{S^q_{U_1U_2}(f)}{S^c_{U_1U_2}(f)}\right)$$

so that

$$S_{U_1U_2}(f)=\sqrt{S_{U_1U_1}(f)S_{U_2U_2}(f)}\,Coh(f)exp(i\varphi(f)) \tag{3.3}$$

When a wind velocity field is desired, whose points are placed on a vertical line (for example, when slender line-like structures are examined), the following expressions for coherence and phase angle can be employed:

$$Coh(f)=exp\left(-2\frac{fC_z|z_1-z_2|}{\bar{u}(z_1)+\bar{u}(z_2)}\right)\quad ;\quad \varphi(f)=-2\frac{fC_\varphi(z_1-z_2)}{\bar{u}(z_1)+\bar{u}(z_2)} \tag{3.4}$$

For more complex velocity fields, the expression of phase angle is not known yet, but the hypothesis is introduced that it usually approaches zero, so that the power cross-spectral density function reduces to

$$S_{U_1U_2}(f)=\sqrt{S_{U_1U_1}(f)S_{U_2U_2}(f)}\,Coh(f) \tag{3.5}$$

where

$$Coh(f)=exp\left(-2f\frac{\sqrt{C_x^2(x_1-x_2)^2+C_y^2(y_1-y_2)^2+C_z^2(z_1-z_2)^2}}{\bar{u}(z_1)+\bar{u}(z_2)}\right) \tag{3.6}$$

of which (3.4) is a special case.

4. THE COMPARISON BETWEEN STATIONARY PARAMETRIC AND MULTI-REGRESSIVE MODELS

The stationary parametric model was compared to multi-regressive models of various orders, specifically of 2nd, 4th and 6th order for the simulation of wind velocity histories.

The comparison was carried out on the simulation of wind speed in four points placed in the vertexes of a hexahedròn whose sides were 20 m. long, placed parallel to the reference axes, so that the coordinates of the four points were (in meters):

$$P_1=\begin{pmatrix}0\\0\\20\end{pmatrix}\ ;\ P_2=\begin{pmatrix}20\\0\\20\end{pmatrix}\ ;\ P_3=\begin{pmatrix}0\\20\\20\end{pmatrix}\ ;\ P_4=\begin{pmatrix}0\\0\\40\end{pmatrix}$$

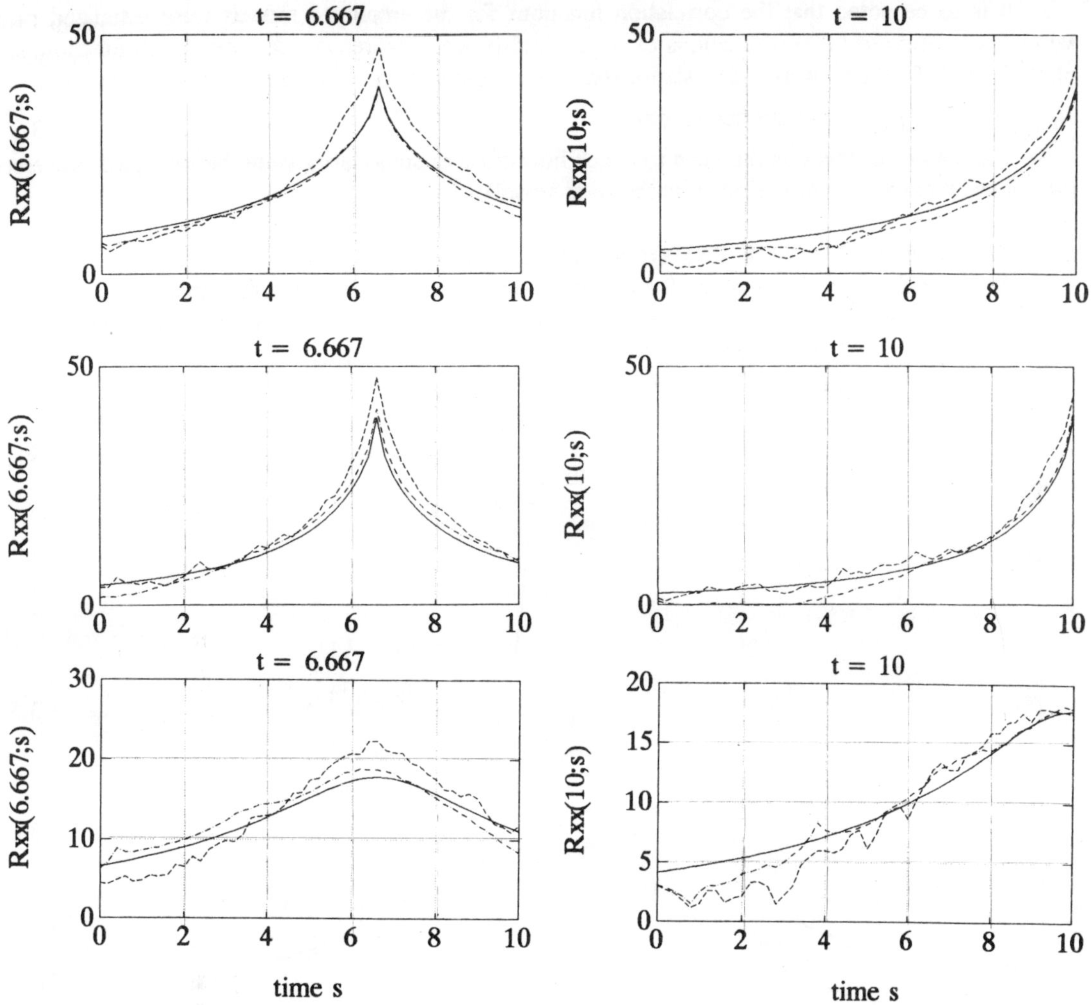

Fig. 1: In these plots, the target correlation function is marked by a continuous line, while the dashed line is the estimated correlation function for the multi-regressive model, and the dash-dot line is the estimated correlation function for the parametric model; the first two plots are relative to $R_{U_2U_2}(t;s)$, the second are relative to $R_{U_4U_4}(t;s)$ and the last ones to $R_{U_2U_4}(t;s)$.

The x axis was parallel to mean wind speed; the mean wind speed profile was logarithmic and the reference was $\bar{u}(10) = 30\ \mathrm{ms}^{-1}$; the roughness coefficient was taken $k = 0.25$ and finally $C_x = 6$, $C_y = 16$, $C_z = 10$.

The power spectral density functions were integrated to obtain the correlation functions between the wind velocities at different sites, and that was the greatest deal of calculation that had to be carried out; the remaining was straightforward enough, as both methods employ memory and time saving algorithms.

It is to be noted that the correlation functions for the employed models were estimated on a sample population of 1000 realizations of the desired process; moreover, the correlation functions are plotted as if they were non-stationary, i.e. comparisons of $R_{U_1U_2}(t,s)$, and not of $R_{U_1U_2}(\tau) = R_{U_1U_2}(t-s)$ have been plotted.

The reason for this was the need to check how close to stationarity were the simulated processes (which, we remind, are not stationary in the wide sense!).

5. THE SIMULATION OF THE QUADRATURE COMPONENT

As we said earlier, expressions for the phase angle are available only for those special cases in which a simulation of the wind speed is desired on a vertical line, these cases are nevertheless quite important, as they arise in the study of the dynamic behavior of line-like structures, such as skyscrapers, broadcasting towers, and so on.

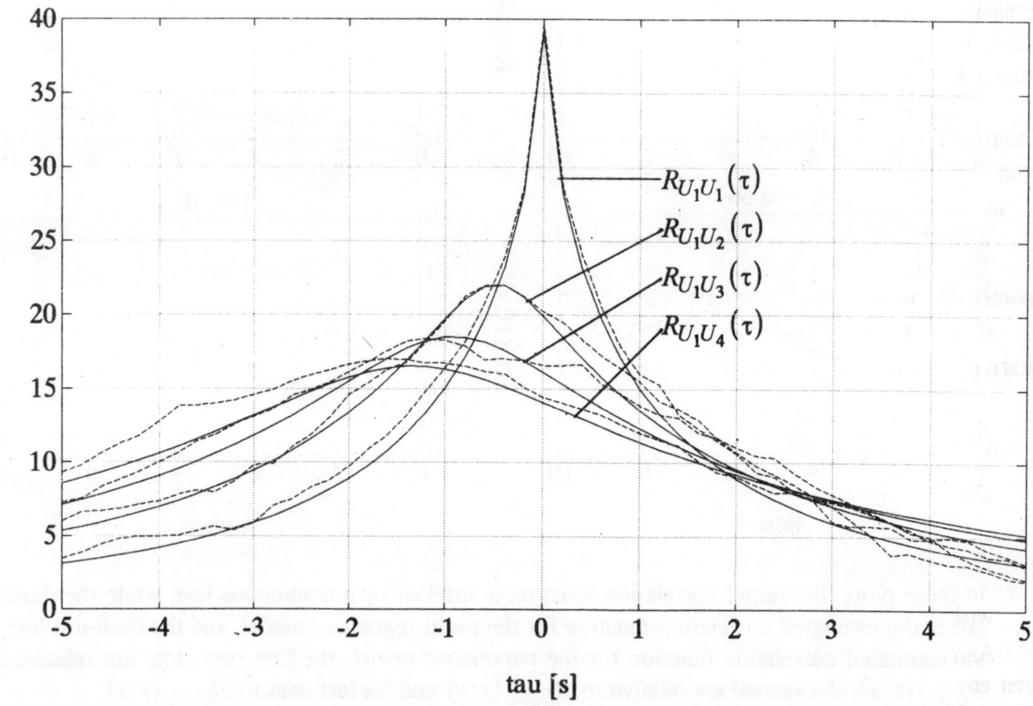

Fig. 2: Plots of correlation function of wind speed at different points; the solid line is the target correlation; the dashed lines are the estimated correlation functions.

Upon substituting (3.4) into (3.3) one obtains

$$S_{U_1U_2}(f) = \sqrt{S_{U_1U_1}(f)S_{U_2U_2}(f)}\,exp\left(-f\frac{C_z|\Delta z|}{\bar{u}_m}\right)exp\left(-if\frac{C_\varphi \Delta z}{\bar{u}_m}\right) \quad 5.1$$

where $\Delta z = z_1 - z_2$ and $\bar{u}_m = \frac{1}{2}(\bar{u}(z_1) + \bar{u}(z_2))$.

This one can be taken as a general expression for the power spectral density, as, if $P_1 \equiv P_2$, then $\Delta z = z_1 - z_2 = 0$ and the whole expression reduces to the power spectral density for the wind speed at P_1.

Here, $C_z = C_\varphi = 10$.

Four points were taken into consideration again, but now they were arranged on a vertical line, at increasing heights (in meters):

$$P_1 = \begin{pmatrix} 0 \\ 0 \\ 10 \end{pmatrix} \; ; \; P_2 = \begin{pmatrix} 0 \\ 0 \\ 20 \end{pmatrix} \; ; \; P_3 = \begin{pmatrix} 0 \\ 0 \\ 30 \end{pmatrix} \; ; \; P_4 = \begin{pmatrix} 0 \\ 0 \\ 40 \end{pmatrix} \qquad 5.2$$

Again, the results were compared to those obtained neglecting the quadrature component of the spectral density.

In particular, a time shift of the peak correlation arises, due to the presence of the phase angle, and consequently of the quadrature component: it is evident from the plots that the time shift grows for increasing height difference of the two points taken into consideration, and it can attain considerable values, while the peak value does not change appreciably; this situation was noted by other Authors also e.g. Peil and Nölle (1990) in a full scale experiment.

This is the only difference between the two approaches, because in the auto correlation functions of the wind speed in the same point, the height difference vanishes, and thus vanishes the quadrature component as well.

6. THE NON STATIONARY PARAMETRIC MODEL

This model can well be employed to simulate general non stationary models, as the sampling theorem can be extended in two-dimensional spaces.

If the double Fourier transform of a given correlation function of two instants vanishes out of a square $Q \equiv \left[-\bar{f};\bar{f}\right] \times \left[-\bar{f};\bar{f}\right]$, then this function can be expanded as

$$R_{UU}(t;s) = \sum_{k,l=-\infty}^{\infty} R_{UU}(kT;lT)\alpha_k(t)\alpha_l(s) \qquad 6.1$$

with the same meaning of the symbols employed; as before, the simulation has the form

$$X_N(t) = \sum_{k=-N}^{N} A_k \alpha_k(t) \qquad 6.2$$

with $E\left[A_k A_l\right] = R_{UU}(kT;lT)$. 6.3

As an example, we tried to simulate a stationary wind speed, at different points, modulated by a given envelope of time; this result can be obtained by simply generating a stationary noise and then multiplying it by the desired envelope.

In the present work, however, we wanted to generate a non stationary vector field, starting from its correlation function; in other words, the non stationary correlation function was estimated for the wind speed, and the non stationary signal was directly generated from it.

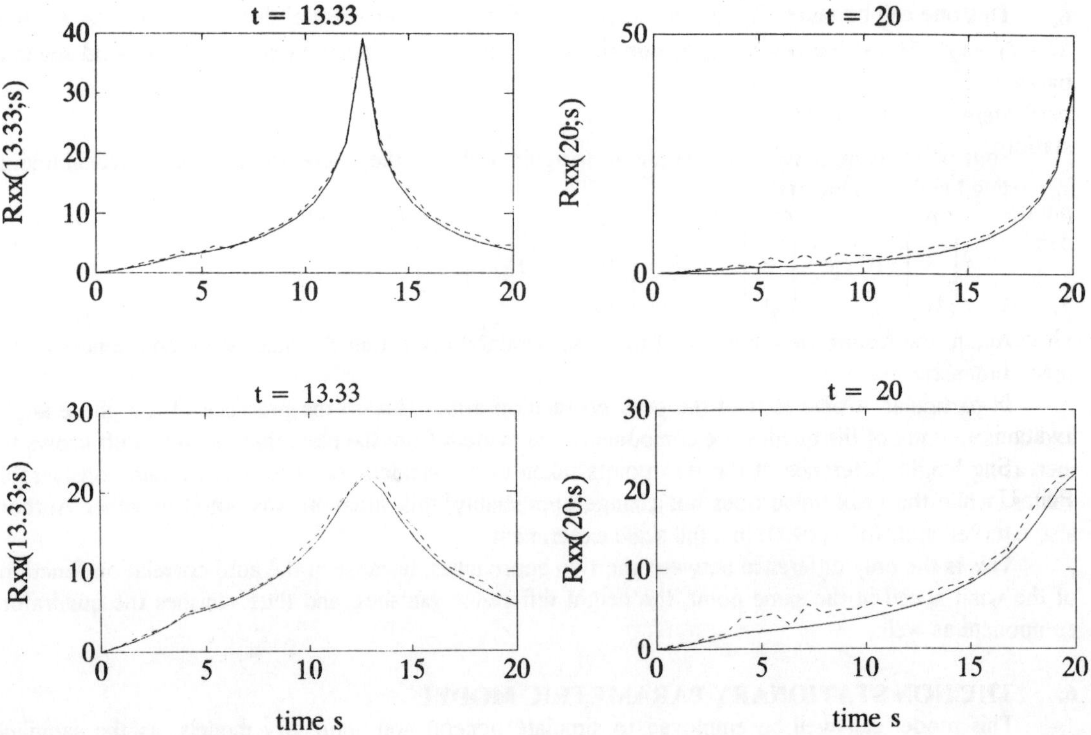

Fig. 3: The non stationary correlation function for a modulated stationary noise is plotted here: the solid line indicates again the target correlation, while the dashed one indicates the evaluated correlation function; the first two plots are relative to the autocorrelation $R_{U_1U_1}(t;s)$, while the second ones are relative to the cross-correlation $R_{U_1U_4}(t;s)$.

If we denote with $\chi_i(t)$ the time envelope of the fluctuating part of the wind speed at point P_i, then the non stationary correlation function can be put in the form:

$$R_{U_iU_j}(t;s) = \chi_i(t) R^{(s)}_{U_iU_j}(t-s)\chi_j(s) \qquad 6.2$$

As before, four cross-correlated histories were simulated with a time envelope given by:

$$\chi(t) = \begin{cases} 0 & \text{if } t \le 0 \\ t/t_0 & \text{if } 0 < t \le t_0 \\ 1 & \text{if } t_0 < t < T_{max} \end{cases} \qquad 6.3$$

which was common for each point; the points were the same as before, as every other parameter used in the simulation.

6. CONCLUSIONS

Two different simulation procedures are presented here; if a non stationary noise is desired, the parametric model proves more useful and flexible than the multi-regressive one, giving good performance, especially in the time intervals where the desired process shows pronounced non stationary characteristics.

If a stationary process is desired, it is our opinion that the multi-regressive model is faster than the parametric one, but gives a bit less accurate results; if a good approximation isn't needed, this method can be employed quite effectively.

7. REFERENCES

Grigoriu, M. and Balopoulou S. (1992), " A Simulation Method for Stationary Gaussian Random Functions Based on the Sampling Theorem" Technical Report NCEER-92-0015, 11 June 1992, National Center for Earthquake Engineering Research, State Univ. of New York at Buffalo.

Iwatani, Y., (1982), "Simulation of Multi-dimensional Wind Fluctuations Having any Arbitrary Power Spectra and Cross-spectra", Journal of Wind Engineering, vol. 11, 1982, pp. 5-18.

Peil, U., & Nölle, H.,(1990), "Measurement of Wind Load and Response of a Guyed Mast", Proc. of the European Conference on Structural Dynamics, Eurodyn 90, 5-7 June 1990, Bochum, Germany, vol. 2, pp. 1161-1198.

AN EMPIRICAL AUTOREGRESSIVE MODEL FOR WIND DATA

SMITH, D.A.[1] and MEHTA, K.C.[2]

[1] *Research Associate Dept. of Civil Engineering, Texas Tech University, Lubbock, Texas, 79409 USA*
[2] *P.W. Horn Professor of Civil Engineering, Texas Tech University, Lubbock, Texas, 79409 USA*

Abstract

AutoRegressive Moving Average (ARMA) models were previously used to investigate the wind speed and u-component speed processes measured at the Wind Engineering Research Field Laboratory at Texas Tech University. AR(3) models were found to be appropriate for this data. The AR(3) model parameters are highly linearly correlated, thus enumerating one of the parameters, for example ϕ_1, specifies the entire AR(3) model. In this paper, the model parameter, ϕ_1, is correlated with physical parameters using regression techniques. Empirical models are proposed to specify the AR(3) model parameters. Results of time domain simulations using AR(3) model parameters computed from these models are presented and compared to field data and theoretical models.

1. INTRODUCTION

Understanding and describing the near ground wind field in which a structure is submersed is the first step in predicting wind-induced pressures on low-rise buildings. The Texas Tech University Wind Engineering Research Field Laboratory (WERFL) provides a unique opportunity to study both the wind field and the associated wind induced pressures on a low-rise building in the field. The facility is located in flat, open terrain on the campus of the university in Lubbock, Texas and is described in detail by Levitan et al. (1992). The objective of this work is to provide an empirical univariate stochastic description of the surface layer wind field at the WERFL site. A description of surface layer winds is given by Cermak et al. (1993).

The data used for this investigation is the wind speed and the longitudinal component speed (u-component speed) collected on the 49 meter high meteorological tower. Wind speed data is collected at a rate of 10 Hz for a continuous 15-minute period. One hundred and eleven 15-minute time series from twelve Mode 15 records are used to develop this empirical univariate stochastic description of the wind field. Digitized observations of the wind data can be considered as a sequence of random variables; thus time series analysis techniques termed AutoRegressive Moving Average (ARMA) modelling are considered appropriate (Brockwell et al., 1987; Box et al., 1976).

The ARMA model parameters for the WERFL data are highly linearly correlated (Smith et al., 1993a, 1993b; Smith, 1993). This correlation implies that the wind process at WERFL is a constant process. That is, even though two time histories may have very

different characteristics as indicated by summary statistics, there is not a significant difference between the underlying turbulence processes for the two time histories. Furthermore, it implies that if one ARMA model parameter can be described in terms of physical
parameters, then an empirical time domain description of the wind data can be developed. Regression methods are used to establish the relationship between the ARMA model parameters and physical parameters.

This empirical model to specify the AR(3) model parameters is used with Monte Carlo simulation to generate wind speed and u-component speed time histories. Three simulations consisting of 10 repetitions each are obtained and compared to field data and theoretical models. These simulated time histories have uses in nonlinear dynamics problems where a time domain solution is required and in computational fluid dynamics modelling for input boundary flow conditions.

2. ARMA MODELLING

The general ARMA model is given by the expression:

$$\phi(B)X_t=\theta(B)Z_t \tag{1}$$

where $\phi(B)$ and $\theta(B)$ are the characteristic polynomials evaluated at B for the autoregressive (AR) and the moving average (MA) part of the model, respectively. X_t is the time series observation at time t, Z_t is the random variable at time t (Z_t~Empirical$(0,\sigma^2)$), and B is the backshift operator.

The autoregressive characteristic polynomial evaluated at B is given by:

$$\phi(B)=1-\phi_1B^1-\phi_2B^2-...-\phi_pB^p \tag{2}$$

where p is the order of the autoregressive part of the model. An autoregressive model of order p is termed an AR(p) model.

The moving average characteristic polynomial evaluated at B is given by:

$$\theta(B)=1+\theta_1B^1+\theta_2B^2+...+\theta_qB^q \tag{3}$$

where q is the order of the moving average part of the model. A moving average model of order q is termed an MA(q) model.

The backshift operator, B, operates on the time index set of the random variables X_t and Z_t. The backshift operation performed on the random variable X_t is given by:

$$BX_t=X_{t-1} \quad and \quad B^2X_t=X_{t-2} \quad etc. \tag{4}$$

Performing the backshift operations indicated by the characteristic polynomials, the general ARMA model is given by the expression:

$$X_t-\phi_1X_{t-1}-\phi_2X_{t-2}-...-\phi_pX_{t-p}=Z_t+\theta_1Z_{t-1}+\theta_2Z_{t-2}+...+\theta_qZ_{t-q} \, . \tag{5}$$

When a model consists of an autoregressive component of order p and moving average component of order q then the model is termed an ARMA(p,q) model.

Differencing the data provides a method of eliminating stochastic nonstationarity from the data. The dth difference of a data set is defined as:

$$(1-B)^dX_t \tag{6}$$

where d is the degree of differencing. The first difference of a data set, d=1, is given by:

$$(1-B)^dX_t=(1-B)^1X_t=X_t-X_{t-1} \quad . \tag{7}$$

ARMA models where the data is differenced to eliminate stochastic nonstationarity are referred to as ARIMA(p,d,q) models. After differencing, the modelling process is as described below for undifferenced data.

ARMA modelling is an iterative three-step process. The first step in the modelling process is to determine the appropriate model type and model order for the time series. The autocorrelation function (ACF) and the partial autocorrelation function (PACF) are used to estimate the model type and model order. The ACF is prescriptive of the moving average component of the model and the PACF is prescriptive of the autoregressive component of the model.

Once the model type and order have been estimated, the second step is the estimation of the model parameters. Parameter estimation is accomplished using the Yule-Walker equations, maximum likelihood methods, or least squares techniques (conditional and unconditional). Unconditional least square techniques are used for parameter estimation in this study since the parent and residual distributions are nongaussian.

The final step is to establish the adequacy of the ARMA model. For a model to be adequate, the residuals of the fitting process must be independent and identically distributed. Model diagnostics can be performed in the time domain or the frequency domain. In the time domain, the Portmanteau test is used (Brockwell et al., 1987). For model diagnostics performed in the frequency domain, the residuals are transformed from the time domain into the frequency domain and the Kolmogorov-Smirnov test and/or the Fisher's Kappa test used to check the assumption that the residuals are independent and identically distributed (Fuller, 1976). For this work, the model diagnostics are performed in the frequency domain using the Kolmogorov-Smirnov test and the Fisher's Kappa test.

If the selected model is not adequate, then a competing model is selected for trial, the parameters are estimated, and model diagnostics performed again. This three step process is iterated until an adequate model is achieved. Evên after an adequate model is achieved, investigations of competing models is performed until the optimum model is obtained.

3. ARMA MODELLING OF WERFL DATA

Previous investigations of 111 wind speed and u-component speed time series collected at WERFL indicated an AR(3) model is appropriate for this data (Smith et al., 1993b; Smith 1993). An AR(3) model is given by the expression:

$$X_t - \Phi_1 X_{t-1} - \Phi_2 X_{t-2} - \Phi_3 X_{t-3} = Z_t \quad . \tag{8}$$

The AR(3) model parameters for the u-component speed at the 10 m height are given in Smith et al. (1993b) and for all the time series in Smith (1993). A typical u-component speed time history from record M15N342 and its associated ACF and PACF are shown in Figures 1 through 3, respectively.

The previous investigations used 10 records, 15 minute duration each, from three different windstorms. Summary statistics for the wind speed at 3.7 m above ground are given in Smith et al. (1993a). From the 10 records chosen for analysis, 111 wind speed and u-component speed time histories out of 120 were available for analysis. Nine time histories were unavailable due to instrument malfunctions. Of these 111 time histories, 66 were wind speed time histories and 45 were u-component speed time histories. Both stationary and nonstationary records were included in the analysis. Stationarity as used

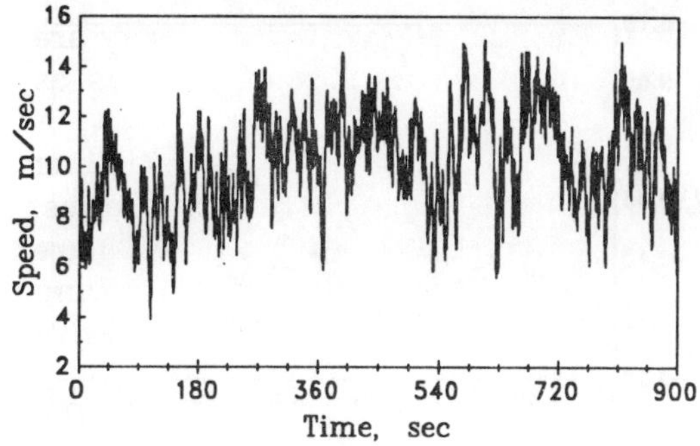

Figure 1. u-component speed at 10 m.

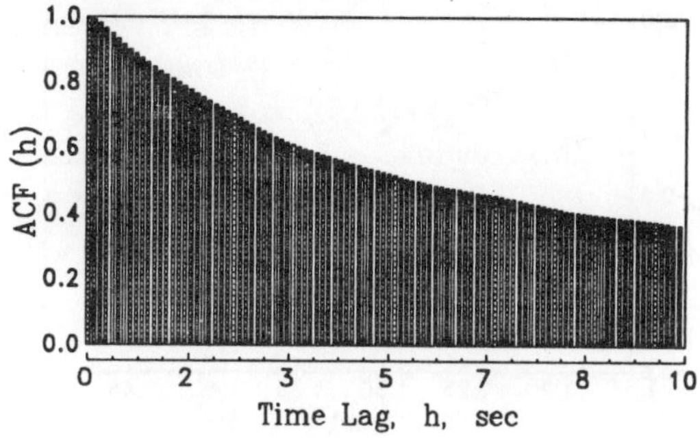

Figure 2. ACF for u-component speed at 10 m.

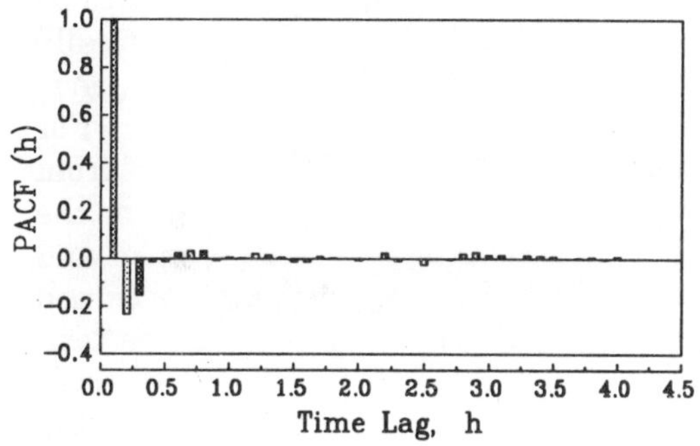

Figure 3. PACF for u-component speed at 10 m.

here refers to the classification of records using nonparametric run and reverse arrangement tests (Bendat et al., 1986). The AR(3) model, given above, was found to be appropriate for both nonparametrically classified stationary and nonstationary WERFL wind data. Furthermore, the AR(3) models were all stationary based on the AR(3) model criteria (Smith, 1993). The residual distributions from the modelling process for both stationary and nonstationary records are nongaussian (Smith et al., 1993b; Smith, 1993).

The AR(3) model parameters, ϕ_i's, were found to be highly linearly correlated (Smith et al., 1993a; Smith, 1993). The linear relationship is graphically shown in Figures 4 through 6 for uvw instruments. This linear relationship was used to study differences in nonparametrically classified stationary and nonstationary wind data (Smith et al., 1993a).

4. RELATIONSHIP OF MODEL AND PHYSICAL PARAMETERS

The linear correlation of the model parameters shown in Figures 4 through 6 implies that a single model parameter can be used to describe the process, i.e., once one

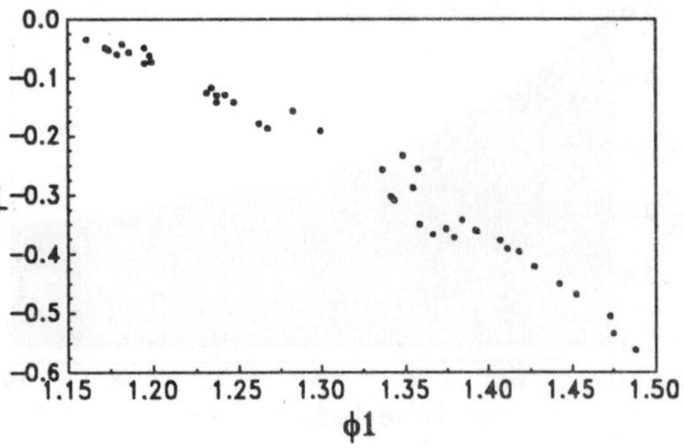

Figure 4. Plot of ϕ_2 versus ϕ_1 for u-component speed.

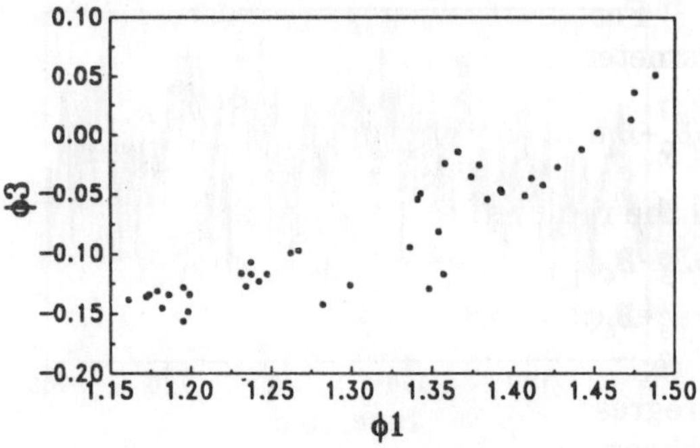

Figure 5. Plot of ϕ_3 versus ϕ_1 for u-component speed.

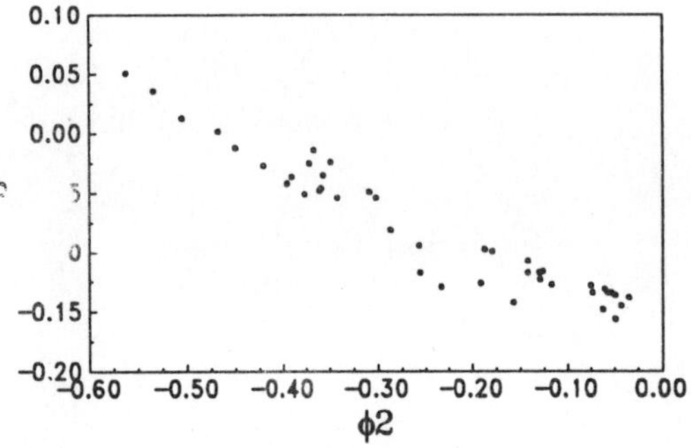

Figure 6. Plot of ϕ_3 versus ϕ_2 for u-component speed.

model parameter is established the other two model parameters can be computed using linear relationships. It also implies that if one of the AR(3) model parameters can be expressed in terms of physical parameters then a general time domain representation of wind turbulence is possible (Smith, 1993).

Investigation of the relationship between physical and model parameters is accomplished using linear regression techniques. Nondimensional physical parameters such as the Reynolds number, turbulence intensity, and height above ground divided by roughness length, z/z_0, as well as dimensional parameters such as shear velocity, u_*, and height above ground, z, were used for independent variables to describe the AR(3) model parameters (dependent variables). Stepwise regression was used for the analysis to insure that all the independent variables in the final model were significant (Draper et al., 1981). The residuals from the regression analysis were typically normally distributed.

For u-component speed this investigation shows that ϕ_1 is related to physical parameters in the form:

$$\phi_1=B_0+B_1u_*\ln\left(\frac{z}{z_0}\right)+B_2z\ ; \tag{9}$$

and the remaining two model parameters are computed as:

$$\phi_2=B_3+B_4\phi_1\ ;\ and, \tag{10}$$

$$\phi_3=B_5+B_6\phi_1+B_7\phi_2\ ; \tag{11}$$

where:
B_i=regression coefficients;
u_*=shear velocity;
z= height above ground; and,
z_0=roughness length.
The variance of the residual distribution, σ^2 (termed the white noise variance, WNV) is established using:

$$\sigma^2=B_8+B_9u_*\ . \tag{12}$$

The regression coefficients, B_i's, for the wind speed and u-component speed models (measured using uvw anemometers at the 3.7 m and 10 m heights) are given in Table 1. It is interesting to note that the regression coefficients, and thus the AR(3) model parameters, are nearly identical for the wind speed and the u-component wind speed.

Table 1. Regression Coefficients, B_i, and Multiple Correlation Coefficients, R^2, for Equations 9 Through 12.

Model	Constant	Time Series Type	
		Wind Speed	u-Component Speed
ϕ_1	B_0	1.030	1.010
	B_1	0.121	0.121
	B_2	-0.036	-0.033
	R^2	0.966	0.961
ϕ_2	B_3	1.784	1.820
	B_4	-1.549	-1.572
	R^2	0.978	0.974
ϕ_3	B_5	1.026	1.278
	B_6	-1.030	-1.247
	B_7	-0.996	-1.141
	R^2	0.998	0.967
σ^2	B_8	-0.071	-0.068
	B_9	0.202	0.195
	R^2	0.921	0.927

This is the case for smooth open terrain where there is not a large difference in the turbulence intensities of wind speed and u-component wind speed. It is anticipated that as the terrain becomes rougher, the difference between the AR(3) model parameters for the two processes will become larger.

The multiple correlation coefficient, R^2, values for the empirical models are also shown in Table 1. A value of R^2 close to 1 or -1 indicates a good fit of the model to the data. The R^2 values for the empirical models, which exceed 0.9, indicate that Eqs. 9 through 12 provide a good description for the AR(3) model parameters. The regression coefficients and multiple correlation coefficients for the cup-vane instruments at other heights are given in Smith (1993).

Ideally, the model parameter, ϕ_1, should be expressed in terms of nondimensional quantities. A number of nondimensional numbers were used as independent variables during the regression analyses. Unfortunately, these models resulted in unacceptably low R^2 values. The search for a nondimensional model for ϕ_1 based on the principles of fluid mechanics is continuing.

5. MONTE CARLO SIMULATION OF U-COMPONENT SPEED RECORDS

The AR(3) model parameters computed using Eqs. 9 through 12 with their associated nongaussian residual distributions can be used to simulate wind speed or u-component speed time histories using Monte Carlo methods. Three simulations of 10 repetitions each were performed. The three simulations represented u_* and z_0 values which are on the lower bound, middle, and upper bound of the data used to develop the empirical models. The u_* and z_0 values used for the simulations are shown in Table 2.

The physical parameters given in Table 2 were used to compute the AR(3) model parameters using Eqs. 9 through 12. These values are given in Table 3 along with the model parameters obtained from field data that have identical u_* and z_0 values. Monte Carlo simulation was performed using the residual distributions from the u-component speed at 10 m from records M15N080, M15N342, M15N438. Using these residual distributions results in a slight mismatch between the variance of the residual distribution (WNV or σ^2) used for the simulations and the one computed using Eq. 12. It is not anticipated that the slight difference in the WNV significantly affects the simulation results presented below. These residual distributions used for the simulations are nongaussian. Simulations 1 and 2 have identical ϕ_i's but have different variances of the residual distributions (WNV).

Table 2. Shear Velocities, u_*, and Roughness Lengths, z_0, Used for Simulations.

Parameter	Simulation		
	1	2	3
u_*, m/sec	0.558	0.624	0.977
z_0, cm	0.549	1.341	2.286

Table 3. AR(3) Models Obtained from Field Data and from Equations 9 Through 12.

Parameter	Simulation 1		Simulation 2		Simulation 3	
	M15N080	Model	M15N342	Model	M15N438	Model
ϕ_1	1.172	1.187	1.195	1.187	1.347	1.401
ϕ_2	-0.049	-0.046	-0.049	-0.046	-0.357	-0.383
ϕ_3	-0.136	-0.150	-0.156	-0.150	-0.035	-0.033
WNV	0.051	0.040	0.053	0.056	0.128	0.123
I%	17		19		17	

The simulations produced both stationary and nonstationary wind time histories. An example of a stationary and nonstationary u-component speed time history from simulation 2 at the 10 m level is shown in Figures 7 and 8, respectively.

Summary statistics for the turbulence intensities, I, and longitudinal integral scale of turbulence, L_U^X for the three simulations are given in Table 4 along with field measured quantities. For the simulations, L_U^X is a decreasing function of increasing z_0 and u_*. This is consistent with results cited by Simiu et al. (1986). As expected, the turbulence intensity for the simulations shows an increasing value with increasing z_0 and u_*, except for simulation 3. This apparent inconsistency is a result of the residual distribution used for the simulation. The residual distribution used for simulation 3 has a high variance as compared with the residual distributions used for simulations 1 and 2. This implies that turbulence intensity, I, is a decreasing function of the residual distribution's variance, σ^2.

The power spectral density for the ensemble of the 10 repetitions for simulations 1 through 3 is shown in Figures 9 through 11, respectively. The results show good agreement with the theoretical power spectral density function proposed by Tieleman (1991) for flat, smooth, uniform terrain. In general, the simulation spectral density

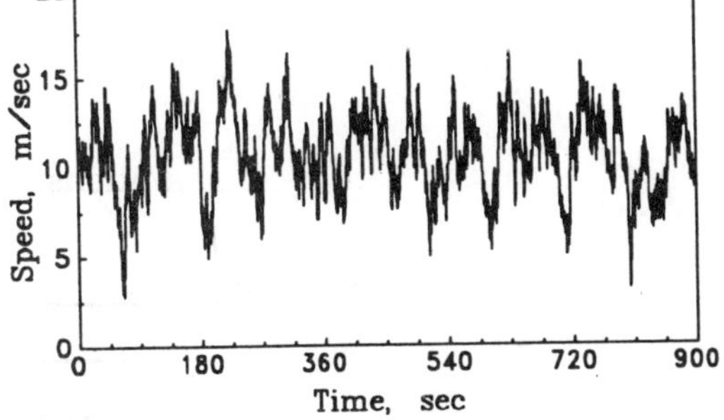

Figure 7. Stationary time history.

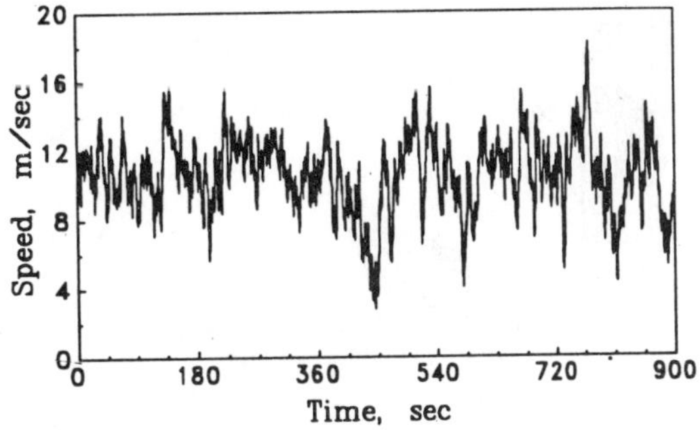

Figure 8. Nonstationary time history.

Table 4. Turbulence Intensities and Integral Scales for Simulations and Field Data.

Parameter	Simulation			Field Data		
	1	2	3	Mode A	Mode 15[1]	Mode 15[2]
	Turbulence Intensity, I%					
Mean	19.7	20.4	16.8	17	18.3	20.0
Min	18.1	17.9	15.4	12	12.7	17.0
Max	21.2	22.97	18.4	18	29.5	26.0
	Longitudinal Integral Scale, L_U^X, m					
Mean	92	78	52	145	107	122
Min	56	54	35	55	53	63
Max	189	107	61	256	255	291

1. I is based on 411 records, L_U^X based on 14 records.
2. I is based on 12 records selected for analysis, L_U^X computed using exponential fit (Teunissen, 1980).

functions match well in the low and middle frequency range of the spectrum. In the high frequency range, the deviation from the theoretical spectrum is due to the frequency response characteristics of the instrumentation. Spectra from Mode 15 u-component speed data are similar to the simulation results.

6. CONCLUSIONS

Autoregressive models of order 3, AR(3) models, are appropriate to model the wind speed and u-component (longitudinal) speed processes measured in the surface layer at

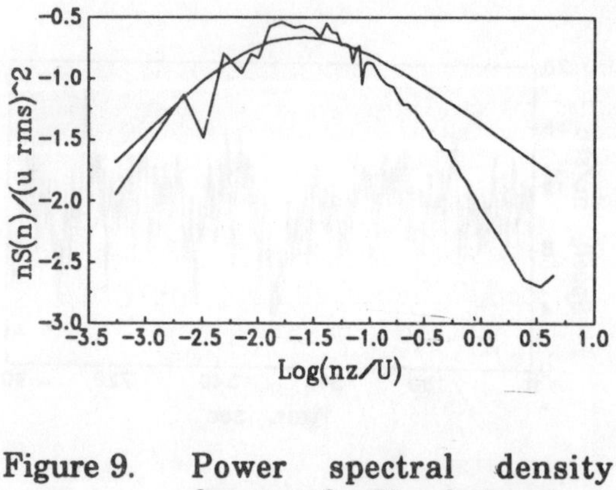

Figure 9. Power spectral density function for Simulation 1.

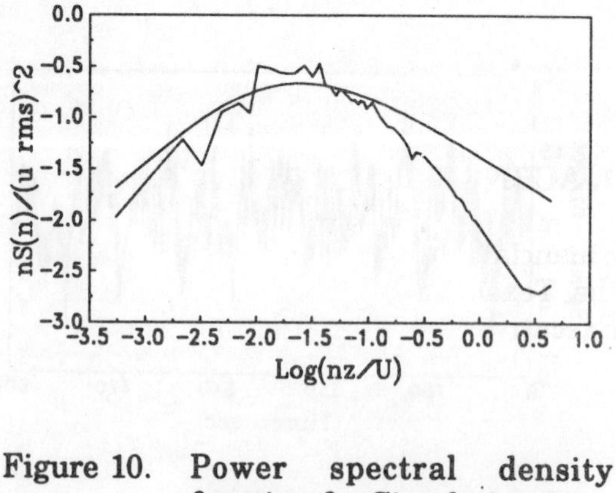

Figure 10. Power spectral density function for Simulation 2.

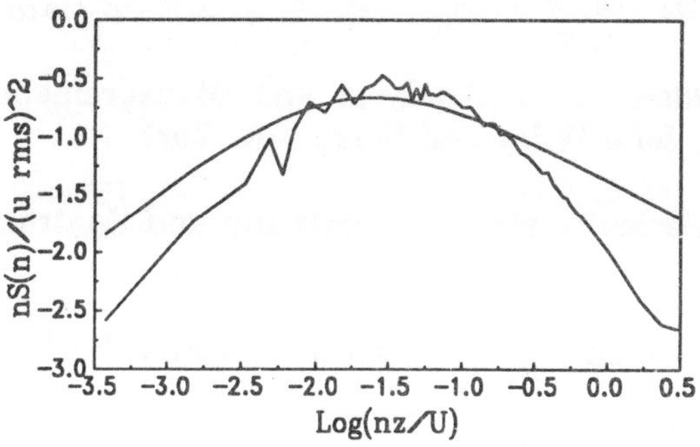

Figure 11. Power spectral density function for Simulation 3.

WERFL (height less than 49 m). These models for the wind speed and u-component speed process are adequate for both nonparametrically classified stationary and nonstationary records (Smith et al., 1992).

The AR(3) model parameters have been related to physical parameters using regression techniques. This correlation implies the resulting empirical models could be used for rougher terrains. Further work is planned to develop these relationships from the principles of fluid mechanics.

In general, the three simulations of 10 repetitions each showed the expected trends for increasing values of shear velocity and roughness length. The longitudinal integral scales of turbulence decreased with increasing u_* and z_0. Turbulence intensities increased with increasing u_* and z_0 (with the exception of 1 simulation). Turbulence intensity appears to be affected by the variance of the residual distribution (WNV). The simulations produced nonparametrically classified stationary and nonstationary time histories similar to those observed in the field.

Future work includes:

- A search for nondimensional independent variables to express ϕ_1;
- Developing a general expression in terms of physical parameters for the probability density function for the residuals; and,
- Expanding the models to a full 2-dimensional case which can be used for computational fluid dynamics input boundary conditions.

7. ACKNOWLEDGMENTS

Financial support from the National Science Foundation (Grant BCS-8821163) and from the Texas Tech University Graduate Student Stipend originated from the Office of Executive Vice President and Provost to pursue this research is acknowledged.

8. REFERENCES

Bendat, J.S., and Piersol, A.G. (1986), Random Data, Analysis and Measurement Procedures, 2nd Ed., Wiley Interscience, John Wiley and Sons, New York.

Box, G.E.P., and Jenkins, G.M. (1976), Time Series Analysis: Forecasting and Control, Rev. Ed., Holden Day, San Francisco.

Brockwell, P.J. and Davis, R.A. (1987), Time Series: Theory and Methods, Springer-Verlag, New York.

Cermak, J.E., Cochran, L.S., and Leffler, R.D. (1993), "Wind Tunnel Modelling of the Atmospheric Surface Layer," Proceedings, Third Asia-Pacific Symposium on Wind Engineering, Hong Kong, China, Vol. II, pp 873-878, December 13-15.

Draper, N.R., and Smith, H. (1981), Applied Regression Analysis, 2nd Edition, John Wiley and Sons, New York.

Fuller, W.A. (1976), Introduction to Statistical Time Series, John Wiley and Sons, New York.

Levitan, M.L., and Mehta, K.C. (1992), "Texas Tech Field Experiments for Wind Loads Part I and II," Journal of Wind Engineering and Industrial Aerodynamics, Vol. 43, pp. 1565-1588.

Simiu, E., and Scanlan, R.H.(1986), Wind Effects on Structures, 2nd Edition, Wiley Interscience, New York.

Smith, D.A., and Mehta, K.C. (1993a), "Investigation of Stationary and Nonstationary Wind Data Using Classical Box-Jenkins Models," Journal of Wind Engineering and Industrial Aerodynamics, Vol. 49, pp 319-328.

Smith, D.A., and Mehta, K.C. (1993b), "Stochastic Models for Stationary and Nonstationary Wind Data," Proceedings, 7th U.S. National Conference on Wind Engineering, Los Angeles, California, June 27-30.

Smith, D.A. (1993), Stochastic Analysis of Wind Data, a dissertation submitted in partial fulfillment for the degree of Doctor of Philosophy, Texas Tech University, May.

Teunissen, H.W. (1980), "Structure of Mean Winds and Turbulence in the Planetary Boundary Layer over Rural Terrain," Boundary Layer Meteorology, 19, pp 187-221.

Tieleman, H.W. (1991), "The Atmospheric Surface Layer and its Simulation for Wind-Load Studies on Low-Rise Structures," College of Engineering, Virginia Polytechnic Institute and State University, Blacksburg, Virginia, Publication VPI-E-91-04, March.

THEME II
Extreme Winds

9ICWE, 1995, New Delhi, India

A STUDY OF OPTIMAL EXTREME WIND ESTIMATION PROCEDURES

GROSS, J.L.[1], HECKERT, N.A.[2], LECHNER, J.A.[2] and SIMIU, E.[1]

[1] *Building and Fire Research Laboratory*
[2] *Computing and Applied Mathematics Laboratory*
National Institute of Standards and Technology, Gaithersburg, Maryland 20899, USA.

ABSTRACT

We describe work aimed at improving procedures for the estimation of non-tornadic extreme wind speeds, regardless of their direction, in regions not subjected to hurricanes. Using the Generalized Pareto Distribution (GPD) approach and the Conditional Mean Exceedance (CME) estimation method, we analyze 115 17-year to 52-year sets of largest annual speeds and sets drawn from 48 15-year to 26-year records of maximum daily wind speeds. Based on this analysis we attempt an assessment of the widely held belief that the Gumbel distribution with site-dependent location and scale parameters is a universal model of extreme wind speeds. Some of our results suggest that the reverse Weibull distribution is a more appropriate model. This would result in more reasonable estimates of wind-induced failure probabilities and wind load factors than the corresponding estimates based on the Gumbel distribution. However, our assessment is so far only tentative owing to uncertainties inherent in our results. Future work based on lower thresholds (larger data samples) and alternative estimation methods is planned.

1. INTRODUCTION

Until recently methods for the estimation of extreme wind speeds were based solely on classical extreme value theory (Gumbel, 1958). Although such methods can be used to obtain credible estimates of wind speeds with relatively short mean return periods (50 years, say), questions remain as to their capability to estimate distribution tails reliably.

In the last two decades a novel theory known as the 'peaks over threshold' approach was developed that offers the potential for more realistic estimates of the tails. This would allow the estimation by statistical methods of wind load factors, which have to date been specified in building standards on the basis of engineering guesses passed from one generation of standards to the next. This paper is part of a long-term project aimed at improving estimates of wind speed distribution tails and wind load factors. The 'peaks over threshold' approach rests on the application of the Generalized Pareto Distribution (GPD) to the excess of the extreme variates over a fixed threshold. For terminology and notations, see Gross et al (1994).

Unlike classical methods, the 'peaks over threshold' approach is applicable to the analysis of the set consisting of all data exceeding a sufficiently high threshold. In addition, it is applicable to data taken from sets of epochal extremes (i.e., maxima over samples of fixed size, such as largest annual wind speeds). According to classical theory, in the asymptotic limit a set of epochal extremes must fit the tail of one of the three extreme value distributions. The epochal extremes that exceed a sufficiently high threshold must therefore fit the GPD with $c>0$, $c\to 0$, or $c<0$.

We review briefly in Section 2 the expression for the GPD, the GPD-based estimator used in this work, and the estimation within the framework of the 'peaks over threshold' approach of variates with specified mean return periods. In Section 3 we analyze 115 sets of observed largest annual wind speeds taken from 17- to 52-year records. Section 4 is devoted to analyses of sets taken from 48 15- to 26-year records of largest daily speeds. Section 5 presents our conclusions and outlines future work.

2. GENERALIZED PARETO DISTRIBUTION, AND DESCRIPTION OF ESTIMATORS

We review here the expression for the Generalized Pareto Distribution and the Conditional Mean Exceedance (CME) method for estimating distribution parameters.

Generalized Pareto Distribution (GPD)

The expression for the GPD is

$$G(y) = \text{Prob}[Y \le y] = 1-\{[1+(cy/a)]^{-1/c}\} \quad a>0,\ (1+(cy/a))>0 \tag{1}$$

Equation 1 can be used to represent the conditional cumulative distribution of the excess $Y=V-u$ of the variate V over the threshold u, given $V>u$ for u sufficiently large; c and a are distribution parameters. The cases $c>0$, $c=0$ and $c<0$ correspond, respectively, to Fréchet, Gumbel, and reverse Weibull (right tail-limited) domains of attraction. For $c=0$ the expression between braces is understood in a limiting sense as the exponential exp(-y/a) (Castillo, 1988, p. 215). For $c<0$ the shape parameter of the corresponding distribution is $\gamma=-1/c$ (Smith, 1989).

Cumulative Mean Exceedance (CME) Method

The CME is the expectation of the amount by which a value exceeds a threshold u, conditional on that threshold being attained. If the exceedance data are fitted by the GPD model and $c<1$, $u>0$, and $a+uc>0$, then the CME vs. u plot should follow a line with intercept $a/(1-c)$ and slope $c/(1-c)$ (Davison et al, 1990). The linearity of the plot is an indicator of the appropriateness of the GPD model. Estimates of c and a can be obtained from the slope and intercept of a straight line fit to the CME vs. u plot.

Estimation of variates with specified mean return periods

The mean return period R, in years, of a given wind speed is defined as the inverse of the probability that that wind speed will be exceeded in any one year. In this section we give expressions that allow the estimation from the GPD of the value of the variate corresponding to probability $1 - 1/(\lambda R)$, where λ is the mean crossing rate of the threshold u per year (i.e., the average number of data points above the threshold u per year), and R is the mean recurrence interval in years. We have

$$\text{Prob}(Y < y_R) = 1 - 1/(\lambda R) \tag{2}$$

$$1 - [1 + cy_R/a]^{-1/c} = 1 - 1/(\lambda R) \tag{3}$$

$$y_R = -a[1 - (\lambda R)^c]/c \tag{4}$$

$$V_R = y_R + u \tag{5}$$

where V_R is the R-year wind speed (e.g., V_{50}=50-year speed) and u is the threshold used to estimate c and a. For epochal sets consisting of the largest annual wind speeds, $\lambda=1$. Note that, given u, λ, c, R and V_R, Eqs. 4 and 5 yield the parameter a inherent in the estimation of V_R.

3. ANALYSIS OF LARGEST ANNUAL WIND SPEEDS

Table A1 given in the Appendix shows estimated values of the tail length parameter, $\hat{c}$. The estimates were obtained by applying the CME method to data samples taken from 115 N-year records of observed largest annual wind speeds adjusted to a 10 m elevation above ground ($17<N<52$). Stations where strong winds are predominantly due to hurricanes were not included in Table A1. All wind speeds are given by the Weather Service in terms of fastest miles (1 mph=0.44704 m/s). For convenience these units have also been used in this paper. In order to include only the strongest winds in each set -- the winds most likely to approach the asymptotic condition inherent in the GPD approach -- we used the CME estimator based for each record on a relatively high threshold. We chose this threshold to be equal to the record's median wind speed, V_{med}. All the CME-based results of Table A1 are based on this threshold. In our calculations for observed data a wind speed, V, was defined as exceeding the threshold if $V \geq V_{med}$, that is, the actual threshold is actually smaller (by an infinitesimal amount) than the nominal threshold.

For the threshold V_{med}, the sample average number of exceedances was $E(n_{ex})=21$, and $E(V_{med})=50$, $SD(V_{med})=6.5$ (E and SD denote sample mean and standard deviation). The mean and standard deviation of the estimated values of c listed in Table A1 are:

$$E(\hat{c})=-0.26,\ SD(\hat{c})=0.38.$$

We denote by V_{med-} and V_{med+} the speed preceding V_{med} and the speed following V_{med}, respectively, in the set of ordered speeds of which V_{med} is the median. Using a threshold V_{med-}, $E(n_{ex})=24$ and a threshold V_{med+}, $E(n_{ex})=17$, results not listed in Table A1 yielded

$$E(\hat{c})=-0.24,\ SD(\hat{c})=0.34$$

$$E(\hat{c})=-0.27,\ SD(\hat{c})=0.48,$$

respectively. For lower thresholds $E(\hat{c})$ was found to increase.

The following results were obtained from Monte Carlo simulations. For 500 25-year samples with mean exceedance rate $\lambda=1$ and $\hat{c}$ estimated by the CME method,

$E(\hat{c})=-0.09$, $SD(\hat{c})=0.27$ (population with Gumbel distribution)

$E(\hat{c})=-0.33$, $SD(\hat{c})=0.24$ (population with reverse Weibull distr., $\gamma=-1/c=1/0.275$).

A comparison between the results based on the observed data on the one hand and on the simulated data on the other would suggest that a reverse Weibull distribution with shape parameter $\gamma \approx 5$ ($c=-0.2$) is a more appropriate model than the Gumbel distribution. (The Gumbel distribution can be interpreted as the limit of a family of three-parameter extreme value distributions as the shape paramcter approaches infinity -- see proof in Simiu et al, 1986)

Let us now hypothesize, nevertheless, that the Gumbel distribution is an appropriate universal model of extreme wind speeds, that is, that for every station the true tail length parameter is $c=0$. The results of the Monte Carlo simulations just shown indicate a bias of about -0.1 in the estimation of c, so let us allow for a bias as large as -0.1 in estimating c. Using a binomial distribution model (with mean $n/2=57.5$ and standard deviation $(n)^{1/2}/2=5.36$), one would expect that about half of the 115 estimated values of c would be below -0.1. Actually, 77 estimated values (significantly more than half) are below -0.1; this number is almost four standard deviations higher than the mean, and would lead to a rejection of the hypothesis that the Gumbel distribution is a universal model for the extreme speeds. However, this tentative conclusion may not be warranted. Indeed, each station may have a different true c, and the sample sizes for the various stations differ. Instead of the average $E(\hat{c})$ for the observed data, it would therefore be appropriate to consider a weighted average of c, where each weight is equal to the inverse of the variance of the estimate of c. Standard deviations of these estimates are listed in Table A1 and were obtained by the expression

$$sd(\hat{c}) = \frac{[\Sigma(n-i)]^{1/2}[\Sigma(n-i)(y_i - \text{intercept} - x_i \cdot (\text{slope}))^2]^{1/2}}{(n-3)^{1/2}(1+\text{slope})^2 \cdot \{[\Sigma(n-i)][\Sigma(n-i)x_i^2] - [\Sigma(n-i)x_i]^2\}^{1/2}} \quad (6)$$

where n is the number of data in the set, x_i are the speeds (i=1,2,..,n-1), and y_i are CME values. The weighted mean of c, based on these standard deviations, is close to -0.1, and its standard deviation is about 0.32. Note that the simplifying assumption implicit in Eq. 6 that the errors in the estimation of y_i for various i's are independent is not correct, and the standard deviations of the c estimates are actually larger than those given by Eq. 6 by factors that pilot Monte Carlo simulations suggested can be as high as two or even more. We conclude that owing to the small sample sizes we used, we do not get a sufficiently good estimate of the weighted average of $\hat{c}$, and the inference made earlier on the basis of the binomial distribution cannot be relied upon with confidence.

A comparison between the tabulated values of CME-based estimated V_N's and values of the maximum speeds on record, V_{max}, shows that the performance of the CME estimator of V_N is very good. We note, however, that a worse set of V_N estimates was obtained, where the CME method was applied to data samples in which identical speeds were made distinct by addition of multiples of 0.001. Though the estimates of c were not much affected by this change, this sensitivity of the CME method appears to cast some shadow upon its dependability.

The CME estimates of V_{100000} appear to be worse than those of V_N: in some cases they differ minimally from the estimates of V_N; in others they can be ridiculously large. We also show in Table A1 wind speed estimates based on the Gumbel model. These were obtained by the probability plot correlation coefficient method (PPCC). It is seen that estimates of V_N based on the Gumbel model are comparable to those based on the CME method.

Table A1 also lists CME-based estimated speeds with mean return period 100000 years, $\hat{V}_{100\,000}$, where the parameter a is based on the CME-based estimated value of V_N, as indicated in the remark following Eqs. 4 and 5, and on a specified c=-0.2.

Load Factors

Let R_u denote the mean return period of the ultimate load. If the wind load predominates (i.e., no load combination need be considered), the wind load factor is

$$\phi = (V_{Ru}/V_{50})^2 \quad (7)$$

Table A1 lists estimated values of ϕ based on Eq. 7, where V_{50} was based on the CME estimates of V_N, and V_{Ru}, corresponding to asymptotically large R_u, was based on a parameter a estimated from V_N by using Eqs. 4 and 5, and the specified parameter c=-0.2 Depending upon the site, the estimates of ϕ vary between 1.24 and 1.68. Their average is ϕ=1.42, as compared to ϕ=1.3 specified in the ASCE Standard 7-93 and earlier versions thereof.

Structural Reliability Implications

Consider, for example, the Fresno, CA data set. Under the assumption that the Gumbel distribution best fits the extremes, for $R_u=10^3$ years, 10^5 and 10^6 years, the estimated wind speeds are 59, 77 and 86 mph, respectively (Simiu et al, 1979). Under the assumption that the reverse Weibull with $\gamma=-1/c=1/0.20$ holds, they are 54, 59, and 60 mph, respectively. It is seen that the tail is considerably shorter for the reverse Weibull than for the Gumbel.

Failure probabilities for wind-sensitive structures designed in accordance with U.S. building code requirements (or safety indices reflecting those probabilities) have been estimated on the basis of the Gumbel model. Ellingwood et al (1980) found such estimates to be substantially higher than for other types of structures. Experience shows that the number of structural failures caused by non-tornadic and non-hurricane winds is vastly smaller than those estimates would indicate. One possible flaw of those probability estimates is in our opinion the fact that they are based on the Gumbel distribution which, as suggested by our results, overestimates extreme winds corresponding to long mean return periods.

The result that the upper tail of the extreme wind speed distribution is finite would invalidate the notion that probabilities of failure of a structure subjected only to wind loading, conditional on the structural strength being sufficiently large, are always larger than zero: if the structural strength corresponded to a wind speed larger than the length of the finite distribution tail, then the conditional failure probability would be zero.

4. ANALYSES OF DATA BASED ON SETS OF LARGEST DAILY WIND SPEEDS

In this section we first analyze data sets that reflect not only extreme winds occurring at various sites, but also ordinary winds. The analyses are intended to verify whether such sets can provide information on the parent population of the extremes. Next, we use a GPD-based approach to analyze sets of data that exceed relatively high thresholds.

4.1 Data Selection

From sets of largest daily wind speeds we obtained data samples that: (1) are relatively large so that sampling errors are acceptably small, and (2) have reduced mutual dependence among the data. The procedure for obtaining the data is as follows: Partition the set of daily maxima into small periods of size equal to or larger than the duration of typical storms in days. (A reasonable choice of the length of the period is eight days, but we also use sets based on four-day periods, and compare results of analyses based on the two choices.) Pick the largest value in each period. If the maxima of two adjacent periods are less than half a period apart, replace the smaller of the two maxima by the next smaller value in the respective period which is at least half a period apart from the larger maximum. A data set is thus obtained in which adjacent data are one period apart on the average and never less than half a period apart. We show below the daily maxima at Boise, Idaho in the first six eight-day periods of the year 1965. The periods are separated by vertical bars. The data selected by the procedure just described are in bold type. In the sixth period we underlined the period maximum (26), discarded and replaced by the next largest value (18) because of the proximity to the larger maximum (31) of the adjacent period.

23,32,**35**,20,26,24,24,14 | 13,**16**, 5,11, 5,12,12, 7 | 6, 6, 9, 9,11,12,25,**26** |

15,12,12, 7,15,12,**29**,10 | 7,10,15,20,20,17,24,**31** | <u>26</u>,9,16,14,**18**,16,14,12|

Our investigation attempts to ascertain whether sets of data selected by this procedure from a set of daily maxima could possibly constitute samples from the parent populations of the extremes. Even though small correlations among data might subsist, we refer to a set obtained by the selection procedure just described as an uncorrelated data set based on eight-day (four-day) intervals or, for short, an eight-day (four-day) interval set.

4.2 Analysis of Uncorrelated Data Sets

We considered 48 uncorrelated data sets based on eight-day intervals, with length N ranging from 15 to 26 years. First we analyzed separately the sets of spring, summer, fall and winter data (seasonal data analyses). Next, we analyzed the data sets unsegregated by seasons. In both cases we estimated the best-fitting distributions (i.e., distributions with the largest PPCC) from among a set of seven distributions or families of distributions (normal, double exponential, lognormal, Gumbel, Fréchet, Weibull, and reverse Weibull).

Seasonal data sets

Our goal in performing the seasonal analyses was to attempt to fit to the spring, summer, fall and winter data, respectively, cumulative distributions $P_{sp}(v)$, $P_s(v)$, $P_f(v)$ and $P_w(v)$. Given these distributions, the distribution for all the uncorrelated data is

$$P(v) = P_{sp}(v)P_s(v)P_f(v)P_w(v). \tag{8}$$

We analyzed, for each season, 48 sets based on eight-day intervals. According to our results, for the spring, fall and winter records the best fitting distribution was predominantly reverse Weibull with shape $4 \leq \gamma \leq 30$. However, 29 summer records were better fitted by Gumbel distributions than by the reverse Weibull; the reverse Weibull (for the stations where it fitted the data better than the Gumbel distribution), and the Gumbel distribution (for the other stations), yielded estimated speeds with mean return period N years, V_N, that in most cases underpredicted the maximum speed recorded during N years, $V_{max,N}$. For summer records underpredictions were 15 percent or more for 16 sets, and 8 to 15 percent for 9 sets; there were only two overpredictions, both less than 8 percent. For spring records there were 12 underpredictions by 8 to 18 percent, and only three overpredictions, all less than 5 percent; comparable results were obtained for fall and winter. The results did not depend significantly on whether eight-day interval sets or four-day interval sets were used. From these and additional analyses we concluded that: (1) inferences from seasonal data sets (obtained as was described earlier from samples of largest daily data) do not provide a dependable basis for estimating extremes, but are likely to underestimate the extreme speeds. In other words, those sets are not drawn from populations underlying the extreme winds, but from mixed populations; (2) a similar conclusion applies to the sets consisting of all largest daily data for each season; (3) for these reasons the approach embodied in Eq. 8 appears to be inapplicable if all the data of the 8-day interval sets are considered.

Some researchers have indicated that the Weibull (as opposed to reverse Weibull) distribution best fits the sets of largest daily data. However, our analysis showed that the Weibull distribution fitted the seasonal data best only for less than ten percent of the sets.

Data sets unsegregated by seasons

The analysis of 48 sets based on eight-day intervals showed that the reverse Weibull (with $4 \leq \gamma \leq 22$) was the best fitting distribution for 27 sets, and fitted the data better than the Gumbel distribution for 41 sets. For 25 sets out of these 41 sets, including 12 sets for which it was optimal, the reverse Weibull underpredicted $V_{max,N}$ by 8 to 25 percent. For the 48 sets there were only 4 overpredictions, all smaller than 5 percent. In addition, the availability of largest annual data for periods N_1 ranging from 30 to 49 years allowed us to check the predictive capability of models inferred from sets based on eight-day intervals by comparing the estimated speed with mean return period N_1, V_{N1}, to the maximum speed recorded during an N_1-year period, $V_{max,N1}$, where $30 \leq N_1 \leq 49$. The underpredictions of the N_1-year speeds were more frequent and drastic than those of the N-year speeds. We concluded that estimated distributions of data sets unsegregated by seasons are too affected by the bulk of the non-extreme data to yield satisfactory estimates of extremes. Each of our conclusions for data segregated by seasons were found to be valid for data unsegregated by seasons as well.

Numerical Experiments

The analyses reported in the preceding paragraph showed that even where other distributions best fitted the data, the reverse Weibull was in most cases very close to being the best fitting distribution, i.e., its PPCC differed only in the fourth or even fifth significant figure from the PPCC of the best fitting distribution. We therefore reanalyzed the data based on eight-day intervals by assuming that the populations for all stations have a single reverse Weibull distribution with site-dependent location and scale parameters. This was done by calculating, for each station, the PPCC's based on the assumption that the shape parameter γ is 1,2,3,...50. For samples of data based on eight-day intervals and unsegregated by seasons the mean value of the PPCC's, taken over all the stations, was largest for $\gamma=11$, and the median PPCC was largest for $\gamma=13$. This is an indication that a reverse Weibull population with $\gamma=12$ would explain the results of the analyses. To see whether this is in fact the case, 48 samples of 730 data points each (corresponding to an 18-year record length based on 8-day intervals) were generated from reverse Weibull populations with (1) $\gamma=8$, (2) $\gamma=12$, and (3) $\gamma=16$. The number of simulated sets for which the best fitting reverse Weibull distribution had shape parameters with $\gamma \leq 12$, $13 \leq \gamma \leq 20$, and $\gamma \geq 21$ are shown in Table 1. Also shown in Table 1 are the numbers of observed sets (average sample size 18 years) with $\gamma \leq 12$, $13 \leq \gamma \leq 20$, and $\gamma \geq 21$. The results of Table 1 suggest that a reverse Weibull distribution with $\gamma \approx 12$ is an appropriate model for the population of extreme winds representing data based on 8-day intervals unsegregated by seasons, except for the larger number of samples with $\gamma \geq 21$ among the observed samples than among the simulated samples. We interpret this larger number as reflecting the relatively frequent presence of outliers among the observed samples. In our opinion this interpretation reinforces the point made earlier that, because wind speed populations which include ordinary speeds in addition to extremes are mixed, samples taken from such populations are not a sound basis for inferences on extremes. It is therefore necessary to "let the tails speak for themselves." This is done by applying to the data the GPD-based 'peaks over threshold' approach.

Table 1. Numbers of Sets Best Fitted by Distributions with Various Values of γ

	$\gamma \leq 12$	$13 \leq \gamma \leq 20$	$\gamma \geq 21$
Simulated sets, $\gamma=8$	48	0	0
Simulated sets, $\gamma=12$	27	17	4
Simulated sets, $\gamma=16$	8	24	16
Observed sets	26	12	10

4.3 'Peaks over Threshold' Analyses

In carrying out 'peaks over threshold' analyses it is tempting to use a relatively low threshold in order to increase the number of data and thus reduce sampling errors. However, this introduces in the samples data that are not representative of the extremes and tend to bias the results. So that this does not happen the threshold being selected should be as high as possible, without reducing the size of the sample being analyzed to the point where the sampling errors become too large.

We selected the largest possible threshold subject to the restriction that the resulting sample size of the exceedances not be smaller than 15. Based on this selection, the average number of exceedances for our 48 sets based on 8-day intervals was $E(n_{ex})=16$, and the average threshold was $E(V_T)=45$, that is, less than the average median, $E(M)=50$, for the largest yearly speed samples analyzed in Section 3. For these thresholds we obtained $E(\hat{c})=-0.22$ and $SD(\hat{c})=0.44$. The results were virtually the same for the 48 sets based on 4-day intervals. These results would appear to lend support to the tentative conclusion of Section 3 that the extreme winds are described by a reverse Weibull distribution with shape parameter $\gamma \approx 5$, or perhaps somewhat larger, rather than by a Gumbel distribution. However, the weighted mean of the estimated c's, obtained as was shown for the results of Table A1, was close to zero. In addition, there were about as many estimated c's larger than -0.1 as there were smaller than -0.1. These results would suggest that the Gumbel distribution is appropriate. However, given the very wide confidence bands for our results, we conclude that no statement on whether the Gumbel or the reverse Weibull distribution is more appropriate can be made on the basis of this analysis.

In principle, the approach inherent in Eq. 8 may be based on 'peaks over threshold' analyses. However, given that the records at our disposal are relatively short and the number of data exceeding a sufficiently high threshold for each of the seasons was judged to be too small, no attempt to perform 'peaks over threshold' seasonal analyses was made in this work.

5. CONCLUSIONS

It is currently assumed in engineering loading models that non-hurricane and non-tornadic extreme wind speeds, regardless of their direction, are described by the Gumbel distribution (which corresponds to a shape parameter $\gamma=-1/c$ approaching infinity). The Gumbel distribution has infinite upper tail. The objective of this paper was to gain insights into the question of whether extreme wind speeds can be described by an extreme value distribution with limited upper tail, that is, by the reverse Weibull distribution.

We used in our analyses observed data, consisting of (a) sets of largest annual wind speeds, and (b) sets of largest daily wind speeds from which we extracted subsets suitable for extreme value analysis; and simulated data. Our results appear to suggest that extreme winds are better described by the reverse Weibull distribution than by the Gumbel distribution. However, given the small sample sizes used in our analyses, the superiority of one of the distributions over the other cannot be affirmed with confidence.

The tentative assumption that the extreme wind distributions are reverse Weibull, with shape parameter $\gamma \approx 5$ (GPD tail length parameter $c=-0.2$) and site-dependent location and scale parameters, yields wind load factors with an average value $\phi \approx 1.4$. This assumption, if confirmed, would invalidate earlier approaches to the estimation of the reliability of wind-sensitive structures, which depend on an infinite-tailed model of extreme wind speeds and therefore yield unrealistically high failure probabilities.

The 'peaks over threshold' analyses were based in this paper on the Cumulative Mean Exceedance (CME) approach, which appears to be extremely sensitive to whether identical values of the variate in a set are left identical or modified by the addition to each of a different number much smaller than unity. Future work aimed at verifying the tentative conclusions of this paper will therefore include analyses based on different estimation procedures, including the de Haan procedure (Dekkers et al, 1989). In addition, we plan to perform analyses based on larger data sets, andmore elaborate

Monte Carlo simulations, in which the sets of samples generated by simulation will have the same sizes as the observed data sets being analyzed, rather than having a constant size. Finally, investigations are envisaged into the possibility that the shape parameter of the extreme wind speed distributions is site-dependent. This would be a departure from current practice, in which it is assumed that extreme winds are described by an extreme value distribution with universal shape parameter (that is, by the Gumbel distribution, which corresponds to a GPD tail length parameter $c \approx 0$), and site-dependent location and scale parameters.

6. ACKNOWLEDGMENTS

We thank S. Coles, D. Walshaw and P. Bortot for valuable suggestions and comments.

7. REFERENCES

ASCE Standard A7-93 (1994), Minimum Design Loads for Buildings and Other Structures, American Society of Civil Engineers, New York.

Castillo, E. (1988), Extreme Value Theory in Engineering, Academic Press, New York.

Davison, A.C., and Smith, R.L. (1990), "Models of Exceedances Over High Thresholds," Journal of the Royal Statistical Society, Series B, Vol. 52, pp. 339-442.

Dekkers, A.L.M., Einmahl, J.H.J., and de Haan, L. (1989b), "A Moment Estimator for the Index of an Extreme-value Distribution," Annals of Statistics, Vol. 17, pp. 1833-1855.

Ellingwood, B. R. et al, (1980), Development of a probability-based load criterion for Amer. National Standard A58, NBS Special Publ. 577, Natl. Bureau of Standards, Washington, D.C.

Gross, J. L., Heckert, A.N., Lechner, J.A., and Simiu, E. (1994), "Novel Extreme Value Estimation Procedures: Application to Extreme Wind Data," Proc., International Conference on Extreme Value Theory and Its Applications, J. Galambos et al (eds.), Kluwer , Boston.

Gumbel, E. (1958), Statistics of Extremes, Columbia University Press, New York.

Pickands, J.(1975), "Statistical Inference Using Order Statistics," Annals Stat., Vol.3, 119-131.

Simiu, E., Changery, M. J. and Filliben, J.J.(1979), Extreme Wind Speeds at 129 Stations in the Contiguous United States, NBS BSS 118, Natl. Bureau of Standards, Washington, D.C.

Simiu, E., & Scanlan, R.H.(1986), Wind Effects on Structures, Wiley-Interscience, New York.

Smith, R. L. (1989), Extreme Value Theory, in Handbook of Applicable Mathematics, Supplement (W. Ledermann et al, editors), pp. 437-472, John Wiley and Sons, New York.

APPENDIX

Table A1. Results of Analyses of Sets of Largest Annual Data

Station	N	V_{max}	V_{med}	n_{ex}	$\hat{c}$	sd($\hat{c}$)	V_N Gum.	V_N CME	V_{50} CME	V_{100000} CME	V_{100000} Gum.	V_{100000} c=-.20	ϕ
(1)	(2)	(3)	(4)	(5)	(6)	(7)	(8)	(9)	(10)	(11)	(12)	(13)	(14)
1. BIRMINGHAM, AL	34	62	44	22	-0.495	0.084	64	62	65	69	104	81	1.47
2. MONTGOMERY, AL	34	77	45	17	0.442	0.022	66	65	72	1071	117	95	1.56
3. TUCSON, AZ	39	78	50	21	0.187	0.086	76	71	79	273	121	103	1.54
4. YUMA, AZ	39	65	46	20	-0.701	0.056	65	66	66	68	116	83	1.44
5. FORT SMITH, AR	31	64	45	17	-0.798	0.158	63	64	64	65	113	80	1.43
6. LITTLE ROCK, AR	39	72	44	21	-0.395	0.070	72	71	73	83	136	97	1.59
7. FRESNO, CA	37	47	34	21	-0.204	0.036	47	46	48	59	77	59	1.42
8. RED BLUFF, CA	42	67	49	25	-0.751	0.115	67	69	67	68	114	81	1.38
9. SACRAMENTO, CA	39	63	43	20	-0.668	0.106	62	64	62	65	114	78	1.45
10. SAN DIEGO, CA	48	61	35	27	0.384	0.022	56	50	57	584	85	75	1.57
11. DENVER, CO	33	61	48	17	-0.265	0.112	60	61	61	69	93	72	1.30
12. GRAND JUNCTION, CO	33	70	52	17	0.104	0.065	67	66	70	146	103	86	1.37
13. PUEBLO, CO	43	79	61	25	-0.477	0.060	80	81	80	85	126	95	1.34
14. HARTFORD, CT	44	67	43	28	-0.238	0.028	67	64	68	84	109	87	1.54
15. WASHINGTON, DC	39	66	47	24	-0.325	0.055	67	65	68	77	106	84	1.44
16. ATLANTA, GA	42	76	46	22	-0.034	0.030	74	72	76	134	131	101	1.59
17. MACON, GA	33	64	45	17	-0.271	0.073	64	65	66	79	118	84	1.46
18. BOISE, ID	48	62	47	24	-0.099	0.068	62	61	62	84	92	75	1.35
19. POCATELLO, ID	48	72	53	25	0.019	0.157	75	71	75	132	113	94	1.43
20. CHICAGO MIDWAY, IL	37	63	46	22	-0.128	0.028	63	61	64	86	97	79	1.41
21. MOLINE, IL	44	72	52	25	-0.665	0.082	72	75	73	75	124	90	1.41
22. PEORIA, IL	42	72	50	22	-0.302	0.075	70	72	71	82	122	89	1.43
23. SPRINGFIELD, IL	32	71	54	16	-0.199	0.118	68	69	69	83	110	83	1.32
24. EVANSVILLE, IN	44	61	47	23	-0.130	0.106	63	63	63	82	103	77	1.36
25. FORT WAYNE, IN	46	69	52	23	-0.163	0.063	69	70	70	88	110	85	1.37
26. INDIANAPOLIS, IN	36	93	53	18	0.023	0.064	81	81	85	171	151	113	1.56
27. BURLINGTON, IA	23	72	55	13	-1.751	0.515	69	76	69	69	140	80	1.28
28. DES MOINES, IA	37	80	56	19	0.042	0.034	79	77	81	155	133	102	1.45
29. SIOUX CITY, IA	46	88	57	24	0.063	0.065	85	80	85	180	132	110	1.49
30. CONCORDIA, KS	20	74	56	11	-0.933	0.188	73	76	74	75	145	89	1.35
31. DODGE CITY, KS	41	72	59	22	-0.709	0.048	72	75	72	73	114	83	1.25
32. TOPEKA, KS	34	79	54	17	-0.180	0.160	71	74	73	91	128	90	1.38
33. WICHITA, KS	41	89	57	22	0.218	0.056	83	80	86	325	134	110	1.49
34. LOUISVILLE, KY	39	66	49	21	-0.085	0.055	63	65	64	87	107	77	1.34
35. PORTLAND, ME	45	73	46	23	-0.283	0.088	73	70	73	90	122	97	1.55
36. BALTIMORE, MD	39	71	54	20	-0.274	0.078	70	71	71	81	116	85	1.34
37. BOSTON, MA	50	85	54	26	-0.126	0.039	84	81	84	119	139	109	1.52
38. NANTUCKET, MA	23	71	55	14	-1.402	0.217	71	73	72	72	128	85	1.33
39. DETROIT, MI	46	68	49	25	-0.207	0.091	67	68	67	81	111	82	1.38
40. GRAND RAPIDS, MI	29	67	47	15	-0.930	0.128	69	71	70	71	136	89	1.47
41. LANSING, MI	38	67	51	21	-0.646	0.063	68	69	68	71	110	83	1.36
42. SAULT STE MARIE, MI	47	67	46	24	-0.351	0.059	65	67	65	74	113	82	1.43
43. DULUTH, MN	36	70	49	19	-0.310	0.033	69	70	70	81	122	88	1.44
44. MINNEAPOLIS, MN	42	82	46	27	-0.023	0.048	79	72	81	149	127	108	1.65
45. JACKSON, MS	29	64	44	16	-0.353	0.073	64	62	66	75	109	84	1.48
46. COLUMBIA, MO	35	65	51	20	-0.103	0.075	66	66	67	89	109	81	1.35
47. KANSAS CITY, MO	51	75	49	31	-0.116	0.038	74	72	74	104	120	94	1.49
48. ST.LOUIS, MO	21	66	46	14	0.035	0.075	64	62	70	135	111	89	1.51

Station	N	V_{max}	V_{med}	n_{ex}	$\hat{c}$	sd($\hat{c}$)	V_N Gum.	V_N CME	V_{50} CME	V_{100000} CME	V_{100000} Gum.	V_{100000} c=-.20	ϕ
(1)	(2)	(3)	(4)	(5)	(6)	(7)	(8)	(9)	(10)	(11)	(12)	(13)	(14)
49. SPRINGFIELD, MO	44	71	49	22	-0.127	0.068	68	67	69	93	111	86	1.42
50. BILLINGS, MT	49	84	58	26	-0.051	0.031	84	81	84	130	132	106	1.45
51. GREAT FALLS, MT	44	75	59	24	-0.400	0.078	74	78	74	80	125	87	1.29
52. HAVRE, MT	27	78	57	14	-0.229	0.227	77	76	80	97	135	99	1.41
53. HELENA, MT	48	71	55	25	-0.241	0.114	70	71	70	81	111	83	1.31
54. MISSOULA, MT	43	71	47	22	-0.157	0.050	71	65	72	97	105	92	1.50
55. NORTH PLATTE, NE	31	74	61	16	-0.709	0.044	74	77	75	76	123	87	1.26
56. OMAHA, NE	51	104	50	29	0.294	0.054	92	81	91	640	145	125	1.68
57. VALENTINE, NE	27	74	61	15	-0.574	0.120	74	78	76	78	130	88	1.27
58. ELY, NV	49	70	51	28	-0.191	0.023	70	68	70	87	107	86	1.39
59. LAS VEGAS, NV	20	70	55	12	-0.938	0.362	68	70	69	69	121	80	1.28
60. RENO, NV	45	77	55	24	-0.463	0.053	77	78	77	83	129	96	1.41
61. WINNEMUCCA, NV	38	63	47	21	-1.102	0.071	63	67	64	64	114	77	1.37
62. CONCORD, NH	46	68	41	23	-0.040	0.038	66	63	66	116	111	89	1.58
63. ALBUQUERQUE, NM	52	85	56	26	0.139	0.023	79	78	78	195	124	98	1.41
64. ROSWELL, NM	36	82	57	18	0.088	0.119	81	81	84	190	145	108	1.48
65. ALBANY, NY	46	68	46	29	-0.086	0.081	66	64	67	95	103	83	1.45
66. BINGHAMPTON, NY	35	65	48	18	-0.370	0.081	64	66	65	72	115	80	1.38
67. BUFFALO, NY	44	79	52	22	0.463	0.050	73	69	74	1049	109	94	1.45
68. LA GUARDIA, NY	33	73	57	17	0.314	0.032	71	71	74	344	113	89	1.33
69. ROCHESTER, NY	45	66	52	23	-0.717	0.094	66	69	66	67	106	78	1.30
70. SYRACUSE, NY	45	67	51	23	0.014	0.089	66	65	66	105	100	79	1.33
71. CHARLOTTE, NC	29	65	42	17	-0.481	0.244	65	62	67	73	112	87	1.55
72. GREENSBORO, NC	50	67	41	25	-0.089	0.073	62	62	62	94	109	81	1.51
73. BISMARCK, ND	40	69	58	20	-0.582	0.145	69	72	69	71	107	79	1.23
74. FARGO, ND	45	100	57	25	0.252	0.032	93	86	95	484	151	127	1.60
75. WILLISTON, ND	18	69	56	9	-0.469	0.127	68	68	71	75	116	84	1.31
76. CLEVELAND, OH	35	69	53	19	-0.224	0.080	68	69	69	81	111	82	1.33
77. COLUMBUS, OH	30	61	49	15	-1.236	0.106	60	64	60	60	106	70	1.26
78. DAYTON, OH	41	72	52	24	-0.170	0.072	74	72	75	97	121	94	1.45
79. TOLEDO, OH	45	82	48	24	0.207	0.050	76	72	78	308	125	103	1.57
80. OKLAHOMA CITY, OK	30	69	53	15	-0.135	0.081	67	67	69	89	105	84	1.35
81. TULSA, OK	35	68	49	18	0.019	0.107	63	65	65	106	109	79	1.35
82. PORTLAND, OR	38	88	49	19	0.268	0.077	80	75	85	503	138	116	1.64
83. HARRISBURG, PA	38	64	45	19	-0.599	0.087	63	65	64	67	114	80	1.42
84. PHILADELPHIA, PA	33	62	47	21	-0.507	0.094	63	64	64	67	105	77	1.38
85. PITTSBURGH, PA	18	60	47	11	-0.591	0.132	61	60	63	65	102	75	1.36
86. SCRANTON, PA	33	57	44	17	-0.387	0.092	56	57	57	62	89	68	1.32
87. BLOCK ISLAND, RI	31	86	60	16	-0.182	0.120	82	82	84	107	138	105	1.42
88. GREENVILLE, SC	43	72	46	22	-0.483	0.069	69	75	70	76	143	90	1.50
89. HURON, SD	49	79	59	26	-0.447	0.059	80	82	80	86	132	98	1.38
90. RAPID CITY, SD	43	70	62	22	-0.341	0.096	71	73	72	76	102	80	1.18
91. CHATTANOOGA, TN	35	76	46	18	-0.336	0.066	76	73	78	93	141	105	1.61
92. KNOXVILLE, TN	33	66	50	18	0.007	0.042	66	65	68	111	109	83	1.38
93. MEMPHIS, TN	21	61	45	11	-0.370	0.369	57	58	59	65	103	71	1.34
94. NASHVILLE, TN	34	70	45	17	-0.178	0.092	67	66	70	94	119	91	1.53
95. ABILENE, TX	36	100	54	19	0.550	0.047	79	78	85	2491	141	111	1.54
96. AMARILLO, TX	34	81	62	17	0.201	0.082	77	78	79	214	124	95	1.32
97. AUSTIN, TX	37	58	45	19	-0.189	0.057	57	58	58	69	91	68	1.31
98. DALLAS, TX	32	67	48	17	-0.233	0.043	65	64	67	81	107	83	1.41
99. EL PASO, TX	32	67	55	17	-0.187	0.103	68	67	69	82	99	81	1.28
100. SAN ANTONIO, TX	36	80	46	19	0.236	0.041	70	68	73	328	123	96	1.55

Station	N	V_{max}	V_{med}	n_{ex}	$\hat{c}$	sd($\hat{c}$)	V_N Gum.	V_N CME	V_{50} CME	V_{100000} CME	V_{100000} Gum.	V_{100000} c=-.20	ϕ
(1)	(2)	(3)	(4)	(5)	(6)	(7)	(8)	(9)	(10)	(11)	(12)	(13)	(14)
101. SALT LAKE CITY, UT	46	69	49	28	-0.333	0.038	70	69	70	80	113	88	1.44
102. BURLINGTON, VT	40	66	44	23	-0.232	0.093	66	63	67	84	107	86	1.51
103. LYNCHBURG, VA	44	53	39	22	-0.750	0.154	52	56	52	53	94	63	1.37
104. RICHMOND, VA	33	61	42	20	0.192	0.058	59	56	63	204	95	79	1.48
105. NORTH HEAD, WA	41	104	67	28	-0.250	0.055	105	96	106	130	158	136	1.55
106. QUILLAYUTE, WA	21	45	35	12	-0.184	0.018	45	44	47	58	72	57	1.37
107. SEATTLE, WA	20	59	43	11	0.027	0.069	56	56	62	112	97	77	1.44
108. SPOKANE, WA	47	65	48	24	0.028	0.040	64	64	65	111	101	79	1.37
109. TATOOSH ISLAND, WA	54	86	66	27	-0.290	0.045	86	85	86	98	128	104	1.34
110. GREEN BAY, WI	36	103	54	18	0.431	0.089	85	82	92	1408	153	126	1.63
111. MADISON, WI	41	75	48	21	-0.516	0.058	75	75	76	82	134	100	1.54
112. MILWAUKEE, WI	42	68	54	21	-0.423	0.059	67	70	67	72	109	79	1.29
113. CHEYENNE, WY	46	73	61	24	-0.476	0.051	74	76	74	77	113	85	1.24
114. LANDER, WY	42	80	58	21	-0.621	0.067	77	83	77	80	142	94	1.35
115. SHERIDAN, WY	44	82	61	24	0.125	0.073	82	80	83	184	128	101	1.38
MEAN	38.1	71.9	50.4	20.5	-0.257 (unweighted)								1.42
SD	8.2	10.5	6.3	4.5	0.384 (unweighted)								0.10

Key

N, V_{max}, n_{ex}:	Sample size, maximum observed speed, number of exceedences, respectively.
$\hat{c}$, sd($\hat{c}$):	Estimated c, standard deviation of $\hat{c}$.
Col. 8,9:	Estimated N-yr wind based on Gumbel model and CME method ($\hat{c}$ from Col. 6).
Col. 10:	Estimated 50-yr wind based on CME method ($\hat{c}$ from Col. 6).
Col. 11,12,13:	Estimated 100,000-yr wind based on CME method ($\hat{c}$ from Col. 6), Gumbel model, CME method (c=-0.20), respectively.
Col. 14:	Load factor, ϕ, based on c=-0.20.

REGIONAL DESIGN WIND SPEED ESTIMATION

CHENG, E.D.H. and CHIU, A.N.L.

Professors, Department of Civil Engineering, University of Hawaii at Manoa, Honolulu, Hawaii 96822, U.S.A.

Abstract

In order to use the available short-term and fragmentary wind data to the full advantage in extreme wind regional analysis, a Rayleigh Simulation Model (RSM) is being developed. RSM basically is a process of generating extreme winds by means of estimating tail quantile probability from historical data. The simulated extreme winds as well as those derived from long-term historical records will be used in a regional extreme wind analysis for the state of Texas.

1. INTRODUCTION

In the current state-of-the-art, there are three approaches for estimating extreme wind speeds from short-term records. The first approach is based on the analysis of the largest monthly wind speeds of at least three years' data (Simiu, et al., 1982). The second approach (Grigoriu, 1982) and the third approach (Cheng and Chiu, 1985, 1994) are simulation models based on short-term continuous hourly wind speed records. The proposed Rayleigh Simulation Model (RSM) generates extreme wind from a parent distribution. Using RSM, extreme winds can be estimated even from records shorter than three years. The selection of a proper approach is dependent upon the available data. Using that selected approach, a distribution function of extreme winds at a station can thus be obtained.

2. RAYLEIGH SIMULATION MODEL

2.1 Tail Quantile Probability

Consider U(t) as the wind speed time series at a specific site, and W as the strong wind or extreme wind random variable associated with U(t). In general the distribution of W is unknown. However, W may be identified by estimating its tail quantile probability

$$P(W \leq V^j) = j, \quad 0 < j \leq 1 \qquad (1)$$

in which j has a fixed large value and V^j is the jth quantile of W. Further,

$$F(V) = P(W \leq V) \tag{2}$$

in which V = a realization of W; F(V) = cumulative distribution function of V.

The task of evaluating Eqs. 1 and 2 begin with analyzing: I. The historical data u(t), which is a realization of U(t), and II. the assumed parametric model F(V) for the distribution of W.

2.2 Rayleigh Function

In this paper, the historical data u(t) are the wind roses (at the 21 Air Force stations in Texas). The assumed parametric model for the tail distribution of a wind rose is a Rayleigh type. To perform a point estimate of the tail quantile V^j for the fixed large j at a site, the following steps are observed.

2.2.1 State division

The first step is to divide the entire range of observed wind rose into a finite number of states. This task is performed with reference to the probability histogram derived from the observed wind rose for that site. In principle, the state divisions should be determined according to the shape of the probability histogram or wind rose; therefore, state intervals will result in various nonequal subrange of wind speeds.

2.2.2 State functions

The second step is the selection of the probability density functions (PDF) of wind speeds in various states determined in the previous step. In this paper, three types of PDF are used to fit a wind rose, that is, uniform, linear and the Rayleigh functions. If relative uniform wind speeds were observed within a state or states, constant values were assigned to the state or states. Likewise, if linear wind speed variations were observed in a state, a linear PDF would be assigned to that state. A Rayleigh PDF is exclusively reserved for the last state in order to take care of the extreme winds. For the last state m, the PDF and CDF are Rayleigh type:

$$\text{PDF: } f_m(V) = (2V / b^2)\exp(-V / b^2)^2 \tag{3}$$

$$\text{CDF: } F_m(V) = 1\text{-}\exp(-V / b)^2 \text{ for } V \geq V_m \tag{4}$$

where b = the scale parameter; and V_m = the lower limit of wind speed of the last state.

The initial value of scale parameter b is evaluated from the wind rose. The subsequent adjustment of b will be demonstrated in the "RAYLEIGH SIMULATION ILLUSTRATION" section.

2.2.3 Synthesis of extreme winds

Rayleigh function has been used by Cheng and Chiu (1993) in modeling extreme winds. There is an important advantage of using Rayleigh function to model F(V) - the unknown parent distribution of strong or extreme winds W at a site: The distribution of

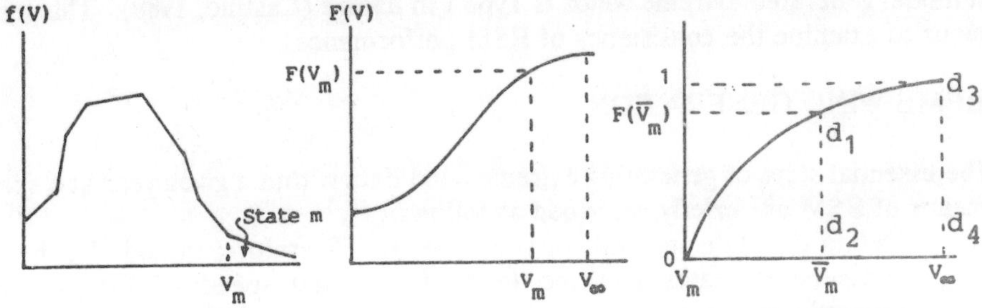

(a) A Hypothetic PDF (b) CDF of (a) (c) CDF of State m

Figure 1. Hypothetical PDF and CDF of wind speeds

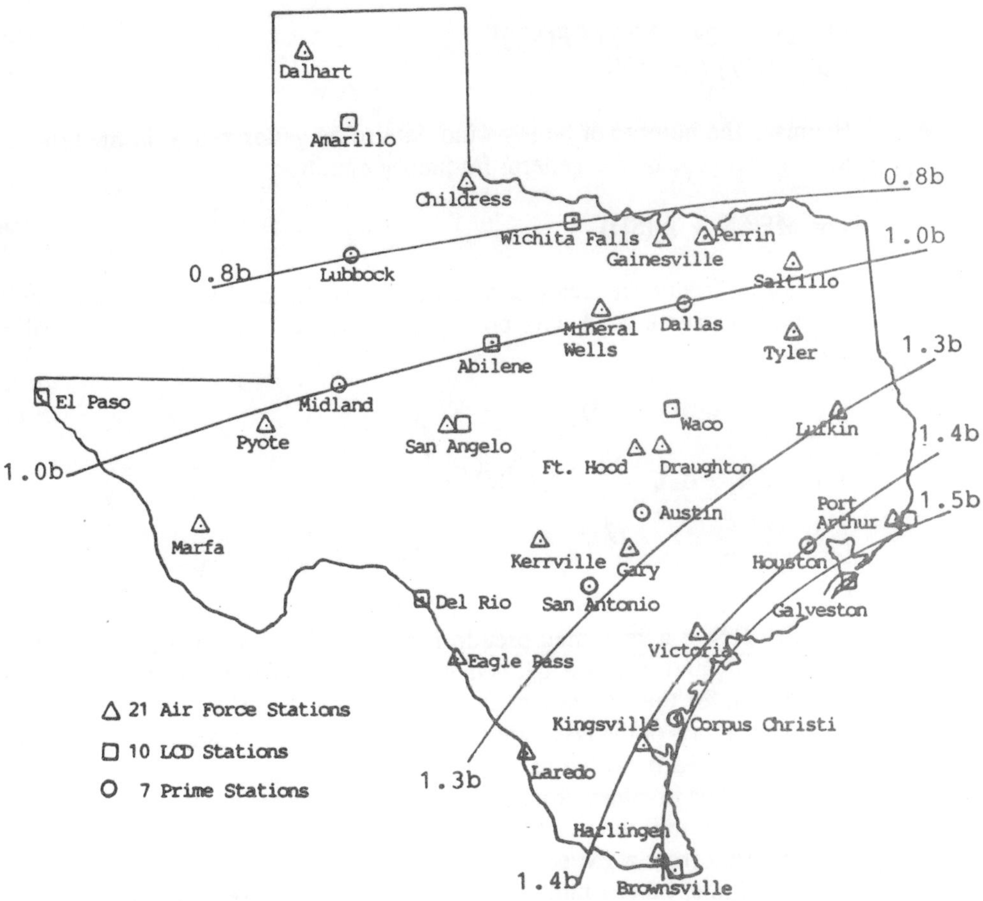

Figure 2. Location of wind stations and Rayleigh Parameters in Texas

Rayleigh model generated extreme winds is Type I in nature (Castillo, 1988). This feature will allow us to examine the consistency of RSM performance.

3. EXTREME WIND GENERATION

The essential steps of generating extreme wind data within a given year at a specific site by means of RSM are briefly described as follows:

a. Define the state intervals of wind speeds. Therefore, m and V_m, the total numbers of states and the lower limit wind speed of the last state, respectively, are known.

b. Determine type of PDF of the state intervals as defined in step a. Therefore, the probability for the last state m, $F(V_m)$, is known (Fig. 1).

c. Determine the coefficient of variance of number of hourly wind data, CV, to be generated in the last state for a given year.

d. Compute the mean (MEAN) and standard deviation (SD) of number of hourly wind data to be generated in the last state for a given year:

$$\text{MEAN} = (24 \times 365)\,(1-F(V_m)) \tag{5}$$

$$\text{SD} = (\text{CV})\,(\text{MEAN}) \tag{6}$$

e. Estimate the number of hourly wind data to be generated, x, in the last state for a given year by the general frequency equation:

$$x = \text{MEAN} + k\,(\text{SD}) \tag{7}$$

where k = frequency factor of assumed distribution of x for a given year.

f. A typical extreme wind data point $\overline{V}_m$ in the last state can be determined from:

$$\text{RN} = \overline{d_1d_2} \,/\, \overline{d_3d_4} = \int_{V_m}^{\overline{V}_m} f_m(V)dV \,\Big/ \int_{V_m}^{V_\infty} f_m(V)dV \tag{8}$$

in which RN = a generated pseudo random number varying uniformly from 0 to 1; $\overline{d_1d_2}$ and $\overline{d_3d_4}$ are defined in Fig. 1; $f_m(V)$ is the PDF of Rayleigh distribution for the last state m and as defined in Eq. 3; V_m and V_∞ are lower and upper limits of the last state m of the wind speed histogram or wind rose, respectively. As indicated in Fig. 1, the CDF of $\overline{V}_m$, $F_m(\overline{V}_m)$, is equal to RN, and $\overline{V}_m$ is thus obtained from the inverse function of $F_m(\overline{V}_m)$.

g. Repeat step f until the number of wind data x in the last state is supposed to be generated from a given year is reached.

h. Repeat steps e, f and g until the desirable number of years of extreme winds is attained.

4. RAYLEIGH SIMULATION ILLUSTRATION

Regional extreme wind analysis is being investigated on wind data collected at 38 weather stations in the state of Texas (Fig. 2). Long-term annual fastest-mile wind speeds exist at 17 of the 38 stations. The 17 stations are: Abilene, Amarillo, Austin, Brownsville, Corpus Christi, Dallas, Del Rio, El Paso, Galveston, Houston, Lubbock, Midland, Port Arthur, San Angelo, San Antonio, Waco and Wichita Falls. For these stations, the annual series of fastest-mile wind speeds were constructed, and Gumbel distributions were fitted. However, the records collected at the remaining 21 Air Force (AF) stations are of short periods and are in wind rose format. The 21 AF stations are: Childress, Dalhart, Draughton, Eagle Pass, Fort Hood, Gainesville, Gary, Harlingen, Kerrville, Kingsville, Laredo, Lufkin, Marfa, Mineral Wells, Perrin, Port Arthur, Pyote, Saltillo, San Angelo, Tyler and Victoria. In order to derive fastest-mile wind speed information from these wind roses, the proposed Rayleigh Simulation Model is applied and described as follows.

4.1 Scale Parameter

In operating RSM, the Rayleigh scale parameters, b, at the 21 AF stations have to be known. They are initially calculated from the wind roses by means of area-moment method, these initial values have to be adjusted for lacking other important weather information such as the effects of thunderstorms to the weather patterns. Therefore, we have decided to derive the Rayleigh scale parameter adjusting information for the 21 AF stations from the seven prime stations: Austin, Corpus Christi, Dallas, Houston, Lubbock, Midland and San Antonio. Our selection of these seven so-called prime stations are primarily because of their geographic locations and the availability of hourly wind data. Upon obtaining wind data at the seven prime stations from the National Climatic Data Center (NCDC) in Asheville, North Carolina, we first examined the thunderstorm effects on the annual series of fastest-mile wind speeds. As indicated in Table 1, about 20% to 52% of the annual fastest-mile wind speeds may be related to thunderstorms at the seven prime stations in Texas. Table 1 was derived from a period of 42 years (except 29 years for Houston) descriptive summary of the day data (TD-3210) obtained from NCDC. Another possible source of thunderstorm information is the expensive hourly present weather data

Table 1. Percentage of Annual Fastest-mile Wind Speed Days with Thunderstrom Occurrences at the 7 Prime Stations

Station	Data Period	o/o
AUSTIN	1950-91	47.6
CORPUS CHRISTI	1950-91	28.6
DALLAS	1950-91	52.4
HOUSTON	1963-91	20.7
LUBBOCK	1950-91	38.1
MIDLAND	1950-91	34.2
SAN ANTONIO	1950-91	38.1

(TD-3280) from NCDC. We tried present weather data on Corpus Christi and was very disappointed. At Corpus Christi, only one out of twelve annual fastest-mile wind speed days (1950-1991) with thunderstorm occurrences was somewhat identifiable from the hourly present weather data. There are no tangible thunderstorm wind speed records available for analyzing this important weather component. Furthermore, by means of the multiple regressional analysis, we found that at the seven prime stations the annual series of fastest-mile wind speeds are not correlated to the annual series of hourly extreme wind speed data obtained from NCDC.

In order to develop a Rayleigh scale parameter spectrum map for Texas, a short-term record based stochastic simulation model developed by Cheng and Chiu (1985, 1994) was modified and applied to the seven prime stations in Texas. The modifications were using the Rayleigh distributions to delineate the last state of the wind speed histograms. The probabilities for V_m, the lower limit of wind speeds of the last state m, used for the seven prime stations vary from 0.993 to 0.997. As indicated in Table 2, the hourly data based used in the stochastic simulations vary from 5-year, 10-year, 15-year and all the hourly data available at the respective stations. At a station, each simulation run begins with the initially obtained Rayleigh scale parameter from the histogram of selected data bases. Both simulated as well as historical annual fastest-mile wind speeds were closely fitted by Type I distributions. Using the Gumbel line obtained from fitting the historical annual fastest-mile wind speeds as the reference, the Rayleigh scale parameter was adjusted and the simulation runs were repeated. The predicted 25-year, 50-year and 100-year fastest-mile wind speeds, which were either stochastically simulated or obtained from historical annual series, are presented in Table 2. In this table, the deviations of simulated 25-year, 50-year or 100-year fastest-mile wind speeds from the reference Gumbel lines were measured by S_v, the standard deviation of the inherent sampling error of the historical records (Gumbel, 1958; Simiu and Scanlan, 1986). As shown in Table 2, the differences between simulated annual fastest-mile wind speeds and values obtained from the Gumbel reference lines at 25-year, 50-year or 100-year recurrence intervals are almost all smaller than one S_v, except five instances at the 25-year recurrence interval. This result is very encouraging. The final Rayleigh scale parameters, b^*, for the seven prime stations in Texas are Corpus Christi - 1.5b, Houston - 1.4b, San Antonio - 1.3b, Austin - 1.2b, Dallas - 1.0b, Midland - 1.0b and Lubbock - 0.8b. Where b is the initially obtained Rayleigh scale parameter from the area moment method of respective wind speed histogram. The values of the finally adjusted Rayleigh scale parameters are independent from the length of the data base used in each of the seven prime stations. Based on the final Rayleigh scale parameters at the seven prime stations, a proposed spectrum of Rayleigh scale parameter for Texas is shown in Fig. 2.

4.2 Frequency Factor

For all the runs in Table 2, the number of hourly wind data generated in the last states for any given year are approximately normally distributed. Consequently Gaussian frequency factors were selected for Eq. 7 during the process of generating fastest-mile wind speed data at the 21 AF stations.

Table 2. Estimated Annual Fastest-Mile Wind Speeds from HistoricalRecords and from Stochastic Simulation Method at the 7 Prime Stations

Data Period	Recurrence Interval (In Years)	V_A	S_V	V_S	$(V_A - V_S)/S_V$
AUSTIN					
1950-1990	25	25.6	1.17		
	50	27.3	1.4		
	100	28.9	1.63		
1/73-12/77	25			27.1	-1.28
	50			28	-0.5
	100			28.8	0.06
1/78-12/82	25			26.7	-0.94
	50			27.7	-0.29
	100			28.8	0.06
1/83-12/87	25			26.7	-0.94
	50			27.9	-0.43
	100			29	-0.06
1/88-12/92	25			26.3	-0.6
	50			27.2	0.07
	100			28.2	0.43
1/73-12/92	25			26.4	-0.68
	50			27.4	-0.07
	100			28.3	0.37
CORPUS CHRISTI					
1950-1990	25	38.1	2.95		
	50	42.3	3.53		
	100	46.4	4.11		
1/81-12/85	25			41.5	-1.15
	50			43.4	-0.31
	100			45.3	0.27
1/86-12/90	25			40.7	-0.88
	50			42.6	-0.09
	100			44.6	0.44
1/81-10/92	25			42	-1.32
	50			44	-0.48
	100			45.9	0.12
HOUSTON					
1983-1990	25	26.6	1.78		
	50	28.8	2.13		
	100	30.9	2.48		
1/73-12/77	25			27.6	-0.56
	50			28.7	0.05
	100			29.9	0.4
1/78-12/82	25			27.8	-0.67
	50			29.1	-0.14
	100			30.3	0.24
1/83-12/87	25			27.4	-0.45
	50			28.4	0.19
	100			29.4	0.6
1/88-9/92	25			27.4	-0.45
	50			28.6	0.09
	100			29.8	0.44
1/73-12/82	25			27.4	-0.45
	50			28.6	0.09
	100			29.7	0.48
1/83-12/87	25			28.1	-0.84
	50			29.4	-0.28
	100			30.7	0.08
1/73-12/87	25			27.6	-0.56
	50			28.6	0
	100			30	0.36
1/73-9/92	25			27.9	-0.73
	50			29.6	-0.38
	100			31.3	-0.16
DALLAS/FORTWORTH					
1950-1990	25	27.7	1.53		
	50	29.9	1.83		
	100	32	2.13		
1/81-12/85	25			28.6	-0.72
	50			29.9	0
	100			31	0.47
1/86-12/90	25			29.2	-0.98
	50			30.3	-0.22
	100			31.3	0.33
1/81-9/92	25			29.3	-1.05
	50			30.4	-0.5
	100			31.4	0.28
LUBBOCK					
1950-1990	25	28.4	1.14		
	50	30	1.36		
	100	31.6	1.59		
1/81-12/85	25			29.1	-0.61
	50			30.1	-0.07
	100			31	0.38
1/86-12/90	25			29.1	-0.61
	50			30	0
	100			30.9	0.44
1/81-12/92	25			29.3	-0.79
	50			30.3	-0.22
	100			31.3	0.19
MIDLAND					
1957-1990	25	30	1.63		
	50	32.1	1.95		
	100	34.2	2.28		
1/73-12/77	25			32.3	-1.41
	50			33.5	-0.72
	100			34.6	-0.18
1/78-12/82	25			32.4	-1.47
	50			33.6	-0.77
	100			34.7	-0.22
1/83-12/87	25			31.6	-0.98
	50			32.7	-0.31
	100			33.8	0.18
1/88-12/92	25			31.6	-0.98
	50			32.8	-0.36
	100			34.1	0.04
1/73-12/92	25			30.9	0.55
	50			31.9	0.1
	100			33	0.53
SAN ANTONIO					
1950-1990	25	26	1.66		
	50	27.7	1.99		
	100	29.4	2.32		
1/81-12/85	25			26.8	-0.48
	50			27.8	-0.05
	100			28.7	0.3
1/86-12/90	25			26.3	-0.18
	50			27.4	0.15
	100			28.4	0.43
1/81-12/92	25			27	-0.6
	50			28.1	-0.2
	100			29.2	0.09

V_A = fastest-mile wind speed from annual series in m/s and 10 m above ground level

V_S = simulated fastest-mile wind speed in m/s and 10 m above ground level

S_V = Cramer-Rao's standard deviation of inherent sampling error of historical records.

4.3 Coefficient of Variance

For all the runs in Table 2, CV, the coefficients of variance of numbers of hourly wind data generated in the last states for any given year are carefully examined. Unfortunately no apparent pattern or patterns of CV are observed. In applying RSM to the 21 AF stations, an average value of CV was used. The probabilities for V_m, the lower limits of wind speeds of the last state m used for the 21 AF stations vary from 0.986 to 0.999. These probabilities were estimated from wind roses. By means of the procedures described in Section 3, the predicted 25-year, 50-year and 100-year fastest-mile wind speeds at the 21 AF stations are presented in Table 3. As shown in Fig. 3, the simulated fastest-mile wind speeds at the 21 AF stations are indeed Type I distributions in nature.

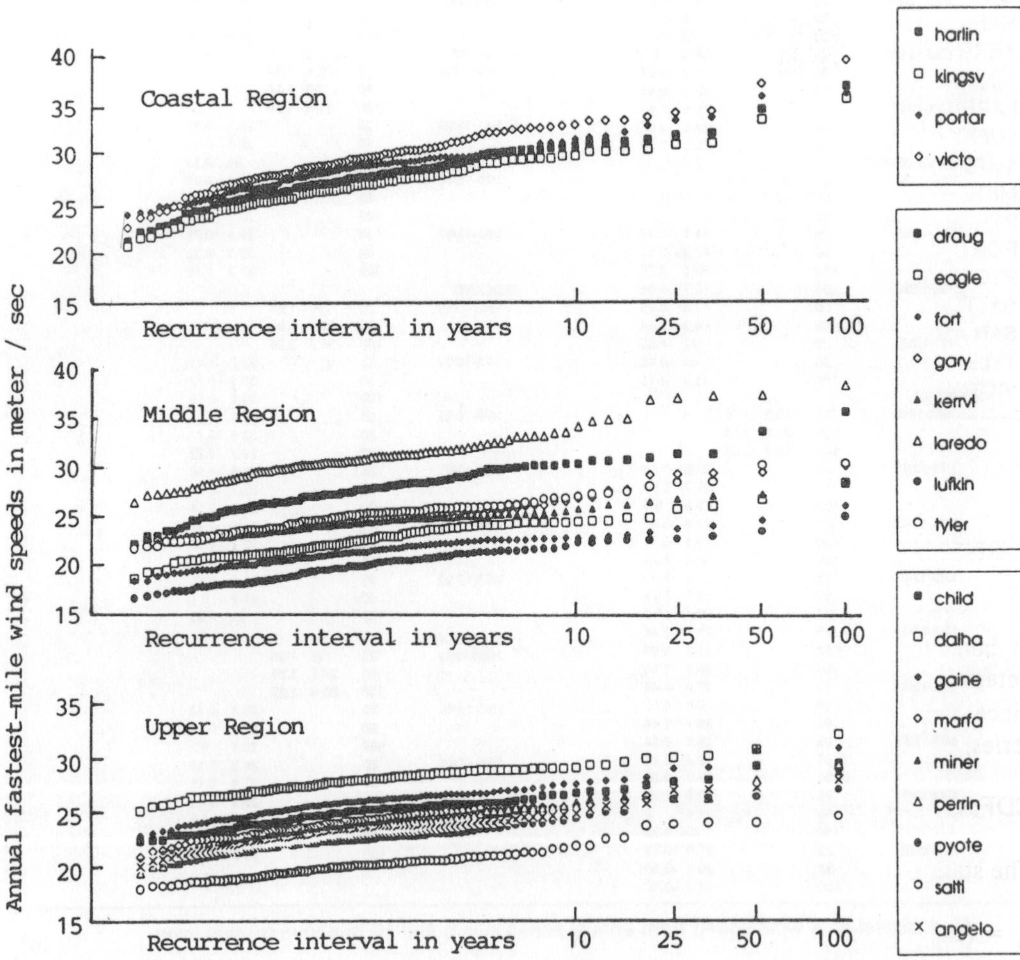

Figure 3. Type I distributions of Rayleigh simulation model generated annual fastest-mile wind speeds at the 21 Air Force stations in Texas

Table 3. Rayleigh Simulation Model Generated Fastest-Mile Wind Speeds (in m/s and 10 m above ground level) at the 21 Air Force Stations

Station	Rayleigh Parameter	$V_{25\text{-yr}}$	$V_{50\text{-yr}}$	$V_{100\text{-yr}}$
CHILDRESS	0.81b	27.4	28.8	29.1
DALHART	0.61b	29.6	30.4	31.7
DRAUGHTON	1.15b	31.1	33.4	35.3
EAGLE PASS	1.24b	25.3	26.5	28.2
FORT HOOD	1.24b	23.5	24.4	25.9
GAINESVILLE	0.89b	28.7	30.3	30.5
GARY	1.33b	28.7	28.9	29.6
HARLINGEN	1.51b	30.9	33.4	35.6
KERRVILLE	1.24b	26.6	27.1	28.4
KINGSVILLE	1.51b	30.1	32.4	34.4
LAREDO	1.42b	36.8	37.1	38.1
LUFKIN	1.33b	22.2	22.9	24.5
MARFA	1.06b	26.3	27.2	28.6
MINERAL WELLS	0.98b	27.6	27.7	28.4
PERRIN	0.89b	26.1	27.6	29.1
PORT ARTHUR	1.51b	32.2	34.6	36.6
PYOTE	0.98b	25.9	26.1	26.7
SALTILLO	0.98b	23.6	23.7	24.3
SAN ANGELO	1.06b	26.5	26.9	27.6
TYLER	1.15b	28.1	29.9	30.1
VICTORIA	1.42b	33.3	36.1	38.4

5. REGIONAL ZONING

The grouping of weather stations into regions or subregions is guided by a sequence of homogeneity tests. For each homogeneity test, permissible ranges of variation are determined by selecting ± 1.0 standard deviation of an appropriate probability distribution. Since Type I distributions are used for representing the annual fastest-mile wind speed series, the Type I distribution in reduced variate w may be expressed as

$$\text{CDF: } G(w) = \exp[-\exp(-w)] \tag{9}$$

The standard deviation of the reduced variate of Type I distribution, S_w, is

$$s_w = (\exp(w) / n^{1/2})[(1-G(w)) / G(w)] \tag{10}$$

where G(w) = CDF of Type I distribution defined in Eq. 9; n = sample size of historical data. Further,

$$G_l(w) = \exp[-\exp(-(w-s_w))] \quad (11)$$

$$G_u(w) = \exp[-\exp(-(w+s_w))] \quad (12)$$

and

$$T_l = [1-G_l(w)]^{-1} \quad (13)$$

$$T_u = [1-G_u(w)]^{-1} \quad (14)$$

where T_l and T_u are lower and upper limits of recurrence intervals, respectively, in year and used in homogeneity test.

In order to identify the inherent tendency of regional characteristics for certain weather stations, all locations with point measurements should be subjected to the initial homogeneity test. Subregions may be defined by successive homogeneity test. In this paper, a 15-year data period was selected for calculating the values of the permissible range of variation in the homogeneity test. All 38 stations in Texas were subjected to a series of homogeneity tests, and three subregions were decided. To construct the dimensionless regional frequency curves for Texas, ratios of fastest-mile wind speeds to the mean annual fastest-mile wind speeds at recurrence intervals of 1.1, 1.5, 5, 15, 50 and 100 years for all 38 stations were calculated and presented in Table 4. The three subregional fastest-mile wind speed frequency curves are plotted in Fig. 4.

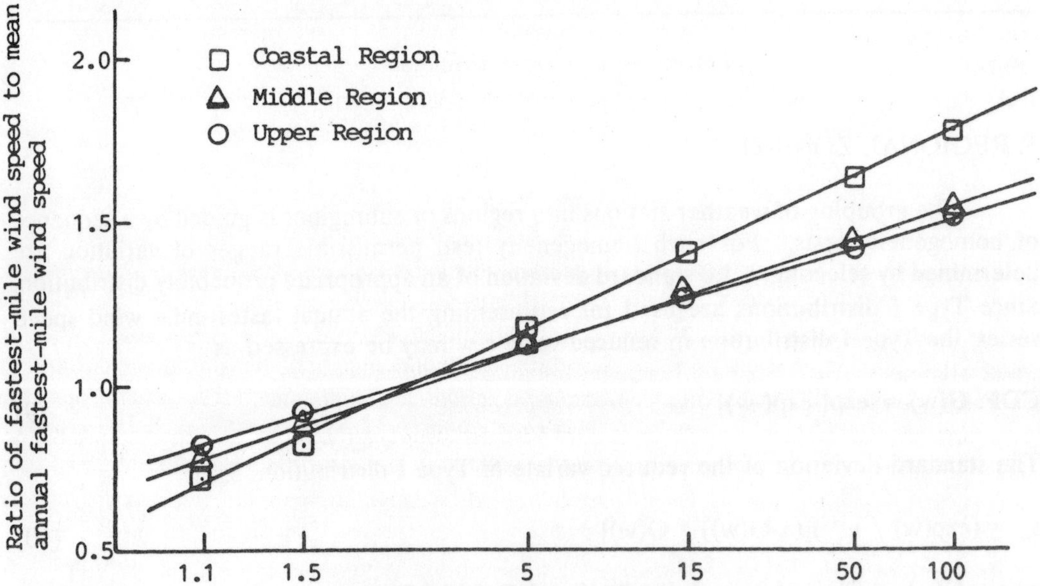

Figure 4. Type I distributions of the three subregional fastest-mile wind speed frequency curves in Texas

Table 4. Extreme Wind Regional Frequency Curves for Texas

Station	V1.1/V2.33	V1.5/V2.33	V5/V2.33	V15/V2.33	V50/V2.33	V100/V2.33
COASTAL REGION						
BROWNSVILLE	0.76	0.89	1.15	1.35	1.55	1.66
CORPUS CHRISTI	0.66	0.84	1.21	1.49	1.77	1.94
GALVESTON	0.73	0.88	1.17	1.39	1.61	1.74
HARLINGEN	0.66	0.84	1.21	1.49	1.77	1.93
HOUSTON	0.81	0.91	1.13	1.29	1.47	1.56
KINGSVILLE	0.67	0.85	1.21	1.47	1.75	1.91
PORT ARTHUR	0.71	0.87	1.18	1.42	1.66	1.79
VICTORIA	0.76	0.89	1.15	1.35	1.55	1.67
Mean	0.72	0.87	1.18	1.41	1.64	1.78
MIDDLE REGION						
AUSTIN	0.84	0.93	1.09	1.22	1.36	1.43
DEL RIO	0.78	0.91	1.14	1.31	1.49	1.61
DRAUGHTON	0.81	0.91	1.12	1.28	1.44	1.53
EAGLE PASS	0.83	0.92	1.11	1.25	1.41	1.48
FORT HOOD	0.85	0.93	1.11	1.22	1.35	1.43
GARY	0.84	0.93	1.09	1.23	1.36	1.43
KERRVILLE	0.87	0.94	1.08	1.18	1.29	1.35
LAREDO	0.77	0.89	1.15	1.34	1.54	1.65
LUFKIN	0.71	0.86	1.19	1.42	1.67	1.81
SAN ANTONIO	0.78	0.89	1.14	1.32	1.49	1.61
TYLER	0.83	0.92	1.11	1.25	1.39	1.47
WACO	0.77	0.89	1.15	1.33	1.53	1.64
Mean	0.81	0.91	1.12	1.28	1.44	1.54
UPPER REGION						
ABILENE	0.77	0.89	1.14	1.33	1.52	1.63
AMARILLO	0.81	0.91	1.12	1.28	1.45	1.54
CHILDRESS	0.87	0.94	1.08	1.18	1.29	1.35
DALHART	0.88	0.95	1.08	1.17	1.27	1.33
DALLAS	0.79	0.91	1.13	1.29	1.46	1.55
EL PASO	0.73	0.88	1.17	1.39	1.61	1.74
GAINESVILLE	0.83	0.92	1.11	1.24	1.38	1.46
LUBBOCK	0.87	0.94	1.08	1.19	1.29	1.36
MARFA	0.81	0.91	1.12	1.27	1.43	1.52
MIDLAND	0.83	0.92	1.11	1.25	1.41	1.48
MINERAL WELLS	0.82	0.92	1.11	1.26	1.41	1.49
PERRIN	0.83	0.92	1.11	1.25	1.39	1.47
PYOTE	0.78	0.89	1.14	1.32	1.51	1.61
SALTILLO	0.82	0.92	1.11	1.26	1.41	1.49
SAN ANGELO	0.79	0.91	1.13	1.31	1.48	1.58
WICHITA FALLS	0.85	0.93	1.09	1.21	1.34	1.41
Mean	0.82	0.92	1.12	1.26	1.42	1.51

6. CONCLUSIONS

The preliminary results obtained from the application of the proposed Rayleigh simulation model to short-term and fragmentary wind data collected at the 21 Air Force stations in Texas are very encouraging. The dimensionless sub-regional fastest-mile wind speed frequency curves may serve as an effective means of estimating fastest-mile wind speeds for prescribed recurrence intervals at an ungauged site in Texas.

7. REFERENCES

Castillo, E. (1988), **Extreme Value Theory in Engineering,** Academic Press, New York, NY.

Cheng, E. D. H. and Chiu, A. N. L. (1985), "Extreme Winds Simulated from Short-Period Records," **Journal of Structural Engineering,** ASCE, Vol. 111, No. 1, January, pp. 77-94.

Cheng, E. D. H. and Chiu, A. N. L. (1993), "Estimating Design Wind Speeds from Short-Term Records," Proceedings of the 3rd Asia-Pacific Symposium on Wind Engineering, Hong Kong, December 13-15, 1993, Vol. I, pp. 519-524.

Cheng, E. D. H. and Chiu, A. N. L. (1994), "Short-Record-Based Extreme Wind Simulation," a paper accepted for publication in the **Journal of Research of the National Institute of Standards and Technology,** U.S. Department of Commerce, Gaithersburg, MD 20899-0001.

Grigoriu, M. (1982), "Estimates of Design Winds from Short Records," **Journal of the Structual Division,** ASCE, Vol. 198, No. ST5, May, pp. 1034-1048.

Gumbel, E. J. (1958), **Statistics of Extremes,** Columbia University Press, New York, NY.

Simiu, E., Filliben, J. J., and Shaver, J. R. (1982), "Short-Term Records and Extreme Wind Speeds," **Journal of the Structural Division,** ASCE, Vol. 108, No. ST11, November, pp. 2571-2577.

Simiu, E. and Scanlan, R. H. (1986), **Wind Effects on Structures,** Wiley-Interscience, New York, NY.

ACKNOWLEDGEMENT

Partial support of this study by the National Science Foundation through Grant BCS-9122224 is gratefully acknowledged.

THEME III
Aerodynamics of Bluff Bodies and Cylinders

QUASI-STEADY MODELS OF AERODYNAMIC LOADS ON TWO CIRCULAR CYLINDERS IN STAGGERED ARRANGEMENT

BIAZIK-BOROWA, E.[1] and FLAGA, A.[2]

[1]*Mr. Sc. Lublin Technological University, Lublin, Poland.*
[2]*Professor of Cracow and Lublin Technological University, Poland*

Abstract

The paper deals with physical and mathematical modelling of wind load on two circular cylinders in staggered arrangement when the downstream cylinder is located in the region of bistable flows between two circular cylinders. The basis of considerations is dimensional analysis of the problem and quasisteady theory approach to the constructing of models of wind load on slender structures Corresponding to the interferential wind load components of wind load on the downstream cylinder as well as on the upstream one, the nonlinear and linear forms of the set of equations of motion of cylinders were determined and analysed. Moreover, the reduced critical velocity of self-induced vibrations of the downstream cylinder was evaluated.

1.INTRODUCTION

Closely spaced structures with cylindrical cross-sections subjected to wind action can be found in numerous applications. Such structures may undergo wind-induced, or wind-structure -interaction-induced vibrations. The oscillation induced by flow-structure interaction of two parallel circular cylinders have attracted much research recently (comp. e.g. Kazakiewicz and Grafskij (1986), Zdravkovich (1991), Ruscheweyh and Dielen (1992)).

For two circular cylinders in staggered or side-by-side arrangements there are three regions i.e. A, B and C (comp. Fig.1)where the downstreamor side cylinder may undergo bistable flow. In these regions two flow regimes can change-over at irregular time intervals. The concomittant change of fluid-dynamic forces can excite and maintain large amplitude oscillations.

Another possibility, which does not exist for a single cylinder, is that the oscillation of one cylinder can strongly influence and synchronize the oscillation of the other one.

The paper aims at modelling of wind load on the downstream cylinder in conditions of aerodynamic interference at following assumptions:

- cross-sections of two cylinders - downstream and upstream one - are circular of identical diameters D;
- cylinders are arranged in a way shown in Fig.2, where $2 < L_x/D < 4$, $5^o < \beta < 15^o$, it means that the downstream cylinder is in the region of bistable flows - region B in Fig.1;
- cylinders are elastically supported, including damping, at the ends in both directions x and y;
- boundary disturbance effects were neglected; the average velocity and fluctuation intensity of the onflowing air were adopted as constant along the whole cylinder length; the problem is treated as a plane one (i.e. two dimensional).

In the region B there may exist some discontinuous changes in the flow between the cylinders. They change rapidly from the flow containing to the flow not containing an additional stream in the gap between the cylinders. Dually established distributions of pressures on the cylinder walls (mainly on the downstream one) changing alternately are the effect of this phenomenon.. It causes additional dynamic loads i.e interferiental loads as a result of which there may occur the vibrations of the cylinders. These vibrations, then, are conductive to the bistable flows. This may, in turn, give rise to synchronization of the cylinder vibration frequency with the frequency of the flow changes. This positive feedback between the interferential wind loads and the response of the cylinders causes a considerable increase of the vibration amplitude of the cylinders (mainly the downstream one).

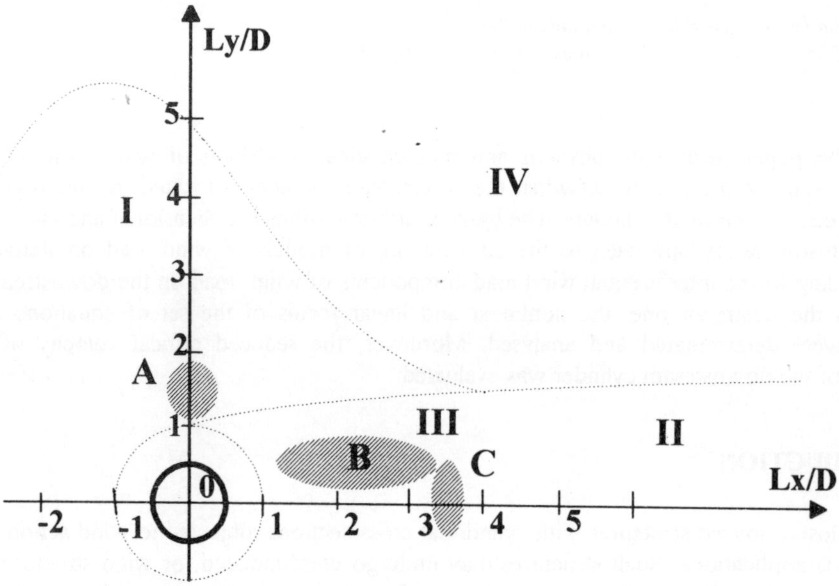

Figure 1. Bistable flow regions after Zdravkovich (1987).

2.MATHEMATICAL MODELS OF WIND LOAD ON THE DOWNSTREAM CYLINDER

Dimensional analysis of the considered problem and quasisteady theory approach to the constructing models of wind load on slender structures were adopted as the basis for determination physical and consequently mathematical models of wind load on the downstream cylinder. This procedure consists of three stages.

2.1.Stage I

In the first stage it was adopted that the flow in front of the cylinders is steady of constant velocity V=const and the cylinders are not subjected to vibrations (Fig.2).The angle of wind attack is constant and equals β. Basing upon investigantions and dimensional analysis following functional relations can be established for dimensionless quantities (i. e. aerodynamic coeficients) of static components of wind load:

$$C_i^I = \frac{W_i^I}{0.5 \cdot \rho V^2 D}\ , \ i = x_1, y_1;\ x_1 = x,\ y_1 = y;\ V = u = \text{const} \qquad (1)$$

$$C_i^I = C_i^I\left(Re, \frac{k_1}{D}, \frac{k_2}{D}, \frac{L_x}{D}, \beta\right) \qquad (2)$$

where: Re - Reynolds number; k_1, k_2 - surface roughness coefficients of the first and second cylinders, respectively; W_x^I, W_y^I - interferential wind loads on the downstream cylinder in directions x and y, respectively.

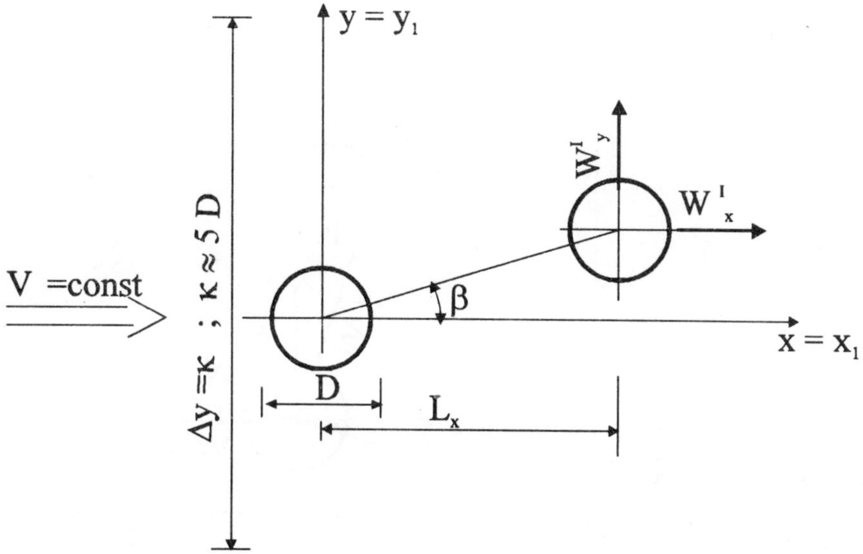

Figure 2. Arrangement of two circular cylinders in stage I.

2.2.Stage II

In the second stage it is adopted that the air onflow is unsteady, but the cylinders do not still vibrate (comp. Fig.3). In this case the spatially averaged angle of oncoming wind γ_s is adopted as the angle of wind attack; it is defined as follows (comp. Fig.3):

$$\gamma_s = \gamma_s(z,t) = \beta - \alpha_s;\ \tan \alpha_s = \frac{v_s}{u_s} \qquad (3)$$

$$\text{where:}\ u_s = u_s(z,t) = \frac{1}{S}\int_S u(x,y,z,t)\,dS\ ;\quad v_s = v_s(z,t) = \frac{1}{S}\int_S v(x,y,z,t)\,dS \qquad (4)$$

The area $S = \Delta y \times 1$ is , as can be seen, the characteristic for the given cylinder arrangement area of spatial averaging of velocity components , located at a distance at least D in front of the cylinders. The dimension Δy of this area is of the order of the "width" of the wake formed behind the cylinders . Mean and fluctuating values of the components $u_s(z,t)$ and $v_s(z,t)$ were obtained as a result of time averaging operation and equal here respecively:

$$\bar{u}_s(z,t)=\bar{u};\ u'_s(z,t)=u_s(z,t)-\bar{u}\ ;\ \bar{v}_s(z,t)=0;\quad v_s(z,t)=v'_s(z,t) \tag{5}$$

In the case of unsteady field of wind velocity in front of the cylinders the wind load on them depends on the fluctuation intensity of the onflowing air I_u . Basing upon dimensional analysis and considering the above remarks and relations the following functional relations was obtained in the second stage:

$$C_i^{II}=C_i^{II}(z,t)=\frac{W_i^{II}(z,t)}{0.5\cdot\rho V_s^2(z,t)D}\ ,\ i=x_2,y_2\ ;\ V_s(z,t)=\frac{u_s(z,t)}{\cos\alpha_s(z,t)} \tag{6}$$

$$C_i^{II}=C_i^{II}\left(Re,\frac{k_1}{D},\frac{k_2}{D},\frac{L_x}{D},\gamma_s,I_u\right) \tag{7}$$

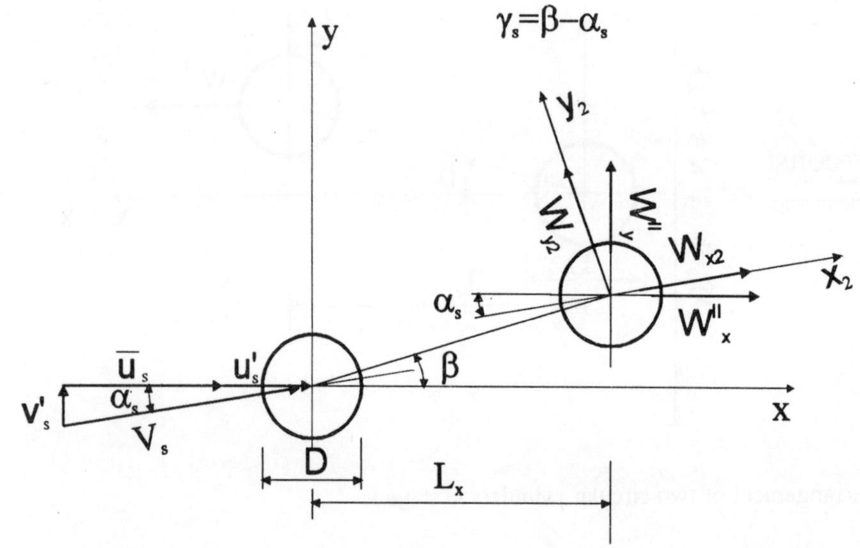

Figure 3. Arrangement of two circular cylinders in stage II.

2.3.Stage III

Apart from variability in time of the onflowing air in the third stage the motion of both cylinders was also considered (comp. Fig.4). Instead of the vector of spatially avereged wind velocity $\vec{V}_s(z,t)$ (as in stage II) the vector of relative velocity $\vec{V}_r$ is now adopted for the downstream cylinder:

$$\vec{V}_r=\vec{V}_r(z,t)=\vec{V}_s(z,t)-\dot{\vec{\eta}}_2(t)-\dot{\vec{\xi}}_2(t) \tag{8}$$

where: $\vec{\dot{\xi}}_2$, $\vec{\dot{\eta}}_2$ - components of motion velocity of the downstream cylinder in directions x and y, respectively. The along-wind component u_r of the relative wind velocity vector $\vec{V_r}$ is given by:

$$u_r = u_r(z,t) = u_s(z,t) - \dot{\xi}_2(t) = \bar{u} + u_s'(z,t) - \dot{\xi}_2(t) \tag{9}$$

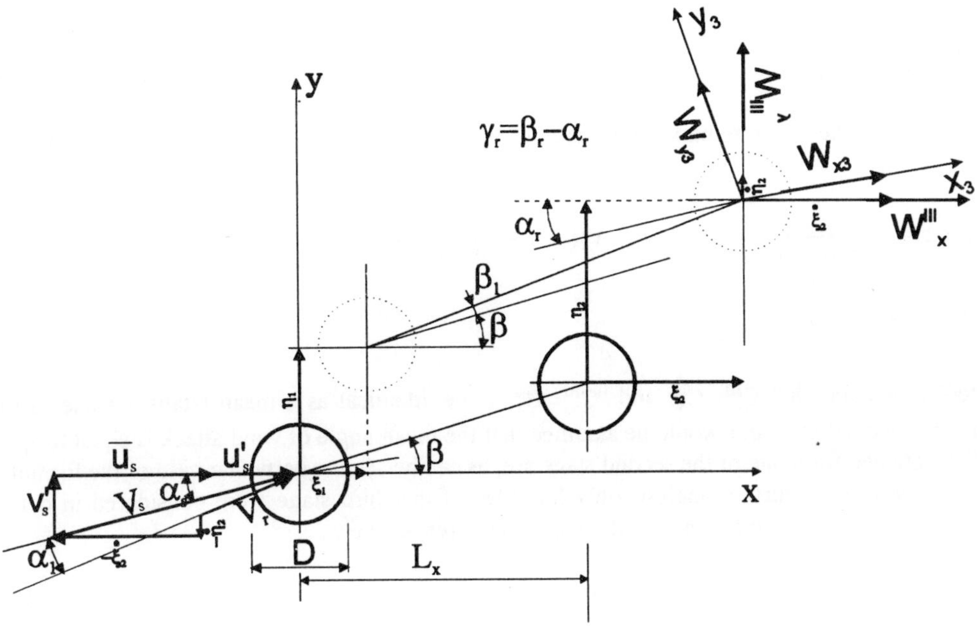

Figure 4. Arrangement of two circular cylinders in stage III.

Instead of the angle of wind attack γ_s (as in stage II) the relative angle of wind attack γ_r was adopted This angle is defined with formulae:

$$\gamma_r = \gamma_r(z,t) = (\beta+\beta_1) - (\alpha_s+\alpha_1) = \beta_r - \alpha_r \; ; \; \alpha_r = \beta_r - \gamma_r \tag{10}$$

$$\gamma_r = \arctan\frac{L_x \tan\beta u_s - L_x \tan\beta \dot{\xi}_2 + \eta u_s - \eta\dot{\xi}_2 - v_s' L_x - v_s'\xi + \dot{\eta}_2 L_x + \dot{\eta}_2 \xi}{L_x u_s - L_x \dot{\xi}_2 + \xi u_s - \xi\dot{\xi}_2 + L_x \tan\beta v_s' - L_x \tan\beta \dot{\eta}_2 + \eta v_s' - \eta\dot{\eta}_2} \tag{11}$$

$$\eta = \eta_2 - \eta_1 \; ; \; \xi = \xi_2 - \xi_1 \tag{12}$$

where: ξ_1, ξ_2, η_1, η_2 - components of upstream and downstream cylinder displacements in direction x and y, respectively.

In this case the relative angle of air onflow α_r is adopted as:

$$\alpha_r = \alpha_s + \alpha_1 \; ; \quad \alpha_r = \arctan\left(\frac{v_s'(t) - \dot{\eta}_2}{u_s - \dot{\xi}_2}\right) \tag{13}$$

As a result of linearization of formulae (11) and (13) we obtain:

$$\gamma_r^1 = \beta + \frac{\eta}{L_x} + \frac{\dot{\eta}_2}{u_s} - \frac{v_s'}{u_s} \tag{14}$$

$$\alpha_r^1 = \frac{v_s' - \dot{\eta}_2}{u_s} \tag{15}$$

Considering the above remarks and relations the following functional relations can be obtained in stage III:

$$C_i^{III} = C_i^{III}(z,t) = \frac{W_i^{III}(z,t)}{0.5 \cdot \rho V_r^2(z,t) D} \; ; \; i = x_3, y_3 ; \quad V_r(z,t) = \frac{u_r(z,t)}{\cos \alpha_r(z,t)} \tag{16}$$

$$C_i^{III} = C_i^{III}\left(Re, \frac{k_1}{D}, \frac{k_2}{D}, \frac{L_x}{D}, \gamma_r, I_u\right) \tag{17}$$

It is adopted, moreover, that $C_{x3}^{III} = C_{x3}^{st}$ and $C_{x3}^{III} = C_{x3}^{st}$, i.e. identical as a mean (static) value of the aerodynamic force coefficient if it would be assumed that the mean angle of wind attack is equal to γ_r.

The obtained formulae of the second stage are, as can be seen, a particular case of the formulae of stage III. Therefore, in further analysis only formulae of the third stage were considered in detail. Components of wind load in directions x and y equal, than, respectively:

$$W_x^{III}(\alpha_r, \gamma_r) = \frac{1}{2} \rho u_r^2 D \frac{1}{\cos\alpha_r}\left[C_{x_3}^{III}(\gamma_r) - C_{y_3}^{III}(\gamma_r) \tan \alpha_r\right] \tag{18}$$

$$W_y^{III}(\alpha_r, \gamma_r) = \frac{1}{2} \rho u_r^2 D \frac{1}{\cos\alpha_r}\left[C_{y_3}^{III}(\gamma_r) + C_{x_3}^{III}(\gamma_r) \tan \alpha_r\right] \tag{19}$$

In above formulae the explicite dependence of particular quantities on parameters: Re, k_1/D, k_2/D, L_x/D, I_u and on dependent variables z and t were neglected. It is:

$$C_x^{III}(\alpha_r, \gamma_r) = \frac{1}{\cos\alpha_r}\left[C_{x_3}^{III}(\gamma_r) - C_{y_3}^{III}(\gamma_r) \tan \alpha_r\right] \tag{20}$$

$$C_y^{III}(\alpha_r, \gamma_r) = \frac{1}{\cos\alpha_r}\left[C_{y_3}^{III}(\gamma_r) + C_{x_3}^{III}(\gamma_r) \tan \alpha_r\right] \tag{21}$$

The above functions can be developed in approximation into Taylor's series aroud the point : $\alpha_r = \bar{\alpha}_r = 0, \gamma_r = \bar{\gamma}_r = \beta$ maintaining only the liner members of development. This can be formulated as follows:

- function described by equation (20)

$$C_x^{III}(\alpha_r, \gamma_r) = C_{x_3}^{III}(\beta) + \left.\frac{\partial C_{x_3}^{III}(\gamma_r)}{\partial \gamma_r}\right|_{\gamma_r = \beta} \cdot (\gamma_r - \beta) - C_{y_3}^{III}(\beta) \cdot \alpha_r \tag{22}$$

- function described by the equation (21)

$$C_y^{III}(\alpha_r,\gamma_r) = C_{y_3}^{III}(\beta) + \left.\frac{\partial C_{y_3}^{III}(\gamma_r)}{\partial\gamma_r}\right|_{\gamma_r=\beta} \cdot (\gamma_r-\beta) + C_{x_3}^{III}(\beta)\cdot\alpha_r \tag{23}$$

In the particular case (stage II) $\alpha_r=\alpha_s$ and $\gamma_r=\gamma_s=\beta-\alpha_s$. Then $\gamma_r-\beta=-\alpha_s$ should be adopted. For the above assumptions it shoud be presumed that:

$$C_{x_3}^{III}(\beta)=C_{x_2}^{II}(\beta)\ ,\ C_{y_3}^{III}(\beta)=C_{y_2}^{II}(\beta)\ ,\ \left.\frac{\partial C_{x_3}^{III}(\gamma_r)}{\partial\gamma_r}\right|_{\gamma_r=\beta} = \left.\frac{\partial C_{x_2}^{II}(\gamma_s)}{\partial\gamma_s}\right|_{\gamma_s=\beta}\ ,\ \left.\frac{\partial C_{y_3}^{III}(\gamma_r)}{\partial\gamma_r}\right|_{\gamma_r=\beta} = \left.\frac{\partial C_{y_2}^{II}(\gamma_s)}{\partial\gamma_s}\right|_{\gamma_s=\beta}$$

Considering the up till now considerations and formulae, the components of wind load on the downstream cylinder W_x^{III} and W_y^{III} for $\gamma_r=\gamma_r^1$ and $\alpha_r=\alpha_r^1$ can be written as follows:

$$W_x^{III} = \frac{1}{2}\rho u_r^2 D\cdot\left\{C_{x_3}^{III}(\beta) + \left.\frac{\partial C_{x_3}^{III}(\gamma_r)}{\partial\gamma_r}\right|_{\gamma_r=\beta}\left(\frac{\eta}{L_x}+\frac{\dot\eta_2}{u_s}-\frac{v_s'}{u_s}\right) - C_{y_3}^{III}(\beta)\cdot\left(\frac{v_s'}{u_s}-\frac{\dot\eta_2}{u_s}\right)\right\} \tag{24}$$

$$W_y^{III} = \frac{1}{2}\rho u_r^2 D\cdot\left\{C_{y_3}^{III}(\beta) + \left.\frac{\partial C_{y_3}^{III}(\gamma_r)}{\partial\gamma_r}\right|_{\gamma_r=\beta}\left(\frac{\eta}{L_x}+\frac{\dot\eta_2}{u_s}-\frac{v_s'}{u_s}\right) + C_{x_3}^{III}(\beta)\cdot\left(\frac{v_s'}{u_s}-\frac{\dot\eta_2}{u_s}\right)\right\} \tag{25}$$

where: $C_{x_3}^{III}(\beta)$, $C_{y_3}^{III}(\beta)$ and their derivatives $\left.\frac{\partial C_{x_3}^{III}(\gamma_r)}{\partial\gamma_r}\right|_{\gamma_r=\beta}$, $\left.\frac{\partial C_{y_3}^{III}(\gamma_r)}{\partial\gamma_r}\right|_{\gamma_r=\beta}$ are determined on the basis of investigations of static components of interferential wind load on the downstream cylinder in stage II (or less precisely in stage I ; the influence of fluctuation intensity I_u on values of these coefficients is then neglected).

3.MATHEMATICAL MODELS OF WIND LOAD ON THE UPSTREAM CYLINDER

The components of wind load on the upstream cylinder could be obtained in the same way as for the downstream one. In the first and second stage substantial difference modelling of wind load don't appear, and I and II stage are particular case of stage III. Therefore, further considerations concern only the last stage.

Basing upon investigantions and dimensional analysis following functional relations can be established for aerodynamic coeficients of static components of wind load on the upstream cylinder (comp. Fig.4):

$$\check{C}_i^{III} = \check{C}_i^{III}(z,t) = \frac{\check{W}_i^{III}(z,t)}{0.5\cdot\rho(\check{V}_r)^2(z,t)\,D}\ ;\ i=x_4,y_4;\quad \check{V}_r(z,t) = \frac{\check{u}_r(z,t)}{\cos\check{\alpha}_r(z,t)}; \tag{26}$$

$$\check{C}_i^{III} = \check{C}_i^{III}\left(Re,\frac{k_1}{D},\frac{k_2}{D},\frac{L_x}{D},\check{\gamma}_r,I_u\right) \tag{27}$$

where: $\check{W}_x^{III}$, $\check{W}_y^{III}$ - interferential wind loads on the upstream cylinder in directions x and y, respectively, in stage III ;

$$\vec{\check{V}}_r = \vec{\check{V}}_r(z,t) = \vec{V}_s(z,t) - \vec{\dot{\eta}}_1(t) - \dot{\xi}_1(t)\ ; \tag{28}$$

$$\check{u}_r = \check{u}_r(z,t) = u_s(z,t) - \dot{\xi}_1(t) = \bar{u} + u_s'(z,t) - \dot{\xi}_1(t) \tag{29}$$

Above functions take into consideration components of motion velocity of the upstream cylinder in directions x and y , i.e. $\vec{\dot{\xi}}_1$, $\vec{\dot{\eta}}_1$ respectively.

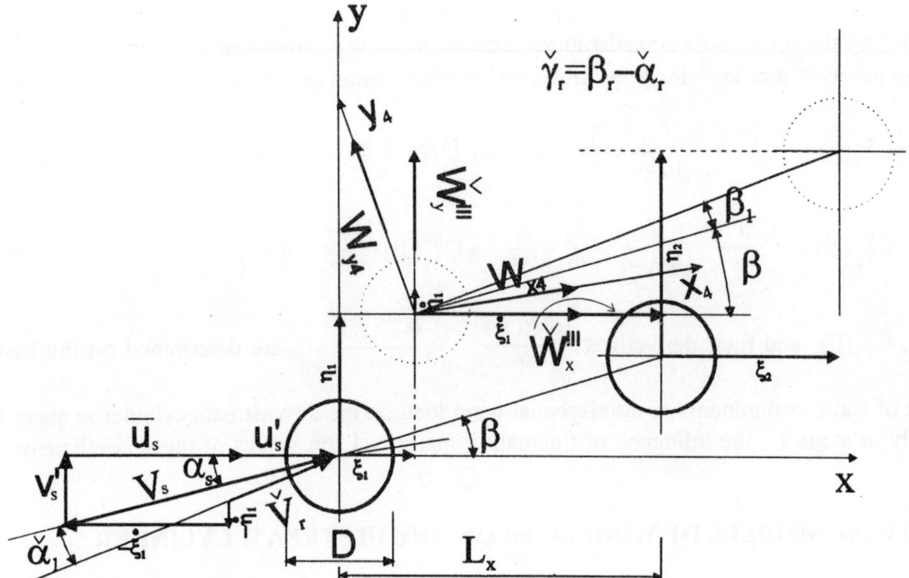

Figure 4. Relation between different quantities for the upstream cylinder.

The relative angles of wind attack $\check{\gamma}_r$ and $\check{\alpha}_r$ can now be given by:

$$\check{\gamma}_r = \arctan \frac{L_x \tan\beta u_s - L_x \tan\beta \dot{\xi}_1 + \eta u_s - \eta\dot{\xi}_1 - v_s' L_x - v_s'\xi + \dot{\eta}_1 L_x + \dot{\eta}_1 \xi}{L_x u_s - L_x \dot{\xi}_1 + \xi u_s - \xi\dot{\xi}_1 + L_x \tan\beta v_s' - L_x \tan\beta \dot{\eta}_1 + \eta v_s' - \eta\dot{\eta}_1} \tag{30}$$

$$\check{\alpha}_r = \arctan\left(\frac{v_s'(t) - \dot{\eta}_1}{u_s - \dot{\xi}_1}\right) \tag{31}$$

As a result of linearization of formulae (30) and (31) one can obtain:

$$\check{\gamma}_r^1 = \beta + \frac{\eta}{L_x} + \frac{\dot{\eta}_1}{u_s} - \frac{v_s'}{u_s} \tag{32}$$

$$\check{\alpha}_r^1 = \frac{v_s' - \dot{\eta}_1}{u_s} \tag{33}$$

Components of wind load on the upstream cylinder in directions x and y , respectively,can be expressed as follows:

$$\check{W}_x^{III}(\check{\alpha}_r, \check{\gamma}_r) = \frac{1}{2}\rho(\check{u}_r)^2 D \frac{1}{\cos\check{\alpha}_r}\left[\check{C}_{x4}^{III}(\check{\gamma}_r) - \check{C}_{y_4}^{III}(\check{\gamma}_r)\tan\check{\alpha}_r\right] \tag{34}$$

$$\check{W}_y^{III}(\check{\alpha}_r, \check{\gamma}_r) = \frac{1}{2}\rho(\check{u}_r)^2 D \frac{1}{\cos\check{\alpha}_r}\left[\check{C}_{y_4}^{III}(\check{\gamma}_r) + \check{C}_{x4}^{III}(\check{\gamma}_r)\tan\check{\alpha}_r\right] \tag{35}$$

In the above formulae the explicite dependence of particular quantities on parameters: Re, k_1/D, k_2/D, L_x/D, I_u and on dependent variables z and t were neglected. It is:

$$\check{C}_x^{III}(\check{\alpha}_r, \check{\gamma}_r) = \frac{1}{\cos\check{\alpha}_r}\left[\check{C}_{x4}^{III}(\check{\gamma}_r) - \check{C}_{y_4}^{III}(\check{\gamma}_r)\tan\check{\alpha}_r\right] \tag{36}$$

$$\check{C}_y^{III}(\check{\alpha}_r, \check{\gamma}_r) = \frac{1}{\cos\check{\alpha}_r}\left[\check{C}_{y_4}^{III}(\check{\gamma}_r) + \check{C}_{x4}^{III}(\check{\gamma}_r)\tan\check{\alpha}_r\right] \tag{37}$$

Similarly as before , the above functions was developed in aproximation into Taylor's series aroud the point : $\check{\alpha}_r = \bar{\check{\alpha}}_r = 0, \check{\gamma}_r = \bar{\check{\gamma}}_r = \beta$. Considering the up till now formulae, the components of wind load on the upstream cylinder for $\check{\gamma}_r = \check{\gamma}_r^1$ and $\check{\alpha}_r = \check{\alpha}_r^1$ can be formulated as follows:

$$\check{W}_x^{III} = \frac{1}{2}\rho(\check{u}_r)^2 D\cdot\left\{\check{C}_{x4}^{III}(\beta) + \left.\frac{\partial \check{C}_{x4}^{III}(\check{\gamma}_r)}{\partial\check{\gamma}_r}\right|_{\check{\gamma}_r=\beta}\left(\frac{\eta}{L_x} + \frac{\dot{\eta}_1}{u_s} - \frac{v_s'}{u_s}\right) - \check{C}_{y_4}^{III}(\beta)\cdot\left(\frac{v_s'}{u_s} - \frac{\dot{\eta}_1}{u_s}\right)\right. \tag{38}$$

$$\check{W}_y^{III} = \frac{1}{2}\rho(\check{u}_r)^2 D\cdot\left\{\check{C}_{y_4}^{III}(\beta) + \left.\frac{\partial \check{C}_{y4}^{III}(\check{\gamma}_r)}{\partial\check{\gamma}_r}\right|_{\check{\gamma}_r=\beta}\left(\frac{\eta}{L_x} + \frac{\dot{\eta}_1}{u_s} - \frac{v_s'}{u_s}\right) + \check{C}_{x4}^{III}(\beta)\cdot\left(\frac{v_s'}{u_s} - \frac{\dot{\eta}_1}{u_s}\right)\right\} \tag{39}$$

where: $\check{C}_{x4}^{III}(\beta$, $\check{C}_{y_4}^{III}(\beta)$ and their derivatives $\left.\frac{\partial \check{C}_{x_4}^{III}(\check{\gamma}_r)}{\partial\check{\gamma}_r}\right|_{\check{\gamma}_r=\beta}$, $\left.\frac{\partial \check{C}_{y4}^{III}(\check{\gamma}_r)}{\partial\check{\gamma}_r}\right|_{\check{\gamma}_r=\beta}$ are determined on the basis of investigations of static components of interferential wind load on the upstream cylinder in stage II (or less precisely in stage I ; the influence of fluctuation intensity I_u on values of this coefficients is then neglected).

4.EQUATIONS OF MOTION OF THE CYLINDERS

Treating the downstream and upstream cylinders as physically linear systems of two degrees of fredom the set of equations of motion of those cylinders can be written as follows:
- upstream cylinder:

$$m\ddot{\xi}_1(t) + C_{\xi_1}\dot{\xi}_1(t) + K_{\xi_1}\xi_1(t) = \frac{1}{H}\int_0^H \check{W}_x^{III}(z,t)\,dz = W_{\xi_1}(t) =$$

$$= \frac{1}{H}\int_0^H \frac{1}{2}\rho(\check{u}_r)^2 D\cdot\left\{\check{C}_{x4}^{III}(\beta) + \left.\frac{\partial \check{C}_{x4}^{III}(\check{\gamma}_r)}{\partial\check{\gamma}_r}\right|_{\check{\gamma}_r=\beta}\left(\frac{\eta}{L_x} + \frac{\dot{\eta}_1}{u_s} - \frac{v_s'}{u_s}\right) - \check{C}_{y_4}^{III}(\beta)\cdot\left(\frac{v_s'}{u_s} - \frac{\dot{\eta}_1}{u_s}\right)\right\}dz \tag{40}$$

$$m\ddot{\eta}_1(t)+C_{\eta_1}\dot{\eta}_1(t)+K_{\eta_1}\eta_1(t)=\frac{1}{H}\int_0^H \check{W}_y^{III}(z,t)\,dz=W_{\eta 1}(t)=$$

$$=\frac{1}{H}\int_0^H \frac{1}{2}\rho(\check{u}_r)^2 D\cdot\left\{\check{C}_{y_4}^{III}(\beta)+\left.\frac{\partial\check{C}_{y_4}^{III}(\check{\gamma}_r)}{\partial\check{\gamma}_r}\right|_{\check{\gamma}_r=\beta}\left(\frac{\eta}{L_x}-\frac{\dot{\eta}_1}{u_s}-\frac{v_s'}{u_s}\right)+\check{C}_{x_4}^{III}(\beta)\cdot\left(\frac{v_s'}{u_s}-\frac{\dot{\eta}_1}{u_s}\right)\right\}dz \qquad (41)$$

- downstream cylinder:

$$m\ddot{\xi}_2(t)+C_{\xi_2}\dot{\xi}_2(t)+K_{\xi_2}\xi_2(t)=\frac{1}{H}\int_0^H W_x^{III}(z,t)\,dz=W_{\xi_2}(t)=$$

$$=\frac{1}{H}\int_0^H \frac{1}{2}\rho u_r^2 D\cdot\left\{C_{x_3}^{III}(\beta)+\left.\frac{\partial C_{x_3}^{III}(\gamma_r)}{\partial\gamma_r}\right|_{\gamma_r=\beta}\left(\frac{\eta_2}{L_x}+\frac{\dot{\eta}_2}{u_s}-\frac{v_s'}{u_s}\right)-C_{y_3}^{III}(\beta)\cdot\left(\frac{v_s'}{u_s}-\frac{\dot{\eta}_2}{u_s}\right)\right\}dz \qquad (42)$$

$$m\ddot{\eta}_2(t)+C_{\eta_2}\dot{\eta}_2(t)+K_{\eta_2}\eta_2(t)=\frac{1}{H}\int_0^H W_x^{III}(z,t)\,dz=W_{\eta_2}(t)=$$

$$=\frac{1}{H}\int_0^H \frac{1}{2}\rho u_r^2 D\cdot\left\{C_{y_3}^{III}(\beta)+\left.\frac{\partial C_{y_3}^{III}(\gamma_r)}{\partial\gamma_r}\right|_{\gamma_r=\beta}\left(\frac{\eta_2}{L_x}+\frac{\dot{\eta}_2}{u_s}-\frac{v_s'}{u_s}\right)+C_{x_3}^{III}(\beta)\cdot\left(\frac{v_s'}{u_s}-\frac{\dot{\eta}_2}{u_s}\right)\right\}dz \qquad (43)$$

where: m - mass per cylinder length unit; $C_{\xi_1}, C_{\eta_1}, K_{\xi_1}, K_{\eta}$, $C_{\xi_2}, C_{\eta_2}, K_{\xi_2}, K_{\eta_2}$ - coefficients of cylinder damping and stiffness of the upstream and downstream cylinders, respectively, in directions x and y; H - cylinder length. The last four differential equations constitute a complex system of stochastic, coupled differential equations, it means, that the vibration of the downstream cylinder, besides many other factors, depends on the motion of the upstream cylinder.

5. EVALUETING OF THE REDUCED CRITICAL VELOCITY

After a number of succesive simplifications and linearizations it is possible to evaluate the reduced critical velocity of self-induced vibrations of the downstream cylinder. Because of the complexity of the problem and on the basis of experimental results Kazakiewicz and Grafskij (1985), Zdravkovich and Medeiros (1991), Ruscheweyh (1983) it is assumped that the vibrations of the upstream cylinder are small in comparison to the vibrations of the downstream one i.e.:

$$\eta_1 \ll \eta_2\,,\ \xi_1 \ll \xi_2 \qquad (44)$$

Then, the resulting damping coefficient C^* can be defined on the base of equation (43) is as follows:

$$C^*=C_{\eta_2}-\frac{1}{H}\int_0^H\left\{\frac{1}{2}\rho u_r^2 D\left(-C_{x_3}(\beta)\cdot\frac{1}{u_s}+\left.\frac{\partial C_{y_3}}{\partial\gamma_r}\right|_{\gamma_r=\beta}\cdot\frac{1}{u_s}\right)\right\}dz \qquad (45)$$

Assuming further simplifications:

$$u_r\cong u_s\,,\quad \bar{u}\cong u_{cr}\,,\quad \frac{1}{H}\int_0^H\frac{u_r^2}{u_s}\,dz\cong\frac{1}{H}\int_0^H u_s\,dz=\bar{u} \qquad (46)$$

equation (45) takes the form:

$$C^{*}= C_{\eta 2}-\frac{1}{2}\rho u_{cr}D\left[-C_{x_3}(\beta)+\frac{\partial C_{y_3}}{\partial \gamma_r}\bigg|_{\gamma_r=\beta}\right] \qquad (47)$$

The necessary condition for self-induced vibrations is in the analysed case condition $C^{*}\leq 0$, then, the reduced critical velocity U_r^{cr} can be wirtten as :

$$U_r^{cr}=\frac{u_{cr}}{f_0 D}=\frac{2\,Sc}{\frac{\partial C_{y_3}^{III}(\gamma_r)}{\partial \gamma_r}\Big|_{\gamma_r=\beta}-C_{x_3}^{III}(\beta)}\,\frac{f}{f_0}=\frac{2\,Sc}{\frac{\partial C_{y_2}^{II}(\gamma_s)}{\partial \gamma_s}\Big|_{\gamma_s=\beta}-C_{x_2}^{II}(\beta)}\,\frac{f}{f_0} \qquad (48)$$

where: f_0 - eigenfrequency of the cylinder; f - frequency of cylinder self - induced vibrations; Sc - -Scruton number ($Sc= 2m\Delta / \rho D^2$).

6.CONCLUSIONS

Formula (48) shows that the reduced critical velocity U_r^{cr} grows proportionally with the increase of the Scruton number Sc and frequency f.This statement is confirmed by experimental results (comp. e.g. Kazakiewicz et al. (1983), Zdravkovich (1987), Knisley and Kawagoe (1988).

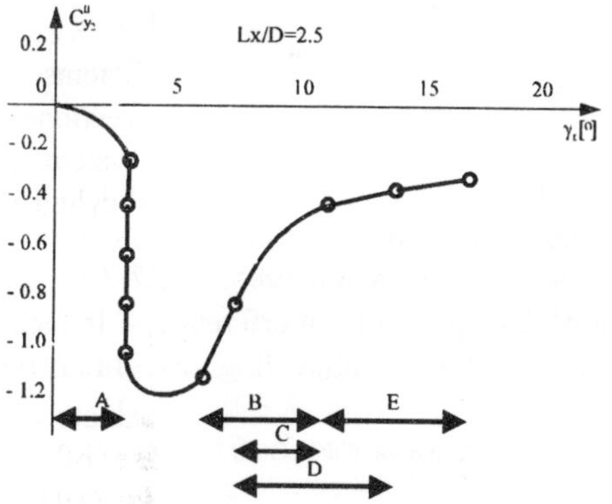

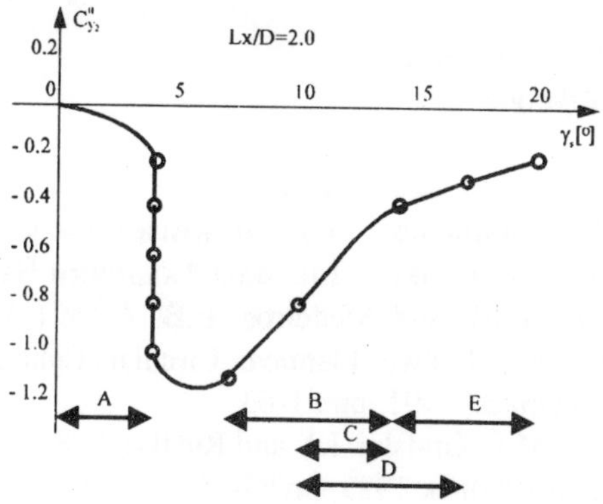

Figure 5. Graphs of the function $C_{y_2}^{II}(\gamma_s$ at $Re = 6 \cdot 10^4$.

Basing upon experimental results obtained by Zdravkovich (1987) two graphs of the function $C_{y_2}^{II}(\gamma_s$ were elaborated and they are shown in Figure 5. Evaluted on this basis values of derivative $\frac{\partial C_{y2}}{\partial \gamma_s}$ and coefficient $\psi = 2/(\frac{\Delta C_{y2}^{II}}{\Delta \gamma_s}-C_{x_2}^{II}(\beta))$ for four different intervals of the angle γ_s A, B, C and D are presented in Table 1. Because of jump character of the function $C_{y_2}^{II}(\gamma_s)$ in the interval A, determination of the derivative $\frac{\partial C_{y2}}{\partial \gamma_s}$ and coefficient ψ in the intervals B and C only seems to be reliable in practice. Values of the coefficient ψ corresponding to these two intervals are less than 0.4 and bigger than 0.2. Values of the reduced critical velocity U_r^{c} evaleated on the base of the simplified formula (48) and obtained in an experiment in natural scale by Studnickova (1993) are of the same order.

Rurther theoretical and experimental works concerning the analysed problem without assumed simplifications (i.e. basing upon full set of differential equatios of motion of the cylinders) are required

Table 1.

	$L_X/D = 2.0$						$L_X/D = 2.5$					
	ΔC^{II}_{y2} [/]	$\Delta\gamma_s$ [rad]	$\Delta C^{II}_{y2}/\Delta\gamma_s$ [/]	C^{II}_{X2} [/]	$\Delta C^{II}_{y2}/\Delta\gamma_s$ $-C^{II}_{x2}$	α [/]	ΔC^{II}_{y2} [/]	$\Delta\gamma_s$ [rad]	$\Delta C^{II}_{y2}/\Delta\gamma_s$ [/]	C^{II}_{X2} [/]	$\Delta C^{II}_{y2}/\Delta\gamma_s$ $-C_X$	α [/]
A	-0.2	0.07	-2.86	-0.2	-2.66	-0.75	-0.2	0.05	-4	0	-4	-0.5
B	0.7	0.12	5.83	-0.2	6.03	0.33	0.65	0.7	0.93	0	0.93	2.15
C	0.4	0.06	6.67	-0.2	6.87	0.29	0.46	0.05	9.2	0	9.2	0.22
D	0.46	0.12	3.83	-0.2	4.03	0.5	0.49	0.1	4.9	0	4.9	0.41
E	0.12	0.11	1.09	-0.2	1.29	1.55	0.08	0.09	0.89	0	0.89	2.25

7.REFERENCE

Kazakiewicz, M.I. and Grafskij, I.J., (1985), "Идентификация ультрагармонических автоколебании при аэродинамическои интерференции тандема круговых цилиндров в скошенном потоке", Доклады Академии Наук Украинскои ССР, Киев 1985, pp 27-30.

Zdravkovich, M.M. and Medeiros, E.B. (1991), "Effect of Damping on Interference - Induced Oscillations of Two Identical Circular Cylinders", Jour. of Wind Engineering and Industrial Aerodynamics, 1991, pp 51-63.

Kazakiewicz, M.I., Grafskij, I.J. and Riedko, C.F. (1983), " Аэродинамика плохообтекаемых тел", Днепропетровск 1983, pp 93-106.

Zdravkovich, M.M. (1987), "Review of Interference-Induced Oscillations in Flow Past Two Parallel Circular Cylingers in Various Arrangements", Seventh International Conference on Wind Engineering, vol.2, Aachen 1987, pp 51-66.

Knisely, C.W. and Kawagoe, M. (1988), "Force - Displacement Measurements on Closely Spaced Tandem Cylinders", A Selection of Papers Presented at the International Colloqium on Bluff Body Aerodynamic and its Applications, Japan, October 17-20, 1988, pp 81-89.

Studnickova, M. (1993), "Galloping of the Leeward Cylinder in Tandem - Application to Bundled Cables ", Czech Technical University in Prague, vol.33,1993 , pp 55-67.

Ruscheweyh,H. (1983), "Aeroelastic Interference Effects Between Slender Structures"; Journal Wind Enginering and Industrial Aerodynamics, vol. 14, 1983, pp 129-140.

Ruscheweyh,H. and Dielen, B. (1992), "Interference galloping-investigations concerning the phase lag of the flow switching", Jour. of Wind Engineering and Industrial Aerodynamics, vol.41-44, 1992, pp 2047-2056.

9ICWE, 1995, New Delhi, India

INTERFERENCE EFFECTS ON THE WIND PRESSURE DISTRIBUTION AROUND A CIRCULAR STRUCTURE DUE TO A RECTANGULAR ELEVATOR

GOWDA, B.H.L.[1], GERHARDT, H.J.[2], KRAMER, C.[2] and JANSER, F.[2]

[1]*Department of Applied Mechanics, Indian Institute of Technology, Madras 600 036, India.*
[2]*Department of Aeronautics, Fachhochschule Aachen, Goethestrasse 1, 56066, Germany.*

Abstract

The wind pressure distribution around a circular structure with a adjoining rectangular elevator block are presented in this paper. The results are obtained at various wind directions and indicate the considerable changes that occur in the lift and drag coefficient of the circular structure. Flow visualisation results are presented which enable a better understanding of the flow field around the structure.

1. INTRODUCTION

The study carried out is on a practical industrial aerodynamic problem concerning the wind pressure distribution on the model of the proposed structure for Star Brewery, Essen, Germany. The structure is circular in shape (32 m in diameter) with a permeable fascade, the gap between the fascade and the main structure being 510 mm. A rectangular block (8.2 m x 5.2 m) is proposed outside the main building which acts as the elevator room. Figure 1 shows the plan view of the building and the elevator block. The height of the circular structure is 120 m and the elevator block extends beyond it to a height of 4 m. It is proposed to have a helipad at the top of the building. It can be expected that the wind pressure distribution and the dynamic response (i.e. response to flow induced vibrations) of the circular shaped main building will get considerably altered due to the presence of the elevator block which has a sufficiently large dimension compared to the main structure. The magnitude of the altered wind pressure distribution is essential for the design purpose and further indicate the tendency of the structure to galloping vibration. Also the study would reveal the local separated regions for various wind directions.

It is proposed for architectural purposes to provide blocks which are approximately conical in shape with a base width of 80 mm and height of 200 mm all round the structure at intervals of 10°. They act as roughness elements in influencing the pressure distribution around the circular building. In the present study the results pertaining to the wind pressure distributon on the outer surface of the structure at various wind directions with and without roughness elements are determined.

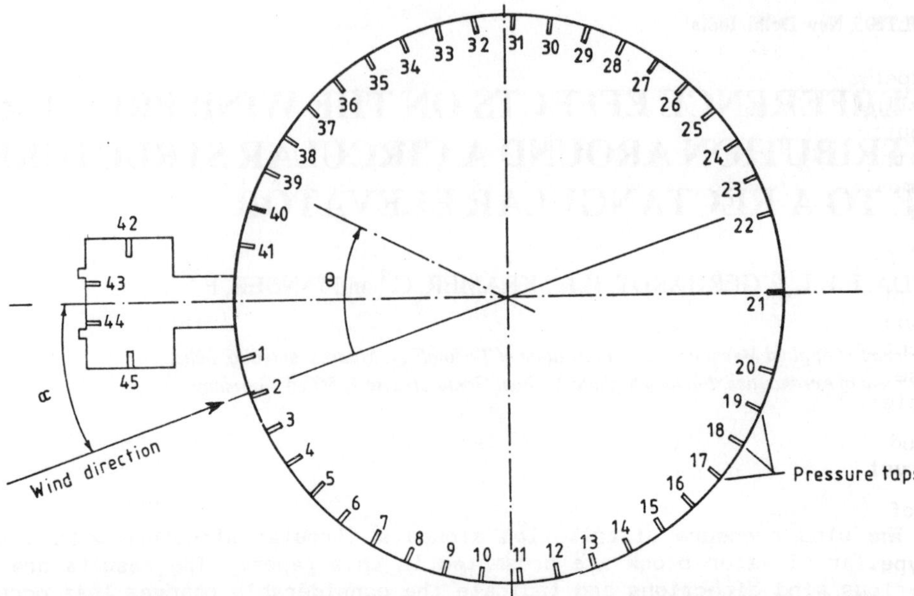

Fig.1 Plan view of the model with pressure tap location

2. EXPERIMENTAL ARRANGEMENT

The wind tunnel tests were carried out on a 1:64 model. A sectional model with three floors (height of each floor 4 m) was utilised for the purpose. The tests were carried out in the open jet, EIFFEL type wind tunnel of the Fachhochschule, Aachen. The exit nozzle has a hexagonal section with a maximum height of 1.41 m and a width of 2 m. The length of the test section is 4 m. For the present study, the test section was modified with a support base plate and a top plate spanning the entire length of the test section. The height between these two plates was 1.26m. The model with the circular cross section and the rectangular elevator block was held at the centre of the test section between the base and the top plates. A turn table with a diameter of 1.9 m is provided which enables the positioning of the model at various angles to the flow. The central portion of the model representing the three floors of the building was made out of PVC and the middle floor was pressure tapped at regular intervals (Fig.1). The rectangular block was hollow and made out of 18 mm thick wooden planks with a smooth outer surface. Pressure taps were provided on all sides of the rectangular block. The pressure taps were connected to a Scani Valve system (Make: Scanivalve Corp. USA; Model J) with 48 ports and a automatic data acquisition system. All the measurements were carried out at a tunnel free stream velocity of 35 m/s which corresponds to a Reynolds number of 1.17×10^6 referred to the diameter of the circular section of the model. The turbulence level in the tunnel is 0.4%. A scaled version of the conical blocks proposed for architectural treatment were provided in the wind tunnel tests. The results pertaining to these are referred to as 'with roughness' in the figures and in the discussion presented.

Flow visualisation results are presented which give a qualitative picture of the flow pattern around the proposed structure and also enable a better understanding of the wind pressure distribution.

3. RESULTS AND DISCUSSION

Experiments were carried out with and without the roughness elements around the cylinder to obtain an idea of the relative changes in the pressure distribution between the two cases. As a first step the pressure distribution around the cylindrical model without the elevator block was determined to check and compare with the pressure distribution around a circular cylinder. The results are shown in Fig.2 where C_p is plotted against θ; here $C_p = (p - p_\infty)/1/2\ \rho U_\infty^2$ where p = pressure at any point on the model, p_∞ and U_∞ are the free stream pressure and velocity respectively. As can be seen they correspond well with the results for a circular cylinder at supercritical Reynolds number (Schlichting, 1960). But it is observed that the maximum negative pressures decrease by nearly a factor of 2 for the case with roughness. The drag coefficient C_D was obtained by integrating the C_p distribution. The values are 0.49 and 0.67 for the case without roughness and with roughness respectively. These values compare well with those given by Achenbach (1977). Further the C_D value of 0.67 for the cylinder with roughness indicates an effective Reynolds number around 10^7, which is very close to the Reynolds number expected in the prototype.

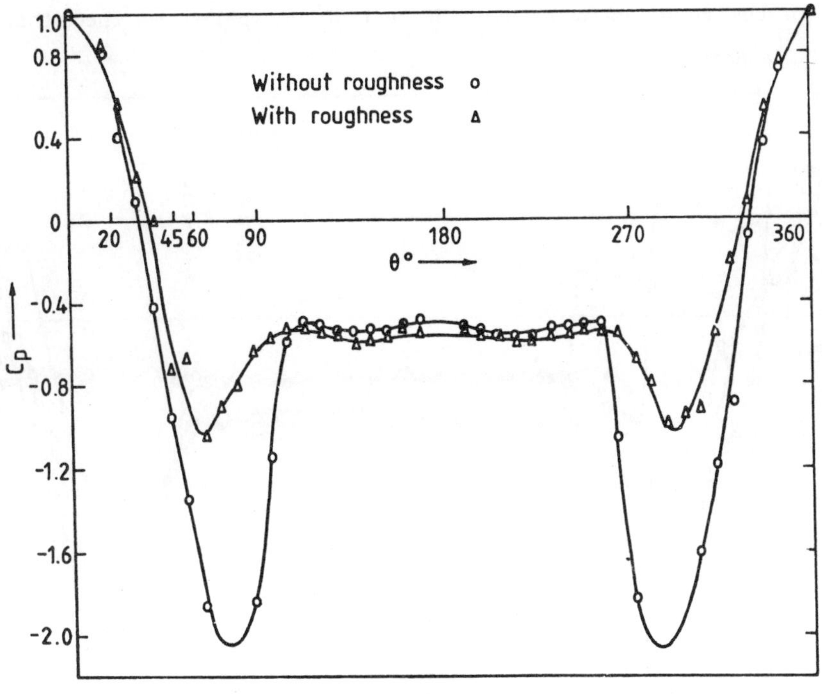

Fig.2 C_p vs θ for cylindrical model (without elevator)

The pressure distribution around the model with the elevator block (Fig. 1) was obtained for α = 0, 5, 10, 15, 20, 30, 45, 60, 90, 120, 150 and 180 degrees for the cases without and with roughness. Results at some typical angles for these cases are shown in Figs. 3 and 4.

The trend in the pressure distribution in Figs. 3 and 4 at corresponding values of α are quite similar particularly in the portion upstream of the rectangular block. Downstream of the block, for $\theta > 270$, higher C_p values (i.e. less negative) occur for the case 'with roughness' compared to that 'without roughness' similar to what is seen in Fig. 2.

Some of the striking features common to both cases seen in Figs. 3 and 4, obviously due to the presence of the elevator block are:

1) As α increases from 0 to 90°, there is a portion on cylinder with $C_p \approx 1$ in the region upstream of the block. The extent over which $C_p \approx 1$ increase with α in this range.

2) For α = 0 to 20°, $C_p \approx -0.45$ over the range $90^o < \theta < 260^o$ and for α = 30 to 90°, $C_p \approx -0.8$ over $90^o < \theta < 260^o$. For $\alpha > 90^o$, the C_p values once again tend towards -0.4 in this range of θ.

Such variations obviously influence the values of C_D. Higher C_D values can be expected in the range of α between 30 and 90° compared to the other values.

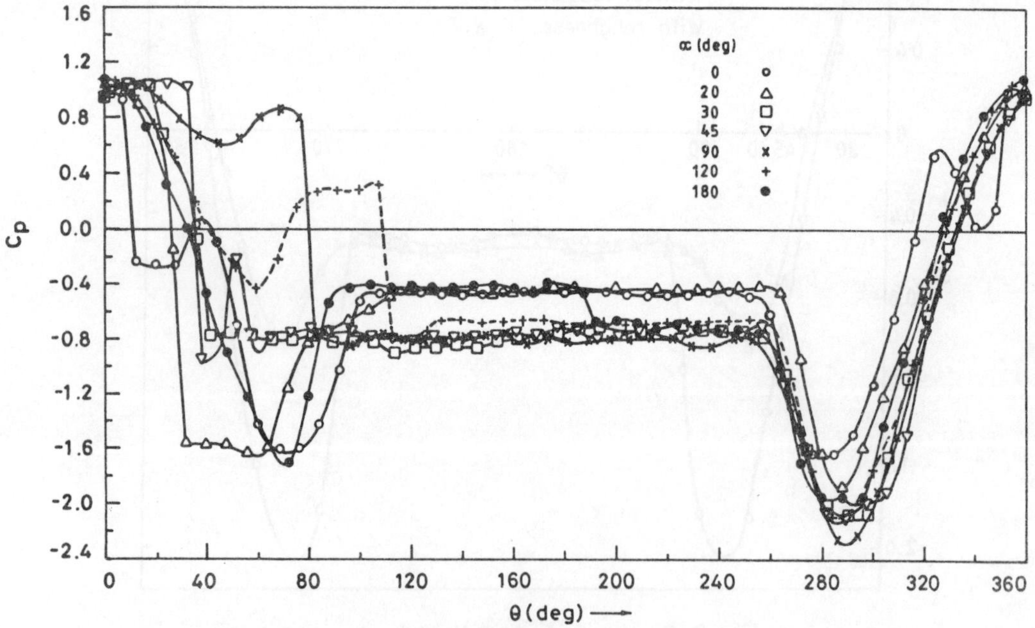

Fig.3 C_p vs θ for various values of α (without roughness)

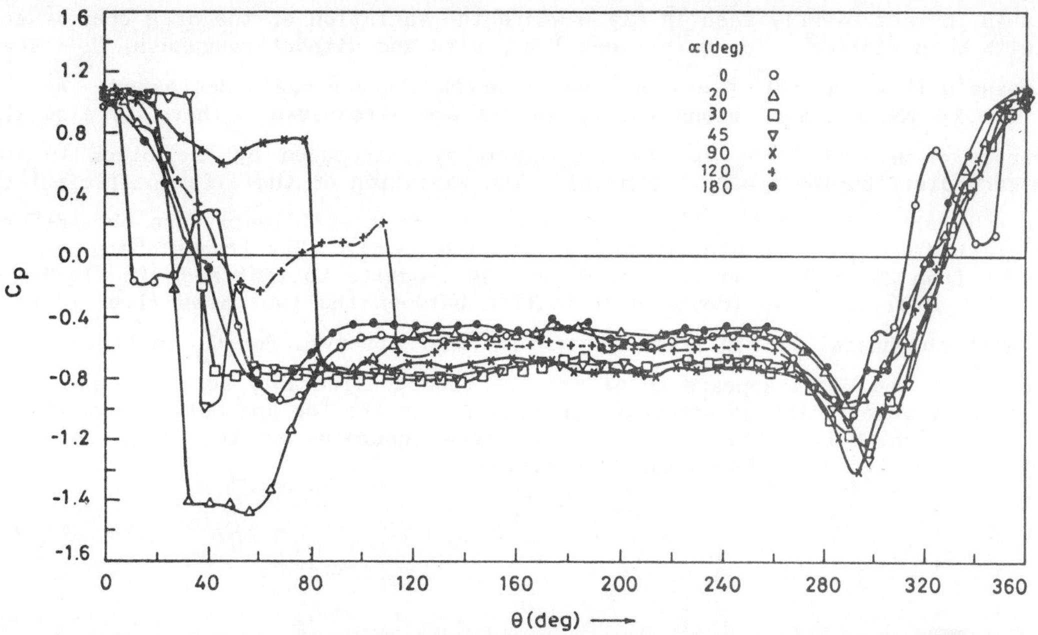

Fig. 4 C_p vs θ for various values of α (with roughness)

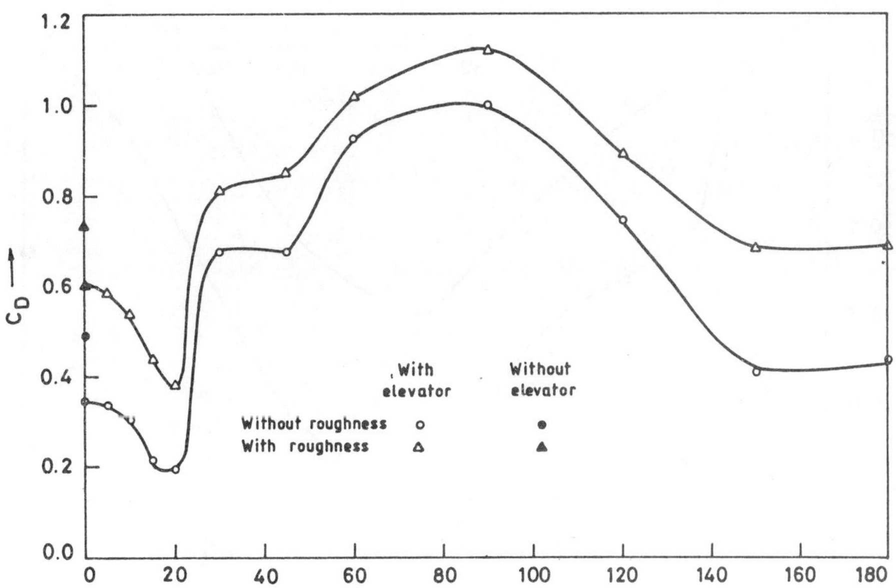

Fig. 5 Variation of C_D with α

This is in fact clearly seen in Fig. 5 where the variation of the drag coefficient C_D with α is plotted. For both cases i.e., with and without roughness, C_D values increase with α over $30^\circ \leq \alpha < 90^\circ$ beyond which it once again decreases. At $\alpha = 0^\circ$, due to the elevator block the C_D values are attenuated. There is also the presence of a lift force due to the asymmetry introduced by the block in the pressure distribution around the model. The variation of the lift coefficient C_L with α is shown in Fig. 6. (Both the drag and lift coefficients were obtained by integrating the pressure distribution around the model). The trends change when α changes from 20 to 30° and at $\alpha = 90^\circ$ but is opposite to that seen in Fig. 5 for C_D. The differences in the C_L distribution between the two cases (i.e. without and with roughness) in Fig. 6 is larger than that observed for C_D in Fig. 5. The presence of the block appears to affect the flow considerably more and give rise to larger asymmetry in the pressure distribution on the top and bottom portions of the model (which contribute to the lift force) compared to the front and rear portions (which contribute to the drag force).

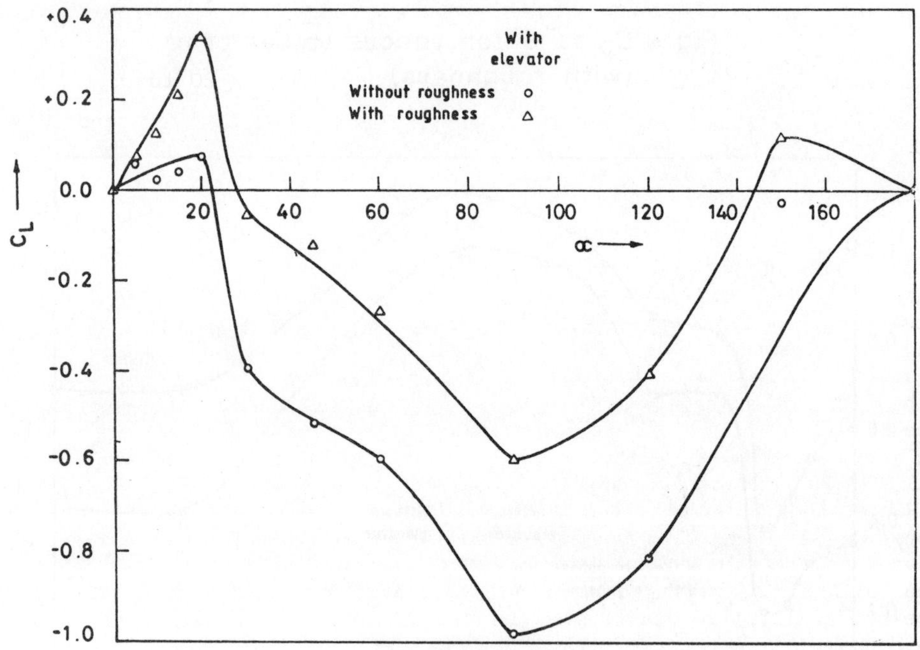

Fig. 6 Variation of C_L with α

To obtain a physical insight into the changes in the flow field that would result around the circular structure due to the presence of the elevator block, flow visualisation studies were carried out on a model. For this purpose flow visualisation tank facility with recirculating water is utilised. Fine aluminium powder is used as the tracer particles. The flow Reynolds number is 10^4 which is much lower than that for the wind tunnel experiments. Nevertheless the results give a qualitative picture of the flow patterns at various values of α as brought out in the discussion which follows. The experiments were carried out at $\alpha = 0$, 10, 20, 30, 45, 60, 90, 120, 150 and 180 degrees. But only results at typical cases are presented in Figs. 7 and 8.

Fig. 7a is for a circular cylinder without the rectangular block. This is to provide a comparison for other cases and bring out the influence of the rectangular block. With the rectangular block, at $\alpha = 0$ (Fig. 7b), there are two small recirculation zones on either side of the block obviously due to the flow separation that occurs from the front edges. The flow separation on the cylinder occurs at almost the same locations as in Fig. 7a. At $\alpha = 20^\circ$, in addition to the localised separation zone around the block, the separation points on the cylinder occur at larger values of θ resulting in a smaller wake width. This would be expected to give rise to a reduced C_D values which is in fact seen in Fig. 5. There is a striking change in the flow pattern when α changes from 20 to 30° (Fig. 7c). Due to the shear layer separating from the rectangular block, separated flow occurs on the entire upper portion of the cylinder. Obviously the pressure distribution can be expected to be much different from that at $\alpha = 20^\circ$. This is seen to be the case in Figs. 3 and 4 where there is considerable variation in the C_p distribution between $\alpha = 20^\circ$ and $\alpha = 30^\circ$. Further there is a change in the trend of variation in the C_D and C_L when α changes from 20 to 30°. The flow patterns at Figs. 7b and 7c provide an explanation for the observed trends. As the angle increases further (Fig. 8a) the base flow almost extends upto the block with obvious influence on the pressure distribution. The flow field at $\alpha = 90$ and 120° are shown in Figs. 8b and c and in Fig. 8d is shown the case when the block is at the rear.

It is clear to see that due to the rectangular block the flow field around the cylindrical structure changes considerably. The block itself might experience severe pressure fluctuations due to the localised separation zones and also the resulting periodic nature of the flow. Another interesting point is the strong recirculating flows that can occur in the vicinity of the proposed structure. Though the flow visualisation results show flow pattern at Re = 10^4 whereas in the prototype Re 10^7, there is correspondence between the two. It is known that periodic vortex shedding around a circular cylinder is suppressed for $3 \times 10^5 \leq Re < 3.5 \times 10^6$, and the shedding is reestablished for $Re > 3 \times 10^6$ (Blevins, 1977). Hence the flow patterns shown in Figs. 7 and 8 reflect particularly conditions for the case with roughness where the equivalent Re is around 10^7 and are relevant to the physical situation. The frequency of vortex shedding could be different in the two cases (i.e. flow visualisation results and those for the prototype) but qualitative similarity can be expected between the results shown in Figs. 7 & 8 and the flow around the prototype.

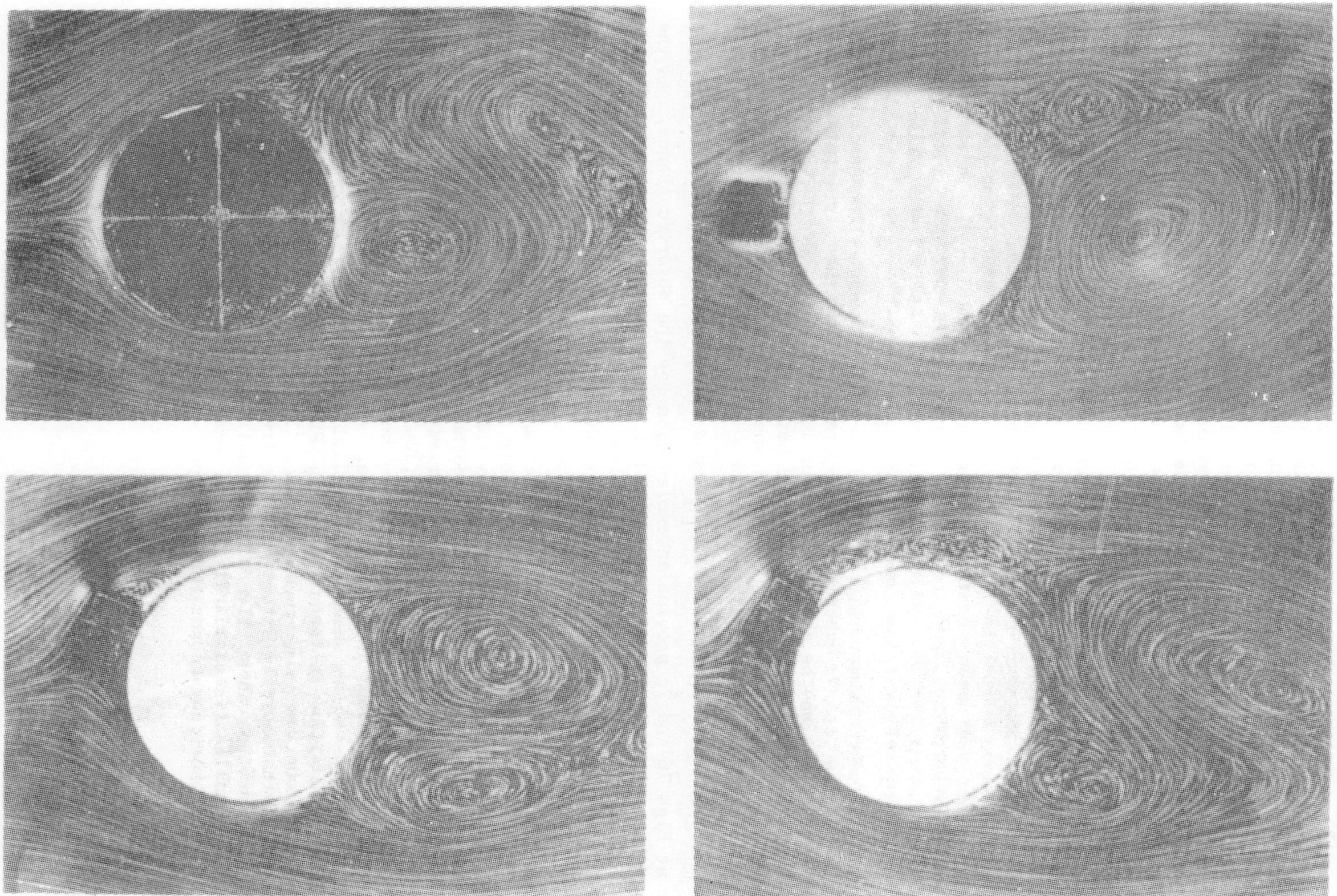

c) α = 20°

d) α = 30°

Fig.7 Flow patterns: $\alpha = 0, 20, 30$ deg

a) $\alpha = 45°$

b) $\alpha = 90°$

c) $\alpha = 120°$

d) $\alpha = 180°$

Fig.8 Flow patterns : $\alpha = 45, 90, 120, 180$ deg

4. CONCLUDING REMARKS

The wind tunnel studies presented have brought out the influence of the elevator block on the wind pressure distribution around the proposed structure. One of the prime effects of the block is seen to be the generation of positive pressures ($C_p \approx 1$) on a considerable portion of the cylindrical shape for wind directions 30 to 90°. The drag and lift coefficients also are altered considerably due to the interference effects of the rectangular block.

The flow visualisation results presented correspond with the wind tunnel results and provide explanation to some of the features observed. The results show that the elevator itself might experience strong periodic forces. Further, the results reveal that there is a possibility of strong periodic flow components in the near wake of the proposed structure.

5. REFERENCES

Achenbach, E. (1977), "The effect of surface roughness on the heat transfer from a circular cylinder to the cross flow of air", Int. Jour. of Heat and Mass Transfer, Vol.20, pp.359-369.

Blevins, R.D. (1977), Flow Induced Vibration, Van Nostrand Reinhold Company.

Schlichting, H. (1960), Boundary Layer Theory, 4th Edition, McGraw-Hill Company.

VORTEX SHEDDING RESPONSE OF FLARE BOOM CIRCULAR MEMBERS

STRØMMEN, E., Aas-JAKOBSEN, K. and HJORTH-HANSEN, E.

University of Trondheim, The Norwegian Institute of Technology, Department of Structural Engineering, 7034 Trondheim, Norway.

Abstract

This paper concerns the dynamic response due to vortex shedding of flare boom circular members. Two series of wind tunnel tests have been carried out. In the first set-up emphasis was put on fulfilment of the similarity requirements for a full-scale member with length 3.5 m and diameter 100 mm, with the cylinder perpendicular to the flow and with pinned supports. The mitigation effect of additional sleeves at various locations were investigated, demonstrating that two short sleeves spanwise located at 1/3- and 2/3-points were effective. For the second series of tests, carried out in smooth as well as in turbulent flow, the set-up was equivalent to a spring-mounted section-model set-up. During these tests the cylinder, with or without sleeves, were tested at various yaw angles, indicating that the sleeves are less beneficial at higher yaw angles.

1. INTRODUCTION

The background to the present wind tunnel investigation was the observations of fatigue cracking which has occurred in flare-boom members (mainly in the struts supporting the flare-boom itself, see Fig.1) on several off-shore platforms in the North Sea. In this particular case it is the Statoil Gullfaks B platform which has been in focus of attention, as the installation of mitigation devices was due to be carried out early summer of 1993 during a short period of production close-down. The fatigue cracking was deemed to be caused by regular vortex shedding effects of individual members, or possibly by a flow interaction effect. Thus, repair would only plaster the problem, and therefore the platform owner decided to make an attempt to eliminate or sufficiently reduce the problem by some sort of non-structural device which could suppress the vortex shedding.

The observed cracks occurred mainly at or in the welded joints between circular hollow section braces and chord members in the lattice struts of the flare-boom. From observations of oscillations and the most typical position of crack initiation, it was concluded that the most critical stress fluctuations were caused by single member vibrations in an across-wind simple bending mode, and not by overall structural vibration effects. Wind buffeting calculations of the overall structure, e.g. as proposed by Wyatt (1992), indicated that such effects were not likely to cause sufficiently large stress fluctuations to explain the fatigue damage which had been observed. The assumption is also supported by simple vortex shedding calculations, which, based on a Strouhal number of 0.2 and a first bending mode eigenfrequency of 30 Hz, would yield a resonant lock-in wind velocity of about 15 m/s for the most relevant members, and again, this agrees well with observations.

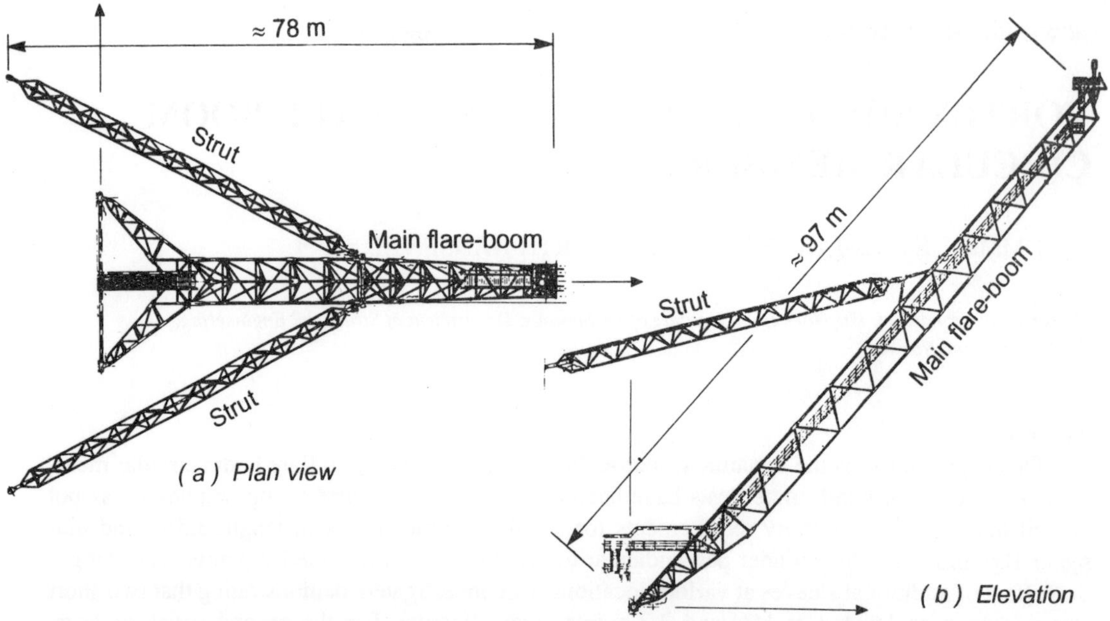

Figure 1 Typical flare-boom design on off-shore production platforms

The most important 'questionmark' to the hypothesis above was the possibilty of any wake induced interference effects, because some damage had also been observed on brace member joints on the top face of the triangular strut, i.e. on diagonal members which are partly shielded by larger diameter chord members. But the main bulk of fatigue damage has occurred in or at the welded joints between the front face diagonal and chord members of the lattice structure, supporting the hypothesis of a regular vortex shedding effect.

The problem of suppressing vortex shedding has been subject to considerable attention by many researchers. A comprehensive review of various means for suppressing vortex shedding has been published by Zdravkovich (1981). However, for this particular case, the most common mitigation devices were deemed unsuitable, either because of uncertainties regarding their effectiveness or because they involved welding or bolts through parent metal, which would damage the coating and thus diminish the corrosion protection. It was therefore decided to perform wind tunnel tests of a mitigation device suggested by Rudge et.al. (1992), where a flow-disturbing device is introduced in the form of short nonstructural sleeves along the member span.

2. FIRST SERIES OF TESTS

Two series of tests have been carried out. In the first test set-up the main emphasis was put on the fulfilment of the appropriate similarity requirements for a typical full-scale member with an overall length of 3.5 meters, an outer diameter of 100 millimeters and a tube wall thickness of 10

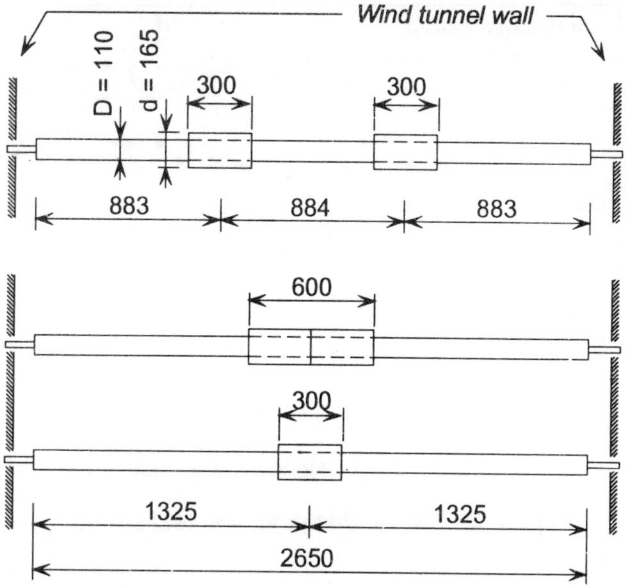

Figure 2 First test set-up and test configurations

millimeters. Its eigenfrequencies were unknown, but calculations indicated that its first bending mode would occur in the range between 25 and 40 Hz, depending on the degree of fixity at its supports. Also damping characteristics were unknown, but for such all-welded structures damping is most likely to be low. Apart for the damping similarity requirements, the appropriate model laws comprise

* mass-scaling, i.e.: $\lambda_m = \lambda_L{}^2$
* correct Reynolds number, i.e.: $\lambda_U = \lambda_{Re}/\lambda_L$
* Cauchy-scaling ('indirect'), i.e.: $\lambda_f = \lambda_U/\lambda_L$

where λ=fullscale/modelscale, and the subscripts m, L, U, f and Re refer to mass (per unit length), length (i.e. dimension), wind velocity, frequency and Reynolds number, respectively.

Since it was desirable to obtain resemblance also between displacement modes, it was for this first test set-up decided to adopt a test arrangement with a steel tube as a simply supported beam spanning across the width of the wind tunnel and perpendicular to the flow. In order to avoid butting against the maximum wind tunnel velocity, a geometric scale λ_L of 0.909 was chosen. With λ_{Re} set equal to 1.0, the velocity and frequency scaling are then equal to 1.1 and 1.21, respectively. The relevant Reynolds number for the full-scale shedding was expected to be somewhere between $0.7 \cdot 10^5$ and $1.5 \cdot 10^5$, i.e. if the prototype member has a sufficiently smooth surface the full-scale vortex shedding is within the sub-critical range but very close to becoming critical.

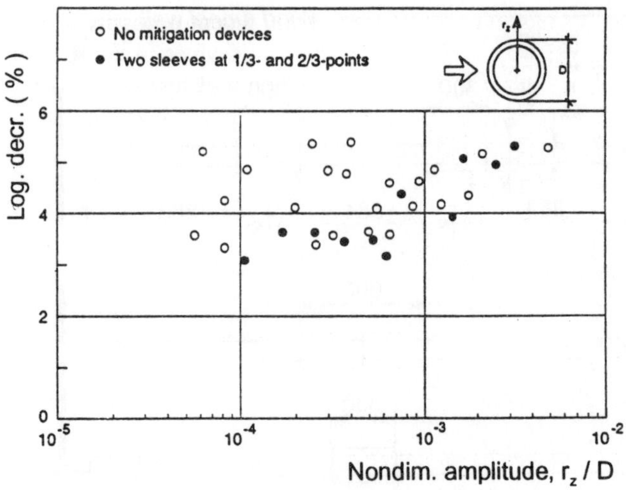

Figure 3 Still-air damping, first test set-up

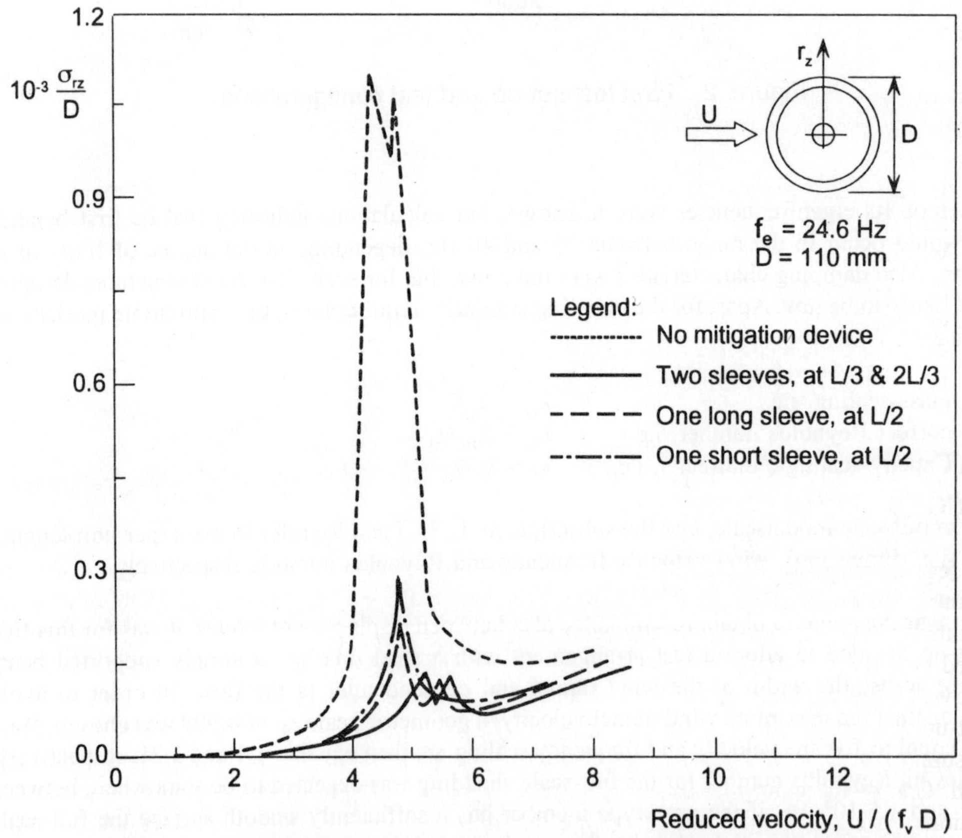

Figure 4 RMS across-wind displacements at midspan, first test set-up ($R_e < 1.5 \cdot 10^5$, $|r_z|_{max} / \sigma_{rz} \approx 2.5$)

This first test set-up was to some extent coloured by the urgent necessity to provide information regarding the possible mitigation effect of flow disturbing devices in the form of additional non-structural short compact (but light-weight) sleeves. The model test set-up is illustrated in Fig.2, together with the chosen configurations of mitigation devices. All testing took place in smooth flow. The aspect of flow interaction with neighbouring members was not studied at this stage.

The model (in-air) eigenfrequencies in the across-wind vertical mode were 24.7 Hz for the single 'naked' cylinder, and about 23.9 Hz for the configurations with two additional sleeves. For the relevant amplitudes of motion, the logarithmic decrement of damping (see Fig.3) was about 4 %, which is likely to be higher than what can be expected for this type of fullscale structure. Damping was not significantly affected by the presence of the additional sleeves.

The test results are shown on Fig.4. Two striking features are apparent. Firstly, from the tests with the 'naked' cylinder there is a distinct vortex shedding response peak at a reduced velocity between 4 and 5.5, which in tests with lower damping would have manifested itself with a larger peak value and been extended over a wider range of wind velocities. But, even at the present level of damping, the vortex shedding response is stresswise considerable. Secondly, for all the other test configurations, the vortex shedding response is considerably reduced. For the configuration with two short sleeves at 1/3- and 2/3-points of the span, which apparently seems to be the most effective mitigation device, the maximum observed RMS-value is reduced to about a tenth of the equivalent values for the 'naked' cylinder. It is likely that these encouraging results may mainly be ascribed to the favourable effects of the following two mechanisms:

* firstly, that vortex shedding from regions along the span with different diameters will occur at different levels of intensity and that the shedding frequency from one region to another will not coincide at various wind velocities,

* and secondly, and probably more important, that the steps between different cross-sectional diameters (due to the sleeves) represent significant disturbances to the vorticity, such that the regularity and the spanwise coherence is reduced.

3. SECOND SERIES OF TESTS

Since few investigations have been published on the vortex shedding response of members at a yaw angle different from zero, and since no tests had been carried out to substantiate any favourable effects of mitigation devices in the form of additional sleeves for members at an inclined position in the flow, it was decided to launch a second series of tests. In these tests the model was a tubular aluminium member with an outer diameter of 90 millimeters, see Fig.5, and the set-up was chosen as a regular spring-mounted 'section-model' test set-up, i.e. the model ends were supported by flexible springs such that its first eigenmode was very close to a rigid body motion. One of the main reasons for chosing a 'section-model' test set-up was to obtain testing conditions with an inherent rig damping as low as possible. The test programme comprise configurations with or without the most effective mitigation devices at yaw angles of 0, 20, 40 and 60 degrees. The tests have been carried out in smooth as well as in turbulent flow.

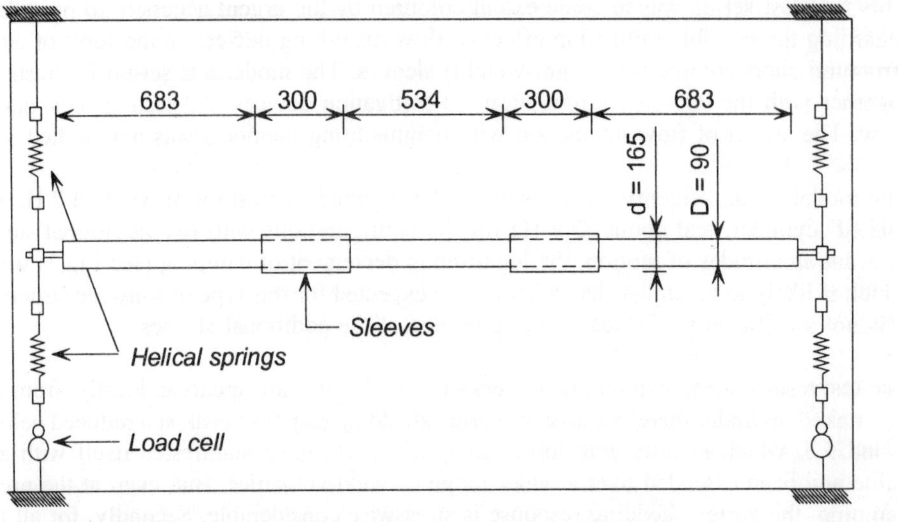

Figure 5 Second test set-up (cross-section of wind tunnel test section, with inner width and height of about 2.7 m and 1.8 m, respectively; dimensions in mm).

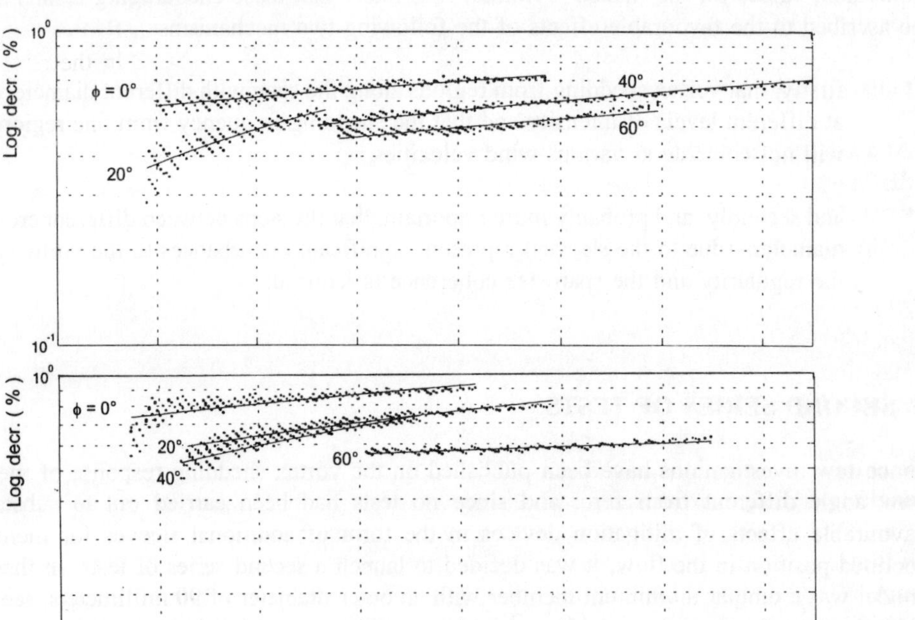

Figure 6 Still air damping (upper diagram: naked cylinder, lower diagram: sleeves included at 1/3- and 2/3-points)

roll were observed between 8.5 and 11.5 Hz (depending on configuration), while the eigenfrequencies for eigen-bending were estimated to be located beyond 25 Hz. Thus, resonant vortex shedding was not only observed in the vertical translational mode, but, at higher velocities, also in the roll mode of vibration.

In the following only displacements associated with the first vertical translational mode will be presented. The test results are shown on Figs. 7 and 8, where the RMS response values, reduced by the outer diameter of the cylinder (D=90 mm), are plotted versus the reduced velocity $U/(f_eD)$.

The interesting range of reduced velocities occur between about 4 and 15. In this range, the Reynolds number is located between about $1.2x10^4$ and $4x10^4$, i.e. the flow is likely to be subcritical. Mass and damping vary from one configuration to another. However, the mass-damping parameter, $m_e\zeta/(\rho D^2)$, is close to 0.4 for all tests (which is relatively low).

The following may be observed:

At a yaw angle of (nominally) 0° resonant vortex shedding oscillations occur in the reduced velocity range between about 4 and 5.5 for the 'naked' cylinder, while hardly any oscillations can be observed for the cylinder with sleeves (spanwise at 1/3- and 2/3-pts.).

At a yaw angle of 20° oscillations occur in the reduced velocity range between about 4.5 and 6.5 for the 'naked' cylinder. For the cylinder with sleeves, oscillations with RMS-values of about a fifth of those which occurred for the 'naked' cylinder can be observed in the reduced velocity range between about 6.5 and 9.

At a yaw angle of 40° oscillations occur in the reduced velocity range between about 5 and 8 for the 'naked' cylinder. For the cylinder with sleeves, oscillations with RMS-values of about half of those which occurred for the 'naked' cylinder can be observed in the reduced velocity range between about 7 and 10.

At a yaw angle of 60° oscillations occur in the reduced velocity range between about 7 and 13 for the 'naked' cylinder. For the cylinder with sleeves, oscillations with RMS-values of the same magnitude as those which occurred for the 'naked' cylinder can be observed in the reduced velocity range between about 9 and 14.

It is thus concluded that the effect of the additional sleeves is gradually diminishing as the yaw angle is increased, from being fully effective at ϕ=0° to becoming ineffective at ϕ=60°.

It is also apparent that the Strouhal number (S), based on the parent cylinder outer diameter (D=90 mm) and on the mean alongwind velocity (U), is decreasing for increasing values of the yaw angle (ϕ). This is illustrated in Figs. 9 and 10, where S is plotted versus ϕ. As can be seen, for the 'naked' cylinder in turbulent flow (Fig.9) S is reduced from about 0.2 at ϕ=0° to about 0.11 at ϕ=60°, and a similar change of S occur for the 'naked' cylinder in smooth flow (Fig.10).

In Figs. 9 and 10, also the dynamic RMS lift coefficients (σ_{CL}) has been plotted versus ϕ. The calculations of these coefficients is based on the assumption that at the maximum oscillations the response is ultra-narrow-banded (which is supported by power spectra plot).

The turbulent flow was generated by five spires at the inlet of the working section of the wind tunnel. The alongwind and the acrosswind vertical turbulence intensities were about 10 and 9%, respectively. The integral lengthscales horizontal and perpendicular to the flow were measured to ${}^{y}L_{u} \approx {}^{y}L_{w} \approx 0.112$ m for the alongwind (u) and the acrosswind (w) turbulence components, indicating that the flow is too fine-grained to be representative for the conditions which can be expected to occur at full-scale. (The non-dimensional velocity spectra of the flow components have their culmination at a reduced frequency fD/U of about 10^{-1}). The tests in smooth flow were carried out 'in empty tunnel', where turbulence is insignificant (i.e. with intensities below 0.5%).

No corrections have been included for possible blockage effects. The blockage ratio was about 5%, and thus, velocity measurements may be slightly to the conservative side. Similarly, no corrections have been included for possible end effects (at tunnel walls).

Eigenfrequencies, damping-ratios and equivalent distributed masses for the various test configurations are given in Table 1. The results of still-air damping recordings are shown on Fig.6. (Since it was the same cross-sectional dimensions on the aluminium model specimens, and it was the same spring supports for all configurations, the eigenfrequencies varies because of different spans, and equivalent masses per unit length varies because of the contributions from suspension rods, springs and trapped and added mass of air).

Table 1 Eigenfrequencies (f_e), damping-ratios (ζ) and masses (m_e) for the second series of tests

Yaw angle ϕ (degr.)		'Naked' cylinder			Incl. sleeves at 1/3- and 2/3-pts.		
Nominal value	Actual value	f_e (Hz)	ζ(*)	m_e (kg/m)	f_e (Hz)	ζ(*)	m_e (kg/m)
0	0.9	7.22	0.0012	3.376	6.03	0.0015	4.756
20	23.7	7.00	0.0012	3.351	5.87	0.0013	4.654
40	40.9	6.30	0.00095	3.273	5.42	0.0013	4.349
60	58.1	4.79	0.00085	3.191	4.27	0.00095	3.903

*) Damping has been read off at amplitudes of about half the magnitude of those which occurred at the maximum response.

As can be seen from Fig.6, damping is low, i.e. the logarithmic decrement is in the range between 0.6 and 0.9%, and the difference between the various test configurations is not insignificant. The level of damping in this test rig is likely to be of the same order as what can be expected for an all-welded steel structure.

The eigenfrequencies given in Table 1 are associated with the first vertical (acrosswind) translational mode of vibration, i.e. close to a rigid body motion. All the test results which are presented in the following are associated with this type of displacements. However, the test set-up contains at least two secondary modes which may be of some interest. These are associated with roll (in which the member end supports are in vertical motion while its midspan is practically at rest), and with eigenbending of the member between its end supports. The eigenfrequencies for

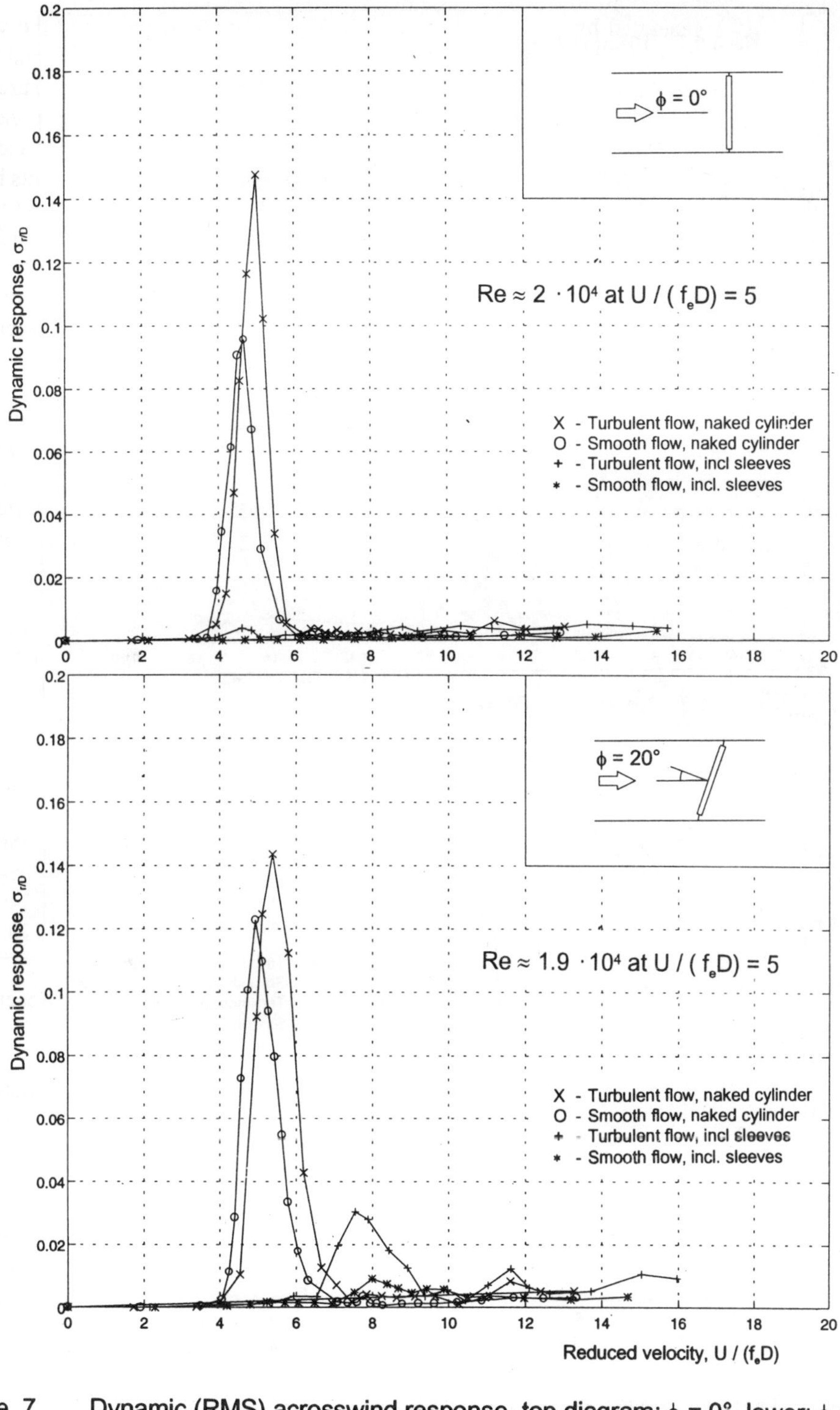

Figure 7 Dynamic (RMS) acrosswind response, top diagram: $\phi = 0°$, lower: $\phi = 20°$

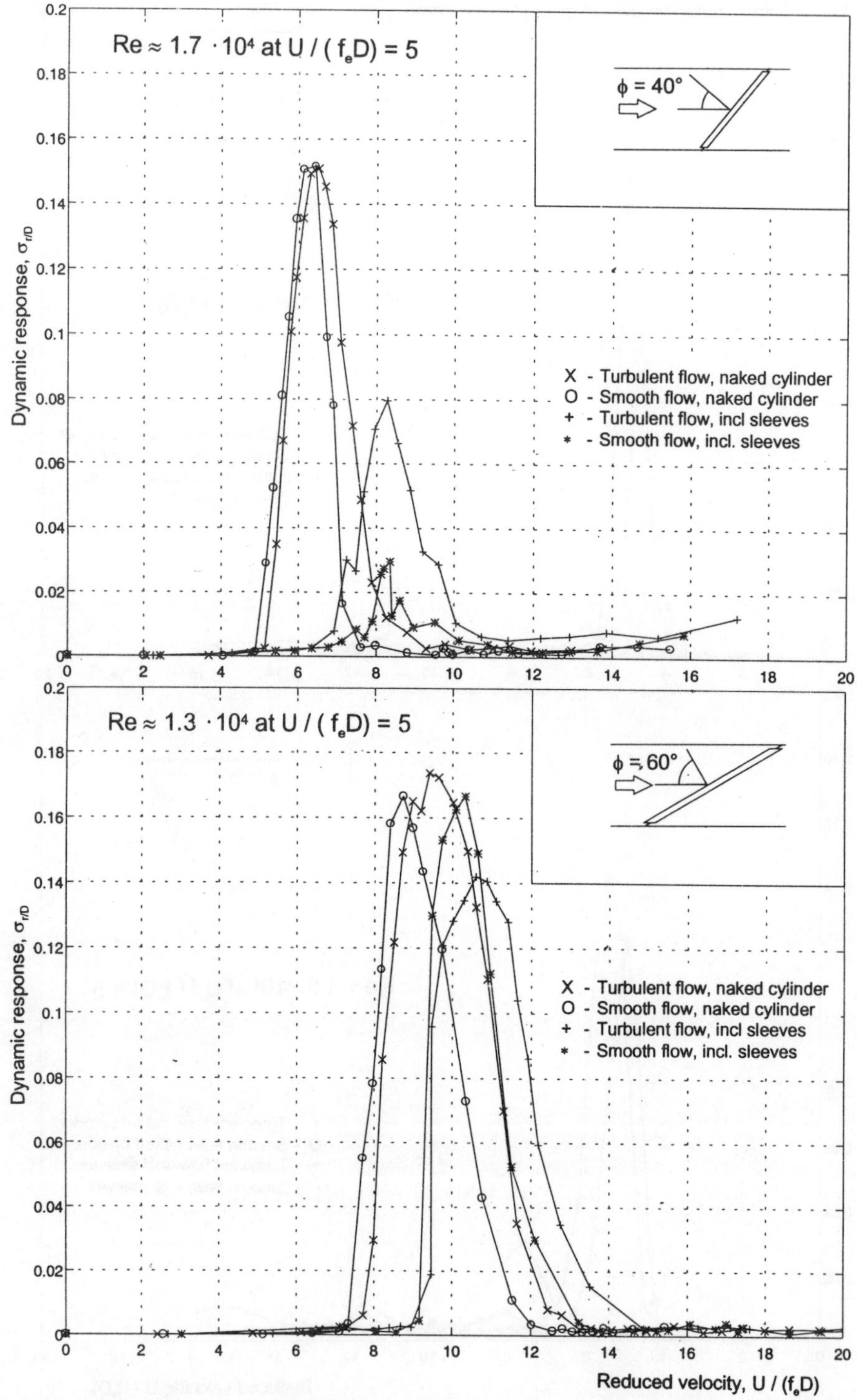

Figure 8 Dynamic (RMS) acrosswind response, top diagram: $\phi = 40°$, lower: $\phi = 60°$

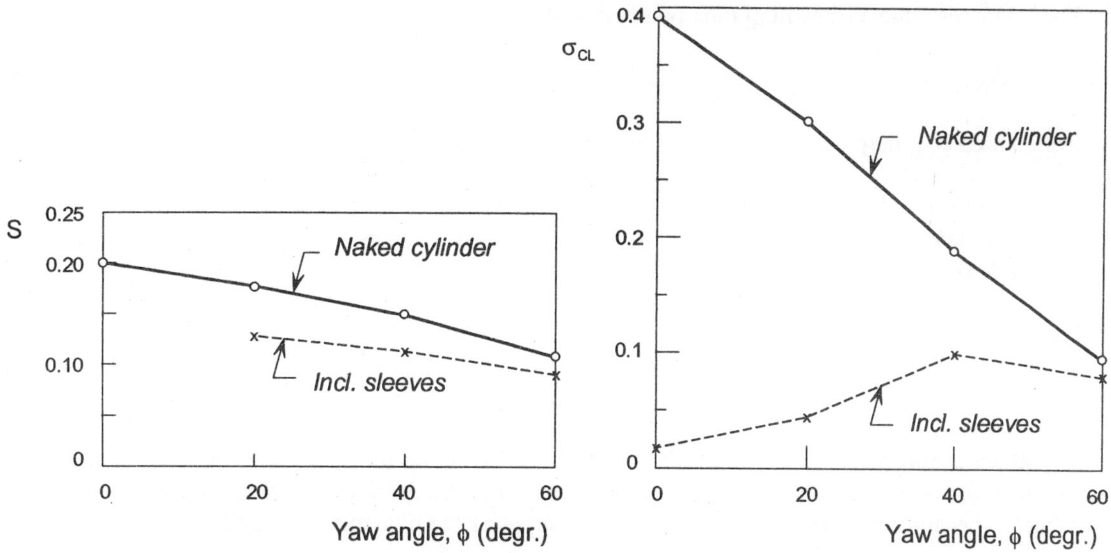

Figure 9 Strouhal number, (read off at peak response) and dynamic RMS lift coefficient, turbulent flow, (D = 90 mm)

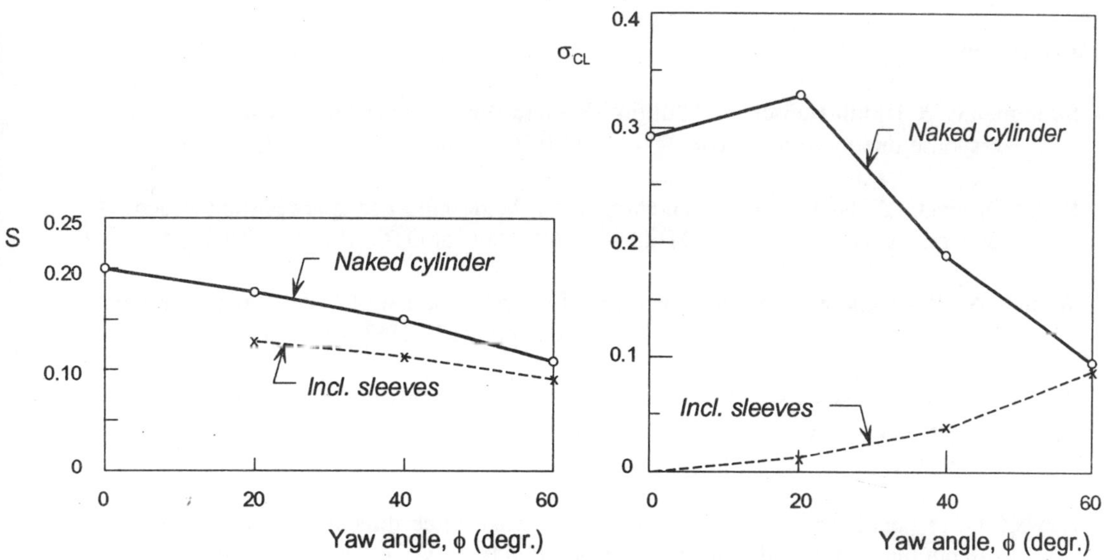

Figure 10 Strouhal number, (read off at peak response) and dynamic RMS lift coefficient, smooth flow, (D = 90 mm)

The RMS response values may thus be expressed by

$$\sigma_{r/D}(\text{max.}) \approx \rho D^2 \sigma_{CL} / (16\pi^2 m_e \zeta S^2) \qquad (1)$$

from which σ_{CL} may be estimated.

For the 'naked' cylinder in turbulent flow (Fig.9) σ_{CL} is almost linearly decaying from about 0.39 at ϕ=0° to about 0.09 at ϕ=60°, while for the cylinder with sleeves σ_{CL} is increasing from about 0.02 at ϕ=0° to a largest value of about 0.10 at ϕ=40°.

For the 'naked' cylinder in smooth flow (Fig.10) σ_{CL} is first increasing from about 0.29 at ϕ=0° to about 0.33 at ϕ=20° and then decreasing to about 0.10 at ϕ=60°, while for the cylinder with sleeves σ_{CL} is continually increasing from approximtely 0 at ϕ=0° to about 0.10 at ϕ=60°.

A somewhat surprising result should be mentioned. As can be seen on Fig.7, the peak dynamic response for ϕ = 0° and 20° is larger in turbulent flow than it is in smooth flow. This was not expected because turbulence is often assumed to suppress the dynamic effects of vortex shedding.

4. CONCLUSIONS

From the present investigation it may be concluded that the Strouhal number and the dynamic lift coefficient for a circular cylinder are strongly yaw dependent, both diminishing with increasing yaw, and that a mitigation device in the form of short sleeves (spanwise at 1/3- and 2/3-points) is effective only at small values of the yaw angle (say below 20°).

References

Strømmen,E. & Hjorth-Hansen,E.: 'Gullfaks B Flare-Boom; Wind tunnel tests to reduce the response due to vortex shedding', SINTEF Rep. No. STF70 F93034, 1993-04-02.

Rudge,D., Fei,C-Y., Nichols,S. & Vandiver,J.K.: 'The design of fatigue-resistant structural members excited by wind', OTC 6902 (24th Annual OTC, Houston, Texas, May 1992).

Wyatt,T.A.: 'Dynamic gust response of inclined towers', Journal of Wind Engineering and Industrial Aerodynamics, Vol. 41-44 (1992) pp. 2153-2163, Elsevier.

Zdravkovich,M.M.: 'Review and classification of various aerodynamic and hydrodynamic means for suppressing vortex shedding', Journal of Wind Engineering and Industrial Aero-dynamics, Vol. 7 (1981) pp. 145-189.

Lewis,C.G. & Gharib,M.: 'An exploration of the wake three dimensionality caused by a local discontinuity in cylinder diameter', Phys. Fluids A, Vol.4, No.1, Jan. 1991, pp. 104-117

ON THREE DIMENSIONALITY OF FLUCTUATING AERODYNAMIC FORCES ON CIRCULAR CYLINDRICAL STRUCTURES

GARG, R.K.[1] and NIEMANN, H.J.[2]

[1] *Structural Engineering Research Centre, Ghaziabad, 201001, India.*
[2] *Building Aerodynamics Laboratory, Ruhr-Universität Bochum, 44780 Bochum, Germany.*

Abstract
The present study aims for more understanding of the fluctuating aerodynamic forces with the help of a study of the pressure field around a cantilevered, stationary circular cylinder. The variation of mean and fluctuating components of pressure, drag and lift, and their correlation / coherence along the height of cylinder of aspect ratio of 8.5 diameter is discussed. The magnitude of the aerodynamic forces as well as their (cross)correlation are higher in 30% to 60% of the upper portion of the cylinder height. The effect of top opening has also been discussed.

1. INTRODUCTION

The flow around the circular cylinders is a quite complex phenomenon. The approaching flow separates from a cylinder not at a fixed point but it depends upon several parameters such as Reynolds number, turbulence characteristics of the approaching flow, body surface texture (roughness), taper, attached appurtenances etc.. The flow has been characterised by Re-regimes as sub critical, critical, supercritical and transcritical for idealised two-dimensional situations. The separated shear layers can roll up when leave the surface in form of vortices (Karman vortices) occurring in certain cellular fashion along the length of the cylinder. The finite aspect ratio, different end conditions, irregularity over the surface, fluctuations in the approaching flow, motion of the body, interaction with wake etc. make the situation more complex and for practical problems the flow around circular cylinder remains that of three-dimensional nature (Niemann et al 1988; West et al 1990; Roshko 1992). This results in a complex nature of induced aerodynamic forces on the cylinder (Vickery/Clark,1972; Basu/Vickery1983; Billah,1989; Steckley/Vickery,1992; Larsen,1993). Although several studies have been conducted on the flow around circular cylinders but still not a clear picture emerges, particularly, for the three-dimensional fluid-structure behaviour. In the case of varying cross-section of the structure, the choice of critical diameter where the across wind excitation force acts is not well addressed in the literature. Also, there remains large coefficient of variation of about 25% between the observed and predicted models of Vickery/Basu and ESDU(Item no.85038).

The present study aims for more understanding of the fluctuating aerodynamic forces with the help of study of the pressure field around a circular cylinder. The nature of along wind and across wind forces, and their variation along the height will be discussed. The correlation of lift force and its variation with height will also be discussed particularly in the light of our experiment as well as recent data from wind tunnel tests (Blackburn et al 1992) which differ from the suggested response prediction model of Vickery/Basu(1983) considering the correlation length of lift force as 1 D. The normal pressure distribution obtained from the wind tunnel (WT) has been analysed to obtain the circumferential correlation of pressure and subsequently the eigen functional analysis of such correlation matrix has also been carried out for various heights of the model. The mean and fluctuating component of along wind as well as across wind forces are deduced by integrating the surface pressures. The vortex shedding frequency has been obtained from the auto-spectrum of the lift force, the correlation of drag and lift forces along the height has been calculated by integrating the correlation coefficients from the point of reference to infinity. The effect of opening and closing of top mouth is also discussed.

2. EXPERIMENTAL SET-UP

The experiment in the wind tunnel was conducted on a parallel sided circular cylinder (stationary) made from wood. The cylinder of desired aspect ratio can be constructed by joining together the wooden sections. Each section is of outer diameter of 70 mm and of height 60 mm. The wind tunnel is open circuit type with test section dimension 1.8 m wide and 1.6 m high (variable upto 1.9 m) yielding a blockage ratio of 1.5 percent. The boundary layer was simulated by using spires (modified Counihan type) and floor surface roughness consisting of rectangular blocks. The simulated mean velocity profile (α =0.17) and turbulence intensities are typical to open country type terrain (Fig.1a). The length scale of longitudinal turbulence (L_{ux}) has been calculated by integration of auto-correlation function upto the first zero crossing in auto-correlation curve (Fig.1b) which is 4D where, D is the outer diameter of the cylinder. Other flow features has been described in another reference by Garg et al. The test Reynolds number based on the velocity at model height (appx. 600 mm) is 0.667E+5. The blockage ratio is less than 1.5 percent and the data has not been corrected for it.

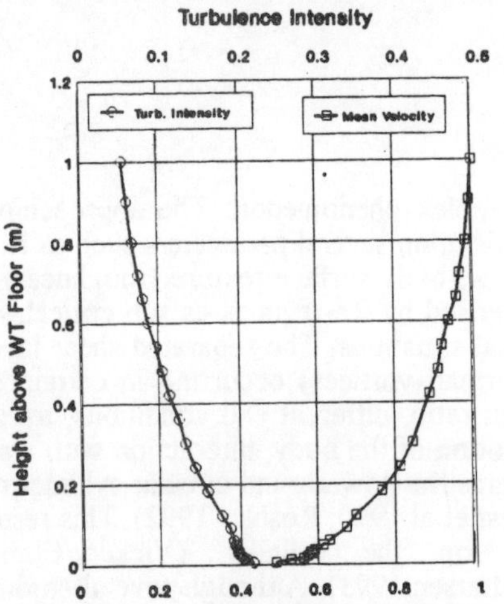

Figure 1a. Mean velocity and turbulence intensity profile.

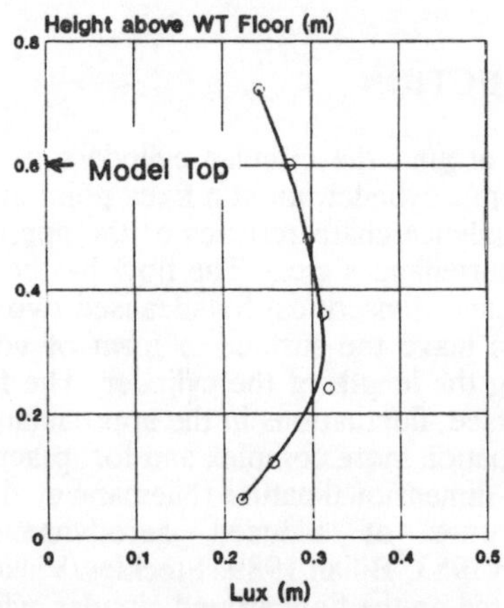

Figure 1b.Longitudinal turbulence length scale profile.

The holes of 0.8 mm diameter were bored at each 15° and at 30° around the circumference in the model. Simultaneous measurement of fluctuating pressure around the cylinder was possible with the help of 800 mm long PVC tubes which were optimised (Holmes/Lewis,1987) for pressure transfer to Honeywell (170PC type) pressure transducers. The pressure tubes consist of two variable diameter tubes with two restrictors. The tube response was quite good upto 100 Hz and after 200 Hz it damps fast as shown in Fig. 2. For estimating the correlation of the forces at two different levels, the tubes were arranged on the pressure transducers such that it measure the difference of pressures between two diametrically opposite points around the circumference and the evaluation process weighted according to their contribution to the lift /drag correlation. The signals were digitised through A/D converter at frequency of 1600 Hz. The sampling time was chosen high enough to have good statistics, which was about 300 Seconds, resulting in more than 6 Million points per set. The digitised data were stored on DDS tapes for further analysis. There are some acoustic disturbances at 8 Hz and at 64 Hz due to a standing wave in wind tunnel, however, their influence on the lift force spectrum is negligible.

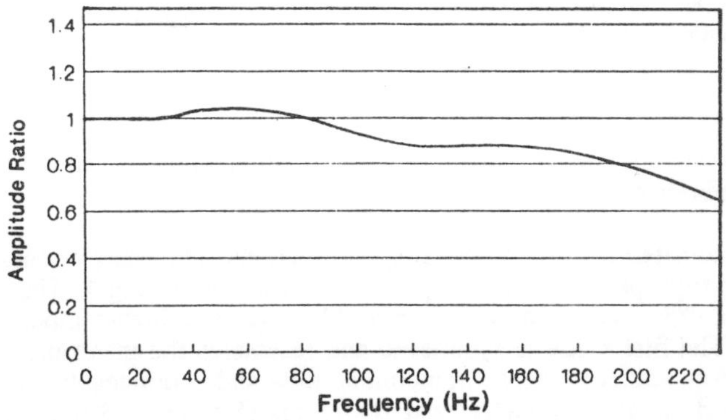

Figure 2. Response of optimised tube system

3. RESULTS AND DISCUSSIONS

3.1 Mean and Fluctuating Pressure :

The mean pressure and fluctuating pressure distribution around the cylinder of aspect ratio (AR) 8.5 is shown in Fig. 3. For determining the pressure coefficients, the mean velocity pressure at the respective level is used. Near base a pressure coefficient of more than unity at the stagnation point is also observed which is possible in turbulent flow and as well near tip a stagnation pressure less than unity has also been observed. The base pressure along the span follows the trend as described by Fox et al(1993) as reflected in the wake width which decreases near top and base as can be seen in mean pressure distribution curves.

The rms pressure normalised by turbulence intensity (Fig.3b) decreases towards the free end and the maximum value occur at 90° from the stagnation. Near stagnation point rms pressures are close to quasi-steady predictions and the deviation is more towards base. However, away from the stagnation point the rms pressures are higher than compared to the quasi-steady prediction which may be because the quasi-steady theory does not take into account vortex shedding components. The correlation of fluctuating pressure around the circumference was also calculated. A negative correlation between stagnation point and wake was observed. There are some anti symmetric correlation curves at 30° and low correlation in the separated zone near the top within one diameter distance which differs from Kareem(1989) but corroborate with Bruun et al(1975). This may be due to secondary local flow, however, in other parts of the span the correlation are similar to as observed by Kareem et al(1989). The eigen function analysis (Armitt,1968; Holmes,1988) of the circumferential pressure correlation is a technique to decompose the pressure fluctuations in temporal and spatial components which satisfy the orthogonality condition. It helps to represent the fluctuating pressure in simpler form apart from

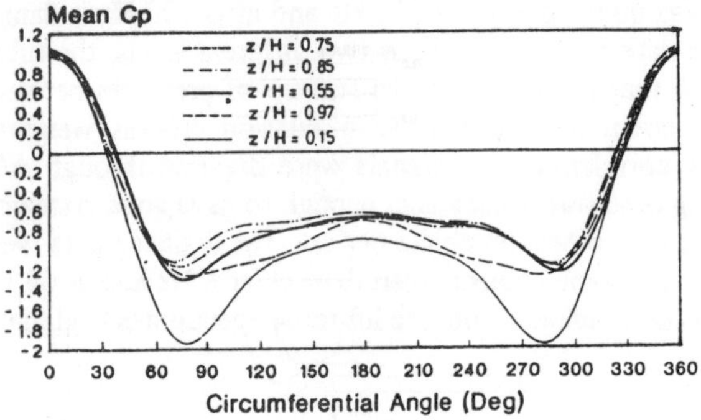

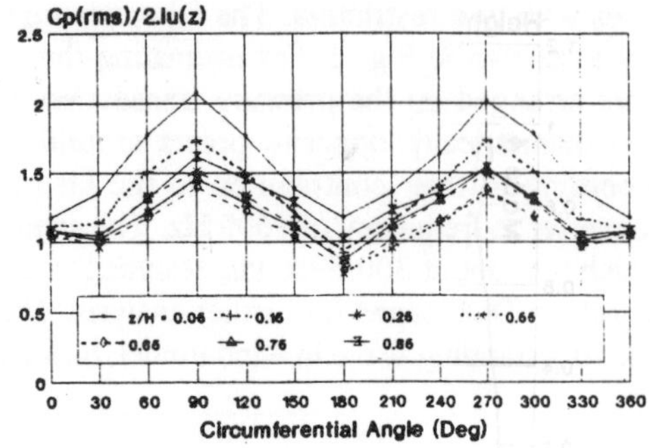

Figure 3a. Mean pressure distribution around circumference.

Figure 3b. Fluctuating pressure distribution

studying physical causes associated with the various eigen functions. The energy contained in the first four modes for the case of top open as well as top closed are shown in Fig.4. The first five modes contain energy of the order 98%, and at the top 98.6%. The sixth and higher modes contain energy individually less than 1%. The first mode is symmetric and resembles the mean pressure distribution which is attributed to the large scale turbulence in incoming flow and is noticeably more in the case of top closed. The second mode is anti symmetric and represents (Kareem,1989) the vortex shedding contribution which in this case is maximum at 0.85H (=1.3D from top), however, the spectrum of lift force does not reflect such maximum. The energy in this mode along the height towards base reduces and over top third portion is more than the rest of the span in the case of top open. First two modes (total energy 80% to 85%) reflect being complementary in terms of energy content i.e. when energy in the first mode increases, the energy content decreases in the second mode. The third mode is also anti-symmetric and looks like the derivative of the first mode or mean pressure, as per the quasi-steady theory is a contribution of lateral turbulence. The fourth mode is symmetric and may be correlated to the moderate scale turbulence in incoming flow. It is observed that the sequence of occurrence of various modes is not quite same as observed by other researchers although the trend is similar.

3.2 Mean and Fluctuating Drag

3.2.1 Mean Drag:

The mean drag varies along the height and is lower than the two-dimensional case except close to the top where it can reach a value corresponding to the two-dimensional conditions as shown in Fig.5 along with the results of Gould et al (AR=8 and 12) and of Etzold (AR=8). Within two diameters from the top the drag is higher in the case of top open, as was also observed by Gould for AR 12, by 10 percent for the top open case within 3.5D from top, however, the trend reverses marginally in the lower height of the cylinder.

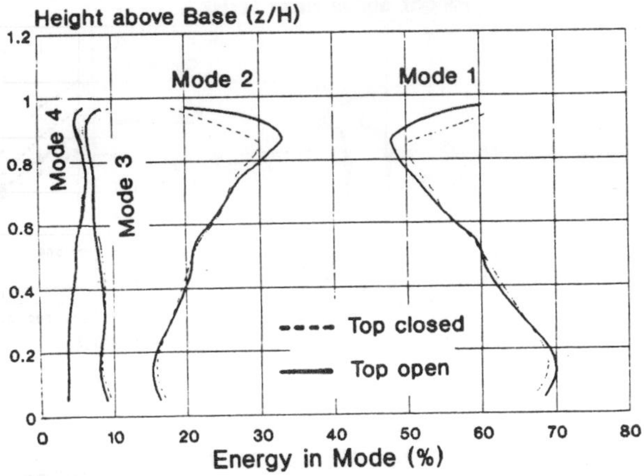

(a). Energy distribution in different modes

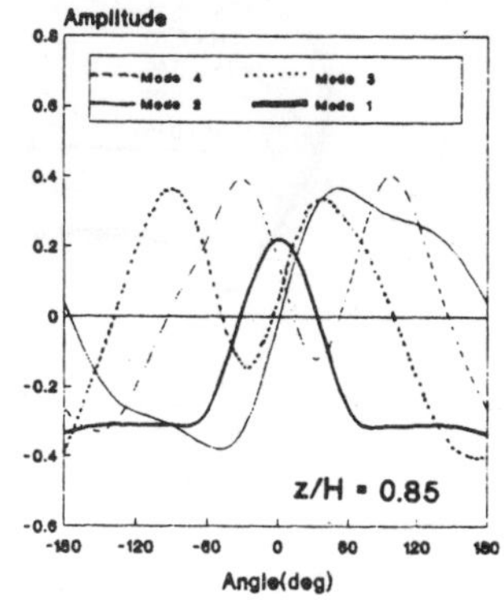

(b) First four eigen forms

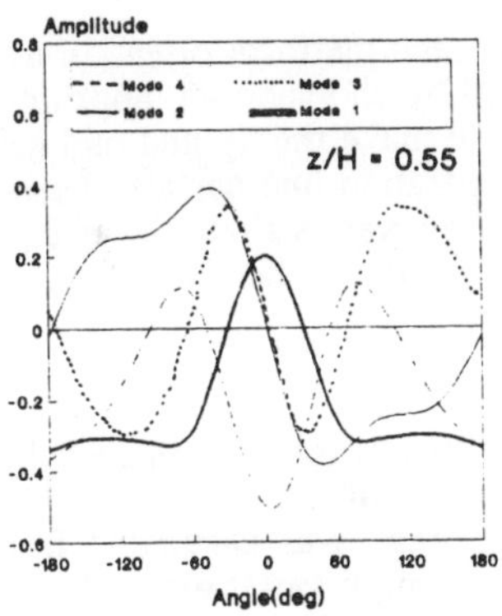

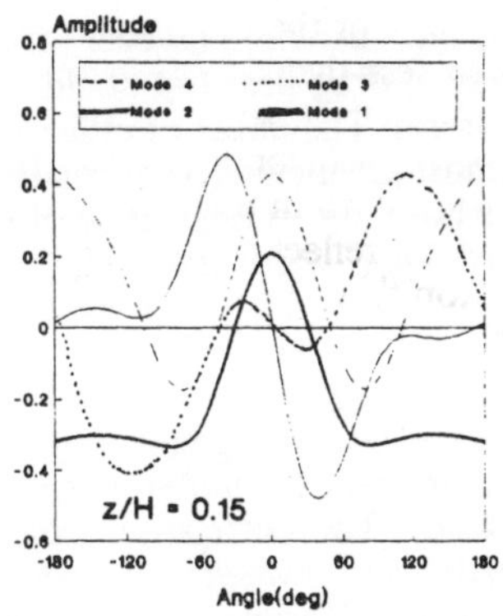

(c),(d) First four eigen forms

Figure 4. Eigen functional analysis of the pressure covariance

The high value of Cd near tip is due to the shedding of trailing vortices from the upper edge of the cylinder which roll up further downwards causing a secondary flow behind the separation line resulting in the local low pressure which causes a local increase in drag. It can be observed that the mean drag observed in present study is larger than the transcritical Re number value of Gould but is lower in the upper half compared to the uniform flow value of Etzold. The complex flow separation associated with the secondary flow near top makes it difficult to generalise the force distribution at least in this region whose influence could be even upto 20 diameter deep along the span as suggested by Fox et al (1993) based on the tests in uniform smooth flow. Only a few standards (ACI) and data units (ESDU) specify drag distribution considering the enhancement near top. Several times there are certain features e.g. platform, efflux velocity etc. near top which can change completely the force distribution. The effect of platform has been studied by Rao/Kumar(1991) and observed the reduction

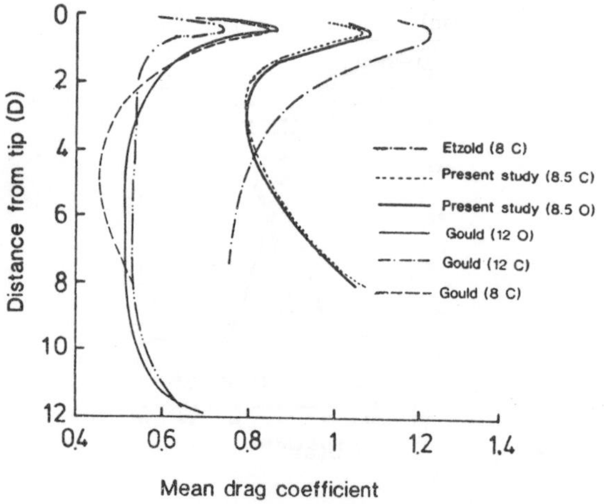

Figure 5. Mean drag distribution

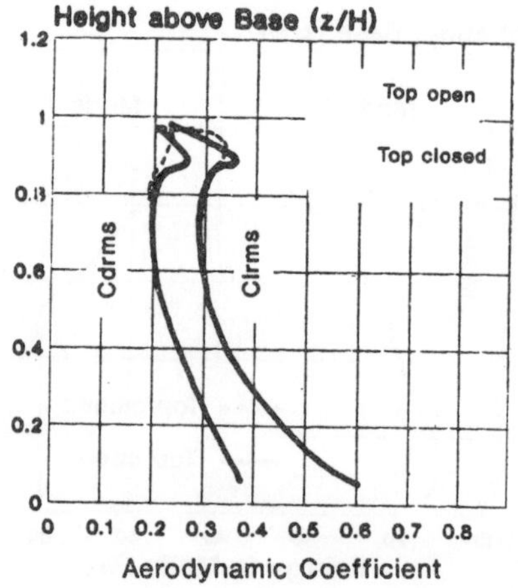

Figure 6. Fluctuating drag and lift

of drag coefficient near tip, however, drag could increase in some platform orientations. They noted that the increase of drag near top could be completely nullified by choosing an appropriate size and position of platform. The effect of efflux velocity (Lawson) is to make the tip and mid height pressure distribution almost identical. The secondary flow along the separation line has also been observed by Etzold(1976) whose directional sense of the mean secondary flow varies along the span, near top it is directed upward, i.e. towards free end, in centre stochastically alternating flow direction, and near base it is directed downwards, i.e., towards base.

3.2.2 Fluctuating drag:

The trend of the fluctuating rms drag distribution along the height follows that of the mean drag, and is slightly more near top within one diameter region for the case of top closed compared to the top open case (Fig. 6). The correlation length of the fluctuating drag was calculated from the cross-correlation coefficient between two levels. The drag correlation length varies along the height (Fig.7), the average value is 4.3D near top at 0.75H and reduces towards base. The drag correlation is stronger going upwards from a reference point compared to going downwards suggesting that the drag is more strongly correlated in upward direction than downward from a reference point. Sonneville (1976) has measured 2D to 2.5D at Reynolds numbers of 1.E+4 to 3.E+4, respectively in smooth flow. The observed drag correlation along the height can be utilised to consider defining the correlated drag.
The coherence (root coherence) curves of fluctuating drag along the height considering the mid height (0.55H) as reference level are given in Fig. 8. In low frequency range the coherence is lower than unity and reduces with increasing separation distance. In the higher frequency range it follows an exponential decay upto a reduced frequency of appx. 0.2 as described by

$$\sqrt{Coh} = e^{\left[-Cz\cdot\left(\frac{f\cdot\Delta z}{U_{av}}\right)\right]} \qquad ...1.$$

The value of the coefficient Cz after fitting to the coherence curves is equal to 7 which reflects similarity with the vertical coherence of longitudinal component of wind velocity fluctuations observed in the wind tunnel.

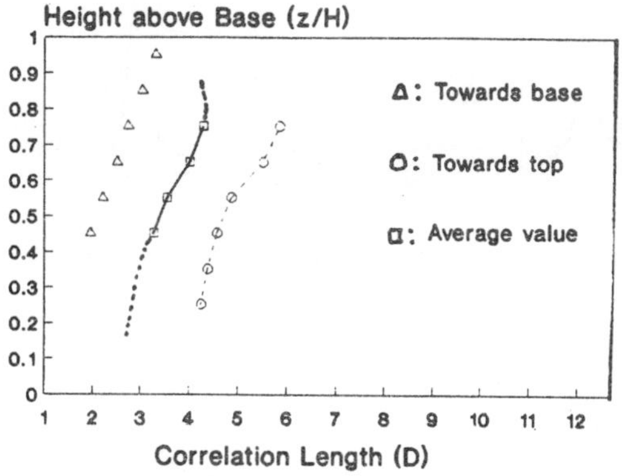

Figure 7. Drag correlation length

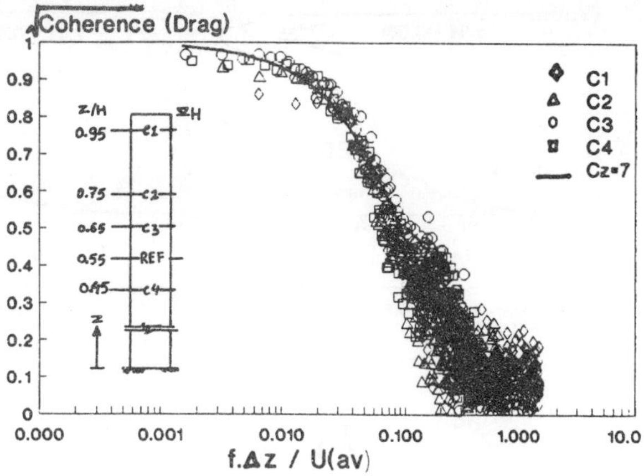

Figure 8. Drag coherence

3.3.3 Fluctuating Lift

3.3.1 General:

The across wind response is caused by the flow separation phenomenon around the body which is influenced by turbulence and velocity gradient in flow, Re, local diameter, amplitude of oscillation frequency, end condition etc. The lift force at various positions over height is obtained after integrating the circumferential pressure around the cylinder. The spectrum of lift force along the height is shown in Fig. 9. Near top within 1D to 2D distance a vortex shedding contribution to the lift is not noticed. The spectra are quite broad banded, sometimes double/multiple peaks occur which may be due to turbulence in flow (Vickery,1972; Fox et al,1990,1993). These curves can be utilised in multi-degree of freedom system dynamical analysis induced by aerodynamic forces on cylinders. The ratio of rms lift to rms drag is approximately 1.45 except near top (1.1) and base (1.6). Also, the ratio of vortex shedding contribution to total lift is of the order of 80 percent.

Due to broad banded nature of spectra, for the purpose of determining the vortex shedding frequency, a frequency near middle of the range of frequencies in the broad peak was considered. The Strouhal numbers, S, calculated based on the wind speed at model height are presented in Fig.10. The cellular pattern of shedding persists near top to almost mid height after which it falls gradually. Based on studies on cylinder of AR of 37 in uniform smooth flow, Budair(1991) described four spanwise vortex shedding zones. In the first region within two diameters from top there exists vortex shedding at lower frequency than the corresponding two-dimensional case. In the next two diameter region it is wide band non-shedding (weak) . In the third region wide band shedding persists upto 3diam. to 13 diam. distance whose central frequency increases moving away from the tip, and in the fourth one is in the rest of the span (except close to base) and compares to the two-dimensional conditions. Almost a similar shedding pattern along the span was observed in the present study (Fig.10) when compared in terms of ratio of the height although with a single cell of shedding extending from 0.8H to 0.4H (3.4D) in the case of top closed, and two cells spanning from 0.85H to 0.67H (1.5D) and from 0.67H to 0.4H (2.3D) in the case of top open. The length of the cell can be interpreted in terms of extent of a region where the vortex shedding is coherent. But the cell formation along the span is attributed to the changes in the fluid due to which the cells are formed and are destroyed (Noack, 1991). Also the number of cellular patterns does not originate simultaneously but one by one as a local event and does not lead to global changes along the span . Noack et al. have proposed a model for the formation of spanwise cells behind slender bodies of revolution in cross flow. The model yields estimate for the position of the cells, their frequencies, amplitude of oscillation along the span and the local shedding angle.

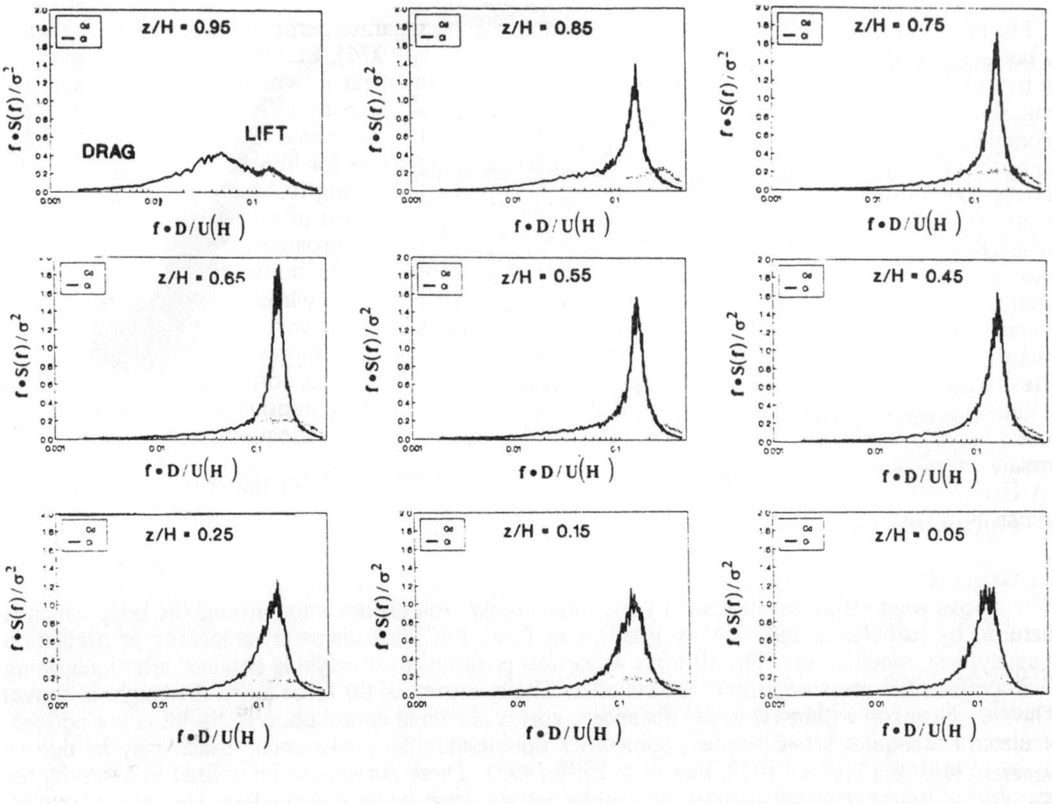

Figure 9. Spectra of the drag and lift force at a level specified ba z/H.

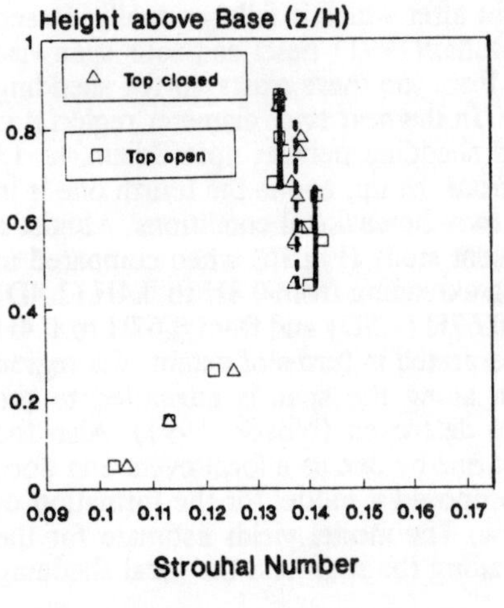

Figure 10. Strouhal number variation

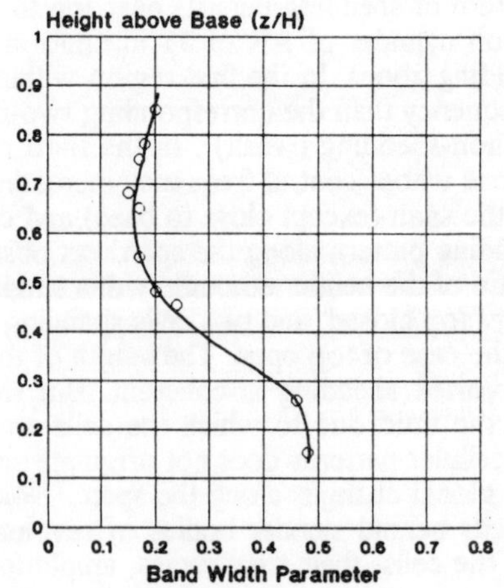

Figure 11. Band width parameter (B).

For finite height cylinders the surface streamlines do not separate perpendicular to the separation line but at an oblique angle as observed by Feng(1968), Stansby(1976), Etzold(1976) etc. This may be due to end conditions, cylinder oscillation and the presence of shear in which the Re number varies along the height. The inclination of vortices was observed (Stansby,1976) depending upon the oncoming shear, and in uniform flow the inclination was less than 10° Feng(1968) in uniform flow has also observed inclined shedding at 7° to 9° for stationary as well as for low amplitude cylinder, and almost parallel shedding for large amplitude cylinder motion. According to Maull and Young(1973) , the primary difference between the uniform and shear flow is presence of vorticity in the shear flow whose vector is normal to the plane of the flow. As the shear flow approaches the body, this vorticity vector is progressively turned until eventually the vorticity near the body is bent into the flow direction. At the trailing edge of the bluff body, vorticity in the B.L. whose vector is parallel to a generator of the body is shed and this rolls into the familiar Karman vortex street. An interaction between this vortex and the free stream vortex takes place. It is also suggested that the longitudinal vortex marks the boundary of each cell and that these longitudinal vortices come from the lorring up of the initial vorticity in the stream. The presence of longitudinal vortex in an otherwise uniform flow can cause a jump in shedding frequency along the span of the body with a consequent variation in the base pressure coefficient.

A Gaussian type of distribution is considered (Vickery/Clark,1972) to fit the vortex shedding peak and can be written as follows:

$$\frac{f\cdot S(f)}{\sigma^2} = \frac{k\cdot f/f_s}{\sqrt{\pi}\cdot B}\cdot e^{\left\{-(1-f/f_s)^2/B^2\right\}}, \qquad \ldots 2.$$

where, *B*, is the bandwidth parameter and *k*, signifies the variance contribution of vortex shedding to the overall variance to lift, i.e., $C_{LV} = \sqrt{k}\cdot C_{L(Total)}$. It may be noted that the estimated C_{LV} includes contribution from lateral turbulence of wavelengths near D/S. The bandwidth parameter calculated is drawn in Fig. 11. The narrowing of bandwidth in upper half of cylinder (excluding top 2 diameter region) is interesting because in this region the cellular form of vortex shedding is observed (Fig. 10) although *B* is influenced by *Lc* and *Iu*. The larger value of *B* in lower half region could be due to presence of relatively larger turbulence as well as shear in flow.

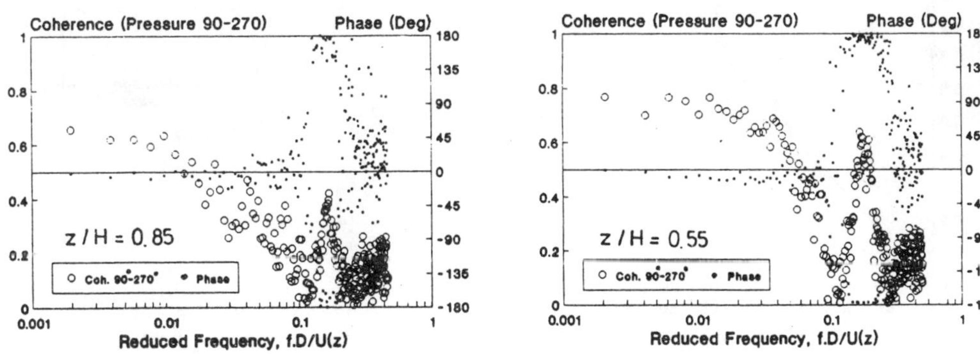

Figure 12. Coherence of the vortex shedding.

3.3.2 Coherence and Correlation Length:

The coherence between side pressures i.e. at 90° and 270° from stagnation point at level z/H=0.55 is shown in Fig. 12. The exponential decay is observed upto the frequencies of vortex shedding peak with almost no phase, however the phase increases and fluctuate between +180° and -180° as vortex shedding range which reflects alternate shedding from the sides of cylinder.

To study the extent of correlation of forces along the span, the correlation coefficient of lift force was obtained and was integrated to obtain the correlation length along the height. The correlation length can be defined as a function of correlation coefficient R_{ij} as follows:

$$L_c = \int_0^{\infty} R_{ij} \cdot dx \, , \qquad \ldots 3$$

This concept is utilised in defining the cross-correlation between forces in spectral methods. The Correlation length normalised with the diameter is calculated for the whole frequency range termed as broad band correlation length and its variation along the height is shown in Fig. 13a. The maximum Lc of 1.47D is observed at 0.85H, after which it decreases gradually. In a similar manner narrow-band correlation length was also calculated (Fig. 13b) but for the entire vortex shedding band-width. For this purpose the lift time series were passed through band pass filter corresponding to the vortex shedding frequency band. Although it is considered that the maximum response occurs near the natural frequency of the structure except when the Scruton number is very low (Blackburn/Melbourne,1992), this type of correlation length seems more representative of vortex shedding excitation, particularly away from resonance as is in the present case. The correlation length varies along the height and depends upon the reference point as well as on the direction of considered upwards or downwards as also observed by Novak/Tanaka. In general, the correlation length is greater upwards indicating that the lift-force is more correlated in upward direction of reference point. The broad band values are for top closed and have been averaged for the spanwise directional effect. The correlation lengths are less than those obtained for the individual pressures. However, the broad band Lc (=1.47D) is close to the value (Lc=1.5D) obtained by Blackburn/Melbourne in high sub-critical turbulent flow at Re=10E5. It is also observed that Lc is higher in the case of top closed as compared to the top open which is evident from the larger length of the cell(s) observed in the case of top closed in Fig. 10.

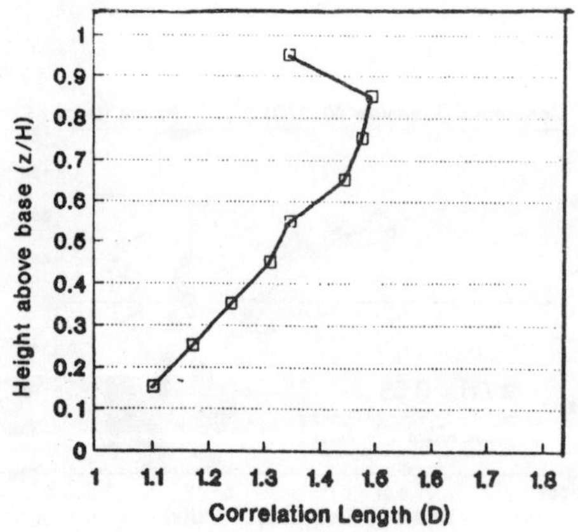

Figure 13(a). Broad band correlation length of lift.

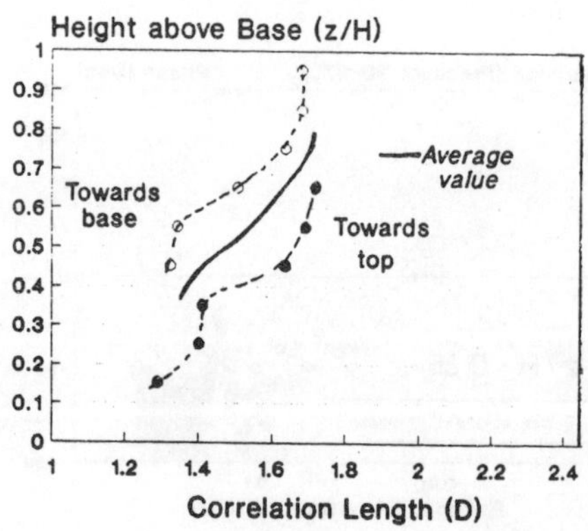

(b) Narrow band (vortex shedding band) correlation length of lift.

4.CONCLUSIONS

The following conclusions can be drawn from the present analysis of experimental data of a stationary circular cylinder of aspect ratio 8.5.

The aerodynamic coefficients as well as their vertical correlation vary along the height. Their largest values are near top extending over the upper 30% to 60% of the cylinder height. The aerodynamic forces are more correlated in the upward direction than downward from a reference point.

The drag near tip is greater than total mean which should be taken into account for base moment calculation, however, consideration should be given to nearby appurtenances e.g. platform in chimneys etc. The coherency (root coherence) of drag along the height is compatible with the vertical coherency of the wind fluctuations and the maximum drag correlation length is 4.3D close to the height of 0.75H.

The cellular shedding also occur at a low aspect ratio of 8.5. The ratio of rms lift to rms drag is appx. 1.45 over the height. The correlation of lift force also varies along the height. The broad band correlation length of lift force in upper quarter height is greater than compared to the rest of the span and the maximum is 1.47D at a level of 0.85H. However, the narrow band (based on vortex shedding band of frequency) correlation length in upper half portion is greater than the rest of the span and the maximum is 1.67 D near level 0.7H.

ACKNOWLEDGMENT

Garg acknowledges the support provided by the DAAD,FRG and CSIR,India, and colleagues at Building Aerodynamics laboratory, Bochum. He is also indebted to Prof.G.N.V.Rao of I.I.Sc. Bangalore, India, to provide the wooden model.

REFERENCES

ACI, 307-88, Standard Practice for the Design and Construction of cast-in-place Reinforced Concrete Chimneys , Detroit, USA.

Basu,R.I. and Vickery,B.J.(1983),"Across-wind Vibrations of Structures of Circular Cross-section.Part I and II.Development of a Mathematical Model for Full-scale Application",Jour. W.E.I.A., Vol. 12, pp75-97.

Billah,K.Y.R.(1989),"A Study of Vortex Induced Vibration",Ph.D. Thesis, Princeton University.

Blackburn,H.M. and Melbourne,B.H.(1992),"Lift on an Oscillating Cylinder in Smooth and Turbulent Flow",8 ICWE, Ontario, Canada, Jour. W.E.I.A,Vol.41-44, pp.79-90.

Bruun,H.H. and Davies,(1975),"An Experimental Investigation of the Unsteady Pressure Forces on a Circular Cylinder in a Turbulent Cross Flow ", Jour. Sound and Vibration, Vol.40(4), pp535-559.

Budair,M., Ayoub,A. and Karamcheti,K.(1991), "Frequency Measurements in a Finite Cylinder Wake at a Subcritical Reynolds Number", AIAA Journal, Vol. 28, No.12,pp-2163-2168.

ESDU Item No. 70014,1970, Fluid Forces Acting on Circular Cylinder for Application in General Engineering, Part II-Finite Length Cylinders, U.K.

ESDU, Item No. 85038, 1985, Circular Cylinderical Structures:Dynamic Response to Vortex Shedding . U.K.

Feng,C.C.(1968),"The Measurement of Vortex Induced Effects in Flow Past Stationary and Oscillating Circular and D-section Cylinders", M.A.Sc. Thesis, The University of British Columbia, Canada.

Fox,T.A, Apelt.C.J. and West.G.S.(1993),"The Aerodynamic Disturbance Caused by the Free-ends of a Circular Cylinder Immersed in a Uniform Flow", Jour. W.E.I.A., Vol.49, pp389-400.

Garg,R.K., Lou,J.X. and Kasperski,M.,"Some Features of Modeling Spectral Characteristics of Flow in Boundary Layer Wind Tunnels", submitted to 9 ICWE, New Delhi.

Gaster,M.(1971),"Vortex shedding from circular cylinders at low Reynolds Number", JFM, Vol.46, pp

Holmes,J.D.(1988),"Analysis and Synthesis of Pressure fluctuations on Bluff Bodies using Eigenvectors", Proc. first BBAA, (Ed. M.Ito et al), pp219-230, Kyoto.

Holmes,J.D. and Lewis,R.E.(1987),"Optimization of Dynamic Pressure Measurement Systems I:Single Point Measurements", Jour. W.E.I.A, Vol. 25, pp.249-273

Kareem,A., Cheng,C.M. and Lu,P.C.,(1989)"Pressure and Force Fluctuations on Isolated Circular Cylinder of Finite height in Boundary Layer Flows", Journal of Fluids and Structures, 3,pp 481-508.

Larsen,A.(1993),"AGeneralised Model for Assessment of Vortex-induced Vibrations of Flexible Structures", Proc. of First I.A.W.E. Afro-Euro. Reg. Conf. on Wind Engineering , Guernsey, pp.357-366.

Maull,D.J and Young,R.A.(1973),"Vortex Shedding from Bluff Bodies in a Shear Flow", Jour. Fluid Mechanics, vol. 60, pp 401-409.

Noack,C.G. et al.(1991),"On Cell Formation in Vortex Streets",Jour. Fluid Mech.,Vol.60,pp 401-409.

Niemann,H.J. and Hölscher,N. (1988),"A Review of Recent Experiments on the Flow Past Circular Cylinders", Proc. BBAA-1, Kyoto, pp 381-393.

Novak,M. and Tanaka,H.(1975),"Pressure Correlations on a Vibrating Cylinder", Proc. ICWE, Heathrow, U.K., pp-227-232.

Rao,G.N.V. and Kumar,S.(1991),"Some Issues Concerning Enhancement of Alongwind and Acrosswind Loads near the Top of Vertical Stacks.", Proc. SMIRT-11, J 07/1, Tokyo.

Roshko,A.(1992),"Perspectives on Bluff Body Aerodynamics", Proc. BBAA-2, Melbourne, Australia.

Sonneville,P.(1976),"Study of the Three Dimensional Structure of Flow Round a Circular cylinder", Doc. Sc. Thesis (in French), Universite Pierre et Marie Curie, Paris.

Steckley,A. and Vickery, B.J. (1992),"Aerodynamic Damping and Vortex Excitation on an Oscillating Prism in Turbulent Shear Flow", Proc. second BBAA, Australia.

Vickery,B.J.and Clark,A.W.(1972),"Lift or Across Wind Response of Tapered Stackes", Jour. of the Structural Div., Proc. ASCE, Vol. 98, pp ST1., pp.1-20

Waldeck,J.L.(1992),"The Measured and Predicted Response of a 300 m Concrete Chimney", J. W.E.I.A., Vol. 44, pp229-240. (presented in 8ICWE, Ontario, Canada).

West,G.S. and Apelt,C.J.(1990),"Measurements of Fluctuating Effects on a Circular Cylinder in a Uniform Flow at Subcritical Reynolds No.:Parts-1,2,3", Report CE 109, Univ. of Queensland, Australia.

9ICWE, 1995, New Delhi, India

EFFECTS OF FREE STREAM TURBULENCE ON FLOW INTERFERENCE BETWEEN TWO CIRCULAR CYLINDERS IN SIDE BY SIDE ARRANGEMENT

ZHANG, H. and MELBOURNE, W.H.

Department of Mechanical Engineering, Monash University, Clayton, VIC. 3168, Australia.

Abstract

This paper presents experimental results on flow interference between two side-by-side arranged circular cylinders. Results of aerodynamic forces and pressures on the cylinders are provided for a range of cylinder spacings and freestream turbulence intensities. The effect of freestream turbulence on the flow interference are also discussed.

1. INTRODUCTION

Interference between two side by side arranged circular cylinders in low turbulence flow has been investigated by a number of researchers, such as, Biermann et al. (1933), Bearman et al. (1973), Zdravkovich et al. (1977), Kiya et al. (1980), Williamson (1985) and Zhang et al. (1991). One of the most intriguing characteristics observed by many researchers is the bistability exhibited in the flow for a range of cylinder spacings: each of the cylinders experiences distinctly different forces despite the symmetric positions of the cylinders in the flow. Based on the fashion of vortex shedding, three interference regimes can be classified in low turbulence flow as summarised by Zhang et al. (1991). In the first regime, the two cylinders are well apart and flow patterns around both cylinders are essentially identical. Vortex shedding can be observed from both cylinders and vortices are shed in an antiphase fashion, that is, vortices from both outer sides (or inner sides) of cylinders are shed simultaneously. In the second regime the strength of vortex shedding becomes weakened and the flow exhibits bistability. In the third regime vortex shedding re-establishes and a single vortex street is formed behind the cylinders.

Although much research on flow interference for side by side arranged cylinders has been conducted, most of the previous investigations was carried out in low turbulence flow. Effects of turbulence on flow interference for this arrangement, which are of practical interest, have received little attention and relevant data are rarely available.

Over the past few years a series of wind tunnel studies focused on the effects of turbulence on flow interference was conducted at the Department of Mechanical Engineering, Monash University. This paper presents some of the results regarding the effects of turbulence for two cylinders in side by side arrangement.

2. EXPERIMENTAL ARRANGEMENT

Experiments were conducted in the 450 kW wind tunnel at the Department of Mechanical Engineering, Monash University. The working section used for the investigation had a width of 1m and a height of 2m. Cylinders used in the experiments were made of aluminium tubes with a wall thickness of 2 mm and a diameter of $D = 100mm$. The aspect ratio of the cylinders was 8. The relative surface roughness, K/D, was of the order of 10^{-6}, where K was the average height of surface roughness.

The cylinders were placed horizontally across the width of the working section. To keep the flow close to two dimensional, a pair of vertical end plates of dimension 1.8 × 1.8m was used for both cylinders.

Freestream turbulence was produced by placing slatted grids upstream of the cylinders. By changing the position of the grid, different turbulence intensities were generated. For the present work, the longitudinal turbulence intensities, $I_u = \sigma_u / \overline{U}$, generated at the model position, were 4.5% and 11.5%, respectively, where $\overline{U}$ and σ_u were the mean and standard deviations of the incident longitudinal velocity. The longitudinal integral scale was 140mm or $1.4D$. For nominal smooth flow a honeycombed channel mesh was used and the longitudinal turbulence intensity generated at the model position was 0.4%. The Reynolds number for all the flow configurations was kept at 1.1×10^5.

A two dimensional force balance was used to measure the global forces on one of the cylinders. The balance could measure the alongwind and crosswind dynamic forces of up to 120 Hz. For pressure measurement, sixteen honeywell pressure transducers were used. The pressure tappings, which were equally spaced circumferentially, were located at the mid span of the cylinders. A pressure transducer tubing system was selected so that the frequency response was flat up to 120 Hz.

Data were collected at a sampling frequency of 1000 Hz. The collected data were then processed by a mini-computer on which mean and standard deviations based on a period of 35 seconds were obtained.

The total blockage ratio of the two cylinders was 10%. The relatively high blockage would affect the force and pressure measurements. However, it would be a daunting task to correct all the measurements for the blockage effect at every cylinder spacing. Furthermore, it is thought that the presence of a constant blockage was unlikely to have a significant influence on both the trend of the variations of interference coefficients and factors with cylinder spacings, and the effects of turbulence on interference. Thus, the results presented in this paper were not corrected for any blockage effects.

3. EXPERIMENTAL RESULTS AND DISCUSSIONS

In order to demonstrate the magnitude of flow interference more effectively, interference force (pressure) coefficients ($I_{\bar{d}\text{ or }\bar{l}}$ and $I_{\bar{p}}$) and interference factors of fluctuating force (pressure) coefficients ($I\sigma_{d\text{ or }l}$ and $I\sigma_p$) were introduced. They are defined as follows:

$$I_{\bar{d}\text{ or }\bar{l}} = C_{\bar{d}\text{ or }\bar{l}} - C_{(\bar{d}\text{ or }\bar{l})_i} \qquad I_{\bar{p}} = C_{\bar{p}} - C_{\bar{p}_i} \tag{1}$$

$$I\sigma_{d\text{ or }l} = \frac{C\sigma_{d\text{ or }l}}{C\sigma_{(d\text{ or }l)_i}} \qquad I\sigma_p = \frac{C\sigma_p}{C\sigma_{p_i}} \tag{2}$$

where

$$C_{\bar{d}\text{ or }\bar{l}} = \frac{\bar{d}\text{ or }\bar{l}}{0.5\rho V_\infty^2 DL} \qquad C\sigma_{d\text{ or }l} = \frac{\sigma_{d\text{ or }l}}{0.5\rho V_\infty^2 DL} \tag{3}$$

and $$C_{\bar{p}} = \frac{\bar{p} - p_\infty}{0.5\rho V_\infty^{\ 2} DL} \qquad C\sigma_p = \frac{\sigma_p}{0.5\rho V_\infty^{\ 2} DL} \tag{4}$$

In the above equations, d, l and p denote the aerodynamic forces, drag and lift, and pressure measured on the cylinders. The overbar and σ represent time mean and standard deviation, the subscript "i" denotes measurements on an isolated cylinder (not subject to interference) under the same flow conditions. V_∞ and p_∞ are the freestream mean wind speed and static pressure, ρ is the air density and L is the length of the cylinders.

Figure 1 shows a sketch of the Cartesian coordinates as well as the definition of the angular position, θ, used for the presentation of the results.

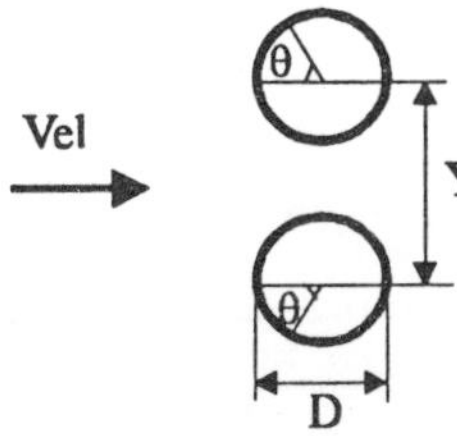

Fig. 1. The definition of cylinder spacing Y and angular position θ. The front inner region is in the range of $270° \le \theta \le 360°$ and front outer region is in the range of $0° \le \theta \le 90°$.

3.1 Low turbulence flow ($I_u = 0.4\%$)

3.1.1 The first interference regime (Y/D>1.9)

Figures 2, 3 and Figures 4, 5 present the interference force coefficients and interference factors, respectively, as functions of cylinder spacing Y/D at different turbulence intensities. Interference force coefficients obtained by Biermann et al. (1933) and Bearman (1973) in low turbulence flow, shown in Figures 2 and 3, are in good agreements with the present results in this regime. Figures 6 and 7 present the mean and fluctuating pressure distributions on one of the cylinders.

It was observed, under the present freestream conditions, ie. $\mathrm{Re} = 1.1 \times 10^5$ and $I_u = 0.4\%$, that the critical spacing for the first interference regime is $Y/D > 1.9$. From Figures 2 to 5 it can be seen that the cylinders have practically no interference at $Y/D = 5$. As Y/D is reduced to below 5, interference becomes more pronounced. The more intense interaction between the two vortex streets, as Y/D is reduced, can be observed in Figure 8, which shows an increase in the cross correlation coefficients of the pressure fluctuations across the cylinders. Consequently an increase in both the fluctuating pressures and global fluctuating forces can be observed. These are shown in Figures 4, 5 and 7. From Figure 8 it can be seen that the correlation between the pressures in the inner sides of the cylinders (tapping E & H) is higher than that in the outer sides (tapping G & F). This indicates that a greater coupling occurs between the "inner" vortex pair than the "outer" vortex pair. This may be responsible for the existence of larger fluctuating pressures in the inner side of the cylinder as indicated in Figure 7 at $Y/D = 2$. In Figure 8, the cross-correlation coefficients obtained by Bearman is also provided for comparison.

3.1.2 The second interference regime

In his theoretical study on the stability of two pairs of infinitely long parallel vortex sheets, Kyoji (1976) concluded that if the distance between two pairs of vortex sheets got below a critical value, the

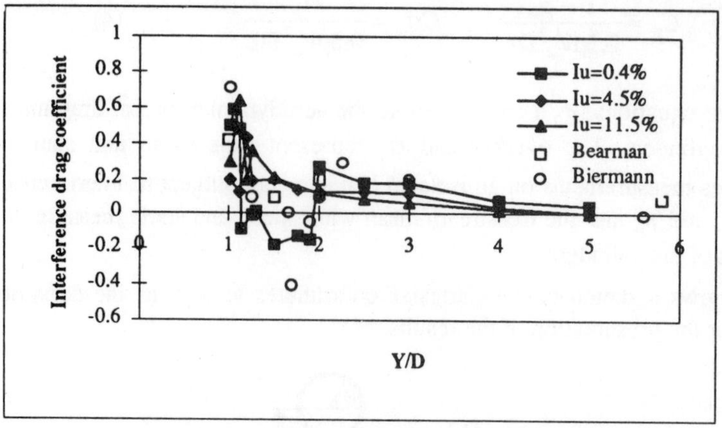

Fig. 2. Interference drag coefficients, $I_{\bar{d}}$, as functions of Y / D.

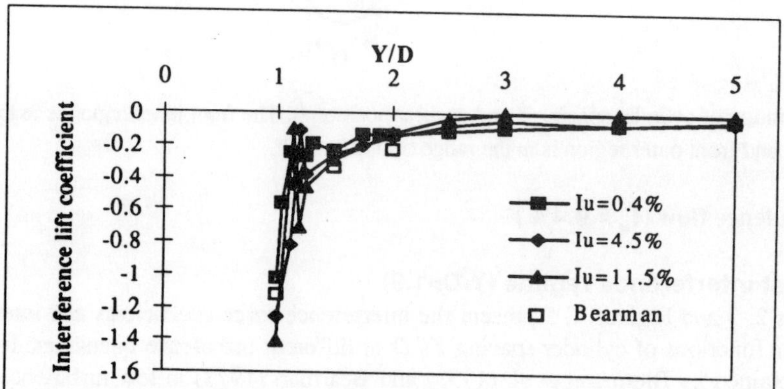

Fig. 3. Interference lift coefficients, $I_{\bar{l}}$, as functions of Y / D.

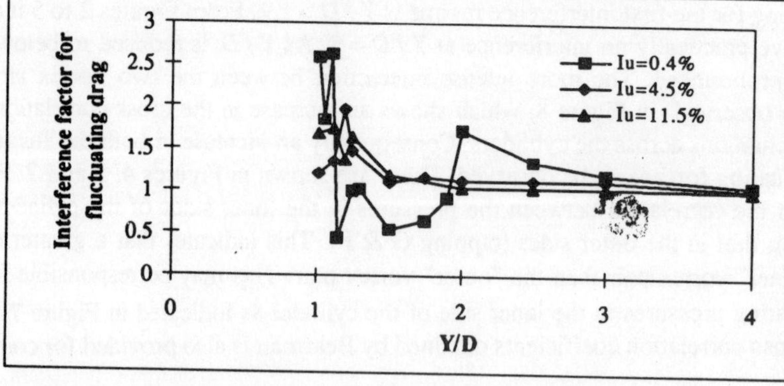

Fig. 4. Interference factors of fluctuating drag coefficients, $I\sigma_d$, as functions of Y / D.

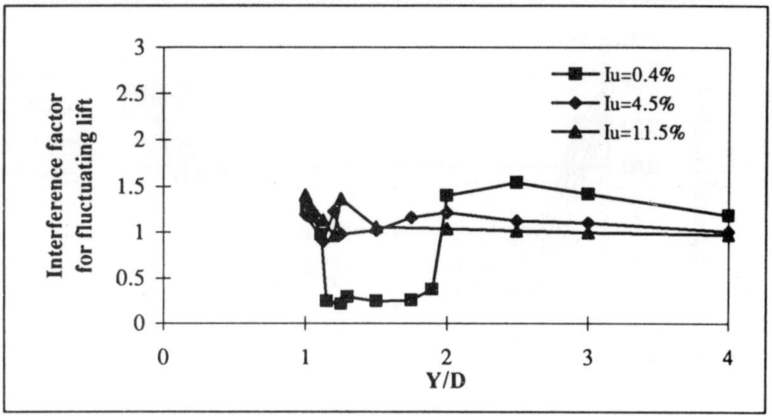

Fig. 5. Interference factors of fluctuating lift coefficients, $I\sigma_l$, as functions of Y/D.

vortex sheets became unstable. His results were based on the application of linear theory to ideal fluid. Applying his results to the four rows of vortices in the wake of side by side arranged cylinders, the onset spacing of instability between the centres of cylinders was found to be two diameters which is surprisingly close to the value of 1.9 observed in the present work. Different from the tandem arrangement for which the transition between the first and second interference regimes takes place over a range of cylinder spacing (see Zhang et al., 1992), the transition for side by side arrangement occurs within a very narrow range of spacings. Figure 9 presents time traces of pressure fluctuations for both $Y/D = 3$ and $Y/D = 1.5$. It can be seen that compared to the trace at $Y/D = 3$, the pressure fluctuation at $Y/D = 1.5$ becomes much more irregular and unsteady. Its spectrum reveals that most of the energy is contained in the lower end of frequency range, as opposed to that at $Y/D = 3$ at which the spectral peak is centred at the regular vortex shedding frequency. Figure 10 presents the difference between the mean base pressure coefficients of the two cylinders in the range of $1 < Y/D < 2$. It can be seen that the base pressures of the two cylinders in this spacing range are distinctly different. This indicates the asymmetry of the flow pattern around the cylinders. The difference in the base pressures will cause the flow through the gap between the cylinders to be biased towards the cylinder wake with the lower base pressure. As shown in Figure 10, the differential base pressure is significantly greater in the range of $1.12 < Y/D < 1.5$ than in $1.5 < Y/D < 1.9$. Thus it is expected that the biased gap flow is less stable in $1.5 < Y/D < 1.9$. This can be seen in Figure 9 in which the pressure trace at $Y/D = 1.75$ displays a more prominent low frequency unsteadiness than that at $Y/D = 1.25$. The low frequency fluctuation is believed to be caused by the swing of the gap flow. This conjecture is supported by Figure 8 which shows that the cross-correlation coefficients in $1.5 < Y/D < 1.9$ are negative in general and appear to have lower values than those in $1.12 < Y/D < 1.5$.

As a result of the biased gap flow, wide and narrow wakes are formed behind the cylinders. Ishgai et al. (1972), among other researchers, observed that the cylinder with a narrower wake has a relatively large Strouhal number of 0.3, whereas the cylinder with a wider wake has a lower Strouhal number of 0.1. He also observed that the biased gap flow was bistable and consequently the cylinders experienced two Strouhal number alternatively. The Reynolds number in his experiments was of the order of 10^3. In the present work, however, only the higher Strouhal number could be observed. This can be seen in Figure 11 which also shows the Strouhal numbers obtained by other researchers in low turbulence flow at the Reynolds numbers of the order of 10^4.

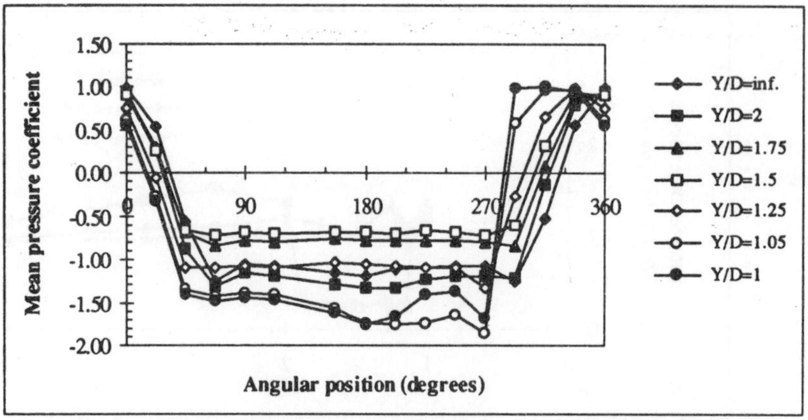

Fig. 6. Mean pressure distributions for various Y / D in flow with $I_u = 0.4\%$.

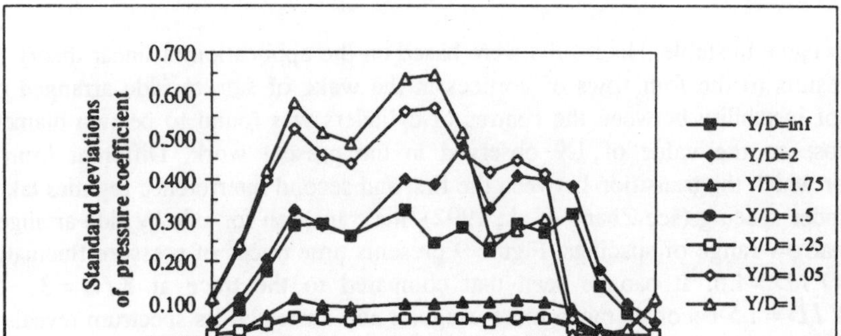

Fig. 7. Fluctuating pressure distributions for various Y / D in flow with $I_u = 0.4\%$.

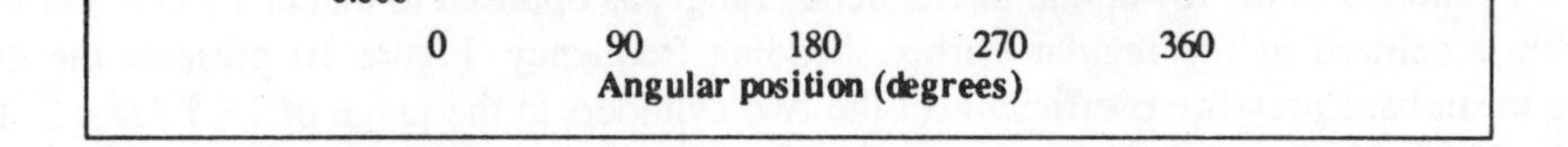

Fig. 8. Cross-correlation coefficients across the cylinders as functions of Y / D at $I_u = 0.4\%$.

Because of the disruption to the regular vortex shedding, the fluctuating pressures and forces in the second regime are small in general, as shown in Figures 4, 5 and 7. From Figures 6 and 7 it can be seen that the pressure distributions become significantly more asymmetric as Y/D is reduced. The mean pressures in the front inner region ($270^\circ \le \theta \le 360^\circ$) are increased and the mean pressures in the front outer region are slightly decreased as Y/D is reduced. Consequently large mean lift force is generated, acting in a direction which separates the cylinders. The disruption in the vortex shedding, as the flow transits from the first to the second regime, creates a large recovery in the base pressure which causes a significant reduction in the mean drag as indicated Figure 2. However, the increase in the base pressure has a negligible effect on the mean lift. Thus no characteristic change can be observed in the mean lift force as the flow interference changes from the first to the second regimes.

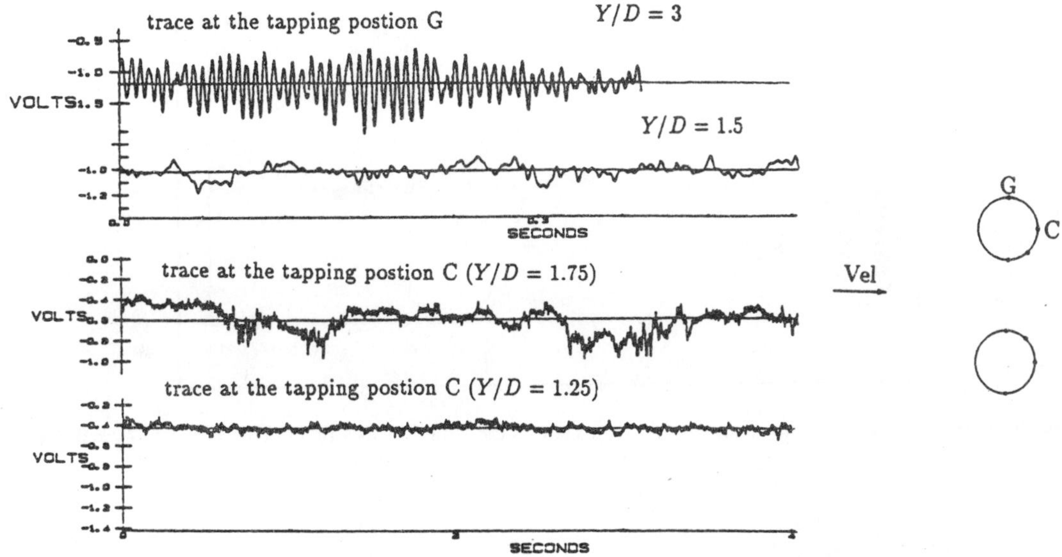

Fig. 9. Comparison of pressure traces at various cylinder spacings in the second interference regime, $I_u = 0.4\%$.

3.1.3 The third interference regime

Similar to the results obtained by the researchers mentioned previously, the two cylinders behave like a single cylinder when the spacing is small enough. Vortex shedding re-establishes and a single vortex street is formed behind the cylinders. The onset spacing observed in the present research is $Y/D = 1.12$. Due to the re-establishing of the vortex shedding, mean and fluctuating pressures as well as the cross-correlation coefficients undergo significant changes as indicated in Figures 6, 7 and 8. Similarly, mean and fluctuating forces on the cylinders also change significantly. As shown from Figures 2 to 5, all the force coefficients reach their extreme values in this regime.

3.2 Turbulent flow

In turbulent flow, the interference undergoes a significant change. The distinctions among various flow interference regimes become much less prominent. This can be demonstrated from Figures 2, 4 and 5, which show that the interference force coefficients and factors in turbulent flow do not display the characteristic changes as observed in low turbulence flow particularly across the first and second regimes. This is largely attributed to the turbulence effect on the vortex shedding process. At the present Reynolds number, the freestream turbulence can severely disrupt the regular vortex shedding of a cylinder, ie. vortex shedding becomes weakened and less correlated. This effect of turbulence on the vortex shedding is expected to have a significant impact on the interaction between the two vortex streets behind the cylinders. The interaction, which is thought to be the main interference mechanism in the first and second regimes, becomes less intense in turbulent flow.

As shown in Figures 2 to 5 the magnitude of interference in turbulent flow is generally reduced significantly compared to those in low turbulence flow and remains fairly small for cylinder spacings larger than 1.5D. As the cylinder spacing gets below 1.2D for $I_u = 4.5\%$, and 1.12D for $I_u = 11.5\%$, vortices were observed to shed periodically to form a single vortex street behind the cylinders. This is similar to that which occurs in low turbulence flow in the third interference regime. As shown in Figures 2 to 5 large mean and fluctuating forces can be observed, in particular the interference coefficient for the lift force reaches a higher magnitude than that observed in low turbulence flow.

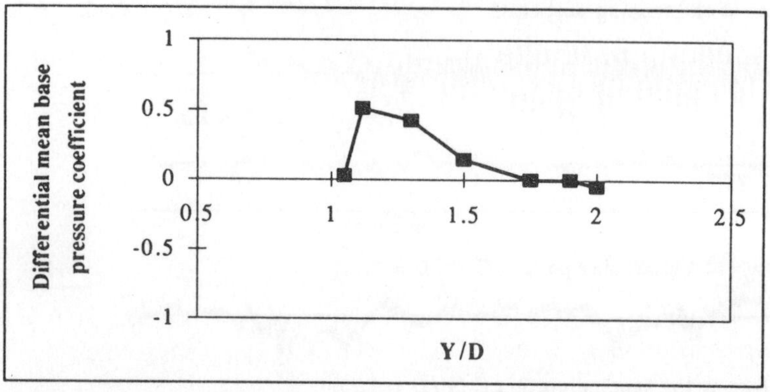

Fig. 10. Difference between the mean base pressures across the cylinders as a function of Y/D at $I_u = 0.4\%$.

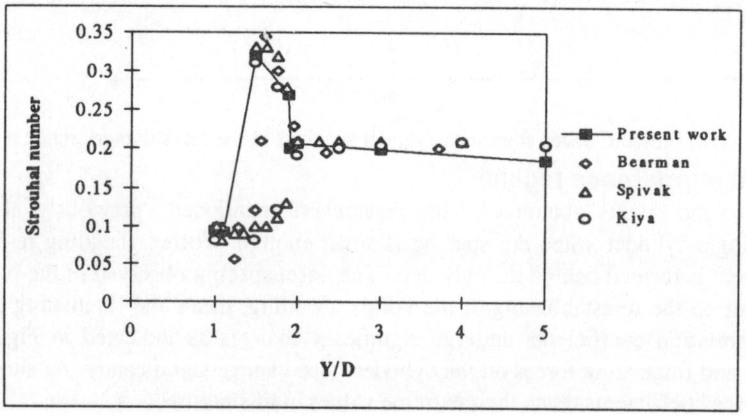

Fig. 11. Variations of Strouhal numbers with Y/D in flows with $I_u = 0.4\%$.

Figures 12 to 15 present the mean and fluctuating pressure distributions on one of the cylinders for a range of Y/D in flows with $I_u = 4.5\%$ and 11.5%. As shown in these figures the mean pressure distributions become more asymmetric as Y/D is reduced. A comparison between the pressure distribution on a single cylinder and those at, say $Y/D = 1.5$ for both turbulence intensities, indicates that the asymmetry is mainly caused by the large reduction in the mean pressures in the outer region around $\theta = 60°$. The asymmetry of the pressure distribution contributes to the large mean lift force observed in turbulent flow for $Y/D < 1.5$.

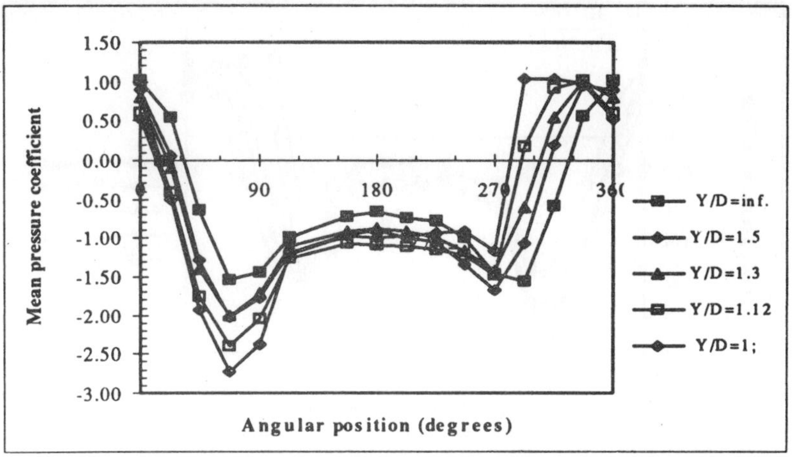

Fig. 12. Mean pressure distributions for various Y / D in flow with $I_u = 4.5\%$.

One of the most interesting phenomena observed near the critical spacing of the third interference regime is the peculiar mean pressure distribution. As shown in Figure 14 the mean pressure features a very non-uniform base pressure distribution at $Y/D = 1.12$ for $I_u = 11.5\%$. However, this was not observed for $I_u = 4.5\%$ as shown in Figure 12. Figures 16 and 17 present mean pressure coefficients measured simultaneously on both cylinders. It can be seen that near the critical spacing the mean pressure on one of the cylinders at $\theta = 225°$ becomes significantly lower than the mean pressure measured at the symmetric angular position on the other cylinder. The value of the lower mean pressure coefficient at $\theta = 225°$ and $Y/D = 1.12$ for $I_u = 11.5\%$, as shown in Figure 17, is almost the same as the pressure coefficient observed in Figure 14 at the same angular position. This suggests that the cylinder with the lower pressure coefficient at $\theta = 225°$ and $Y/D = 1.12$ for $I_u = 11.5\%$ experiences a distorted pressure distribution. The different mean pressure coefficients at $\theta = 225°$ for both cylinders indicate that it is likely that the cylinders experience different types of pressure distributions. While one of the cylinders has the non-uniform distribution, the other cylinder may experience a "normal distribution". This may explain why the non-uniform base pressure distribution was not observed for $I_u = 4.5\%$ as shown in Figure 12. Further investigations on the causes of these distorted distributions are needed.

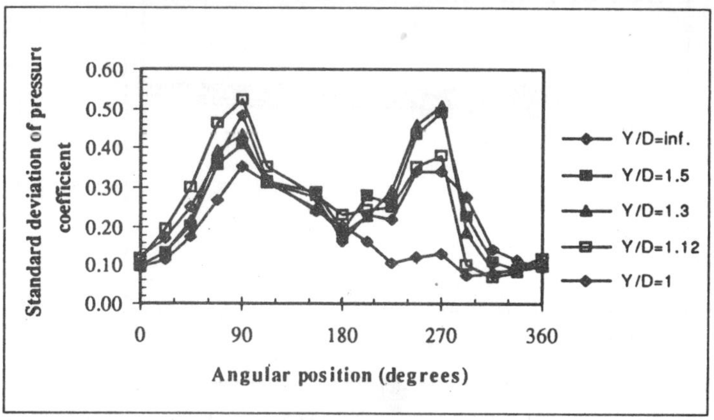

Fig. 13. Fluctuating pressure distributions for various Y / D in flow with $I_u = 4.5\%$.

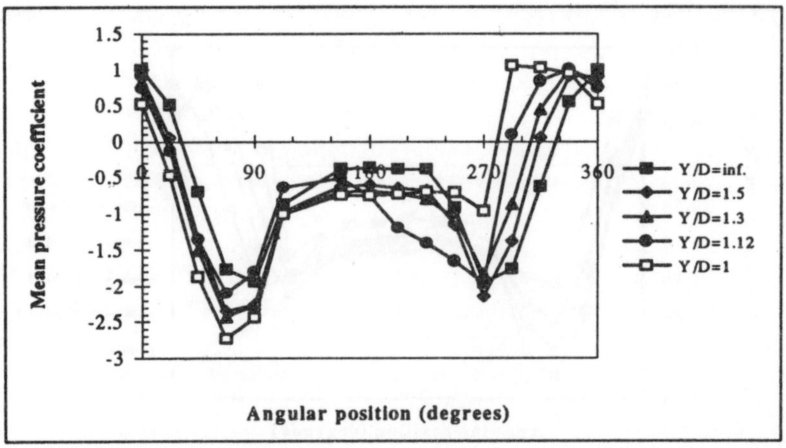

Fig. 14. Mean pressure distributions for various Y / D in flow with $I_u = 11.5\%$.

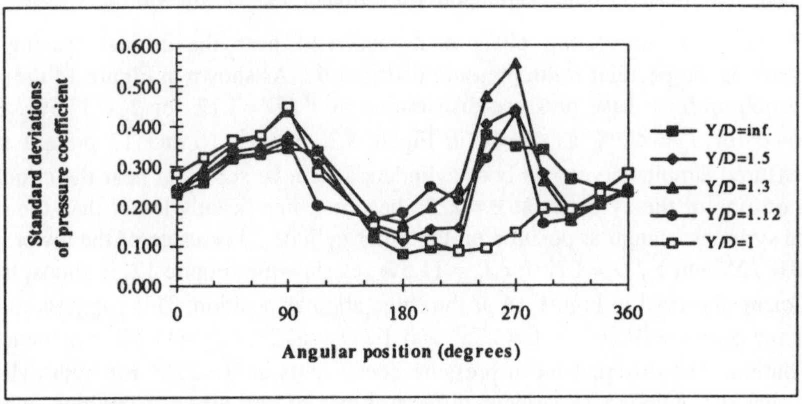

Fig. 15. Fluctuating pressure distributions for various Y / D in flow with $I_u = 11.5\%$.

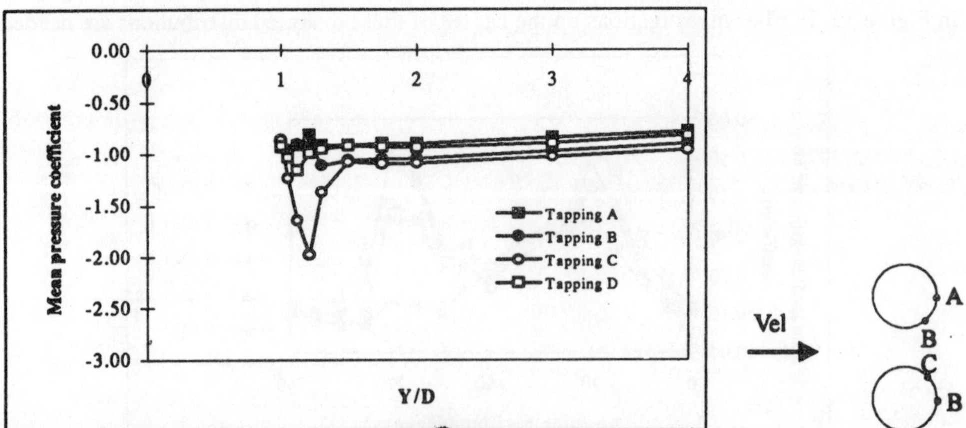

Fig. 16. Variations of mean pressure coefficients measured simultaneously on both cylinders with Y / D at $I_u = 4.5\%$.

One of the main differences between the interference in low turbulence flow and that in turbulent flow is that, in low turbulence flow, interference has a "global effect", whilst in turbulent flow the effect is more localised. This can be observed in Figure 18 which shows the variations of interference factor of fluctuating pressure coefficients at different tapping positions for both $I_u = 0.4\%$ and 11.5%. It can be seen that the variations of fluctuating pressures at different regions of the cylinder display a similar trend in low turbulence flow. However, the same trend cannot be observed in turbulent flow. The existence of a common trend in low turbulence flow suggests that one interference mechanism plays a dominant role, that is the interaction between the separated shear layers and shed vortices, which influences the pressures in most areas of the cylinder in a similar way. However, in turbulent flow, the strength of the vortex shedding is much weakened and the flow is less correlated, which results in the much localised interference effect as shown in Figure 18. The localised effect is thought to be achieved by the interaction between the freestream turbulence and the boundary layer of the cylinders. As the cylinders are closely positioned, the flow field around the both sides of the cylinders are significantly different. Hence, the freestream turbulence is distorted differently on both sides. The different distortions cause different interactions between the freestream and the boundary layers of the cylinders, which is thought to be responsible for the prominent asymmetric pressure distributions observed in turbulent flow.

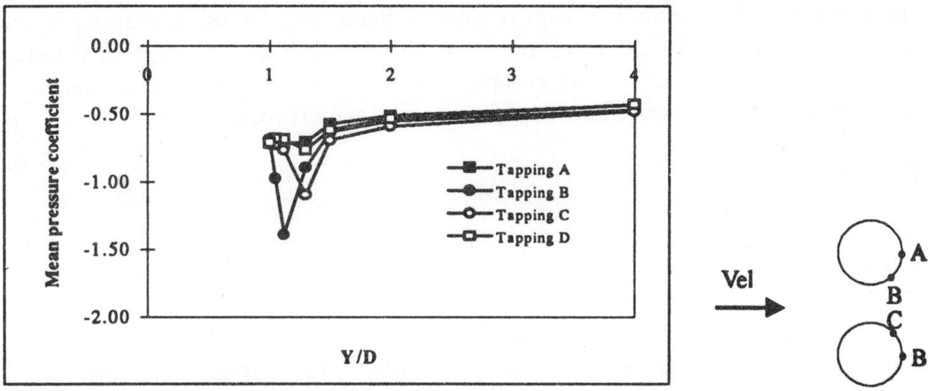

Fig. 17. Variations of mean pressure coefficients measured simultaneously on both cylinders with Y/D at $I_u = 11.5\%$.

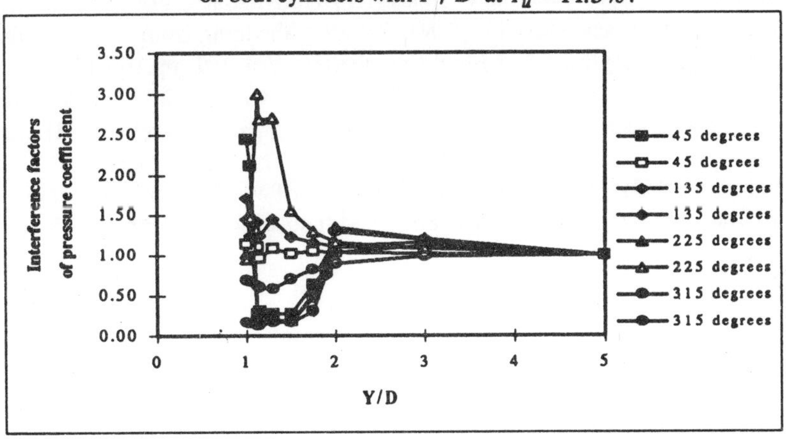

Fig. 18. Variations of interference factors of pressure coefficients with Y/D. The solid symbols are for $I_u = 0.4\%$ and non-solid symbols are for $I_u = 11.5\%$.

4. CONCLUSION

The present investigation on flow interference between two side by side arranged cylinders indicates that the flow interference in low turbulence flow reaches its maximum magnitude in the third regime, $Y/D<1.12$, in which a single vortex street is formed behind the cylinders. In the second regime in which regular vortex shedding is suppressed, it was observed that the flow is more unstable in the range of $1.9<Y/D<1.5$ than in $1.5<Y/D<1.12$. In this regime, despite the severe interference, the magnitudes of interference force coefficients and factors are generally small. In the first regime, no severe interference takes place for $Y/D>5$. For $Y/D<5$, the interaction of the separated shear layers and vortices is strengthened as Y/D is reduced and large magnitude of interference is reached at the end of this regime, ie $Y/D=2$.

The present results show that freestream turbulence can significantly affect flow interference. It is believed that when the turbulence intensity is increased, the interference mechanism changes from that controlled by the interaction of the separated shear layers and shed vortices to that controlled by the interaction between the distorted local turbulence and the cylinder boundary layers. Consequently, in turbulent flow the critical spacing, under which prominent interference occurs, is reduced. Also the magnitude of interference, particularly with respect to fluctuating forces, is reduced in most cases. When the cylinders are almost in contact, the interaction between the distorted local turbulence and boundary layers is thought to be responsible for introducing prominent asymmetric pressure distributions which cause the relative large interference drag and lift coefficients.

5. REFERENCE

Biermann, D and Heerrnstein, W.H.Jr (1933), "The Interference between Struts in Various Combinations", National Advisory Committee for Aeronautics, Technical report (468).

Bearman, P.W and Wadcock, A.J. (1973), "The Interaction between a Pair of Circular Cylinder Normal to a Stream", Journal of Fluid Mechanics, Vol. 61, pp 499-511.

Kyoji, K (1976), "Formation and Interaction of Two Parallel Vortex Streets", Bulletin of the JSME, Vol. 19, pp 283-290.

Zdravkovich, M.M and Pridden, D.L (1977), "Interference between Two Circular Cylinders; Series of Unexpected Discontinuities", Journal of Industrial Aerodynamics, Vol. 2, pp 255-270.

Kiya, M., Arie, M., Tamura, H. and Mori, H. (1980), "Vortex Shedding from Two Circular Cylinders in Staggered Arrangement", Journal of Fluid Engineering, Vol. 102, pp.166-173.

Zhang, H. and Melbourne, W.H. (1991), "Interference between Two Circular Cylinders for Side by Side Arrangement", Proceedings of Australian Wind Engineering Workshop, Huntervalley, NSW.

Zhang, H. and Melbourne, W.H. (1992), "Interference between Two Circular Cylinders in Tandem in Turbulent Flow", Journal of Wind Engineering and Industrial Aerodynamics, Vol. 41-44, pp.589-600.

WAKE INTERFERENCE BETWEEN TWO CIRCULAR CYLINDERS

BRIKA, D. and LANEVILLE, A.

Départment de génie mécanique, Université de Sherbrooke, Sherbrooke (Québec), J1K 2R1, Canada

Abstract

This paper deals with an experimental investigation of the vortex-induced oscillations of a long flexible circular cylinder immersed in the wake of an identical cylinder. In comparison with the hysteretic response of an isolated cylinder, the steady-state response of the cylinder downstream of a stationary one is represented by a single branch and shows a wider synchronization region that decreases with increasing values of the spacing between the cylinders. When the upstream cylinder is free to oscillate, the downstream cylinder response becomes hysteretic. The hysteresis is characterized with two branches and two discontinuities; each branch is associated with a particular vortex pattern and the discontinuities coincide with those of the upstream cylinder. The results also indicate that beyond an inclination β=20° of the bundle, the leeward cylinder behaves as an isolated one.

1. INTRODUCTION

The investigation of the wake interference between the circular cylinders is of great importance, such a configuration being commonly used in many engineering applications. Transmission line bundles, offshore structures, heat exchangers, radar masts, twin chimney stacks and twin struts to support aeroplane wings are typical applications. That a cylinder in the wake of another cylinder does not behave as would an isolated one is a well known fact. This is attributed in part to the reduction of the velocity of the flow impinging on the downstream cylinder. In the relatively simple case of two stationary cylinders, the experimental results, as summarized by Ohya *et al.* (1989) and Zdravkovich (1977), show that the aerodynamic characteristics of the downstream cylinder such as the drag and lift coefficients, the pressure distribution, the Strouhal number and the vortex patterns differ significantly from those of an isolated cylinder. The most important result of this case is the presence of a discontinuity in these characteristics at a critical spacing ratio L/D ranging between 3.5 and 4; L is the distance between the cylinders' centres and D the cylinder external diameter. This discontinuity is attributed to a sudden change from one mode of vortex pattern to another (Zdravkovich 1977).

In the case of a cylinder oscillating in the wake of a stationary cylinder, the interaction between the flow, the wake vortices and the cylinders becomes more complex as reported by Zdravkovivh (1972), Tanida *et al.* (1973), Price (1975), King & Johns (1976) and Bokaian & Geoola (1984) amongst others. The dynamic response of the downstream cylinder is influenced by the spacing ratio L/D in addition to the reduced velocity, the damping ratio and the Reynolds number. Bokaian & Geoola (1984) have observed that, depending on the spacing ratio L/D, the downstream cylinder may exhibit separated or combined vortex-induced oscillations and wake galloping, D being the cylinder external diameter. The wake galloping is observed for L/D$\leq$ 3. King & Johns (1976) have investigated vibrations in the *in-line* and *cross-flow* directions. These vibrations depend on the parameters mentioned previously and are attributed to different vortex patterns of the cylinders' wake. The *in-line* vibrations are present only when the stability parameter $2m\delta/\rho D^2$ is less than 1.2 (m is the mass per unit length of the cylinder, δ, the logarithmic decrement and ρ the fluid density) King & Johns also observed that in general, the vortices shed from the upstream cylinder and the mechanical coupling between the two cylinders reinforce the excitation of the downstream cylinder. Finally, when both cylinders are free to oscillate, the wake interference between the cylinders gains in complexity.

Since damping may have caused some effects on the cylinder response, the purpose of this study is then to investigate experimentally the influence of L/D on the dynamic response of a bundle of two similar cylinders when damping is the least significant. Tandem and staggered arrangements are examined. The downstream cylinder is freely excited by wind and the upstream cylinder is either stationary or wind-excited.

2. EXPERIMENTAL SET-UP

The experimental set-up is described with more details in Brika & Laneville (1993). Figure 1 shows a sketch of a bundle of two cylinders and the different parameters studied, i.e. the separation L between the cylinders' centre and the inclination angle, β, of the bundle with respect to the flow direction. Both cylinders have smooth surfaces, an external diameter D= 33.4 mm and a length l= 3.313 m. Each elastic model has a natural frequency f_n= 17.58 Hz and is suspended horizontally with thin steel flexible blades at the nodes of its first free-free vibration mode. The nodes coincide with the tunnel walls and the part of the model internal to the wind tunnel vibrates in a sinusoidal mode shape. The vibrating system has a low damping ratio, ζ= 0.8x 10^{-4}. The models have a solid blockage ratio of 1.8% and an aspect ratio l_w/D= 52.7, where l_w is the net length of the cylinder exposed to the airflow. In the case of cylinders in tandem, the spacing ratio is varied between 7 and 25, while for a separation L/D= 13 the bundle inclination, β, is varied between 0° and 20°. The Reynolds number ranges between 5100 and 27500.

The displacement of each cylinder is measured with a B&K 4393 accelerometer (mass= 2.4 grams) placed on the part of the vibrating model external to the wind tunnel at a distance of about 45 mm from the nodal point. A hot wire anemometer positioned at 1.5D downstream and 1.5D below the leeward cylinder is used to measure the phase difference Φ between the vortex shedding and the downstream cylinder displacement. The phase Φ and the cylinders' displacement are recorded on a double channel tracer. The results are performed according to two different regimes referred to as the *progressive* and *impulsive* regimes. In the progressive regime, the flow velocity is progressively increased or decreased while the downstream model is vibrating at its stationary amplitude of the previous velocity. In the impulsive regime, for each flow velocity, the downstream cylinder is released either from rest or from a high amplitude initially pumped with a shaker (B&K 4809).

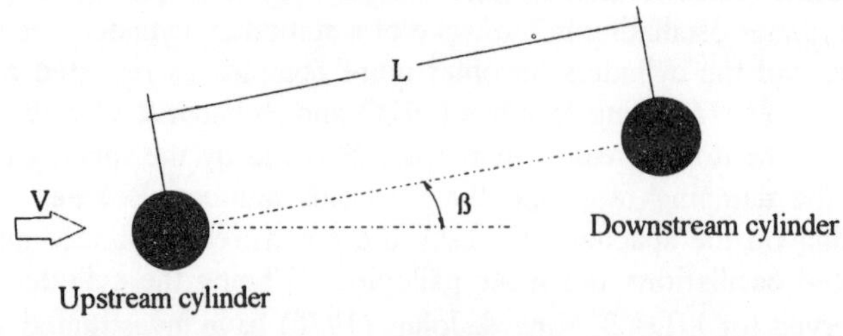

Figure 1. Sketch of a bundle of two cylinders.

3. RESULTS

3.1 Single cylinder

The results obtained with an isolated flexible cylinder are reviewed shortly for comparison purposes. These results are presented with more details by Brika & Laneville (1993). Figure 2 summarizes the results obtained by Brika & Laneville (1993) and presents the adimensional amplitude A/D as a function of the reduced velocity V_r= V/fD, where A is the single amplitude of vibration at the model mid-span, V the mean flow velocity and f, the cylinder vibrating frequency (f≈ fn= 17.58 Hz). The hysteresis phenomenon in the response of the model is characterized with two branches and two discontinuities. Flow visualization (Brika & Laneville 1993) has shown that each branch of the hysteresis loop is associated with a particular vortex pattern in the wake of the cylinder and each discontinuity is attributed to a sudden change from one mode of vortices to another, supporting the hypothesis previously advanced by Williamson & Roshko (1988). The first branch, generally obtained with a progressive increase of flow velocity and ranging between the synchronization onset velocity and the resonance velocity, is associated with the so-called 2S mode (Williamson & Roshko 1988) where a single vortex is shed from each side of the cylinder at every cycle of vibration like the natural Karman vortices. The second branch, ranging between the end of synchronization and V_r= 5.5 and generally obtained with a progressive decrease of the reduced velocity, is associated with another vortex pattern referred to as the 2P mode where a pair of vortices of opposite signs are shed from each side of the cylinder at every cycle of vibration. For convenience, the branch associated with the 2S mode will be called the C2S curve and that associated with the 2P mode will be designated as the C2P curve.

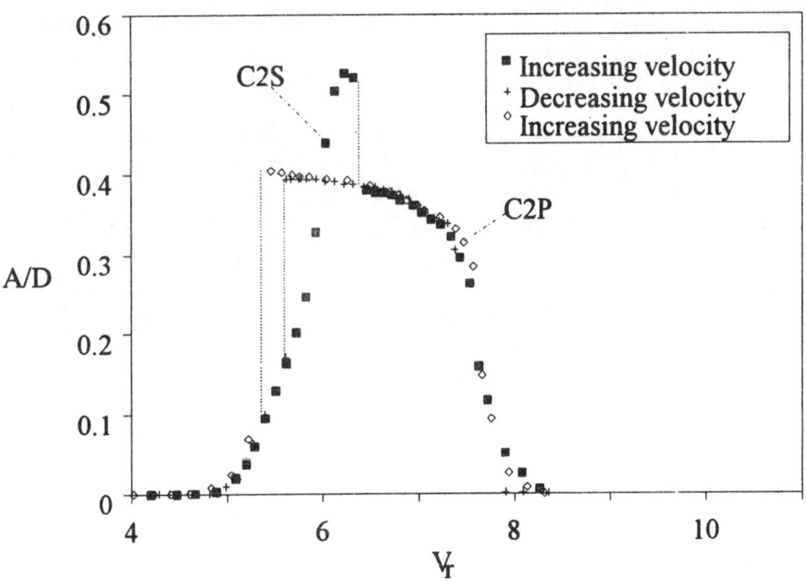

Figure 2. The dynamic response of an isolated cylinder

3.2 Bundle with a Fixed Upstream Cylinder

Tandem arrangement

Figure 3 shows the effect of the spacing ratio L/D on the dynamic response of the downstream cylinder A/D as a function of the reduced velocity V_r in the case of the progressive regime. In comparison with the results of an isolated cylinder (see figure 2), the presence of a stationary upstream cylinder clearly influences the response of the leeward one. The dynamic respcnse is no more hysteretic and the jumps in the oscillation amplitude have completely disappeared. The results of figure 3 also indicate that a) the synchronization onsets at a higher reduced velocity than that of an isolated one, b) the synchronization region is much larger and decreases with increasing value of L/D. For L/D= 10, say, the synchronization region is twice as large as that of an isolated cylinder. The expansion of the synchronization region occurs essentially at higher reduced velocities. The maximum amplitude as well as the velocity for which this maximum occurs (resonance velocity, V_R) also decrease with increasing L/D. Up to the largest spacing ratio tested (L/D= 25), the leeward cylinder response remains different from the one of a single cylinder and the upstream cylinder can be said to still influence the response of the downstream one. The maximum amplitude of osillations is roughly of the same order as that of an isolated cylinder but with higher values at small spacing ratios and lower values at large separations. The results of figure 3 also indicate the presence of a first peak with A/D< 0.06 around the reduced velocity V_r= 5. In the neighbourhood of this velocity, the vortices shed from the upstream cylinder have a frequency approximately equal to the natural frequency of the downstream cylinder and when these vortices impinge on the leeward cylinder they may excite it. As the vortices of the upstream cylinder cannot lock-in to the vibrating frequency of the downstream model, this oscillation amplitude vanishes with increasing value of the reduced velocity.

The results of figure 3 show that for a given mean flow velocity in the range defined by the. synchronization onset velocity and Vr= 6.7, the amplitude of oscillations remains roughly the same for all the spacing ratios examined. For velocities larger than V_R, the effect of L/D can be seen to reduce the

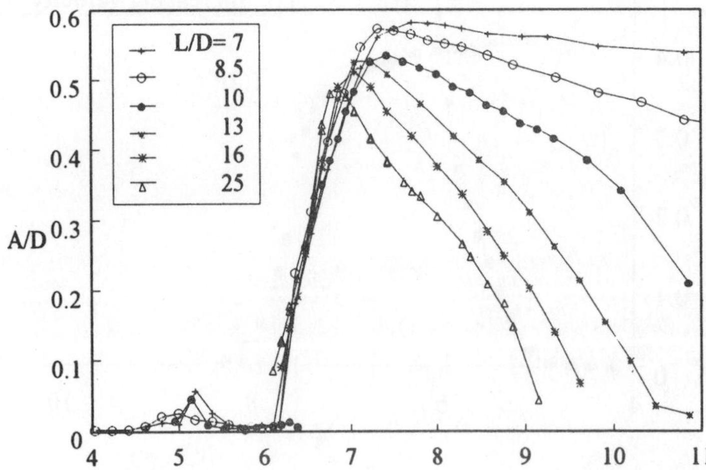

Figure 3. Effect of L/D on the response of the downstream cylinder (tandem arrangement, upstream cylinder is fixed).

amplitude of oscillations at a given velocity. For each spacing ratio, the amplitude A/D increases more rapidly with V_r before resonance and decreases in a more progressive manner past the resonant velocity. A point of inflection, more apparent for L/D= 25 (V_r= 7.25, A/D= 0.41), can be observed just past the resonance. Because of these differences of behaviour prior and past the resonance velocity and because of the results of the isolated cylinder, it can be deduced that two different vortex patterns can be possiblt present in the wake of the downstream cylinder even if no hysteresis phenomenon is observed in its dynamic response. The velocity where a *progressive* jump from one vortex pattern to another occurs is expected to be around the resonance velocity. This hypothesis will be verified with a flow visualization in the near future.

For the smallest separations examined, i.e. L/D= 7 and 8.5, another phenomenon has been observed in addition to the one of vortex-induced oscillations. At higher velocities, the amplitude A/D does not vanish as in the case of the other separations (see figure 3): after reaching a minimum as shown in figure 4, it undergoes a progressive and slight increase with increasing value of V_r. The amplitude seems to reach a maximum of about A/D= 0.56 (L/D= 7) and the cylinder continues to vibrate at higher reduced velocities (at least until V_r= 22). Similar behaviour has been observed by Bokaian & Geoola (1984) but only at very small separations (L/D≤ 3). This phenomenon is sometimes referred to as *wake galloping* and can be inferred to take over from the minimum.

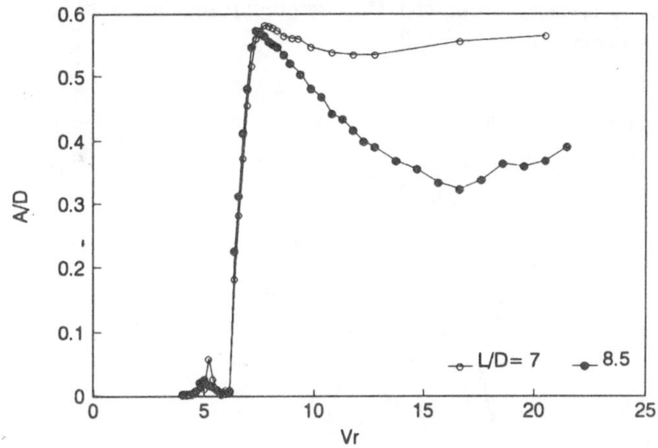

Figure 4. The dynamic response of the downstream cylinder at low spacing ratios (L/D= 7 and 8.5).

Figure 5 shows typical time responses of the downstream cylinder when the model is released from rest (full symbols) and from a high amplitude initially pumped with a shaker (empty symbols). For a spacing ratio L/D= 13, three reduced velocities are chosen: one at the resonance where the maximum amplitude occurs (Vr= 7.25), the second before and the third past the resonance velocity The results of figures 3 and 5 indicate that, for a given flow velocity, the steady-state amplitude is always the same whether the model is released from rest, from a high amplitude or in the case of a progressive increase or decrease of flow velocity. This is in contrast with the single cylinder case for which it was observed that the steady-state amplitude is strongly influenced by the initial conditions of excitation and the history of the cylinder motion (see Brika & Laneville 1993). This behaviour is attributed to the absence of the hysteresis phenomenon for which two states are possible at the same flow velocity. Th results of figure 5 also indicate that the steady-state amplitude is reached more rapidly at the resonance velocity than at any other velocity. This means that the power imparted by wind to the downstream cylinder is maximum at the resonance velocity.

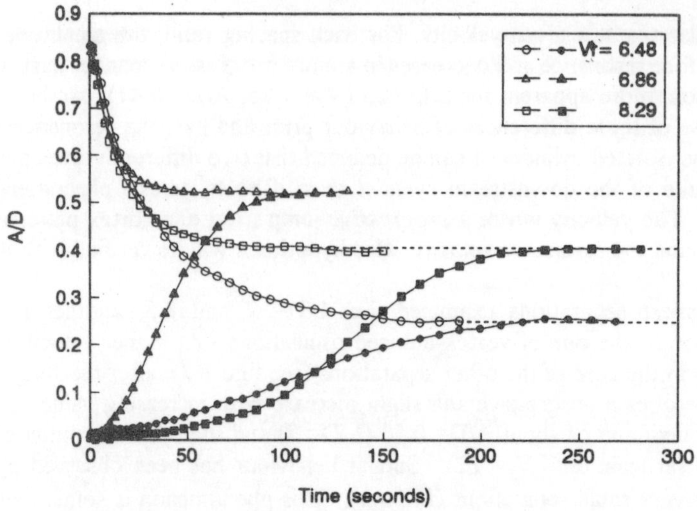

Figure 5. Examples of the time response and the steady-state amplitude of the downstream cylinder when the model is released from rest (full symbols) and when it is released from a high amplitude (empty symbols), L/D= 13, β= 0°.

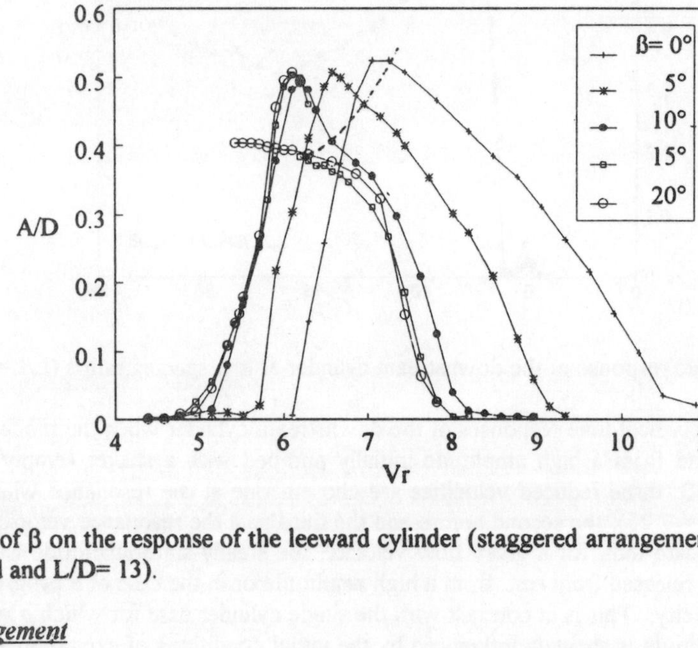

Figure 6. Effect of β on the response of the leeward cylinder (staggered arrangement, upstream cylinder is fixed and L/D= 13).

Staggered arrangement

Figure 6 shows the effect of the bundle inclination angle β on the dynamic response of the leeward cylinder for a separation L/D= 13, typical of the high transmission lines bundles. The results of this figure indicate that the increase of angle β reduces the synchronization region, the synchronization onset velocity, and, to a certain degree, the maximum amplitude of oscillation. The hysteresis phenomenon and

the associated discontinuities observed in the response of an isolated cylinder begin to appear progressively once again with an increasing value of β. For a bundle inclination angle β= 20°, the response of the leeward cylinder is similar to that of a single cylinder. Beyond β= 20°, the downstream cylinder is outside the wake of the fixed upstream cylinder and therefore behaves as an isolated one. Pon *et al.* (1989) also noticed the same behaviour for L/D= 15. A *"kink"* appears in the response of the downstream cylinder at a speed slightly higher than the resonance velocity (see figure 6, dashed line). This kink is more obvious at higher values of β than at lower values, and for β≥ 15°, it shows as a clear discontinuity as in the case of an isolated cylinder. The progressive appearance of this kink with β supports the hypothesis of two different vortex patterns in the wake of the downstream cylinder. Figure 7 shows a comparison between typical records of the cylinder displacement build-up from rest and the associated phase angle variation, Φ, between the hot wire anemometer signal and the vibrating cylinder displacement, for both cases of a single cylinder (a) and a bundle of two cylinders (b and c). Figures 7(b) and (c) are obtained in the case of a bundle inclination angle β= 15° (L/D= 13) and V_r= 6.0 and 6.2 respectively. These figures indicate that when the model is released from rest, its displacement increases progressively and tends to stabilize on a lower steady-state amplitude when it suddenly bifurcates and keeps on increasing to stabilize finally on a higher stable amplitude. This bifurcation is accompanied with a sudden change of the phase angle Φ between the vortices shed from the downstream cylinder and its displacement. Figure 7(a) shows a similar behaviour in the case of an isolated cylinder for a velocity V_r= 5.47. For this last case, the flow visualization results have shown that the bifurcation is attributed to a sudden change of the vortex pattern from the 2P mode to the 2S mode (Brika & Laneville 1993). The bifurcation observed in the downstream cylinder response may then originate from an instantaneous variation of the flow pattern in the wake of the leeward cylinder. It is worthwile to mention here that in the case of an isolated cylinder (for V_r> 5.6), the amplitude always stabilizes on the C2P curve when the model is released from rest and no bifurcation similar to that of figures 7(b) and (c) has been observed. This difference is attributed to the presence of a fixed upstream cylinder and such bifurcation in the downstream cylinder response occurs when the downstream cylinder crosses the wake of the upstream one.

Figure 7(a) also shows another jump (I) in the phase angle Φ at earlier stages of the displacement motion. Since the vortices shed from a cylinder at rest are the natural Karman vortices (2S mode), the discontinuity (I) is then due to a first change of the vortex pattern from the 2S mode to the 2P mode. In the case of a bundle of two cylinders, however, three different jumps of the phase angle Φ can be observed at different stages of the motion of the downstream cylinder when it is released from rest. These jumps in Φ are shown in figures 7(b) and (c) and are referred to with (I), (II) and (III). With a bundle inclination angle β= 15°, the downstream cylinder released from rest is expected to be initially outside the upstream cylinder wake. The presence of the three different jumps can be explained as follows:

i) the discontinuity (I) occurs when the cylinder begins to oscillate and is caused by an abrupt change from the 2S mode to the 2P mode as in the case of an isolated cylinder,
ii) the second discontinuity (II) may occur when the vibration amplitude increases and the lower side of the downstream cylinder approaches sufficiently the wake of the upstream model for which the flow pattern of the 2S type. This mode dislodges and replaces the pair of vortices shed from the lower side of the downstream cylinder. The flow pattern will then be of the (P+S) type, a pair of vortices shedding from the upper side and a single vortex shed from the lower side,
iii) the third dicontinuity (III) occurs when the vibration amplitude keeps on increasing, the upper side entering the wake of the upstream cylinder wake: the remaining pair of vortices changes to a single vortex and the whole wake of the leeward cylinder is of the 2S type.

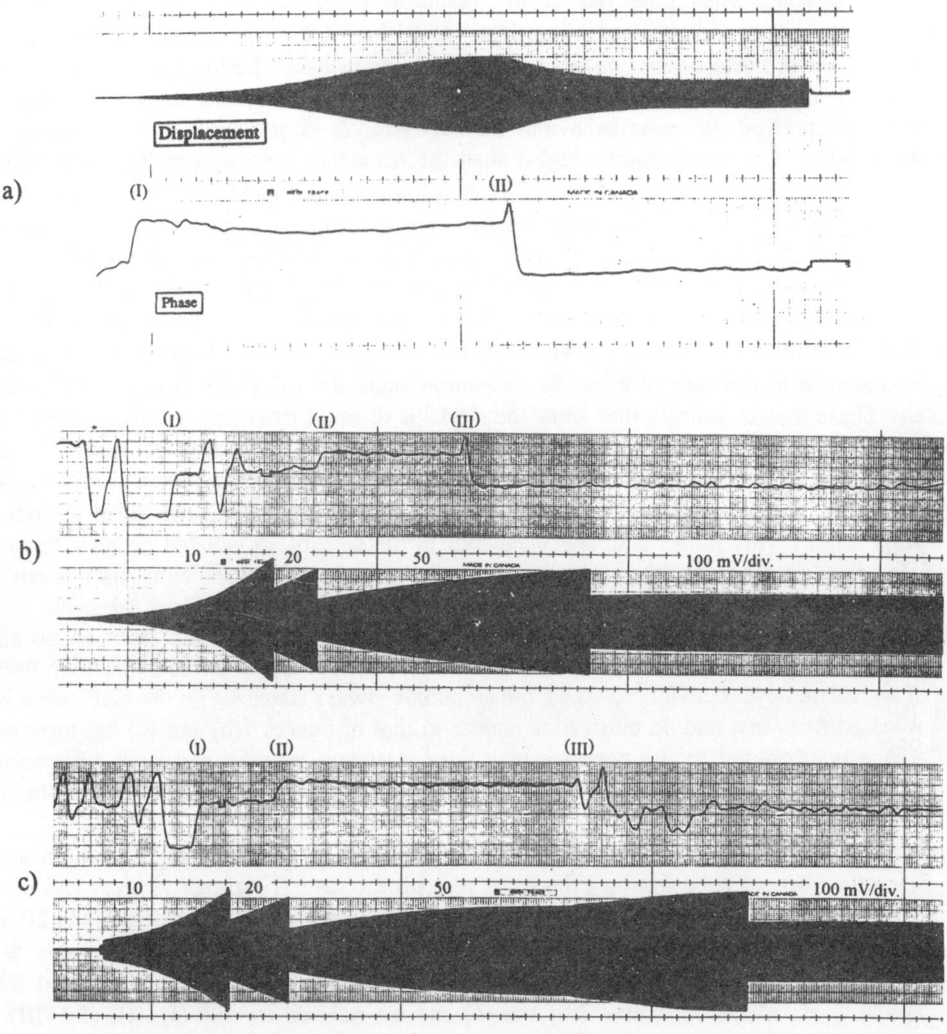

Figure 7. The vibration amplitude and phase angle Φ as a function of time. (a), isolated cylinder (V_r= 5.47); (b) and (c), bundle of two cylinders (stationary upstream cylinder, L/D= 13, β= 15°). (b), V_r= 6.0; (c), V_r= 6.2.

Figure 8 shows examples of the downstream cylinder displacement build-up from rest for different reduced velocities in the case β= 15°. This figure indicates the presence of bifurcations at different amplitudes depending on the reduced velocity. The bifurcations are indicated with arrows in figure 8. For larger values of the bundle inclination angle, say β= 20°, such bifurcations are also oberved in the displacement and phase angle Φ but only at lower reduced velocities (V_r< 5.6). Figure 9 shows the traces obtained for a reduced velocity V_r= 5.34 as the model is released from rest or from a high amplitude initially pumped with a shaker. The records show bifurcations occuring at different amplitudes for the same flow velocity. Each bifurcation is again accompanied with a jump in the phase angle Φ, indicating a sudden variation of the flow pattern from one mode to another. In comparison with the

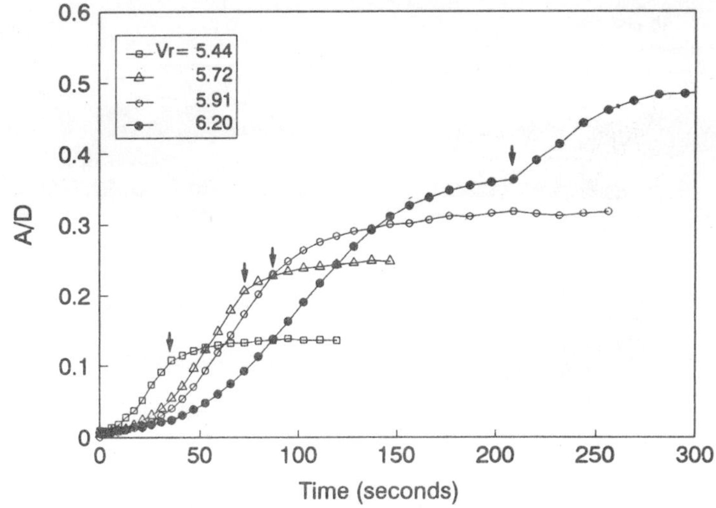

Figure 8. Time response of the downstream cylinder when it is released from rest at different reduced velocities and the accompanied bifurcations (L/D= 13, β= 15°).

records of figures 7(b) and (c) where β= 15°, we can observe the absence of the discontinuity (II) in figure 9(d) and this record is rather similar to that of an isolated cylinder. This confirms once again that beyond β= 20°, the downstream cylinder will be outside the upstream cylinder wake and behaves as an isolated one.

3.2 Bundle with Free Vibrating Cylinders

For cylinders free to oscillate (pure aerodynamic coupling), the wake of the upstream cylinder impinging on the downstream one differs from the one of a stationary cylinder: the vortex shedding frequency of the upstream cylinder is synchronized with its vibrating frequency and the upstream cylinder is now oscillating. Figure 10 shows the results obtained with a spacing ratio L/D= 10. A_1 and A_2 correspond respectively to the vibration amplitude of the windward and the leeward cylinders. The windward cylinder response is similar to the one of an isolated cylinder, while the leeward one differs from both that of a single and that of a cylinder in the wake of a fixed one (figure 3). Figure 10(a) shows the hysteresis phenomenon which is characterized with two branches, each branch being associated with a particular vortex pattern (Brika & Laneville 1993) and delimited by a discontinuity featuring a jump to the other branch.

In the synchronization region of the upstream cylinder, the response of the leeward cylinder (figure 10(b)) is now hysteretic. It is also characterized with two branches and two discontinuities which coincide with those of the upstream one. Each branch is then associated with a particular vortex shedding mode and the discontinuities are attributed to a sudden change from one mode to another. The last discontinuity appearing in the response of the leeward cylinder coincides with the end of synchronization of the windward cylinder (V_r= 7.4). Beyond this value. the upstream cylinder comes to rest and the downstream cylinder response will be similar to that of a cylinder in the wake of a fixed upstream one (see figure 3). The results of figure 10 also show that the synchronization of both cylinders onsets at the same velocity ($V_r \approx 5$) that is different from the one measured in the previous case (V_r= 6 in figure 3).

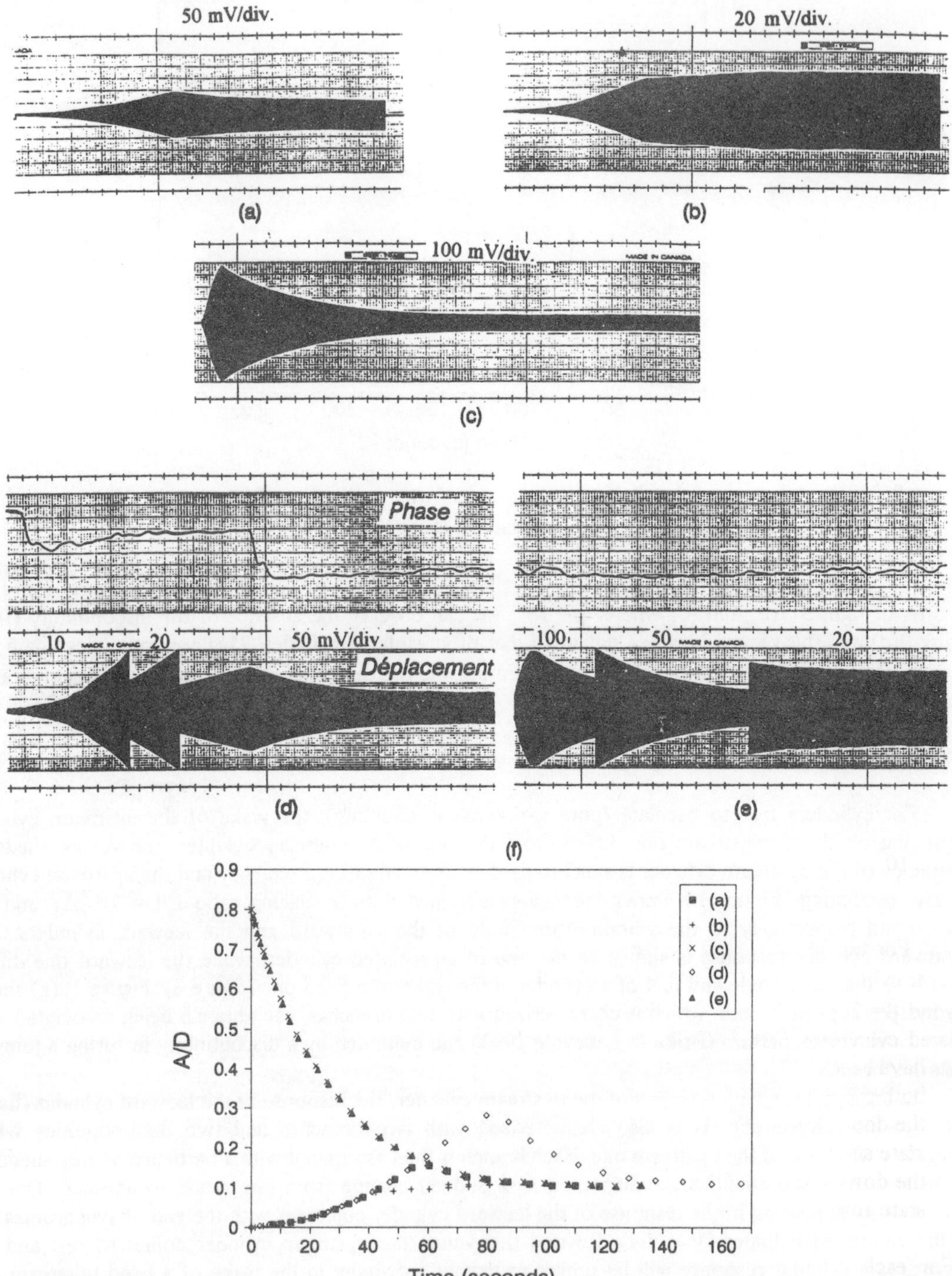

Figure 9. The vibration amplitude and phase angle Φ as a function of time (stationary upstream cylinder, L/D= 13, β= 15° and V_r= 5.34). (a), (b) and (d); from rest, (c) and (e); from a high amplitude.

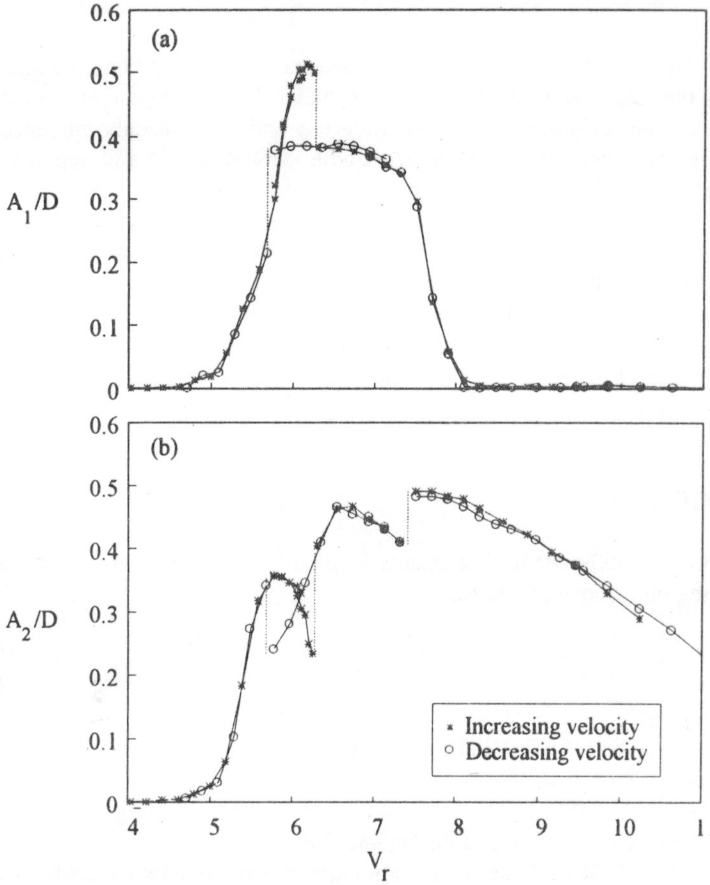

Figure 10. The cylinders' response as a function of V_r (L/D=10, β= 0°), a); upstream cylinder, b); downstream cylinder.

For the staggered arrangement (L/D= 13), the results also indicate that the dynamic response of the rear cylinder tends progressively to that of an isolated one as the bundle inclination angle β is varied. Beyond β= 20°, the downstream cylinder is outside the wake of the upstream one and behaves as an isolated cylinder. In the impulsive regime case, different conditions of excitation have been examined. Tests have been performed according to the following three different procedures:

i) both cylinders are released from rest,
ii) the downstream cylinder is released from rest while the upstream one is vibrating with its steady-state amplitude of the C2S curve (flow pattern: 2S mode),
iii) the downstream cylinder is released from rest while the upstream one is vibrating with its steady-state amplitude of the C2P curve (flow pattern: 2P mode).

For each case, the downstream cylinder response is different and when the upstream cylinder undergoes a bifurcation in its displacement build-up from rest for Vr< 5.6, a similar and simultaneous bifurcation is observed in the displacement of the downstream cylinder. This shows the strong effect of the aerodynamic coupling between the two cylinders even at separations as large as L/D= 25.

4. CONCLUSION

The results of the present study show that the presence of an upstream cylinder and its vibration have great effects on the response of the downstream cylinder. When the windward cylinder is stationary, the response of the leeward cylinder is no more hysteretic and it is strongly influenced by the spacing between the cylinders; the synchronization decreases with increasing L/D and appears on a wider range of the reduced velocity. In spite of the absence of the hysteresis, it seems that two different flow patterns may exist on both sides of the resonance in the wake of the downstream cylinder. When the upstream cylinder is free to oscillate, the downstream cylinder response becomes hysteretic. The hysteresis is characterized with two branches and two discontinuities, each branch is associated a particular vortex pattern and these discontinuities coincide with those of the upstream cylinder. An additional discontinuity in the response of the downstream cylinder is observed as the upstream cylinder ceases to be synchronized. The results also indicate that beyond an inclination $\beta = 20°$ of the bundle, the leeward cylinder behaves as an isolated one.

5. ACKNOWLEDGEMENT

The authors would like to acknowledge the financial support of Hydro-Québec and the Natural Science and Engineering Research Council of Canada.

6. REFERENCES

Bokaian, A. and Geoola, F. (1984) "Wake-induced galloping of two interfering circular cylinders", Journal of Fluid Mechanics, Vol. 146, pp. 383-415.

Brika, D. and Laneville, A. (1993) "Vortex-induced vibrations of a long flexible circular cylinder", Journal of Flüid Mechanics, Vol. 250, pp. 481-508.

King, R. and Johns, D.J. (1976) "Wake interaction experiments with two flexible circular cylinders in flowing water", Journal of Sound and Vibration, Vol. 45, No 2, pp. 259-283.

Ohya, Y., Okajima, A. & Hayashi, M. (1989) "Wake interference and vortex shedding (chapter 10)" *in* Encyclopedia of fluid mechanics (ed. N. Cheremisinoff), Gulph Publishing Co., Vol. 8, pp. 322-389.

Pon, C.J., Currie, I.G. and MacDonald, R. (1989) "Aeolian vibration excitation of bundle conductors", Canadian Electrical Association, Technical report, Vol. I & 2.

Price, S.J. (1975) "Wake-induced flutter of power transmission conductors". Journal of Sound and Vibration, Vol. 38, pp. 125-147.

Tanida, Y. Okajima, A. and Watanabe, Y. (1973) "Stability of a circular cylinder oscillating in uniform flow or in a wake". Journal of Fluid Mechanics, Vol. 61, pp. 769-784.

Williamson, C.H.K. and Roshko, A. (1988) "Vortex formation in the wake of an oscillating cylinder" Journal of Fluids & Structures, Vol. 2, pp. 355-381.

Zdravkovich, M.M. (1972) "Flow induced vibrations of two cylinders in tandem and their suppression". *In* Proc.Intl. Symp. Flow-Induced Structural Vibrations, Karlsruhe, Paper E7, pp. 631-639, Springer.

9ICWE, 1995, New Delhi, India

EFFECTS OF TURBULENCE ON SURFACE PRESSURES OF A FLAT PLATE AND RECTANGULAR CYLINDERS IN SEPARATED AND REATTACHING FLOWS

LI, Q.S. and MELBOURNE, W.H.

Department of Mechanical Engineering, Monash University, Clayton 3168, Australia.

Abstract

Wind tunnel experiments were conducted to study the effects of the intensity and scale of free-stream turbulence on the streamwise surface pressure fields of two-dimensional blunt flat plate and rectangular cylinders with different afterbody lengths exposed to grid turbulence and smooth flow. The results of the experiments demonstrate that not only turbulence intensity but also turbulence scale have significant effects on pressure fluctuations in separated and reattaching flows.

1.INTRODUCTION

Numerous wind tunnel studies have been carried out to investigate the effects of turbulence intensity and scale on the surface pressures on flat plates and rectangular cylinders over the last three decades. However, much of these previous studies have involved extensive measurements in the reattchment zone where the maximum r.m.s. pressure occurs. Saathoff and Melbourne(1989) thus concentrated their investigation on the peak pressures which occur in the forward part of the bubble and which are of primary concern in wind engineering, as fluctuating pressures in this region have been recognized as a major cause of roof failure on low-rise buildings and of cladding failure on tall buildings. But the largest ratio of turbulence integral scale to the model thickness in their study was less than 2.1, which is smaller than the typical values of the turbulence scale ratio in the natural wind around buildings. More work is thus required over a much larger range of turbulence scale to be relevant to the wind engineering field. Although Nakamura and Ozono (1987) investigated the effects of turbulence on streamwise pressures over a large turbulence scale range, only mean pressures were measured in their study. Flow visualization experiments by Cherry et al. (1984) and Saathoff (1988) have shown that the instantaneous flow field in a separation bubble is significantly different from that obtained by mean flow measurements.

This paper describes an experimental study on the effects of the intensity and scale of turbulence on streamwise surface pressures of two-dimensional blunt flat plate and rectangular cylinders with different afterbody lengths exposed to grid turbulence over a large range of turbulence scale by focussing attention on the fluctuating and very low peak pressures measured near separation.

2.EXPERIMENTAL ARRANGEMENTS

The experiments were conducted in a 450kw wind tunnel with a working section of 2.0m square and 15m long at Monash University. The desired free-stream turbulence was generated by using bi-planar wooden grids. Seven different grids were used. The ratio of

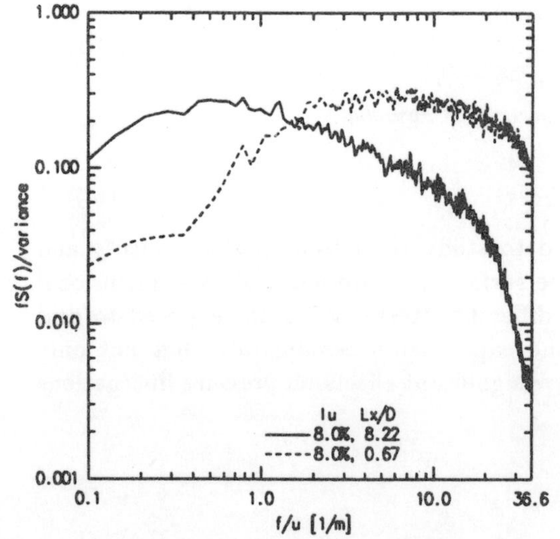

Figure 1. Longitudinal Velocity Power Spectra Obtained at Model Position For Different Grid Configurations.

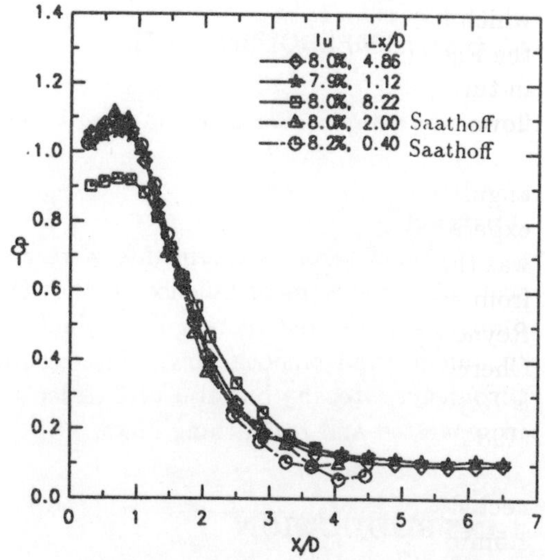

Figure 2. Streamwise Mean Pressure Distributions On The Blunt Flat Plate In Turbulent Flows

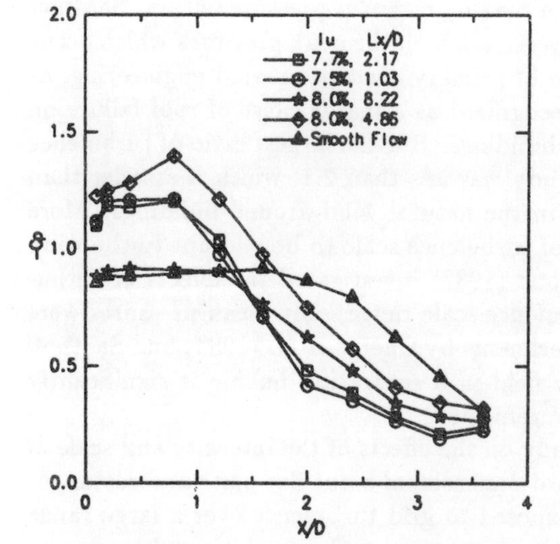

Figure 3. Streamwise Mean Pressure Distributions On The Cylinder With H/D=4 In Turbulent & Smooth Flows

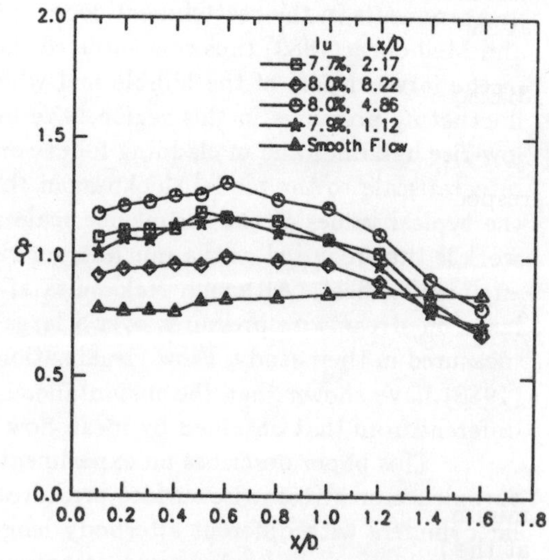

Figure 4. Streamwise Mean Pressure Distributions On Cylinder With H/D=2 In Turbulent & Smooth Flows

mesh size to bar widths was approximately 4.0 for each grid. The intensity and scale of free-stream turbulence could be altered by changing the grids or the distance from the grids to the experimental model. The experiments were carried out over a wide range of the ratio of turbulence integral scale to the model thickness, $L_x/D = 0.35$ to 8.22 for an approximately constant turbulence intensity, while the turbulence intensity varied from 8% to 25%. Smooth flow has a turbulence intensity of 0.8%. Figure 1 shows a typical example of longitudinal velocity spectra for different grid configurations. The spectra of the two turbulent flows which have approximately the same turbulence intensity but different scales are presented in the Figure 1. For the small scale turbulence the grid is very close to the model. The difference in turbulence energy distributions in the frequency domain in large and small scale turbulent flows can be seen.

A blunt flat plate with a rectangular cross-section, a square cylinder and two rectangular cylinders with depth/thickness ratio,H/D, of 2 and 4, respectively, were used as experimental models which had the same thickness, D, of 50mm. The spanwise dimension was the same, 1.6m.The models were mounted horizontally between endplates located 200mm from each side wall of the tunnel to insure the uniformity of the flow around the models. The Reynolds number based on the thickness of models was kept approximately at 4.5×10^4. Cherry et al. (1984) found that the mean pressure distribution on a blunt flat in smooth flow is Reynolds number independent in the range of $3.0 \times 10^4 < Re < 6.0 \times 10^4$. Nakamura et al. (1984) also drew a similar conclusion for various sizes of afterbody rectangular cylinders.

Pressure data on the models were collected using Honeywell 163pc transducers connected to pressure tappings with 60mm lengths of PVC tubing with an internal diameter of 1.5mm. Restrictors placed in the tubing provided a flat frequency response within 10% up to 250Hz. Streamwise pressure distributions were measured using a row of tappings on the centrelines of the bottom surfaces on the models. A spanwise row of tappings located near the leading edge on the bottom surface of each model was used to measure lateral cross-correlations of fluctuating pressure. All the measurements presented in the paper were made at zero incidence.

3.EXPERIMENTAL RESULTS

Mean, fluctuating and peak pressure coefficients in the data presentation are defined, respectively, as:

$$C_p = \frac{\overline{p} - p_0}{1/2\rho\overline{u}^2} \tag{1}$$

$$C_{\sigma_p} = \frac{\sigma_p}{1/2\rho\overline{u}^2} \tag{2}$$

$$C_{\check{p}} = \frac{\check{p} - p_0}{1/2\rho\overline{u}^2} \tag{3}$$

where: $\overline{p}$ is the time mean pressure, σ_p is the standard deviation of pressure, $\check{p}$ is the minimum pressure, p_0 is the static pressure at the model location, $\overline{u}$ is the mean wind velocity at the model location, ρ is the density of air.

In the experiments, pressure data were recorded and analyzed by using a Perkin-Elmer computer. The sampling frequency was 1000 Hz. A large amount of data were recorded to

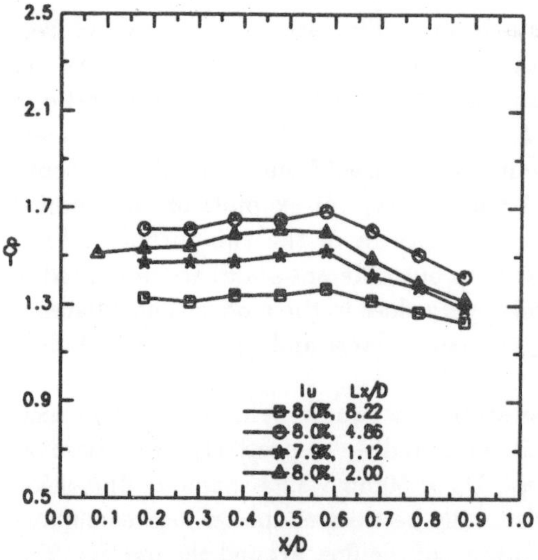

Figure 5. Streamwise Distributions of Mean Pressure Coefficient On The Square Cylinder In Turbulent Flows

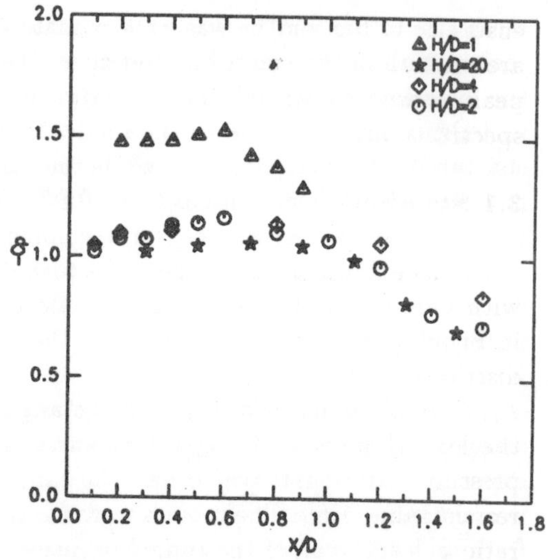

Figure 6. Distributions of Mean Pressure Coefficient On The Four Models In Turbulent Flows(Iu=8%,Lx/D=1.12)

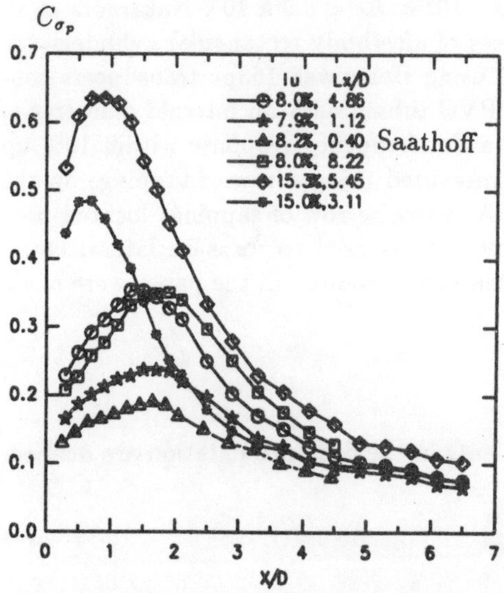

Figure 7. Distributions of Fluctuating Pressure Coefficient On The Blunt Flat Plate In Turbulent Flows

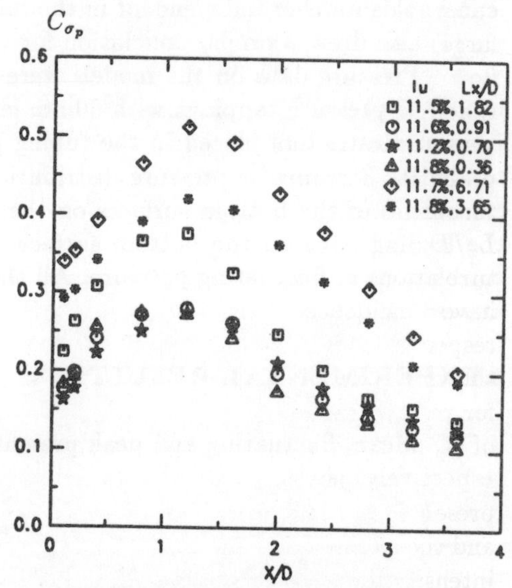

Figure 8. Distributions of R.M.S. Pressure Coefficient On The Cylinder With H/D=4 In Turbulent Flows

ensure that random nature of the data can be statistically estimated. Values of C_p, C_{σ_p}, are ensemble averages from 400 seconds of data collection. The $C_{\hat{p}}$ is the average of 100 peak measurements from consecutive 4 second samples. 65536 data points were recorded for spectral analysis.

3.1 Streamwise Pressure Distributions

Distributions of the mean pressure coefficient, C_p, on the flat plate in turbulent flows with approximately the same intensity (8.0%) but different turbulence scales are presented in Fig.2. The data presented in Fig.2 show little effect of turbulence scale on mean pressure distributions up to L_x/D=4.86. However, with further increase in turbulence scale ratio to L_x/D=8.22, it can be seen that the mean pressure coefficients, which were measured near the leading edge in particular, are scale-dependent. Fig.3, Fig.4 and Fig.5 show the mean pressure distributions on the rectangular cylinders with depth/thickness ratio of 4, 2 and 1, respectively. The results obtained on models with $H/D = 2, 4$ indicate that as turbulence scale ratio changes from 1.03 to 2.17, little scale effect on mean pressure distribution is evident. With further increase in turbulence ratio to 4.86, the values of C_p become more negative. However, when the scale ratio reaches to 8.22, the magnitude of C_p are becoming smaller. As can be seen in Fig.3, the mean pressure coefficients measured near the leading edge in a very large scale turbulent flow ($L_x/D = 8.22$) are close to the values of C_p obtained in smooth flow. This is in agreement with Nakamura and Ozono's conclusion (1987) which suggested that turbulence of very large scale is equivalent to a flow with slowly fluctuating velocity, and hence it can no longer influence the bluff body mean flow effectively. The measured mean pressure coefficients on the square cylinder model shown in Fig.5 are apparently changing with turbulence scale in all the ranges of turbulence scale tested. The values of Cp measured at $L_x/D = 4.86$ are more negative than data obtained at $L_x/D = 1.12$. With further increasing turbulence scale ratio to 8.22, it can be found that the magnitude of mean pressure coefficients have decreased approxmately 18% and 10% than those obtained at $L_x/D = 4.86$ and 1.12, respectively. The distributions of C_p on the four models tested for a given turbulent flow are shown in Fig.6. It can be seen that the mean pressure coefficients measured near separation for rectangular cylinders with $H/D > 1$ demonstrate a similar distribution while the values of C_p obtained on the square cylinder are more negative.

Streamwise distributions of r.m.s. fluctuating pressure, C_{σ_p}, on the flat plate are presented in Fig.7. An increase in turbulence intensity causes fluctuating pressures to increase and the location of maximum C_{σ_p} to move closer to the leading edge. An increase in turbulence intensity from 8% to 15% moves the position of maximums C_{σ_p} upstream from 1.5D to 0.7D. As shown in Fig.7, C_{σ_p} increases with increasing scale from 0.40 to 4.86. The magnitude of this effect in this case is about 100%. With further increase in turbulence scale ratio to 8.22, the maximum value of C_{σ_p} does not keep increasing and are about the same as measured at L_x/D=4.86. It can be seen in Fig.8 and Fig.9 that the profiles of C_{σ_p} measured on the cylinders with $H/D = 2$ and 4 are not sensitive to changing turbulence scale for $0.36 < L_x/D < 0.91$. However, with further increase in scale ratio to 1.82, turbulence scale has a significant effect on the distributions of C_{σ_p}. The distributions of C_{σ_p} in turbulent and smooth flows measured on the square cylinder are presented in Fig.10. The fluctuating pressures at all points on the streamwise surface are larger in the absence of free-stream turbulence than those obtained

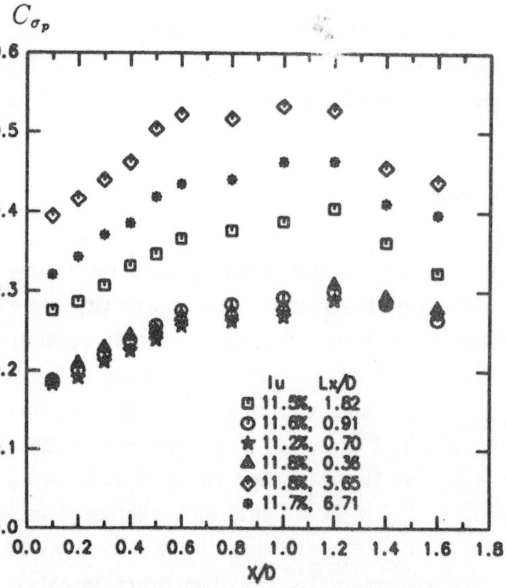

Figure 9. Distributions of R.M.S. Pressure Coefficient On The Cylinder With H/D=2 In Turbulent Flows

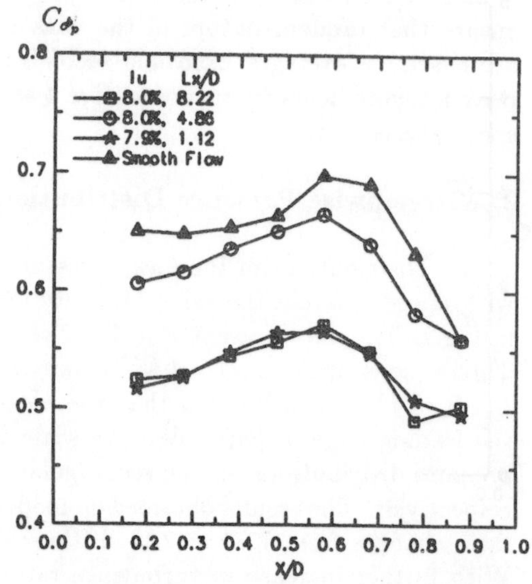

Figure 10. Distributions of Fluctuating Pressure Coefficient On The Square Cylinder In Turbulent and Smooth Flows

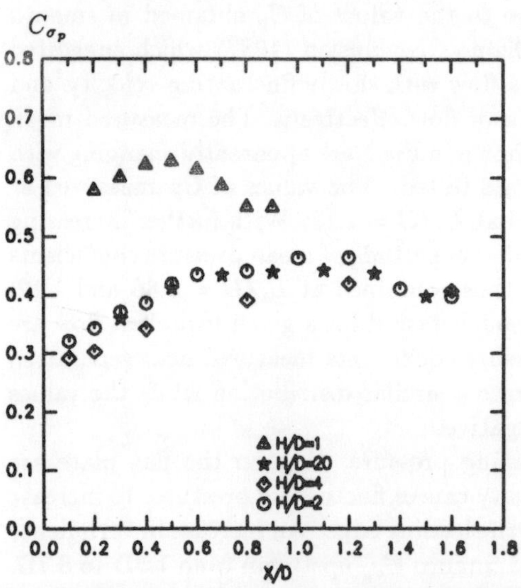

Figure 11. Distributions of Fluctuating Pressure Coefficient On The Four Models In Turbulent Flows(Iu=8%,Lx/D=1.12)

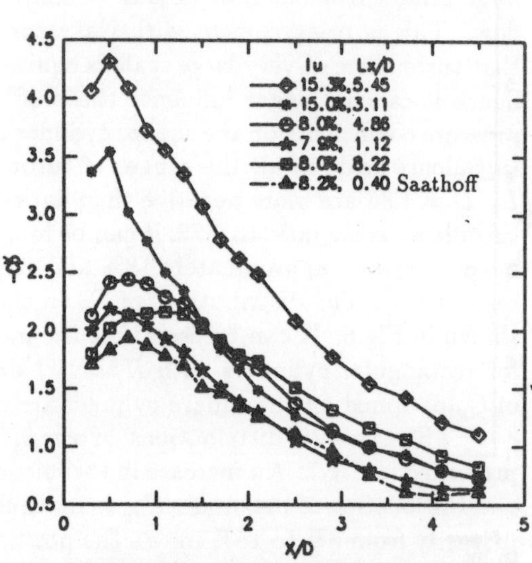

Figure 12. Negative Peak Pressure Coefficient Distributions On The Flat Plate

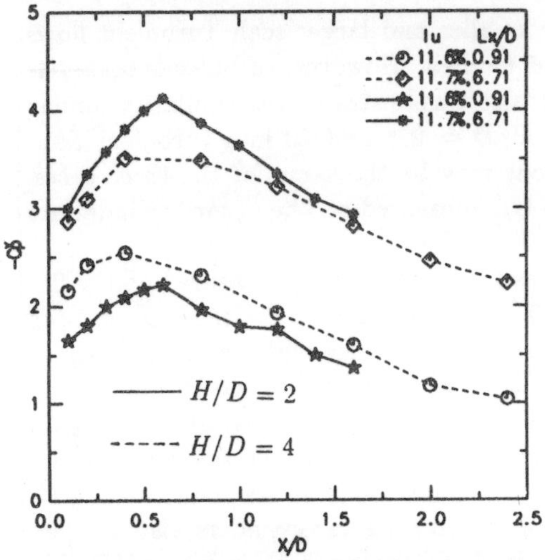

Figure 13. Negative Peak Pressure Coefficient Distributions On The Cylinders With H/D=2,4

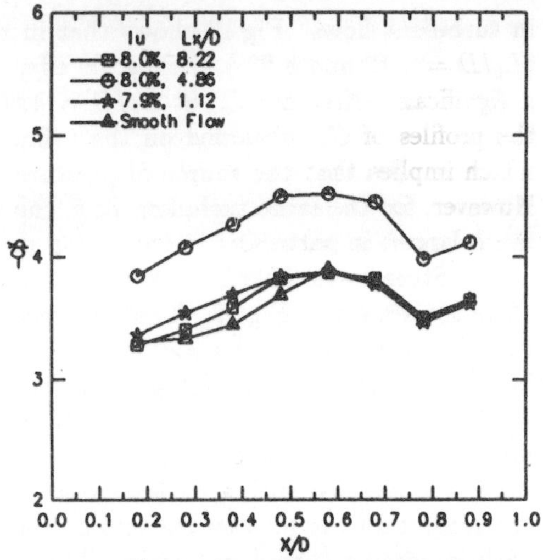

Figure 14. Distributions of Peak Pressure Coefficient On The Square Cylinder In Turbulent and Smooth Flows

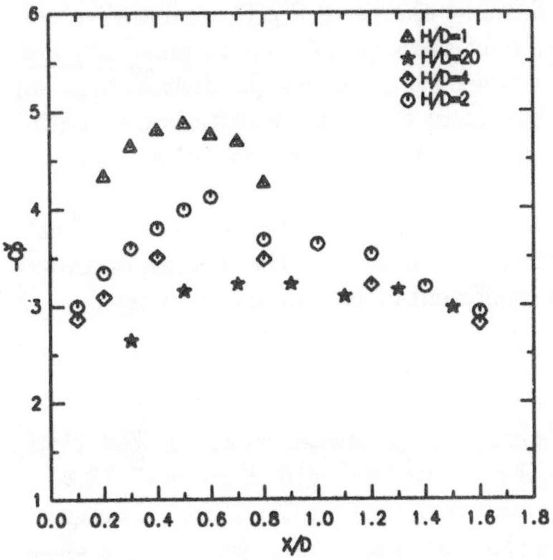

Figure 15. Distributions Of Peak Pressure Coefficient On The Four Cylinders In Turbulent Flows(Iu=12%,Lx/D=6.7)

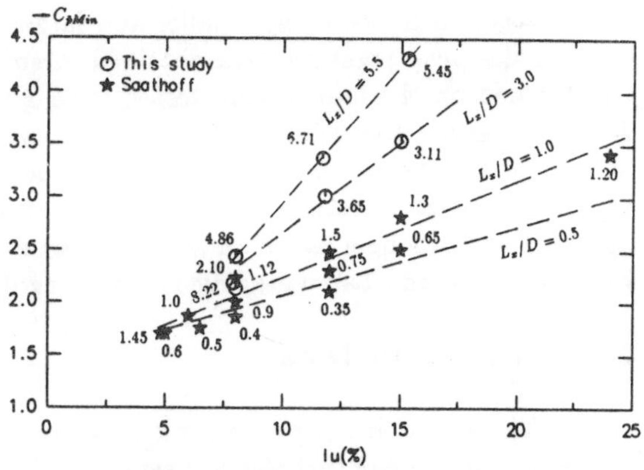

Figure 16. Minimum Negative Pressure Coefficients On The Flat Plate(Values of Lx/D Are Shown Next to Each Data Point)

in turbulent flows. Fig.10 shows that in relatively smaller and larger scale turbulent flows ($L_x/D = 1.12$ and 8.22), little scale effect on C_{σ_p} is evident. However, turbulence scale has a significant effect on C_{σ_p} at $L_x/D = 4.86$. Fig.11 also demonstrates the similarity among the profiles of C_{σ_p} obtained on the cylinders with $H/D = 2, 4$ and 20 in a turbulent flow, which implies that the source of pressure fluctuations may be the same for the three cases. However, for the same turbulent flow, the values of C_{σ_p} measured on the square cylinder are much larger, in particular in the region near the leading edge.

Streamwise distributions of negative peak pressure coefficient,$C_{\check{p}}$ measured on the flat plate are shown in Fig.12. The peak pressure coefficients also show a dependence on both turbulence intensity and scale. In particular, the effect of turbulence scale on peak pressures becomes greater as turbulence intensity increases. It is interesting to note in Fig.12 that the magnitude of $C_{\check{p}}$ increases as L_x/D from 0.40 to 4.86. However, the values of $|C_{\check{p}}|$ have decreased at a scale ratio of 8.22. It can be found in Fig.12 that a 1.75:1 increase in turbulence scale is associated with a 24% increase in the maximum value of $|C_{\check{p}}|$. As shown in Fig.13, the magnitudes of $|C_{\check{p}}|$ on the streamwise surface of the rectangular cylinders also increase with increasing turbulence scale. A 7.4:1 increase in L_x/D is found to increase 38% and 86% in the maximum magnitude of $C_{\check{p}}$ for the cylinders with $H/D = 4$ and 2, respectively. Therefore, turbulence scale has more significant effect on $C_{\check{p}}$ measured on the cylinder with shorter afterbody length. The effect of turbulence scale on $C_{\check{p}}$ obtained on the square cylinder becomes more apparent at $L_x/D = 4.86$ for $I_u = 8.0\%$ as shown in Fig.14, and the values of $C_{\check{p}}$ measured in relatively smaller and larger turbulent flows ($L_x/D = 1.12$ and 8.22) are close to the peak pressure data obtained in smooth flow. Fig.15 shows the distributions of $C_{\check{p}}$ obtained on the four models tested for a given turbulent flow. As the afterbody length, H/D, decreases, the values of $C_{\check{p}}$ become more negative. As discussed above, an increase in turbulence intensity increases the magnitude of $C_{\check{p}}$. Therefore, an increase in the level of turbulence intensity causes the distribution of $C_{\check{p}}$ for a short model to be equivalent to that obtained with a longer model in low turbulence. This is consistent with the conclusion drawn by Courchesne and Laneville (1984) regarding drag coefficient of rectangular cylinders.

3.2 Peak Pressure Data

The minimum negative peak pressure coefficient, $C_{\check{p}Min}$, measured on the flat plate model as a function of turbulence intensity and scale, is presented in Fig16. Figures 17, 18, and 19 summarize the effect of the turbulence intensity and scale on the $C_{\check{p}Min}$ for rectangular cylinders with different afterbody lengths. The results obtained on the four models show that the magnitude of $C_{\check{p}Min}$ increases with increasing turbulence scale, and the scale effect becomes more pronounced at higher values of turbulence intensity. It is worth noting that turbulence intensities significantly greater than 15% are frequently encountered by buildings and structures in boundary layer wind flows. Moreover, peak pressure coefficients are widely used as design pressure coefficients.on buildings. Therefore, it is important to correctly simulate free-stream turbulence scale to provide the largest magnitude of design pressures on buildings.

Fig.20 shows $C_{\check{p}Min}$ as a function of L_x/D for $I_u = 8.0\%$ measured on the four models. It can be seen that as the afterbody length ratio,H/D is increased, the values of $C_{\check{p}Min}$ become progressively less negative, in particular in larger turbulent scale flows. It is interesting to

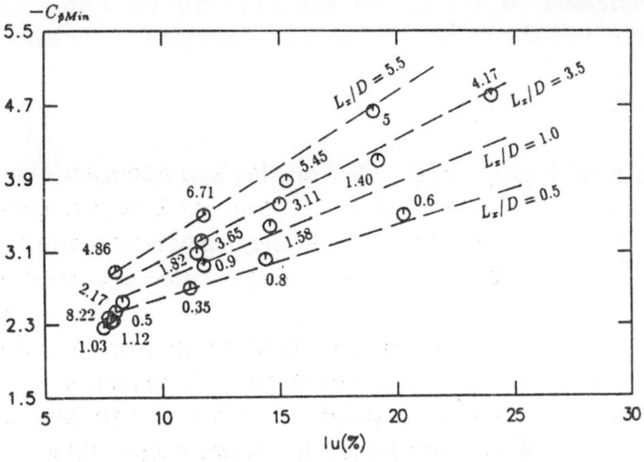

Figure 17. Minimum Negative Peak Pressure Coefficients On The Cylinder With H/D=4(Values of Lx/D Are Shown Next to Each Data Point).

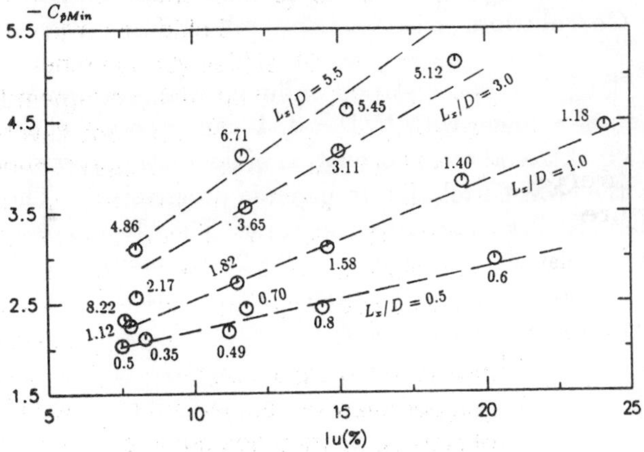

Figure 18. Minimum Negative Peak Pressure Coefficients On The Cylinder With H/D=2(Values of Lx/D Are Shown Next to Each Data Point).

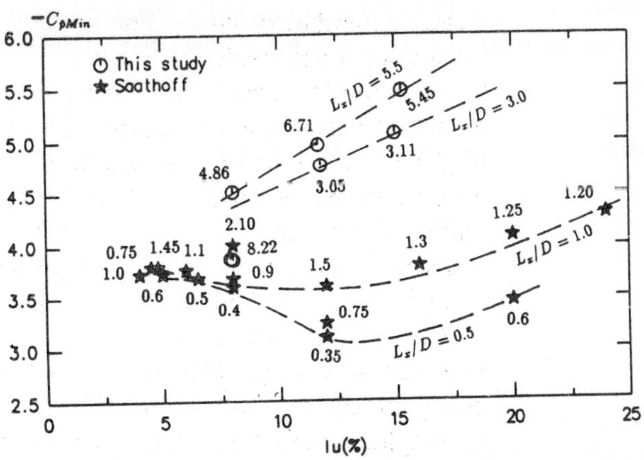

Figure 19. Minimum Negative Peak Pressure Coefficients On The Square Cylinder(Values of Lx/D Are Shown Next to Each Data Point)

note that as L_x/D is increased up to 8.22, the effect of turbulence scale on minimum peak pressures is not significant and the values of $C_{\check{p}Min}$ approaches that in smooth flow.

3.3 Spectrum Analysis

Pressure spectra measured at $X/D = 0.1$ on the four models are displayed in Fig.21 for a turbulent flow and the turbulence spectrum is also presented. A high-amplitude spike in pressure spectrum corresponding to the Strouhal frequency in a narrow band is evident in the case of the square cylinder, which illustrates that pressure fluctuations near the leading edge of a square cylinder at this turbulence level are dominated by vortex shedding at the Strouhal frequency. The pressure spectra obtained on the flat plate and rectangular cylinders with $H/D = 2$ and 4 have a similar distribution with the longitudinal velocity spectra except in the high-frequency range where the decay is fast. This implies that longitudinal turbulence is responsible for the majority of pressure fluctuations near the leading edge of rectangular cylinders with $H/D = 2$ or larger.

3.4 Lateral Pressure Correlation

Fig.20 shows the lateral cross-correlation of fluctuating pressures near separation (at $X/D = 0.2$) measured on the cylinder with $H/D = 4$. It can be seen that the turbulence scale dominates the distribution of lateral cross-correlation of fluctuating pressures. An increase in turbulence scale causes the cross-correlation to increase progressively. The results obtained on the other three models also draw the same conclusion. The effect of depth/thickness ratio on the lateral pressure correlation can be seen in Fig.21 by comparing the data measured at the same turbulence configurations for the models tested with different afterbody lengths. The spanwise correlation measured on the bluff 2:1 section is higher than that obtained on the slender 4:1 section.

4.CONCLUSION

Experimental results indicated that mean pressure distributions are strongly dependent on free-stream turbulence intensity but not significantly affected by turbulence scale over a range of turbulence scale ratio for $L_x/D < 4.9$ for a blunt flat plate and $L_x/D < 2.1$ for rectangular cylinders with $H/D = 2$ and 4. However, the measured mean pressure coefficients on the square cylinder model are apparently changing with turbulence scale in all the ranges of turbulence scale tested. Fluctuating pressures and negative peak pressure coefficients are increased with both increasing turbulence intensity and scale, as noted previously. However, with further increase in L_x/D (e.g. $L_x/D = 8.22$, $I_u = 8.0\%$), the magnitudes of C_{σ_p} and $C_{\check{p}}$ were found to be smaller than those measured at the same turbulence intensity but with lower scale ratios. An increase in the level of turbulence intensity causes the distribution of $C_{\check{p}}$ for a short model to be equivalent to that obtained with a longer model in low turbulence. As the afterbody length ratio,H/D is increased, the values of $C_{\check{p}Min}$ become progressively less negative. But as L_x/D is increased to a larger value (e.g. $L_x/D = 8.22$), the effect of turbulence scale on $C_{\check{p}Min}$ is not significant and the values of $C_{\check{p}Min}$ approaches that in smooth flow. Spectrum analysis shows that longitudinal turbulence is responsible for the majority of

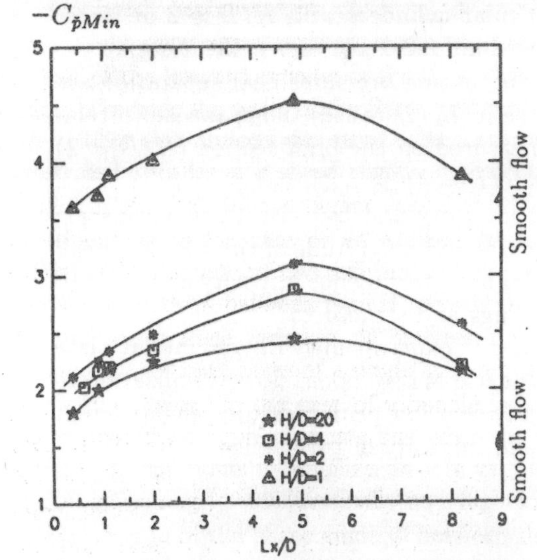

Figure 20. Minimum Negative Peak Pressure Coefficients On The Four Models As a Function of the Turbulence Scale(with Iu=6%)

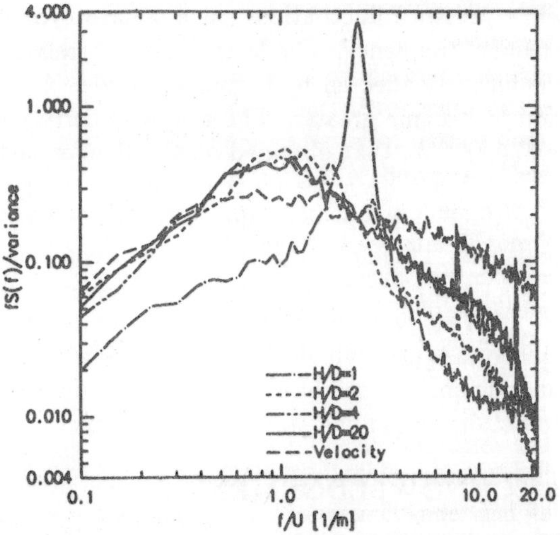

Figure 21. Logitudinal Velocity and Pressure Spectra Measurd On The Four Models Near The Leading Edge(Iu=8%,Lx/D=4.86)

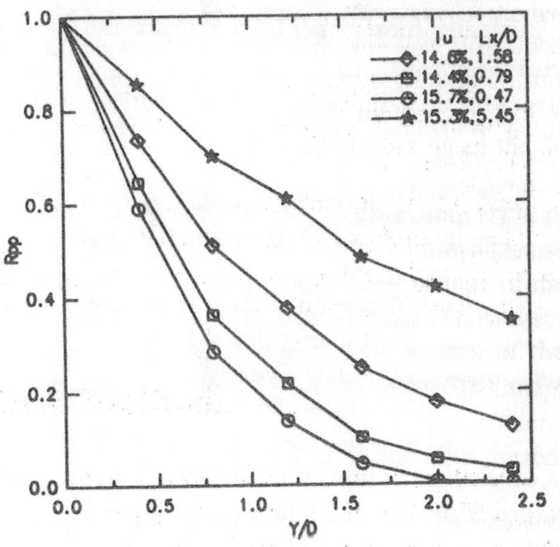

Figure 22 Cross-Correlation of Fluctuating Pressure On The Cylinder With H/D=4 As a Function of Lateral Displacement.

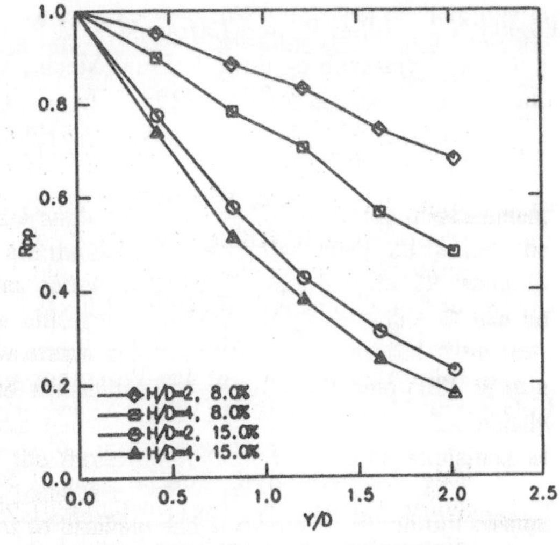

Figure 23. Cross-Correlation of Fluctuating Pressure Measured On The Cylinder With H/D=2,4.

pressure fluctuations near the leading edge of rectangular cylinders with $H/D = 2$ or larger while pressure fluctuations on a square cylinder are dominated by vortex shedding at the Strouhal frequency. The lateral cross-correlation of fluctuating pressures near separation are significantly affected by turbulence scale. An increase in L_x/D causes the spanwise correlation to become greater. The spanwise correlation measured on the bluff section ($H/D = 2$) is higher than that obtained on the slender section ($H/D = 4$).

It has been shown that not only turbulence intensity and but also turbulence scale have significant effects on the r.m.s. fluctuating pressures and magnitude of the very low peak pressures. Therefore, it is necessary to correctly model both turbulence intensity and scale to predict the highest value of design pressures on buildings and structures.

This paper presents quantitative results obtained in experiments with a blunt flat plate and rectangular cylinders. Future work will involve the use of high-speed flow visualization to provide greater insight into the mechanisms causing the very low peak pressures, in particular to obtain some understanding of the effect of scale and intensity of turbulence.

5.ACKNOWLEDGEMENT

The support of the Australian Research Council is gratefully acknowledged.

6.REFERENCE

Cherry,N.J., Hillier,R. and Latour,M.E.M.P.(1984), Unsteady measurments in a separated and reattaching flow, J. Fluid Mech., Vol.144, pp.13-46.

Courchesne,J. and laneville,A.(1984), An experimental evaluation of drag coefficient for rectangular cylinders exposed to grid turbulence, ASME, J. Fluids Eng., Vol. 104, pp.523-527.

Li,Q.S. and Melbourne,W.H.(1993), A study of turbulence effect on surface pressures in separated/reattaching flows, Proc. of Third Workshop on Wind Engineering, Brisbane, ppB5-B8.

Li,Q.S. and Melbourne,W.H.(1994), Effects of free-stream turbulence on surface pressures fluctuations of a rectangular cylinder in separated and reattaching flows, to be submitted to Proc. of 4th Workshop on Wind Engineering, Oct. Sydney.

Li,Q.S. and Melbourne,W.H.(1994), An experimental investigation of the effects of free-stream turbulence on streamwise surface pressures in separated and reattaching flows, J. Wind Eng. Ind. Aerodyn., Special issue of Third Asia-Pacific Symposium on Wind Eng..

Nakamura,Y.and Ohya,Y.(1984), The effects of turbulence on the mean flow past two dimensional rectangular cylinds, J. Fluid Mech.,Vol.149 pp.255-273.

Nakamura,Y.and Ozono,S.(1987), The effects of turbulence on a separation and reattaching flow, J. Fluid Mech.,Vol.178 pp.477-490.

Saathoff,P.J.(1988), Effects of free-stream turbulence on surface pressure fluctuation in separated/reattaching flows, Ph.D Theis, Monash University.

Saathoff,P.J. and Melbourne,W.H.(1989), The generation of peak pressures in separated and reattaching flow, J. Wind Eng. Ind. Aerodyn., Vol.32, pp.121-134.

9ICWE, 1995, New Delhi, India

UNSTEADY WIND FORCES ON CYLINDERS WITH H-SHAPED CROSS SECTIONS

LAM, K.M., TO, A.P. and KWAN, S.H.

Department of Civil and Structural Engineering, The University of Hong Kong, Pokfulam Road, Hong Kong

Abstract

We measured in the wind tunnel the mean and fluctuating wind forces on a member of H-section cylinders of square envelope, with an aim to investigate the effect of the size and geometry of the recessed cavities on the wind loading of the cylinders. This paper presents some limited results which show that the recessed cavities indeed lead to modifications of the wind forces on the cylinders.

1. Introduction

While wind codes give loading coefficients for buildings with regular shapes in plan, irregular-plan tall buildings are almost the rule rather than the exception in most commercial and residential developments. Irregular plans usually feature re-entrant corners such as in "X"-shaped plans, and recessed bays such as in "H"-shaped plans. For a H-plan building, it is general believed that mean stagnant flow is found inside the recess bays so that the overall loads tend to be the same as those on an unrecessed building of the same external rectangular envelope (Cook 1990). However, supporting quantitative wind-tunnel data from parametric studies are lacking.

On the other hand, unsteady dynamic wind loads are equally important, especially for flexible and slender tall buildings. Little information is available on the unsteady loads on H-plan buildings. Nakamura & Nakashima (1986) studied flow over prisms with elongated H-shaped cross-sections and found that the dynamics of shear layers separated from the upstream sides of the recessed bays play an important role in vortex excitation of the prisms. For H-plan buildings with near-square envelope, it is not known whether flow interactions occur between the external flow and the "mean stagnant" flow inside the recess bays; and if interactions occur, how the fluctuating loads on the buildings be modified.

We address ourselves to these problems by carrying out wind-tunnel tests on two-dimensional cylinders with different shaped H-cross-sections with a square envelope. While aiming at obtaining some quantitative loading data, we also hope to explore the complex flow interactions through measurements of mean and unsteady lift and drag as well as velocity data in the wake.

2. Experimental Arrangement

Experiments were carried out in a recirculating wind tunnel in the Department of Civil and Structural Engineering at the University of Hong Kong. The working section was 3 m wide and 1.8 m high. The model assembly with a width of 1.2 m was mounted at the centre of the test section. A number of wire meshes were installed at the entrance of the working section to produce a reasonably smooth flow at the test section. All experiments were conducted with wind speed $U_o = 10$ m/s.

The H-section cylinder models had a square envelope of breath B = 10 cm. Recessed bays on the cylinders were of the same size on the two sides. The depth d and width w of the bays varied over a range shown in Table 1. All models were 1.2 m long and mounted between two end plates of 1 m by 0.6 m. The model support on the either end of the model was designed to be a strain gauge type force balance which measured the lift and drag. Velocities in the cylinder wake were measured with a hot wire.

Table 1. Model number and dimensions of cavities.

Model type	Model no.	w/B	d/B	w/d
No Cavity	Ref.	0	0	----
	D1	0.18	0.40	0.46
Deep	D2	0.20	0.30	0.67
Cavity	D3	0.18	0.20	0.9
	D4	0.30	0.30	1.0
	D5	0.35	0.25	1.5
Shallow	S1	0.25	0.10	2.7
Cavity	S2	0.30	0.10	3.4
	S3	0.40	0.08	5.0

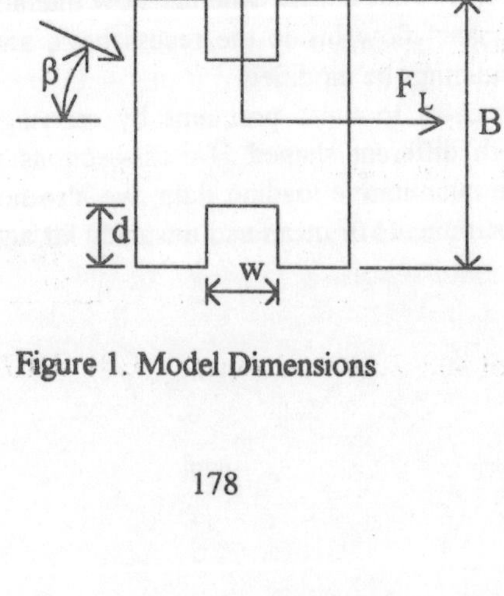

Figure 1. Model Dimensions

3. Results and Discussion

3.1 Drag for normal incidence

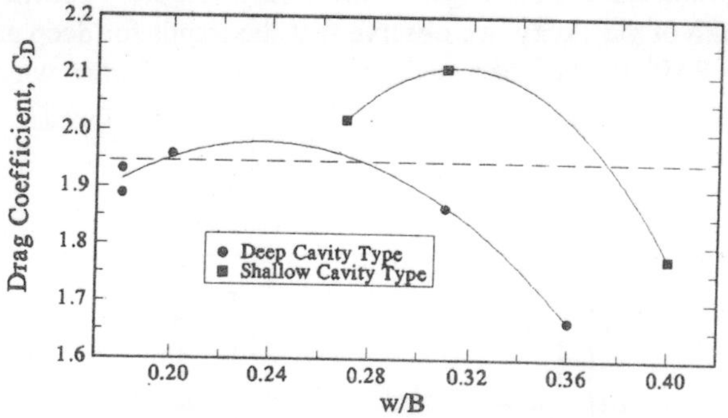

Figure 2. Drag coefficients of models

In Figure 2, the drag coefficients are plotted against the width of the cavity. We observe that, for both shallow and deep cavity type cylinder, the drag on the cylinder decreases with the width of the cavity after a little increase. However, the trend of deep cavity type cylinder starts to drop with a smaller value of the width of the cavity than that of shallow cavity type. In addition, the cylinders have deep cavities experience smaller drag. Some of them even experience drag smaller that the of the reference model. (The level of drag of the reference model is plotted with dash line.)

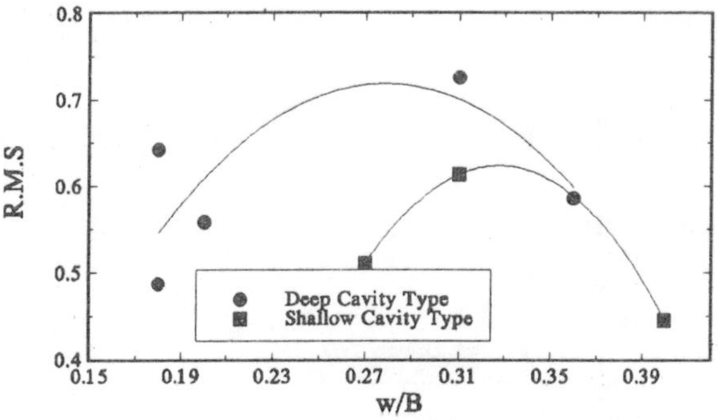

Figure 3. RMS drag fluctuations

Figure 3 shows the RMS drag fluctuations with the width of the cavity, we observe that the drag r.m.s. exhibits a similar trend as that of the drag coefficient. Also, the trend of the deep cavity type models starts to drop at a smaller w than that of the shallow cavity type. However, the r.m.s of the deep cavity type models are higher in this case.

3.2 Lift for normal incidence

Lift forces on the models are found to be very small as expected.

Although the lift are nearly zero, this is not the case in RMS . Since vortices are shedded alternately from the top and the bottom edges of the model. Figure 4 shows the trends of RMS lift on the models with the width of the cavity, we observe that the trends for deep and shallow cavity type are similar. However, the RMS of the deep cavity type model are higher which discloses that they are exerted a higher unsteady force than that of the shallow cavity type models.

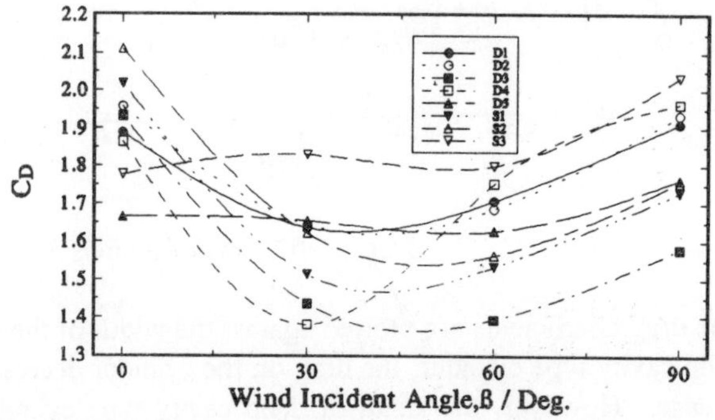

Figure 4. RMS lift fluctuations

3.3 Drag for oblique incidence

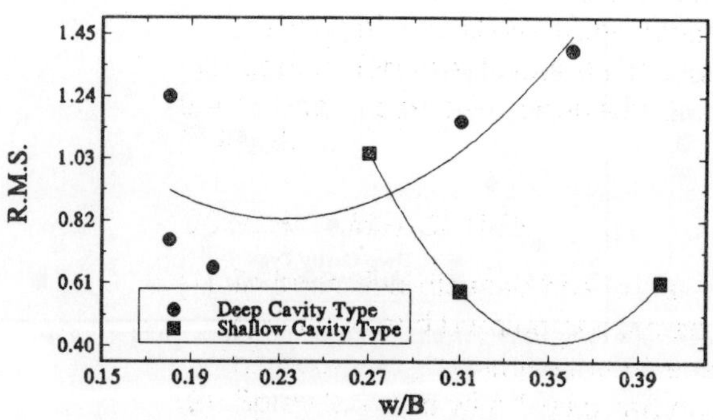

Figure 5. Drag coefficients of models

In Figure 5, drag coefficients are plotted against the wind incident angle. It is found that the trends of variation are quiet similar for different models of different w/d ratios except the model D5 and S3. In turn, their trends are similar to each other. For these two models, their w are very larger, both of them are having w/B exceed or equal to 0.35. This is an evidence that the width of the cavity plays an important role in drag with different wind incident angle.

3.4 Wake Characteristics

In Figure 6, each curve represents normalized velocity profile behind the rear surface of model by 100mm at wind incident angle of 0°, where the models were placed at the elevation of z/B = 2.1. These curves are symmetric as expected, with minimum value at the centre. It is found that wake sizes do not give a good indication of how drags vary.

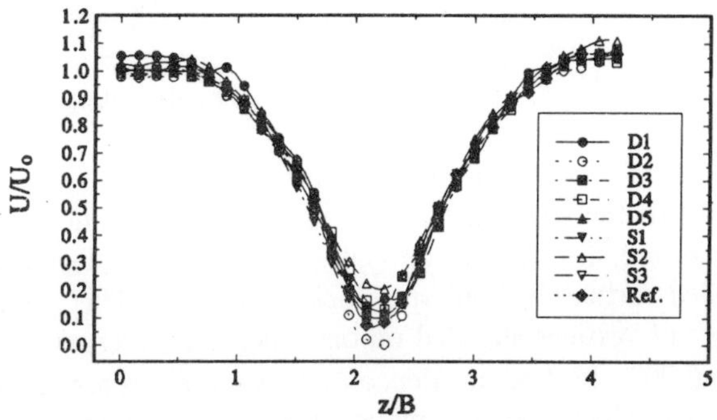

Figure 6. The velocity profiles

Location of vortex formation is also found, which is plotted in Figure 7, it is discovered that the locations of vortices formation of different models do not vary so much. Therefore, it seems that there is no relationship between the existence of cavity and the location of vortex formation at the wind incident angle of 0°

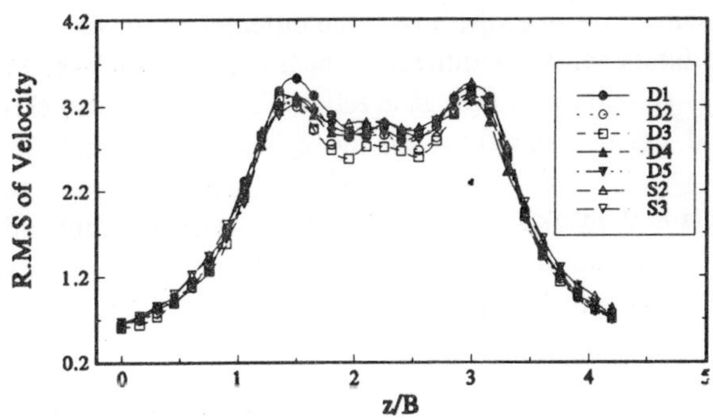

Figure 7. Location of formation of vortex

The shedding frequencies of vortex are found to be inside the range of 13 Hz to 15.5 Hz in all cases. Strouhal number is found to be 0.13 when the wind incident angle of 0° and 90° for all models. This is consistent with what Bearman (1982) found. Figure 8 shows how strouhal numbers, which are calculated with the peak frequencies of drag spectra, varies with the flow incident angle.

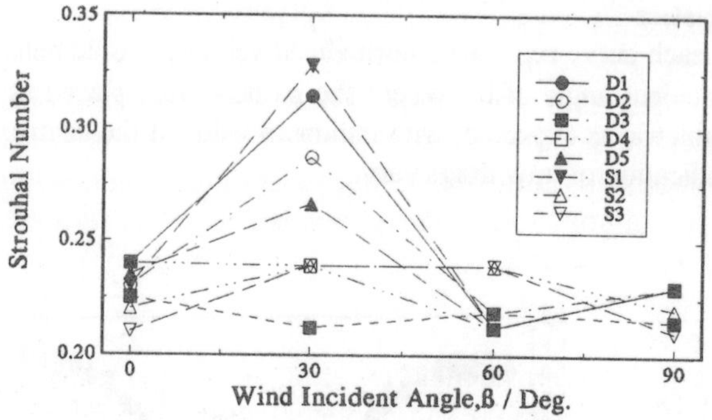

Figure 8. Strouhal numbers

It is expected that the strouhal numbers in drag should double that of the vortices shedding frequency, since vortices are shedded on top and bottom side of a rectangular block alternately. The suction from the shedded vortices will pull the block backwards. This implies that model will be pulled twice times the number of vortices shedded on one side. However, this is only observed at the wind incident angle of 0° and 90° In these incidences, nearly all model are subjected to a strouhal numbers in drag of around 0.23 and which is approximately twice of the strouhal numbers of vortices shedding.

However, in case of the wind incident angle of 30° the strouhal numbers of drag do not follow the above explanation. These values are scattered, therefore, cavity influences the peak frequency at this orientation.

4. Concluding Remarks

Mean and fluctuating wind forces on a number of H-section cylinders of square envelope were measured. The results show that the differences exist between wind forces on H-cylinders and these on unrecessed cylinders of the same envelope. Although the differences are not large in magnitudes, they seem to be caused by some fundamental modifications of unsteady flow over the cylinders due to the presence of the recessed bays and complex flow interaction. The results also show that wind forces on cylinders with deep recessed cavities and these on the cylinders with shallow cavities fall into two different regiemes.

Limited measurements in the wake suggested that these exist negligible differences in wake size, shedding frequency and vortex location among the H-cylinder models. An exception is for a wind incidence of 30°, at which some differences were observed.

5. References

Bearman et al, "An experimental study of pressure fluctuations on fixed and oscillating square-section cylinders" Journal of Fluid Mechanics, 1982, 119, 297-321.

Cook, N. J., Designer's Guide Book to Wind Loading of Building Structures, Part II, Butterworths, 1985.

Nakamura et al, "Vortex excitation of rectangular cylinders with a long side normal to the flow". Journal of Fluid mechanics, 1987, 180, 171-191.

9ICWE, 1995, New Delhi, India

PRESSURE FIELD AND VORTICAL STRUCTURE AROUND THREE-DIMENSIONAL PRISMS

TANIIKE, Y.[1] and OKUDA, Y.[2]

[1] *Associate Professor, Disaster Prevention Research Institute, Kyoto University, Uji, Kyoto, 611, Japan*
[2] *Research Associate, Osaka City University, Osaka 558, Japan*

Abstract

An experimental investigation on the behavior of pressure field generating around square section prisms placed in a turbulent boundary layer is described. Measurements of the body and floor surface pressures, including the simultaneous fluctuating pressure measurements of several points, and the flow visualization around the prisms are presented. The generating process of vortices produced by separating shear layers emerging from the leading edges of the prism, and the structure of the vortices shed into the near wake have been examined. The effects of the variation of the aspect ratio on vortex formation around the prism, the approximate estimation of the vortex strength along height, and vortex formation and decay processes in the wake are described and compared to those for a two-dimensional bluff body.

1. INTRODUCTION

The relationship between the vorticity field induced by the shear layer separated from two-dimensional bluff bodies and the associated pressure-and-velocity fields has been investigated and clarified by many authors, e.g. Bearman(1965), Vickery(1966), Lee(1975), and Davies(1976).

On the other hand, the flow around a three-dimensional bluff body contains the vertical velocity component so that the vortices produced by the separated flows from the top and side surfaces and the corresponding pressure field will show more complex and different characteristics compared to those of two-dimensional flow. Castro and Robins(1977) measured the surface pressures on cubic bodies and the mean and fluctuating velocities within the wake and demonstrated the effects of upstream turbulence on the reattachment of the separated shear layers and the decays of vorticity and velocity deficits in the wake. Sakamoto(1985) measured the aerodynamic forces acting on square section prisms of various heights in a boundary layer and clarified the effects on the forces due to variations of the aspect ratio and parameters characterizing the boundary layer.

However, in spite of those substantial results, information on the vortical structure generated by separated shear layers from three-dimensional bluff bodies has not been sufficiently obtained owing to the difficulty of direct measurements on instantaneous flow velocities with three components in the turbulence wake.

The present study is designed to examine the pressure field on and around three-dimensional prisms in detail and to evolve the vortical structure in the near wake. Pressure measurements on the square section prisms of various aspect ratios and on the floor surfaces around them, including the simultaneous fluctuating pressure measurements of several points, have been presented in a turbulent boundary layer developed along a smooth floor. The dry-ice mist flow visualization technique was utilized to clarify the position of vortex formation, the vortex street spacing and the convection velocity for the prisms of various heights. Experimental data were collected to examine the generating process of vortices which will occur immediately after flow separation at the leading edges of the prisms and to speculate the lower part of the vortices shed into the turbulent wake by transforming the pressure field on the floor surface into the vorticity field using a viscous vortex representation.

2. EXPERIMENTAL PROCEDURE

The experiments were carried out in the DPRI boundary layer wind tunnel with a working section 2m high, 2.5m wide and 21m long. The incident flow simulated a natural wind over open country terrain with a power law component of 0.14 and a low turbulence intensity distribution (Taniike1991).

Square section prisms of three different aspect ratios (H/B=1, 2 and 4; height H=7, 14 and 28cm) with a constant width (B=7cm) were prepared. For measurements of the mean and fluctuating pressures acting on the prism with one face normal to the flow and the base floor around it, a large number of pressure tappings were connected to pressure transducers (Setra model 239) through pressure tubes of 1.5mm in internal diameter and 40cm in length. Eight pressure transducers with the same capacity were used for simultaneous pressure measurements. Pressure measured by the pressure tubing system had a flat frequency response up to 40Hz. The outputs from the pressure transducer were analyzed with a sampling time of 2ms through low pass filtering at 40Hz by a wave form analyser. The free stream velocity U_0 outside the boundary layer was kept at 4.35m/s giving the Reynolds number of $Re=2.1x10^4$ based on the width of prisms.

Flow visualization around the prisms was performed by a dry-ice mist technique at a fixed free velocity of U_0=0.8m/s corresponding to the Reynolds number of $Re=3.8x10^3$. The dry-ice mist was exhausted from a slit on the base floor located at about 1.4m upstream the prism and was carefully adjusted to prevent interference against the flow (Taniike 1992).

3. PRESSURE DISTRIBUTIONS ON THE BODY SURFACE AND THE BASE FLOOR

Figure 1 presents the mean and fluctuating r.m.s. pressure coefficients, C_p and C_p', on the prisms with aspect ratios of H/B=1 and 2, where the results of H/B=2 were similar to those of H/B=4. Figure 2 shows the corresponding results on the base floor around the prisms of H/B=1, 2 and 4. The pressure coefficients are normalized using the dynamic pressure $q_0(=\rho U_0^2/2$; ρ is the air density) and the reference static pressure in the free stream.

The distributions of mean pressure coefficient C_p on the side surface were almost uniform along the height for each case, whereas the r.m.s. coefficient C_p' became larger towards the base. The fact that the values of C_p were constant along height suggests that the mean velocity U_b of the separating flow at the leading edge must be constant along height according to the following relationship:

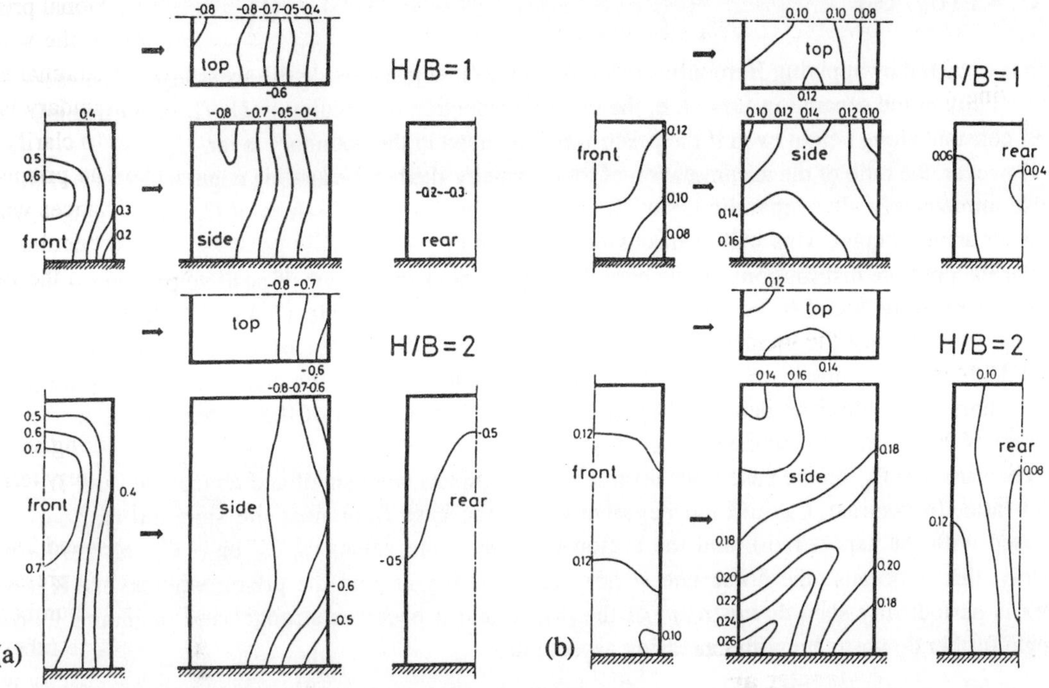

Figure 1. Distributions of (a) the mean pressure coefficient C_p and (b) the r.m.s. coefficient C_p' of pressure fluctuations on the prism surface.

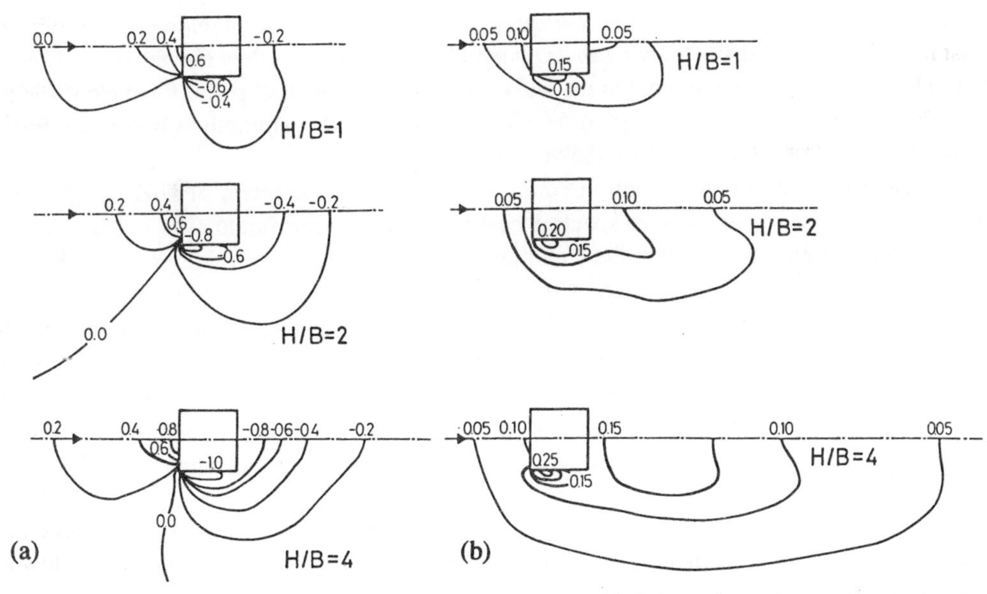

Figure 2. Distributions of (a) the mean pressure coefficient C_p and (b) the r.m.s. coefficient C_p' of pressure fluctuations on the base floor around the prisms.

$$C_p = 1-(U_b / U_0)^2 \quad (1)$$

which is obtained by applying Bernoulli's equation to the flow just outside the shear layer. Therefore, the total vorticity in the separating flow, i. e. the rate of production of circulation $U_b^2/2$, is considered to be almost constant along height even if the prism was immersed in the boundary layer.

However, the ratio of the accumulation of total vorticity shed at the leading edge, which is considered to arise immediately after separation, vary with the height so that the values of C_p' on the side surface will differ at each height. This will be discussed in section 6.

For the pressure distributions on the base floor, the region and values of positive pressures upstream of the prisms were found to be expanding with the aspect ratio. This seems to be related with the blocking effect against the incident flow which increases with the front surface area of the prisms. The abrupt decrease of C_p upstream of the prism until C_p reaches 0.4 is associated with the position of the primary horseshoe vortex as described by Baker(1980). The position of the vortex was seen to move upstream of the prism with increasing aspect ratio.

The values of C_p' on the base floor upstream of the prisms were small and almost the same for each aspect ratio. In contrast, C_p' and the negative C_p on the base floor near the side and rear surfaces increased with the aspect ratio, and the region with the same values of C_p' or ($-C_p$) also enlarged. Possibly, this is because the horseshoe vortex stays steadily around the prism, whereas the Karman vortex is periodically shed downstream of the prism, and it becomes stronger and maintains its own strength further downstream with increasing aspect ratio.

4. FLOW VISUALIZATION AND VORTEX FORMATION

Figure 3 shows photographs of the streak pattern around the lower part of the prisms. The alternate vortex formation is clearly seen close to the rear surface of the prism of H/B=4. However, clouds of the dry-ice mist rapidly became indistinct owing to its own diffusion property as they moved away. Therefore, two pairs of clouds could be seen at most in the wake, and only one and half pairs (three clouds in total) for H/B=2. This result seems to correspond to the actual decay of vortex strength as is obvious from the distributions of C_p and C_p'on the base floor shown in figure 2.

On the other hand, no vortex formation could be seen behind the prism of H/B=1. This is also recognized from the absence of distinct spectral peaks in the pressure fluctuation on the base floor around the prism (see figure 4), and that the simultaneous pressure drop distributions on the base floor behind the prism differed from those formed by the vortex formation (see figure 7 in section 5).

For the prism of H/B=4, the predominant spectral peak of the pressure fluctuation on the base floor behind the prism, which corresponds to the vortex shedding frequency f_s , coincided with the sharp spectral peak of the pressure fluctuation on the base very close to the side surface (see figure 4). This could be due to the fact that the 'origin' or 'source' of the Karman vortex being shed into the wake has been formed in the region between the shear layer and the side surface immediately after separation at the leading edge (see figure 3a).

Assuming that the cloud center of the dry-ice mist corresponds to that of shedding vortices, the positions of each vortex center by display readings of the video tape at 1/3-second intervals are shown in figure 5 in which the same symbol designates the positions of each vortex at a given time.

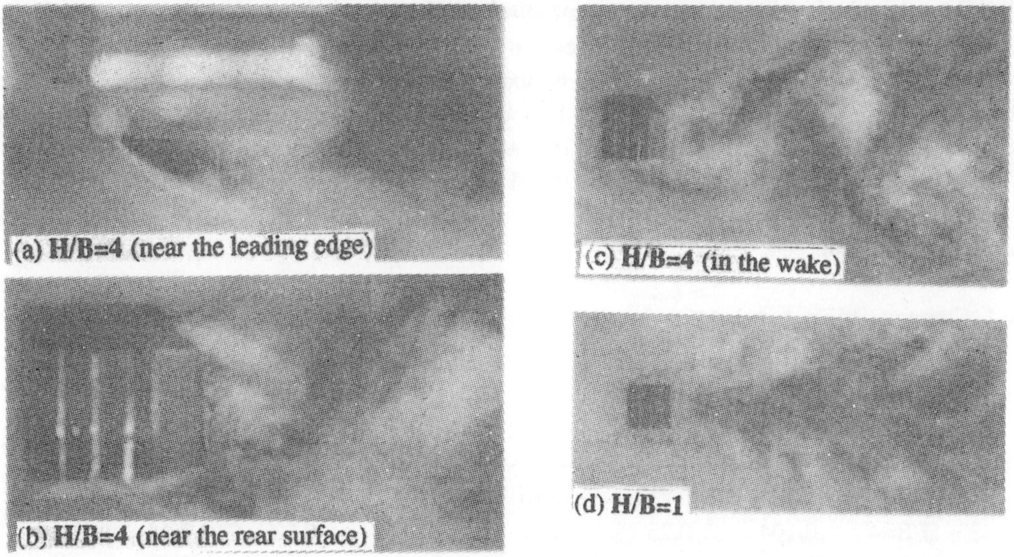

Figure 3. Streak patterns of flow (from left to right) around the lower part of the prisms.

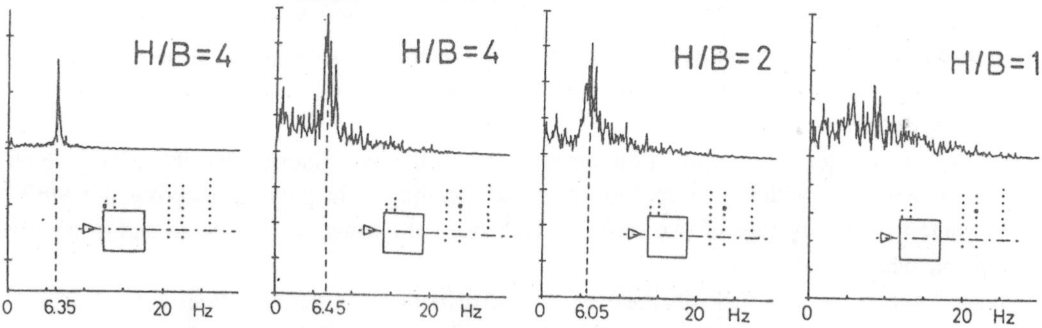

Figure 4. Power spectra of pressure fluctuations on the base floor (at closed square) close to the leading edge or behind the prism.

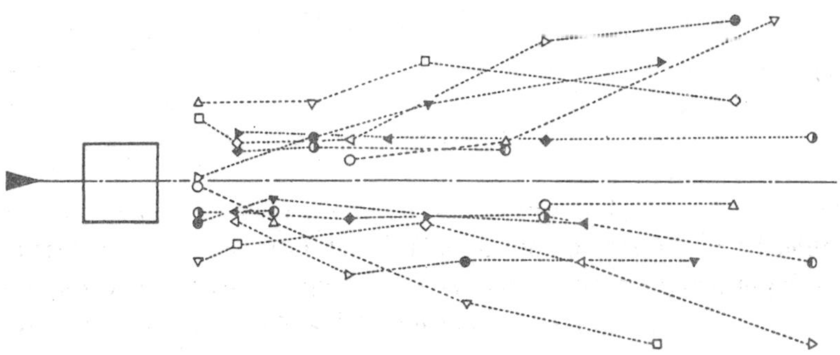

Figure 5. Paths of the center of vortex clouds for the prism of H/B=4.

Both the lateral separation b and the longitudinal separation l of vortex streets gradually expanded as the streets moved downstream. The convection velocity U_c of vortices increased with the vortex flow-down. The velocity varied from $0.33U_0$ at about 2B downstream from the prism center, through $0.52U_0$ at about 4B, to $0.92U_0$ at about 6B. This could be because the decay of individual vortices reduced the windward velocity induced by the interaction of mutual vortices.

The average spacing ratio b / l, the normalized mean convection velocity U_c / U_0 and the Strouhal number $S_t(=f_sB / U_0)$ are indicated in table 1 compared with those of a two-dimensional D-shape cylinder (Davies 1976).

Table 1. Characteristic parameters for the vortical structure in the wake for prisms of various aspect ratios.

	St	b / l	U_c / U_0	$\tilde{\Gamma}_0$	$\tilde{R}_0$	ε
2-D	0.125	0.21	0.80	2.39	1.0	0.26
H/B=4	0.104	0.25	0.59	1.25	0.23	0.13
H/B=2	0.097	—	—	0.95	0.24	0.10

5. VORTEX STRENGTH AND ITS SIZE NEAR THE BASE FLOOR

Figure 6 shows the time records of the simultaneous fluctuating pressures near the leading edges of the top and side surfaces and on a lateral line of the base floor across the wake where the separating flow rolls up into vortices (see figure 3b).

Here let us consider the pressure drop distributions on the base floor behind the prism, which are thought to be associated with the vortex formation and sheddings. The pressure drop $\hat{p}$ is defined as the negative value from the average of two peaks on both sides of a trough to the trough and is normalized as $\widehat{C_p} = \hat{p} / q_0$, where $q_0 = \rho U_0^2 / 2$.

Figure 7 shows the simultaneous pressure drop distributions on the base floor behind the prisms. Closed circles represent the average of pressure drops and error bars denote the spread of the measured values. The vertical chain line represents the wake center, the abscissa indicates the lateral line along the measuring points (closed squares) with separation distance 1.25cm (=0.18B), each of which is located respectively at 7.5cm (=1.07B), 10cm (=1.43B) and 15cm (=2.14B) downstream from the prism center.

Based on the pressure drop distributions, let us investigate the lower part structure of vortices shed into the wake, using the following viscous circular vortex model, which is obtained from the solution for a line vortex diffusing under the action of viscosity:

$$U_\theta = (\Gamma_0 / 2\pi r)\left[1 - \exp\{-1.26\,(r / R_0)^2\}\right] \qquad (2)$$

$$\partial p / \partial r = \rho U_\theta^2 / r \qquad (3)$$

where r is the radial distance from the vortex center, U_θ the tangential velocity with a maximum at the core radius R_0 of the vortex.

Using equations (2) and (3), the pressure p(r) in the vortex is given by

$$p(r) = \int_0^r (\rho\Gamma_0^2 / 4\pi^2 r^3)\left[1 - \exp\{-1.26\,(r/R_0)^2\}\right]^2 dr + p_0 \qquad (4)$$

where p_0 is the pressure at the vortex center. When r approaches infinity, we get

$$\Gamma_0^2 = 4\pi^2(p(\infty) - p_0) \Big/ \int_0^\infty (\rho / r^3)\left[1 - \exp\{-1.26\,(r/R_0)^2\}\right]^2 dr \qquad (5)$$

Therefore, the vortex strength (circulation) Γ_0 can be numerically calculated if $p(\infty)$, p_0 and R_0 are given. Here, let us apply equations (2) and (3) to the pressure drop distributions shown in figure 7. First, let p_0 be equal to the experimental value $\widehat{p}(0)$ (or $\widehat{Cp}(0)$ in the form of the pressure drop coefficient) and the upper limit of integral in equation (5) be given by r=100mm instead of r=∞. Next, R_0 and $p(\infty)$ were chosen as varying parameters of the matching procedure, varying respectively with intervals of 0.5mm and $0.01q_0$. Using p(r) obtained, R_0, $p(\infty)$ and Γ_0 were determined in the way that the square derivation of $\sum_{i=1}^{8} \{p(r_i) - \widehat{p}(r_i)\}^2$ was reduced to minimum.

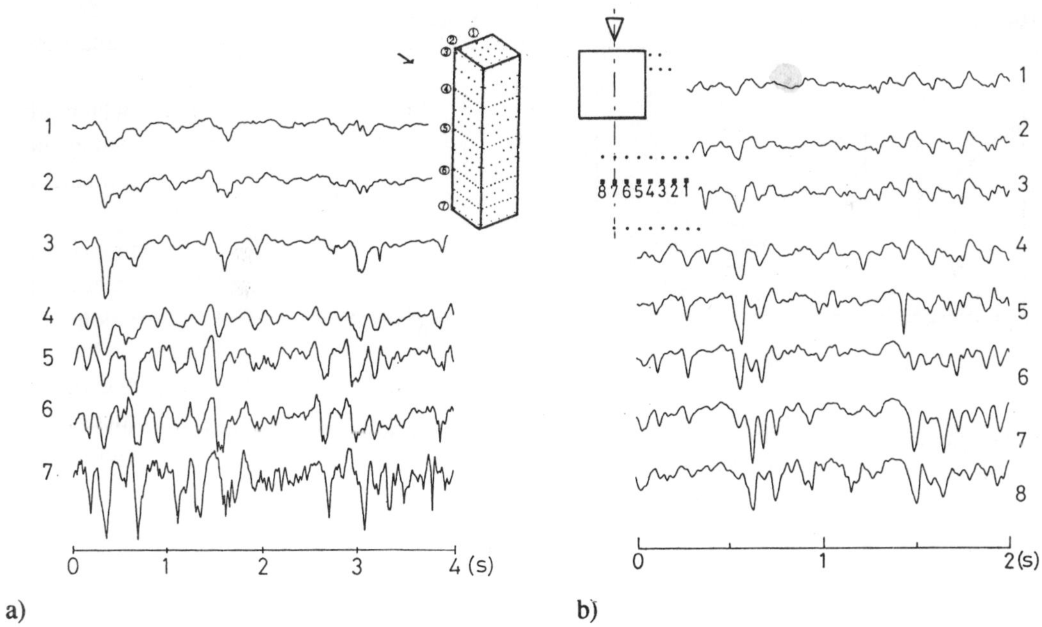

Figure 6. Time records of pressure fluctuations (a) near the leading edges on the top and side surfaces and (b) on the nearest lateral base line to the rear surface for the prism of H/B=4.

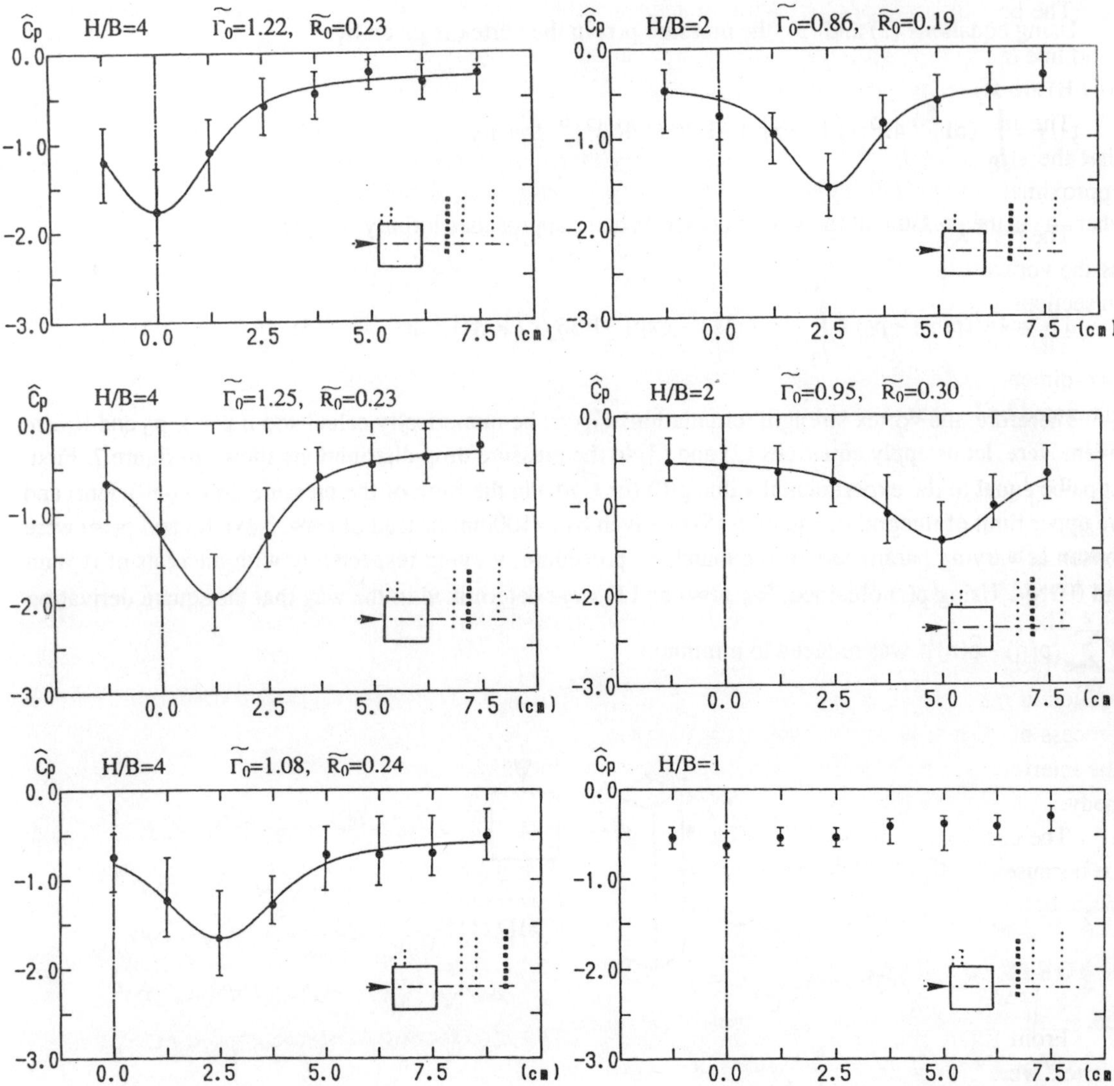

Figure 7. Distributions of the pressure drop coefficient $\widehat{Cp}$ on the base floor in the near wake and the fitting curves based on equations (4) and (5).

The best-fit curve of $\widehat{C_p}(r)$ with the most suitable combinations of $p(\infty)$ and R_0 was plotted as the solid line in figure 7. The corresponding normalized vortex strength $\widetilde{\Gamma_0}(=\Gamma_0 / BU_0)$ and core radius $\widetilde{R_0}(= R_0 / B)$ are also indicated in the figure and table 1.

The analytical result demonstrates a favorable agreement with the experimental results, and shows that the structure of the lower part of roll-up vortices behind the three-dimensional prisms can be approximately expressed by the two-dimensional viscous vortex model.

The results that $\widetilde{\Gamma_0}$ increased with the aspect ratio, and that $\widetilde{R_0}$ enlarged and conversely $\widetilde{\Gamma_0}$ decreased as the vortex was convected downstream, substantiate the inference on the vortical structure referred to in sections 3 and 4.

The values of $\widetilde{\Gamma_0}$ for H/B=4 and 2 are almost equal to or less than half of the value of the two-dimensional body (see table 1). That is connected with the following vorticity accumulation ratio ε estimated by Roshko(1954):

$$\varepsilon = \Gamma_0 f_s / \frac{1}{2}(1-C_p)U_0^2 \tag{6}$$

where f_s is the vortex shedding frequency and C_p the mean pressure coefficient near the leading edge of the lower part of the side surface.

The resulting values of ε for the three-dimensional prisms were almost half of the two-dimensional value (see table 1). Therefore, the vortex strength $\widetilde{\Gamma_0}$ will become smaller with roughly the same reduction ratio as that of the vorticity accumulation ratio ε . The result suggests that the accumulation process of vorticity involved in a separating flow from the side surface becomes deteriorated owing to the interference effect of the separating flow from the top, compared to that of a two-dimensional bluff body.

The core radius $\widetilde{R_0}$ was also considerably small compared to the two-dimensional value, that could be because the vortices diffused and decayed owing to the diffusive action due to high turbulence in the wake before they fully grew up and entrained the separating flow from the opposite side surface.

6. VORTICAL STRUCTURE ALONG HEIGHT

From figure 6a, wave forms which would suggest the formation of the 'source' of the Karman vortex were seen near the leading edges in the middle and lower parts of the side surface, and the amplitudes became larger towards the base.

Figure 8 shows the simultaneous pressure drop distributions near the leading edges of the top and side surfaces and along the center line of the rear surface. Same symbols in the figure designate values of $\widehat{C_p}$ occurring roughly at the same time and the solid lines indicate empirical curves of the ensemble average of $\widehat{C_p}$, which can possibly be expressed by the following exponential relationship:

$$\hat{p}(z) \propto \exp(-\gamma z/H) \tag{7}$$

where z/H is the vertical distance from the base normalized by each prism height H and γ is a constant.

For the $\widehat{C_p}$ decreasing exponentially along height, the following two reasons may be considered: 1) the accumulation of total vorticity shed at the leading edge of the side surface decreases with height due to the interference effect by the separating flow from the top surface; 2) the core radius of the accumulated vorticity, which is the 'source' of the Karman vortex, expands with height owing to be free from the boundary condition by the base floor.

Now, consider the limiting case of the above assumption (1) under the condition where the core radius is uniform along height. A pressure drop $\widehat{p}$ at the edge of the side surface at height z may be connected with the vortex strength Γ as follows:

$$\widehat{p}(z) \propto \Gamma(z)^2 \tag{8}$$

Using the relations of (7) and (8), the vortex strength Γ decreasing with height is roughly estimated as;

$$\Gamma(z/H)=\Gamma_0 \exp(-0.98\ z/H) \quad \text{for } H/B=4 \tag{9}$$
$$\Gamma(z/H)=\Gamma_0 \exp(-1.19\ z/H) \quad \text{for } H/B=2$$

where Γ_0 is the vortex strength on the base floor.

Next, consider the limiting case of the other assumption (2) under the condition where the accumulated vorticity involving in the vortex is constant along height. The pressure drop $\widehat{p}$ may be newly connected with the core radius R as follows:

$$\widehat{p}(z) \propto R(z)^{-2} \tag{10}$$

which gives the following estimation for the core radius R expanding with height:

$$R(z/H)=R_0 \exp(0.98\ z/H) \quad \text{for } H/B=4 \tag{11}$$
$$R(z/H)=R_0 \exp(1.19\ z/H) \quad \text{for } H/B=2$$

where R_0 is the core radius on the base floor.

The above results suggest that the vortex strength for three dimensional prisms becomes weak along height whereas the radius of the vortex core expands along height, and both of the decreasing and expansion ratios along height for the vortex strength and the vortex core become larger with decreasing aspect ratio. The phenomenon that the core radius expands along height was also seen in a previous paper (see figure 9a; Okuda and Taniike 1993), which visualized the separating flows over the side surface of a three dimensional prism.

On the other hand, the pressure drop distribution on the rear surface is supposed to correspond to the strength and expansion distribution along the height of the roll-up vortices shed into the wake. The decreasing ratio of $\widehat{C_p}$ along height on the rear surface was a little larger than that on the side surface. The coefficients in equations (9) and (11) were -1.31 and 1.31 respectively for the prism of H/B=4. This could be because the vortex core was slightly expanding along height in the shedding process of the 'source' vortex due to the diffusive action caused by increase of eddy viscosity with height (see figure 9c; Okuda and Taniike 1990).

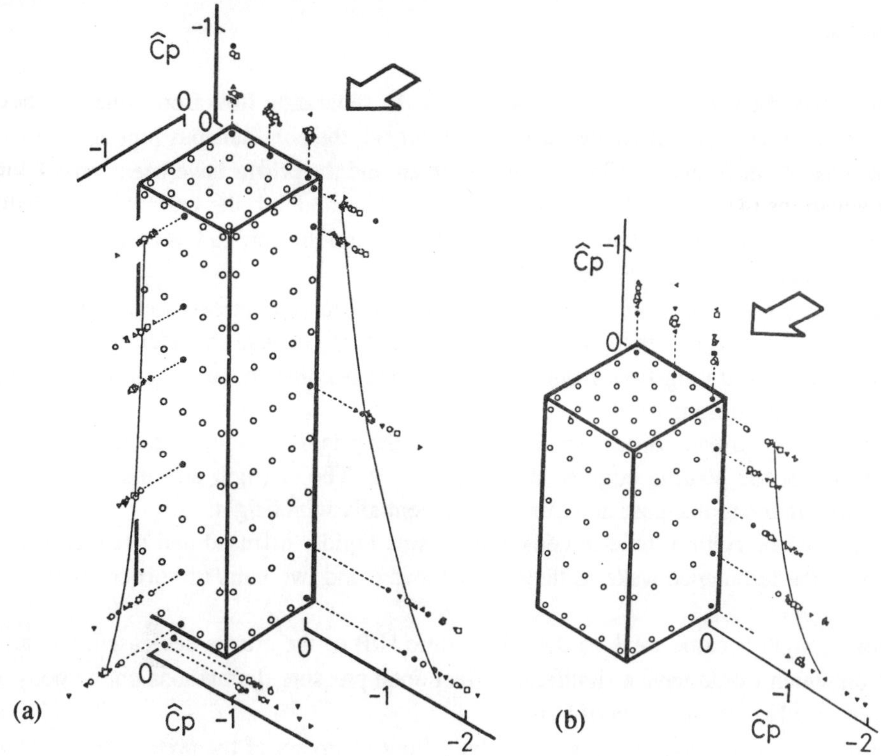

Figure 8. Distributions of the pressure drop coefficient $\widehat{Cp}$ near the leading edges of the top and side surfaces and along the center line of the rear surface.

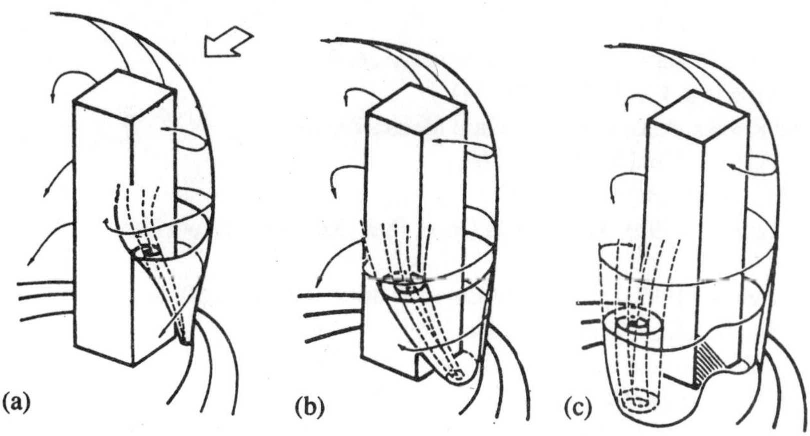

Figure 9. Sketch of the expansion of the 'source' vortex along height and its shedding process in to the wake (a→b→c).

7. CONCLUSIONS

In order to clarify the vortical structure produced by the separating flow from square section prisms of various aspect ratios in a low turbulence wind environment, the simultaneous pressure measurements on the body and base surfaces and the flow visualization around the prisms have been carried out.

Applying equations (4) and (5) based on a viscous vortex model to the pressure drop distributions on the base floor behind the prisms, the strength (circulation) and the core radius of vortices periodically shed into the near wake were precisely obtained.

The vortex strength and the core radius for the three-dimensional prisms were almost half or less than half of those for a two-dimensional body. This could be because the accumulation process of vorticity involved in a separating flow from the side surface becomes deteriorated owing to the three-dimensional boundary condition.

The 'source' of the Karman vortices shedding behind the prisms could be formed immediately after the flow separation at the leading edge of the side surface. The strength and the core radius of the 'source' vortex were found to decrease and expand exponentially with height.

The roll-up vortices behind three-dimensional prisms rapidly diffused and decayed owing to the diffusive action in the turbulence wake as they moved away, and two pairs of vortices survived at most in the wake.

The vortex strength became weak as the aspect ratio H/B of the prisms decreased, and any sign of vortex formation which would have a significant effect upon pressure fluctuations on the body and floor surfaces was not seen behind the prism of H/B=1.

From the mean pressure distributions on the base floor upstream of the prisms, the blocking effect on the incident flow, which increased with the aspect ratio, was found to move the position of the primary horseshoe vortex to further upstream of the prism.

REFERENCES

Baker, C.J. (1980), J. Wind Engng. Industr. Aerodyn., 6, 9.
Bearman, P.W. (1965), J. Fluid Mech., 21, 241.
Castro, I. P. and Robins, A. G. (1977), J. Fluid Mech., 79, 307.
Davies, M. E. (1976), J. Fluid Mech., 75, 209.
Lee, B.E. (1975), J. Fluid Mech., 69, 263.
Okuda, Y. and Taniike (1990), Y., Annuals Disas. Prev. Res. Inst., Kyoto Univ., 33B-1, 297 (in Japanese).
Okuda, Y. and Taniike (1993), Y., J. Wind Engrg. Industr. Aerodyn., 50, 163.
Roshko, A. (1954), N.A.C.A. Tech. Note, No.3169.
Sakamoto, H. (1985), J. Wind Engrg. Industr. Aerodyn., 18, 131.
Taniike, Y. (1991), J. Engrg. Mech., ASCE, 117(3), 457.
Taniike, Y. (1992), J. Wind Engrg. Industr. Aerodyn., 41-44, 1073.
Vickery, B.J. (1966), J. Fluid Mech., 25, 481.

9ICWE, 1995, New Delhi, India

UNSTEADY WIND FORCES ON A SQUARE PRISM IN A TURBULENT BOUNDARY LAYER

NISHIMURA, H.[1] and TANIIKE, Y.[2]

[1] *Chief Researcher, General Building Research Corporation, 5-8-1, Suita, Osaka 565, Japan*
[2] *Assoc. Professor, Disaster Prevention Research Institute, Kyoto University, Uji, Kyoto 611, Japan*

Abstract

The study about aeroelastic instability of tall buildings is very important in wind resistant structure design. This paper describes the characteristics of unsteady aerodynamic forces acting on a square sectioned tall building in a turbulent boundary layer. Wind tunnel experiments were conducted by using a forced oscillation method. Aeroelastic instability regions for a square sectioned tall building was clarified.

1. INTRODUCTION

In order to study aeroelastic instabilities of prismatic high-rise buildings in strong winds (so called 'vortex induced oscillation' or 'galloping'), unsteady aerodynamic forces must be investigated as the external forces acting on buildings. From this point of view, wind tunnel experiments were performed to obtain information about the unsteady cross wind moments as modal forces on buildings, which are subjected to the forced harmonic oscillation with a fundamental linear mode in a turbulent boundary layer.

Several measurements have been made concerning the aerodynamic forces acting on the two-dimensional model by Otsuki et al.(1974), Nakamura et al.(1975) and Washizu et al.(1978). Recently, Taniike(1984, 1989) and Steckley et al.(1990) reported the characteristics of unsteady aerodynamic forces acting on a three-dimensional building model by using the forced oscillation technique. However, these studies had been restricted to buildings immersed in a relatively low turbulent flow. The objective of this paper is to investigate the unsteady aerodynamic base moments as modal forces of a high-rise building model in a highly turbulent boundary layer and to clarify their charactcristics.

2. EXPERIMENTAL PROCEDURE

Wind tunnel experiments were carried out to obtain the information on the aeroelastic instability of a tall square building in a turbulent boundary layer. The incident flow showed in Fig.1 simulated natural wind conditions over an urban area with a power law exponent of 0.25 in horizontal velocity profile and a high turbulence intensity distribution. Unsteady aerodynamic forces acting on a square section prism were measured, which was harmonically oscillating with a fundamental linear mode in the cross wind direction.

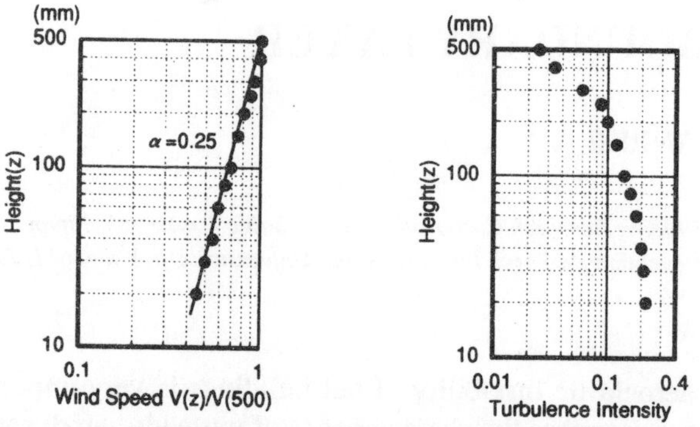

Fig.1 Wind profile and turbulence intensity

Figs.2 and 3 show the illustration of the forced oscillation instrument and the square section model. Two models with the identical mass and shape were mounted on the upper and lower driven beams, the former (called the active model) was placed inside the test section and the latter (the dummy model) outside. The two models were simultaneously forced to give a harmonic oscillation with a fundamental linear mode, i.e., a rocking mode in the cross wind direction by the shaker. The outputs from gauges cemented on the upper and lower beams yielded the signals proportional to bending moments. These two signals were digitalized by computer and the subtractions eliminating the inertial moment of the model were obtained, which indicate the unsteady cross wind moments

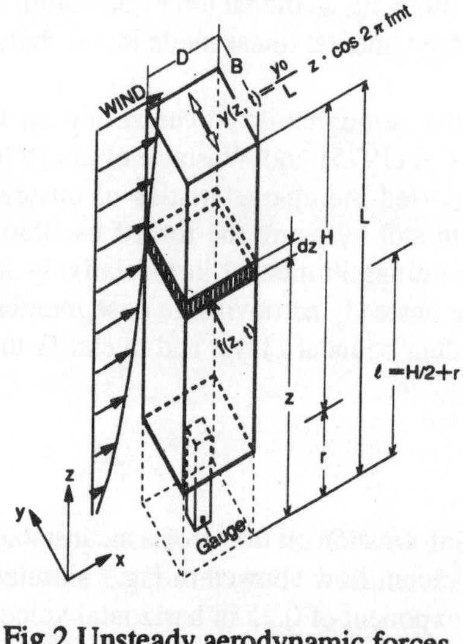

Fig.2 Unsteady aerodynamic forces on an oscillating building model

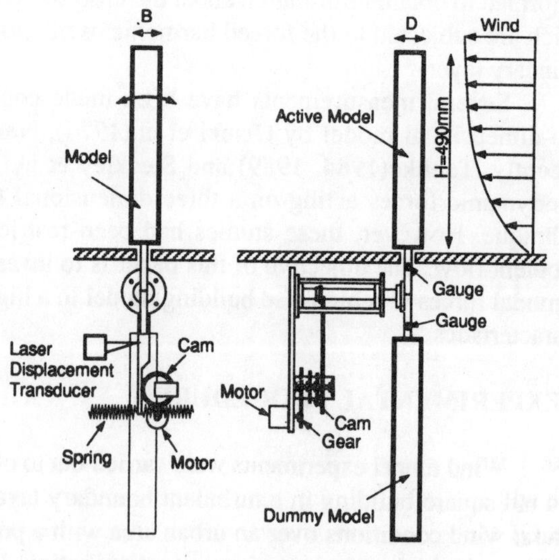

Fig.3 Forced oscillation instrument

acting on the active model. The mean wind velocity was varied discretely from 1.0m/s to 11.0m/s while the frequency of the forced oscillation was kept at a constant value of 6.8Hz.

3. ANALYSIS METHOD

The oscillation form of the model is considered as a rocking of one-degree of freedom with a harmonic oscillation as shown in Fig.3. In this experiment, the rotation angle θ (t) was small enough to be expressed as

$$\theta(t) = \frac{y_0}{L}\cos 2\pi f_m t = y_r\, q(t) \qquad (1)$$

where L(=H+r) is the length from the center of rotation (at the location of the gauge) to the top of the model, H the height of the model, r the distance between the gauge and the bottom of the model, $y_r(=y_0/L)$ the nondimensional amplitude and $q(t)(=\cos 2\pi f_m t)$ the generalized coordinate. Here, the values of H, r, and L were respectively 490, 60 and 540mm and the nondimensional amplitudes y_r were taken equal to 0.002, 0.004, 0.008, 0.016 and 0.032, where y_0 is the amplitude at the top of the model.

$$Q(t) = \int_r^L f(z,t)\, z\, dz \qquad (2)$$

where f(z,t) is the unsteady lift force per unit length at height z.

The unsteady aerodynamic forces Q(t) is composed of a random component due to turbulence in the incident flow, the Strouhal component induced by regular vortex sheddings, and the frequency component $Q_m(t)$ caused by the motion of the model. The frequency response component $Q_m(t)$ among the above components involved in Q(t) mainly contributes to the harmonic oscillation with a frequency of f_m. This component is obtained by taking the same frequency part as f_m out of the Fourier series expanding Q(t).

$$Q_m(t) = |Q_L|\cos\left(2\pi f_m t + \beta_Q\right) = Q_R \cos 2\pi f_m t - Q_I \sin 2\pi f_m t \qquad (3)$$

$$Q_R = \frac{1}{T}\int_{-T}^{T} Q(t)\cos 2\pi f_m t \cdot dt \quad , \quad Q_I = \frac{-1}{T}\int_{-T}^{T} Q(t)\sin 2\pi f_m t \cdot dt \qquad (4)$$

$$|Q_L| = \sqrt{Q_R^2 + Q_I^2} \quad , \quad \beta_Q = \tan^{-1}\left(\frac{Q_I}{Q_R}\right) \qquad (5)$$

where T is the analysis duration of 24 sec, Q_R is in phase with the model displacement while Q_I is in phase with the model velocity, and β_Q is the phase angle between $Q_m(t)$ and the transverse displacement. Here, Q_R and Q_I are nondimensionalized as

$$C_R = Q_R / \tfrac{1}{2}\rho V^2 D H \ell \ , \ C_I = Q_I / \tfrac{1}{2}\rho V^2 D H \ell \qquad (6)$$

where ρ is the air density, D the depth of the model and ℓ =r+H/2, where r<<H/2. The aerodynamic coefficients C_R and C_I are described as functions of the reduced wind velocity Vr(=V/f_mD) and the nondimensional amplitude y_r(≅y_0 /H). The aerodynamic moment coefficient C_I is related to the damping of the oscillating system. In case of C_I >0, the works done by the unsteady moment become negative, i.e., the negative aerodynamic damping effect is generated in the oscillating system. Therefore, self-excited oscillations may occur in the condition that C_I >0 and the magnitude of the negative aerodynamic damping exceeds that of the structural damping of the oscillating system.

4. RESULTS

Unsteady aerodynamic coefficient C_I

The results of coefficient C_I for a tall square section building with the aspect ratio of $H/\sqrt{BD}$ =7 are shown in Fig.4. The ordinate is C_I and the abscissa Vr. The arrow on the abscissa indicates the reduced wind velocity for vortex resonance Vres. The curves in Fig.4(c) will be discussed later.

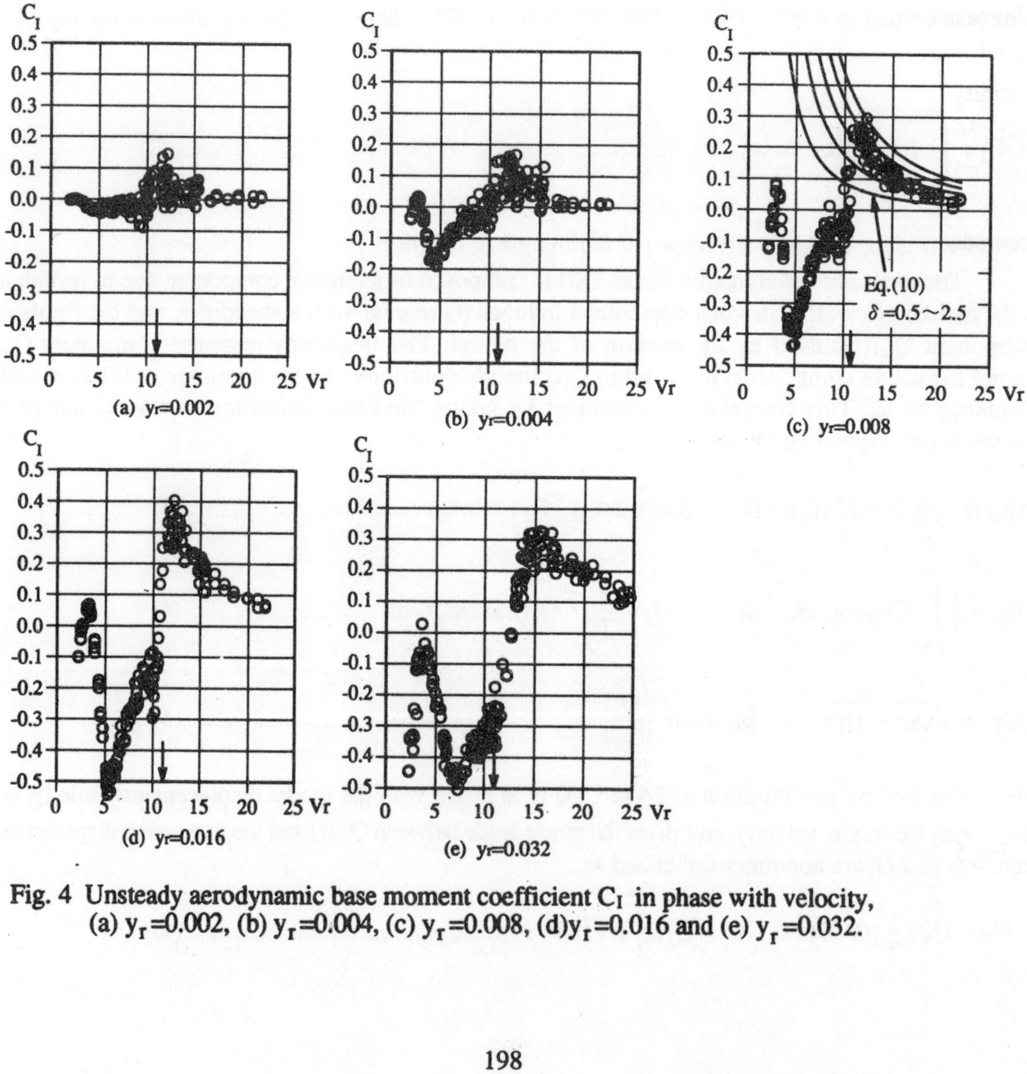

Fig. 4 Unsteady aerodynamic base moment coefficient C_I in phase with velocity, (a) y_r=0.002, (b) y_r=0.004, (c) y_r=0.008, (d)y_r=0.016 and (e) y_r=0.032.

When y_r is small, the values of C_I scatter around zero near the resonance wind velocity Vres, so that the negative aerodynamic damping effect does not clearly appear. When y_r becomes large, two positive peaks of C_I were obviously observed. The wind velocity corresponding with each peak is found to increase with y_r, and the wind velocities which change their signs from negative to positive are also increased with y_r. Each positive peak will not exceed a certain value even if y_r become large, consequently the oscillation will be limited and show a peak response.

Unsteady aerodynamic coefficient C_R

The aerodynamic moment coefficient C_R shown in Fig.5 is related to the spring constant of the oscillating system. In case of $C_R>0$, the unsteady moment acts on the oscillating system to weaken the stiffness of the spring and to make the frequency lower, and vice versa. C_R shows positive values in wide velocity range below Vres, and abruptly reduces over Vres. The magnitude of Q_R is almost an order of magnitude smaller compared to the stiffness of oscillating system, so that the frequency change of the oscillating system will hardly be observed in practice.

Phase angle β_Q

The examples of β_Q are shown in Fig.6. The phase angle β_Q abruptly rises up to 90° in the vicinity of Vres, and a positive peak of C_I appeared. Values of β_Q were around 90° at velocity ranges over Vres. In the case where β_Q is nearly 90°, the quasi-steady theory on galloping analysis can be applied. On the other hand, β_Q decreases with wind velocity in the lower velocity range less than the Vres. When the Vr=3, β_Q became 90° and the positive peak of C_I once again appeared, because it decreased from -180° to -270°.

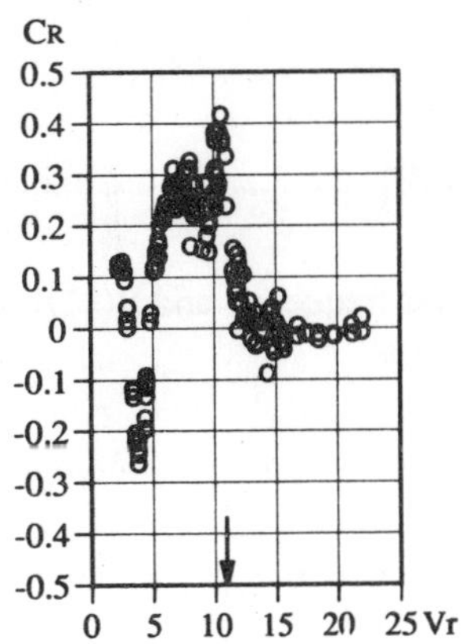

Fig.5 Unsteady aerodynamic base moment coefficient C_R in phase with displacement (y_r=0.008)

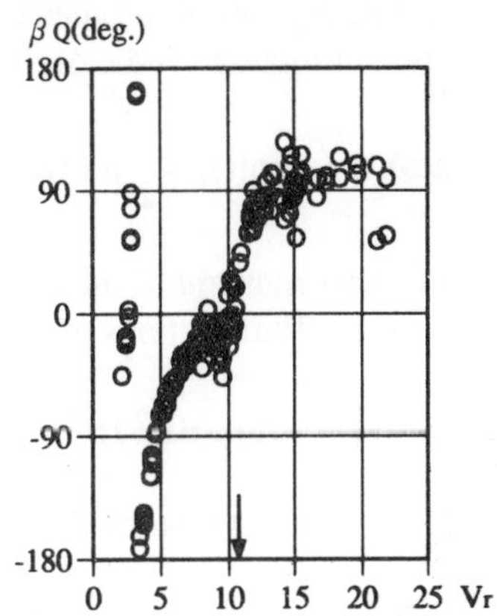

Fig. 6 Phase angle β_Q between unsteady base moment and transverse displacement (y_r=0.008)

Power spectra

Power spectra of the unsteady aerodynamic moments are shown in Fig.7. The ordinate is the nondimensionalized power spectrum and the abscissa is the frequency in Hz. The characters 'm' and 's' in the figure indicate the frequency f_m of the forced oscillation, and the frequency fs is the shedding of vortices specified by the Strouhal number, respectively. The peak at fs is found to increase with the wind velocity, while the peak at f_m is kept constant. The locked-in phenomenon that fs coincides with f_m was observed near Vres.

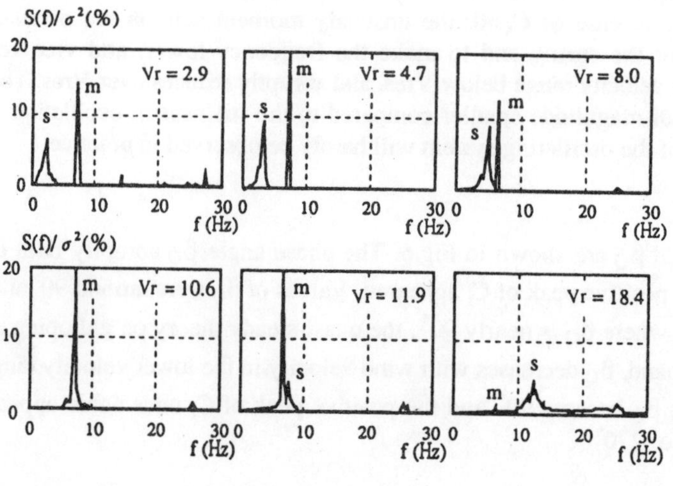

Fig. 7 Power spectra of unsteady aerodynamic base moment

Response curves and regions of aeroelastic instability

The equation of motion for the oscillating model is written as

$$y_r I\left\{\ddot{q}(t) + 2\mu\omega\dot{q}(t) + \omega^2 q(t)\right\} = \frac{1}{2}\rho V^2 BH \ell \left\{C_R q(t) + \frac{C_I}{2\pi f_m}\dot{q}(t)\right\} \quad (7)$$

where I is the total mass moment of inertia, ω the natural circular frequency and B(=D) the breath of the model, respectively. The solutions of the equation are written as

$$4\pi^2 \cdot I\, y_r \left\{(f_o/f_m)^2 - 1\right\} = (1/2)\,\rho BD^2 H \ell V_r^2 C_R(y_r, V_r) \quad (8)$$

$$8\pi^2 I \eta\, y_r (f_o/f_m) = (1/2)\,\rho BD^2 H \ell V_r^2 C_I(y_r, V_r) \quad (9)$$

where f_0 the natural frequency of the oscillating system. Assuming that the model oscillates with the natural frequency, the condition for transverse oscillations caused by the negative aerodynamic damping effect is

$$C_I \geq 32\pi^2 \frac{H}{D} \delta\, y_r \frac{1}{V_r^2} \quad (10)$$

where $\delta(=\eta M\rho BDH)$ is the mass-damping parameter for a building with a linear mode and uniformly distributed mass, where M is the generalized mass. The curves in the Fig.4(c) indicate the right side values of equation (10) varied with the mass-damping parameter δ when y_r=0.008. When the value of C_I is equal or more than the value of the curve, the oscillation becomes steady or developing, respectively.

The limit cycles of transverse oscillation, namely the response curves showing the relations between y_r and Vr are obtained by prescribing values of δ and comparing the magnitudes of both sides in equation (10). The result is shown in Fig.8. Some remarks may be derived from Fig.8. If the mass-damping parameter is extremely small, the oscillation showing peak response may occur near Vr=3 and the fairly strong motion regarded as the composite actions of vortex excitation and the galloping may be also seen near Vres. If the parameter becomes larger, i.e. $\delta \geqq 1.5$, only the vortex excitation with small amplitude will be shown near Vres in this case the galloping will disappear.

The region of the aeroelastic instability at the amplitude of y_r=0.008 obtained from the response curves in Fig.8 is shown in Fig.9. The aeroelastic instability region is divided into two parts; one is situated near Vr=3 and it appear only when δ becomes extremely small, while another is seen nearer to, and greater than the Vres. The occurrence velocity of aeroelastic instability near Vres is found to shift toward higher velocity range with an increase of δ. As the mass-damping parameter

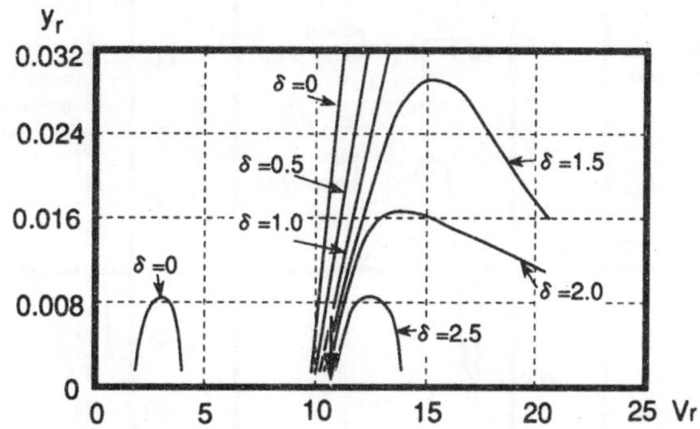

Fig.8 Response curves with mass-damping parameter

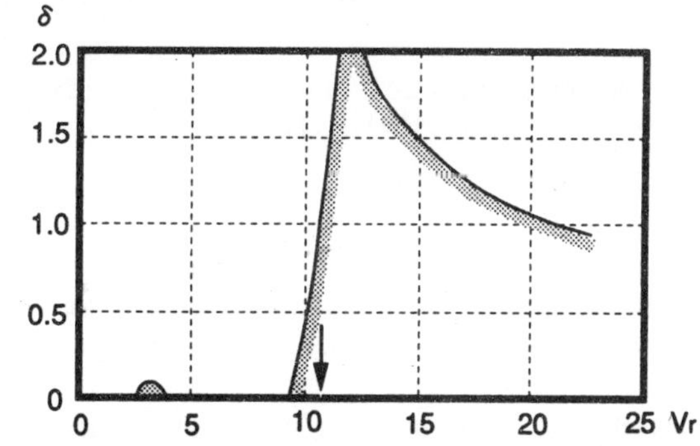

Fig.9 Region of the aeroelastic instability when y_r=0.008

for tall buildings exceeds δ=0.4, the aeroelastic instability for a tall square section building will appear at a velocity region beyond the resonance velocity.

Effect of the Karman vortex on the unsteady aerodynamic forces

In order to investigate the effect of the Karman vortex on the aeroelastic instability, the unsteady aerodynamic forces were also measured on the condition that the splitter plate was set up behind the model. There is a good agreement found between the unsteady aerodynamic force coefficients with and without the splitter plate in the lower velocity region, as shown in Fig.10. For the model with a splitter plate the positive peak of C_I, the abrupt decrease of C_R and the abrupt jump of β_Q could not be observed near the Vres, while these phenomena could clearly be seen in the case of a model without the splitter plate as shown before. From the results, the Karman vortex obviously affected the vortex induced oscillation near the Vres. A remarkable result is that the positive values of C_I near the Vres appear even if the Karman vortex had disappeared using the splitter plate behind the model. This suggests that the occurrence of aeroelastic instability does not depend on the Karman vortex.

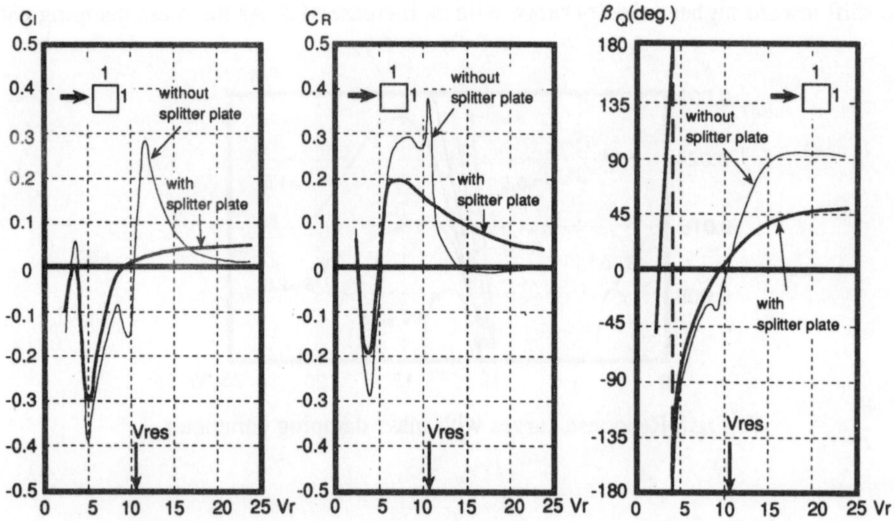

Fig.10 Unsteady aerodynamic base moment with and without splitter plate when y_r=0.008. (a) C_R, (b) C_I, (c) β_Q

5. CONCLUSIONS

Wind tunnel experiments were performed to obtain information of unsteady cross wind base moments on a square sectioned tall building oscillating transversely with a fundamental linear mode in a turbulent boundary layer.

The aeroelastic instability for a tall building was found to occur even in a highly turbulent boundary layer.

The unsteady aerodynamic force predominantly consists of forces due to Karman vortices and the vortices induced by model oscillation.

The frequency response component $Q_m(t)$ of the unsteady cross wind moments (which is related to the onset of self-excited oscillations) showed a non-linear behavior as for the oscillation amplitude and the reduced wind velocity.

The Karman vortex affects on the unsteady aerodynamic forces near the resonance velocity. However, in the case where the Karman vortex is not existent, the negative aerodynamic damping effect occurs in the velocity range nearer to, and greater than the resonance velocity.

The response curves of the aeroelastic cross wind oscillations were obtained with the product of the mass and damping of a structure varying as a parameter by substituting the unsteady aeroelastic moments for the external force term in the equation of motion.

The region of aeroelastic instability related to the mass-damping parameters of a square section building were obtained, which seem to be of great use for wind resistant design of high-rise buildings.

REFERENCES

Otsuki, Y., et al. (1974), "A note on the aeroelastic instability of a prismatic bar with square section", J. Sound and Vibration, 34.

Nakamura, Y., et al. (1975), "Unsteady lifts and wakes of oscillating rectangular prisms", J. Engrg. Mech. Div., ASCE, 101 (EM6).

Washizu, K., et al. (1978), "Aeroelastic instability of rectangular cylinders in a heaving mode", J. Sound and Vibration, 59.

Taniike, Y. (1984), "Unsteady aeroelastic forces on a three dimensional model with square section", J. Wind Engineering, JAWE, No.21.

Taniike, Y. (1989), "Unsteady aeroelastic base moments on a high-rise building", Proc. of the Sessions Related to Design, Analysis and Testing at Structures Congress '89.

Steckley, A., et al. (1990), "On the measurement of motion induced forces on models in turbulent shear flow", J. of Wind Engineering and Industrial Aerodynamics, Vol.36.

FLUID DYNAMICS OF FLAT PLATES AND RECTANGULAR PRISMS IN THE PRESENCE OF MOVING SURFACE BOUNDARY-LAYER CONTROL

MODI, V.J.[1], MUNSHI, S.R.[1] and YOKOMIZO, T.[2]
Professor *Graduate Fellow* *Professor*

[1] *Department of Mechanical Engineering, University of British Columbia, Vancouver, B.C., Canada V6T 1Z4.*
[2] *Department of Mechanical Engineering, Kanto Gakuin University, Matsuura, Kanazawa, Yokohama, Japan 236.*

Abstract

Fluid Dynamics of two-dimensional flat plates and rectangular prisms is studied in the presence of Moving Surface Boundary-layer Control (MSBC) employing three complementary approaches: (i) extensive wind tunnel test program; (ii) finite element based numerical integration of the complete Navier-Stokes equations; and (iii) flow visualization. The surface pressure distribution, resultant forces and wake characteristic results suggests significant delay in the boundary-layer separation, reduction in drag, and increase in the Strouhal number indicating effective decrease in the bluffness. The numerical analysis clearly shows progressive narrowing of the wake with an increase in the momentum injection. The flow visualization study in a water channel using slit lighting in conjunction with polyvinyl chloride tracer particles dramatically substantiate effectiveness of the MSBC. Implication of the study to alleviate wind induced instabilities is apparent.

1. INTRODUCTION

Wind effects on bluff bodies with the resulting vortex resonance and galloping type of instabilities have been a subject of considerable study. A vast body of literature accumulated over years has been reviewed by several authors including Cermak (1975), Welt (1983, 1988) and Modi et al.(1994). A wide variety of devices have been developed aimed at arresting the vibrations. They may be classified as:

(a) Spoilers which tend to change fluid dynamic characteristics of structures in such a way as to interfere with and weaken the excitation force. Helical strakes, shrouds, slats, fairings, etc. belong to this category.

(b) Dampers, which provide a mechanism for dissipation of energy or its transfer from one part of the system to the other. Here we have hydraulic dashpots; stockbridge, tuned-mass and nutation dampers; etc.

Zdravkovich (1981, 1984); Every, King and Weaver (1982); as well as Welt and Modi (1992) have summarized this aspect of the literature. However, potential of the momentum injection to alter fluid dynamics of bluff bodies, as in the case of Moving Surface Boundary-layer Control (MSBC), has received relatively less attention. Modi et al. reported successful application of the MSBC with reference to high performance airplanes (1987, 1991) as well as tractor-trailer truck configurations (1990, 1993). Kubo et al. also investigated application of the concept to two-dimensional square prisms representing tall structure (1992, 1993). The present investigation builds on this background to study fluid dynamics of two fundamental geometries, two dimensional flat plates and rectangular prisms, using three different approaches providing comprehensive understanding of the problem. To begin with, a planned wind tunnel test-program is described which provides information on the effect of MSBC on the boundary-layer separation, drag reduction and Strouhal number. This is followed by the finite element analysis of the Navier-Stokes equations to assess effect of the MSBC on the wake geometry. Finally, a flow visualization study is undertaken which shows, rather spectacularly, effectiveness of the MSBC in suppressing the vortex shedding phenomenon.

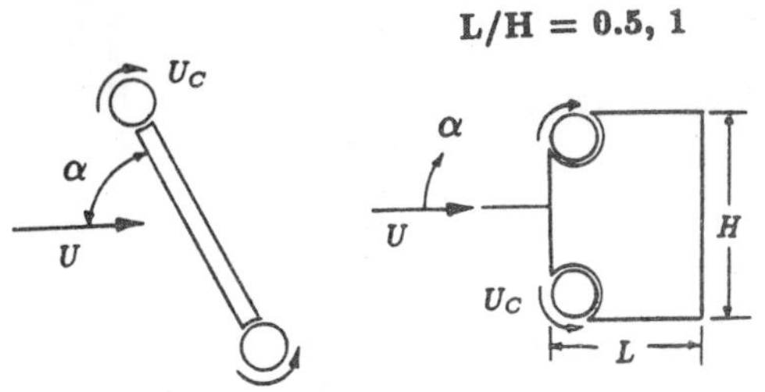

Figure 1 Bluff body geometries studied using the wind tunnel tests, finite element numerical approach and flow visualization.

2. WIND TUNNEL TEST RESULTS AND DISCUSSION

A family of flat plate and rectangular prism models (Figure 1), fitted with end plates and rotating cylinders, spanned the tunnel test-section, 0.91×0.68 ×2.6 m, to create essentially two-dimensional condition. Typically, a model was supported by an Aerolab six-component strain guage balance and tested in a low-speed, low turbulence, return-type wind tunnel, where the airspeed can be varied from 1 – 50 m/s with a turbulence level less that 0.1%. The spatial variation of velocity in the tunnel test-section is less than 0.25%.

The smooth surface rotating cylinders were mounted between high-speed bearings housed in the brackets at either end of the model. They were driven by 1/8 h.p. variable speed motors, located outside the tunnel, through standard couplings. The models were provided with pressure taps, suitably distributed over the circumference, to yield detailed information concerning the surface loading. The pressure distribution was recorded using a highly sensitive ($6.9\times10^{-2}N/m^2$) capacitance type Barocel transducer system. The lift and drag data were obtained through integration of the pressure distribution as well as direct measurement of the forces (sensitivity 0.5 gm/mV). Their correlation substantiated two-dimensional character of the flow.

The tests were conducted over an extended range of the angle of attack ($\alpha = 0 - 105^o$) with an increment of 5°. The cylinder rotation speed was varied to give $U_C/U = 0, 1, 2, 3, 4$ (U_C, cylinder surface velocity; U = freestream velocity) with the Reynolds number (Re) in the range $3 - 5 \times 10^4$ (Re based on the freestream velocity and the plate or the prism height H).

The maximum blockage ratio during the tests was around 20%. A satisfactory procedure applicable to a bluff body in a flow with separating shear layers is still not available. With rotation of the cylinders, the problem is further complicated. As shown by the pressure data and confirmed by the flow visualization, the unsteady flow can be separating and reattaching over a large portion of the model surface. In absence of any reliable procedure to account for wall confinement effects in the present situation, the results are purposely presented in the uncorrected form. Note, this is of little consequence, as the main objective is to assess the effect of momentum injection under the same blockage condition.

2.1 Flat Plate

Variations of the lift (C_L) and drag (C_D) coefficients with the angle of attack (α) and the momentum injection parameter U_C/U are shown in Figure 2. With the two cylinders at rest, the plate stalls at around 10° (Figure 2a). In general, for a given angle of attack, an increase in U_C/U affects the lift adversely, primarily due to the rise in pressure on the top surface induced by the downstream cylinder. However, the focus here is on the drag which should decrease due to narrowing of the wake and increase in the wake pressure. This is shown quite vividly in Figure 2(b). Note, for $\alpha = 90^o$, a decrease in the drag coefficient from 1.85 at U_c/U=0 to 0.47 at U_C/U=3

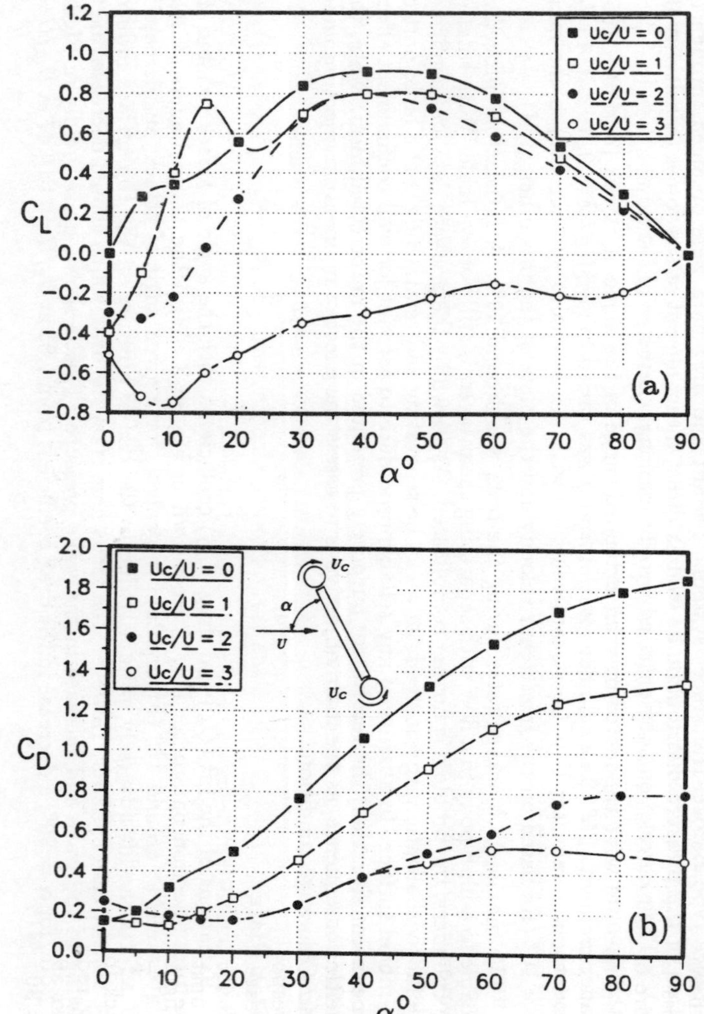

Figure 2 Plots showing the effect of momentum injection and angle of attack on the lift and drag characteristics of a two dimensional flat plate.

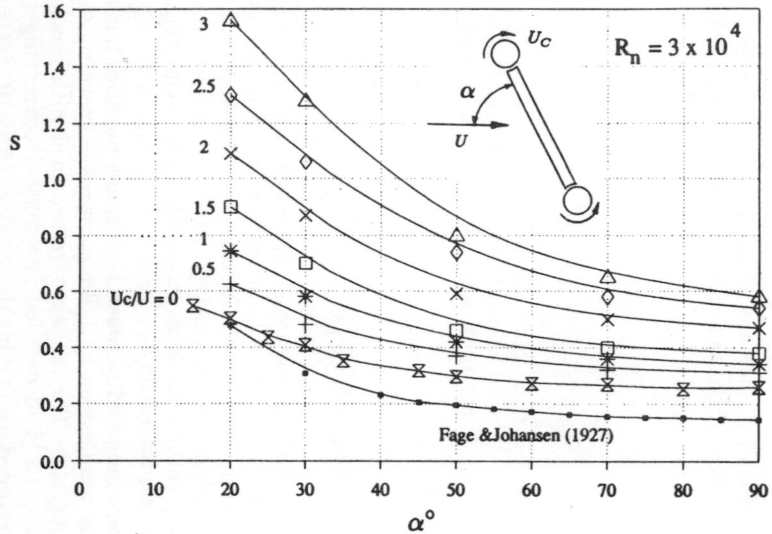

Figure 3 Effects of momentum injection on the Strouhal number, associated with a flat plate, over a range of the angle of attack.

represents a reduction of around 75%.

Figure 3 shows variation of the Strouhal number (S) with the angle of attack and U_C/U at a fixed Reynolds number of 3×10^4. Classical results of Fage and Johansen are also included for comparison. A remarkable increase in the Strouhal number with cylinder rotation at a given α is apparent. It suggests an increase in the shedding frequency corresponding to narrowing of the wake. Obviously this will delay onset of the vortex resonance.

2.2 Rectangular Prisms

Detailed pressure distribution results were obtained for both a square (H/L = 1) and a rectangular prism (H/L = 0.5). A systematic variation of the cylinder speed, its direction of rotation and angle of attack resulted in a large amount of information. Only a small sample of representative results useful in establishing trends are presented here.

Figure 4 shows typical pressure plots as affected by the momentum injection and angle of attack for a square prism. At the outset, presence of a suction peak near the downstream cylinder (S/R = 0.8), increasing with the momentum injection, is apparent. Obviously this would contribute to the overall reduction in the drag. Furthermore, in general, with an increase in U_C/U, the wake pressure also increases, which, in turn would reduce the drag. Of interest is the movement of the stagnation point with the angle of attack, from S/R = 0 at α=0 to ≈ -1 at 45°, i.e. near the upper (downstream) cylinder. With further increase in the angle of attack, the stagnation point crosses over to the lower face, progressively moving downstream ($\alpha = 75^o - 105^o$). Note the presence of pressure peaks in the vincinity of the lower (upstream) cylinder. They are associated with the complex flow near the momentum injection region and suggest the transient separation- reattachment condition. As can be expected, the transient character accentuates with an increase in U_C/U as the flow tries to reattach. The upper (downstream) cylinder, being in the wake (for α greater than around 60°) does not show such pressure fluctuations. Pressure plots for a rectangular prism (H/L = 0.5) showed similar trends with minor local variations (Figure 5).

To take advantage of the momentum injection, the direction of the lower cylinder should be changed from counterclockwise to the clockwise sense when crossover of the stagnation point across the rotational element takes place. The corresponding plots are shown for $\alpha = 90^o$ in Figure 6. To help assess relative changes, pressure plots with the two different rotation strategies for the upstream cylinder are presented togather. Note, a significant rise in suction at both the cylinder locations. Furthermore, the wake pressure also rises. As explained later this results in a considerable reduction in the drag.

A word concerning behaviour of the separation point would be appropriate. Even with the momentum injection, there is a local separation bubble due to a sharp corner downstream of the cylinder. As can be expected, its size depends on the level and direction of the injection. However, it must be emphasized that, with an increase in U_C/U, the streamlines do come closer to the prism surface resulting in a smaller wake as shown by the flow visualization.

Corresponding drag results for the prisms are presented in Figure 7. For the square prism, the peak drag occurs at with $\alpha = 40^o$ both with and without the MSBC (Figure 7a). Furthermore, the effect of momentum injection is also relatively small resulting in a drag reduction of around 8.8%. On the other hand, for α=0, the change in C_D from 1.2 ($U_C/U = 0$) to 0.5 ($U_C/U = 3$) corresponds to 58.3% reduction in the drag. This is also reflected in larger suction peaks with an increase in U_C/U as observed in Figure 4. Note, at $\alpha = 90^o$, there is actually an increase in drag ($\approx$ 9.4%). This is due to outward rotation of the upstream cylinder which, obviously, is opposing the upstream flow. The effect of imparting the desired downstream rotation is also indicated in the figure by a dashed line. Now the reduction in drag was found to be 41.1%.

For the rectangular prism, the peak drag values are associated with different angles of attack for without ($\alpha = 10^o$) and with ($\alpha = 45^o$) rotation as shown in Figure 7(b). The maximum drag reduction (61.7%) occurs at $\alpha = 0$, where the pressure plots show relatively large suction peaks as well as increase in the wake pressure with the MSBC. On the other hand, in the range of $\alpha = 90^o - 105^o$ the drag increases by around 13% due to outward rotation of the front cylinder.

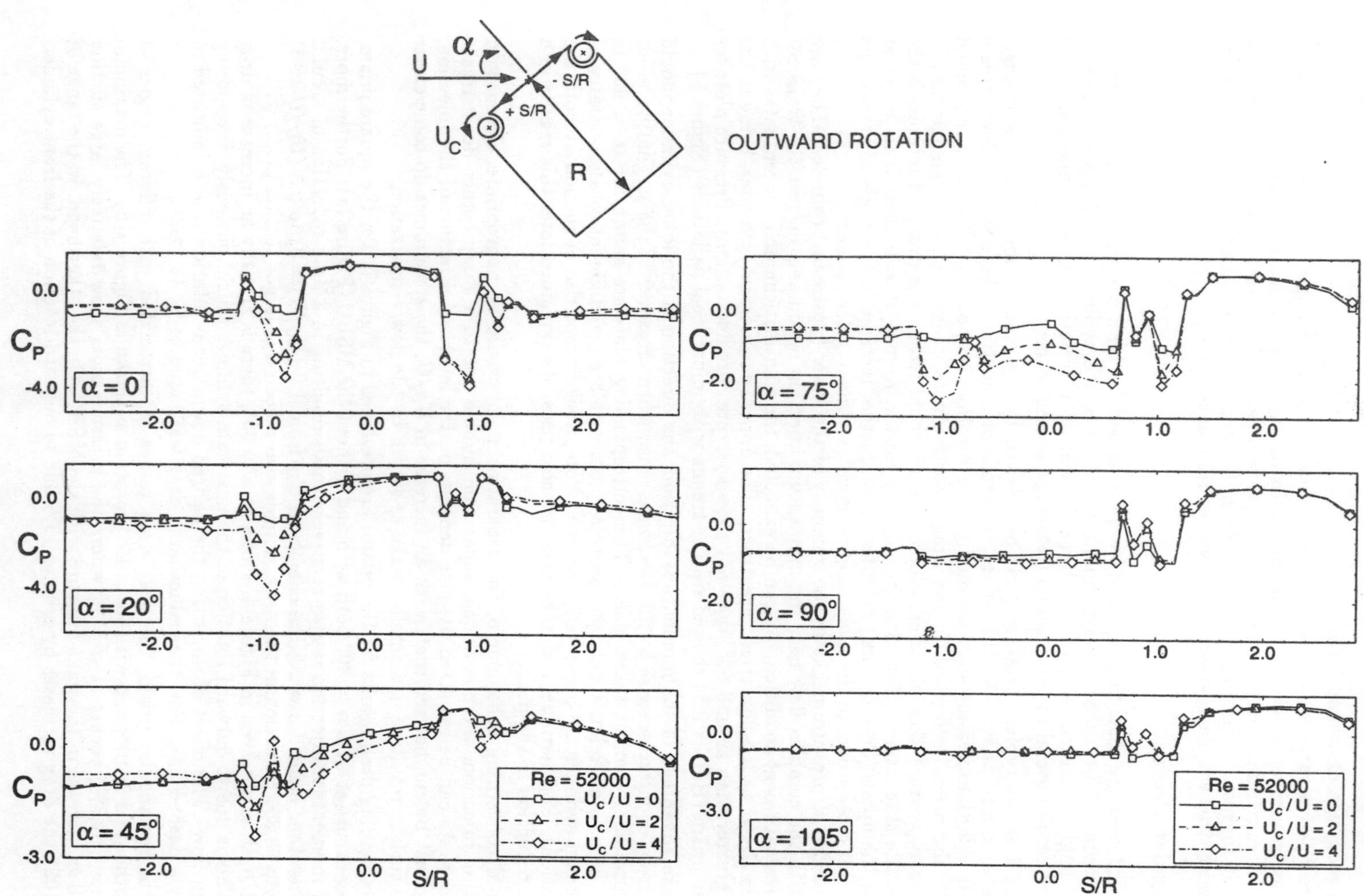

Figure 4 Pressure plots for a square prism showing effects of the angle of attack and momentum injection.

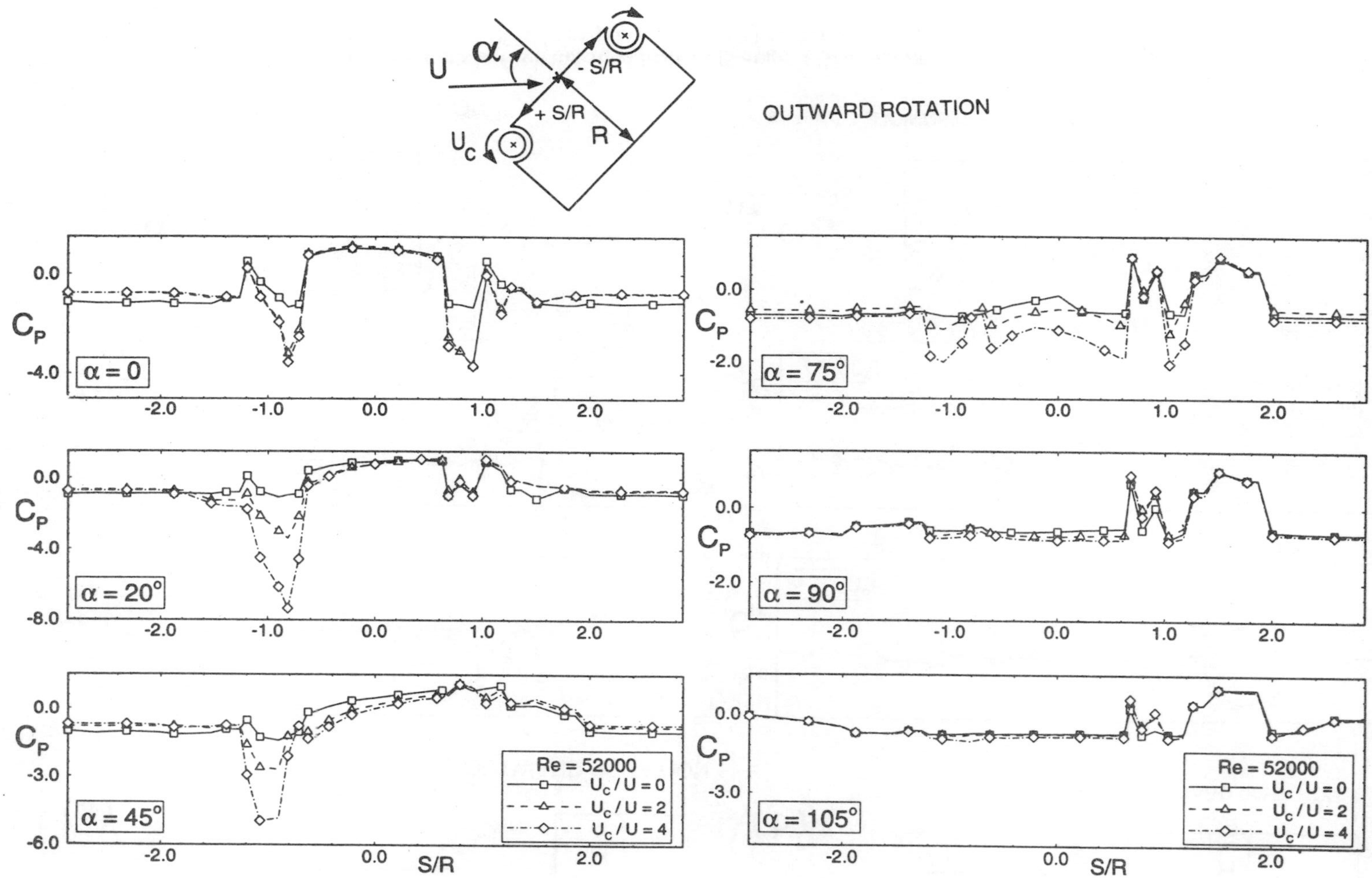

Figure 5 Effects of the angle of attack and MSBC on the surface pressure distribution for a rectangular prism (H/L = 0.5).

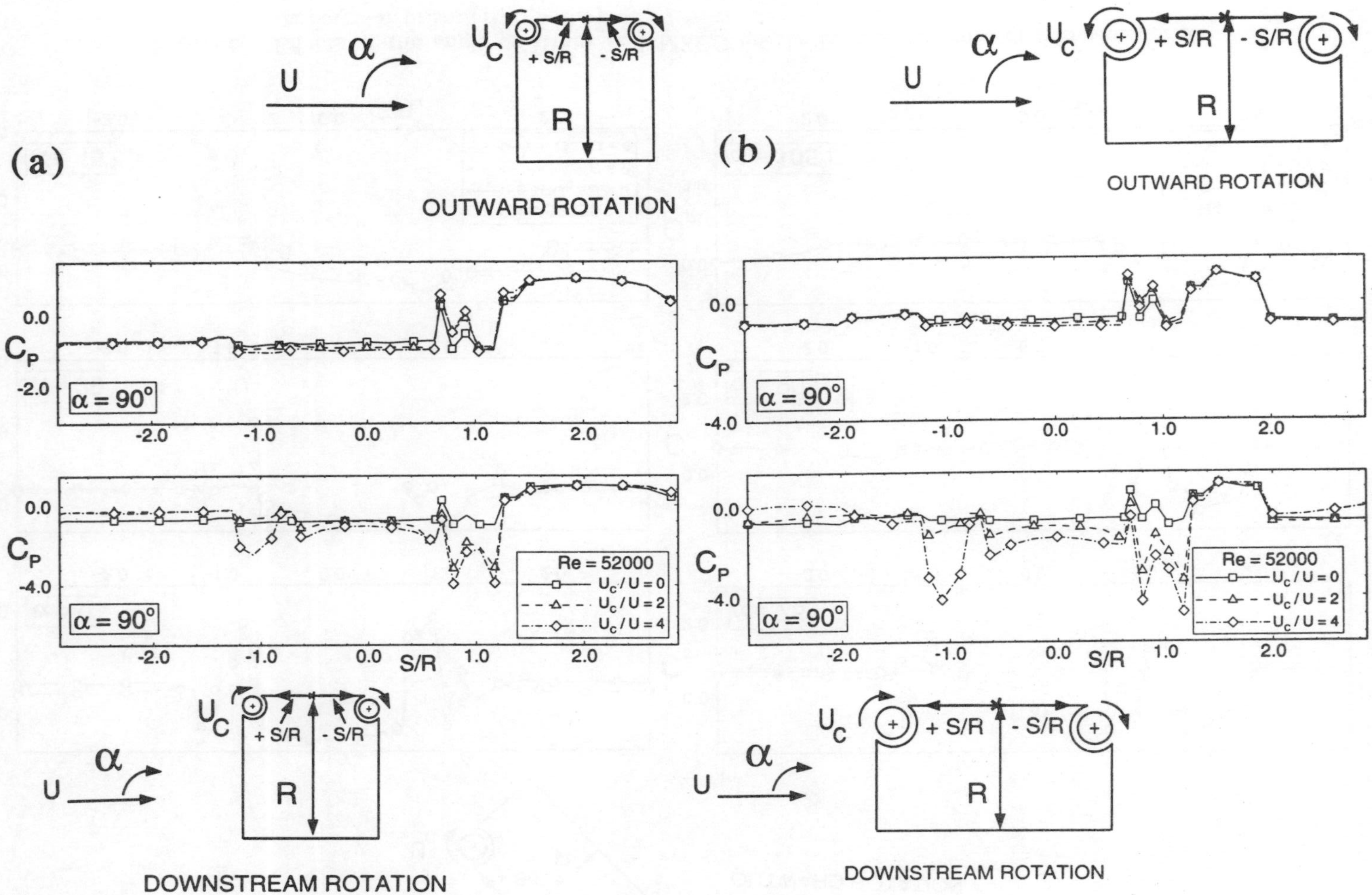

Figure 6 Pressure plots showing the effect of change in the direction of rotation of the upstream element, $\alpha = 90^o$: (a) square prism; (b) rectangular prism (H/L = 0.5).

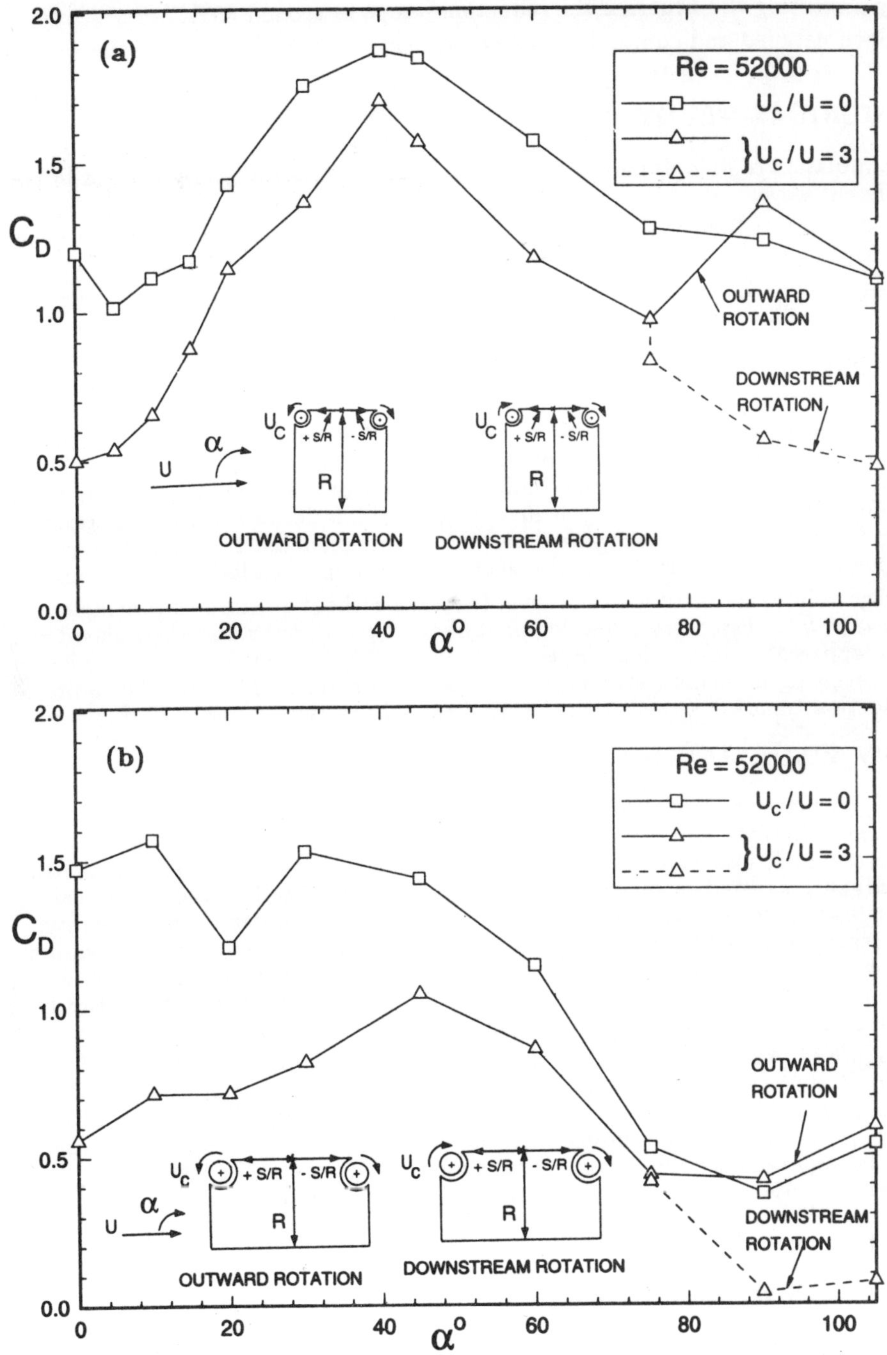

Figure 7 Influence of momentum injection on the drag: (a) square prism; (b) rectangular prism (H/L = 0.5).

Note, by correcting the situation the reduction jumps to around 85%.

Such detailed and comprehensive experiments with bluff bodies in the presence of MSBC are not reported in the literature.

3. NUMERICAL STUDY

Simulation of fluid dynamical problems has been classically approached in two fundamentally different ways:

(a) modelling of the physical character of a phenomenon, as approached by Prandtl, through insight into the physics of the problem;

(b) simulation of the governing equations of motion as accomplished through finite element or finite difference schemes. As can be expected, each has its advantages and limitations. Here the second procedure was used to study the complex problem of a multielement bluff bodies with momentum injection using the finite element integration of the Navier-Stokes equations.

The procedure employs stream function-vorticity form of the equations in conjunction with the variable grid-size, finite element discretization ($\approx$ 3000 nodes, Figure 8). Such a numerical solution of the two and three element bluff bodies with momentum injection has not been investigated before. The parametric analysis involving a systematic variation of the speed ratio, angle of attack and the Reynolds number gave detailed information about the separation condition and the time dependent wake (Figure 9). It also showed, rather spectacularly, effectiveness of the MSBC.

Figure 10(a) shows time evolution of the wake without any momentum injection for a square prism at $\alpha = 90^o$. Progressive development of the twin vortices and their shedding in the wake are quite apparent. Note, with U_C/U=4 (Figure 10b), the generation of vortices is essentially arrested resulting in smooth streamlines and a narrow wake. This matches quite well with the drag reduction trends observed with the wind tunnel test-data.

4. FLOW VISUALIZATION

To get better appreciation as to the physical character of the complex flow field as affected by the angle of attack and momentum injection parameters, extensive flow visualization study was undertaken. It also gave useful information about relative importance of the various parameters and hence assisted in planning of the experiments as well as the numerical analysis. The flow visualization tests were carried out in a closed circuit water channel facility. The models were constructed from Plexiglas and fitted with rotating cylinders driven by variable speed d.c. motors. A suspension of fine polyvinyl chloride powder was used in conjunction with slit lighting to visualize streaklines. Both angle of attack and cylinder speed were systematically changed and still photographs as well as video movies taken.

Figure 11 presents a typical set of flow visualization pictures, in this case for a square prism at $\alpha = 0$. It shows, rather dramatically, effectiveness of this form of boundary-layer control. In absence of any momentum injection the flow separates near the leading edge resulting in a long and wide wake with alternately shedding vortices. However, with progressive increase in the U_C/U, the wake becomes shorter and narrow. This clearly substantiates the trends indicated by the wind tunnel test–results and the numerical analysis. Effective reduction of the bluffness led to high frequency oscillation of the wake which can be better appreciated through the video, which forms a part of the presentation.

5. CONCLUDING REMARKS

Based on the wind tunnel tests, complemented by numerical and flow visualization studies, the following general comments can be made:

(i) The Moving Surface Boundary-layer Control (MSBC) promises to reduce drag of bluff

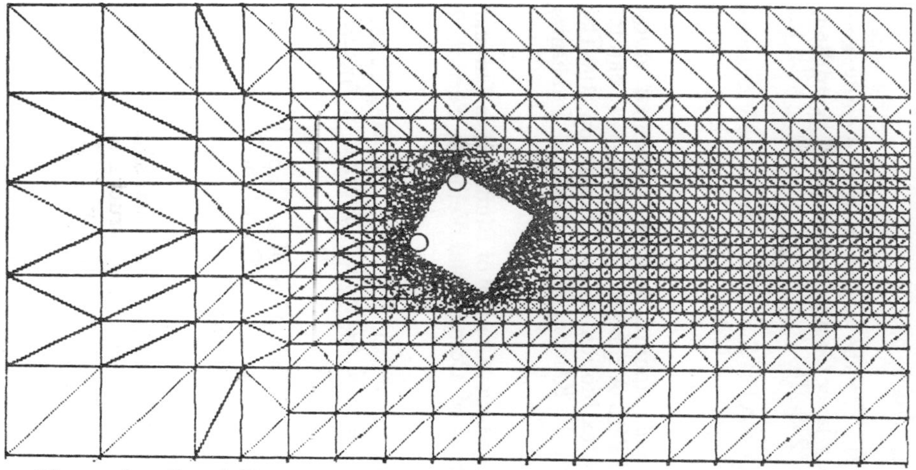

Figure 8 Spatially varying grid with triangular elements used in the numerical analysis of flat palates and rectangular prisms with the Mcving Surface Boundary-layer Control (MSBC).

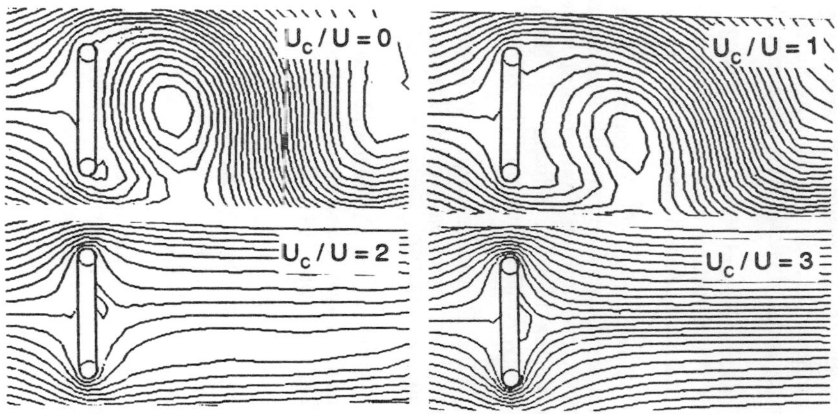

Figure 9 Finite element analysis of flow past a two-dimensional flat plate showing effectiveness of the MSBC. Note, for $U_C/U = 2$, the flow is beginning to approach the potential character.

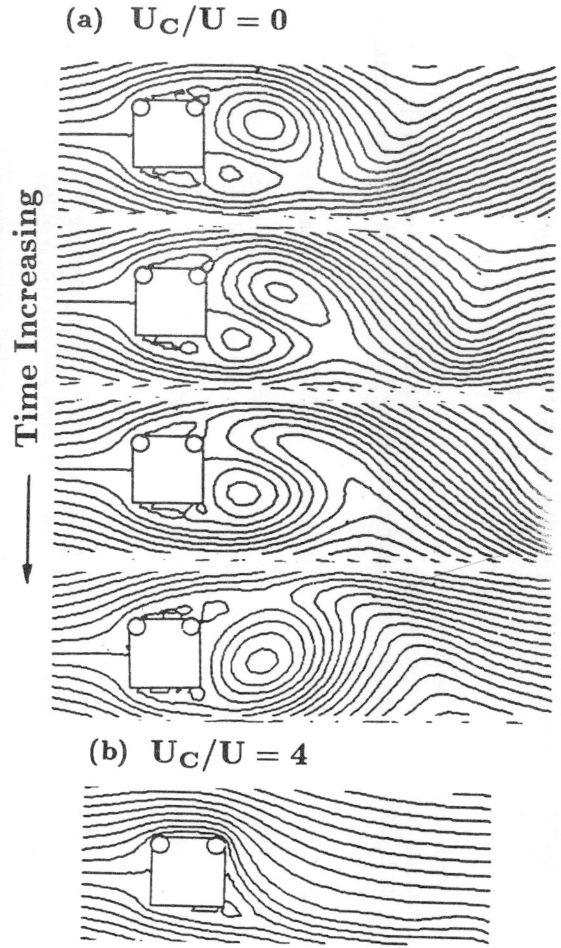

Figure 10 Numerically obtained flow patterns showing progressive evolution of wake for a square prism at $\alpha = 90^o$: (a) $U_C/U = 0$; (b) $U_C/U = 4$.

bodies by a significant amount. For a two-dimensional flat plate normal to the flow it was around 75%. In the case of two-dimensional prisms studied here, the peak reduction varied, approximately, between 60% and 85% depending on the direction of the cylinder rotation.

(ii) Numerical approach to the problem, even with a modest level of discretization, is able to predict character of the flow field with a remarkable accuracy. Considering highly complex character of the flow, this promises significant saving in time, effort and computational cost.

(iii) Flow visualization study confirms effectiveness of the MSBC quite dramatically.

(iv) The concept of the MSBC presents several avenues of promising applications which are under study. They include performance improvement of a delta wing and its control surfaces; drag reduction of bluff bodies such as tractor-trailer truck configurations; and suppression of vortex resonance and galloping type of wind induced instabilities.

REFERENCES

Cermak, J. E. (1975), "Application of Fluid Mechanics to Wind Engineering," Freeman Scholar Lecture, Journal of Fluids Engineering, Transactions of the ASME, Vol.97, No.1, pp. 9-38.

Every, M. J., King, R. and Weaver, D. S. (1982), "Vortex Excited Vibrations of Cylinders and Cables and Their Suppression," Ocean Engineering, Vol.9, pp. 135-157.

Kubo, Y., Modi, V. J., Yasuda, H. and Kato, K. (1992), "On the suppression of Aerodynamic Instabilities through the Moving Surface Boundary-layer Control," Journal of Wind Engineering and Industrial Aerodynamics, Vol.41, pp. 205-216.

Kubo, Y., Yasuda, H. and Kotsubo, C. (1993), "Active Control of Super-tall Structure Vibrations under Wind Action by a Boundary-layer Control Method," Journal of Wind Engineering and Industrial Aerodynamics, Vol.50, pp. 361-372.

Modi, V. J., Mokhtarian, F., Yokomizo, T., Ohta, G. and Oinuma, T. (1987), "Bound Vortex Boundary-Layer Control with Application to V/STOL Airplanes," Proceedings of the IUTAM Symposium on Fundamental Aspects of Vortex Motion, Tokyo, Japan, pp. 196-199; also Vortex Motion, Editors: H. Hashimoto and T. Kambe, North-Holland Publishing Company, Amsterdam, pp. 225-230.

Modi, V. J., Ying, B. and Yokomizo, T. (1990), "Boundary-Layer Control of Bluff Bodies Through Momentum Injection," SAE Transactions, Journal of Commercial Vehicles, Section 2, Vol.99, pp. 778-794.

Modi, V. J., Mokhtarian, F., Fernando, M. S. U. K. and Yokomizo, T. (1991), "Moving Surface Boundary-Layer Control as Applied to Two-dimensional Aerofoils," Journal of Aircraft, Vol.28, No.2, pp. 104-112.

Modi, V. J., St. Hill, S. and Yokomizo, T. (1993)., "Drag Reduction of a Cube-Type Truck Configuration Through Boundary-Layer Control: Experiments and Prototype Road Tests," 7th International Pacific Conference on Automotive Engineering, Phoenix, Arizona, U.S.A., Paper No. 931893.

Modi, V. J., Welt, F. and Seto, M. L. (1994), "Control of Wind Induced Instabilities through Application of Nutation Dampers: A Brief Review," Engineering Structures: Auxiliary Damping Systems for Vibration Mitigation, Special Issue Guest Editor: J. D. Holms, in press.

Welt, F. (1983), "A Parametric Study of Nutation Dampers," M.A.Sc. Thesis, University of British Columbia, Vancouver, Canada.

Welt, F. (1988), "A Study of Nutation Dampers with Application to Wind Induced Oscillations," Ph.D. Thesis, University of British Columbia, Vancouver, Canada.

Welt, F. and Modi, V. J (1992), "Vibration Damping Through Liquid Sloshing: Part 2 - Experimental Results," Transactions of ASME, Journal of Vibration and Acoustics, Vol.114, No.1, pp. 17-23

Welt, F. and Modi, V. J (1992), "Vibration Damping Through Liquid Sloshing: Part 2 - A Nonlinear Analysis," Transactions of ASME, Journal of Vibration and Acoustics, Vol.114, No.1, pp. 10-16.

Zdravkovich, M. M. (1981), "Review and Classification of Various Aerodynamic and Hydrodynamic Means for Suppressing Vortex Shedding ," Journal of Wind Engineering and Industrial Aerodynamics, Vol. 7, pp. 145-189.

Zdravkovich, M. M. (1984), "Reduction of Effectiveness of Means for Suppressing Wind-Induced Oscillaions," Engineering Structures, Vol. 6, pp. 344-350.

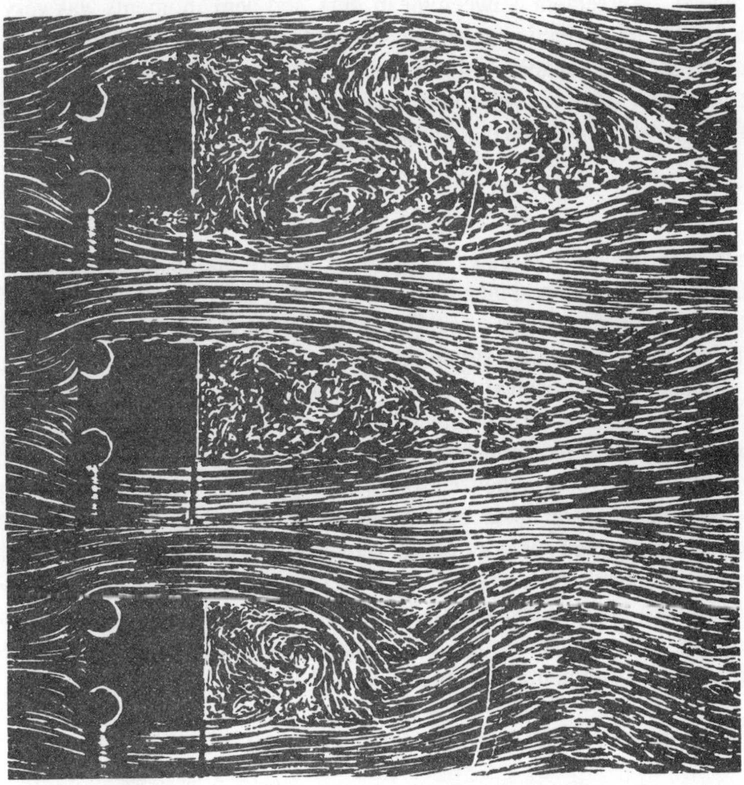

Figure 11 A typical series of flow visualization pictures showing effectiveness of the MSBC as applied to a two-dimensional square prism. Reduction in the wake at $U_C/U = 4$ is rather spectacular, suggesting a significant reduction in the drag. The wind tunnel results for the case showed a reduction in the drag of around 62%.

9ICWE, 1995, New Delhi, India

FORCE AND PRESSURE MEASUREMENTS ON A ROUGH CONE IN TURBULENT SHEAR FLOW

FLAY, R.G.J.[1] and VICKERY, B.J.[2]

[1] *Senior Lecturer, Mechanical Engineering Department, University of Auckland, Auckland, New Zealand*
[2] *Professor, Boundary Layer Wind Tunnel Laboratory, University of Western Ontario, London, Canada*

Abstract

A detailed wind tunnel test has been carried out on a rough cone using a 256-channel electronic pressure transducer system. The results from a reduced data set consisting of weighted integrations of pressure signals at each level to form time series of lift and drag at each level are compared with results from two balance tests. Agreement between mean, rms. and peak moments was very good. Spectra of the pressure data for the drag direction showed good agreement with the wind spectra at each level, and with the von Karman spectral equation. Motion induced forces were examined by oscillating the model sinusoidally. Aerodynamic damping was found to be negative only when the forcing frequency was close to the shedding frequency, and only then at the lower sections of the model.

1. INTRODUCTION

The results described in this paper are part of a very complete programme of measurements aimed at understanding the nature and magnitude of the mean and dynamic loads on a 1:500 scale model of a proposed tower / building. The model employed was a fairly accurate representation of the prototype. It was of conical form with structural elements spiralling around the outside. The top piece was perforated. Electronically scanned pressure taps were used to get surface pressure information. The model was mounted on a special oscillator plate which could oscillate the model about its base sinusoidally at prescribed amplitude and frequency. This paper discusses the pressure results with the model stationary, and also the motion induced forces arising when the model was oscillated.

Force measurements were made in separate tests using two balances, a stiff relatively high frequency balance which enabled spectra of shear and overturning moments to be measured, and a low frequency aero-elastic balance which enabled the model response to be determined. Moment and shear coefficients are compared in the paper from all three sources, and good agreement is shown. Further details concerning the tests and more data are contained in Flay and Vickery (1993a). A detailed analysis of the pressure measurements is given in Flay and Vickery (1993b).

2. MODEL, INSTRUMENTATION, AND DATA RECORDING DETAILS

The model was constructed at a linear scale of 1:500, and can be seen in Fig. 1. It is of conical shape, 1600 mm high, on a 258 mm base The lowest 1200 mm was roughened with plastic pieces about 4 mm square glued to the outside in a spiralling pattern to simulate structural members in the prototype,

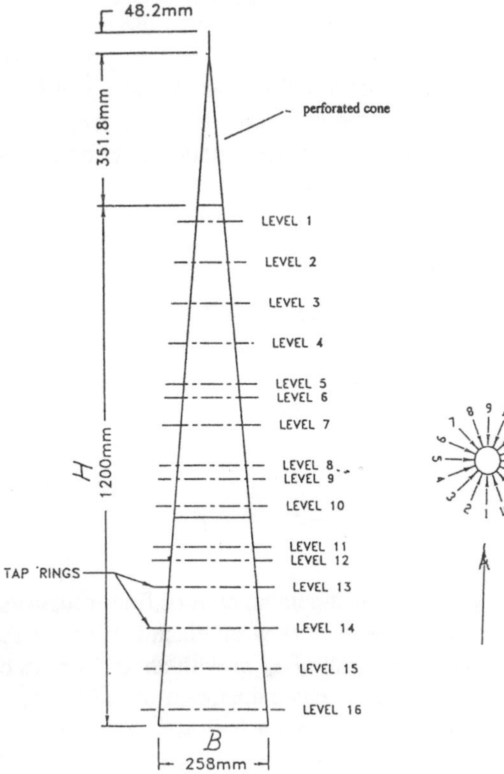

Fig. 1 Principal dimensions of model and locations of pressure taps.

and introduced considerable roughness to the surface - at a height of 1200 mm the relative roughness k/D is 7%, reducing to 1.5% at the base. The external roughness elements were not continued above a height of 1200 mm in the model tested. The apex of the cone was made of perforated acrylic.

The pressure data were obtained using 256 electronically scanned pressure channels arranged with 16 pressure taps equi-spaced at 22.5^0 intervals at each of 16 levels as shown in Fig. 1. At each level the taps were plumbed in a clockwise direction with tap number 1 facing directly upstream. Each transducer module also consists of 16 channels, which meant that one module was used per level. The pressure tubing was made up of a 610 mm length of standard 1.35 mm bore plastic tubing with a 50 mm length of smaller 0.86 mm bore tubing inserted into one end of the larger tubing a depth of about 6 mm and used to connect to the small transducer tubes. The other end was pushed through holes drilled in the skin of the model, glued, and then cut off flush with the model surface. No restrictors were used in the tubing, and calibration showed that the frequency response of the tubing was good up to 50 Hz.

Most of the pressure results discussed herein were recorded with the model stationary with a wind speed of about 12 m/s recorded by a pitot-static probe at a height of 1486 mm above the wind tunnel floor. The sampling frequency was 200 Hz and the sampling period was 120 seconds.

Normal pre-processing was carried out to minimise the effects of zero and gain drifts in the instrumentation, and then files of pressure coefficients

$$Cp = \frac{P - P_\infty}{q_\infty} \tag{1}$$

were formed where P_∞ and q_∞ are the static and dynamic pressures respectively, as recorded by the reference pitot-static probe, and P is a pressure reading. For the most part, data at each level were combined together to give time histories of lift and drag, in order to reduce the bulk of the data, using the following 2 equations

$$C_D = \frac{\pi}{16}\Sigma_{i=1}^{16} Cp_i \cos(22.5 * (i-1))$$
$$C_L = \frac{\pi}{16}\Sigma_{i=1}^{16} Cp_i \sin(22.5 * (i-1)) \tag{2}$$

where i is tap number.

On the stiff strain gauged balance, measurements were made of the mean, rms. and the spectra of both the base shear and base moment in the along-wind (drag) and across-wind (lift) directions. The balance is of simple design, consisting of 4 strain-gauged posts between top and bottom plates. The bottom plate was rigidly attached to the wind tunnel turntable, and the top plate was attached to the base of the model. The same model was used as in the pressure tests, but the pressure tubes were removed.

Two degree of freedom "straight stick" aero-elastic measurements were made by mounting the same model on a strain gauged 50 mm diameter Al alloy tube. The tube was 800 mm long. The top end was rigidly attached centrally in the model, and the bottom end was attached to a "basket" which was in turn attached to the turntable. A smaller tube with a block of foam at its lower end ran through the balance tube, and was immersed in a container of oil to provide the appropriate level of damping. The minimum achievable damping with no oil was 0.6%. The design was tuned to have a natural frequency of 9.5 Hz, and response measurements in the lift and drag directions were made for relative damping levels of nominally 0.5, 1.0 and 2.0 %. Data were acquired for each damping case through the speed range 4.88 m/s to 18.3 m/s at 0.61 m/s increments. At each speed, four blocks of 40 seconds of data were recorded at 80 Hz, with low pass filters set at 40 Hz.

3. APPROACH FLOW

The shear flow employed in the model study was generated by a combination of floor roughness and spires. For the bulk of the test programme the aim was to produce flow similar to that over the ocean. To achieve this, the floor was roughened uniformly over a length of 32 m with about 78 rows of 13 mm high blocks. Three 1.5m high spires spaced 1.0m apart were located upstream of the blocks at the test section entrance. For most tests, data were recorded for two approach flows. However, the paper deals only with data recorded in the rougher "Exposure 2" flow. Results from hot-wire and pitot-static traverses at the model position are given in Figs. 2 and 3 and are compared with full-scale measurements from Shiotani, (1975) assuming a mean speed of 80 m/s at a height of 30m. The mean wind speed profile is well described by a power-law profile:

$$\frac{U}{U_{ref}} = \left(\frac{Z}{Z_{ref}}\right)^{\alpha} \qquad (3)$$

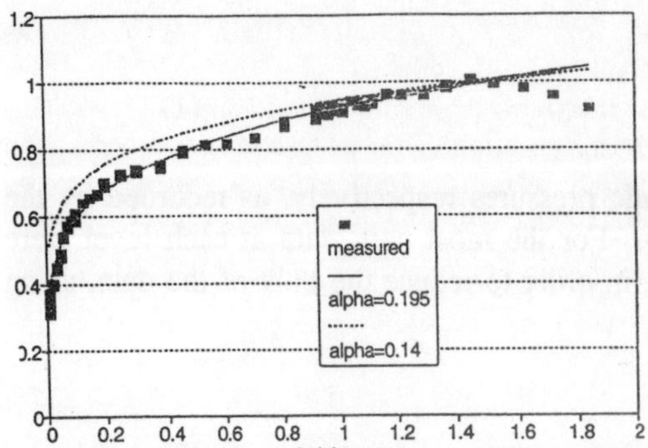

Fig. 2 Normalised velocity profiles, linear axes.

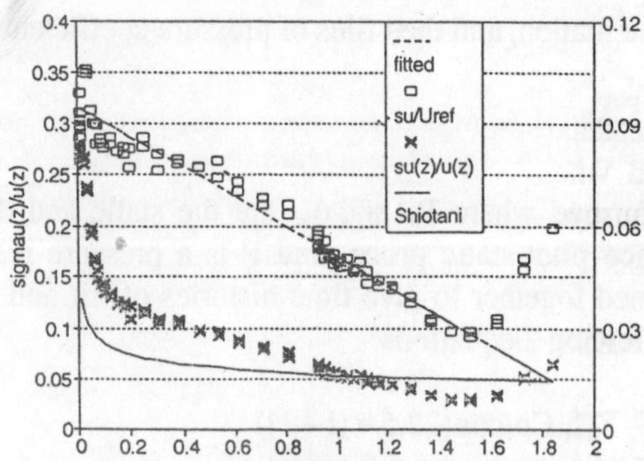

Fig. 3 Turbulence variation with height.

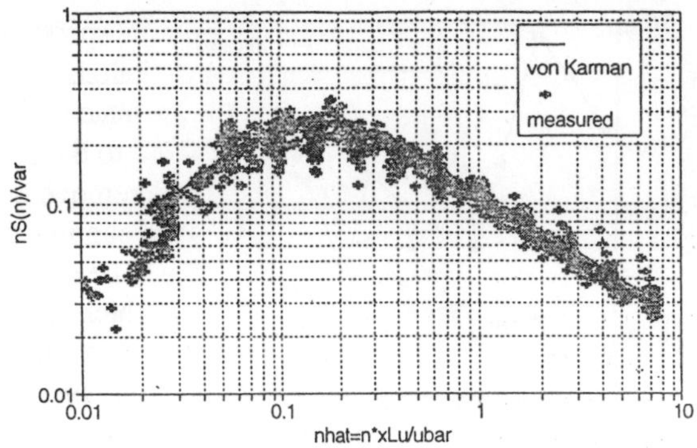

Fig. 4 84 measured spectra compared to von Karman spectrum with ${}^xL_u$ defined by equations (5).

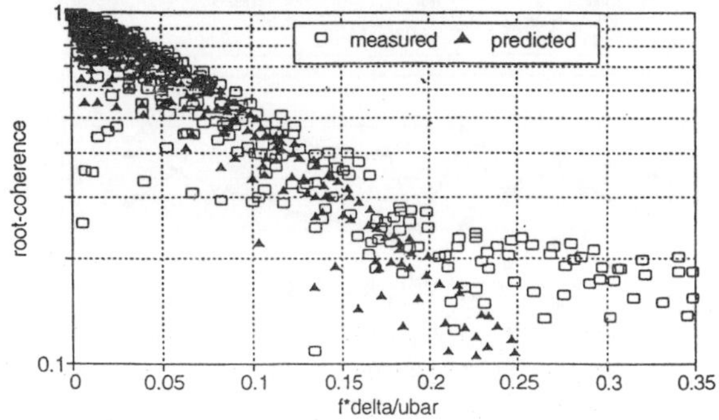

Fig. 5 Comparison of measured and predicted root coherence (all data) versus f*Δ/ubar semi-log axes.

with α = 0.195, and Z_{ref} =1.47m, for heights below 1.5 m. The normalised standard deviation of the wind tunnel velocity data show a roughly linear decrease with height up to 1.5m described by

$$\frac{\sigma_u}{U_{ref}} = 0.095 - 0.044 * z \qquad (4)$$

Eighty-four wind auto-spectra measurements at the model position were obtained. By using an optimising routine to fit von Karman equations to the measured spectra at all levels (Fig. 4), it was found that the integral length scale was given approximately by

$${}^xL_u = 0.95 * z^{0.25}; z < 0.5\text{m}, {}^xL_u = 0.81 - 0.39(z-0.5)^2; \geq 0.5\text{m}. \qquad (5)$$

Simultaneous hot-wire measurements for different heights and separations enabled the root-coherence to be evaluated for 42 different combinations. These results were used in an optimising procedure to produce the best fit of the equation

$$\gamma(f) = \exp\left[-\left\{\left(a\frac{\Delta}{{}^xL_u}\right)^2 + \left(\frac{f\Delta}{\bar{U}}\right)^2\left(b + c\frac{\Delta}{\bar{z}}\right)\right\}^m\right] \qquad (6)$$

to the data by adjusting the coefficients a, b, c and m appropriately. Relatively good agreement was found when a=1.052, b=7.099, c=3.204, m=0.653, as can be seen in Fig. 5. Furthermore, it was demonstrated that, as expected, this equation was also a good fit to the normalised co-spectrum of the wind.

4. PRESSURE RESULTS FOR THE DRAG DIRECTION

This section considers the integrated pressure data at each level resolved into the drag direction. Fig. 6 shows that the resulting drag coefficients at each level show an increase with height, and that the maxima and minima are about 3.5 standard deviations away from the means (the level 10 results are in

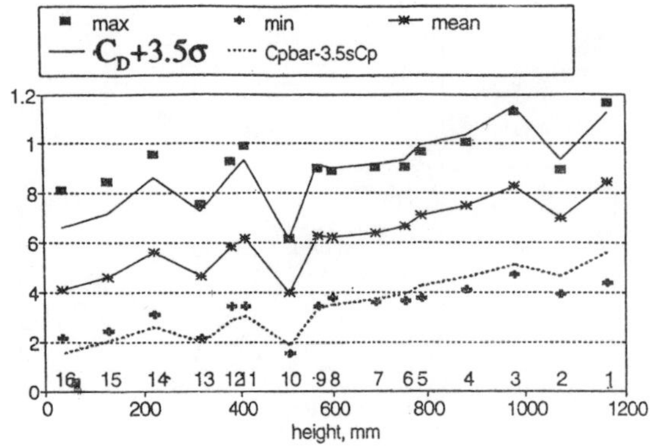

Fig. 6 Maximum, minumum, mean, and mean +/- 3.5σ for local drag coefficients along model.

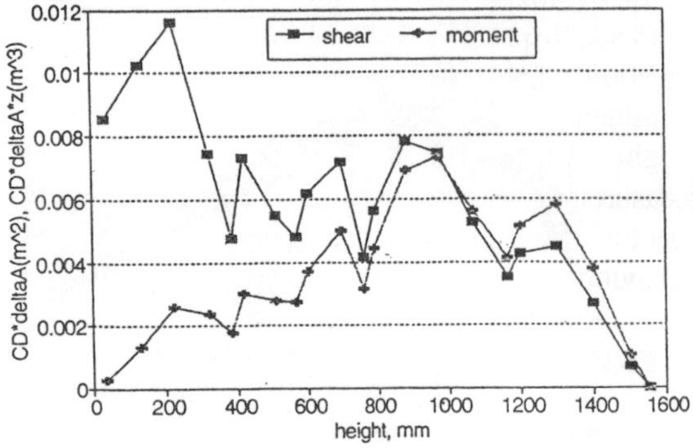

Fig. 7 Variation with height of contribution to total drag shear and moment.

error). A nominal Reynolds number for these tests is $Re_D = 10^5$. For a smooth circular cylinder in smooth flow, a drag coefficient $C_D = 1.2$ would be expected for this value of Re_D.

Turbulent onset flow reduces the Reynolds number dependency of the drag coefficient, and increasing surface roughness causes emulation of higher Reynolds number behaviour. In the present case where the roughness varies from 1.5% at the base to 7% at a height of 1200 mm, we might expect the drag coefficient to be around C_D =1.0+/-0.1. Clearly the values in Fig. 6 are a lot less than this, for which there are two main reasons. Firstly, the model is conical in form, not cylindrical, so the flow will be more three-dimensional than for flow around a cylinder, and secondly the cylinder is very rough and is embedded in turbulent flow.

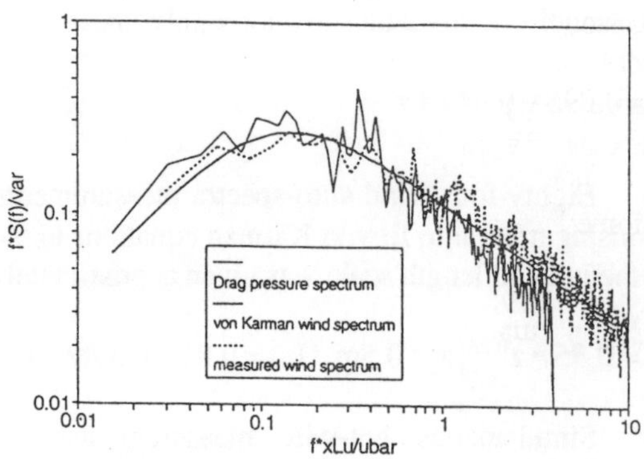

Fig. 8 Pressure and wind spectra for level 8, z=0.6 m, ${}^xL_u$=0.81 m.

The means can be used to determine the contribution to the total drag shear and moment by using appropriate areas and heights. This is shown in Fig. 7. The bottom two-thirds is the most important for shear and the top two-thirds for moment. Integration of these two curves gives values for base shear and moment coefficients (ignoring the contribution from the exterior roughness elements) which have been worked out to be C_S=0.39, and C_M=0.20, using a reference area of 0.258*1.2m^2 and a reference length of 1.2m (see Fig. 1). It is worth noting that the reference area is twice the actual area, and this has the effect of artificially making the coefficients appear smaller than for a cylinder. These values are discussed later in regard to the balance results.

The variance is made up of fluctuations over a range of frequencies, in this case from 0.2 to 50 Hz, and an example spectrum for level 8 which is about half-way up the model is shown in Fig. 8. It can be seen that there is fairly good agreement between the von Karman and drag pressure spectrum, although the pressure spectrum reduces with increasing frequency more quickly that the von Karman equation. This effect was more noticeable at lower elevations and must be due to the aerodynamic admittance falling below unity. The figure also shows a wind spectrum measured at almost the same height. It shows better agreement with the von Karman equation than the drag spectrum. Spectra measurements from individual taps on the front of the model showed good agreement with the wind spectra. Spectra from individual taps on the rear of the model showed increasingly organised wake behaviour at lower heights evidenced by a substantial peak in the spectra.

5. PRESSURE RESULTS FOR THE LIFT DIRECTION

This section considers the integrated pressure data at each level resolved into the lift direction. Spectra were obtained as for the drag results, and example spectra for levels 3 and 14 are shown in Figs. 9a and 9b. The appearance of a peak in the rear drag spectra at lower heights was mimicked in the lift data where the spectra were broad for the upper levels but peaked at low levels. It can be seen that the spectrum for level 3 at z=0.98 m is moderately peaked whereas for level 14 at z=0.22 m it is very peaked. This result was unexpected and indicates that shedding from the model becomes more organised at lower levels, and is contrary to the behaviour observed by Vickery and Clark (1972). This effect is probably the result of the rather rough surface. Near the top the high value of k/D=7 %, almost as large as the 10 % figure for the Scruton spirals on chimneys, would virtually destroy any organised vortex shedding.

It was found that the lift spectra could be adequately fitted by the forced damped harmonic motion equation

$$\frac{fS(f)}{\sigma^2} = k\frac{\frac{f}{f_s}\frac{4\zeta}{\pi}}{\left(1-\left(\frac{f}{f_s}\right)^2\right)^2+4\zeta^2\left(\frac{f}{f_s}\right)^2}. \tag{7}$$

k is a measure of the energy in the peak, and was usually very close to unity; ζ is the damping ratio, and f_s is frequency. Numerical values of these variables were found using an optimising procedure. f_s can be

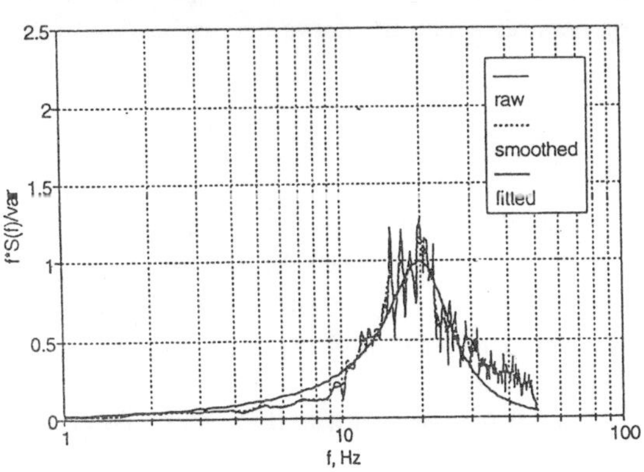

Fig. 9a Measured and fitted (FDHM) lift spectra for level 3.

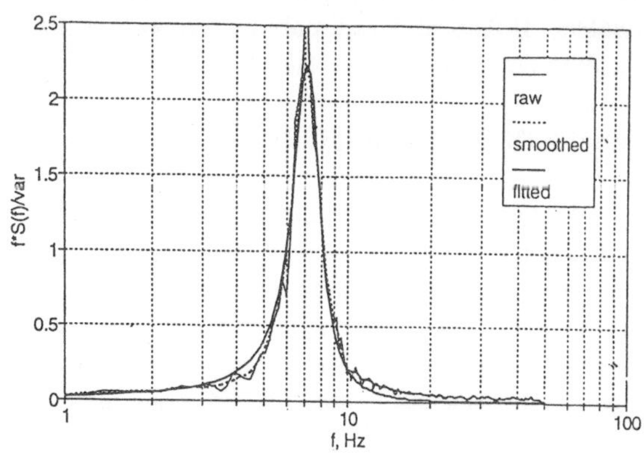

Fig. 9b Measured and fitted (FDHM) lift spectra for level 14.

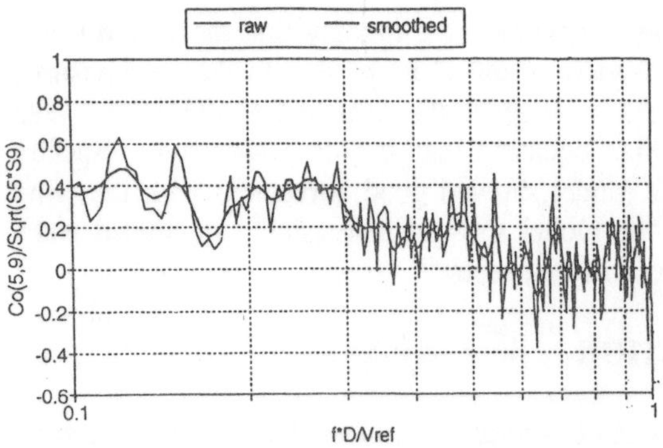

Fig. 10 Normalised co-spectrum for lift, levels 5 and 9, Z_5=0.788 m, Z_9=0.569 m.

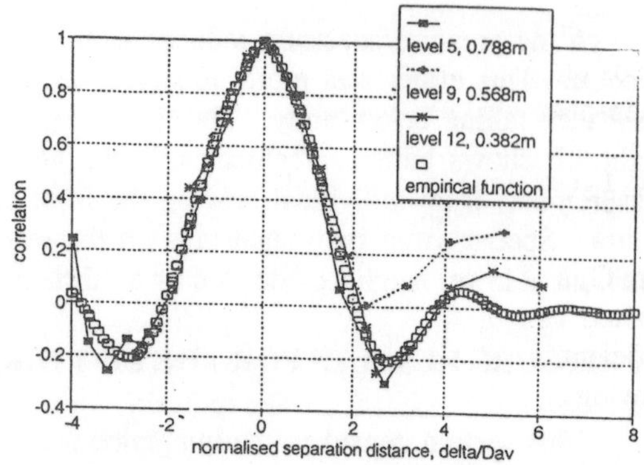

Fig. 11 Correlation of normalised co-spectra, evaluated at the shedding frequency of the reference level, versus Δ/D_{av}.

interpreted as the shedding frequency, while the half power spectral width at f_S is very approximately equal to 2ζ. The variation with height of these parameters, including the Strouhal number were determined, and are given in Flay and Vickery (1993a, 1993b).

One of the objectives of this research was to determine the response of the model by modal analysis from the spectra and co-spectra of the wind forces, and so the results have been fitted by empirical curves to enable the required integrations to be carried out. Prediction of the lift response requires the normalised co-spectra to be determined. These were determined from the lift data using 42 pairs of levels, and a typical example of a co-spectrum is given for levels 5 and 9 in Fig. 10. Normalised co-spectra results enable the correlation to be evaluated for any desired frequency. In this work levels 5, 9, and 12 have been used as references, and from lift spectra, their shedding frequencies have been estimated to be 13.33, 9.45, and 7.75 Hz respectively. These frequencies were used to determine the narrow-band cross-correlation coefficients as a function of separation distance up and down the cone from the reference levels. The results were found to collapse extremely well when plotted as a function of the ratio of separation distance to average local cone diameter as shown in Fig. 11. The correlation falls to zero at twice the average cone diameter, falling further to about -0.2 at a separation of 3 diameters. It then rises again above zero at four diameters, beyond which the curve becomes somewhat difficult to define with these data.

6. MOTION INDUCED FORCES

Motion induced lift forces were measured at sixteen levels by integrating the sixteen pressure signals at each level appropriately during forced motions of the model at 8, 9, 10, 11,12, 14, 16, 18, 20 Hz. In all cases the amplitude of the forced motion at height z was (for rigid body rotation about the base) z/1000. The rms. displacement at 1200 mm was 0.85 mm which corresponds to 0.42 m at prototype scale. While the chosen amplitude was reasonable in terms of anticipated prototype behaviour, it was small in absolute terms and the motion-induced forces were difficult to extract from the force records.

The model was forced to oscillate in the cross-wind direction sinusoidally at frequency ω_0, i.e. $y=y_0\cos\omega_0 t$, and it was then supposed that the resulting forces per unit length F(t) at each height would have in-phase and out-of-phase components described by

$$F(t) = \tfrac{1}{2}\rho V^2 D(C_R\cos\omega_0 t - C_I\sin\omega_0 t) \tag{8}$$

where C_R and C_I were coefficients to be extracted from the time history records. The C_R coefficient was extracted from the data by using a signal from an accelerometer located in the model, and the C_I coefficient by integrating the acceleration signal to produce a velocity signal, and then using the following equations

$$C_R = \tfrac{2}{T}\int_0^T F(t)a(t)dt,\ C_I = \tfrac{2}{T}\int_0^T F(t)v(t)dt \tag{9}$$

where a(t), and v(t) are unit amplitude acceleration and velocity signals. C_R and C_I can be used to obtain other coefficients of interest. The force per unit length variation at the forced oscillation frequency can be represented by added mass, m_a, and aerodynamic damping, c_a coefficients acting in-phase with the acceleration and velocity respectively, i.e.

$$F(t) = -m_a\ddot{y} - c_a\dot{y} \tag{10}$$

and since nominally $y=y_0\cos\omega_0 t$, we get

$$F(t) = m_a y_0\omega_0^2\cos\omega_0 t + c_a y_0\omega_0\sin\omega_0 t \tag{11}$$

Comparing equations (8) and (11), it is evident that

$$m_a y_0\omega_0^2 = \tfrac{1}{2}\rho V^2 DC_R \text{ and } c_a y_0\omega_0 = -\tfrac{1}{2}\rho V^2 DC_I \tag{12}$$

We now introduce the added-mass coefficient C_m, where

$$C_m = \frac{\text{added mass/unit length}}{\text{mass of displaced fluid/unit length}} = \frac{m_a}{\rho\frac{\pi}{4}D^2} \tag{13}$$

and the aerodynamic damping parameter K_a, where

$$K_a = \frac{\text{mass of structure/ unit lengtb}}{\text{mass of fluid in square of side D/unit length}} * \text{damping ratio} = \frac{m}{\rho D^2}\zeta \tag{14}$$

and since $\zeta = \frac{c_a}{2m\omega_0}$ (note that it is legitimate to define ζ in this way because the forcing frequency ω_0 is simulating the natural frequency of the structure),

$$K_a = \frac{m}{\rho D^2}\frac{c_a}{2\rho D^2\omega_0} = \frac{c_a}{2\rho D^2\omega_0}. \tag{15}$$

Similarly $C_m = \frac{2}{\pi}\frac{V^2}{\omega_0^2 D y_0} C_R$

(16)

If we now let $y_0=\alpha D$, where α is a constant, and since in this experiment $y_0=z/1000$, it follows that $\alpha=z/1000D$. Substitution for y_0 and α in (15) and (16) yields the following pair of equations which have been used to show the results.

$$C_m = \frac{2000}{\pi}\frac{D(z)}{z}\left(\frac{V(h)}{\omega_0 D(z)}\right)^2 C_R \tag{17}$$

$$K_a = 250\frac{D(z)}{z}\left(\frac{V(h)}{\omega_0 D(z)}\right)^2 C_I \tag{18}$$

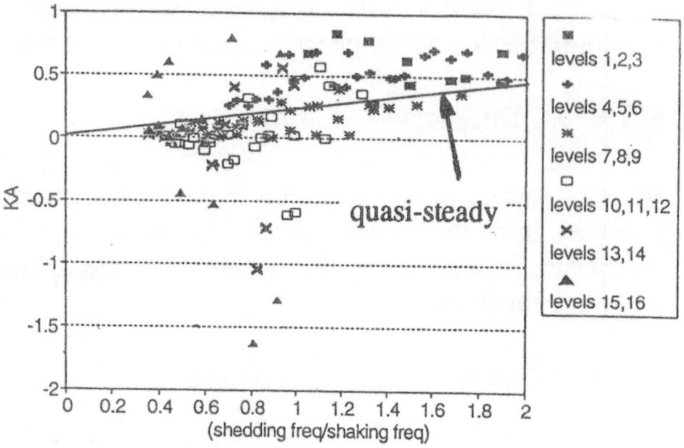

Fig. 12 K_a versus f_s/f_0 for all levels 1 - 16.

Fig. 12 is a composite plot of K_a as a function of f_s/f_0 i.e. shedding frequency at height in question divided by forcing frequency, for all levels. Because of the low signal levels there is considerable scatter, but there are a number of significant conclusions that can be drawn. There is no indication of negative damping at level 9 or above. This is attributed to the high relative roughness levels (above 2%). For roughness levels below 2 % (levels 10 to 16) the aerodynamic damping is negative in the vicinity of $f_s/f_0=1$ and the minimum values for these levels range from about -0.6 to -1.6 which is broadly consistent with other observations on turbulent cylinders in turbulent shear flow. The observation that the damping is negative only near $f_0=f_s$, and only in the lower sections of the model suggests that the overall damping is likely to be positive or weakly negative for all speeds.

Quasi-steady assumptions lead to the following expression for aerodynamic damping coefficient due to cross-wind motion

$$c_a = \frac{1}{2}\rho V D\left(\frac{\partial C_L}{\partial \alpha} + C_D\right). \tag{19}$$

Since the planform is circular, the lift term in (19) disappears, and introducing the Strouhal number, S, and using (15) we can write

$$K_a = \frac{C_D}{8\pi S}\left(\frac{f_s}{f_0}\right) \tag{20}$$

Equation (20) is shown in Fig. 12 using nominal values $C_D=1.0$, and $S=0.20$. This relationship would be expected to be valid at large values of $\frac{f_s}{f_0}$, i.e. for $\frac{V}{f_0 D}$ greater than 5 say. It can be seen that this expression seems to correctly follow the trend in the data of K_a increasing positively as $\frac{f_s}{f_0}$ increases.

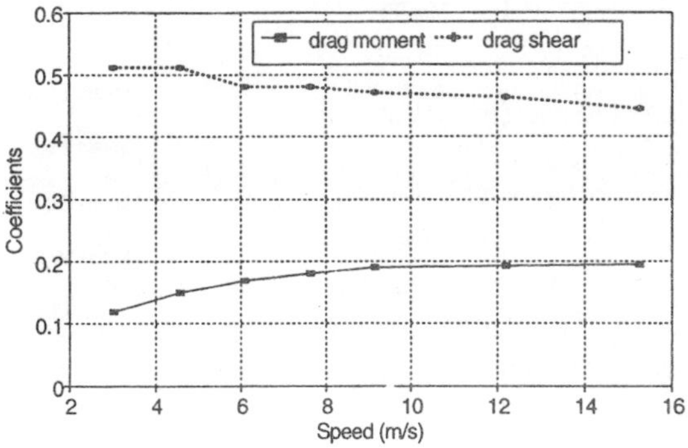

Fig. 13 Mean base shears and moments for the drag direction.

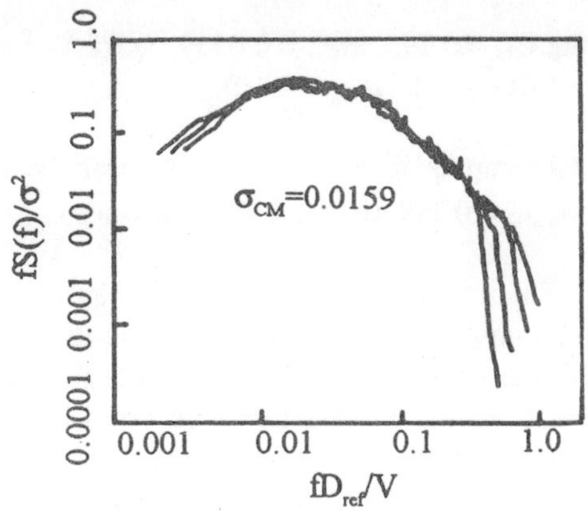

Fig. 14a Spectra of drag base moment.

7. BASE BALANCE RESULTS

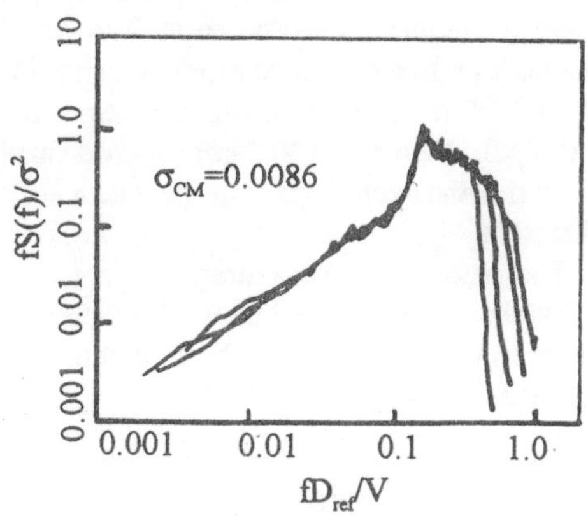

Fig. 14b Spectra of lift base moment.

Plots of mean base balance shears and moments are given in Fig. 13 and show some Reynolds number dependency up to a test speed of around 10 m/s. With roughness as large as on this model (1.5 - 7 % of diameter), one would expect the loads to settle to a steady value at Re≈10^5 , which corresponds to a test speed of 12 m/s based on the mean model diameter. It should be noted that the coefficients for these mean coefficients are in excellent agreement with those in Section 4 calculated from pressure integrations. Plots of the base moment drag and lift spectra are given in Figs. 14a and 14b. The overlaid plots are for the speeds of 7.6, 9.1, 12.2, and 15.2 m/s and are in good agreement, which supports the earlier conclusion that Reynolds number dependency is weak.

The peak base moment coefficient for a dynamic structure can be estimated from the data presented and the following approximate relationships

$$\hat{C}_M = \bar{C}_M + g\sigma_{C_M}\left[1 + \frac{\pi}{4\zeta}\frac{f_0 S(f_0)}{\sigma^2}\right]^{\frac{1}{2}} \qquad (21)$$

where g is the peak factor with an approximate value of 3.5, σ_{CM}=rms. base moment coefficient, and $S(f_0)$ is the moment spectrum in Figs. 14a and 14b. For example, if the reference speed was 50 m/s and the natural frequency f_0 was 0.08 Hz, the value of f_0D_{ref}/V is 0.08*129/50=0.206. The relevant quantities from Figs. 14a and 14b for the computation of $\hat{C}_M$ are then:

for Drag $\bar{C}_M = 0.195$, $\sigma_{C_M} = 0.0159$, $f_0S(f_0)/\sigma^2 = 0.05$,
and for Lift $\bar{C}_M = 0$, $\sigma_{C_M} = 0.0086$, $f_0S(f_0)/\sigma^2 = 0.6$.

For total damping of 1% of critical, the peak dynamic moment coefficients are:
Drag: $\hat{C}_{MD} = 0.319$ and for Lift: $\hat{C}_{ML} = 0.21$.

The resonant component of the base moment will be related to the accelerations. For drag the resonant component is $C_{MDR} = 0.11$, and for lift it is $C_{MLR} = 0.205$. These simple calculations indicate that, for the values of f_0, V, and z chosen in the example, the total base moment due to drag exceeds that due to lift but the situation is reversed for the resonant component. The strength requirements are therefore determined by drag loading, but the accelerations are primarily due to across-wind loading. These coefficients are compared to results obtained from the aero-elastic tests in the next section.

8. AERO-ELASTIC BALANCE RESULTS

Moment coefficients for drag and lift, using a reference area of D*H (see Fig. 1) were measured for a range of reduced velocities from 2 to 7.5. Figs. 15a and 15b show the base drag and lift moment coefficients as a function of reduced velocity, for damping nominally 1 % of critical. Note f_0 =9.5 Hz, and D_{ref}=0.258 m. There is a slight increase in the peaks with increasing speed, and a large peak drag result at V_r=3. Data at 0.5 % damping confirmed that something is occurring at this speed. The figures also show that the excursions from the mean are about twice as large in the lift direction compared to the drag direction.

The results give peak drag moment coefficients of about 0.3 for virtually the whole range of reduced velocities tested, and peak lift moment coefficients of about 0.2. There is thus amazingly good agreement in the predictions of the moments from the two balance sources - in fact the two balance tests were carried out by different people.

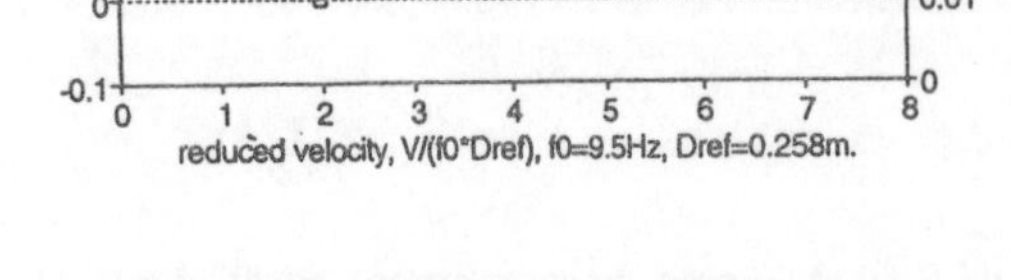

Fig. 15a Aero-elastic balance base moment coefficients for drag.

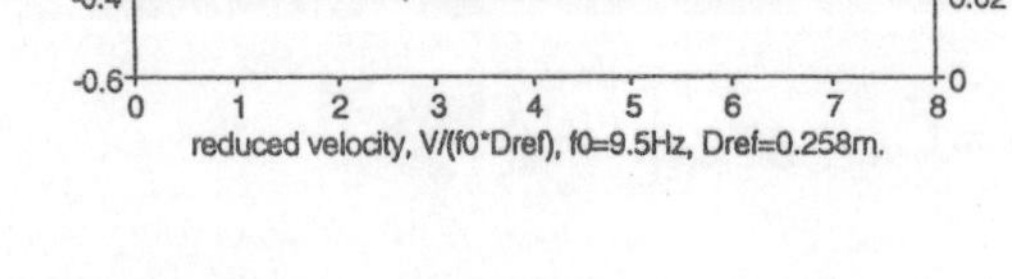

Fig. 15b Aero-elastic balance base moment coefficients for lift.

9. CONCLUSIONS

A large quantity of data have been collected on this model using different techniques. The results from the different sources are self consistent, and provide a means of verifying present analysis techniques. Mean results from integrating the pressure data gave moment coefficients which are in good agreement with the mean balance results. The two sets of balance results are also in good agreement for both the means, and the peaks.

The roughness on this cone has made it behave in quite a different way to previously observed behaviour of an unroughened cone. It is believed that this behaviour has occurred because of the relatively large roughness elements over the surface of the cone breaking up any organised wake behaviour.

10. REFERENCES

Flay, R.G.J. and Vickery, B.J., (1993a) "A wind tunnel study of the pressures, forces, and response of a rough cone in turbulent shear flow", Boundary Layer Wind Tunnel Laboratory Report BLWT-1-1993, University of Western Ontario, London, Ontario, Canada, N6A 5B9.

Flay, R.G.J. and Vickery, B.J. (1993b) "Pressure correlations on a rough cone in turbulent shear flow", 7th US National Conf. on Wind Engineering, Univ. of California, Los Angeles, USA, June 27 - 30.

Shiotani, M., (1975) "Turbulence measurements at the sea-coast during high wind", J. Met. Soc. Japan, Vol. 53, No. 5, pp 340 - 354.

Vickery, B.J. and Clark, A.W. (1972) "Lift and across-wind response of tapered stacks", J. of the Struc. Div. ASCE, Vol. 98, No. ST1, Proc. Paper 8634, pp 1 - 20, Jan.

9ICWE, 1995, New Delhi, India

EFFECT OF INTERFERENCE ON THE PRESSURE DISTRIBUTION ON THE GROUND PLANE AROUND A LOW-RISE AND A TALL BODY IN TANDEM ARRANGEMENT

GOWDA, B.H.L.

Fluid Mechanics Laboratory, Department of Applied Mechanics, Indian Institute of Technology, Madras 600 036, India.

Abstract

In this paper the pressure distribution on the ground plane around a low-rise prismatic body with height to width ratio (h/b) of 1 and a tall body (h/b = 5) in tandem arrangement are presented. Two cases are considered: 1) When the shorter body is in front and the tall body at the rear, 2) when the tall body is in front and the short one at the rear. In each case the gap between the two is systematically varied. The results indicate that in both cases it is the tall body which controls the pressure and the flow field and due to the resulting interference effects very severe flow conditions can be expected around the low-rise body.

1. INTRODUCTION

Prismatic bodies with square cross section and height-to-width ratio (h/b) of 1 and 5 are typical shapes of low-rise and tall buildings. A clear understanding of the flow field around such three-dimensional bluff bodies would be helpful in estimating the forces and analysing the fluid motion around buildings. The dispersion of affluents released, the movement of waste material, and the pedestrian comfort are some of the other features which could be better understood and controlled by analysing the flow around these bluff bodies (Sharan, 1975; Beranek, 1979; Hunt et al , 1978; Gowda et al, 1985). Another important aspect which decides the general wind environment is the interference effects of one body on another (Gowdà et al, 1990). It would be both interesting and useful to study such effects on the flow field as buildings invariably occur in combination. The flow around buildings near the ground surface would be very much influenced by the pressure distribution on the ground plane. Such pressure distributions would reveal the regions of favourable and adverse pressure gradients which in turn decide the direction of ground surface flow - towards or away from the buildings. In the present study, the pressure distribution on the ground plane around prismatic bodies with square cross section and differing height to width (h/b) ratio is considered. A low-rise body with h/b = 1 and a tall body with h/b = 5 are considered in tandem arrangement. Two cases are investigated: 1) when the short body (h/b = 1) is in front and the tall body (h/b = 5) is at the rear ii) when the tall body is in front and the short one at the rear. The gap (g) between the bodies is systematically varied in each case. The results presented are applicable to understand flow around buildings in practical situations. But there are some limitations in the applicability of the results which are pointed out at a later stage in section 3.3.

2. EXPERIMENTAL ARRANGEMENT

The experiments are carried out using the wind tunnel facility shown in Fig. 1. Air from the blower passes through a set of screens and a settling chamber which terminates in a nozzle conforming to DIN 1952 standards. The dimensions of the nozzle exit are 295 mm x 365 mm (365 mm being the height). The velocity at exit is very uniform, the variation both in the horizontal and vertical planes (taken at the central section of the nozzle exit) being less than 0.5 per cent.

A stand designed and fabricated out of mild steel angles is used to support the surface plate on which the bodies are placed. A smooth hylam sheet 12 mm in thickness with a length of 1140 mm and a width of 655 mm is used as the surface plate. Pressure tappings (1 mm in diameter) are provided all along the length of the plate (on the central line) with a centre to centre distance of 6 mm. There is provision for levelling the surface plate and also varying its position in the vertical direction. The plate is positioned in front of the nozzle so that its leading edge abuts with the lower edge of the nozzle as shown in Fig. 1. The position of the pressure tappings coincide with the middle of the nozzle exit. The pressure taps are connected through PVC tubing to the scanning boxes type FC091 (Make: Furness Controls, U.K.). The boxes are in turn connected to micromanometer Type FC012 (Make: Furness Controls, U.K.). The accuracy of the manometer is 0.01 mm water column and by the use of the scanning boxes the pressure at several stations could be monitored quickly and accurately.

Prismatic bodies with width b equal to 30 mm made out of fine teak wood and finished to have smooth surface and sharp edges (to eliminate Reynolds number effect) are utilised. Measurements are carried out (along the centre line) with individual bodies and with bodies in tandem arrangement to study interference effects. Results are obtained when the single body is placed at the leading edge

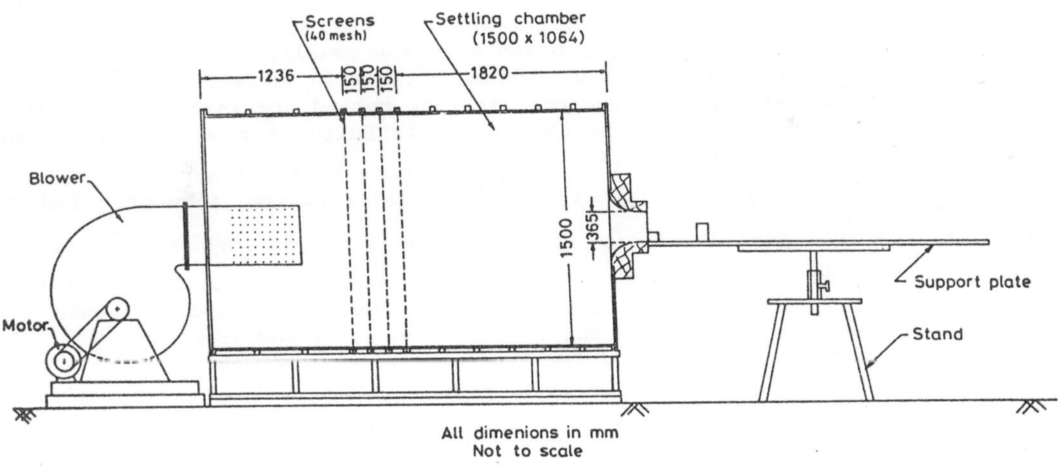

Fig.1. Experimental set-up

and at several distances from the leading edge of the plate in order to obtain a comparison between such results and the pressure distribution obtained when two bodies are simultaneously placed on the plate in tandem arrangement. Both when the bodies are kept at the leading edge and in the boundary layer, there is sufficient length of the surface plate behind the bodies. The uniform pressure distribution on the surface plates without the bodies confirmed the proper alignment of the surface plate.

All measurements are carried out at a exit velocity of 15.5 m/s which gives a Reynolds number (referred to the width of the body) of 30000. The results can be used for analysing practical situations as they are independent of Reynolds number. (The limitations are pointed out at a later stage). The pressure distributions on the surface plate is obtained without the bodies and this is utilised as the reference pressure for the calculation of C_{pw}. $C_{pw} = (p - p_r)/1/2\ \rho U_r^2$, where p = pressure at any point with the body; p_r = pressure at the same point on the surface plate without the body; ρ = density of air; U_r = the velocity in the free stream approaching the bodies.

3. RESULTS AND DISCUSSION

Two cases are considered: i) when the short body (h/b = 1)is at front and the tall body (h/b = 5) at the rear; ii) when the tall body is at front and the short one at the rear. For each case the results are obtained for g/b = 1, 2, 3, 4, 5, 6, 7 and 8. In all the figures where the pressure distribution is presented the results obtained for the relevant single body cases without interference are also included for comparison purposes.

3.1. Pressure distribution for case (i)

The front body is always located at the leading edge and the rear body located at various positions to obtain the different g/b ratios. The results are presented in Figs.2 to 7 for g/b = 1, 2, 3,4, 6 and 8. (The results for g/b = 5 and 7 have not been given as they are typically similar to those in other figures).

Considering first the pressure distribution behind the rear body, it is seen that in all the figures (2 to 7) the C_{pw} distribution behind the tall body is considerably altered due to the interference effects of the front body. The point of $(C_{pw})_{max}$ at the rear would indicate the point of flow reattachment. When there is no well defined peak in the C_{pw} distribution, a tangent can be drawn from the point of inflexion and the intercept of this tangent on the horizontal axis would then approximately indicate the location of the reattachment point (Gowda et al, 1985). Such tangents have been drawn in Fig.2 and it is seen that the reattachment point decreases from about x/b = 4 for the no-interference case to about x/b = 2.5 with interference. Such a decrease is evident at other values of g/b also (Figs.3 to 7). This decrease is most probably due to the turbulence generated by the front body. Further the $(C_{pw})_{min}$ with interference occurs close to the body in all cases (Figs.2 to 7) indicating that the saddle point (Gowda et al, 1985) moves closer to the body with interference. The interference effects behind the rear body appears to be maximum for g/b = 2; but even at the large gap ratio of g/b = 8 (Fig.7) the effects persist. There is in general an increase in the base pressure due to interference.

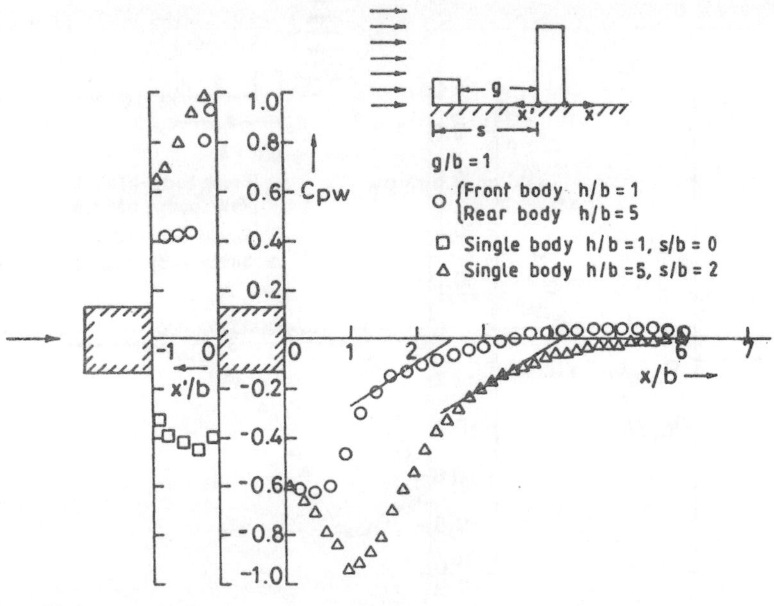

Fig.2. Interference effects on pressure distribution: g/b=1

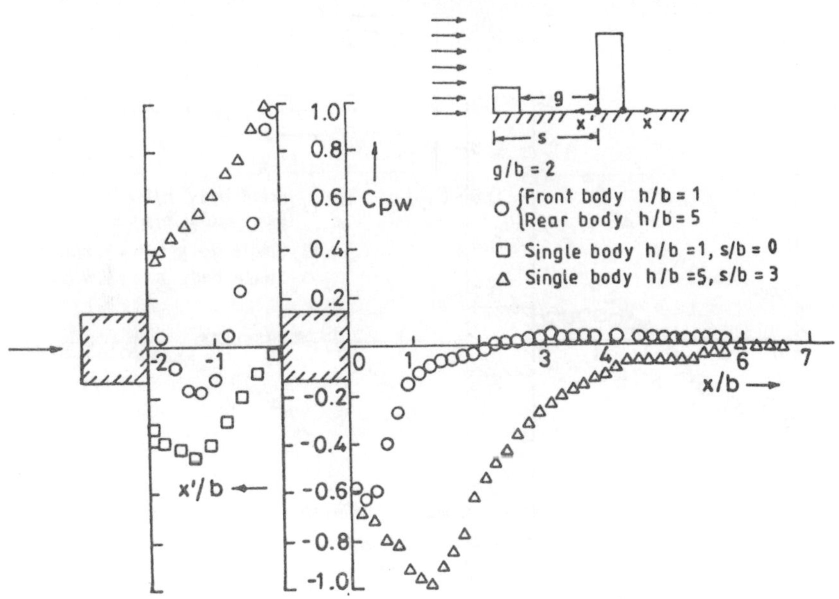

Fig.3. Interference effects on pressure distribution: g/b = 2

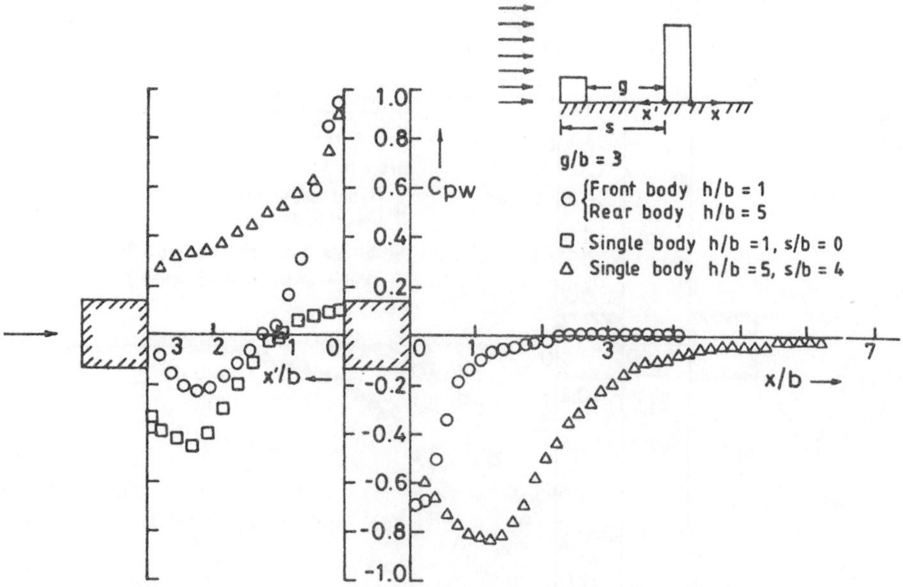

Fig.4. Interference effects on pressure distribution: g/b = 3

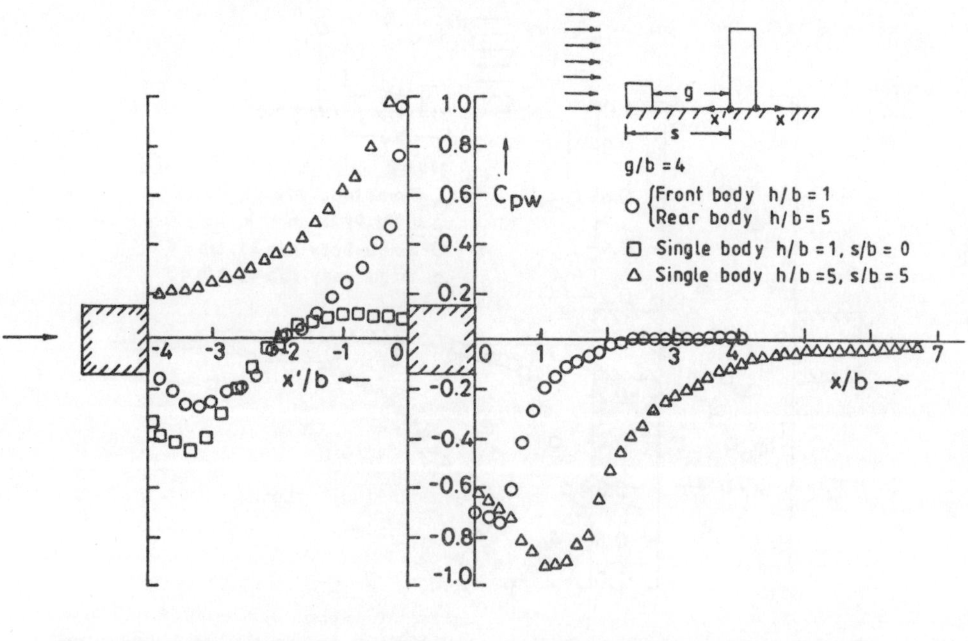

Fig.5. Interference effects on pressure distribution: g/b = 4

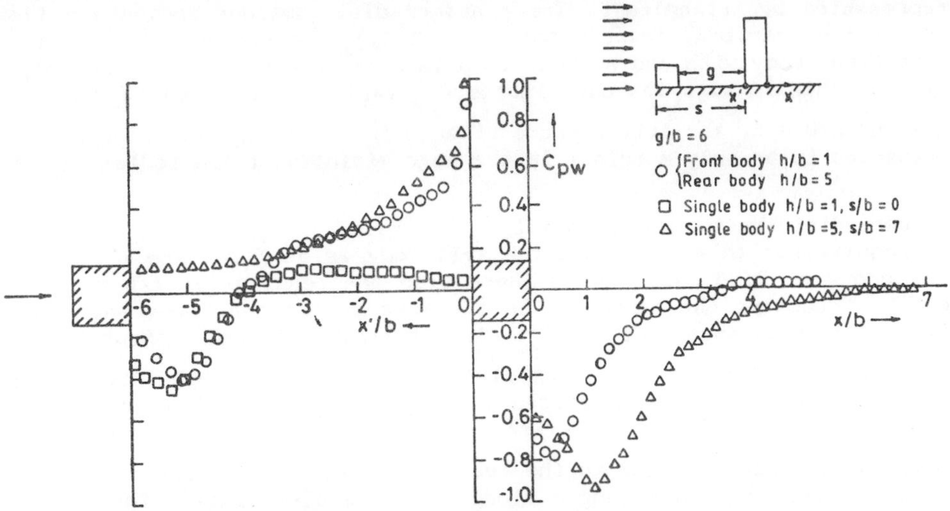

Fig.6. Interference effects on pressure distribution : g/b = 6

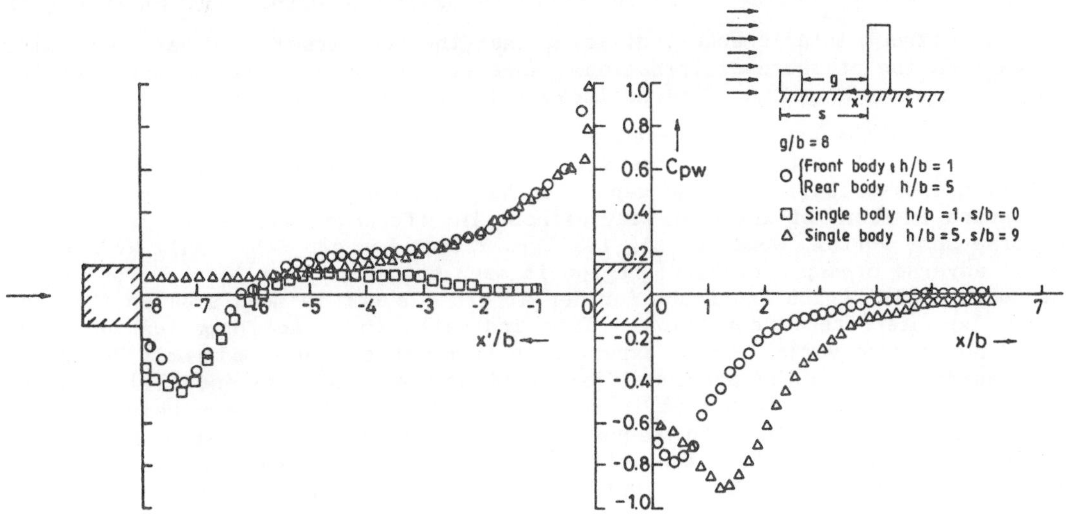

Fig.7. Interference effects on pressure distribution : g/b = 8

Focussing the attention now on the pressure distribution in the gap between the bodies it is quite interesting to observe the strong influence of the tall body on the gap flow, specially at low values of g/b (Figs. 2 and 3). Positive pressures occur in the gap though attenuated compared to the values for the no-interference case (represented by triangles). The pressure distributions indicate a flow from the front of the rear body towards the front body. The pressure distribution just behind the front body with and without interference, become nearly same only at g/b = 6 (Fig.6). This is also the case in the C_{pw} variation in front of the rear tall body. Beyond g/b = 6, the interference effects as far as the gap flow is concerned can be expected to be only marginal in the near vicinity of the bodies.

3.2. Pressure distribution for case (ii)

The results for this case when the tall body is at front are shown in Figs.8 to 13 for g/b = 1, 2, 3, 4, 6 and 8 (here also the results for g/b = 5 and 7 are not shown). Focussing the attention on the wake of the rear body and considering Fig.8 (g/b = 1) it is seen that the C_{pw} distributions with and without interference are nearly the same for $x/b < 2$. But it is to be noted that at this position the short rear body is completely submerged in the wake of the tall front body. Hence the C_{pw} values observed with interference are in effect due to the wake of the front body rather than that due to the rear body. At other values of g/b i.e. ≥ 2 (Figs.9 to 13) the near wake length along the flow direction of the rear body is considerably reduced. Also the base pressure increases due to interference similar to case (i).

Considering next the C_{pw} variation in the gap between the bodies, the strong influence of the tall body in this case also is quite clearly seen. For example, at g/b = 1, the C_{pw} values with interference nearly coincide with those for the tall body without interference indicating that the downstream body has very little influence on the pressure distribution. Whereas, it was the reverse when the tall body was at rear (Fig.2). Some noticeable influence of the rear body on the C_{pw} distribution in the gap is observed at g/b = 2 and 3 (Figs.9 and 10). At higher g/b values (Fig.11 to 13) again there is very little influence of the short body on the pressure distribution in the gap. Further the steep adverse pressure gradient that occurs in front of the rear body without interference (due to the formation of the horse-shoe vortex, Gowda et al, 1985) is absent for $g/b \leq 5$. Only at g/b = 6, a small adverse pressure gradient region is seen but very close to the front of the rear body. This region increases further at g/b = 8 but is much smaller than that for the no-interference condition. This indicates that the formation of a well defined horse-shoe vortex can be expected in front of the rear body only for $g/b \geq 6$.

Considering both the pressure distributions for cases (i) and (ii) presented in Figs.2 to 13, it is seen that in both cases it is the tall body which controls the pressure distribution and hence the flow pattern particularly in the gap between the bodies. Another interesting feature noticed is that in case (ii) (i.e. when the tall body is in front), upto g/b = 3, the pressure in front of the rear body is lower than that at its base indicating a possible flow from the base towards the front and to the possibility that there could result a thrust force rather than a drag force on the rear body. (The confirmation of the later requires measurement of pressure distribution on the bodies).

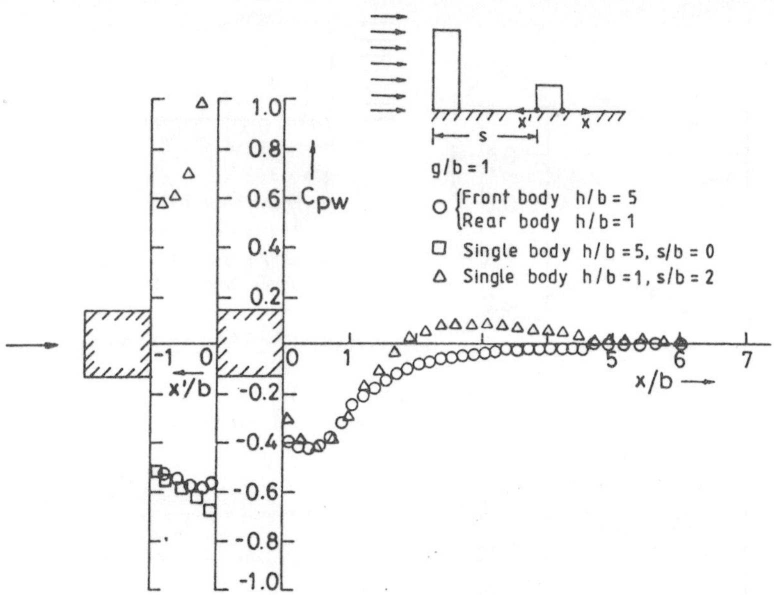

Fig.8. Interference effects on pressure distribution : g/b = 1

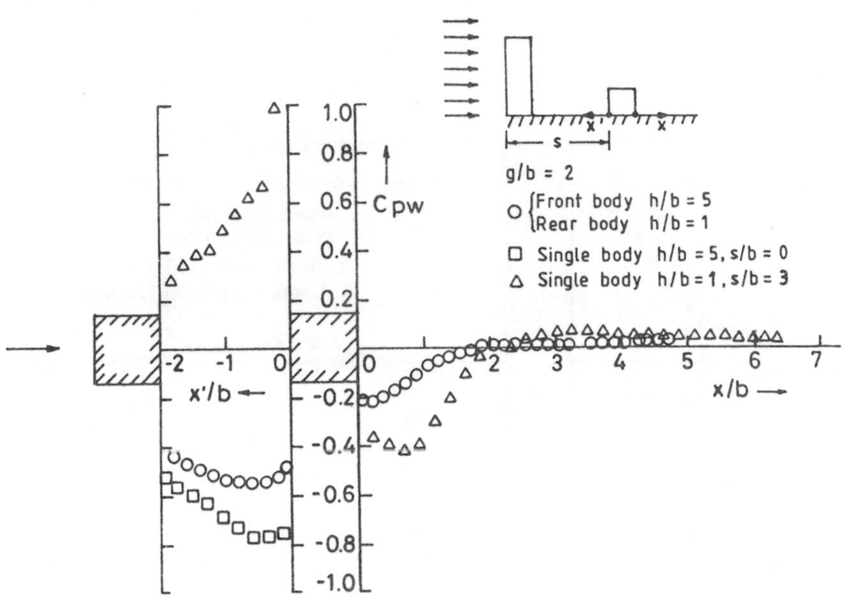

Fig.9. Interference effects on pressure distribution : g/b = 2

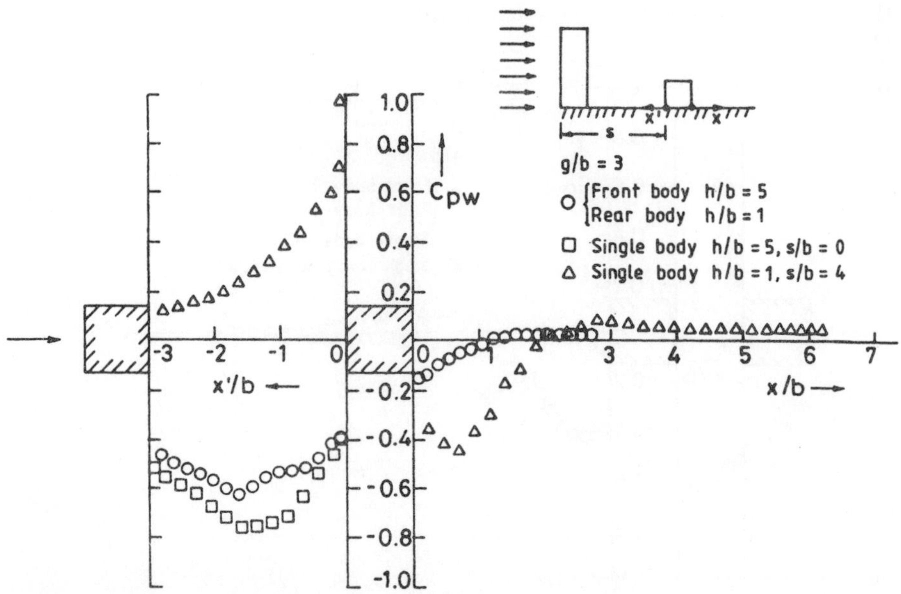

Fig.10. Interference effects on pressure distribution : g/b = 3

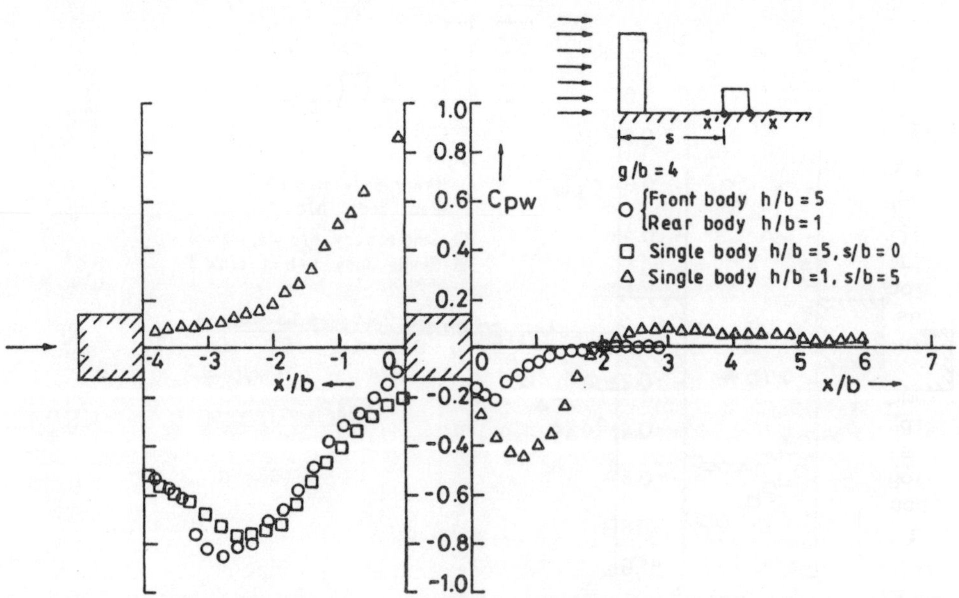

Fig.11. Interference effects on pressure distribution : g/b = 4

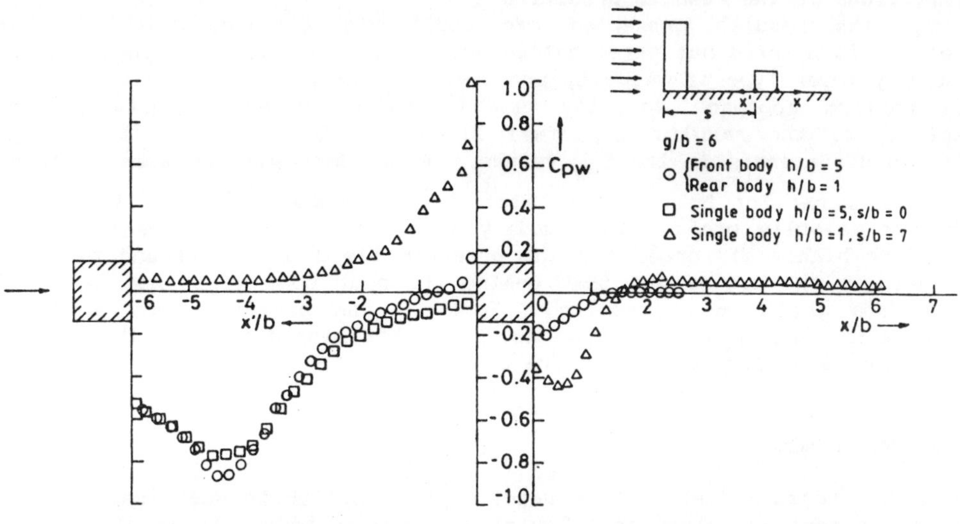

Fig.12. Interference effects on pressure distribution : g/b = 6

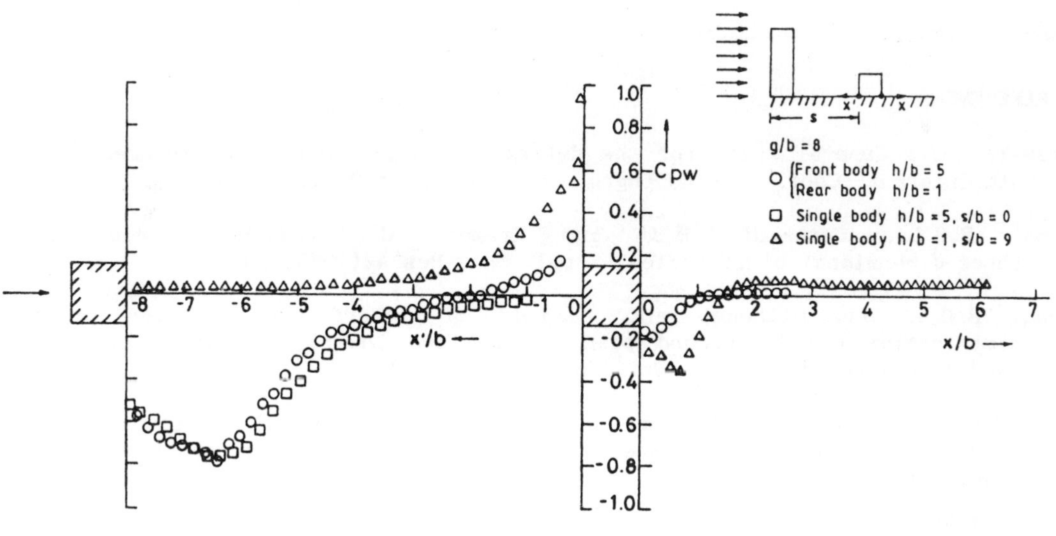

Fig.13. Interference effects on pressure distribution : g/b = 8

3.3. Limitations of the results presented

Though the results presented are applicable to understand flow around buildings, it is pointed out here that in actual practice the buildings are exposed to a boundary layer type of velocity profile. As the present results are obtained with an uniform approach velocity profile there is some limitation on the applicability of the results to actual situation; for example, there could be quantitative differences in the C_{pw} values. Also there will be some influence of the Reynolds number for buildings in atmospheric boundary layers. But even when the approach velocity profile is boundary layer type, the flow behind the front body would be highly disturbed and the pressure field on the ground plane within the gap between the bodies and that behind the rear body might not be very much different from the present results. Further, the present study provides the necessary basic results which can be used for obtaining the comparative influence when other approach flow conditions are used.

4. CONCLUDING REMARKS

From the pressure distribution results presented it is seen that when a tall body is at the rear and close to a low-rise prismatic body, the resulting pressure gradient causes a flow towards the short body and strong vortical flows can occur around instead of at the base of the later. The HSV around the rear tall body will be formed only for $g/b > 4$. A low-rise body behind a tall body would distort the base flow behind the tall body only marginally. The short body would be completely enveloped by the wake of the tall body at low values of g/b. Some distortion of the pressure field in the wake of the tall body can be expected at $g/b = 2$ and 3. In all cases, the wake behind the rear body gets shortened probably due to the turbulence generated by the upstream body.

It is to be remarked here that the pressure measurements presented have been confined to the plane of symmetry only. Results at other planes away from the plane of symmetry are desirable.

5. REFERENCES

Beranek, W.J.: General rules for the determination of wind environment. Proc. 5th Internat. Conf. on Wind Engrg. Colorado: Fort Collins (1979).

Gowda, B.H.L., Gerhardt, H.J. and Kramer, C.: Pressure field around three-dimensional bluff bodies. Appl. Sci. Res. 42(1985) 245.

Gowda, B.H.L. and Sitheeq, M.M.: Effect of interference on the pressure distribution on the ground plane around two low-rise prismatic bodies in tandem arrangement. Appl. Sci. Res. 47(1990) 129.

Hunt, J.C.R., Abell, C.J., Peterka, J.A. and Woo, H.: Kinematical studies of the flow around free and surface mounted obstacles applying topology to flow visualisation. J. Fluid Mech. 86 (1978) 179.

Sharan, V.Kr.: On characteristics of flow around building models with the view of simulating the minimum fraction of theboundary layer. Internal. J. Mech. Sci. 17 (1975) 557.

9ICWE, 1995, New Delhi, India

A SEMI-EMPIRICAL WAKE MODEL FOR TWO-DIMENSIONAL (SHARP EDGED) BLUFF BODIES

HANGAN, H. and VICKERY, B.J.

Research Assistant *Professor & Research Director*

Boundary Layer Wind Tunnel Laboratory, University of Western Ontario, London, Ontario, Canada N6A 5B9.

Abstract

A two-dimensional wake model for turbulence profiles is developed for (sharp edged) bluff bodies based on similarity (self preservation) and universality (different shapes). The method employs spectral filtering around the shedding peak and normalization by the local maximum longitudinal turbulence (Ium) and by the corresponding lateral coordinate (ym).

1. INTRODUCTION

A universal wake model for bluff bodies has significant scientific and engineering applications. In wind engineering such a model would simplify the approach to a large variety of problems such as self induced vibrations , buffeting, turbulent diffusion of pollutants and warm air, or dense gas dispersion in the presence of obstacles . At present these problems are solved either by means of specific experimental modelling, or (incompletely) by CFD simulation. Most of the classical wake studies have been concerned with the mean velocity description of the wake, usually in terms of velocity deficit profiles. Similarity (self preservation) and universality (different bodies) have already been established for mean velocity deficits in the wake. There are, however, no complete practical wake models from the turbulence point of view. The fact that turbulence develops a structural similarity in free turbulence flows was first argued by Hinze and Townsend (1959) and demonstrated by Tennekes and Lumley (1972) but only in the far wake. Counihan, Hunt and Jackson (1974) presented a theoretical wake model that predicts self-preservation of Reynolds stress for 2D fence configurations. Practical applicability of the two powerful theoretical concepts of similarity and universality , their range of validity and eventual restrictions has not yet been established.

LIST OF SYMBOLS

Cd = drag coefficient based on transverse body dimension
Cpw = base pressure coefficient
d = transverse body dimension (here d=d square=0.03m)
dw = wake width defined as the maximum distance between streamlines
f = frequency
fs = shedding frequency
Iu = longitudinal turbulence intensity (by free stream velocity)

Ium = maximum longitudinal turbulence intensity for each transverse profile
Iv = transverse turbulence intensity
k = "cavity number" = Uw/Uo
Uo = free stream velocity
Uw = free stream-line velocity at separation
uv = Reynolds stresses
uvm = maximum Reynolds stress for each transverse profile
St = body Strouhal number
Stu = universal Strouhal number
Su = "u" velocity component power spectral distribution
Sv = "v" velocity component power spectral distribution
x = longitudinal coordinate
y = transverse coordinate
ym = transverse position of the maximum longitudinal turbulence intensity (Ium)

2. WAKE CHARACTERIZATION

The objective of the present study is to define, by a semi-empirical approach, the turbulent structure of the wake of a sharp edged 2D bluff body in smooth and turbulent flow.

Similarity and universality of turbulence intensity profiles are established and a simple wake model is constructed. Primarily the model employs a dimensionless measure of turbulence intensity and coordinates by normalizing by the maximum longitudinal turbulence *Ium* at every downstream station and by the lateral position corresponding to this value (y_m). This approach based on wake and not on body parameters was suggested by Davies (1976) and used by Vickery (1981) in a study of wake buffeting, by Raju et al (1988) and B.S. Shiau (1992) for porous fence studies. Also a wake width (dw) based on universal Strouhal number, as defined by Roshko (1960), is employed for the longitudinal dimensionless coordinate (see Fig. 1a). Secondly the model focuses on the vortex shedding as basic wake coherent structure. This is done by defining turbulence intensities and Reynolds stresses in terms of rms value of the spectral peak defined for a suitable frequency band. From this point of view a certain similarity exists between this experimental model and Direct Vortex Simulation as a CFD method.

3. EXPERIMENTAL PROCEDURE

The test program was designed to allow a wake mapping in terms of narrow band turbulence intensities of the longitudinal (I_u) and lateral (I_v) components. Three series of tests were carried out on 2D shapes: square, rhombus, hexagon 1, (face perpendicular to the flow), hexagon 2, (rotated at 30 degrees), U shape and flat plate.

3.1 Preliminary Smooth Flow Tests

The preliminary tests were carried out in smooth flow in a small (0.46 x 0.46)m section open circuit wind tunnel with a contraction ratio of 10/1. The reference wind speed was of 11 m/s and the turbulence levels were of 0.77% for the longitudinal component and approx. 0.2% for the transverse one. The Reynolds number for the 4 shapes tested (square, rhombus, and the two hexagons) was of about 2.2×10^4. A cross hotwire connected to a TSI 1051-6 anemometer measured the two component velocity field.
A HP Structural Dynamics Analyzer 5423A was employed for spectral measurements. The shedding peak signal was "empirically" filtered observing the spectral density plot for each station.

The measurements allowed the acquisition of wind data on half of the wake region for six longitudinal stations: x/d = 4, 6, 7, 8, 10, 12 and 12 transverse stations: y/d = 0.25, 0.5, 0.75, 1, 1.25, 1.5, 1.75, 2, 2.5, 3, 3.5 or a total of 72 measuring points (where d refers to the square edge length of 0.03m and x, y are longitudinal and transverse coordinates - see Fig. 1).

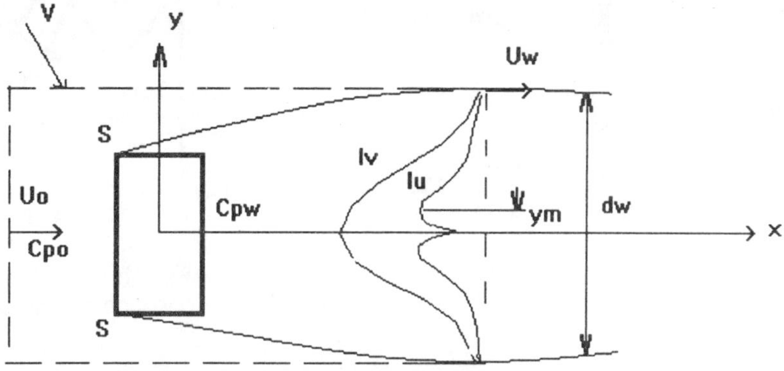

Fig. 1 : Wake parameters

3.2 Secondary Smooth Flow Tests

These tests were carried out in a larger, open circuit, wind tunnel (BLWT1) with a (1.70 x 2.45)m section. The reference wind speed was about the same as for the small tunnel (preliminary) tests but the turbulence levels were slightly different: 1.3% for the longitudinal component and 0.46% for the transverse one.

Time series, each corresponding to a wake mapping position (as described in #3.1), were measured for the two cross hot-wire channels. A FFT routine was employed to process the data and obtain the two wind component spectra and the corresponding cross-spectrum. Preliminary spectral analysis (see # 3.1) showed that a bandwidth of ±8% around the shedding frequency defines well the vortex spectral signature. A short routine was written to detect the shedding peak and compute the rms integral for the established bandwidth.

3.3 Turbulent Flow Tests

Free turbulence tests were carried out in BLWT1. Large scale turbulent flow was developed using an upstream grid of 0.138m bar width and 0.55m spacing. Turbulence levels of 11.5% longitudinal and 8.67% transverse were obtained at the 2D body position, 5.5m downstream from the grid. This is in good agreement with Vickery results (1968) for a similar plane grid. Based on these results the lateral length scale should be of 0.076m or approx. 2.5d. The longitudinal length scale was of approx. 6d which is in agreement with previous results by Makita and Sassa (1987). Such large turbulence scales are typical for wind engineering problems. The energy containing region of the grid turbulence is centred near fs*d/Uo = 0.025 which is well removed from the shedding value of about 0.14 to 0.2.

The reference wind speed was of approx. 9.5m/s. Again time series were measured for every wake position and for four 2D shapes. Similar data processing routines as for #3.2 but wider spectral window (±25%) for rms of peak were employed. Typical spectral plots for smooth and turbulent flows are shown in Fig. 2a and 2b.

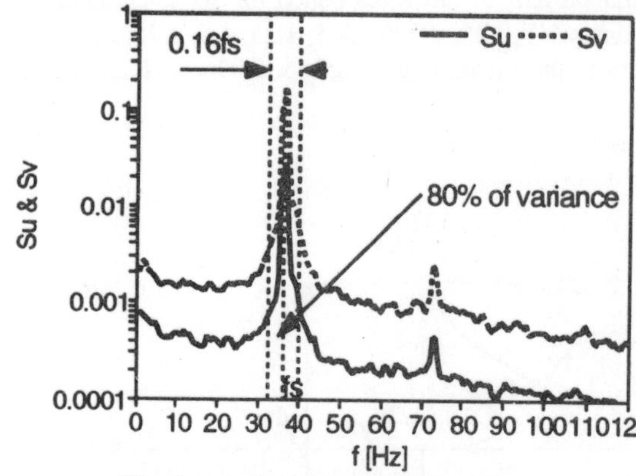

Fig.2a : u&v Velocity Spectra
U Shape-Smooth:x/d=6,y/d=1.75

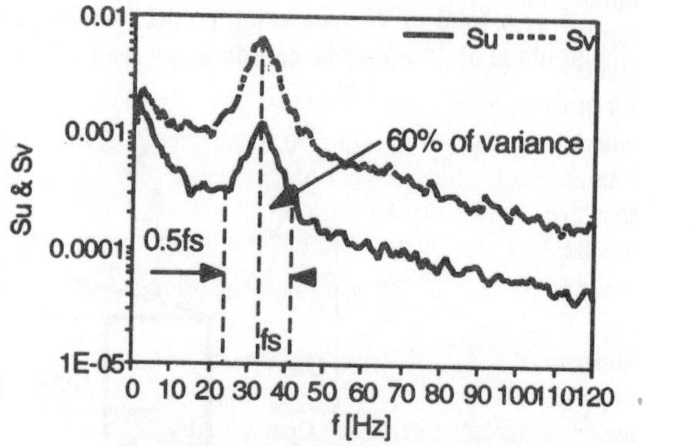

Fig.2b : u&v Velocity Spectra
Hexagon-Turbul.:x/d=6,y/d=1.75

4. RESULTS

4.1 Smooth Flow Results

Based on these experiments, similarity and universality of turbulence profiles was found. Similarity for longitudinal and transverse turbulence profiles is shown in Fig. 2a, and 2b for different 2D shapes. A Gaussian regression for the lateral component $I_v/I_{um} = 1.8*exp[-0.2*(y/ym)^2]$ was found and holds for downstream profiles (*x/d>6*) and for different shapes. Similar regressions were used by Hah and Lakshminorayana(1982) for an airfoil case.

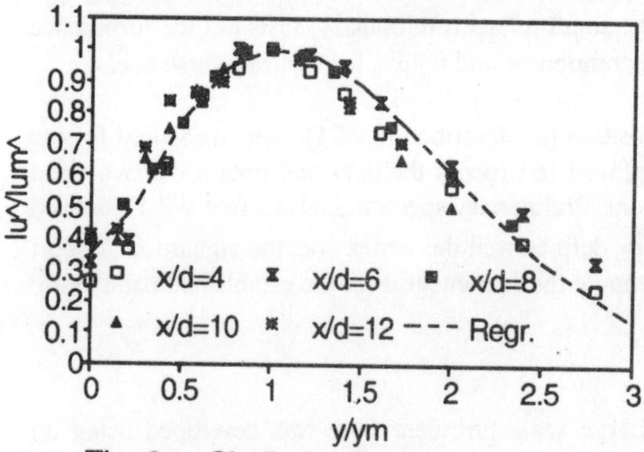

Fig. 3a : Similarity-Smooth
U Shape-Long. Turbulence

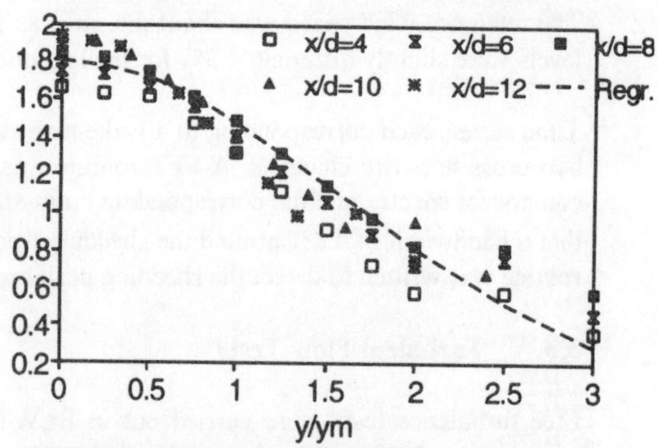

Fig.3b : Similarity-Smooth
Hexagon1-Transverse Turbulence

The form of the longitudinal nondimensional turbulence profiles $\left(I_u/I_{um}\right)$ and Reynolds stresses (*uv/uvm*), where Ium and uvm are the maximum values and ym is the transverse location of Ium, is different and double Gaussian regressions defined as:

$$Iu/Ium = 1 - 0.65 * \exp[-4.25 * (y/ym)^2] \quad \text{for } y/ym < 1$$
$$= \exp[-0.5 * (y/ym - 1)^2] \quad \text{for } y/ym > 1$$

and

$$uv/uvm = 1 - 0.7 * \exp[-7 * (y/ym)^2] \quad \text{for } y/ym < 0.75$$
$$= \exp[-0.55 * (y/ym - 0.75)^2] \quad \text{for } y/ym > 0.75$$

are considered. We mention here the difficulties in measurements for the middle wake region where a double shedding frequency of the longitudinal component was observed, a result of the influence in this region from both sides of the wake. Following the coherent (vortex) signature of the wake flow we switched between the double shedding frequency peak at the wake axis and the usual shedding frequency for transverse distances y/ym>0.25. This and the use of a flow based nondimensional coordinate y/ym, where ym refers to the maximum longitudinal turbulence position, result in a collapse of data points for the internal and external wake boundaries. The use of the double Gaussian regression is based on our belief that two wake regions can be defined: an internal wake region for y/ym<1 where a deficit turbulence mechanism (similar to the deficit velocity and hence of Gaussian type) can be assumed, and an external wake region where dispersion laws are dominant.

Universality of turbulence profiles is shown in Fig. 4a and 4b. The present data is in good agreement with experiments on a D shape made by Davies (1979). The difficulties for the middle wake region measurements which have been mentioned are the possible cause of the data scatter for y/ym<0.5. Similarity and universality for Reynolds stresses and regressions are shown in Fig. 5a and 5b.

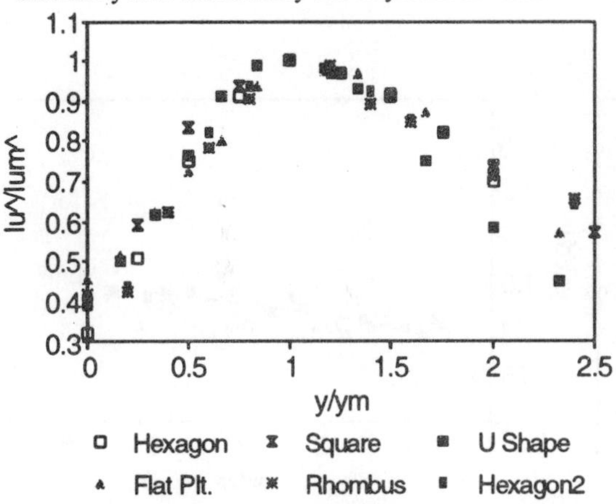

Fig. 4a : Universality-Smooth
Longitudinal Turbulence x/d=8

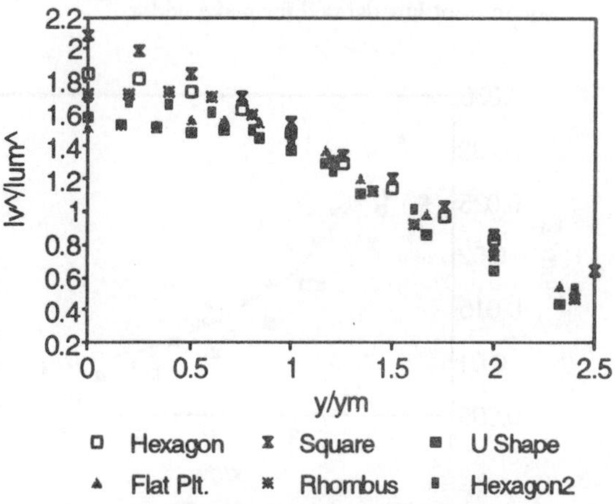

Fig. 4b : Universality-Smooth
Transverse Turbulence x/d=8

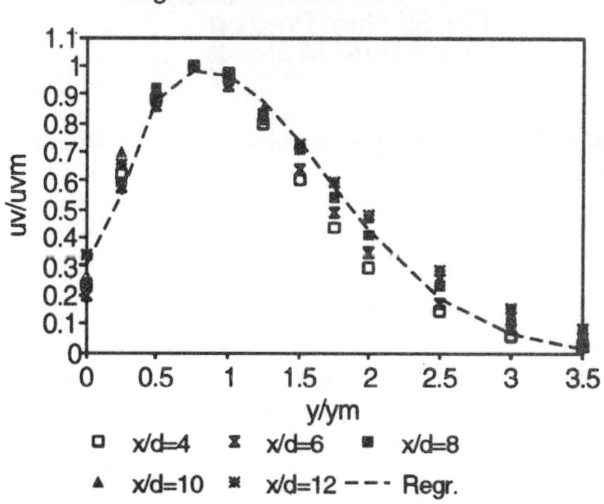

Fig. 5a : Similarity-Smooth
Square-Reynolds Stress

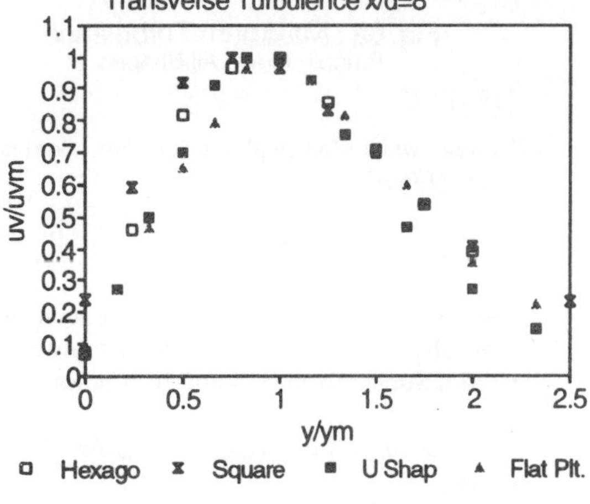

Fig.5b : Universality-Smooth
Reynolds Stress x/d=8

Again universality is manifested for the maximum longitudinal component (I_{um}) as a function of downstream distance $\left(x/d_w\right)$, see Fig. 6a. Fig. 6b marks the position of this maximum *(y_m/dw)* as a function of downstream distance *(x/dw)* and shows a collapse of the data for the different cross sectional shapes. Also maximum Reynolds stresses decay is similar to the longitudinal turbulence decay.
Power-law regressions were found for these two plots. Decreasing of the maximum turbulence and maximum Reynolds stresses are well described by inverse power laws :

$$I_{um}^2 = 0.055 * \left(x/d_w\right)^{-1}$$

and

$$uvm/Uo^2 = 0.075 * (x/d_w)^{-1}$$

which is in agreement with experimental results by Perara (1979) and Castro (1983). The corresponding maximum turbulence position fits well the regression $y_m/d = 0.56 * \left(x/dw + 1.2\right)^{1/2}$ which is similar to the square root law defined for wake width.

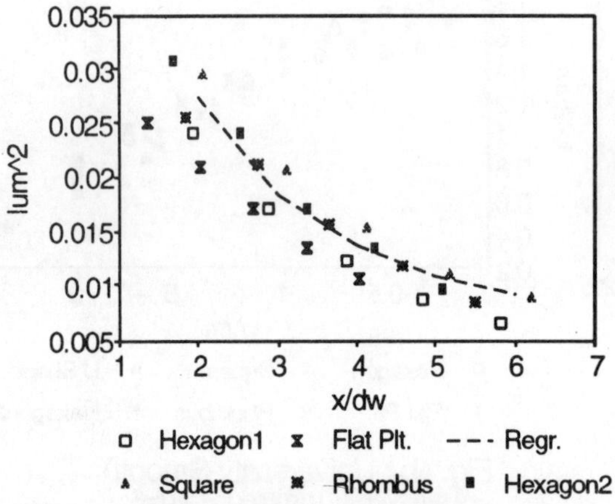

Fig. 6a : Maximum Turbulence
Smooth Flow - All Shapes

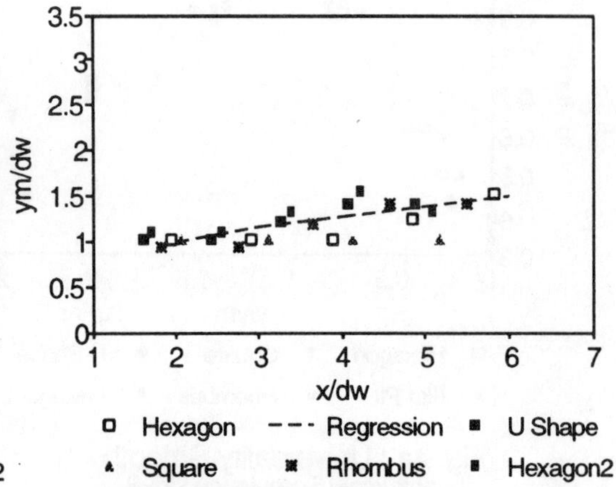

Fig. 6b : Ium Position
Smooth Flow - All Shapes

The wake width (dw) employed for x/dw coordinate is based on universal Strouhal number as defined by Roshko (1960)

$$Stu = fs * dw/Uw$$

where Uw is the free stream-line at separation and fs the shedding frequency. From experimental analysis on circular cylinders, wedges and flat plates at subcritical Reynolds numbers Roshko deducted an universal Strouhal number Stu=0.16. The wake width (dw) for the present study was computed from :

$$St/Stu = d/dw * (Uw/Uo) = d/dw * (1 - Cpw)^{1/2} = d/dw * k$$

or from :

$$dw = \frac{Stu}{fs} * kUo$$

where Cpw is the base pressure coefficient and $k = Uw/Uo$. Details regarding this wake width computation and a description of aditional pressure tests which were employed are shown in Hangan and Vickery (1994).

4.2 Turbulent Flow Results

Analogous general results regarding similarity and universality of turbulent profiles were obtained for large scale turbulent flow. Similarity tendency is shown in Fig. 7a and 7b. The procedure was the same as for smooth flow except for the bandwidth for the rms peak integration which was changed from approx. ± 8% of the shedding frequency to ±25%. This is due to the broader spectral peak in turbulent flow compared to smooth flow.

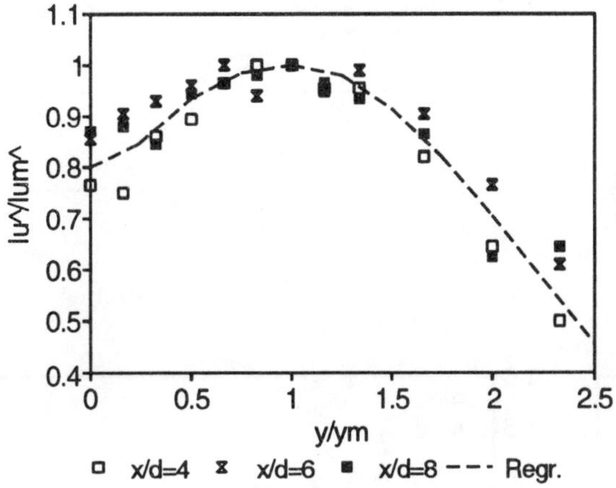

Fig. 7a : Similarity-Turbulent
Flat Plate-Long.Turbulence

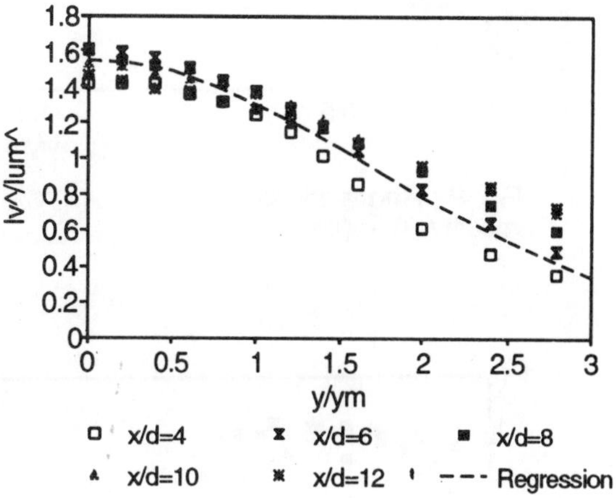

Fig. 7b : Similarity-Turbulent
U Shape-Transverse Turbulence

Two observations can be made: First it was evident that the coherent structure (vortex) in turbulent flow "dies" faster than in smooth flow and it is probable that for a smaller scale turbulent flow this fact would be more evident. This is the reason why investigation of similarity was limited at x/d<10. Secondly there was no evidence of the "frequency switch" in the middle wake region as it was the case for smooth flow. Also a relative important low frequency spectral signal due to the large scale free turbulence was observed. The corresponding turbulence regressions for similarity plots (Fig. 7a and 7b) and Reynolds stresses are:

$$
\begin{aligned}
Iu/Ium &= 1 - 0.2 * \exp[-4.25 * (y/ym)^2] \quad \text{for y/ym} < 1 \\
&= \exp[-0.35 * (y/ym - 1)^2] \quad \text{for y/ym} > 1
\end{aligned}
$$

$$
Iv/Ium = 1.55 * \exp[-0.17 * (y/ym)^2]
$$

$$
\begin{aligned}
uv/uvm &= 1 - 0.75 * \exp[-5.4 * (y/ym)^2] \quad \text{for y/ym} < 1 \\
&= \exp[-0.65 * (y/ym - 1)^2] \quad \text{for y/ym} > 1
\end{aligned}
$$

It can be observed that these expressions are close to those for smooth flow except for the inner and outer wake regions where the profiles have to adjust to the free, external flow, turbulence levels. This behaviour is ilustrated in Fig. 8 in which the corresponding regressions for smooth and turbulent flow are plotted.

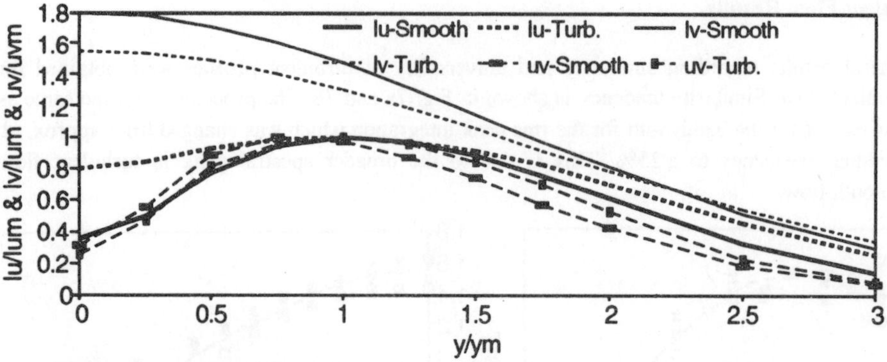

Fig. 8 : Regressions
Smooth & Turbulent Flow

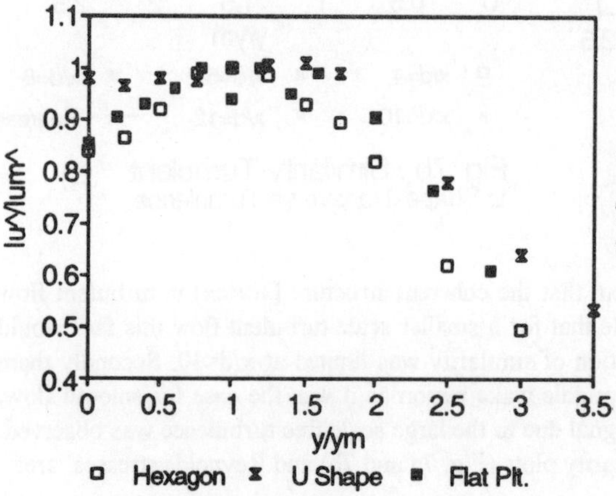

Fig. 9a:Universality-Turbulent
Longitudinal Turbulence, x/d=6

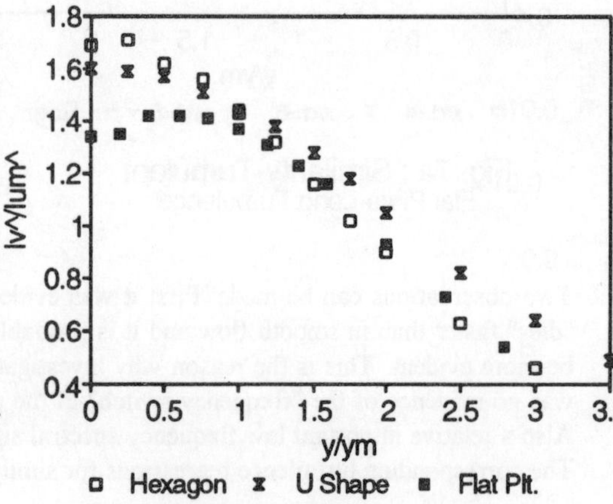

Fig. 9b:Universality-Turbulent
Transverse Turbulence, x/d=6

Universality is shown in Fig. 9a and 9b. A slight scatter of the data points compared to the smooth flow case is observed mostly for the inner wake region. Similarity and universality of Reynolds stresses and the corresponding regressions are introduced in Fig. 10a and 10b. Maximum turbulence decay (Fig. 11a) is well fitted by a similar inverse power low as for smooth flow. The maximum turbulence position data (Fig. 11b) collapse well for different cross sections and follow almost a constant law.

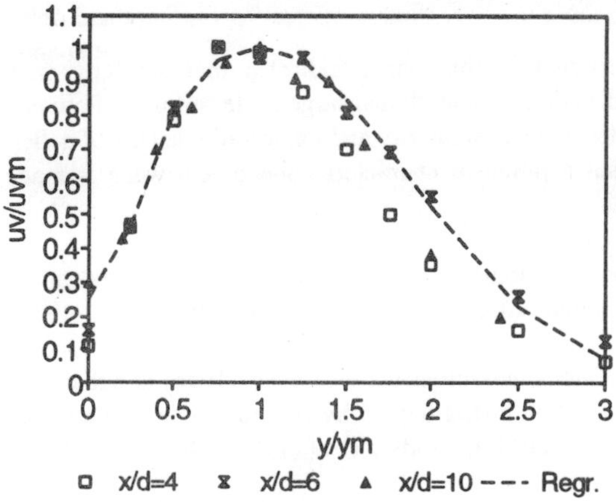

Fig. 10a : Similarity-Turbulent
Hexagon1-Reynolds Stress

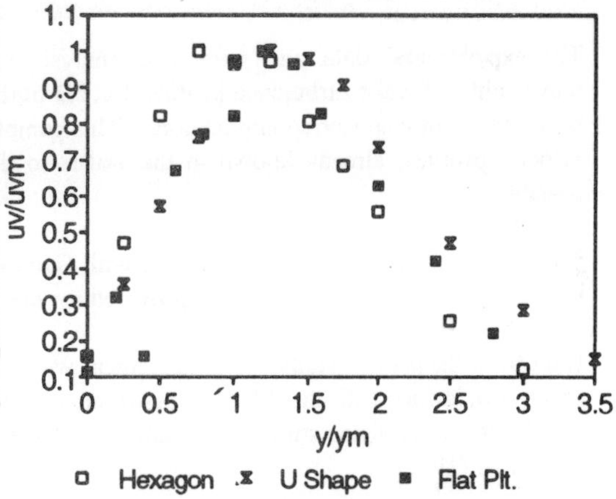

Fig. 10b:Universality-Turbulent
Reynolds Stress, x/d=6

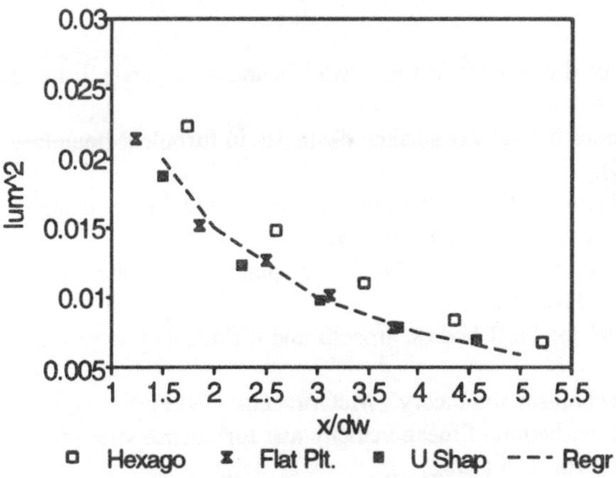

Fig. 11a : Maximum Turbulence
Turbulent Flow - All Shapes

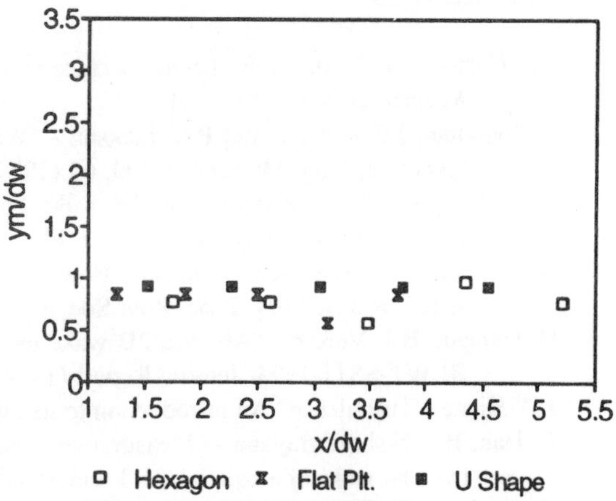

Fig. 11b : Ium Position
Turbulent Flow - All Shapes

4.3 Reattachment, smooth flow

Particular attention was paid to the rectangular shape for which two spectral peaks were identified and this was attributed to the flow reattachment due to the longer afterbody of this type of shape.

The same experimental procedure and data processing as for #3.2 were employed focussing on the first spectral peak Similarity and universality of longitudinal and transverse turbulence profiles and of Reynolds stresses were found for practical enginnering purposes and details are introduced in Hangan and Vickery (1994).

CONCLUSIONS

The experimental data and numerical analysis reported in this paper indicates that similarity and universality of wake turbulence profiles for 2D bluff bodies in smooth and large scale turbulent flow can be assumed for engineering applications. This complements the similarity and universality aspect of deficit velocity profiles, already known in the wake, so that a practical characterization of the wake is made possible.

Turbulence intensities profiles are fitted with Gaussian distributions so that semi-empirical models can be built and power law regressions are provided for maximum turbulence and its location in the wake.

Data from the present study are in agreement with previous measurements by Davies (1976) and there are strong similarities with the different cases such as that of an airfoil treated by Hah and Lakshminarayana (1982) or porous fence studies by Castro (1979), Perara (1981), Bradley et al (1983), Raju et al (1988) and Shiau (1992)

Smooth and large scale turbulent flow cases were studied. Particularities of flow reattachment are underlined. Applicability of the wake model is found for all these different flow scenarios.

REFERENCES

J.P. Castro - "Relaxing wake behind surface-mounted obstacles in rough-wall boundary layers", J. Fluid Mechanics, Vol. 93 (1979).

J. Counihan, J.C.R. Hunt and P.S. Jackson - "Wakes behind 2D surface obstacles in turbulent boundary layers", J. Fluid Mechanics, Vol. 64 (1974).

M.E. Davies - "A comparison of the wake structure of a stationary and oscillating body, using a conditional averaging technique", J. Fluid Mechanics, Vol. 75 (1976).

A.Fage and F.C.Johansen- "On the flow of air behind an inclined flat plate of infinite span", A.R.C.R.&M.1104, Proc. Roy. Soc. a, Vol. 116 (1927)

H. Hangan, B.J. Vickery- "About a 2D wake model for bluff bodies, smooth and turbulent flow results", BLWT-SS11-1994- Interim Report (1994)

J.D. Hinze -"Turbulence. An introduction to its mechanism and theory", McGraw Hill (1959).

C. Hah, B. Lakshminarayana - "Measurement and prediction of mean velocity and turbulence structure in the near wake of an airfoil", J. Fluid Mech., Vol. 115 (1982).

K. Kushioka,T. Saito and A. Honda - "Simulation of turbulent flows by Direct Vortex Method", J.Wind Eng. and Industrial Aerodynamics, Vol. 46 -47 (1993).

M.D.A.E.S. Perara - "Shelter behind 2D solid and porous fences", J. Wind Eng. and Industrial Aerodynamics, 8 (1981).

K.G. Ranga Raju et al - "Experimental study on characteristics of flow past porous fences", J. Wind Eng. and.Industrial Aerodynamics, 29 (1988).

A. Roshko - "On the drag and shedding frequency of 2D bluff bodies", NACA TN No. 3169.

B.S. Shiau - "The turbulence structure behind the multiple wind breaks across-wind", J. Wind Eng. and Industrial Aerodynamics, 41-44 (1992).

H. Tennekes and Lumley, J.C. - "A First Course in Turbulence", MIT Press (1972).

B.J. Vickery - "Across Wind Buffeting in a Group of Four In-Line Model Chimneys", J. Wind Eng. and Industrial Aerodynamics, 8 (1981).

B.J. Vickery - "Wind loads on buildings", Ph.D Thesis, University of Sydney, Australia (1968)

9ICWE, 1995, New Delhi, India

EFFECTS OF THE WIND VELOCITY PROFILE ON THE AEROSTATIC FORCES

NODA, M.[1], UTSUNOMIYA, H.[1], NAGAO, F.[1], and TAKENAKA, K.[2]

Graduate Student *Professor* *Associate Professor*

[1] *Tokushima University, Tokushima, Japan 770.*

[2] *Omoto Construction Company, 1-1-13, Yamashita, Okayama, Japan, 700*

Abstract

The effects of the turbulent flows on the aerostatic forces were experimentally examined by wind tunnel test. The test models were two-dimensional rectangular cylinders of B/h=1.0, 4.0 placed on the floor normal to flow direction. Some different turbulent flows (such as turbulent boundary layer flows, grid turbulent flows and uniform flow) were investigated. The behavior of the aerostatic forces (such as lift force coefficient, C_L, base pressure coefficient, C_{pb} and the pressure distribution on the roof) and the flow properties around model were examined under some different conditions of turbulent flows, and it became clear that the aerostatic forces and the flow properties around model depended strongly on the non dimensional velocity gradient at the height of model and the past assessments of the turbulence effect on the aerostatic forces in turbulent boundary layer flows include the effect of the velocity gradient.

1. INTRODUCTION

Many researchers investigate the effect of the wind characteristics of the turbulent flows on the aerostatic force, because the effect of the wind characteristics of the turbulent flows on the aerostatic forces is important problem for an estimate of the values of the aerostatic forces on a wind resistant design. For two-dimensional bluff bodies in uniform turbulent flows, Lee [1] and Petty [2] examined by paying attention to the turbulence intensity and the turbulence scale of the approaching flow. As their result, it became clear that the base-pressure coefficient turned down with increase of the turbulence intensity but the turbulence scale had no effect for the base-pressure coefficient. For bluff bodies in turbulent boundary layer flows, the effects of the wind characteristics of approaching flows on the aerostatic forces are investigated by many researchers. Most of the researchers pay attention to the local turbulent factors (such as the turbulence intensity and the turbulence scale) at the height of model, like under the uniform turbulent flows. Their result was that the aerostatic force caused by the separation (such as lift forces and the distribution on the roof) depended strongly on the turbulence intensity (turbulence energy) [3], [4]. Certainly, we can measure them easily. However, the explanations by the local turbulent factors are not sufficient, because many kinds of the factor of the turbulent boundary layer flows are distributed vertically and the source of the turbulence of the approaching flow is the vertical gradient of the mean velocity in turbulent boundary layer flows. Therefore, we consider the effects of the vertical profile of mean velocity which is main source of the turbulence in turbulent boundary layer flows, and the non dimensional mean velocity gradient defines by

$$\gamma = (d\tilde{U}/d\tilde{z})_{\tilde{z}=h/\delta} \qquad \text{where } \tilde{U} = U_z/U_G,\ \tilde{z} = z/\delta, \tag{1}$$

U_z is the mean velocity at the height of model, U_G is the mean velocity outside boundary layer, z is height, δ is the boundary thickness and h(=6.0cm) is the height of model.

In this study, the effects of the velocity gradient on the aerostatic forces and on the flow properties around model were investigated by using two-dimensional rectangular cylinders placed on the floor normal to flow direction under the several kinds of turbulent flows.

2. EXPERIMENTAL CONDITION

The purpose of this study was not to clarify the effects of the turbulence intensity but to clarify the effects of the vertical profile of mean wind velocity on aerostatic forces. The two dimensional models were examined to simplify the distribution of aerostatic forces caused by the separation from the leading edge of the roof. Furthermore the turbulence intensity should be changed without changing the vertical profile of wind velocity for the investigation of the effect of only the turbulence intensity. Therefore, the turbulence intensity was changed under the constant vertical profile of wind velocity by using the grid turbulent flows. In this chapter, models and the characteristics of the approaching flows used here were explained precisely.

2.1 Models

The two-dimensional rectangular cylinders of the cross sections which have ratio width to height B/h=1.0 for Model A and 4.0 for Model B were used. Each model was placed on the floor normal to flow direction as shown in Figure 1. Models have the pressure taps at the center of the span. The lift force was calculated by the integration of the difference pressures between the surface pressures of roof and the reference static pressure. The lift coefficient, C_L and the surface pressure coefficients were normalized by the mean wind velocity at the height of the model U_h by following equation;

$$C_L = L/q_h A \tag{2}$$

$$C_p = (p - p_\infty)/q_h \tag{3}$$

where L is lift force, q_h is dynamic pressure at the height of model which was calculated by measured mean velocity at the height of model (=½ρU_h), p is surface pressure, and p_∞ is reference static pressure which was measured by Pitot tube at the upwind point of model above the boundary layer.

2.2 Approaching flows

The flows used here were turbulent boundary layer flows (TBLF), grid turbulent flows with thin thickness of boundary layer (GTF), and the uniform flow with very thin boundary layer (UF) as shown in Table 1. The TBLF were generated by roughness mats and spires designed by Irwin's method [5]. The vertical profiles of wind speed of TBLF, GTF and UF at the position placed the model are shown in Figure 2. In this figure, the wind velocities were normalized by the mean velocity at the height of model. The wind velocity gradient, γ was given 0, if the height of model, h was higher than the thickness of boundary layer, δ. Figure 3 represents the spectrums of fluctuating wind velocity of the turbulent boundary layer flow (T-1) and grid turbulent flow (G-1) which agree with Von Karman's spectrum.

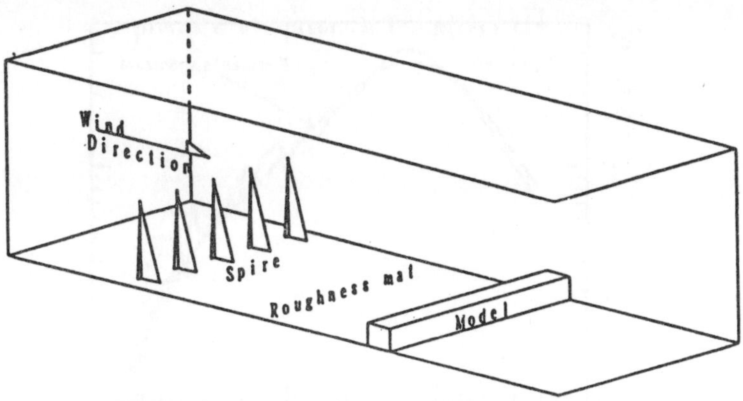

Figure 1 General view of wind tunnel test

Table 1 Properties of used flows

Case	T–1	T–2	T–3	T–4	T–5	T–6	G–1	G–2	G–3	UF
I_u (%)	15.3	10.2	6.6	7.4	9.5	10.4	14.2	6.7	5.7	0.2
δ cm	30	30	30	10	20	40	20	6.0	6.4	4.5
α	0.30	0.18	0.11	0.22	0.20	0.17	0.05	0.06	0.10	0.19
γ	0.93	0.67	0.46	0.33	0.52	0.79	0.16	0.06	0.11	0.00
L_x^u/h	0.9	0.9	1.1	0.3	0.8	0.9	0.9	0.9	0.8	---

where, turbulence intensity, I_u and turbulence scale, L_x^u were measured at the height of model.

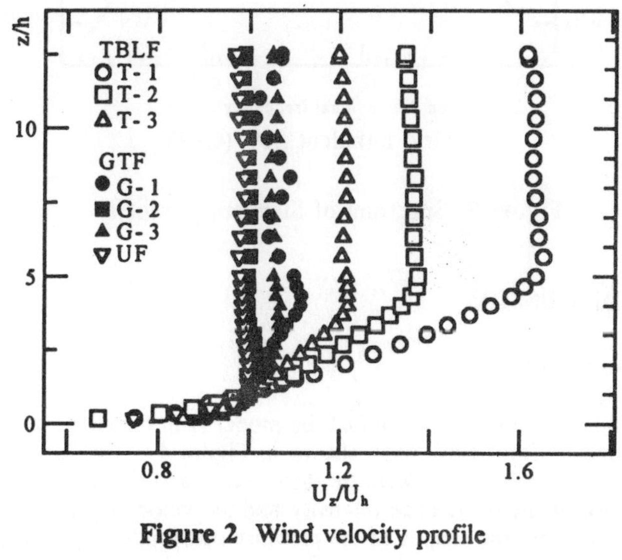

Figure 2 Wind velocity profile

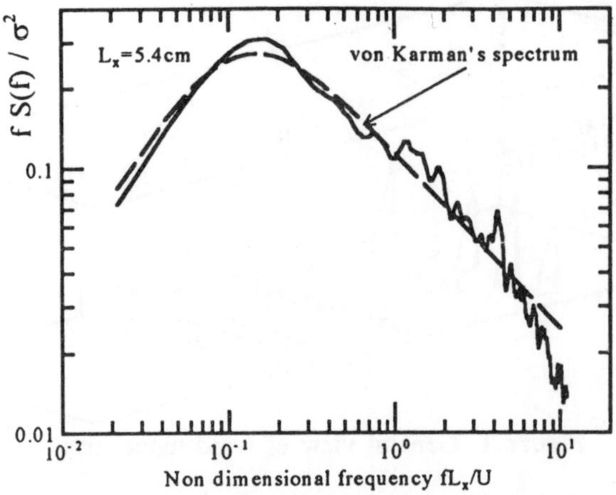

(a) Turbulent boundary layer flow (T–1)

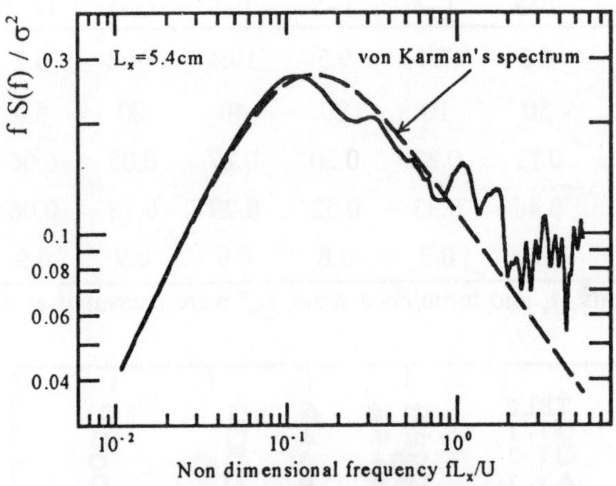

(b) Grid turbulent flow (G–1)

Figure 3 Spectrum of fluctuating velocity

3. RESULTS AND DISCUSSION

3.1 Pressure distribution on roof

(1) Model A (B/h=1.0)

The pressure distributions on the roof of the model A in turbulent boundary layer flows, grid turbulent flows and uniform flow were shown in Figure 4. In turbulent boundary layer flows, it is found that the shapes of the pressure distribution and the peak values of the pressure change with the increase of the turbulence intensity and the velocity gradient. From only this result, it can be conducted that the shapes of the pressure distribution and the peak values of the pressure depend on the turbulence intensity. If the turbulence intensity at the height of model is the main cause of the variation of the shapes of the pressure distribution and the peak values of the pressure, the result under grid turbulent flows must be the same result as under turbulent boundary layer flows. However, the shapes of the pressure distribution and the values of pressure hardly change with the drastic increase of the turbulence intensity from I_u=0.2% to 14.2% in grid turbulent flows and uniform flow. Apparently the explanations for these results by turbulence intensity are deficient. But these results could be explained by the increase of the velocity gradient at the height of model. Therefore, it is found that the shapes of the pressure distribution on the roof depend on not the turbulence intensity, but the velocity gradient at the height of model.

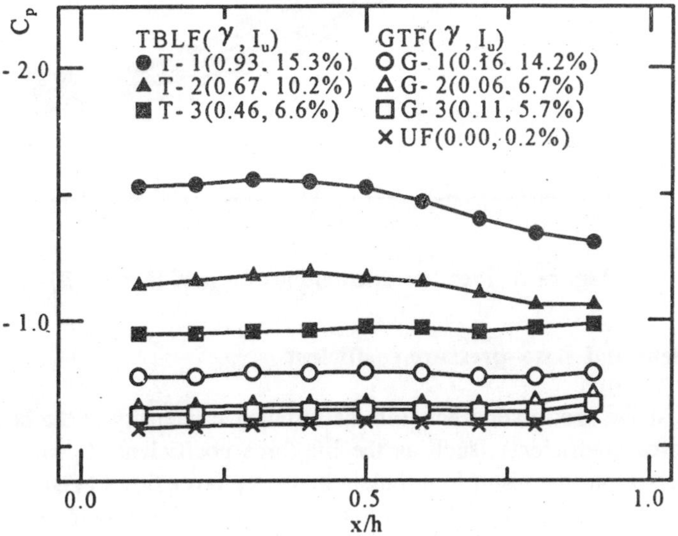

Figure 4 Pressure distributions on roof (Model A)

(2) Model B (B/h=4.0)

Figure 5 shows the effects of the characteristics of the approaching flows on the pressure distribution on the roof of Model B. In the case of Model B, the shapes of the pressure distribution on the roof show the reattachment flow pattern, where the separated shear flow from leading edge induced high negative pressure near the leading edge and pressures recover close to the trailing edge. In comparison with the results of T–1 and those for G–1, where the turbulence intensities were almost the same (approximately I_u=15%), high negative pressures occurred in T–1 near leading edge, on the other hand, negative pressure was not so large in G–1. Therefore it is found that the change of the shape of the pressure distribution on roof are not caused by only change of the turbulence intensity and the effect of the velocity gradient are included these results. In the case of small vertical gradient of mean wind speed, that is the result in the grid turbulent flows and uniform flow, the magnitudes of negative pressures near leading edge increase with the increase of turbulence intensity. Therefore, in the case of reattachment flow patterns occurred in the roof of Model B, the pressures on roof are influenced by both turbulence intensity and the velocity gradient at the height of model.

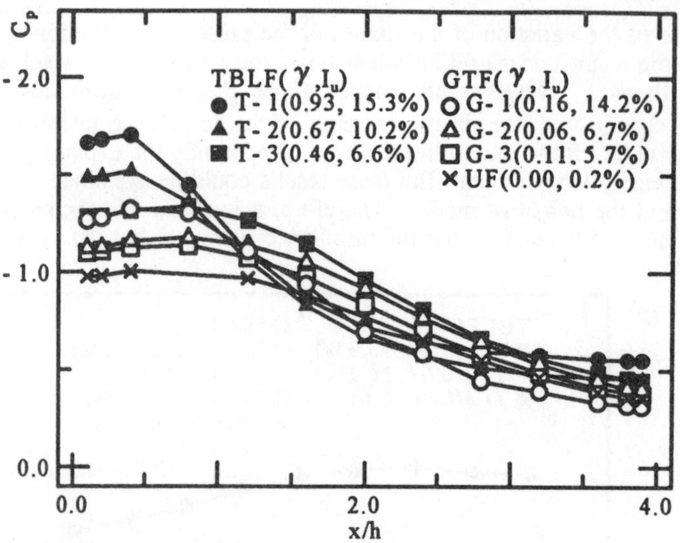

Figure 5 Pressure distributions on roof (Model B)

3.2 Lift coefficient and Base-pressure coefficient

(1) Model A (*B*/*h*=1.0)

Figure 6 shows the relation between the turbulence intensity at the height of model and the aerostatic force coefficients (such as the lift force coefficient, C_L and the base-pressure coefficient, C_{pb}). From the result in turbulent boundary layer flows (TBLF) and uniform flow (UF, I_u=0.2%, γ=0), it is clear that the aerostatic force coefficients increase with the increase of turbulence intensity. Furthermore, it is found that the aerostatic force coefficients depended on turbulence intensity in turbulent boundary layer flows. However, if the turbulence intensity would induce the increase of the aerostatic force coefficients, the result under turbulent boundary layer flows and the grid turbulent flows gave the same values, but the aerostatic force coefficients in grid turbulent flows do not coincide with those in turbulent boundary layer flows. Therefore, it can be judged that the turbulence intensity is not the main cause of the increase of the aerostatic force coefficients.

Figure 7 indicates the relation between the velocity gradient at the height of model and the aerostatic force coefficients. It is found that the aerostatic force coefficients correlate closely with non dimensional velocity gradient where the correlation coefficients are given 0.98 for C_L and 0.97 for C_{pb}. Therefore, it is found that the aerostatic force coefficients caused by completely separated flow pattern (such as Model A are influenced strongly by the velocity gradient) · and the behavior of the aerostatic force coefficients are explained by the non dimensional velocity gradient at the height of model.

(2) Model B (*B*/*h*=4.0)

Figure 8 shows the relation between the aerostatic force coefficients and turbulence intensity at the height of model, and Figure 9 indicates the relation between the aerostatic force coefficients and the non dimensional velocity gradient at the height of model. From Figure 8, it is found that the aerostatic force coefficients increase with the increase of turbulence intensity in turbulent boundary layer flows, but they are almost the same magnitudes for the change of

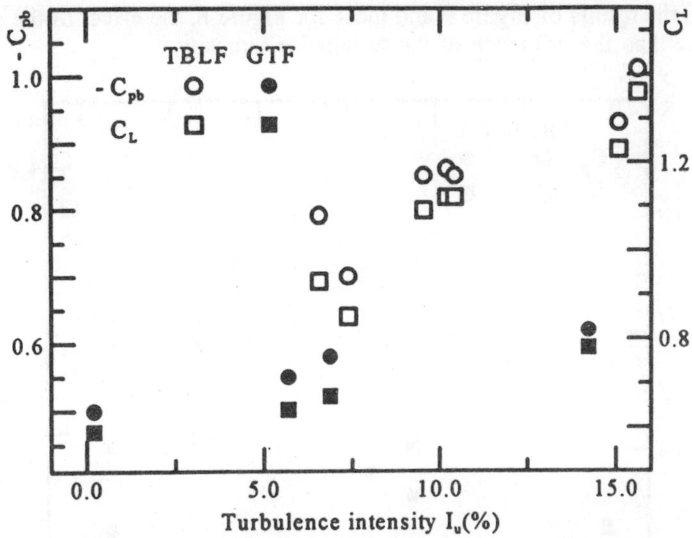

Figure 6 Turbulence intensity v.s. aerostatic force coefficients (Model A)

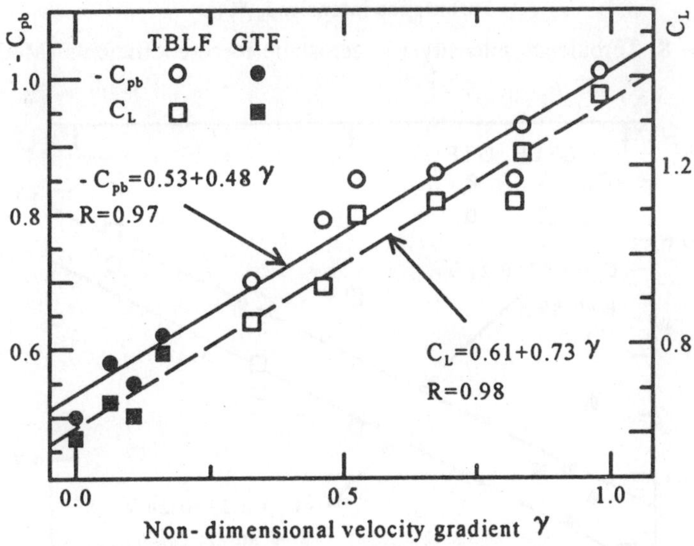

Figure 7 Velocity gradient v.s. aerostatic force coefficients (Model A)

turbulence intensity in the grid turbulent flows. Therefore, it is hard to say that turbulence intensity is the main cause of the change of the aerostatic force coefficients.

On the other hand, the relation between the non dimensional velocity gradient at the model height and the aerostatic forces shown in Figure 9 are almost linear; the correlation coefficients are 0.89 for C_L and 0.96 for C_{pb}. The dispersion of values is wider than the result of Model A given in Figure 7. It can be estimated that the variation of the aerostatic force coefficients are influenced by both turbulence intensity and the velocity gradient. However, in comparison with the results of Figure 9 and those for Figure 8, the effect of the velocity should be greater than the influence of the turbulence intensity.

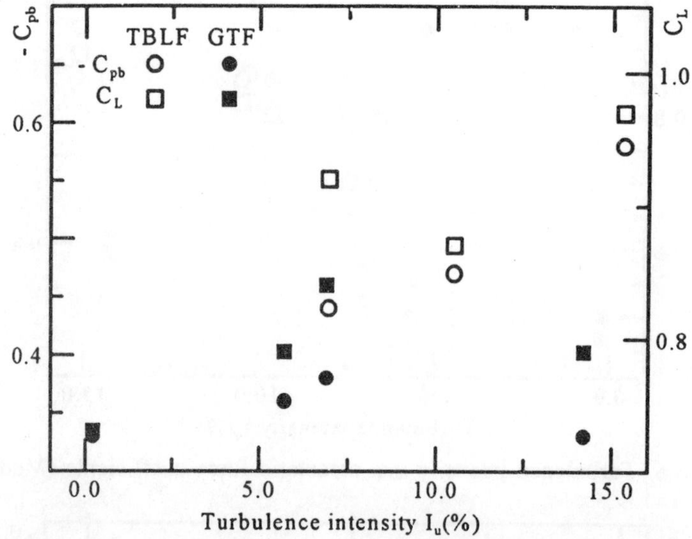

Figure 8 Turbulence intensity v.s. aerostatic force coefficients (Model B)

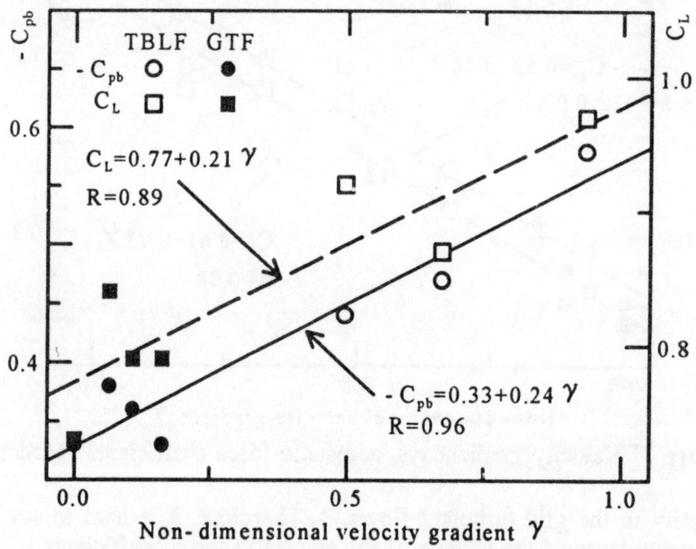

Figure 9 Velocity gradient v.s. aerostatic force coefficients (Model B)

3.3 Flow properties around model

In section 3.1 and 3.2, it became clear that the effect of the velocity gradient on the aerostatic force was stronger than that of the turbulence intensity. In this section, the effect of the velocity gradient on the flow properties around model is discussed. The velocity distribution and the flow direction were measured by x-type hot-wire anemometer. Mean velocities were normalized by measured mean velocity at the height of model.

(1) Model A (B/h=1.0)

Figure 10 indicates the vector of the mean velocity above Model A. In Figure 10 (a), the arrows represented by solid line indicate the result in grid turbulent flow (G–1) and the arrows given by broken line show the result in turbulent boundary layer flow (T–1). These flows have almost the same turbulence intensity and different velocity gradient. From Figure 10 (a), the inclined angle of flow in T–1 is smaller than that in G–1. In Figure 10 (b), the arrows represented by solid line indicate the result in G–1 and the arrows given by broken line show the result in uniform flow (UF). These flows have almost the same velocity gradient and different turbulence intensity. In spite of the very different turbulence intensity, flow patterns around model are almost the same. From these results, it can be found that the flow around model is influenced strongly by the velocity gradient.

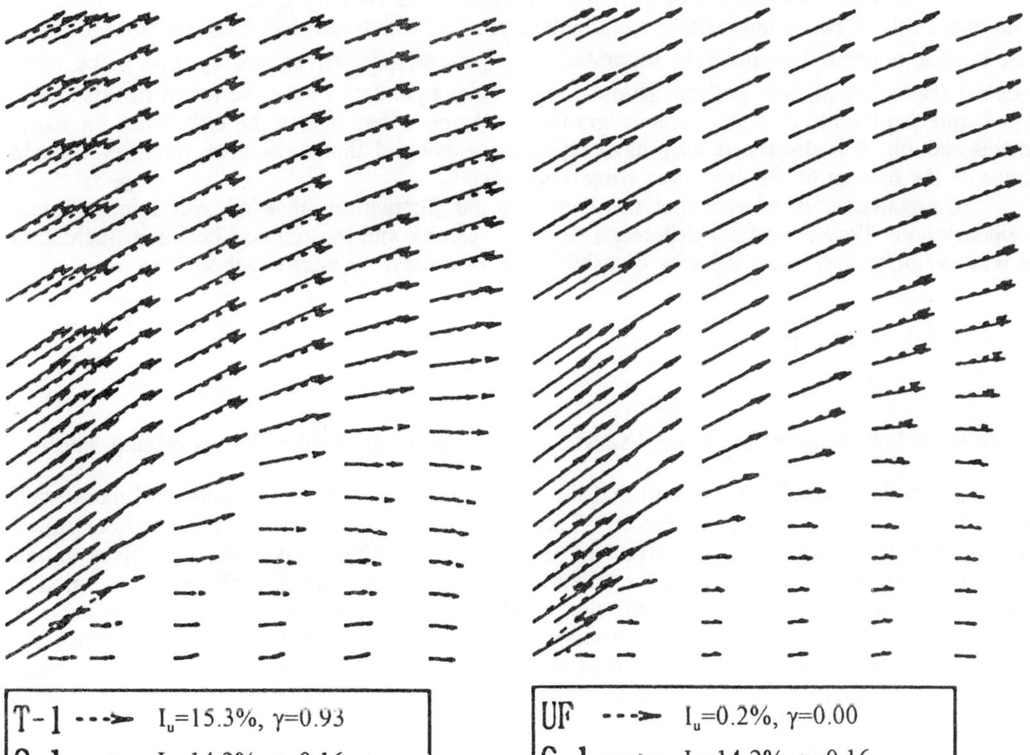

(a) Turbulent boundary layer flow (T–1) v.s. Grid turbulent flow (G–1)

(b) Uniform flow (UF) v.s. Grid turbulent flow (G–1)

Figure 10 Flow pattern above Model A

(2) Model B (B/h=4.0)

Figure 11 indicates the vector of the mean velocity above Model B. In Figure 11 (a), the arrows represented by solid line indicate the result in grid turbulent flow (G–1) ant the arrows give by broken line show the result in turbulent boundary layer flow (T–1). In Figure 11 (a), it can be found that the angle of flow near the leading edge in T–1 is also smaller than that in G–1 and the separating bubble in T–1 is smaller than that in G–1. The arrows represented by solid line indicate the result in G–1 and the arrows given by broken line show the result in uniform flow (UF) in Figure 11 (b). From Figure 11 (b), the flow patterns are almost the same with the drastic increase of turbulence intensity. Therefore, it can be found that the flow patterns above Model B were influenced by the velocity gradient, however, the effects of the velocity gradient on flow patterns for Model B are meaner than those for Model A.

3.4 Discussion

(1) The relation between the turbulence intensity and the non dimensional velocity gradient

In this study, the effects of turbulence intensity and velocity gradient were investigated with two kinds of two–dimensional rectangular cylinder, which induced different flow patterns, that is, one generated completely separated flow pattern; [B/d=1, square cylinder], the other induced reattachment flow pattern; [B/d=4, rectangular cylinder]. As the result of this study, it is found that the effect of the velocity gradient is stronger than that of the turbulence intensity. In this section, it is discussed why most researchers asserted that turbulence intensity is main cause of the change of the aerostatic force coefficients.

Generally, it is known that the cause of the fluctuation of wind speed in turbulent boundary layer flow is vertical difference of wind velocity and the relation between fluctuation of wind velocity and mean velocity distribution is expressed by next equation.

$$u' = \bar{u}(z+l') - \bar{u}(z) \fallingdotseq \frac{d\bar{u}}{dz} \tag{4}$$

where u' is fluctuation of main wind speed, $\frac{d\bar{u}}{dz}$ is velocity gradient and l' is mixing length.

From this formula, it is found that the root mean square of the fluctuation of main wind speed depend on the velocity gradient. Therefore, the turbulence intensity is also controlled by the velocity gradient and the effect of the velocity gradient looks like the effect of the turbulence intensity. However, the result of this study shows that the effect of the velocity gradient is stronger than that of the turbulence intensity, and it can be inferred that the most of all assessments of the effect of the turbulence intensity in turbulent boundary layer flows are indirect judgements for the investigation of the effect of the velocity gradient. It doesn't mean that the velocity gradient is equivalent to the turbulence intensity. On the other hand, it is true that the effect of the turbulence intensity is in existence on the cross section with reattachment flow pattern like Model B. The effect of the turbulence intensity on aerostatic forces have to appraise independently the effect of the velocity gradient.

(2) The mechanism of the effect of the non dimensional velocity gradient

The mechanism of the effects of the velocity gradient on the aerostatic forces and the flow properties around model must be made clear. However, the results of this study were not enough to do it. Two factors would be important to explain the mechanism of the effect of the velocity gradient are arisen. One is the vorticity in approaching flow which affects the separation from leading edge and the separated shear layer flow. The other is the vertical difference of the momentum of the approaching flow. In the near future, these two factors must be examined minutely.

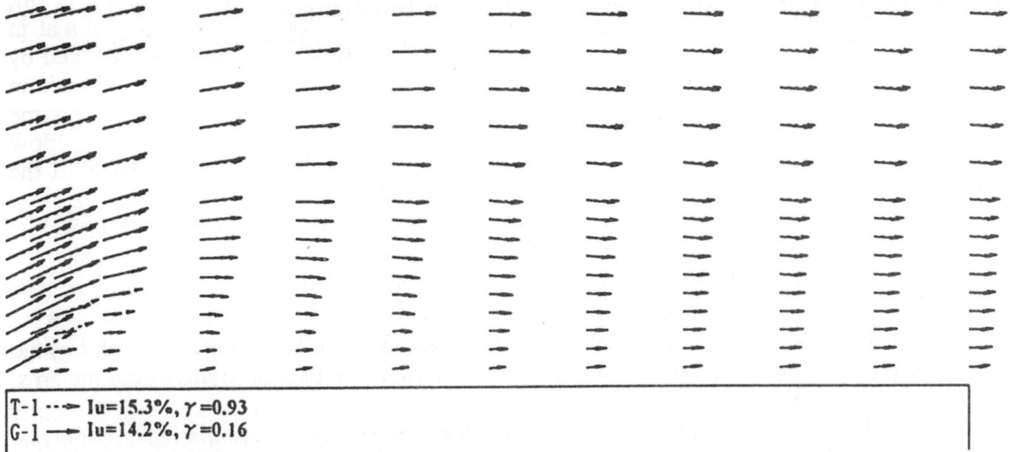

(a) Turbulent boundary layer flow (T–1) v.s. grid turbulent flow (G–1)

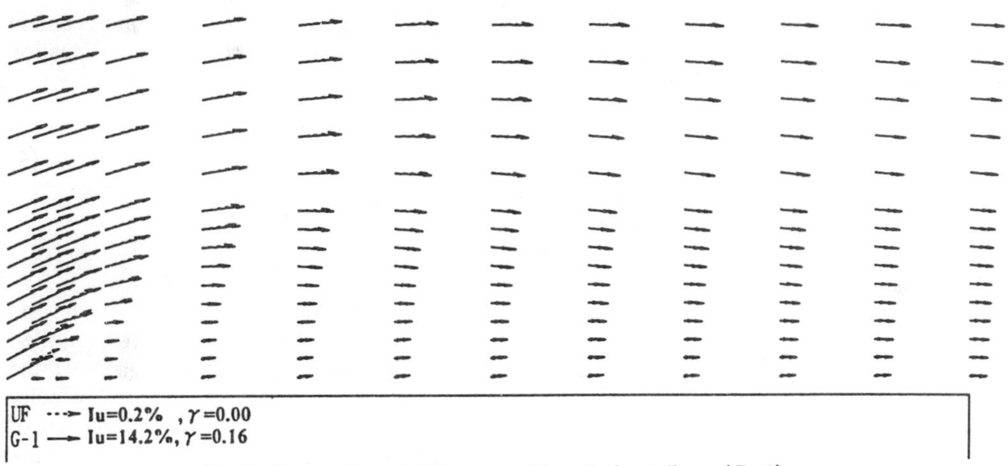

(b) Uniform flow (UF) v.s. grid turbulent flow (G–1)

Figure 11 Flow pattern above Model B

4. CONCLUSION

The following results are obtained:

1) The pressure coefficients on the roof and the base pressure coefficient are influenced by the wind characteristics of the approaching flow, especially, the effects of the non dimensional velocity gradient are much greater than those for turbulence intensity.

2) It is inferred that the effects of the non dimensional velocity gradient originate in the vertical difference of the momentum or the vorticity of approaching flow. The relation between the non dimensional velocity gradient and the momentum or the vorticity must be explained in the future.
3) The effect of the velocity gradient, the turbulence intensity and three dimensional flows on the aerostatic forces of three-dimensional models must be examined in the future.

5. REFERENCE

1. Lee B. E.; "Some effects of turbulence scale on the mean forces on a bluff body", J. of Industrial Aerodynamics, Vol. 1, pp.361-370, 1975.
2. Petty D. G.; "The effect of turbulence intensity and scale on flow past square prisms", J. of Industrial Aerodynamics Vol. 4, pp. 247-252, 1979.
3. Ueda H., Tamura Y. and Fuji K., "Effects of turbulence of approaching wind on mean wind pressures acting on flat roofs", J. of Struct. Constr. Engng, AIJ, No. 425, 1991.
4. Tielman H. W., Pnzer E. C. et. al.; "Effect of Free-Stream Turbulence on Peak Suctions observed below Coner Vortices", Proc. of Third Asia-Pacific Symposium on Wind Engineering, Vol. II, pp.723-728, 1993.
5. Irwin P. A.; "Design and Use of Spires for Natural Wind Simulation", Laboratory Technical Report LTR-LA-233, National Research Council Canada.

9ICWE, 1995, New Delhi, India

VISCOUS FLOW SIMULATION AROUND BLUFF BODY BY VORTEX METHOD

SHIRATO, H., MATSUMOTO, M. and SHIRAISHI, N.

Department of Civil Engineering, Kyoto University, Kyoto, Japan

Abstract

The VIC (Vortex-In-Cell) method is employed in the discrete vortex method in order to make direct treatment of viscosity. The numerical viscosity induced by VIC execution is quantified by comparing its amount of vorticity scattering to the surrounding grid points with that of single Lamb vortex diffusion. The modified VIC method which aims to reduce the numerical viscosity is newly proposed. Its performance is examined by the Reynolds number dependence of flow pattern and Strouhal number of a 2-D stationary rectangular cross section.

INTRODUCTION

With the aid of computer advance, the flow field itself as well as the aerodynamic phenomena of a body in fluid come to be investigated also numerically. Although many sorts of sophisticated CFD codes are published, their performance is much dependent on specification of a computer. The users, who only want to handle these softwares as one of tools, often have to suffer from the hardware requirement such as memory size, CPU speed, and corresponding running cost, etc.

The discrete vortex method (DVM) is one of the CFD techniques to save the difficulties in computation. It is in principle pressure-independent, based on the transport equation of vorticity. The conventional DVM is even mesh-free, therefore, is easy to apply to arbitrary configurations. On the contrary, its disadvantage such as: 1)incapability of direct treatment of turbulence, 2)difficulty of taking into account of the diffusive motion of each vortices, 3)difficulty of vortex convection near body surface due to overestimation of induced velocity has been pointed out. [1],[2],[3]

The diffusive motion of vortices have been taken into account by introducing the Rankine vortex with its core growth, or time decay of the circulation. These are, of course, based on some physical characteristics of fluid motion, and give fairly good agreement with the observed phenomena especially in higher Reynolds number regime. However, these are more or less indirect treatment to the diffusion process of vorticity field. In addition, there is a statement that the superposition of velocity field, which is used in the conventional DVM, cannot give the exact solution of multiple vortices field, because of the nonlinearity of the Navier-Stokes equation [1].

The VIC (Vortex-In-Cell) method [4],[5], which gives the vorticity and velocity at the mesh points surrounding a point vortex, is one of the alternatives to save the above problems 2) and 3). The VIC method also saves CPU time in comparison with the conventional velocity calculation by the Biot-Savart law, when large number of vortices are treated. Furthermore, the diffusion process can be simulated directly. On the contrary, it may bring the additional viscosity to the solution, since its procedure contains the spatial scattering of vorticity to the surrounding grid points.

In this study, the additional numerical viscosity due to VIC method execution is investigated by comparison with the diffusion of single Lamb vortex. The modified VIC method is then proposed to reduce the additional viscosity. In order to examine the proposed method, flow around a stationary 2-D rectangular cross section is simulated, and the Reynolds number dependence of its Strouhal number is discussed.

COMPUTATIONAL PROCEDURE

Certain number of vortices are placed on the body surface, whose strength is determined by the condition of zero-normal velocity at the mid of each two adjacent vortices on the body surface together with the Kelvin's theorem [6]. All surface vortices are released simultaneously to the surrounding fluid in every time step and convected with the flow velocity at each vortex position. The procedure is based on the "time-splitting method" [7], which decomposes the original transport equation of vorticity into two sub-step, convection and diffusion, respectively. Once the vorticity at each mesh point is determined by the VIC method (in detail described later), the diffusion process of vorticity field is then evaluated by solving the following equation:

$$\frac{\partial \omega}{\partial t} = \nu \nabla^2 \omega \qquad (1)$$

where, ω: vorticity, ν: kinematic viscosity of fluid

Here, the exact diffusive motion of vortices should be described not only by eq.(1) but also by changing the circulation of each vortices. This may be completed if new vorticity $\omega(x, y)$ at the mesh point is reinterpolated back to the new circulation of vortices around the mesh points in proper way. This is, however, an indeterministic problem. The circulation of each vortex is kept constant in this study. Reinterpolation by Graham's method [8], difference vorticity field, $\omega^{(n+1)} - \omega^{(n)}$ (the superfix means the time step at (n+1), (n), respectively), onto the original vortex field with the conservation of the first three moments of vorticity, is one of the successful solution to overcome this problem.

The velocity components at each mesh point are then calculated by solving the Poisson equation:

$$\nabla^2 \psi = -\omega \qquad (2)$$

where, ψ: stream function

The convection velocity of a vortex is evaluated by the conventional area-weightening manner (see Fig.1):

$$\vec{u}_k = \sum_{i=1}^{4} \vec{u}_i A_i / \Delta x / \Delta y \qquad (3)$$

where, $\vec{u}_k$: velocity component at the center of vortex k, $\vec{u}_k = (u_k, v_k)$, $\vec{u}_i$: velocity component at mesh point i, $\vec{u}_i = (u_i, v_i)$

The location of each released vortex in the next time step is approximated by taking the rate of velocity change into account:

$$\vec{x}_k^{(n+1)} = \vec{x}_k^{(n)} + \left\{\frac{3}{2}\vec{u}_k^{(n)} - \frac{1}{2}\vec{u}_k^{(n-1)}\right\}\Delta t \tag{4}$$

where, $\vec{x}_k^{(n)}$: position of vortex k at time $t = n\Delta t$, $\vec{x}_k^{(n)} = (x_k^{(n)}, y_k^{(n)})$

NUMERICAL VISCOSITY EVALUATION AND MODIFIED VIC METHOD

The conventional VIC method is to convert the circulation of one point vortex Γ to the vorticity of the surrounding four-grid points, $\omega_1, \omega_2, \omega_3, \omega_4$ by the area-weightening technique. The assigned voticity, ω_1, in Fig.1 by VIC method is expressed as:

$$\omega_1 = \Gamma(1 - \frac{\xi}{\Delta x})(1 - \frac{\eta}{\Delta y})/\Delta x/\Delta y \tag{5}$$

It is pointed out that this method induces numerical viscosity to some extent. This can be approximately quantified by comparing with the diffusive expansion of a single Lamb vortex. Assume one Lamb vortex at (ξ, η) in a cell (see Fig.2), whose vorticity distribution is:

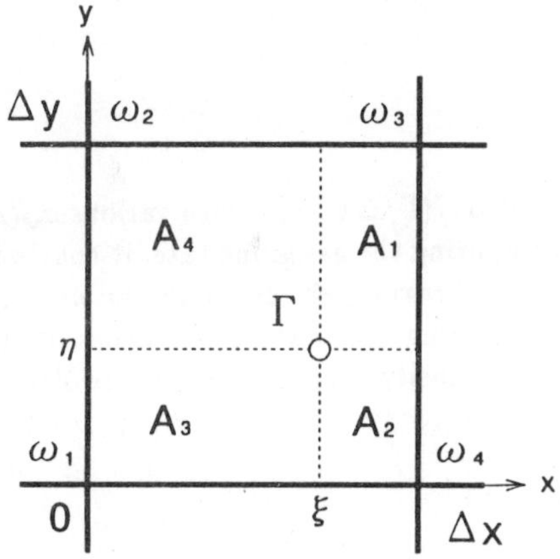

Fig.1 Area-weightening for a vortex in a cell

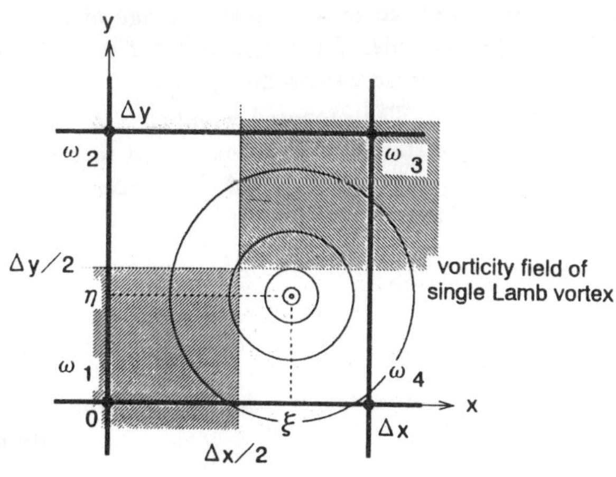

Fig.2 Lamb vortex in a cell

$$\omega(r,t) = \frac{\Gamma}{4\pi\nu t}\exp\left\{\frac{r^2}{4\pi\nu t}\right\} \tag{6}$$

where, t: time after its origin

The vorticity origined at $t = 0$ diffuses according to the spatial growth of the standard deviation:

$$\sigma_w(t) = \sqrt{2\nu t} \tag{7}$$

At $t = \Delta t$, the most appropriate way to convert $\omega(r, \Delta t)$ field to the vorticity at the surrounding four grid points would be assigning each fraction of vorticity field divided by the line $x = \Delta x/2, y = \Delta y/2$ to the nearest grid point (see Fig.2). Namely, for ω_1:

$$\omega_1 = \int_{-\infty}^{\Delta x/2}\int_{-\infty}^{\Delta y/2} \omega(r, \Delta t) dA/\Delta x/\Delta y \tag{8}$$

here,

$$\sum_{i=1}^{4} \omega_i = \iint \omega dA/\Delta x/\Delta y = \Gamma/\Delta x/\Delta y \tag{9}$$

Fig.3 plots the isoline of the assigning rate for ω_1, namely $\omega_1/(\Gamma/\Delta x/\Delta y)$, with various $\sigma_w(\Delta t)/\Delta x$ together with that of VIC method (eqs.(5),(8)). By comparing the assigning rate, it concludes that the numerical viscosity by the VIC method per Δt is approximately equal to the diffusion of a single Lamb vortex during the standard deviation σ_w growing up to about 0.3 times of the mesh size Δx, when $\Delta x = \Delta y$. Consequently, the numerical viscosity ν_{num} per Δt by the VIC method can be evaluated as:

$$\nu_{num} = \frac{\sigma_w^2}{2\Delta t} = \frac{0.09(\Delta x)^2}{2\Delta t} \tag{10}$$

In order to minimize ν_{num} for a given Δt, one has to reduce mesh size Δx, or to increase Δt under given Δx. The latter approach is, however, confined by the optimal range of Δt in the convection substep according to the references. (for example, [6]) If ν_{num} is equal to ν of actual fluid, then the inviscid calculation may represent the realistic viscous flow pattern.

Following the same approach in the previous section, The proposed modified VIC method is derived by use of an *inviscid Lamb vortex*, in which the standard deviation is kept to zero. That is, the vorticity field is expressed by a delta function, $\omega(x, y) = \Gamma\delta(x - \xi)\delta(y - \eta)/\Delta x/\Delta y$. Using eq.(8), the assigned vorticity, ω_1, can be formulated as follows (see Fig.4):

$$\omega_1 = \Gamma\{1 - h(\xi - \Delta x/2)\}\{1 - h(\eta - \Delta y/2)\}/\Delta x/\Delta y \tag{11}$$

where, $h(x) = 0(x < 0), \quad 0.5(x = 0), \quad 1.0(x > 0)$

Eq.(11) represents the assigned vorticity ω_1 for an inviscid vortex, which gives the minimum numerical viscosity.

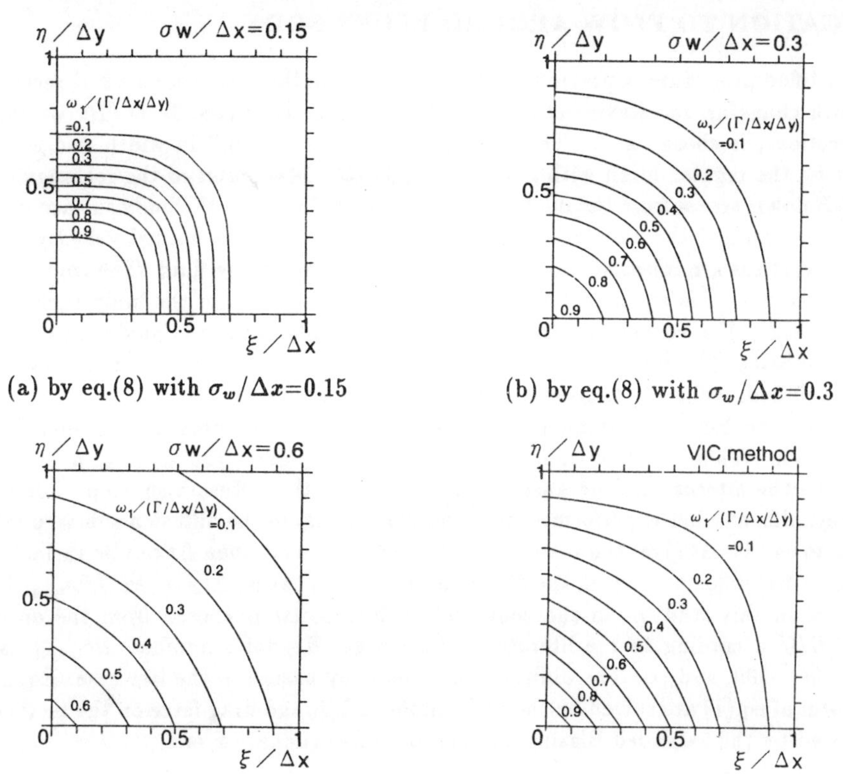

(a) by eq.(8) with $\sigma_w/\Delta x$=0.15

(b) by eq.(8) with $\sigma_w/\Delta x$=0.3

(c) by eq.(8) with $\sigma_w/\Delta x$=0.6

(d) by conventional VIC method (eq.(5))

Fig.3 Isoline of assigning rate for ω_1 ($\omega_1/(\Gamma/\Delta x/\Delta y)$

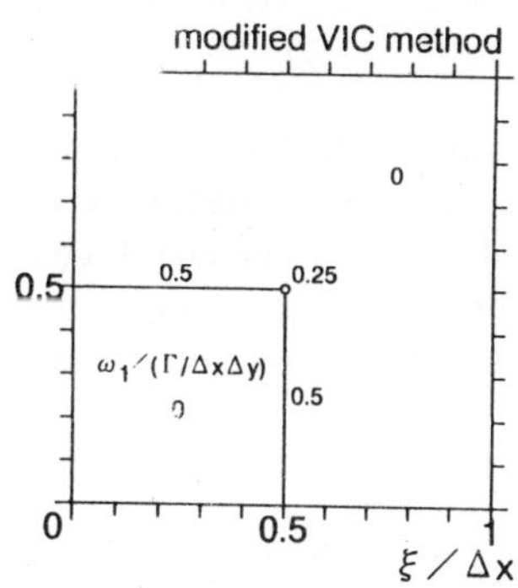

Fig.4 Isoline of assigning rate for ω_1 ($\omega_1/(\Gamma\Delta x/\Delta y)$ by proposed modified VIC method

APPLICATION TO FLOW AROUND BLUFF BODY

The modified procedure in previous section is applied to the flow around a 2-D rectangular cross section with changing the Reynolds number. The slenderness ratio, B/D (B: width, D: depth of cross section), is chosen as 2. The computational domain, $20D$ in width and $21B$ in length, is covered by the regular mesh with $\Delta x = \Delta y = 0.125D$. The center of the rectangular section is placed $5.5B$ downstream from the inlet. Vortices are distributed on the body surface with the pitch Δx. After determined the circulation of these vortices by the zero-normal velocity condition, all the surface vortices are released simultaneously to the surrounding fluid. The convecting velocity of the released vortex is equivalent to the velocity being apart from the body surface with small distance, $0.025D$. If a vortex penetration through the body surface is predicted during the next convection substep, the motion is modified so that the velocity component normal to the body surface is nullified, while the tangential component is maintained. [9]

To change the Reynolds number, the rearrangement of the mesh size is required in proper FDM approach. When the time-splitting technique is employed, the usage of another value of ν or Δt can be the alternatives, because ν appears only in the diffusion substep. Computation for various Reynolds number is performed by manipulating Δt_d in the diffusion sub-step in this study. According to eq.(8), Δt gives the same order of effect to σ_w as ν, then Δt can be variable depending on the target Reynolds number in the diffusion sub-step. That is, $\Delta t_d = (Re_o/Re_{target})\Delta t_c$ ($Re_o = UD/\nu = 300$ in this study). In the convection sub-step, Δt is chosen from the optimal range, $\Delta t_c = 0.1B/U$ according to the literature. The target Reynolds number, Re_{target}, is chosen as 100, 300, 800, 3000, and ∞. The diffusion sub-step is by-passed in the last case. Eq.(2) and fully implicit form of eq.(1) are solved by the SOR method. Lift and drag force on the rectangular body are evaluated by the extended Blasius formula for non-stationary flow:

$$Lift = -Imag\left\{\frac{i\rho}{2}\oint_C\left(\frac{df}{dz}\right)^2 dz + i\rho\frac{\partial}{\partial t}\oint_C \bar{f}d\bar{z}\right\} \tag{12}$$

$$Drag = Real\left\{\frac{i\rho}{2}\oint_C\left(\frac{df}{dz}\right)^2 dz + i\rho\frac{\partial}{\partial t}\oint_C \bar{f}d\bar{z}\right\} \tag{13}$$

where, f: complex potential, $f = \phi + i\psi$, ϕ: velocity potential, $z = x + iy$, $\bar{(\ )}$: conjugate operator

Fig.5 shows the flow patterns around the section in Re=100 and ∞. Each dot represents the individual released vortex. Streamlines are drawn together in these figures. Vortex shedding seems to be rather unstable in both cases, forming irregularly larger or shorter vortex spacing. More frequent reattachment to the body side surface may be observed in Re=100 than in $Re=\infty$.

The wind tunnel experiments support the above flow properties. It is reported that the vortex shedding modes, classified into three categories, are varying irregularly from one to another[10], and that the unstable wake flow pattern is easily affected by the external perturbation in a certain frequency range.[11] It is known that the wake pattern of a rectangular cross section is sensitively modulated by the separated flow reattachment, and correspondingly, the discontinuity of Strouhal number (St, $St = f_s D/U$, f_s:vortex shedding frequency) exists at B/D=2.8 and 6.0. Even in case of a single cross section with fixed B/D, the St discontinuity is also recognized by changing Re. As for B/D=2.0, the flow changes from the reattachment type to the fully separated type in $Re \geq 500$

approximately, and St jumps at $Re \doteq 500$.[12] So the entire wake flow can be devided into two patterns by the boundary of $Re \doteq 500$. And the computed flow patterns in Fig.5 for Re=100 and $Re = \infty$ belong to the reattachment type and the fully separated type, respectively. Fig.6 shows the flow visualization in water tank for B/D=2 in Re=8000[11]. The inviscid case in Fig.5(b) seems to give better agreement with this visualization result, because both (Fig.5(b) and Fig.6) belong to the fully separated type.

Fig.7 plots the Strouhal number (St) for various Re together with the values obtained from the experiment[12]. St by the computation is evaluated from the time history of lift force. Power spectra of the lift force for each Re are also shown in the figure. The spectral peak corresponding to the dominating vortex shedding frequency component is not so sharp as the observation[12]. This may be caused by the unstable wake flow characteristics of B/D=2 section. Although most of the St by the computation gives a bit of smaller value than that from experiments, the St jump around Re=500 is reproduced well by the computation.

Consequently, it can be conclude that the effect of viscosity is properly appeared to the computed wake flow pattern.

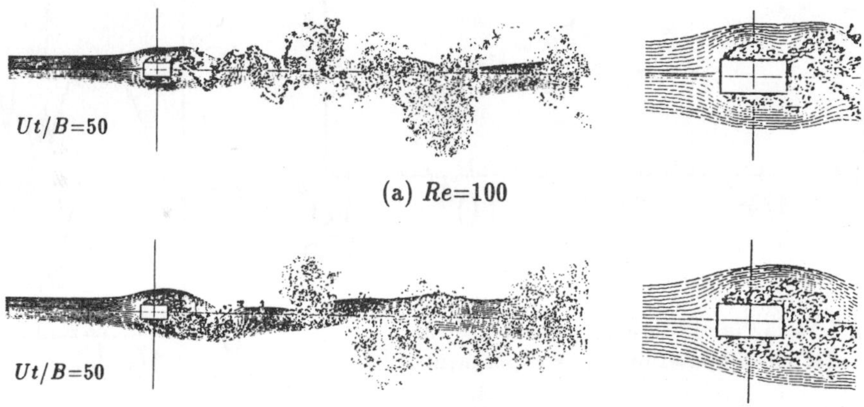

(a) Re=100

(b) inviscid flow

Fig.5 Computed flow pattern around B/D=2 rectangular section

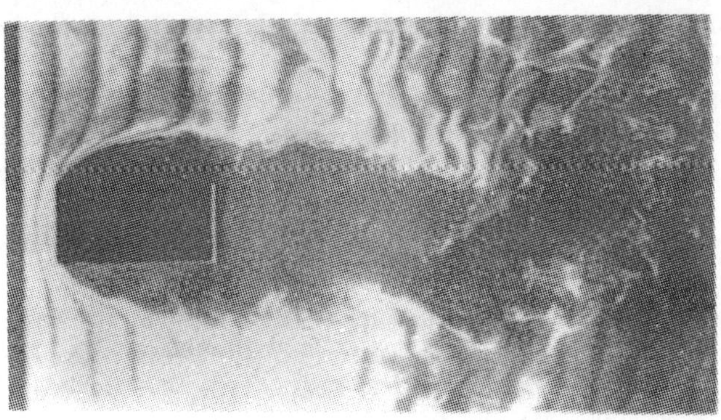

Fig.6 Flow visualization around a rectangular section (Re=8000)[11]

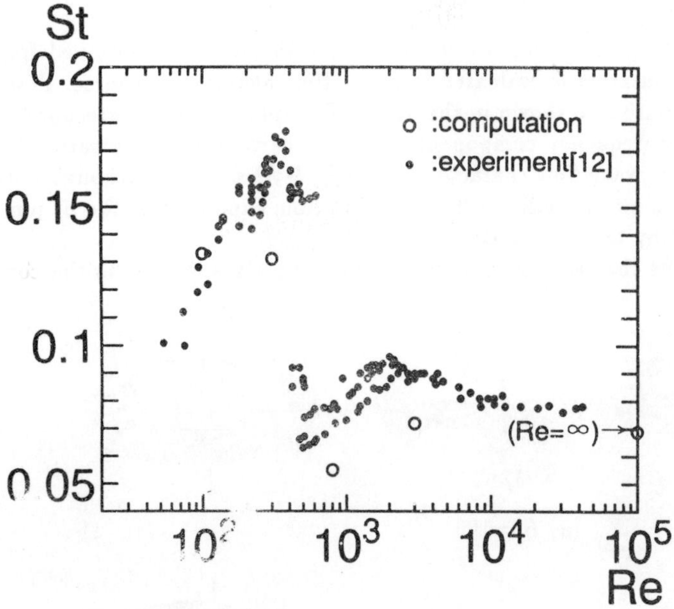

Fig.7 Variation of Strouhal number with *Re*
(•: by computation, o: by experiments[12])

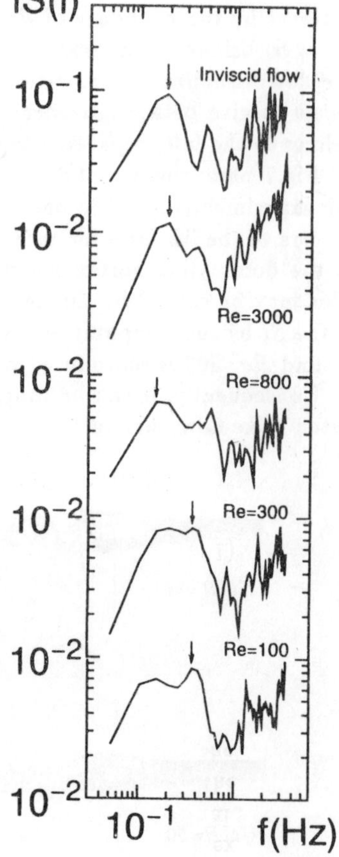

Fluctuating velocity components in the computational flow are monitored. Figs.8, 9 are the time histories and their power spectra of transverse velocity fluctuation (v-component) at several points along the line parallel to the approaching flow and $D/2$ apart from the body side surface. Both are for Re=100 and $Re=\infty$, respectively.

In both cases, the velocity fluctuation develops its periodicity in the downstream wake, which doesn't, however, coincide with the Strouhal frequency. According to careful observation of u- and v- velocity components, it should be emphasized that the Strouhal frequency component dominates not in far downstream but near side surface (eg. point No. 3, 4, 5). As for the reattchment type, Re=100, the power spectra near the body side surface seems to contain another peak which specifies the cahracteristic frequency of movement induced type excitation, f_m, ($f_m = 0.6U/B$) [13]. This is because the flow near the body side surface is bounded by the separated flow, which interferes the direct interaction to the Karman vortex in downstream. On the other hand, no spectral peak at $f = f_m$ are recognized in the fully separated type, $Re=\infty$, since the flow near the body side surface interacts directly with the downstream wake fluctuation.

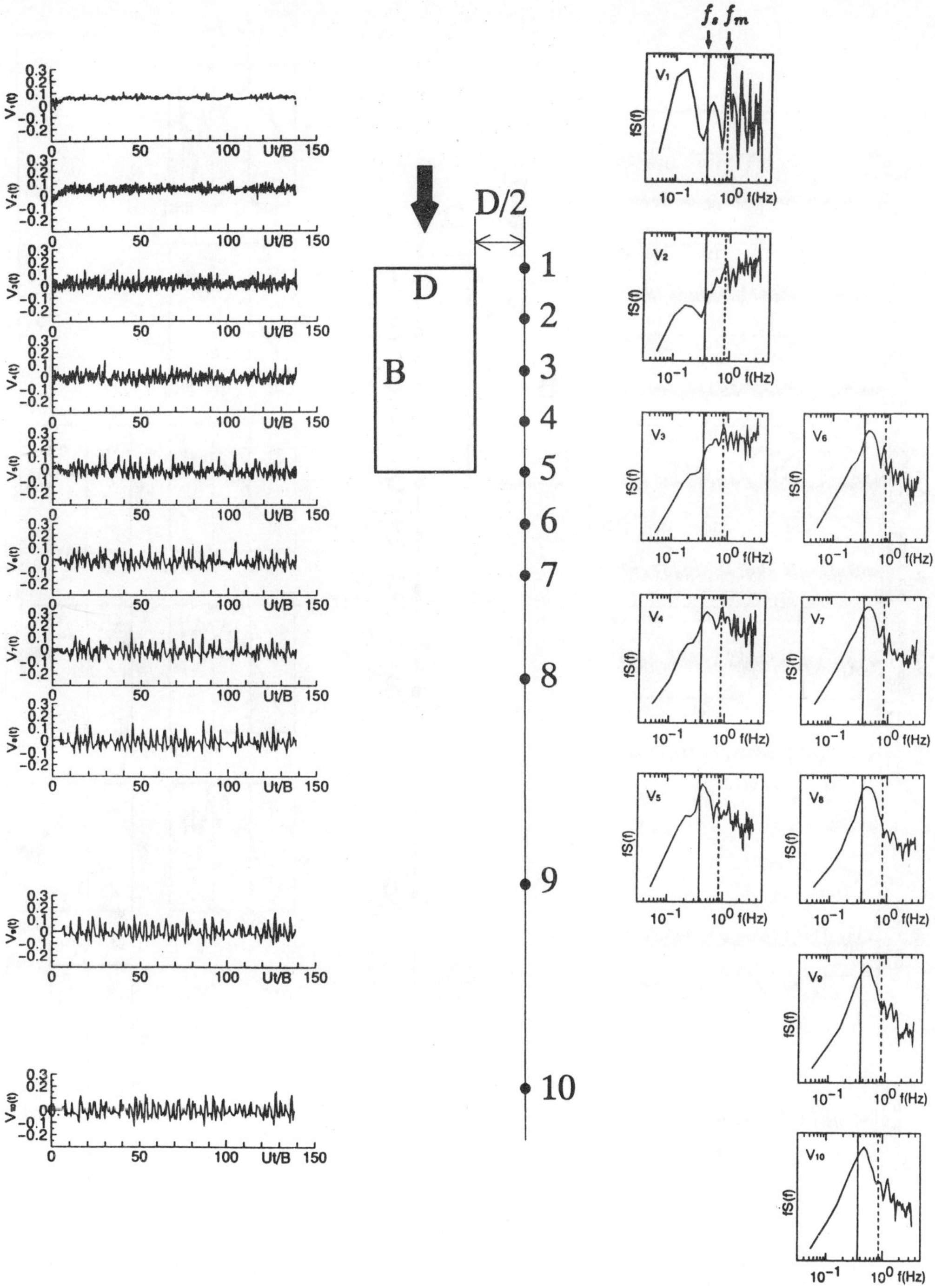

Fig.8 Velocity fluctuation near body and downstream (Re=100)
(f_s: Strouhal frequency, f_m: characteristic frequency of movement induced excitation)

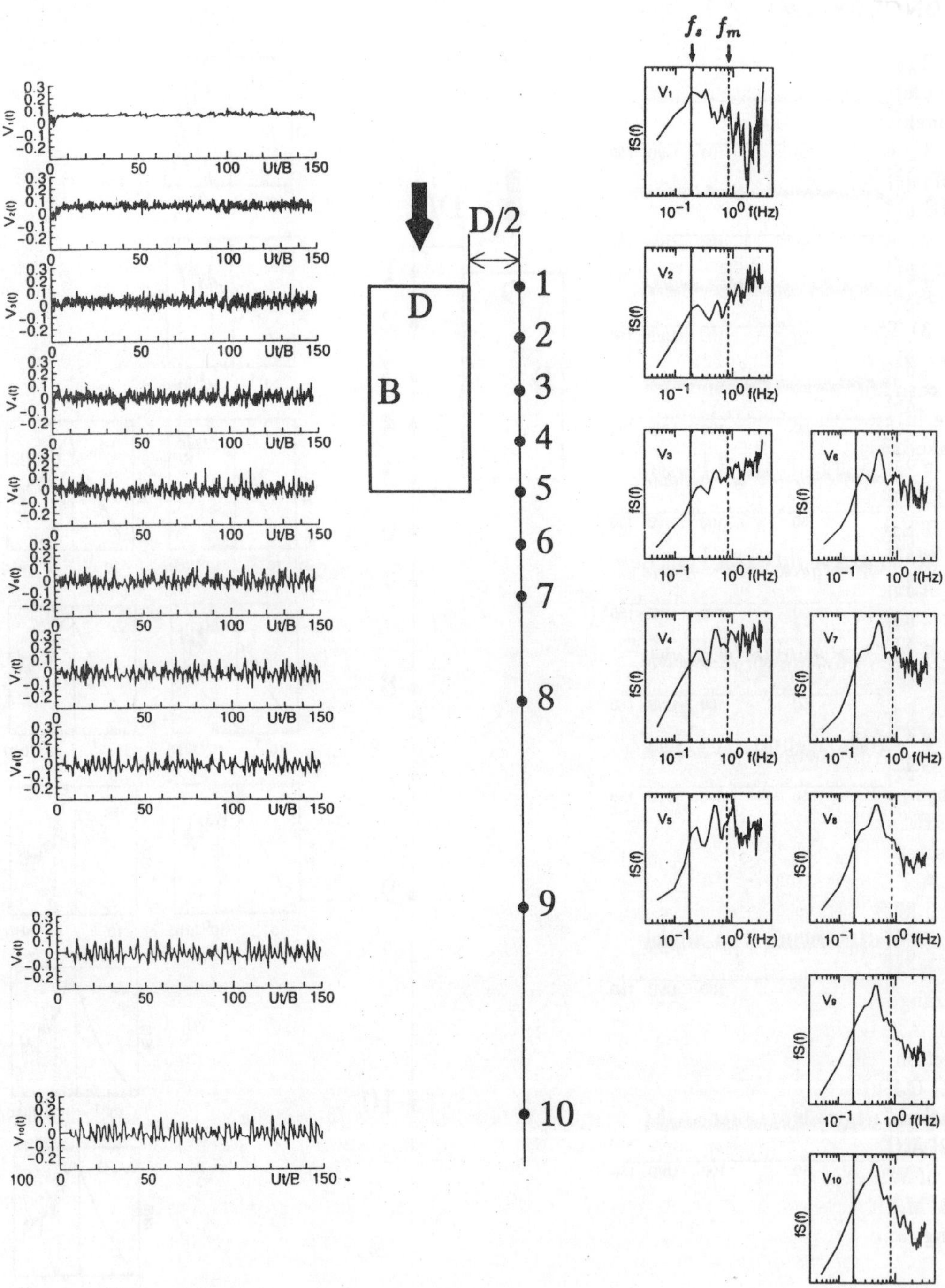

Fig.9 Velocity fluctuation near body and downstream ($Re=\infty$)
(f_s: Strouhal frequency, f_m: characteristic frequency of movement induced excitation)

CONCLUSION

The discrete vortex method with the modified vortex-in-cell (VIC) scheme is performed to simulate the flow around a stationary 2-D rectangular cross section (slenderness ratio B/D=2). Concluding remarks are shown in the following:

1) The numerical viscosity induced by the VIC method is evaluated by comparing with the diffusion of a single Lamb vortex. By use of an inviscid vortex, the modification is proposed to the VIC method, which can reduce the numerical viscosity.

2) Applying the proposed method to the flow around a 2-D rectangular cross section, the change of flow pattern due to reattachment of separated flow is reproduced. The discontinuity of the Strouhal number at around Re=500 gives fairly good agreement with the experimental result.

3) There is significant difference in the velocity spectra around the body between Re=100 and $Re=\infty$. The frequency component corresponding to the movement induced excitation is recognized in case of the reattachment type flow pattern (Re=100). This result supports the discussion that the interaction of velocity fluctuation between at the body side surface and at the downstream wake is signmificantly controlled by the separated flow reattachment.

REFERENCES

[1] T.Sarpkaya, "Computational Methods With Vortices - The 1988 Freeman Scholar Lecture", J. of Fluids Eng., Trans. of ASME, vol.111, 1989

[2] A.Leonard, "Review - Vortex methods for flow simulation", J. of Comp. Phys., vol.37, 1980 , pp.289-335

[3] H.Shirato, M.Matsumoto, N.Shiraishi, "Unsteady aerodynamic force characteristics on 2-D oscillating bluff body", CWE92 Computational Wind Engineering (Ed. S.Murakami), 1993

[4] J.P.Christiansen, "Numerical Simulation of Hydrodynamics by the Method of Point Vortices", J. of Comp. Phys., vol.13, 1973

[5] G.R.Baker, "The "Cloud in Cell" technique applied to the roll up of vortex sheets", J. of Comp. Phys., vol.31, 1979

[6] H.Sakata, T.Adachi, R.Inamuro, "Simulation method for unsteady flow with separation by discrete vortex model", Proc. of JSME, vol.49, No.440, 1983

[7] A.J.Chorin, "Vortex sheet approximation of boundary layers", J. of Comp. Phys., vol.27, 1978

[8] J.M.R.Graham, "Computation of viscous separated flow using a particle method", Numerical Methods for Fluid Dynamics III, 1988

[9] N.Arai, K.Taguchi, T.Tani, "Flow simulation around twin rectangular prisms in side-by-side arrangement by discrete vortex method", Proc. of JSME, vol.53, No.486, 1987

[10] A.Okajima, "Aerodynmamic charatcteristics of a rectangular prism and an angle-structural section (Effects of Reynolds numbers)", J. of Wind Eng., No.59, 1994

[11] C.Knisely, M.Matsumoto, F.Menacher, "Rectangular cylinder in flows with harmonic perturbations", J. of Hydr. Eng., ASCE, vol.112, No.8, 1986 , pp.690-704

[12] A.Okajima, "Flow around a rectangular cylinder with a section of various width/height ratios", J. of Wind Eng., No.17, 1983

[13] M.Matsumoto, N.Shiraishi, H.shirato, "Bluff body aerodynamics in pulsating flow", J. of Wind Eng. and Ind. Aerod., vol.28, 1988, pp.261-270

THEME IV
Wind Tunnel Techniques, Studies

9ICWE, 1995, New Delhi, India

DEVELOPMENT OF A TWISTED FLOW WIND TUNNEL FOR TESTING YACHT SAILS

FLAY, R.G.J. [1], LOCKE, N.J. [2], and MALLINSON, G.D.[3]

Senior Lecturer *Postgraduate Student* *Associate Professor*

Mechanical Engineering Department, University of Auckland, Auckland, New Zealand

Abstract

This paper presents the results obtained in the course of developing a special wind tunnel aimed at reproducing the variation in the onset wind speed and direction with height up the mast as seen by yachts as they sail in the boundary layer above the sea. Earlier work has shown that the variation in the wind speed with height can easily be modelled by using standard boundary layer wind tunnel techniques. The variation in wind direction with height is a more difficult problem with which to deal. The paper shows results from a pilot wind tunnel study which investigated two concepts to produce sheared twisted flow. Both of these concepts were successful, but the concept using a series of twisted vanes gave the better solution as the area of good flow was larger than that produced by a dual wind tunnel concept, and, it is a more economical solution from the point of view of both running and initial capital costs.

1. INTRODUCTION

The motion of a yacht when it sails in the boundary layer above the sea produces an apparent wind velocity which results from the vector addition of the wind and yacht velocities. Since the wind speed increases with height, this means that the apparent wind speed incident onto a yacht also increases with height, and in addition its direction changes, rotating away from the yacht's heading with increased height. This rather complicated onset flow causes difficulty when analysis using computational fluid dynamic codes is used, and also, when modelling in the wind tunnel is carried out. Requirements for wind tunnel testing have been reported on by Flay and Jackson (1992), and Flay (1992), where it has been noted that the variation in wind speed with height can be modelled in the wind tunnel in a similar way to the procedures used for conventional wind engineering testing. However, the twist in the wind direction with height is a more difficult problem to deal with for a stationary wind tunnel model, because the true and apparent wind speeds are the same. This problem is considered in the paper.

2. WIND STRUCTURE OVER THE SEA

The velocity profile over a sea surface over the height range of importance can be described by the simple logarithmic law, with a typical roughness length of 0.2 mm, i.e.

$$\bar{U} = \frac{U_*}{k} \ln \frac{Z}{Z_0} \qquad (1)$$

where U_* is the friction velocity, k is von Karman's constant, Z is height above the sea surface and Z_0 is the roughness length. For flow over the water Z_0 varies with wind speed because wave heights increase with wind speed (and fetch). A correlation for Z_0 over water as a function of wind speed is provided by Cook, (1985), namely

$$Z_0=5*10^{-5}*U_{10}^2/g. \tag{2}$$

When evaluated using U_{10}=8 m/s these give Z_0=0.3 mm, which is a very small roughness length, smaller than that obtained for the smooth surface of desert areas. In relatively low winds, "rounded" waves, generally translating in the direction of the wind, obviously offer little drag to the wind compared to the stationary generally sharp-edged obstacles on land.

The flow over the sea is turbulent, except in stable stratified conditions. In neutrally stable conditions, with a roughness length of 0.3 mm, the turbulence intensity will be about 10% over the height of typical masts. This is much lower than that appropriate to building studies where the roughness length is typically around 0.2 - 2 m, and the turbulence intensity is 15 - 20%.

Yacht model scales are often in the range 1:6 to 1:40, (cf. 1:100 to 1:1000 for buildings), and this means that potentially it would be very difficult to properly model the large low frequency eddies important to building aerodynamics. However, it has been argued earlier (Flay and Jackson, 1992), that it is not necessary to model the complete wind spectrum for wind tunnel testing sails. It is important to model the high frequency end of the spectrum, as this affects the way the sail boundary layer develops, and the curvature of any separated shear layers. The low frequency end of the spectrum can be considered to be quasi-static. The resulting turbulence intensity for modelling has been shown to be

$$Tu_{Z,app} = \frac{0.45}{\ln\left(\frac{Z}{Z_A}\right)} \tag{3}$$

where Z_A is the apparent roughness length, and is given by

$$\frac{Z_A}{Z_0} = \frac{1}{f}\text{ (upwind)}; = f\text{ (downwind), where } f = \left(\frac{h}{Z_0}\right)^{\frac{V_S}{V_h}} \tag{4}$$

h is a reference height, e.g. the height of the mast, V_h the wind speed at this height, and V_S the boat speed. Thus the roughness is typically diminished by a factor of 300 when sailing upwind, and increased by the same factor downwind. This means that for the downwind case when the yacht is sailing with the wind, the turbulence intensity is somewhat enhanced, whereas the reverse is true when sailing upwind.

3. APPARENT WIND STRUCTURE - TARGET FLOW

The direction the yacht sails with respect to the mean wind direction affects the shape of the apparent velocity profile and the amount of twist, as illustrated in Fig. 1. For the upwind case (a), the amount of twist, γ, between the two levels is smaller than for the downwind case (b). Case (b) represents the more challenging wind tunnel modelling situation, and is also the one which is least amenable to CFD investigation. It is this more difficult case which is the main focus of this paper. Case (b) is also of some importance for America's Cup racing yachts, as they spend considerable time sailing downwind.

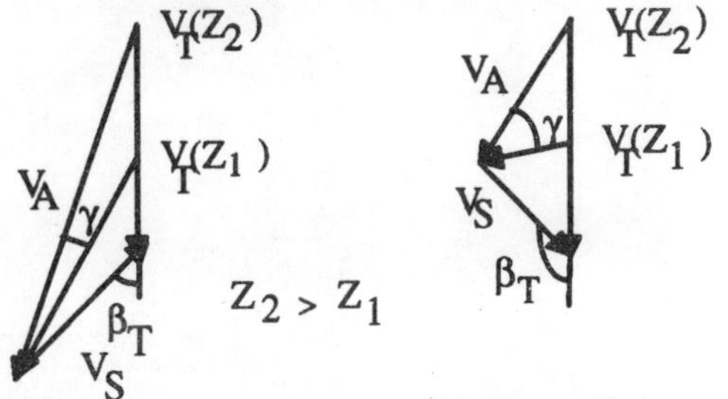

(a) upwind (b) downwind

Figure 1 Illustration of wind twist. Two true wind vectors of different lengths lie on top of each other with their heads aligned. The angle γ is the apparent-wind "twist" angle.

4. DESIGNS FOR PRODUCING WIND TWIST

A further look at Fig. 1 reveals possible solutions to the modelling problem. A model yacht could be fired at an angle β_T through the boundary layer flow generated in a wind tunnel. This would give the correct aerodynamics, but measuring the forces during its short "transit" through the test section would be difficult. Such an approach has been taken by the vehicles group at Nottingham, (Humphreys and Baker, 1992), to obtain force measurements on trains and lorries, but many repetitions of each configuration have been necessary to obtain statistically reliable results. The practical difficulties associated with this approach for yachts, with their flexible sails, etc. seem enormous, and provided sufficient reasons for the authors not to pursue this approach further.

A rather different approach, also suggested by Fig. 1, was taken in the present work for one of the test configurations. The basic idea was to keep the model stationary, and to add a velocity of $-\vec{V}_S$ to the velocity profile, $\vec{V}_T(Z)$, and hence introduce twist into the onset flow. Conceptually one can see that this could be done by building a boundary layer wind tunnel and having porous walls so that an appropriate amount of uniform flow can be introduced through one side, and the same amount withdrawn through the other side. A means would have to be available to alter the angle β_T to that desired for a particular test. Unfortunately, it appeared difficult to the authors to implement this idea in a practical fashion for the general case, and so it was decided to restrict β_T to 90^0, and then to test a practical realisation of this particular case by measuring the flow produced in the jet of two wind tunnels intersecting at 90^0. A very useful feature of such a configuration is that the amount of twist would be expected to change as the relative speeds in the two wind tunnels changed, and thus changing the wind twist would be as easy as altering one fan speed.

A second idea was also of interest. This was to use a set of twisted turning vanes at the outlet of a boundary layer wind tunnel. The boundary layer wind tunnel would be used to develop the shear flow, as is normally done, and the twisted vanes would be used to develop appropriate twist based on velocity triangles produced from yacht performance velocity prediction programs. A drawback of this idea was that potentially the twisted vanes would be difficult and expensive to make. Also, they might prove difficult to adjust and thus not be set to different angles as often as they should be, thus reducing the productivity of the facility.

Figure 2 Model wind tunnel in closed circuit mode.

Figure 3 Model wind tunnel in open circuit mode.

A pilot facility comprising two separate wind tunnel circuits, a straight Boundary Layer Wind Tunnel (BLWT) and a return circuit Uniform Flow Wind Tunnel (UFWT), whose flows intersected at 90° in a common test-section as shown in Figure 2 was built. The return leg of the UFWT was removable which allowed the converging flow concept to be tested with a closed-loop or open jet UFWT. The entire UFWT was removed to test the twisted vane concept, although the flow was still constrained horizontally in the test-section by extended floor and roof sections that were used in all test cases to constrain the jet flows to the horizontal direction. Both wind tunnels had constant cross-sections of height 300 mm and width 450 mm. The wind speed in each wind tunnel was variable with a maximum of around 7.1 m/s in the UFWT and 8.7 m/s in the BLWT.

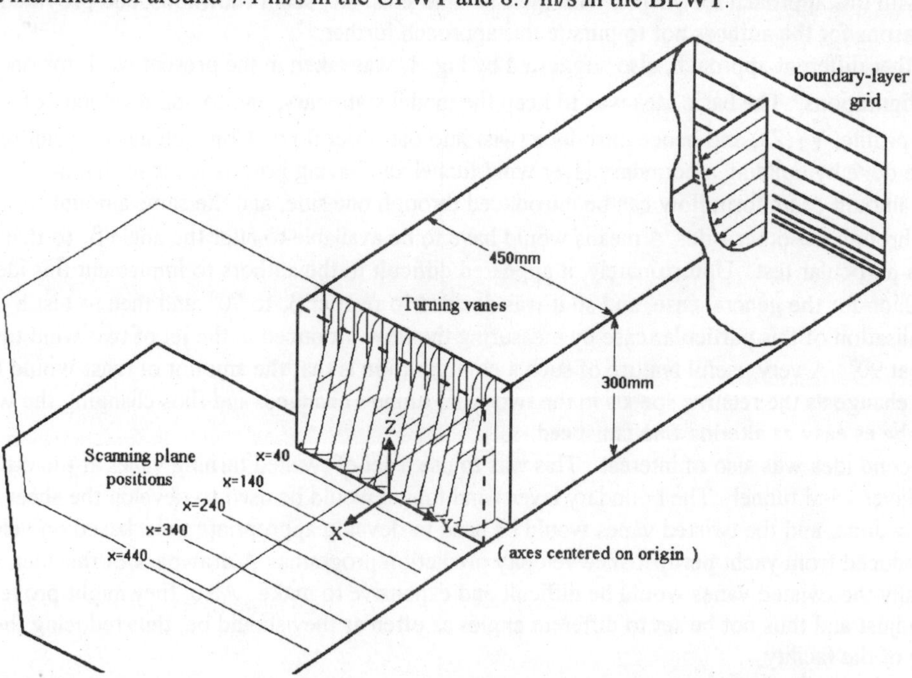

Figure 4 Schematic diagram of the twisted turning vane experimental setup.

Twisted vanes were designed using the ideas of conventional turning vanes in the corners of wind tunnels. An array of 15 turning vanes was constructed from 0.5 mm polycarbonate plastic, with chords of 60 mm, and lateral spacings of 30 mm. The leading edges were held straight into the flow by horizontal steel strips (combs). Horizontal wires were attached to the trailing edges at six heights and "guitar string" tensioners used to twist all the trailing edges in parallel.

5. COMPUTATIONAL FLUID DYNAMICS MODELLING

In order to investigate extrapolating the experimental results to other configurations, a computational fluid dynamics (CFD) investigation was also carried out. The CFD modelling was done using a package called CFDS-FLOW3D marketed by AEA Technology Harwell. CFDS-FLOW3D solves the equations governing the motion for a viscous fluid using a finite volume method to create the numerical approximations. Two wind tunnel configurations were modelled with the intention of representing the physical prototypes as closely as possible. The CFD simulations were compared with the physical measurements using the SeeFD visualisation system written at the University of Auckland.

For both configurations, the fluid was assumed to be incompressible with a viscosity equal to that of air. The mesh spacing was set at 0.045 m in the horizontal dimensions and 0.03 m in the vertical direction. Mesh refinement tests and comparison with the physical measurements indicated that the mesh resolution was adequate. The k, ε turbulence model was used.

A plan view of the region modelled for the dual wind tunnel is shown in Figure 5(a). The boundary conditions for the two merging streams were set 0.45 m upstream from where they entered the test section. The blocked regions delineate the wind tunnel walls on which the no-slip boundary condition was imposed. At the two outlet surfaces, the pressure was specified to be constant. CFDS-FLOW3D automatically ensures that overall mass conservation is satisfied. The incoming uniform flow was set to be 75 m/s. The shear flow was set to match measurements from the wind tunnel. When the turbulence model was used, the turbulence energy was set at 2% of the kinetic energy of the flow.

Figure 5(b) shows the geometry used to model the single wind tunnel with vanes. The flow was specified to correspond to that measured at the exit of the vanes, and this served as the inlet boundary condition for the test section. Otherwise, the conditions correspond to those described above for the dual wind tunnel.

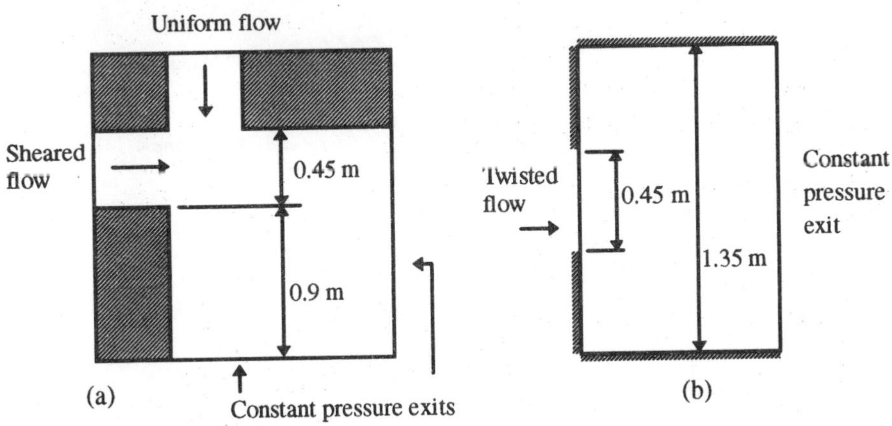

Figure 5 Plan views of the regions modelled for (a) the dual wind tunnel and (b) the single wind tunnel with inlet vanes which generate a twisted flow.

6. DUAL WIND TUNNEL CONCEPT

6.1 Comparison of Experimental and CFDS-FLOW3D results

The purpose of the flow investigations was to determine the extent and uniformity of the twisted flow region to ascertain its usefulness for testing yacht models in a full-size version of the wind tunnel. The first flow investigations were carried out very simply and quickly, using a grid tufted with cotton, in case there was no point to making more quantitative investigations. It was clear from these investigations that the flow was indeed twisted in the manner expected. Further investigation using this simple tufted grid into the extent of twisted flow in both the closed and open circuit configurations showed conclusively that there was a larger extent of uniformly twisted flow in the open circuit mode (Figure 2) compared to the closed circuit mode (Figure 1). Thus further investigation was confined to the open circuit configuration only.

Detailed measurements using a hemispherical head 5-hole flow angle probe (Locke, 1994, Locke et al, 1994) attached to a computer controlled traversing rig were then obtained. The velocities at the exit of each wind tunnel (test-section inlet) were measured with the other wind tunnel off, and the average resulting velocity profiles were vectorially added to give an 'expected' flow twist. From this, a twist of 19° in the flow was calculated between the heights of 10 mm and 110 mm. Six planes of area 600 mm * 250 mm normal to the boundary layer wind tunnel flow were scanned over a grid at 25 mm spacing in both directions with both wind tunnels running. A plot of the velocities measured in the pilot wind tunnel test-section at heights of 35, 110 and 235 mm is given in Figure 6. The twist in the flow was calculated relative to the velocity vectors at $z=110$ mm as these vectors were reasonably uniform over the horizontal plane making them a good reference. Figure 7 shows the region in the test section where the experimentally measured twist between the heights of 10 mm and 110 mm was greater than or equal to 19° and clearly show the useful testing region lies within the BLWT side of the coalesced jet. However, within this region the flow is not uniform. Plots of the velocity and twist profiles show the flow is more uniform towards the upper right corner of the figure (i.e. over scanning planes 4 and 5 defined in Figure 6).

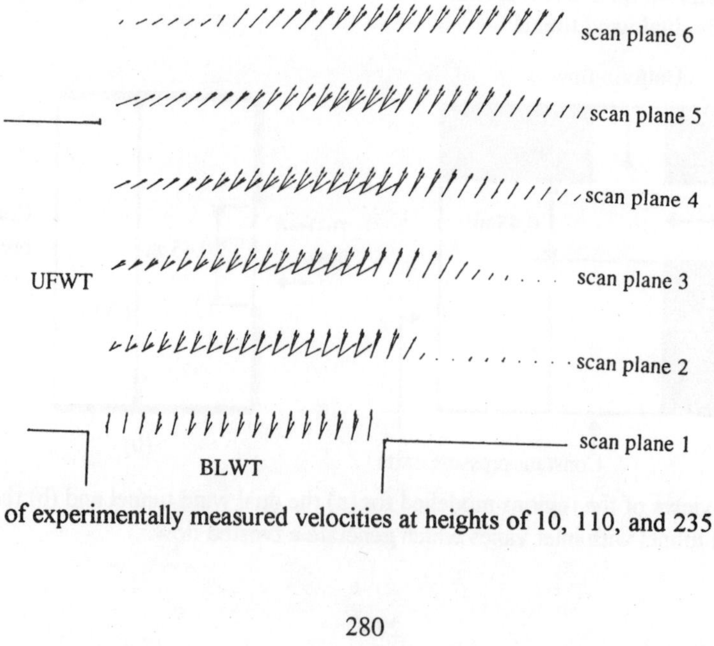

Figure 6 Plot of experimentally measured velocities at heights of 10, 110, and 235 mm.

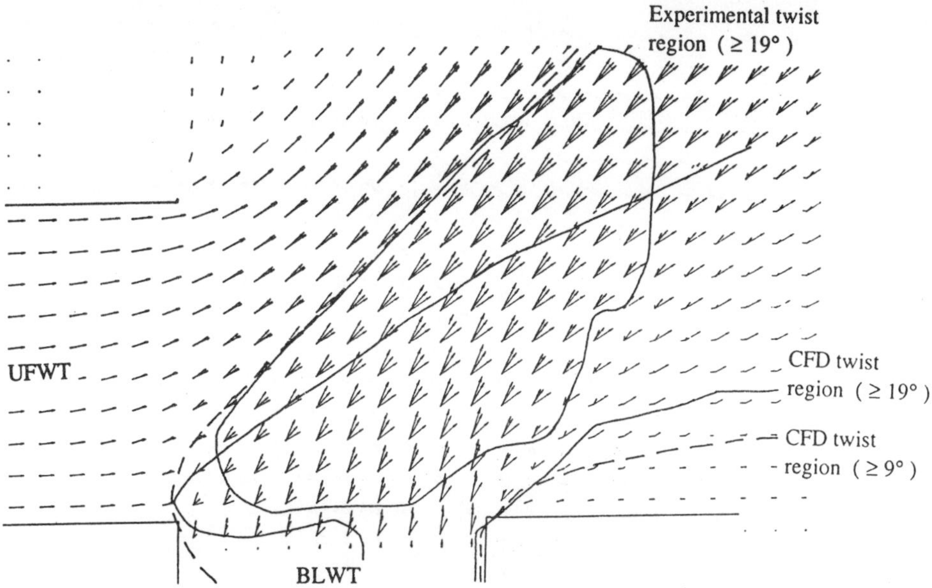

Figure 7 CFDS-FLOW-3D solution and twist regions from CFD and experimental results (19^0 is the expected twist from vector addition of the inlet flows).

CFDS-FLOW3D solutions were found using laminar and turbulent (k-ε) models and an increase in the effective Reynolds number from 10^4 (pilot wind tunnel scale) to 10^5 (full scale) was also modelled. Both the laminar / turbulent and different Reynolds number solutions were very similar so only the turbulent results with $Re=10^5$ are discussed hereafter. As with the experimental measurements, the twist in the CFDS-FLOW3D flow is calculated relative to the velocities in the z=110 mm plane. The CFDS-FLOW3D predictions compare well with the experimental measurements, especially over the boundary layer mixing region at the right hand side of the coalesced jet (compare Figs. 6 and 7).

The region where CFDS-FLOW3D predicts 19° of twist or more, shown in Figure 7, is much smaller than the corresponding experimental region. This is expected as the twist in the CFDS-FLOW3D results was more uniform vertically compared to the experimental measurements where nearly all the twist occurred between z=10 and 110 mm. The relatively small region of useful flow for testing, compared to the width and length of the wind tunnels, is a drawback of this design.

6.2 Extrapolation using CFDS-FLOW3D to other conditions

In order to increase the useful test area, relatively wider BLWT and UFWT's with respect to their heights were proposed that were 600 mm wide thus increasing the wind tunnel aspect ratio (height : width) from 1:1.5 to 1:2. The effect of increasing the aspect ratio was investigated using CFDS-FLOW3D, as it was much easier to alter the numerical model than the physical model. The twist distribution across the test-section was shown to scale proportionally to the width. Thus an increase in the useful test area is obtained by using wind tunnels with larger relative widths.

A change in the relative speeds of the wide wind tunnel configuration (effectively changing the simulated sailing conditions) was modelled using CFDS-FLOW3D, by halving the BLWT speed while

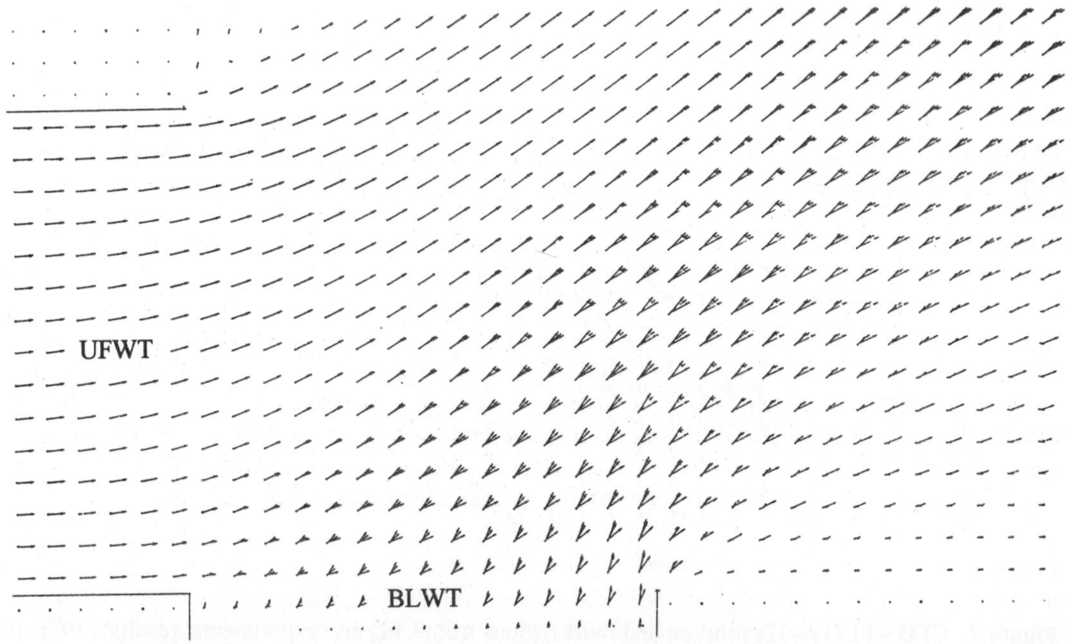

Figure 8 CFDS-FLOW3D solution for wide wind tunnel configuration (300 mm H x 600 mm W) with BLWT running at half speed. Velocity vectors plotted at z=35, 110, and 235 mm.

keeping the UFWT speed constant. The resulting velocities are given in Figure 8 and clearly show a reduction in the flow twist compared with Figure 7. Tufted grid experimental observations and CFD also showed that the useful testing region moved significantly when the relative speeds of the two wind tunnels were changed. This result is important as it means that identifying the optimum test position of a model would be difficult. The ease of altering the wind twist by adjusting a fan speed would be negated if the model / balance position had to be moved to keep it in a good region of the flow at the same time.

6.3 Dual Wind Tunnel Conclusions

1. The foregoing has shown that useful measurements were obtained from a tenth scale pilot wind tunnel with consistent flow angle and speed results.
2. The CFDS-FLOW3D results showed good agreement with physical results.
3. The model and CFD results proved the idea that a region of twisted flow suitable for testing yacht sails could be developed. Unfortunately the region of uniform speed and twist was relatively small compared to the width of the two wind tunnels.
4. CFD results showed that the useful testing region could be increased by increasing the wind tunnel width.
5. Physical and CFD investigations verified that the amount of twist could be altered by changing the speed of one of the wind tunnel flows. An unfortunate consequence of this was movement of the useful twisted flow test region.
6. The main conclusion from this work is that although the idea is sound in principle, in practice it is not a convenient solution for testing yacht models in twisted wind flow.

7. SINGLE WIND TUNNEL / TWISTED TURNING VANE CONCEPT

The pilot wind tunnel was re-configured without the UFWT and a set of twistable vanes, as described in section 4, was attached to the outlet of the BLWT. They were then set to the largest twist angle that could conveniently be achieved, giving a component of the flow in the positive y direction at the top, and the negative y direction at the bottom (see Figure 4 for the definition of the axes). The flow was surveyed as before using the tufted grid and 5-hole flow angle probe attached to the computer controlled traversing rig.

7.1 Experimental and CFDS-FLOW3D results

The experimental measurements were taken from the pilot BLWT setup with a set of twisted turning vanes installed at the wind tunnel exit as illustrated in Figure 4. The experimental and CFDS-FLOW3D velocities, plotted in Figures 9 and 10 respectively, show excellent agreement in both speed and twist over most of the test-section, except at the edges of the jet in the shear layer where the experimental velocities decay slightly faster. More important is the horizontal (x, y directions) uniformity of the flow throughout the useful test region. Figures 11 and 12 show the uniformity in both the twist and the speed in the downstream direction. There is also excellent agreement between the vane exit angle and the twist induced in the flow making it easy to predict and implement changes in the twist. As expected, the useful test area is triangular in shape, decreasing with distance downstream of the vanes as the jet boundaries converge towards each other - one way at the top, and the other at the bottom.

In the scanning plane immediately downstream of the model turning vanes the flow was non-uniform. It is unclear if the disturbances were caused by the vanes or poor flow quality upstream of the vanes, but the influence of the vanes on the local flow was apparent and thus it is important to design the vanes and their supporting structure to attempt to minimise the aerodynamic interference. The non-uniformity decayed away by the next wake survey plane at x=140 mm.

7.2 Quality of expected simulation

An analysis of the velocity triangles at different levels for a sailing yacht easily shows that most of the twist occurs at the lower levels. To keep the flow straight at say 40% mast height means that it would not be untypical to wish to have flow twisted at $+6^0$ at the top, and -30^0 at the bottom of the mast. The experimental and CFD results showed good agreement, and verified that the twisted vane wind tunnel concept can produce this kind of desired onset flow. The results thus support the idea of developing this idea into a full size wind tunnel capable of producing the desired flows onto large model yachts. In such a wind tunnel the shear and turbulence would be developed upstream of the vanes in the BLWT as for conventional wind engineering testing. The vanes would then simply twist this flow to the appropriate wind angle. Since the flow has been shown to basically continue on in the same direction as the trailing edge of the vanes, it is anticipated that it will be quite easy to set the vanes to the required angles.

A good quality flow simulation for yacht sail investigations is expected with this wind tunnel design.

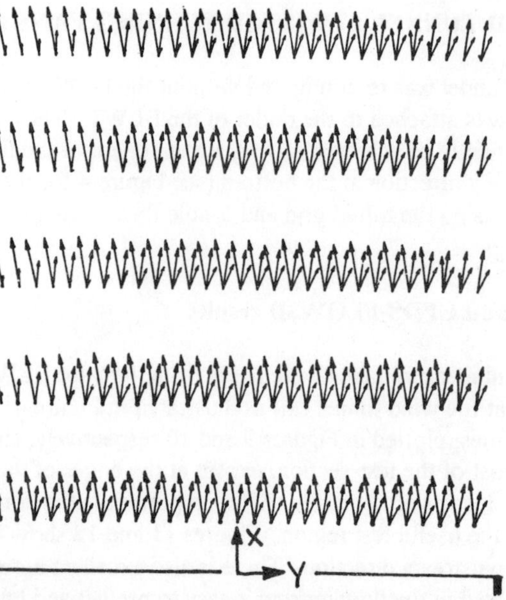

Figure 9 Experimentally measured velocities plotted at heights z=35, 110, and 235 mm.

Figure 10 CFDS-FLOW3D velocity solutions plotted at heights z=35, 110, and 235 mm.

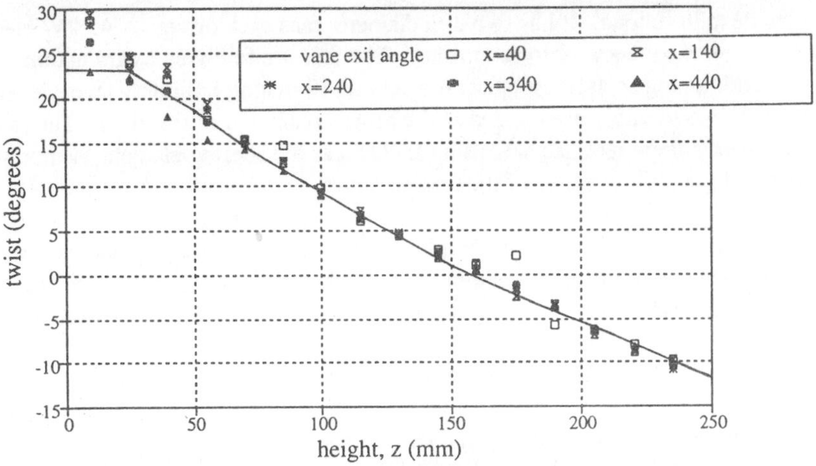

Fig. 11 Twist profiles along the test-section centreline, x=40 to 440 mm, y=0 (twist at each point is the average of twists from y=+/- 60 mm).

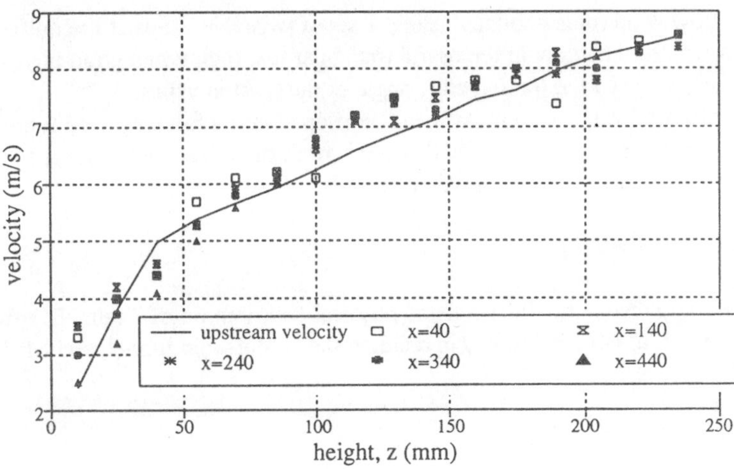

Figure 12 Velocity profiles along the test-section centreline, y=0, x=40 to 440 mm (velocity at each point is the average of velocities from +/- 60 mm).

7.3 Full scale design concept

In view of the foregoing promising results, the twisted vane option was developed into a full scale wind tunnel. In addition to the better flow quality expected using vanes based on the results from this study compared to the dual wind tunnel concept, the cost of the facility is less as only one wind tunnel is required, and the space requirement is therefore reduced. A height to width ratio of 1:2 was chosen for the full scale wind tunnel to maximise the useful testing area. The final dimensions of the wind tunnel are

3 m high x 6 m wide x 29 m long. It has two 3 m diameter fans each driven by 45 kW electric motors controlled by a variable frequency motor controller. The fans are followed by transitions, a honeycomb, a screen, a flow conditioning section for spires etc. and an 18 m long boundary layer development section. The special twisted vanes are located at the downstream end of this duct. The resulting top speed is 11 m/s. A six-component balance, using LVDT's as the sensing element, is mounted centrally 1.5 m downstream of the trailing edges of the turning vanes.

The wind tunnel is a general purpose facility and the twisted vanes can be removed for conventional wind tunnel testing such as for building aerodynamics investigations. It is now complete and has been used successfully in some preliminary investigations. A detailed survey of the flow is still being undertaken at the time of writing; these commissioning results will be reported on in due course.

8. CONCLUSIONS AND RECOMMENDATIONS

The dual wind tunnel concept has been shown to work, and to produce a region of twisted flow similar to that desired to test yacht sails, but the useful test region was smaller than desired. The useful test region was shown to move when the relative speeds of the two wind tunnels was changed. This effect reduces the attractiveness of this wind tunnel concept for sail aerodynamics investigations because the ease of changing the twist is offset by the need to move the model and balance when the twist is changed.

The twisted flow wind tunnel concept using a set of twistable vanes at the outlet of a BLWT showed good performance. The flow in the useful triangular test region had good uniformity, and the flow direction was very close to the trailing edge angle of the twisted vanes.

The BLWT / twisted vane concept has been developed into a full size wind tunnel design and has been built. Initial commissioning results show that the flow in the full scale facility is twisted as expected and that the wind tunnel is very useful for developing sails.

9. ACKNOWLEDGEMENTS

The authors acknowledge the financial and physical contribution of Team NZ towards this research project, the work of Mr Jim Hefkey in building the model wind tunnel, and Mr Greg Collecutt in building the model twisted vanes.

10. REFERENCES

Flay, R.G.J. (1992), "Wind tunnel tests on a sixth-scale laser model", Ship Science Report No. 55, Department of Ship Science, Faculty of Engineering and Applied Science, University of Southampton, Highfield, England, UK.

Flay, R.G.J., and Jackson, P.S. (1992), "Flow simulation for wind tunnel studies of sail aerodynamics", J of Wind Eng. and Ind. Aerod., 41-44, 2703-2714.

Humphreys, N.D., and Baker, C.J. (1992), "Forces on vehicles in cross winds from moving model tests, J of Wind Eng. and Ind. Aerod., 41-44, 2673-2684.

Locke, N.J. (1994), "Lift and drag distributions of sails from wind tunnel wake surveys", M.E. Thesis, Mech. Eng. Dept., Univ. of Auckland, Auckland, NZ.

Locke, N.J., Flay, R.G.J., and Jackson, P.S. (1994), "Design of a wind tunnel traversing rig and application to aerofoil wake measurements, IPENZ Ann. Conf., Nelson, New Zealand, Feb.

DEVELOPMENT OF HYBRID EXPERIMENTAL SYSTEM COMBINED WITH RANDOM RESPONSE ANALYSIS FOR UNSTEADY AERODYNAMIC VIBRATION OF STRUCTURE

KANDA, M.[1], MARUTA, E.[1], HOMMA, Y.[2] and UEDA, K.[1]

[1] *Department of Architecture and Architectural Engineering, College of Industrial Technology, Nihon University, Izumi-cho 1-2-1, Narashino-city, Chiba, 275, Japan.*
[2] *Obayashi Corporation Technical Research Institute, Shimokiyoto 4-640, Kiyose, Tokyo, 204, Japan.*

Abstract

The new type hybrid testing method, which simulates vibration of structure for unsteady aerodynamic force, is presented in the present paper. This method is referred to as "the hybrid vibration method". The system, which can simulate the rocking vibration of a single degree of freedom system, is developed on the basis of the concept of the present method. The out line of the system, especially including the vibration device, the mechanism of control and measurement and the numerical integration algorithm, is described. In order to verify the validity of the present method, free vibration, buffeting vibration and orthogonal vibration were simulated. It was confirmed that the present method can simulate the vibration for aerodynamic forces including unstable aerodynamic phenomena such as galloping and also can predict the critical value of viscous damping accurately when the unstable aerodynamic phenomena occur.

1. Introduction

With the advent of large scale structures such as high rise buildings or long span bridges with long natural period and low viscous damping, it becomes increasingly important to understand response characteristics of structures against external wind forces from a standpoint of wind resistant design. In general, response characteristics of structures against external wind forces depend on characteristics of unsteady aerodynamic forces as well as mechanical properties of structures.

In the past researches, an attempt has been made to estimate the unsteady aerodynamic forces for various type of structures. The characteristics of unsteady aerodynamic forces are highly complicated, since an effect of the interaction between structural response and external wind force is inevitably included in those characteristics, in addition to the effects of the geometry of structures and the properties of natural winds. Especially, it is noteworthy that such interaction is one of sources which causes unstable aerodynamic vibration of structures.

At present, there are two major methods to evaluate unsteady aerodynamic forces by means of the wind tunnel test. One is the free vibration method and the other is the forced vibration method. The free vibration method is the testing approach that measures unsteady aerodynamic forces induced when the model of structure is freely vibrating under the flow generated in the wind tunnel. In this case, the interaction with structural response is correctly involved in the measured unsteady aerodynamic forces. However, the unsteady aerody forces have to be identified by applying the rather complicated procedure to the mea values. Thus, it often happens that the identified values are not sufficiently accurate. The forc vibration method is also the testing approach that measures unsteady aerodynamic forces induced when the model of structure is forcedly vibrating under the flow generated in the wind tunnel. An advantage of the forced vibration is that unsteady aerodynamic forces can be measured directly. However, the interaction between structural responses and unsteady aerodynamic forces may not be exact, since the model of structure is forced to continue a steady vibration.

Taking the best characteristics of both methods, it would be ideal to develop such a method that could directly measure unsteady aerodynamic forces including an exact interaction with structural responses.

The objective of this paper is to present "hybrid vibration method" as an ideal method to be applied and to verify validities of this method. The hybrid vibration method is the procedure that combines the analysis with the experiment in order to measure unsteady aerodynamic forces and also to simulate unsteady aerodynamic vibration. The hybrid vibration method has several advantages, in addition that it is possible to measure directly unsteady aerodynamic forces including an exact interaction with structural responses. Those advantages will be also discussed in this paper.

2. Principle, advantages and limitations of hybrid vibration method

2.1 Principle of hybrid vibration method

The hybrid vibration method to be presented is the mean that simulates unsteady aerodynamic vibration of structures and also measures unsteady aerodynamic forces directly by applying the response analysis method to the equation of motion. In this method, the term of external forces to be appeared in the equation of motion is experimentally evaluated through the wind tunnel test. That is, the unsteady aerodynamic forces, to be measured by the wind tunnel test, acting on the model of structure are regarded as the external forces. Parameters involved in the vibrational system such as mass, stiffness and damping are numerically set up in the computer. First, the response value of structural model is calculated in the computer. Then, the response value is forced to be given by means of the loading device to the model on which external wind force is to be measured. The concept and the flow chart of the hybrid vibration method are shown in Fig.1 and 2, respectively. The present hybrid vibration method is the on−line computer control testing method. The procedure of the hybrid vibration method is proceeded as follows.

① The equation of motion is assumed to be expressed as follows;

$$M\ddot{X}+C\dot{X}+KX=F(\dot{X},X,t) \tag{1}$$

Where, M, C and K indicate the mass, damping and stiffness matrices, $\ddot{X}$, $\dot{X}$ and X indicate the vectors of acceleration, velocity and displacement, $F(\dot{X},X,t)$ indicates the vector of external force, and t indicates the time.

② The mass, damping and stiffness matrices; M, C and K in Eq.(1), are numerically set up in the computer.
③ The response values; $\ddot{X}$, $\dot{X}$ and X, are calculated in the computer by applying the step by step integration method.
④ The response value to be calculated in the process ③ is forced to be given by means of the loading device to the model of structure on which the external force is to be measured.
⑤ The external force; $F(\dot{X},X,t)$ in Eq.(1), is determined by measuring the unsteady aerodynamic force acting on the model of structure which was arranged in the wind tunnel apparatus.
⑥ The processes from ③ to ⑤ are repeated in a real−time.
If these processes could be executed accurately, it is possible to simulate the unsteady aerodynamic vibration of structure and also to measure directly the unsteady aerodynamic forces including the interaction with structural responses.

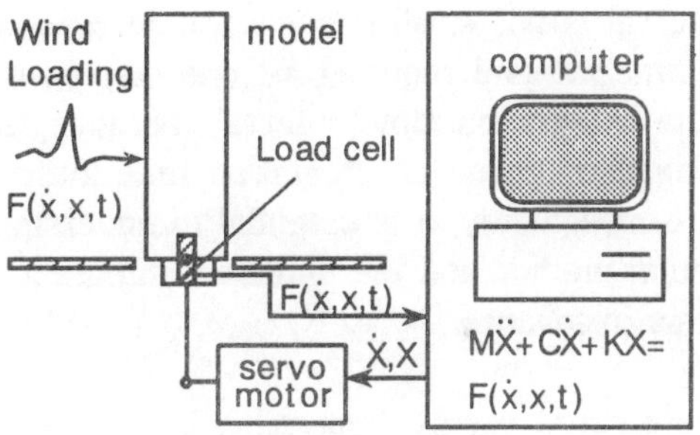

Fig.1 Concept of the System

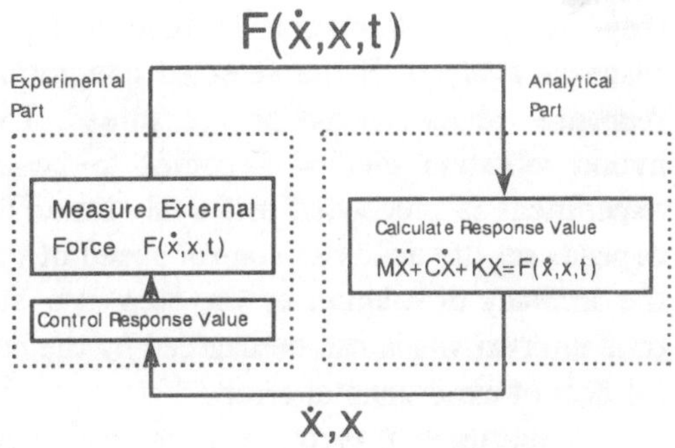

Fig.2 Flow Chart of Hyblid Vibration Method

2.2 Advantages and limitations of hybrid vibration method

The hybrid vibration method has the following advantages;

Ⅰ. The unsteady aerodynamic forces including the exact interaction with structural responses can be directly measured.

Ⅱ. The aerodynamic phenomena including the unstable vibration such as vortex−induced oscillation, galloping and flutter can be simulated.

Ⅲ. It is easy to set up the vibrational parameters of structure, because they is set up numerically in the computer. When the hybrid vibration method is applied to a multi−degrees of freedom system, it is possible to set up an orthogonal damping such as the Rayliegh damping if desired.

Ⅳ. It is possible to obtain coupled response values for two vibrational systems, which are not actually coupled, by coupling the responses for two systems to be evaluated independently for both systems in the computer. For example, if a mathematically formulated model for damper is implemented in the computer and this model is coupled with a vibrational system to be tested numerically, then the response of the model with damper can be experimentally obtained without using a real damper.

As stated in the above, the hybrid vibration method has several advantages such as Ⅰ, Ⅱ, Ⅲ and Ⅳ. Furthermore, this method also can be extended to another useful applications.

It should be mentioned that the hybrid vibration method is not the all−round approach for simulating the unsteady aerodynamic vibration of structure. As a matter of course, there are several limitations.

For example, the following limitations may be pointed out.

1)Limitation for modeling of structure

In the hybrid vibration method, the structure to be simulated has to be replaced with the discretized spring−mass model. Thus, it is questionable to apply the hybrid vibration method to such structures that do not behave as the spring−mass system but behave as the continuum. Hence, the validity of the model has to be discussed enough.

2)Effect of discretization of time on accuracy of responses

The hybrid vibration method predicts response values of structure by means of the step−by−step integration method using the discretized time as well as the conventional response analysis. If the selected integration time interval is not small enough, the accuracy of response values can not be guaranteed. The control time interval required for one step in the hybrid vibration method becomes to be equal to the integration time interval, because the experiment is proceeded in a real−time. Thus, the minimum value of integration time interval depends on the ability of control speed of the system. Consequently, it is essential to understand the accuracy of solution by the step−by−step integration method and the minimum integration time interval which can be handled by the control system in advance.

3)Effect of experimental errors

Experimental errors are inevitably involved in measured values. Especially, in the hybrid vibration method, since the external wind forces to be measured experimentally are utilized to calculate the response values, the effect of experimental errors shall be included in the response values. In order to clarify the reliability of results evaluated by the hybrid vibration method, it is needed to investigate intensively whether an effect of experimental errors on responses may be allowed or not.

3. Outline of system

In this chapter, the system actually manufactured on the basis of the concept of the hybrid vibration method is described. Furthermore, the accuracy of the system is rigorously investigated. In order to investigate the aerodynamic vibration of building by the hybrid vibration method, the developed system is applied to the high rise building which can be modeled by a single degree of freedom system with the rocking vibration.

3−1. Vibration device

The developed device, which measures external wind forces acting on the model and controls the calculated response value to the model, is shown in Fig. 3. This device has the revolving shaft to reproduce the rocking vibration of the single degree of freedom system. The testing model can be installed at the upper side of the shaft. The servo motor; that is, AC 200V and 400W, is installed at lower side of the shaft. This servo motor works as the loading source and is utilized to impose vibration on the model forcedly so that the response values predicted in the computer are reproduced. The target location of the model, which corresponds to the response value, shall be controlled by using the magnetic scale with the accuracy of 1/1000mm. Furthermore, the load−cell, which is feasible to measure the resultant forces and moments; Fx, Fy, Mx, My and Mz, is installed at the center of rotation of the shaft to measure external wind forces acting on the model.

3−2. Numerical integration algorithm

The response value is predicted on the basis of the measured external wind force in the hybrid vibration method. In this system, the explicit Newmark method, which is one of the step−by−step integration methods, is utilized to predict responses in the computer. The explicit Newmark method is expressed as;

$$\mathbf{X}_{n+1} = \mathbf{X}_n + \Delta t\dot{\mathbf{X}}_n + \Delta t^2/2\ddot{\mathbf{X}}_n \quad (2-a)$$

$$\dot{\mathbf{X}}_{n+1} = \dot{\mathbf{X}}_n + \Delta t/2(\ddot{\mathbf{X}}_n + \ddot{\mathbf{X}}_{n+1}) \quad (2-b)$$

$$\ddot{\mathbf{X}}_{n+1} = \mathbf{M}^{-1}\cdot\{\mathbf{F}_{n+1}(\dot{\mathbf{X}}_{n+1}, \mathbf{X}_{n+1}, t) - \mathbf{C}\dot{\mathbf{X}}_{n+1} - \mathbf{K}\mathbf{X}_{n+1}\} \quad (2-c)$$

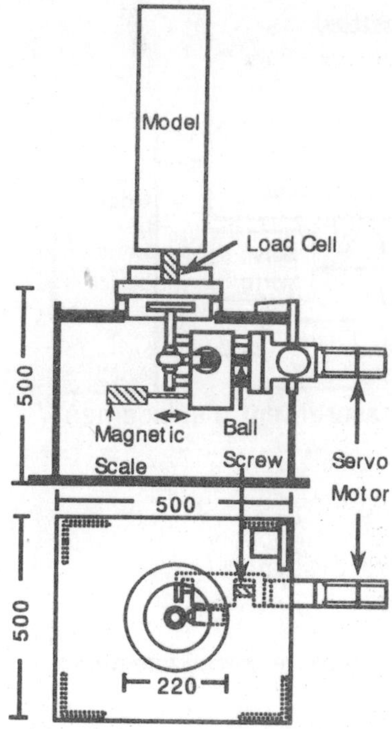

Fig.3 Vibration Device

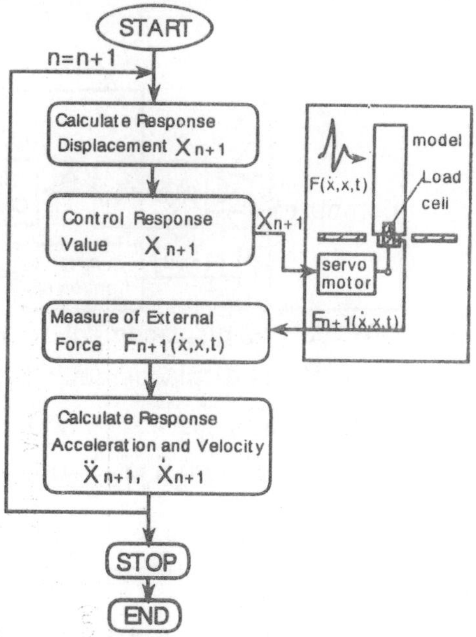

Fig.4 Flow Chart of This System

Where, Δt indicates the integration time interval, the subscripts n and n+1 indicate the step number in the discretized time. The flow chart of this system with the explicit integration method is shown in Fig.4. The explicit Newmark method has an advantage that the response value in the next step can be explicitly predicted. When the hybrid vibration method is applied, it is required to predict the response value in the next step explicitly as indicated in Fig.2. Thus, the explicit Newmark method shall be utilized in this method.

3–3. Control and measurement methods

When the hybrid vibration method is applied, the simulation has to be proceeded in a real–time. In other word, the control time required for one step becomes to be equal to the integration time interval. Hence, the control of response values and the measurement of external wind forces have be carried out with very high speed. In this system, the control and measurement methods as shown in Fig.5 are used in order to make the control and measurement with high speed feasible. The property of this mechanism is that the control of response values is performed not by the computer directly but by the controller. In other words, if the controller instead of the computer is utilized, then the high speed control of response values becomes possible. The control location for the target response value can be confirmed by feedbacking the measured value by the magnetic scale to the controller. The measured external wind forces are sent to the computer in the form of the digital signals converted from the measured analog signals after the location of model is settled

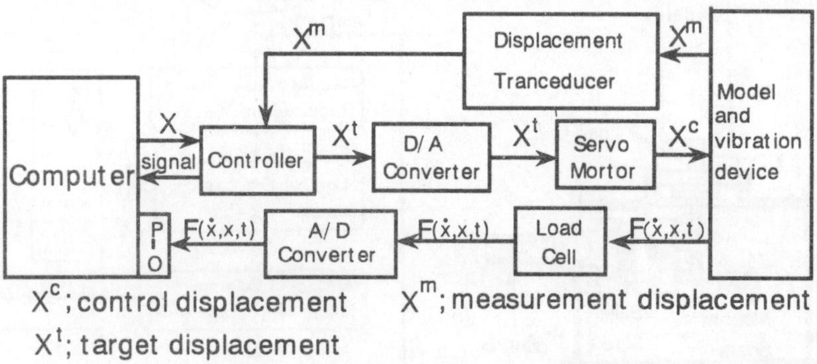

Fig.5 Mechanism of Control System

4. Verification on accuracy of measurement and control values by forced vibration

The hybrid vibration method is the procedure that the response values are predicted based on the measured external wind forces. Hence, the reliability of the hybrid vibration method highly depends on the accuracy of measured external wind forces. In order to ensure the accuracy of measured external wind forces, it is essential not only to be able to measure the external wind forces precisely but also to make the control so that the calculated response value can be imposed on the model accurately. Because the aerodynamic force has the interaction with the response of structure. Before going to the random response simulation under external wind force, in this chapter, forced vibration test under sine wave is conducted, and the accuracy of control values and the magnitude of noises involved in the measured values are investigated. The spectrum of control displacement errors induced under the forced sine wave is shown in Fig.6. The error; e, can be estimated by the following equation.

$$e=\sum_{K=1}^{N}(X^{T}_{K}-X^{C}_{K})/N \tag{3}$$

where, X^T and X^C indicate the target and controlled displacements and N is the total number of steps actually controlled. The superscript T and C indicate the target value and the control value, respectively. The amplitudes of the sine waves generated are 1/175, 1/350 and 1/1750 and the frequency is varied from 3Hz to 12Hz. The control time interval adopted for one step is 4msec, which is the fastest control speed for the system. In all cases that the amplitude of displacement angle is set equal to 1/1750, 1/350 or 1/175, the displacement errors tend to become larger with an increase in the frequency. This may be due to the fact that the magnitude of displacement to be controlled during one step becomes larger as the frequency becomes higher under the sine wave with fixed amplitude. However, it can be said that the accuracy of this control device is satisfactory, because the errors induced are at most 1/1000 even in the case of the highest frequency of 12Hz and the largest amplitude of 1/175.

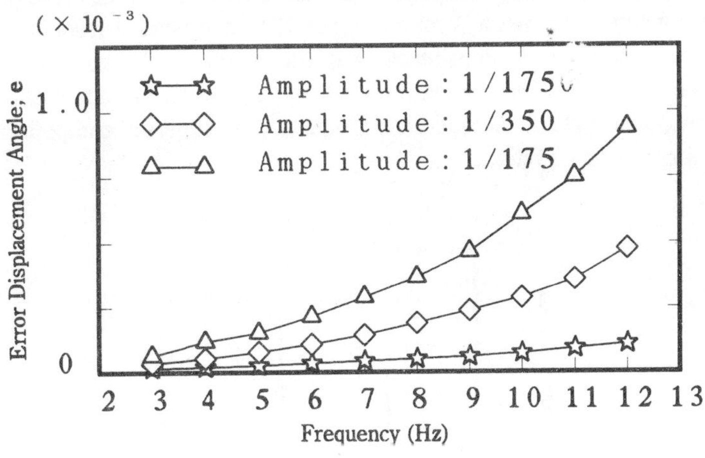

Fig.6 Control Displacement Error Spectrum

Figure 7(a) and (b) shows the comparisons between the target and control values for the displacements and velocities. The amplitude and frequency of sine wave are 1/350 and 6Hz, respectively. The control value of velocity; $\dot{X}^{c}_{n+1}$,is estimated by the following equation.

$$\dot{X}^{c}_{n+1}=(X^{c}_{n+1}-X^{c}_{n})/\Delta t \tag{4}$$

It is seen that both control values for velocities and displacements are consistent with the target values satisfactorily.

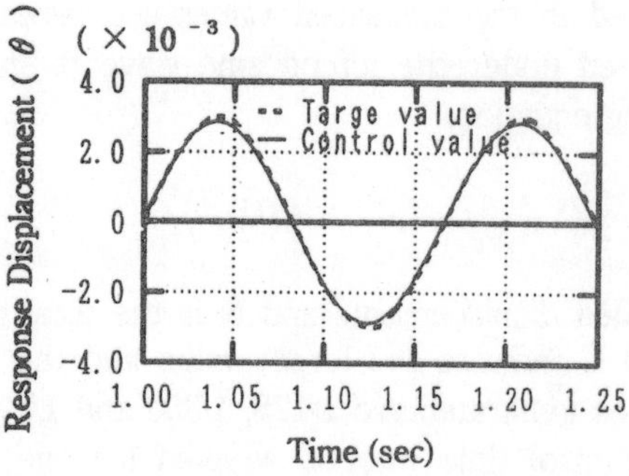

Fig.7(a) Time History of Displacement

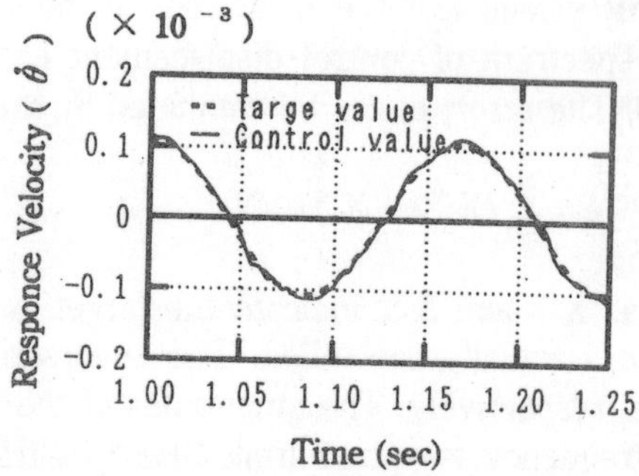

Fig.7(b) Time History of Velocity

Figure 8 shows the Fourier spectrum of noises involved in the measured values by the load−cell under the forced sine wave with constant amplitude. In this case, the amplitude of the sine wave was set to be equal to 1/250 and the frequency was set to be equal to 6Hz. The growing component near 50Hz may be generated by the electric noises or the natural frequency of the experimental apparatus. This component has no effect on the calculated response values since the frequency of the component is much higher than the natural frequency of the vibration system. On the other hand, it is considered that the superior component near 6Hz may be caused by the forced vibration. The noises induced by the forced vibration have some influence on the calculated response values, because the frequency of superior component is nearly equal to the natural frequency of the system.

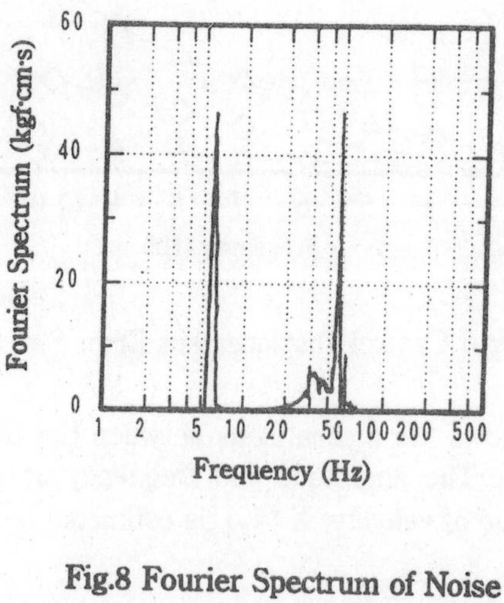

Fig.8 Fourier Spectrum of Noise

5. Simulation of free vibration by impulse force

Before simulating the random vibration against external wind forces, the simulation of the free vibration by impulse force is conducted in order to investigate an accuracy of responses calculated by this system. In this simulation, the parameters to be input to the computer are as follows; the rotational inertia $I=3.53kgfs^2/cm \cdot cm^2$, the natural frequency of the vibration system $f_o=6Hz$ and the integration time interval $\Delta t=4msec$. The damping factors are parametrically varied; 0%, 1%, 2% and 3%. The prismatic model with the dimension of $100 \times 100 \times 500mm^3$ is set up on the load-cell of vibration apparatus. After the impulse force is

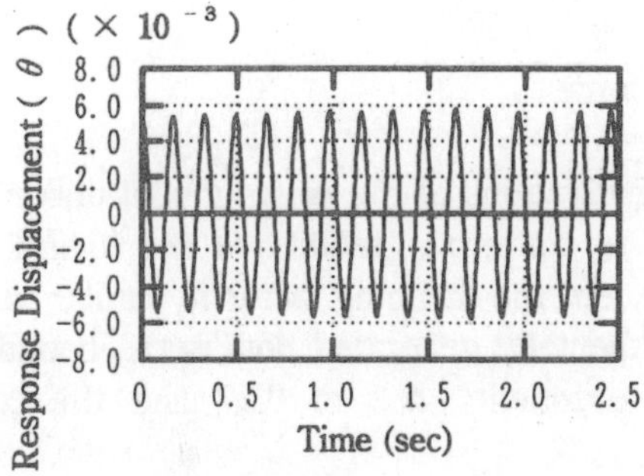

Fig.9(a) Time History of Response Displacement by Impulse Force (h=0%)

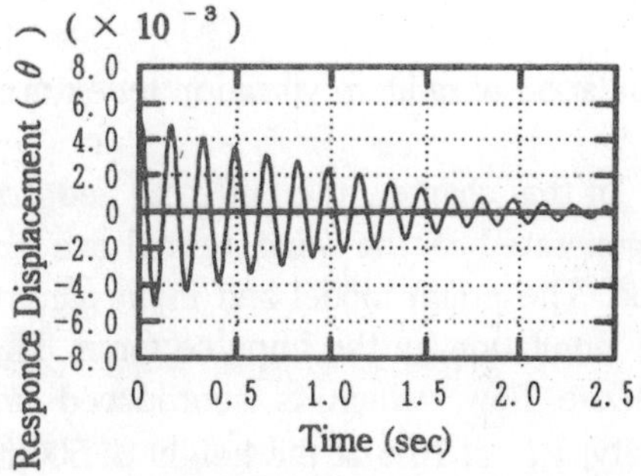

Fig.9(b) Time History of Response Displacement by Impulse Force (h=2%)

Table.1 Comparison with Indentified Value and Input Value

input value (f, h)	indentified value (f', h')
f=6.00 (Hz) h=0.00	f' =6.10 (Hz) h' =0.00
f=6.00 (Hz) h=0.01	f' =6.10 (Hz) h' =0.0094
f=6.00 (Hz) h=0.02	f' =6.10 (Hz) h' =0.02
f=6.00 (Hz) h=0.03	f' =6.10 (Hz) h' =0.032

applied once, the response value is estimated. The time histories of response displacements are shown in Fig.9(a)and(b). The natural frequencies and damping factors, which were identified from the simulated results by applying the nonlinear least square method, are listed in Table 1 as well as the input quantities. The identified values for the natural frequency agree well with the input values although there is a little difference between them; that is, 0.07 ~ 0.1Hz. The identified value for the damping factor agrees with the input value in the case of h=0%. However, the identified values for the cases with h=1%, 2% and 3% are roughly 0.2% ~ 0.8% smaller than the input values. Judging from these results, it is considered that the effect of noises or other factors on the response values can be permitted only in the experimental environment similar to the simulations stated in the above.

6. Simulation of random vibration by external wind force

In this chapter, the buffeting and orthogonal vibrations of the prism model under the wind flow generated in the wind tunnel are simulated to confirm a validity of the hybrid vibration method. The prism model and input parameters except the damping factor is similar to the case of the simulation by the impulse force. The experimentally generated flow is the boundary layer turbulence flow, which is reproduced with the exponent; α , of 0.27 and the turbulence intensity; I_{t} , of 10% at the height of 500mm.

6−1. Simulation of buffeting vibration

Figure 11 shows the time history of response displacement for the case of h=1%. Figure 11 clear illustrates the situation that the model vibrated around the mean value with constant frequency and the control value against the target value is accurate. The buffeting vibration, which was simulated by the present system, gives satisfactory results except the responses in the low range of reduced wind velocity for the case of h=0%.

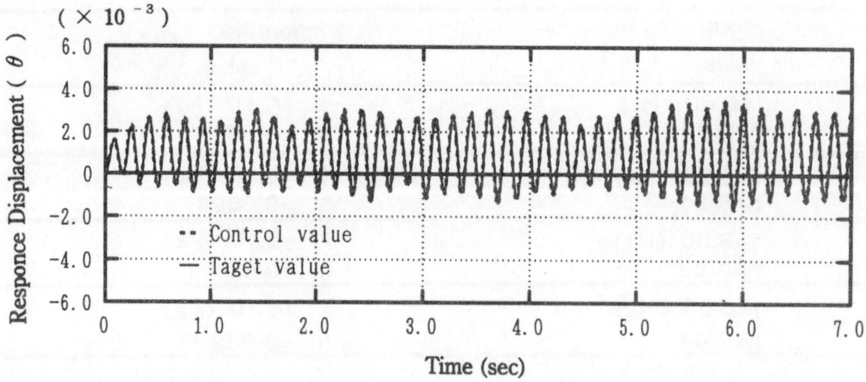

Fig.10 Time History of Displacement (Buffeting Vibration)

6-2. Simulation of orthogonal vibration

Figure 11 shows the relationship between the RMS value of displacement angles in the direction of orthogonal vibration and the reduced wind velocity. Figure 13 shows the time history of displacement angles for the case of 0% and $V/f_o \cdot B=10$. It is seen from Fig.11 that the RMS value increases in proportional to the reduced wind velocity up to $V/f_o \cdot B=10$, and on the other hand the rate of increase in the RMS values becomes gentle for $V/f_o \cdot B>10$ in comparison with that for $V/f_o \cdot B<=10$. However, the remarkable divergence phenomenon, which would be considered as the galloping, can be observed at $V/f_o \cdot B=10$ when h is less than 0.4%. Figure 14 clearly illustrates the situation that the divergence phenomenon is occurring. The present result concerning the above phenomenon agrees with the experimental result; Kawai (1992), qualitatively. The divergence phenomenon in the orthogonal direction occurred when h is less than 0.4% but it did not occur when h is greater than 0.5%. Thus, it can be concluded that there

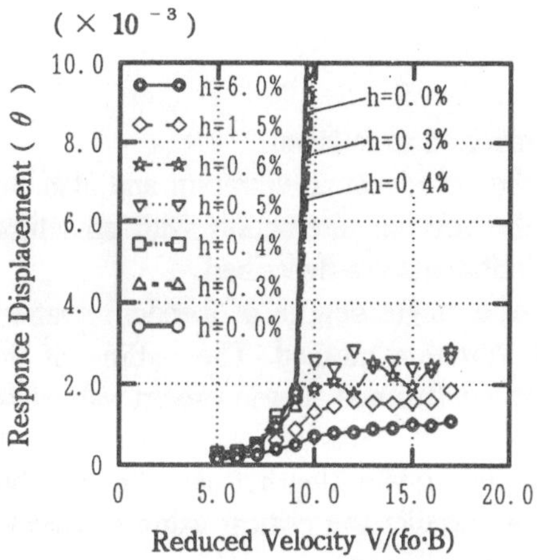

Fig.11 Othogonal Wind Response
(Hybrid Vibration Method)

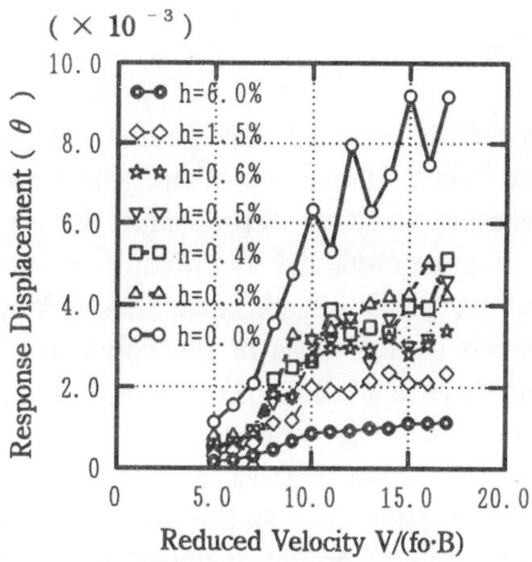

Fig.12 Othogonal Wind Response
without Controling Response Value
(Hybrid Vibration Method)

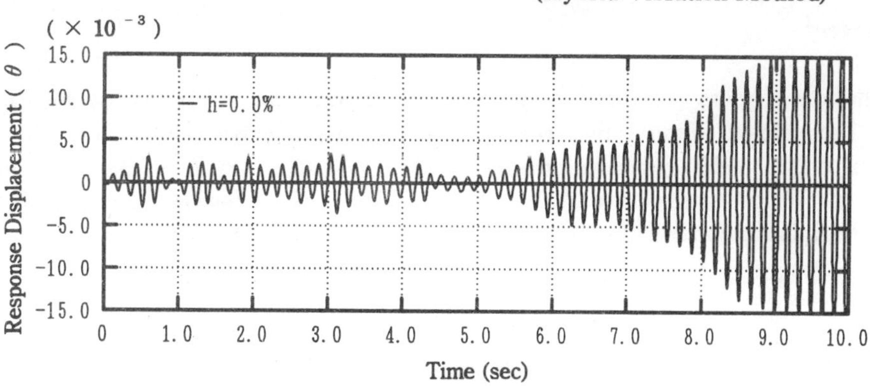

Fig.13 Time History of Control Displacement (h=0.0%)

may be a critical value of viscous damping between h=0.4% and h=0.5%. It is possible to estimate the critical value of viscous damping accurately, when the divergence phenomenon occurs, by the hybrid vibration method.

In the next place, in order to verify that the hybrid vibration method is an effective mean to simulate unstable aerodynamic phenomena, another type of simulation on wind responses is conducted. The simulation to be conducted is the hybrid vibration method, but the process to control the model on the basis of predicted response values is not actually performed. Figure 12 shows the results estimated by the present simulation under the similar condition to the case of Fig.11. For all damping factors h adopted, the RMS values increase with an increase in the converted wind velocity. Thus, the divergence phenomenon was not observed. Consequently, it can be said that the hybrid vibration method can simulate unstable aerodynamic phenomena caused by the interaction between the structural responses and the external wind forces.

7. Conclusion

The results obtained through this study can be summarized as follows;

1. The new hybrid vibration method, which can simulate the aerodynamic vibration and also can directly measure unstable aerodynamic forces including the precise interaction with structural responses, is presented, and its principle, advantages and limitations are described.
2. The system, which can simulate the rocking vibration of a single degree of freedom system, was developed on the basis of the concept of the hybrid vibration method. The outline of the system was described and furthermore the accuracy of the control and measurement values by the present system was verified.
3. The hybrid vibration method can simulate not only the buffeting vibration but also the unstable aerodynamic vibration such as the galloping. The method can predict the critical value of viscous damping accurately when the unstable aerodynamic phenomenon occurs.
4. The noises, to be generated by the forced vibration, have a significant effect on the calculated response values. It must be needed to mitigate the effect of noises on responses to improve the reliability of the hybrid vibration method.

Acknowledgement

The authors express their sincere gratitude to Assoc. Prof. N.SHIRAI of the College of Science and Technology, Nihon Univ., Tokyo, Japan, for his valuable suggestions. The authors also wish to thank under graduate students, Mr. T.Otsuka, Miss. Y.Takahashi and Mr. M.Maeshiba of Wind Engineering Laboratory, College of Industrial Technology, Nihon University, Chiba, Japan, for assisting them in carrying out the experiments.

Reference

Kawai,H.(1992), "Vortex Induced Vibration, Galloping and Flutter of Tall Building ", Proceeding of 12th National Symposium on Wind Engineering, Tokyo, Dec, 1992, pp267−272 (in Japanese).

SOME FEATURES OF MODELLING SPECTRAL CHARACTERISTICS OF FLOW IN BOUNDARY LAYER WIND TUNNELS

GARG, R.K.[1], LOU, J.X.[2] and KASPERSKI, M.[2]

[1] *Structural Engineering Research Centre, Ghaziabad, 201001, India.*
[2] *Building Aerodynamics Laboratory, Ruhr-Universität Bochum, 44780 Bochum, Germany.*

Abstract

The influence of the choice of form of spectra to calculate the integral length scale of the longitudinal component of wind velocity is discussed. Fichtl-McVehil form of spectrum fits better to the measured data. Subsequent influence on other spectral quantities and their estimation including the decision of geometric scale is also discussed, and the same has been applied to the wind velocity data obtained in wind tunnel as well as in nature. Full description of the spectral characteristics of simulated atmospheric boundary layer in WT will help in comparing the results obtained from different sources (WT tests) for identifying the influences of various flow/body parameters on the wind induced effects and for further (mathematical) modeling.

1. INTRODUCTION

A proper wind tunnel (WT) experiment, to estimate accurately the mean and fluctuating forces on a structure, requires to simulate the mean wind profile as well as turbulent properties such as turbulent intensities, spectrum and integral length scales typical to ABL (Atmospheric Boundary Layer), Small scale turbulence (SST) etc. (ASCE,1990; Melbourne,1979; Surry,1982; Tieleman,1992 etc.). There are several methods to simulate wind characteristics in wind tunnels (Cermak,1987; Cook), however, a big problem remains to estimate accurately the simulated spectral properties.

The spectrum of fluctuating wind and consequently the integral length scales directly affect the response of the structure not only dynamic (Lee,1975; Nakamura,1983; Kareem,1987 etc.) but the static response also, thus warranting an accurate estimation of the integral length scales of the simulated wind. There are several methods to estimate the simulated wind properties e.g. by assuming a functional shape in form of spectrum or, by direct method of integrating the data in time domain. By direct method one need to define the upper bound of integration because of impracticability of integrating to infinity. It warrants to have a very long time series for accurate estimations (Kaimal,1989). One can also use integration with window techniques, but all these have practical difficulties/limitations (Schrader,1993). However, the assumption of spectrum may lead to different results for different choice of spectra (Harris,1986; Solari,1987; Flaga,1987), the difference in estimation could be even of the order of 2 (Teunissen,1980), thus leading to systematic errors in estimation of the integral length scales and the related statistical quantities.

Apart from the concept of average size of eddy (Integral length scale) small size eddies (SST) are believed to play a role in vortex dynamics requiring attention to simulate the upper frequency for a given geometric scale (Laneville/Gartshore,1975; Melbourne,1979; Tieleman, 1990,1992, Farell/Iyenger, 1993) such that the test lies in the inertial sub-range. These parameters are also used to determine the model scale in WT (Cook,1977). The modeling of the spectral properties are also desired in accurate formulation of the mathematical model (Schueller et al,1983; Kareem,1987, Kasperski/Niemann,1993) of fluid structure interaction based on WT test results.

In the present paper, the parametric influences of choice of the form of spectrum on Lux and their influence on other statistical quantities will be discussed. Other important spectral characteristics to be simulated in WT and influence on geometric scale/time scale will also be discussed. The analysis is carried out after obtaining the long time series of wind velocity data in WT as well as in nature. Also, the discussion provided helps to fit the data to the spectral quantities.

2. EXPERIMENTATION

2.1 Wind tunnel data:

The wind tunnel is of open circuit type with cross-section of 1.8 m wide and 1.6 m high (variable upto 1.9m). The boundary layer (BL) flow is achieved by employing a castellated barrier wall of height 0.425m/0.225m and the spires (Counihan,1969) followed by the floor roughness in form of rectangular blocks of various sizes having dimension: length, breadth and height in mm as 18*36*36; 36*36*36; 36*36*16 (Fig. 1). The simulated boundary layer (Fig. 2) represents an open country terrain α=0.17.

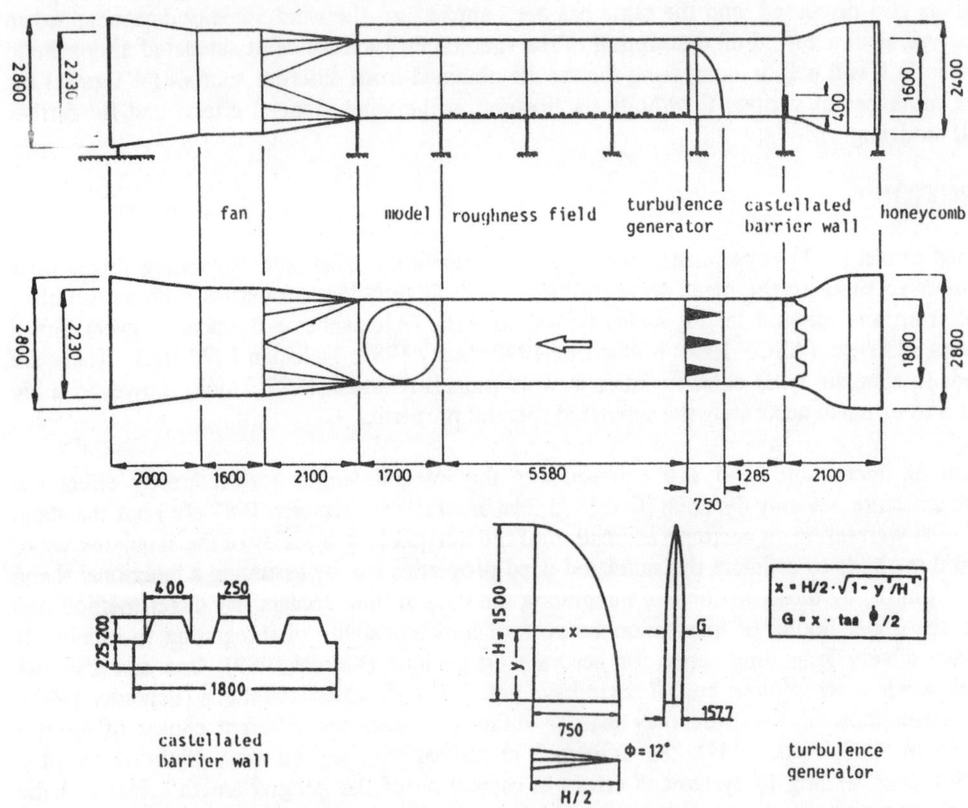

Figure 1. Wind tunnel layout to simulate atmospheric boundary layer.

The velocity data was obtained through hot-wire anemometer at various heights of WT at a sampling rate of 6400 Hz. The sampling time was sufficiently long of 300 seconds duration to have reasonably good statistics representing several hours of data in a typical full scale condition. The digitised data from A/D converter was stored on DDS tapes for further processing.

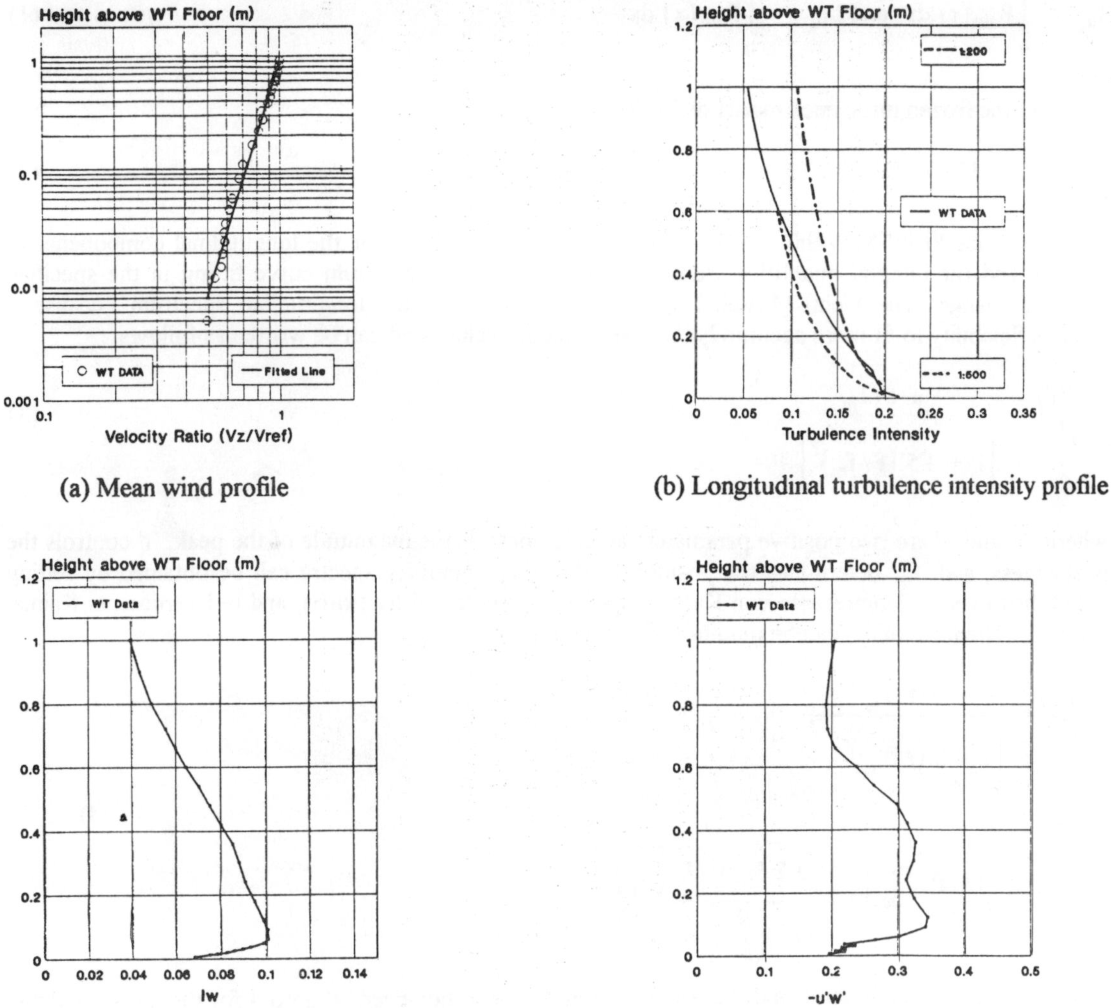

(a) Mean wind profile

(b) Longitudinal turbulence intensity profile

(c) Vertical turbulence intensity profile

(d) -u´w´ distribution

Figure 2. Simulated boundary layer in wind tunnel

2.2 Full scale data:

The full scale wind velocity data was available from a 344m high mast (Peil/Noelle,1992 etc.) measured with the help of cup anemometers installed at the end of 7.5 m long cantilevered arms. The sampling rate was 10 Hz and was low pass filtered before storing on PC. The data was resolved in two orthogonal directions with the help of direction of wind available at a level, however, the direction was interpolated at other levels of measurements where the wind direction was not measured.

3. METHOD OF ANALYSIS

Considering the wind as stationary random process the integral time / length scale can be defined as:

$$T_{u_x} = \int_0^\infty R_{uu}(\tau)\, d\tau, \text{ and } L_{u_x} = \int_0^\infty R_{uu}(x)\, dx \quad (1)$$

and from the frozen turbulence model of Taylor hypothesis

$$L_{u_x} = T_{u_x} \cdot U \quad (2)$$

Here, we use Fichtl-McVehill spectrum mode' (Fichtl et al,1970) for the longitudinal componenet of velocity and the integral quantities are computed by means of spectrum curve fitting in the specified frequency range. The Fichtl-McVehill spectrum is chosen for the longitudinal spectrum because it possess flexibility to fit more accurately the given spectral values and can be written as follows:

$$\frac{f \cdot S(f)}{\sigma^2} = \frac{a \cdot f / f_m}{\left[1 + 1.5 \cdot (f / f_m)^r\right]^{\frac{5}{3r}}} \quad (3)$$

where, `a' and `r' are two positive parameters and, `a' controls the magnitude of the peak, `r' controls the peakedness, and `f_m' determines the position of the peak. Panofsky spectra can be obtained by setting r=5/3, however, r=2 represents von Karman spectra as suggested by Harris, and r=1 represents Kaimal spèctra. The form of (3) can be modified by noting $f_m = a/(4.T_{ux})$;

$$\frac{f \cdot S(f)}{\sigma^2} = \frac{4 \cdot f \cdot T_{u_x}}{\left[1 + b \cdot (f \cdot T_{u_x})^r\right]^{\frac{5}{3r}}} \quad (4)$$

$$\text{where } b = \frac{1.5 \cdot 4^r}{a^r} \quad \text{and } a = \frac{1.5^{1/r} \cdot r \cdot \Gamma\left(\frac{5}{3r}\right)}{\Gamma\left(\frac{2}{3r}\right) \cdot \Gamma\left(\frac{1}{r}\right)}$$

Here, Γ is gamma function and 0.8 > r > 2.5 for WT data, however, `r' upto 4 for the full scale data. The data obtained from WT, $S_d(f_i)$ is fitted to the form $S(f_i)$, as in expression (4) by varying parameter `r' such that the error function Err is minimum over the range of frequency e.g. here limited upto 2000Hz. However, the full scale data was fitted for the frequency range 0.0048 Hz to 0.80 Hz.

$$Err = \sum_{i=1}^{k} \left[\frac{S_d(f_i)}{S(f_i)} - 1\right]^2 \quad (5)$$

L_{ux}, as well as variance is computed from a considered functional form of spectrum after varying `r'.

4. RESULTS AND DISCUSSIONS

4.1 Wind tunnel data:

Figure 3, describes the spectrum of longitudinal component of wind velocity as obtained in WT at the height of 0.54 m and mean wind speed (U) is 24.44 m/S, also curves are drawn corresponding to the least square fit of spectrum (LSFS) for r=1.15, vonKarman (r=2) and Panofsky (r=5/3). The LSFS curve of Fi-Mc spectrum fits well to the measured WT data, however, Karman and Panofsky spectra underestimate spectral values at low frequencies as well as deviation near peak which causes difference in Lux based on these spectra. The same trend has been observed for the spectra at other heights in WT. Figure 4, shows the auto-correlation curve corresponding to the spectrum of Figure 3, and having Tux, 0.013145 Seconds.

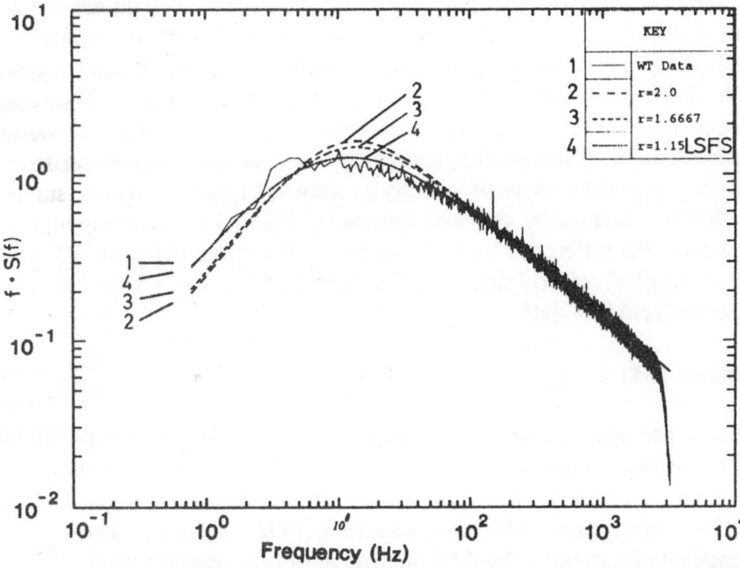

Figure 3. Spectrum of the longitudinal component of wind in wind tunnel

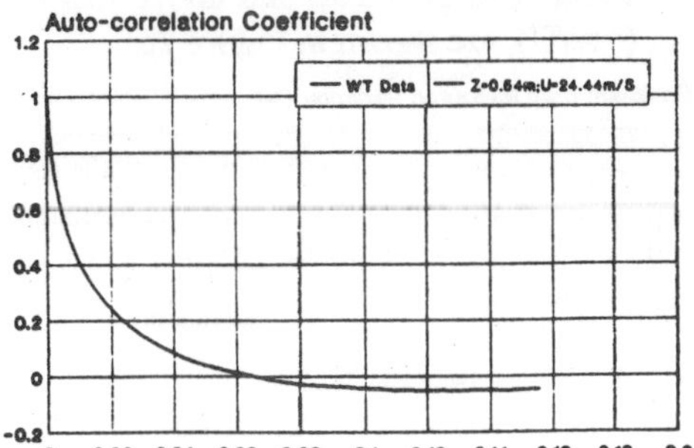

Figure 4. Auto-correlation function corresponding to spectrum in Figure 3.

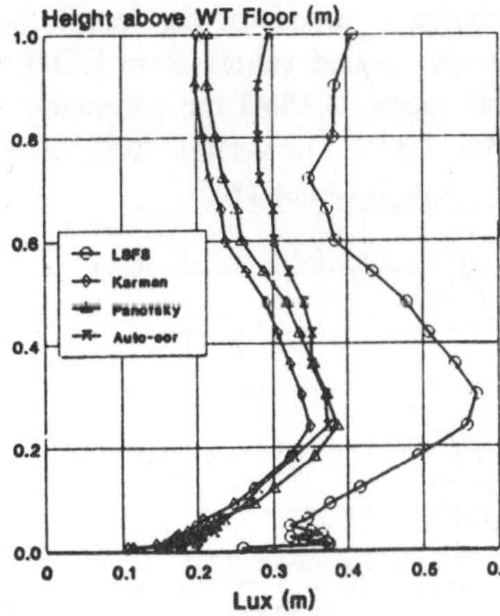

Figure 5. Comparison of L_{ux} based on chosen spectrum in WT

Figure 5, describes the Lux obtained with height in WT for various forms of spectrum chosen namely Panofsky, von Karman/Harris and from direct integration of the Auto-correlation curve upto first zero crossing, the negative lobe has been ignored There is considerable difference between the Lux obtained from various spectral form chosen.. Lux obtained from the LSFS is higher and is appx. of the order of two times as compared with the corresponding Karman spectrum. This difference may be due to poor fit by Karman and Panofsky curves in low frequency region of the spectrum (Teunissen 1979; Waldeck 1990 etc.). A higher value near 0.3 m height could be due to castellated barrier upstream of the spires. The variation of parameter `r' in Fichtl-McVehil spectrum (LSFS) with height is shown in Figure 6. The variation of `r' is quite regular except at greater heights and close to the WT floor where it is very low. The value of `r' ranges from 0.85 to 1.25.

Other length scales are important in certain situations depending upon the flow field around structure e.g. for horizontal structures (bridge) vertical components are more important as compared to tall structures. The other length scales have also been measured as $L_{uy} = L_{uz} = L_{vy}$ and on average are 0.45 times Lux (Hölscher,1993); the Reynolds stress parameter, -u'w', distribution is shown in Figure 2d.

For the purpose of comparison of influence of spectra on related statistical quantities the variance has been calculated from the fitted spectra within the frequency range considered. The comparison is made with the variance obtained directly from time series of velocity as shown in Figure 7, which suggests that there is underestimation of velocity variance by choosen spectra by 4% to 10 % corresponding to r=2 and 2% to 7% for r=5/3, however, the difference by LSFS is upto 3% (within 0.8m of WT height and difference increases with greater heights), clearly depicting the superiority of LSFS with Fichtl-McVehil spectrum in fitting to the spectrum from WT data.

4.2 Effect on Geometric Scale in WT:

4.2.1 General:

Although the geometric scale is decided by the mean velocity profile i.e. Jensen number (H/zo etc.) (Martin/Jensen; Roy/Holmes,1988) and consequently by comparison with the turbulence intensity to be achieved, due consideration is to be given to other modelled spectral quantities such as turbulence scales- large, small (ASCE,1990; Melbourne,1979; Tieleman 1990,1992), Reynolds stress distribution (Cermak et al) and the frequency range typical to the ABL inertial sub-range (Farell/Iyenger,1993).

4.2.2 Turbulence intensity:

The turbulence intensities for longitudinal fluctuations achieved in WT are compared with the correspondingly scaled values from ESDU for the similar B.L. profile (Zo=0.05m) in Figure 2B. At lower heights (appx. 0.15m) the geometric scale can be considered to be 1:200 and above this height (upto 0.5m) as 1:350. The vertical turbulence intensities, I_w, (=w'/U), are shown in Figure 2c.

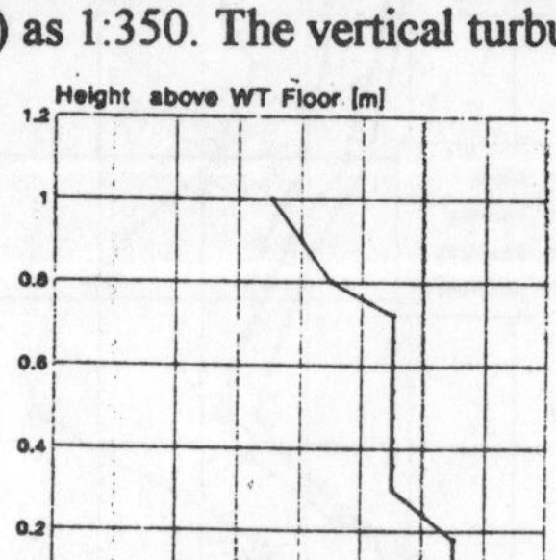

Figure 6. Variation of parameter 'r' in Fi-Mc spectrum in WT

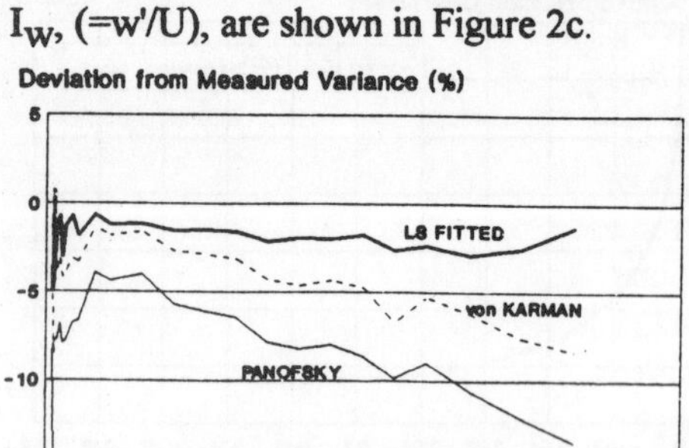

Figure 7. Effect of choice of spectrum on variance

4.2.3 Integral length scale:

The Lux normalised with the corresponding height estimated in WT are compared to the scaled values from ESDU in Figure 8. The ESDU values are corresponding to auto-correlation estimates. The comparison of these values with Lux obtained from WT based on auto-correlation estimates yields a geometric scale of appx. 1:1000, however, the Lux based on LSFS estimates corresponds to a scale of 1:700. If a mismatch of 3 (ASCE,1990) is allowed a geometric scale of 1:350 can be accepted for this ABL.

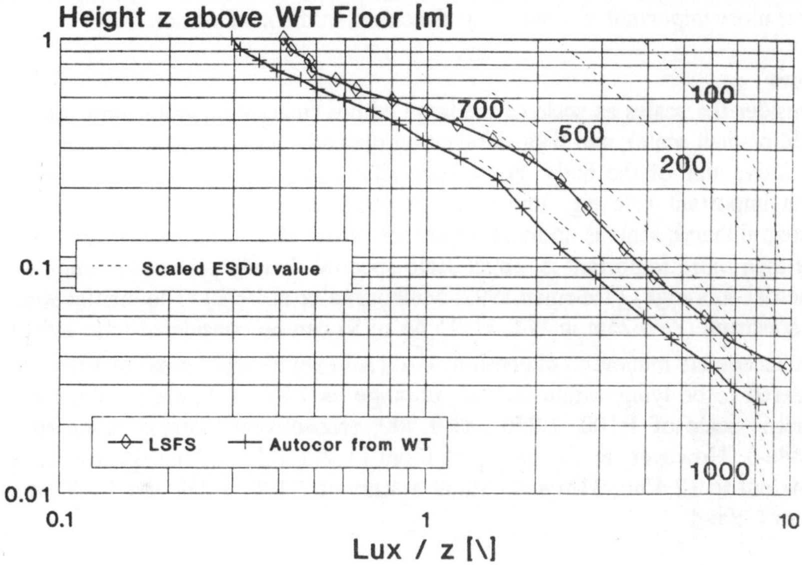

Figure 8. Comparison of scaled ESDU Lux with WT Lux

4.2.4 Inertial subrange:

The lower frequency limit (fl) of the inertial subrange at height Z for longitudinal component of wind velocity fluctuations can be considered as 0.5U/Z (MacCready quoted in Monin/Yaglom,1975) and for vertical or lateral components as 1.U/Z. Experiments by Saddoughi and Veeravalli (quoted by Bradshaw, 1994) suggest that the microscale Reynolds number of more than 1000 is required to achieve a convincing locally isotropic inertial subrange. The upper limit of frequency of inertial subrange i.e. start of the dissipation range (f_2) can be considered as $f_2 = 0.125U/(2\pi\ \eta)$, where, η, is the micro scale defined in terms of energy dissipation rate (ε) and is equal to $(\nu\ 3/\varepsilon)1/4$. The dissipation rate, ε , is function of dissipation length scale $L\varepsilon$, and can be estimated as $0.8\ U_{rms}{}^3/L_{ux}$, based on Batchelor's measurements of integral scales for a wide range of decay time and Reynolds number (Townsend, 1976). At the height of 0.25m in WT, the lower frequency limit, f_1, is 31.3 Hz, and the upper frequency limit , f_2, is 4235 Hz corresponding to U=15.65 m/S, ε=115.70, η=0.735 mm , Lux taken is corresponding to auto correlation curve (rather than LSFS due to similar basis of calculation); the corresponding lower and upper wavelengths λ_1 (=U/f) and λ_2 (=U/f) by using Taylor hypothesis are 0.5m and 3.7mm, respectively. However, at the height of 0.54m in WT (U=24.44m/S) the frequency limits are f_1=22.63 Hz, and f_2=4790 Hz (here, ε=31.81, η=0.1015 mm); the corresponding lower and upper wavelengths λ_1 and λ_2 are 1.08m and 5.1mm, respectively, which correspond to the atmospheric/WT values estimated by Tenneks/Lumley.

The above mentioned analysis suggests that the upper frequency limit of inertial subrange in this WT flow situation is close to 2000 Hz and does not reach the upper range at full speed as discussed in earlier paragraph. The situation is similar for the data obtained at lower speed in WT which yields these frequency limits as f_1=25.63 Hz and f_2=2000 Hz at Z=0.24 m (U=12.305 m/S) while f_2 observed in WT at upper range of the inertial subrange as obtained from the corresponding spectrum is nearly 1000 Hz. The upper frequency limit may vary depending upon the types of turbulence generators and should not be looked as a strict value to be simulated, however, small scale effects by considering appropriate geometric scale is more important to simulate as described in the next section.

4.2.5 Small scales

One could consider the scales as scale of the large eddies (macro/integral length scale) or scale of the small eddies (micro/small scale), and their domination depends upon the type of flow mechanism around the body and relative size of the body. For certain situations where separated shear layer (separation bubble) plays an important role e.g. low pressure peaks associated with separated shear layer, the simulation of integral length scale is no more important provided it is bigger than twice of the dimension of the structure then more important is small scale spectral density parameter (implies that the test is conducted in inertial subrange) (Tieleman,1992; Melbourne et al,1979). The inertial subrange as can be seen from the spectrum (Z=0.25m in WT, U=15.65 m/S) can be considered upto a frequency of 2000 Hz which gives acceptable minimum dimension, $\sqrt{A}$ (A being exposed area) or width of the (full scale) structure considered to be lying within inertial subrange as 7.82m, 20.m and 40m for a WT model if scaled at geometric scale of 1:100, 1:250 and 1:500, respectively, corresponding to SST parameter ($f.S/U^2$) as 1.564e-3. However, at the height of 0.6m in WT (U=24.91m/S) these minimum structure dimensions come out to 12.45m, 31m and 62m at a scale of 1:100, 1:250 and 1:500, respectively, with SST parameter as 1.29e-4.

4.3 Study of full scale data:

To study the effect of choice of spectrum the full scale data was analysed and a spectrum at a height of 156 m, U=29.64 m/S, is given in Figure 9. The spectrum beyond frequency 0.8Hz seems to be influenced by the disturbances such as inertia of cup anemometer etc. (Peil/Nölle,1992). Figure 10, shows the corresponding auto-correlation curve. The variation of L_{ux} with height obtained after fitting several spectral forms is shown in Figure 11. Large values of Lux at heights 246m, 282m and 300m were observed from the method LSFS as 796m, 719m and 976m, respectively, and are not shown in Figure 11 due to less reliability of full scale data at these heights. The parameter `r' corresponding to LSFS varies irregularly with height as can be seen in figure 12. The three large values of Lux converge for low values of `r', as 0.85 to 0.95. For a qualatitive study the Lux distribution with height as obtained from LSFS is compared with the WT data based on LSFS as well as corresponding to Karman spectrum, although the B.L. parametres (here, α =0.22) are not same in both the cases.

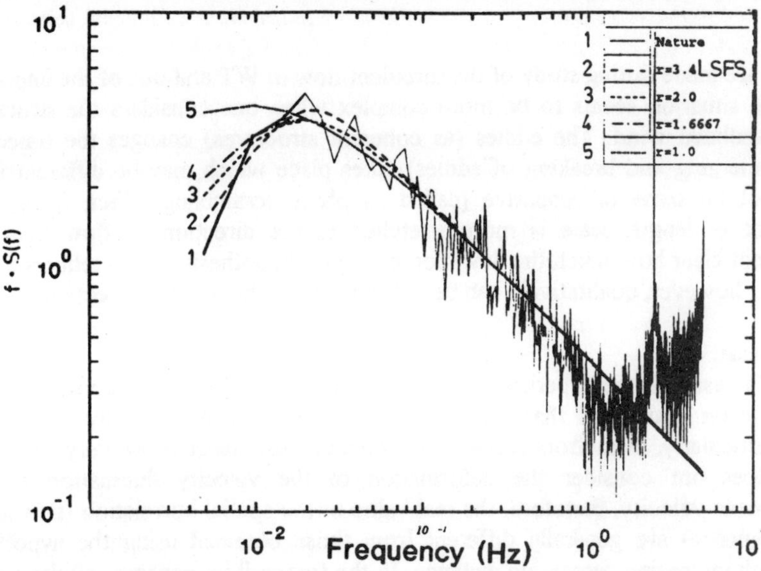

Figure 9. Spectrum of longitudinal component of atmospheric wind (Z=156m, U=29.64m/S)

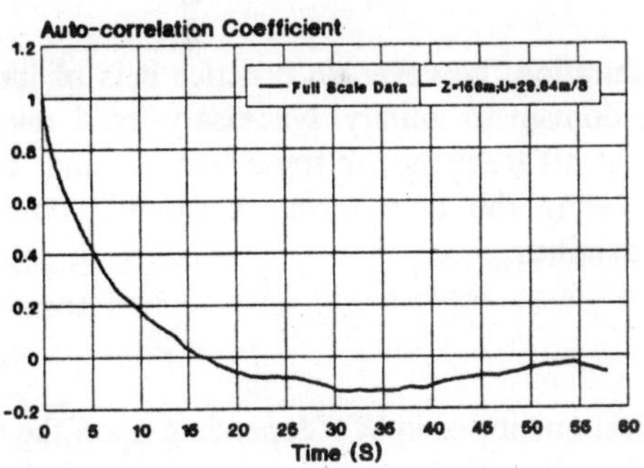

Figure 10. Auto-correlation curve corresponding to spectrum in Figure 9.

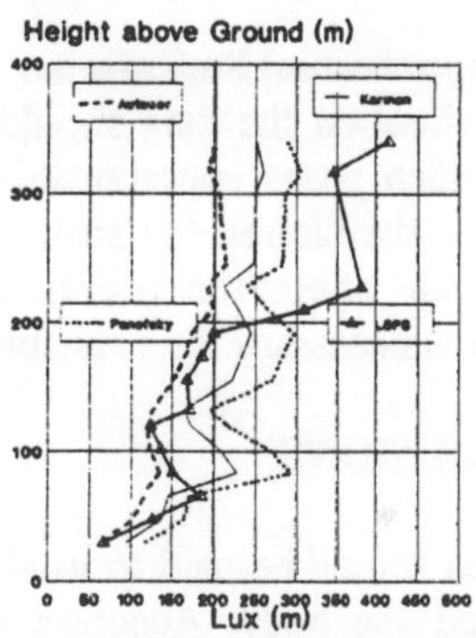

Figure 11. Comparison of Lux based on chosen spectra in ABL.

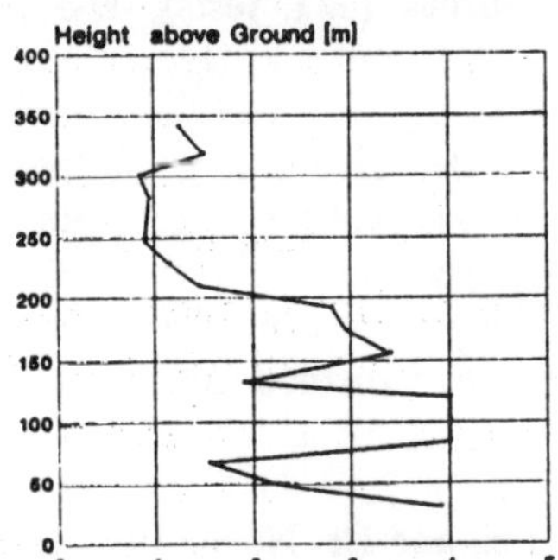

Figure 12. Variation of parameter 'r' in Fi-Mc spectrum for ABL

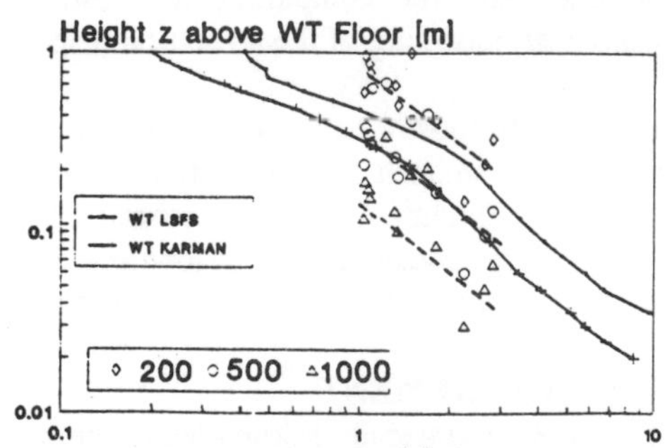

Figure 13. Comparison of ABL scaled Lux with WT Lux

4.4 Other effects

4.4.1 General:

Several assumptions are made during study of the turbulent flow in WT and one of the important is the Taylor hypothesis. The situation seems to be more complex when one considers the structure of the eddies contained in simulated wind. The eddies (as coherent structures) changes the trajectories and distortion (including merging and breaking of eddies) takes place which may be different in different directions and may not be same or repetitive (called as phase scrambling effect, Yule,1980). The longitudinal component of length scale is more stretched in the direction of flow than the other components. It is still not clear how much the deviation of Taylor hypothesis makes influence on various parameters e.g. on Lux, however, qualitatively can be described as in the following section.

4.4.2 Taylor's hypothesis:

Cenedese et al (1991) based on the experiments in uniform turbulent flow on a rectangular duct using Laser Doppler anemometer concluded that the Taylor's hypothesis is substantially correct for small separation distances particularly, away from surface and when the turbulence is not very large. However, Taylor's hypothesis does not consider the deformation of the velocity fluctuation field moving downstream with the mean velocity, therefore, the real value of the spatial correlation (thus longitudinal integral scales of turbulence) are generally different from those obtained using the hypothesis. This difference increases with increasing separation distance. In the frozen flow concept all the eddies travel at the mean flow velocity, however, in reality each eddy travels at a velocity depending on its frequency.

4.4.3 Computational limitations:

In ideal situation the data should be almost of infinite duration, however, in practice it is of limited duration which poses impracticability to integrate in time domain to infinity. Necessary trend removal was done to the filtered full scale data. Consideration of cut off frequency in trend removal and block length greatly affect the results particularly while calculating the time scale. A careful choice of parameters is necessary for a reliable estimate of spectral quantities.

5. CONCLUSIONS

Attention should be paid to model the appropriate spectral quantities in WT depending upon the flow field around the body. Accurate estimation of integral length scales or parameters associated with large/small eddies, which varies along the height of WT, is highlighted. This is necessary for accurate simulation of wind characteristics in WT, determining the appropriate geometric/time scale, for interpreting and for comparing the results from different sources (WT tests), and for further (mathematical) modeling of the structural response to wind.

The methods mostly utilised to estimate the integral length scales give a large difference in the estimated value, in such situations, the method discussed here provides tool for an accurate description of the spectral (Fichtl-McVehil spectrum) characteristics e.g. the spectrum, integral length scales and variances, of the simulated wind in low as well as high frequency range, however, the caution should be observed while comparing data from various sources due to difference in method used.

ACKNOWLEDGEMENT

The authors gratefully acknowledge the support from Prof. Peil and Dr. Noelle from University of Karlsruhe for providing the full-scale wind velocity data, and Prof. Niemann for encouraging and for discussions during preparation of this paper. RKG acknowledges the support of DAAD,FRG and CSIR,India for his present stipendium at RUB.

REFERENCES

ASCE Standard(1990), Minimum Design Loads for Buildings and other Structures Revised ANSI).

Bradshaw,P.(1994), "Turbulence:the Chief Outstanding Difficulty of our Subject", Exp. in luids, Vol.16, pp203-216.

Cenedese,A., Romano,G.P. and Felice,F.Di. (1991), "Experimental Testing of Taylor's Hypothesis by LDA in Highly Turbulent Flow", Experiments in Fluids, Vol.11, pp.351-358.

Cermak,J.E.(1987),"Advances in Physical Modeling for Wind Engineering", Jour of Engg. Mech., Proc. ASCE, Vol. 113, pp.737-756.

Cook,N.J.(1977/78),"Determination of the Model Scale Factor in Wind Tunnel Simulations of the Adiabatic Atmospheric Boundary Layer",Jour. W.E.I.A., Vol. 2, pp.311-321.

Counihan,J(1969),"An Improved Method of Simulating an Atmospheric Boundary Layer in a Wind Tunnel", Atm. Env., Vol. 3, pp.197-214.

ESDU Data Item 85020 (1985) "Characteristics of Atmospheric Turbulence near the Ground (Amendment E)", ESDU International, U.K.

Farell, C. and Iyengar, A. (1993),"Experiments on the Simulation of a Neutral Urban Atmospheric Boundary Layer", Proc. 7th U.S. Natl. Conf. Wind Engg., Los Angeles.

Fichtl,G.H. and McVehil,G.E.(1970),"Longitudinal and Lateral Spectra of Turbulence in the Atmospheric Boundary Layer at the Kennedy Space Center",Jour. of Applied Meteor., Vol.9, pp51-63

Flaga,A. and Wrana,B.(1987),"Analysis of Empiric Formulae of Power Spectral Densities of three Wind Velocity Vector Components", 7ICWE,Aachen, Pre-print, Vol.1, pp.11-20.

Harris,R.I.(1986),"Longer Turbulence Length Scales", Jour. W.E.I.A.,Vol. 24, pp.61-68.

Hölscher,N,(1993),"Ein multivariater Ansatz für die aerodynamische Übertragungsfunktion der Winddrücke in atmosphärischer Grenzschichtströmung", Ph.D.Thesis, Ruhr University, Bochum, FRG.

Kaimal,J.C., Clifford,S.F. and Lataitis,R.J.(1989),"Effect of Finite Sampling on Atmospheric Spectra", Boundary Layer Meteorology, Vol.47, pp.337-347.

Kareem,A.(1987),"Effect of Parametric Uncertainties on Wind Excited Structural Response", Jour.W.E.I.A.,Vol. 30, pp.233-241.

Kasperski,M and Niemann H.J.(1993),"The Fully Correlated Gust:A New Concept for Separating a High-frequent Gust Loading Process and a Low-frequent, Spatially Correlated Gust Load", Proc. ICOSSAR'93, Innsbruck, pp.1715-1718.

Lee,B.E.(1975/76),"Some Effects of Turbulence Scale on the Mean Forces on a Bluff Body", Jour. W.E.I.A., Vol.1, pp.361-370.

Melbourne,W.H. (1979),"Turbulence Effects on Maximum Surface Pressures- A Mechanism and Possibility of Reduction", Proc. Int. Conf. on Wind Engg., Vol.1, Ed. Cermak,J.E., Colorado, pp.541-551.

Monin,A.S. and Yaglom,A.M.(1975), Statistical Fluid Mechanics:Mechanics of Turbulence, Vol. 2, The MIT Press.

Nakamura,Y., Ohya,Y. and Ozono,S.(1988),"The Effects of Turbulence on Bluff-Body Mean Flow",Jour. W.E.I.A., Vol. 28, pp.251-259.

Peil,U. and Noelle H.(1992), "Guyed Masts under Wind Load", Jour.W.E.I.A., Vol. 41-44, pp.2129-2140.

Roy,R.J. and Holmes,J.D. (1988),"The Effect of Scale Distortion on Total Wind Loads on a Low-rise Building model", Jour. W.E.I.A., Vol.29, pp.273-282.

Schrader,P.(1993),"Die Statistische Stabilität Gemessener Integraler Längenmaße und Anderer Windparameter", Ph.D.Thesis, KIB, Ruhr University, Bochum, FRG.

Schueller,G.I., Hirtz,H. and Booz,G.(1983), "The Effects of Uncertainties in Wind Load Estimation on Reliability Assessments",Jour. W.E.I.A., vol. 14, pp.15-26.

Solari, G (1987),"Turbulence Modeling for Gust Loading", Jour. of Structural Engineering (Proc. ASCE), Vol.113, No.7, pp.1550-1569.

Surry,D.(1982),"Consequences of Distortions in the Flow Including Mismatching Scales and Intensities of Turbulence", Proc. Int. Workshop on Wind-Tunnel Modelling for Civil Engineering Applications, Ed.- T.A.Reinhold, pp.137-185.

Teunissen,H.W.(1980), "Structure of Mean Winds and Turbulence in the Planetary Boundary Layer over Rural Terrain", Boundary Layer Meteorology, Vol. 19, pp.187-221.

Tieleman, H.W.(1992),"Problems Associated with Flow Modelling Procedures for Low-rise Structures", Jour. W.E.I.A., Vol.41-44, pp.923-934.

Yule,A.J. (1980),"Phase Scrambling Effects and Turbulence Data Analysis", Proc. Sec. Intl. Symp. on Turbulent Shear Flows, (Ed..BradburgLJS), Vol.2, Springer Verlag.

Waldeck, J.L (1989),"Measurement of Wind Effects on a 300m Concrete Chimney", Jour. W.E.I.A., Vol. 32, pp.199-210.

WIND TUNNEL SIMULATION OF THE ADIABATIC FLOW AROUND A SET OF SINUSOIDAL HILLS

FERREIRA, A.D. and VIEGAS, D.X.
Assistant *Associate Professor*

Grnpo de Mecâmica dos Fluids, Department of Mechanical Engineering, University of Coimbra, 3000 Coimbta, Portugal

Abstract

A wind tunnel study of the adiabatic flow around a set of two or three similar two-dimensional hills of sinusoidal cross-section is presented. The Reynolds number of the flow, based on the velocity outside the boundary layer and on the hill model height varied between 1.8×10^4 and 2.5×10^5. Particular attention was given to the flow close to the surface of the models, that was always smooth in the present study, namely to surface pressure and shear-stress distributions and flow visualization.

1. INTRODUCTION

The knowledge of the wind flow over complex topography is important for many practical situations. We may refer, for example, wind energy studies, pollutant dispersion, microclimatic and fire propagation studies. This last field was the main motivation for this work, included in a wider research program on physical aspects related to forest fires, that is being carried out at the University of Coimbra.

In most cases the orography of a region exhibits groups of various hills or mountains. There are some instances in which those hills have a shape and a relative distribution that allows, with a reasonable degree of approximation, the consideration of a two-dimensional situation. Such is the case when the cross section of each hill is practically constant and their axis are parallel, with wind blowing perpendicular to this direction.

In this work the adiabatic flow around a set of two or three similar two-dimensional hills of sinusoidal cross-section is studied experimentally. This work includes surface pressure measurements and flow visualizations, for various situations corresponding to several distances between hills, and for different wind profile characteristics.

2. METHODOLOGY

2.1 Experimental set-up

For the present work, the cross section of the hills had a geometrical shape defined by the following equation:

$$z = \frac{H}{2}\left[1 + \sin\left(\frac{x_1 - L}{2L} \times \pi\right)\right] \tag{1}$$

where H is the height of the hill and L is its length, measured in the flow direction, for z = H/2 (figure 1).

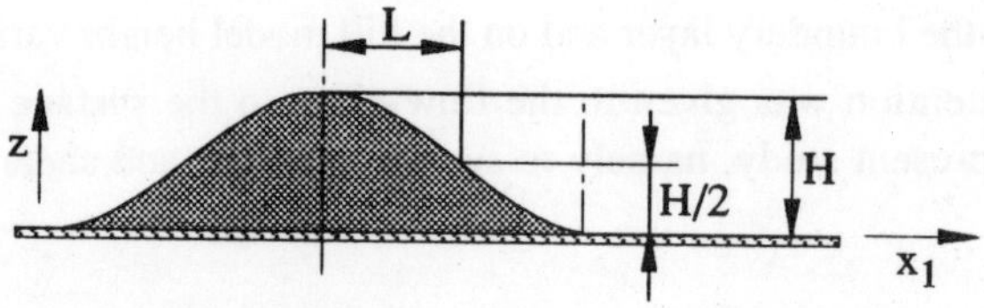

Figure 1 - Cross section of the sinusoidal hill.

In a previous work, Ferreira et al (1991), four different shapes of the cross section of a singular hill were considered, with values of the ratio H/L equal to 0.75, 1, 2 and 4. In this work only the model shape with H/L=1, with H=200mm, was considered, but with two or three interfering hills. Each model had a span of 1500 mm across the flow and end plates were mounted in the upwind model to avoid three dimensional interference between the hills. All measurements were made in the plane of symmetry of the flow. The surface of the models was hydraulically smooth.

The tests were performed in the Atmospheric Boundary Layer Wind Tunnel of the University of Coimbra. This wind tunnel has an open circuit, with an open working section of 4×1 m^2, with a length of 9 m, and with a maximum velocity of 20 m/s. A value of 12% was measured for the turbulence intensity of the undisturbed flow. The technique proposed by Counihan (1973) was used in order to obtain a thick boundary layer in the working section (figure 2). Different types of incident boundary layers were simulated, with several values of the thickness (δ) and exponent (α) of the velocity profile, by the use of various combinations of vortex generators and roughness elements.

The shear-stress sensors were of the type proposed by Irwin (1981). They consist essentially of two concentric pressure taps, one of them flush with the surface and the other protruding around 1 mm from the surface. A pressure difference is created by the flow that can be related by calibration to the wall shear-stress. The sensors used in this research were locally built and calibrated in a wall jet flow, and also in a rectangular duct. The use of this type of probes in a pressure gradient flow, like the one considered in this work, is subject to errors, as shown by Monteiro (1993). No corrections were made to the values obtained with a calibration in a zero pressure gradient flow. This type of sensor is insensitive to the flow direction and therefore only the modulus of the shear-stress is measured.

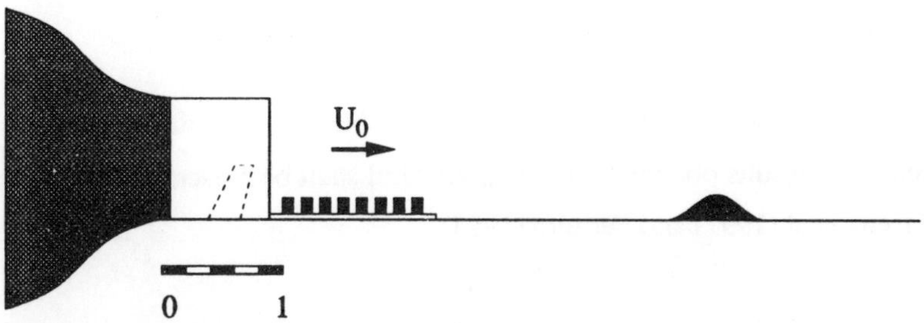

Figure 2. Schematic representation of the wind tunnel.

2.2 Model geometry and definitions

The geometry of the model for two or three similar hills is shown in figure 3, with the coordinate systems used in the work. A basic reference system (x,z) has its origin in the center of the leading edge of the working section, local systems (x_1,z_1), (x_2,z_2) and (x_3,z_3) were considered for the description of the flow properties around each hill. S1 and S2 are, respectively the distances between the first and the second hill, and the second and the third hill. The first hill was placed at x=10.2 H.

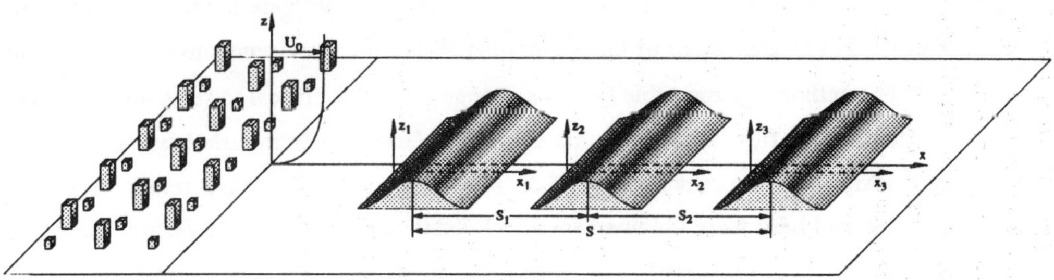

Figure 3 - Model geometry for three sinusoidal hills

Average local static pressure and shear-stress coefficients, C_p and C_f, defined by equations (2) and (3), respectively, were evaluated for various sets of the parameters S1/H and S2/H, where p is the local static pressure, p_0 the undisturbed flow static pressure, τ_w the wall shear-stress, ρ the specific mass of the fluid and U_0 the reference velocity of the undisturbed flow.

$$C_p = \frac{p - p_0}{\frac{1}{2}\rho U_0^2} \tag{2}$$

$$C_f = \frac{|\tau_w|}{\frac{1}{2}\rho U_0^2} \tag{3}$$

Most of the results presented in this work were obtained for δ=0.55 m and α= 0.24. Due to space restrictions only results obtained for the upwind hill shall be presented here, further results can be found in Ferreira et al (1991) and Ferreira (1993).

3. RESULTS

3.1 Flow similarity

In this type of flow, where thermal effects are absent, it is easy to show that the main similarity parameter is the Reynolds number of the flow. In order to determine if there was any significant effect of this parameter in the tests and also to evaluate the feasibility of applying the present results in a natural scale, the variation of the drag coefficient for a singular hill with the Reynolds number ($Re=U_0H/\nu$) was studied. The drag force was obtained by integration of the pressure on the surface of the hill. The shear-stress was not considered as its contribution is less than 2%, as was shown by Ferreira (1993).

The results presented in figure 4 show that there is a marked decrease in the value of C_D for $2\times10^4 < Re < 8\times10^4$. For larger values of the Reynolds number the flow seems to be independent of this parameter. The authors interpret this fact as a change from sub-critical to super-critical flow, like the one that occurs in the flow around circular cylinders. Given the high turbulence intensity of the flow in the boundary layer, the critical value of the Reynolds number for this geometry is around 10^5. All results presented here were obtained for $Re=1.6\times10^5$ and therefore the results should be applicable to full scale conditions without major corrections.

3.2 Pressure distribution

The distribution of the pressure coefficient around the first hill for the case of two hills placed at various distances S1/H, is presented in figure 5. As can be seen in this figure, the presence of the second hill changes the pressure field, specially in the leeward slope. It can be observed also that for increasing values of S1/H the pressure distribution becomes similar to the isolated hill case.

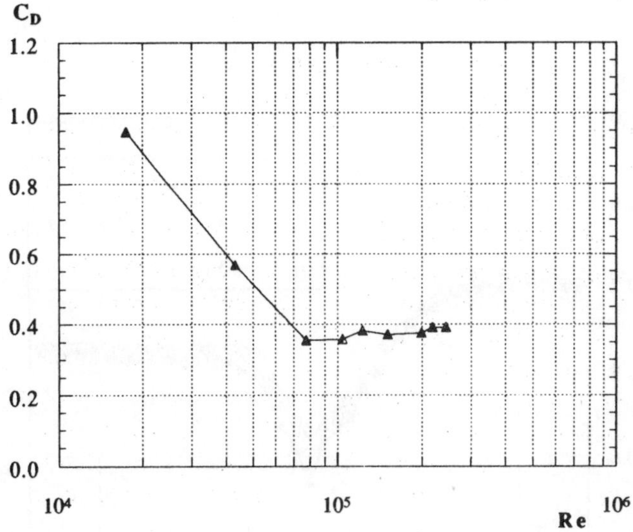

Figure 4 - Evolution of the drag coefficient of a single hill with the Reynolds number.

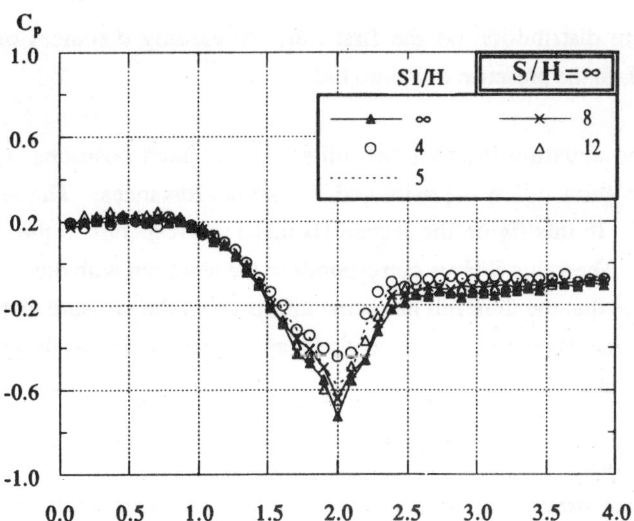

Figure 5 - Pressure distribution around the first hill for the case of two hills.

When a third hill is introduced the results are different and depend on the relative position of the three hills. Two different situations were considered in this study.

In the first one the first and the third hills were at fixed positions, x/H=10.2 and 22.2, respectively, and the second hill was placed at various locations. The results obtained are shown in figure 6. From the analysis of this figure we can conclude that the presence of the third hill modifies the flow properties, but the influence of the second hill is less marked, in comparison with the previous case.

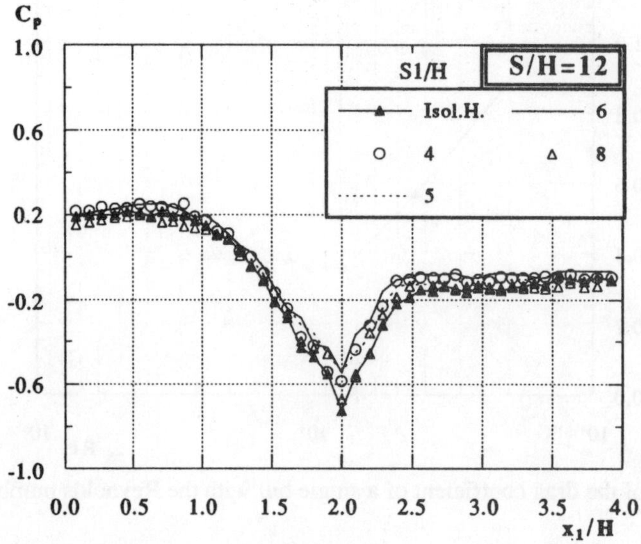

Figure 6 - Pressure distribution on the first hill, for varying distances of the second hill, in the presence of a third hill.

In the second situation the first two hills were at fixed positions, x/H=10.2 and 14.2, respectively, and the third hill was positioned at various distances. The results obtained are presented in figure 7. In this figure the legend (Isol.H.) corresponds to the isolated hill, that is shown as a reference. The value S/H=∞ corresponds to the situation with only two hills. From this figure we can conclude that the third hill has some influence on the pressure distribution around the first hill, but it seems that its position is not very relevant in this context, as the curves for S/H=8, 10 and 12, are all very similar.

3.3 Wall Shear-stress

The local shear-stress coefficient on the first hill for the same situation that was considered in figure 7, is shown in figure 8. Once again it can be observed that the distribution of C_f is very similar for the finite range of S/H considered in this study.

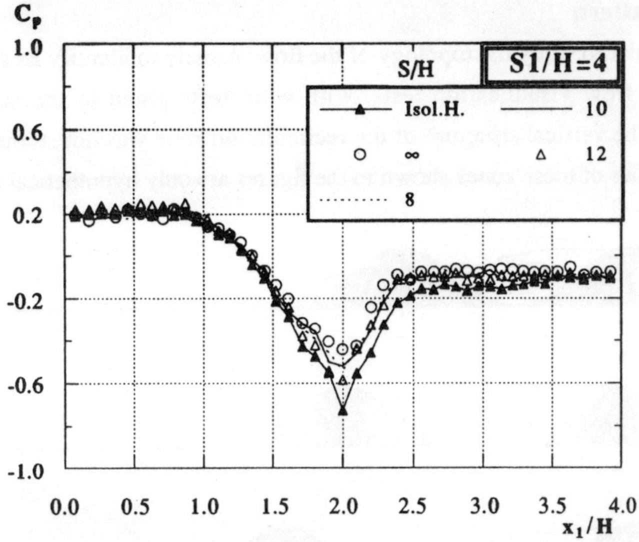

Figure 7 - Pressure distribution on the first hill, with a fixed second hill, for various locations of the third hill.

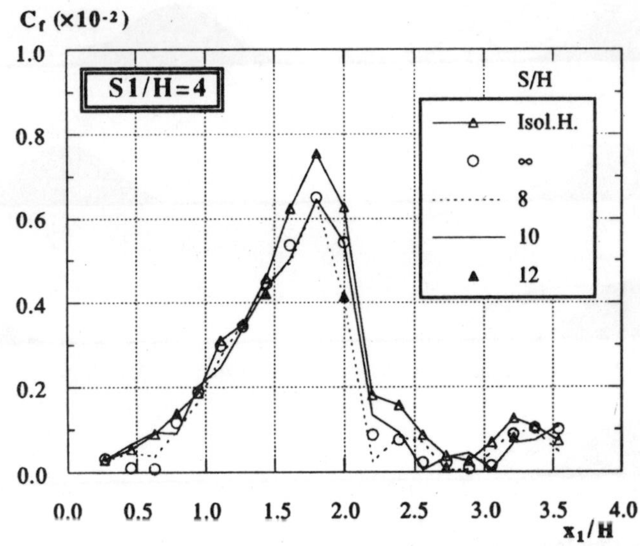

Figure 8 - Wall shear-stress distribution on the first hill, with a fixed second hill, for varying distances of the third hill.

3.4 Flow Pattern

In order to study the topology of the flow, namely to identify its separation and reatachment points, some flow visualisation tests, with wool tufts glued to the surface of the model were performed. The vertical structure of the recirculation zone was not studied in detail in all cases, so the limiting lines of these zones shown in the figures are only hypothetical average streamlines.

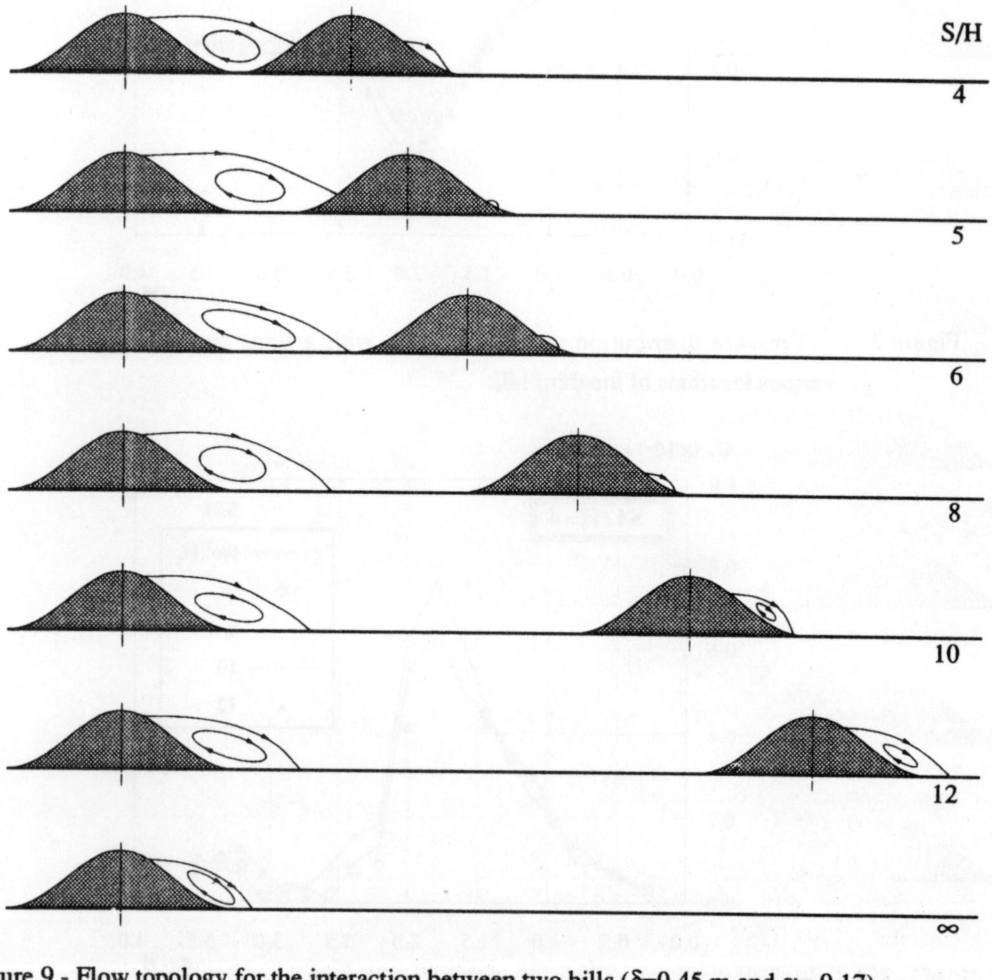

Figure 9 - Flow topology for the interaction between two hills (δ=0.45 m and α= 0.17)

The flow topology for two hills is shown in figure 9, for a boundary layer with δ=0.45 m and α=0.17. We can observe that the recirculation zone in the leeward face of the first hill changes its shape for varying values of S/H. Its length decreases for increasing values of S/H. For S/H=12 the presence of the second hill still affects the extension of this zone. It is interesting to notice that the recirculation in the leeward face of the second hill decreases when S/H varies between 4 and 5, and then increases to a size similar to that of an isolated hill.

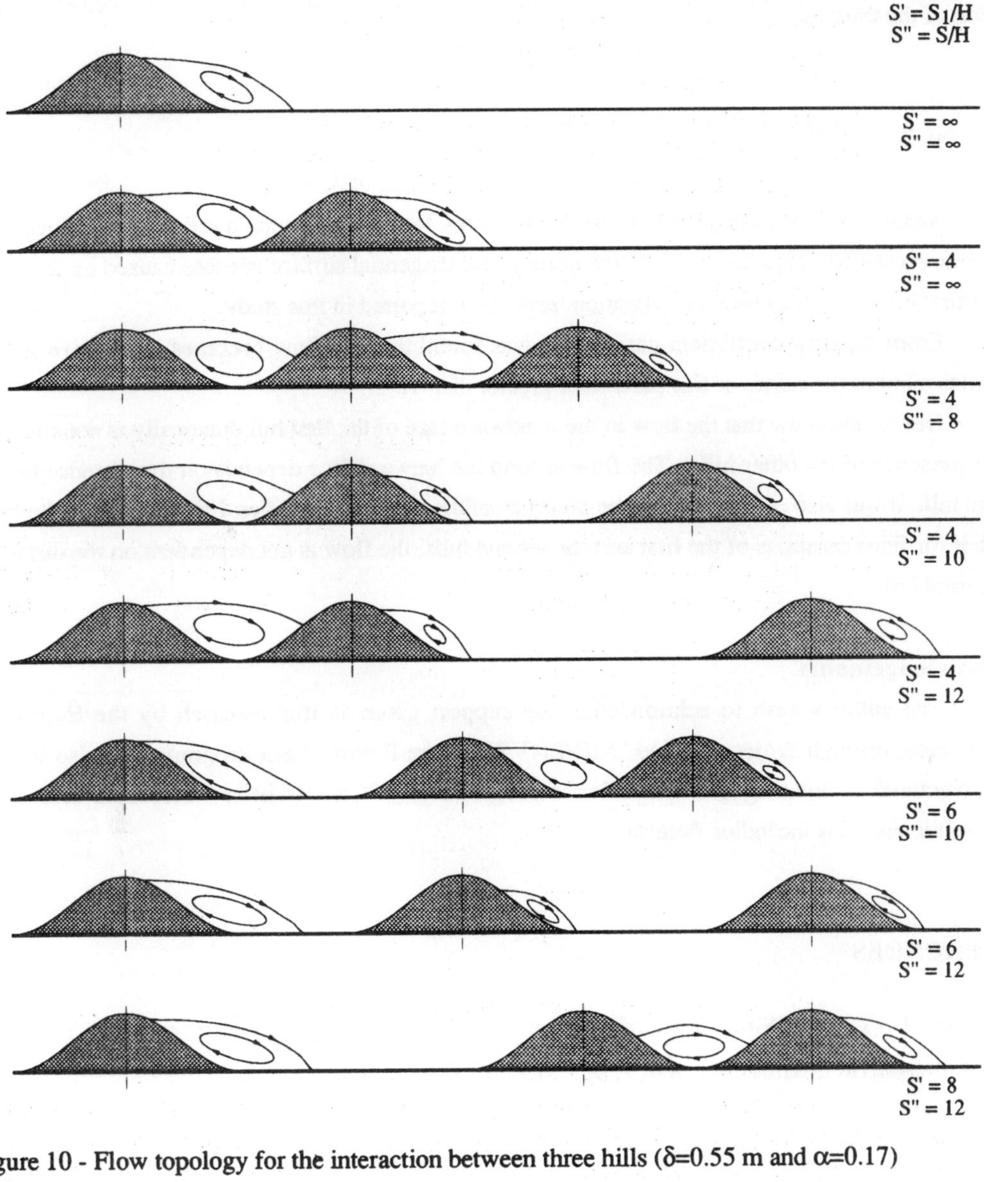

Figure 10 - Flow topology for the interaction between three hills (δ=0.55 m and α=0.17)

The flow topology around three hills was also studied for δ=0.55 m and α=0.24. The results are shown in figure 10. This figure confirms the comments about the pressure distribution that were made for this situation. The influence of the third hill on the first one is very slight, the separation point remains practically fixed, while the reattachment point varies very little with the position of the third hill.

4. CONCLUSION

Some of the average properties of the boundary layer flow around a set of two dimensional hills were presented. Measurements of the normal and tangential surface stresses caused by the flow on the first hill, as well as flow visualization tests, were reported in this study.

From the drag coefficient variation it was found that the flow becomes insensitive to the Reynolds number for values of this parameter greater than 10^5.

Our results show that the flow in the windward face of the first hill practically is not affected by the presence of the other hills. The flow around the leeward face depends on the distance to the second hill. It was also concluded that the presence of the third hill modifies the flow around the first one, but for fixed positions of the first and the second hills, the flow is not dependent on the distance to the third hill.

Acknowledgements

The authors wish to acknowledge the support given to this research by the European Commission, through Project entitled "MINERVE", of the Environment Program, and also by the Junta Nacional de Investigação Científica, through Project "Características físicas da eclosão e desenvolvimento dos incêndios florestais".

REFERENCES

Counihan, J. (1973), "Simulation of an Atmospheric Boundary Layer in a Wind Tunnel", Atmospheric Environment, Vol.7, pp. 673-689.

Ferreira, A.D. (1993), "Contribuição para o estudo do escoamento turbulento adiabático em torno de colinas bidimensionais", Master Thesis, University of Coimbra.

Ferreira, A.D., Silva, M.C.G., Viegas, D.X., Lopes, A.G. (1991), "Wind tunnel simulation of the flow around two-dimensional hills", Journal of Wind Engineering and Industrial Aerodynamics, nº 38, pp. 109-122.

Irwin, H.P.A.H. (1980), "A Simple Omnidirectional Probe For Wind Tunnel Studies of Pedestrian Level Winds", Journal of Wind Engineering and Industrial Aerodynamics, nº 7, pp. 219-239, 1980.

Monteiro, J.M.P. (1993), " Contribuição para o estudo do comportamento das sondas de Irwin e de Preston em escoamentos turbulentos com gradientes de pressão", Master Thesis, University of Coimbra.

Schlichting, H. (1979), *Boundary Layer Theory*, McGraw-Hill Book Company.

9ICWE, 1995, New Delhi, India

EXPERIMENTAL STUDY OF DYNAMIC WIND LOADS ON BUILDINGS USING HFBB TECHNIQUE

LIU, SP.[1], XIE, C.T.[1], ZHANG, D.K.[1], WANG, Q.Z.[1], LEE, B.E.[2]

[1] *Senior Engineers, China Aerodynamics Research and Development Center, P.O. Box 211, Mianyang, Sichnam P.R. China*
[2] *Professor, Dean, Faculty of Engineering, University of Portsmouth, UK*

Abstract

This paper describes a joint China-UK project on the experimental study of dynamic wind loads on buildings being carried out at the China Aerodynamics Research and Development Center. The study strategies, objectives and experimental arrangement are presented, and some typical results are discussed as well.

1. INTRODUCTION

The last decade has seen a significant growth in the demand for dynamic wind loading information for the design of tall, wind sensitive buildings. At the same time there has been a major improvement in the ability of wind engineering consultants to supply this information, from wind tunnel tests and from semiempirical calculation methods. However, whether mathematical or physical models are used, they are only very primitive reflections of the complexities of the real world. It has been established previously that a number of simplifying assumptions are required in order to construct either model above of the relationship between the wind structure and the building response for buildings of given shape in known locations. Since these simplifying assumptions are treated differently by different wind tunnel labs and the authors of different numerical prediction methods, considerable uncertainty may surround the results yielded for any one-off test or calculation. Whilst wind engineering consultants can assure a client of the accuracy and repeatability with which a particular result has been obtained, they are still unable to indicate the degree of certainty of reliability which can be placed on the result itself. That implies that the level of confidence which the structural engineer has when asked to use the results of a single test is low, leading to uncertainties in the safety factor associated with the design, or alternatively, a conservative but costly structural solution.

The structural engineer using such results, or the wind engineer supplying them, would be in a far better position to exercise judgement in their interpretation if more information were available regarding their potential variability as functions of the parameters which influence the simplifying assumptions referred to earlier.

The purpose of a joint China-UK project on the measurement of dynamic wind loads on buildings is to help establish the confidence limits with which dynamic wind loading design data can be used in practice.

2. STUDY STRATEGIES

In order to establish a data base which will help to determine the confidence limits which should be associated with the wind tunnel assessment of dynamic wind loads, it is necessary to carry out a parametric test series of wind tunnel experiments.

The relationship between wind characteristics and building response is a complex one involving an aerodynamic transfer function relating the gust structure to the load function, and a mechanical transfer function relating the load function to the response. Well-known full aeroelastic modeling techniques enable the designer to reproduce all the elements of this process in a wind tunnel. However, the cost of this full modeling procedure for any particular building has tended to be prohibitively high in all but exceptional cases.

The other conseqence of the use of the full aeroelastic modeling method is that it only enables the building response for a particular set of structural characteristics to be determined. Thus the intermediate step in the transfer process, i. e. the load function, is not usually characterized. If the load function itself were capable of independent determination as a function of only the wind structure and the aerodynamic shape of the building, then this would greatly assist our understanding of the nature and origins of fluctuating wind loads. Substantial efforts have been made to meet this objective in recent years in many laboratories in the world. Balances were designed for the measurement of the load function in the form of a quasi-static force spectrum which is independent of the dynamic characteristics of the model (Tschanz, 1983). To obtain a reliable measurment of the fluctuating forces and moments on a stationary model mounted on a force balance, the resonant frequency of the entire system must be significantly higher than any expected forcing frequency of interest. At the same time the transducer should have a high sensitivity. Such a High Frequency Base Balance (HFBB) technique is used in the project, for the determination of modal loads corresponding to the in-wind, cross-wind and torsional response for a wide range of building characteritics and geometries.

Comparisons will be made with calculation methods applied over the same range of parameters as well as with available full-scale data and wind tunnel studies from other labs.

3. PROJECT OBJECTIVES

Specific series of tests will be concerned with the influence on modal loads of the following:

(i) different terrain roughness for buildings of a fixed shape and size,

(ii) change of terrain roughness at different fetch lengths upstream for a given building,

(iii) change of building shape,

(iv) overall size of a given shaped building in a given terrain situation,

(v) incident wind angle for buildings of simple shape, and

(vi) increasing incident mean wind speed.

Objectives (i), (ii) and (iii) are straightforward and can lead to the sensitivity analysis.

Objective (iv) has a multiple variant built into it, and the objective may be illustrated as follows. If a building of a given size, say height 100m, is to be built in a specified terrain location, then it may be modelled at a selected scale ratio, say 1/500, to yield a model size, 200 mm high, in a boundary layer whose linear scale ratio is also 1/500. Modal forces can then be scaled up by the relationship.

$$S_{nm}/S_{nf.s.} = (V_m/V_{f.s.})^3(D_m/D_{f.s.})^5$$

to yield full scale modal forces. If a 1/250, 1/200 or even 1/100 model of the same building (i. e. model of 400mm, 500mm and 1000m) is then tested in the same boundary layer, the results may also be scaled up using the equation above to yield full scale values. Any lack of agreement may then be considered to be a function of the "scaling mismatch" between the model and the boundary layer. Such an exercise provides valuable information concerning the limitations and validity of physical simulation. However, if an identicallly shaped building of a different size, say 200m high, were to be built in the same terrain location, a true 1/500 scale model would now yield a model size of 400mm, i. e. the same size as the 1/250 scale model of the 100m building—fomerly a "scaling mismatch", but now correct scale for the larger building. It will be significant to determine from these tests whether or not the correct scale wind tunnel tests for buildings of different sizes bear the same relationship to each other as that indicated by the results of different empirical calculation methods applied to buildings of different sizes. This exercise will be a test of the spatial integrity of the turbulence characteristics of the terrain simulation as much as of the turbulence descriptions built into the calculation methods.

Objective (v) is important because it will help to clarify the validity of calculation methods as well as to establish the wind directional ranges required for tests on isolated tall buildings. The calculation methods of modal loads on rectangular planform buildings are usually only valid for normal wind angles. If the model tests confirm that the loads are at a maximum when the wind is normal to a face and decrease for oblique wind, then the finding will lend confidence to use calculation methods as conservative results for most wind directions.

4. EXPERIMENTAL ARRANGEMENTS

4.1 Wind tunnel

The experiments were carried out in the Low Speed Aerodynamics Institute,

CARDC which is the biggest aerodynamic research center in China with a comprehensive wind tunnel complex.

The CARDC FL-11 low speed wind tunnel is an open circuit type with a close test section measuring 1.4m wide, 1.4m high and 2.8m long, originally designed for aeronautical testing.

Modifications for wind engineering purpose consist of extending the existing test section floor towards both the contractor and diffuser (Fig. 1), resulting in a new boundary layer test section, 1.4m×1.4m×6.0m. A 0.8m high simulated atmospheric boundary layer may be generated in the model area, 4.9m from the entrance, and the ratio of fetch to boundary layer height is 7.5. The model is to be mounted on a turntable.

4.2 Simulated atmospheric boundary layer

The simulation hardware includes fence, spire and roughness elements. The profiles of mean longitudinal velocity were produced, with power law exponents $\alpha=0.16$, 0.30 and 0.40, corresponding to typical rural, suburban and urban terrain respectively. Based on spectral scaling, the length scale of the simulated boundary layer flow varies from 1:270 to 1:345 depending on altitude. Thus the flow scale 1:300 was chosen as a compromise.

4.3 Models

There is clearly an advantage to be gained by modeling the buildings which have themselves been the subject of full scale dynamic wind load investigations. Three such tall buildings were considered, Sheffield University Arts Tower (78m×36m×20m), U.K., Hume Point Building (67m×23m×18m), U.K., and Yokohama KN Building (68m×48m×36m), Japan. Also the data available for comparison are from wind tunnel tests. A good generic wind tunnel model of a high rise building is the CAARC model (182.9m×45.7m×30.5m), which has been subject to widespread testing with aeroelastic models, fluctuating pressure measurement on rigid model, and HFBB technique as well.

Two types of models have been designed for the project: tall slender type based on the CAARC model (labeled as "C-" series) and cuboid type based on the Arts Tower model approximately ("A-" series), as well Hume Point building ("H-") and Yokohama building ("Y-").

Table 1 indicates that each of C- and A-series has 5 different plan aspect ratios for the same height, which illustrated the objective 3. As for the objective 4 of the size variation, models A4 (80m×40m×40m) and A5(80m×40m×20m) include three different scales, 1/300, 1/600 and 1/150, where 1/300 has been recognized as flow scaling match and the other two as mismatch. Total 16 models were tested. For the objective 1, each of five models (C11, C21, A51, H11 and Y11) has been tested in the three simulated flows corresponding to rural, suburban and urban terrains. It is seen from Tab. 1 that the objectives 5 and 6 are covered to change the incident wind angle and mean speed. Also interesting is that the same model C21 with same balance was tested in both CARDC and Guangdong Provincial Building Research & Design Institute wind tunnels to evaluate the error source from facilities.

Table 1. wind tunnel test program of the project

series	f. s. size (m) H	f. s. size (m) D_y	f. s. size (m) D_x	aspect H/D_x	aspect D_y/D_x	scale	model	terrain⑤ Su.	terrain⑤ Ur.	terrain⑤ Ru.	parameter wind angle⑥	parameter mean speed⑦
		30			1.0		C11	O	O		O	O
		45			1.5		C21①	O	O	O	O	O
tall slender	180	60	30	6	2.0	1/300	C31	O			O	O
		90			3.0		C41	O			O	O
		120			4.0		C51	O			O	
			80	1.0	1.0		A11	O			O	
		80	40	2.0	2.0	1/300	A21	O			O	
			20	4.0	4.0		A31	O			O	
						1/300	A41	O			O	O
cuboid	80	40	40	2.0	1.0	1/600	A42	O			O	O
						1/150	A40	O			O	
						1/300	A51②	O	O	O	O	O
			20	4.0	2.0	1/600	A52	O			O	O
						1/150	A50	O			O	
Hume.	67	23	18	3.7	1.3	1/300	H11③	O	O	O	O	O
Yoko.	68	48	36	1.9	1.3		Y11④	O	O	O	O	O

① CAARC model approx. (182.9×45.7×30.5)
② Arts Tower approx. (78×36×20)
③ Hume Point Building, (UK)
④ Yokohama KN Building, (Japan)
⑤ folw scale 1/300, Su. =Suburban, Ur. =Urban, Ru. =Rural
⑥ change wind angle 0-90°
⑦ fixed wind angle at 0° or 90°

All the models were made of a light laminated wood shell and the resulting frequency of model-balance system is about 100Hz.

4. 4 High frequency base balances

Five component high frequency base balances of strain gage type were designed and manufactured by CARDC who has rich experiences in strain gage balance design. To meet the wide range of wind loading on smaller and larger models, four balances (Chen et al, 1993) were made and used in th project, including plate-type UWO version and stem-plus-plate types.

Tables 2 and 3 describe the balance specifications and natural frequency. Two typical balance versions are shown in Fig. 2.

Table 2. Specifications of four balances from static calibration

Balance	Configuration	Parameters	F_x	F_y	M_x	M_y	M_z
JZW001	Stem Plus Plate	Design load ($\frac{N}{N.m}$)	160	80	60	60	30
		Sensitivity ($\frac{N}{N.m}$)	0. 5	0. 5	0. 1	0. 1	0. 6
		Precision (%)	0. 07	0. 14	0. 07	0. 03	0. 55
JZW002	Stem Plus Plate	Design load ($\frac{N}{N.m}$)	100	100	50	50	50
		Sensitivity ($\frac{N}{N.m}$)	0. 33	0. 34	0. 081	0. 079	0. 21
		Precision (%)	0. 16	0. 21	0. 71	0. 08	0. 17
JZW003	Plate	Design load ($\frac{N}{N.m}$)	10	10	5	5	5
		Sensitivity ($\frac{N}{N.m}$)	0. 022	0. 019	0. 0049	0. 0048	0. 002
		Precision (%)	0. 092	0. 167	0. 153	0. 232	0. 325
JZW004	Plate	Design load ($\frac{N}{N.m}$)	100	75	35	35	4
		Sensitivity ($\frac{N}{N.m}$)	0. 1	0. 1	0. 02	0. 02	0. 001
		Precision (%)	0. 05	0. 18	0. 02	0. 11	0. 16

Table 3. Natural frequencies measured on three balances (Hz)

	F_x	F_y	M_x	M_y	M_z
JZW001	228	265	872	854	567
JZW003	90	126	232	340	142
JZW004	215	287	552	552	287

4. 5 Data acquisition and reductions

The balance signal through multichannel amplifier ($\times 250$) is fed into a signal separator to pick up the mean and fluctuating component. The later one is fed into a secondary amplifier ($\times 10$) and a low-pass filter (cut-off frequency about 50Hz), then is calculated to obtain a spectral power density function by FFT program in a personal computer.

The sample period used was 6. 67 sec, equivalent to 2000 sec in full scale. Over this period 1000 samples were averaged. The sample length of 3072 points at a rate of 280Hz was used for unsteady measurement.

For data normalisation, the mean wind speed U_H measured at model height was used as the reference speed.

$$C_{\overline{M}} = \frac{\overline{M}}{1/2\rho U_H^2 D_y H^2}$$

$$C\sigma_M = \frac{\sigma_M}{1/2\rho U_H^2 D_y H^2}$$

where D_y is the wide dimension of the model plan and σ_M, the rms values of the fluctuating parts, were obtained by integration of spectral power density functions. All the data are provided in the body-fixed coordinate system. The force normal to the wide and narrow faces of the model is denoted by F_x and F_y, and base overturning moments generated by F_x and F_y are denoted by M_x and M_y, respectively.

5. RESULTS AND DISCUSSIONS

Due to limited paper space, only selected typical results are shown here, and the completed results will be presented and discussed in separate paper.

5. 1 Influence of terrain on wind loads (objective 1)

The example in Fig. 3 presents both mean and dynamic overturning moment coefficients for an 1/300 Hume Point model in rural, suburban and urban terrain, respectively. The galloping trend may be seen in mean moment C_{Mr} near 0° or C_{Mx} near 90° wind angle.

It seems that in most cases the influence of boundary layer flow over different terrain on mean loading was contrary to on dynamic loading. It would be obvious if we consider that the profile shape of mean wind speed in the trends as height is contrary to that of turbulence intensity. For instance, in the roughest urban area near ground the mean speed decreased more rapidly than in rural area (then lowest mean load) and the turbulence intensity was more strongly (higest dynamic load).

5. 2 Influence of scaling effect on wind loads (objective 4)

The example in Fig. 4 presents both mean and dynamic overturning moment coefficients for the model A4 (80m × 40m × 40m) with scales 1/300 (A41, "scaling match") and 1/600 (A42, "scaling mismatch") in suburban terrain.

These and other limited results available have shown a "disagreement", which remind the wind engineering consultants of the importance to include the flow (or eddy) scale into the similarity parameters in physical modeling. To clarify this finding more and accurate testing informations are required.

5. 3 Influence of building shape (objective 3)

The example in Fig. 5 is the mean and dynamic overturning moment coefficients for five 1/300 C-series models in suburban terrain (Liu et al, 1993). C21, as mentioned above, is CAARC standard model approximately. Five models have same height 180m and narrow side 30m, while the aspects D_y/D_x are varied as 1. 0 (square), 1. 5, 2. 0, 3. 0 and 4. 0 (slender rectangular).

The mean cross-wind bending moments decrease with D_y/D_x due to stronger separated wake flow. The galloping trends also decrease with D_y/D_x and only for $D_y/D_x < 2.0$ the significant galloping phenomena could be seen.

Comparing Fig. 5c and 5d, it was found that the dynamic wind loads in cross-wind are larger than in along-wind. Note that $C_{\sigma my}$ at 90° is the along-wind dynamic load when the wind is normal to the narrow face (180m × 30m), and it could be seen that the dynamic wind loads decrease with D_y/D_x though all narrow faces have equal area. Such trends are similar as for cross-wind loads ($C_{\sigma mx}$ at 90°).

6. CONCLUSIONS

The initial results of a China-UK joint project are presented to indicate the sensitivity of wind loads for various parameters. The examples of the results presented demonstrated the importance of the objectives of the joint project. More informations should be gathered in order to approach the confident trends of the wind loads with respect to various parameters. In this case, the HFBB technique is a powerful tool.

7. REFERENCES

Chen Feng, Hu Guofeng and Lee, B. E. (1993), "Design and Calibration of High Frequency Base Balances Used for Dynamic Wind Load on Building Models", Proceedings of Third Asia-Pacific Symposium on Wind Engineering, Hong Kong, December 13-15, 1993, Vol. II, pp 923-928.

Liu Shangpei, Wang Qizhi and Lee, B. E. (1993), "An Assessment of Dynamic Wind Loads on a Series of CAARC-Like Slender Buildings", same Proceedings as above, Vol. I, pp 261-266.

Tschanz, T. (1983), "The Base Balance Measurement Technique and Applications to Dynamic Wind Loading of Structures", BLWT-5-1983, The University of Western Ontario.

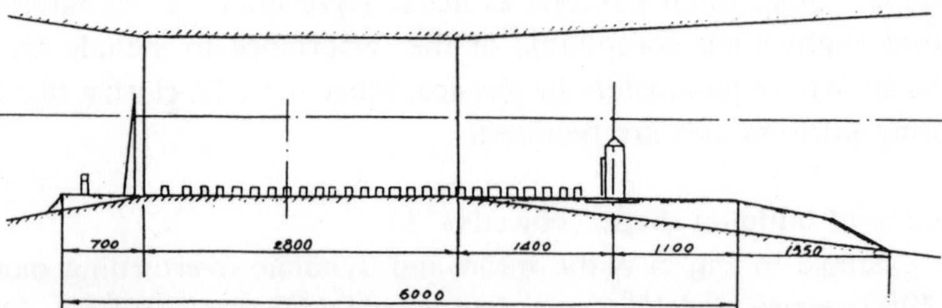

Fig. 1 CARDC 1. 4m×1. 4m×6. 0m boundary layer test section

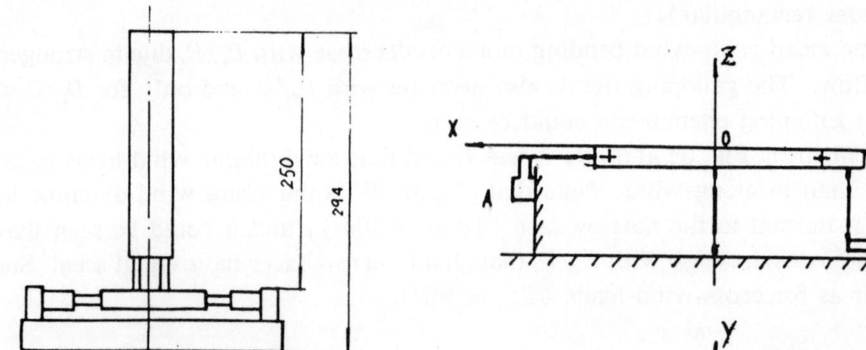

Fig. 2 Schemes of stem-plus – plate (left) and plate version balances

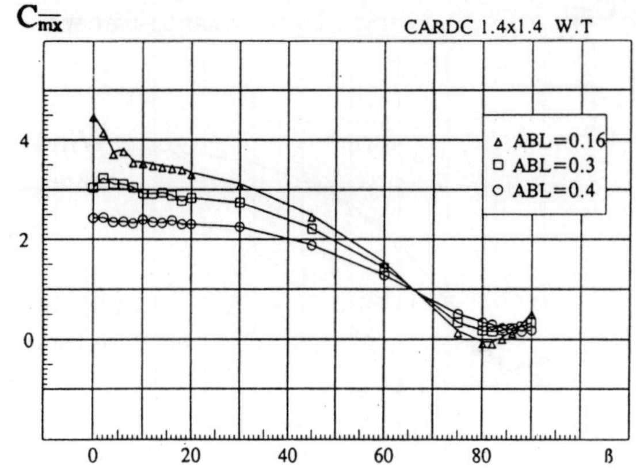

(a) $C_{\overline{mx}} \sim \beta$

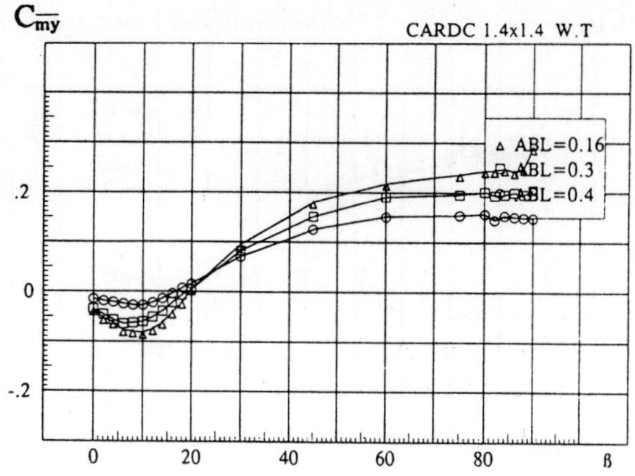

(b) $C_{\overline{my}} \sim \beta$

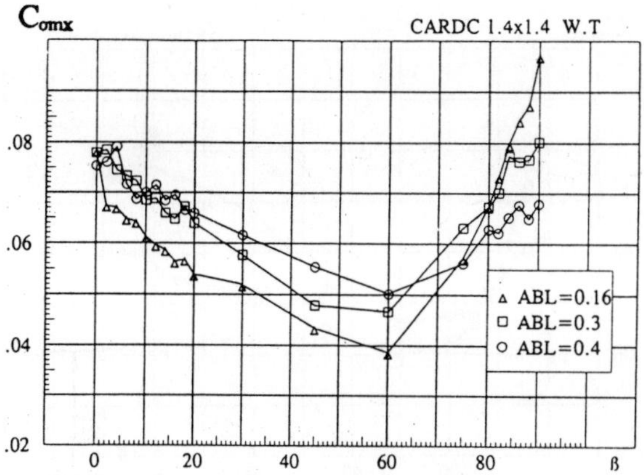

(c) $C_{\sigma mx} \sim \beta$

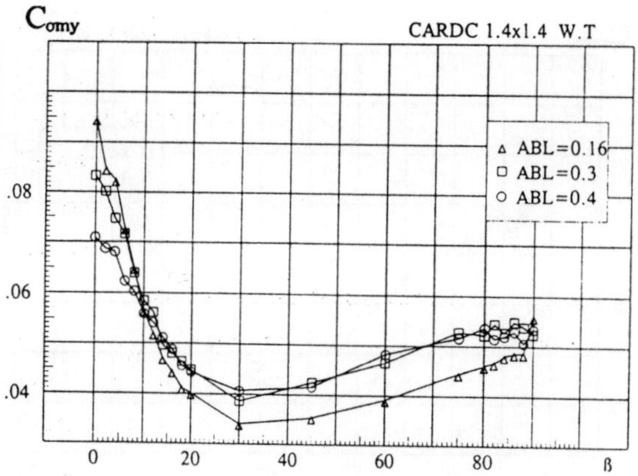

(d) $C_{\sigma my} \sim \beta$

Fig. 3 Influence of terrains on mean and fluctuating bending moments (H11 model, scaling match 1/300)

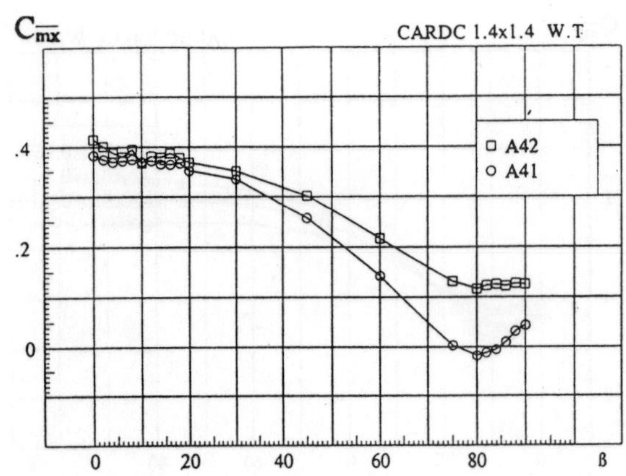

(a) $C_{\overline{mx}} \sim \beta$

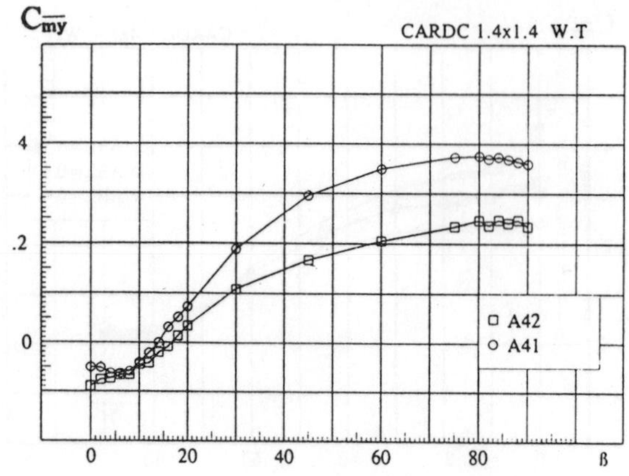

(b) $C_{\overline{my}} \sim \beta$

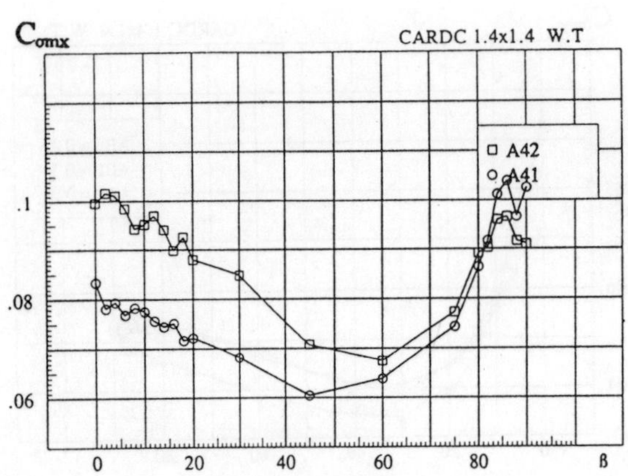

(c) $C_{\sigma mx} \sim \beta$

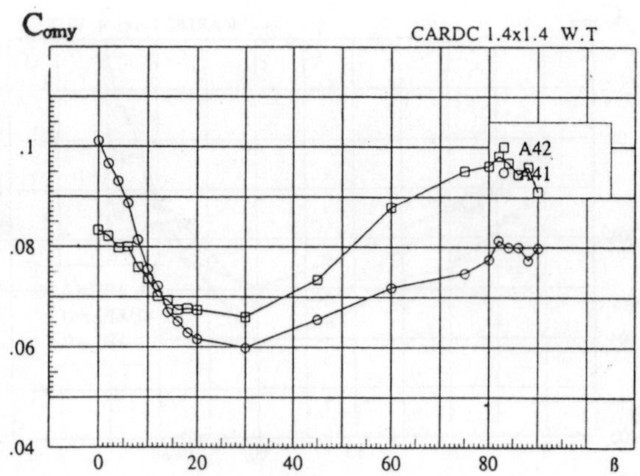

(d) $C_{\sigma my} \sim \beta$

Fig. 4 Scaling effects (A4 model, suburban flow 1/300, A41 1/300, A42 1/600)

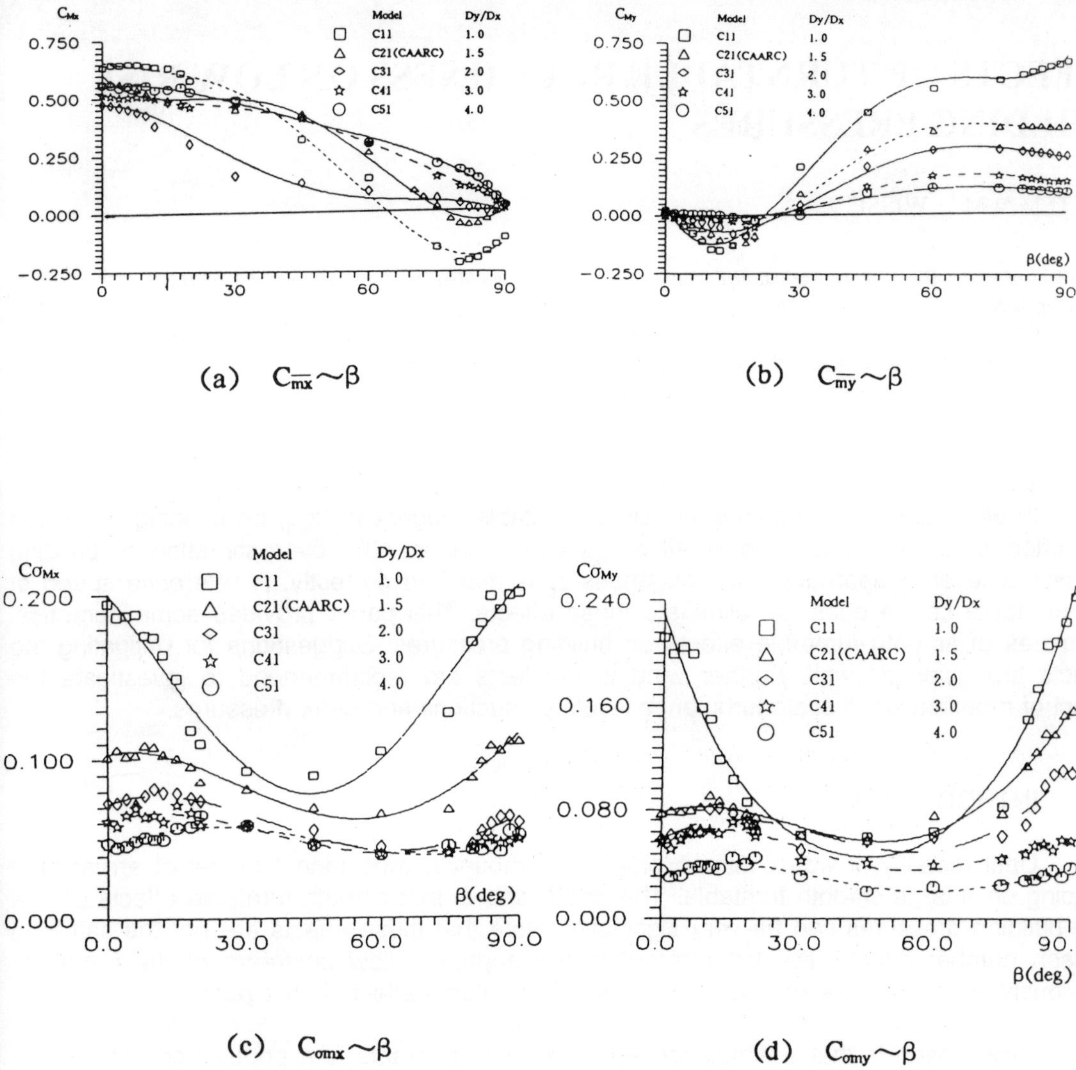

Fig. 5 Influence of model shape on wind loads (C-series model, same height and narrow face with varied aspects, in suburban terrain)

9ICWE, 1995, New Delhi, India

EFFECTS OF TURNTABLE ROUGHNESS ON LOW RISE BUILDING PRESSURES

BOTTEMA, M.[1], WISSE, J.A.[2]

[1] *Post-doctrol fellow, Wageningen Agricultural University (WAU), Group Meteorology, Duivendaal 2, 6701 AP Wageningen, NL*
Present affiliation: Ecole Centrale de Nantes, Division Méc., Energ. et Environnement Lab. de Méc. des Fluides, 1 Rue de la Noë, 44072 Nantes, Cedex 03 France
[2] *Professor in Building Physics, Dept. of Architecture and Building Technology, Group FAGO, Eindhoven University of Technology, P.O. Box 513, 5600 MB Eindhoven, The Netherlands*

Abstract

In wind tunnel experiments, effects of turntable roughness ($z_{o,tt}$) on building pressures are often underestimated. Too small a $z_{o,tt}$ can result in 40% overestimation of building pressures, even if approach flow roughness z_o is matched correctly. A reference speed at the turntable centre does not eliminate these effects. This paper provides some numerical estimates of smooth turntable effects on building pressures. Suggestions for mitigating the effects are given as well. Further wind tunnel tests are recommended to investigate the effect of modified small scale turbulence levels on suctions and peak pressures.

1. INTRODUCTION

Until recently, it was common practice to mount a wind tunnel model of an isolated building on a large smooth turntable. This paper shows that smooth turntable effects on low rise building pressures can be very significant. Note that these effects are not the same as Jensen number effects: the latter refers to the approach flow upstream of the turntable. Obviously, there were reasons to neglect smooth turntable effects in the past.

One reason is that a simulation with a smooth turntable is a popular compromise in order to reduce wake effects of upstream roughness elements. The far wake model of Leene et al (see Leene, 1992) shows that wake length (for 10% wind speed reduction) varies from 36H for very wide obstacles to 5H for cubic obstacles, where H denotes obstacle height. Most authors assure that their model is not in the wake of individual upstream roughness elements.

Another reason is that many wind engineers primarily concentrate on wind loading. Kind (1986) suggests that a smooth turntable has only moderate influence on roof suctions. However, our data show up to 40% 'smooth turntable error' in the pressure difference over windward and leeward walls. For e.g. ventilation applications, this is highly significant.

Basic books (e.g. Plate, 1982; Reinhold, 1982) do at best devote a few lines to the topic of smooth turntable effects. As a result, many authors do not specify the dimensions and roughness of their turntable, even when investigating Jensen effects (Borges et al, 1984). Hussain and Lee (1980) are one of the few who explicitly state that they used a smooth turntable.

Recently, the problem of a smooth turntable has been recognized. For example, Richardson and Surry (1992) have increased their turntable roughness. However, their fig. 1 still shows some distortion in wind profile. The first author found only one reference (Tieleman et al, 1978) in which turntable roughness length (denoted as $z_{o,tt}$ or $z_{o,loc}$) is specified. Finally, the best option, a uniform approach flow (Holmes et al, 1990) can be achieved by using carpet roughness.

This paper will present some estimates of magnitudes and trends of the smooth turntable effect. This is to state the problem, and to provide some first approximate data for correction of earlier results: e.g. data that are used for building codes, or wind tunnel data for full scale comparisons (and intercomparisons; e.g. Vickery et al, 1986).

Furthermore, this paper will discuss how to mitigate smooth turntable effects. This may contribute to the development of future, improved wind tunnel techniques.

2. FLOW OVER A SMOOTH TURNTABLE

Flow over a smooth turntable can be treated as a rough to smooth roughness change, to which internal boundary layer (IBL) theory can be applied.

After a roughness change, the mean wind profile can be described by two logarithmic sections, which are linked by a 'kink' at the internal boundary layer height h_{IBL}. Figure 1 shows an example: a wind profile after 250 m of smooth turntable fetch (corresponding to centre of 1 m radius turntable at scale 1:250). Two logarithmic profile segments can be seen, with a 'kink' in between (solid line). The upper segment yields the approach flow z_o (0.2 m), the lower segment the downstream (turntable) $z_{o,tt}$ (0.002 m). The height of the IBL is given by a diffusion equation (Panofsky et al, 1984), which can be approximated by (Simiu et al, 1986):

$$h_{IBL} = 0.3\, z_o^+ \left(\frac{x}{z_o^+}\right)^{0.8} \tag{1}$$

where x is the fetch. Eq. 1 is only valid for the logarithmic (or constant flux) layer.

Figure 1 shows a considerable increase in wind speed (from 25% at 5m to 45% at 2 m: pedestrian height) in the lower part of the turntable IBL. Note that low buildings will be completely submerged in the IBL if $x/H > 15$, where x is the fetch, and H the model height (both in wind tunnel).

The roughness change at the edge of the turntable does not only affect mean wind profiles but also the IBL turbulence. Full scale (one point) data indicate considerable increase in the normalized streamwise and lateral standard deviations σ_u/U^* and σ_v/U^*, but not in the normalized vertical standard deviation σ_w/U^*. This is due to the fact that large, energy containing, eddies do adapt only slowly to new surface conditions (Panofsky et al, 1984; Tieleman, 1990a). On the other hand, small scale turbulence will quickly adapt to the new surface conditions. As negative pressures (especially peaks) are sensitive to small scale turbulence (Tieleman et al, 1990b), they may be affected by smooth turntable effects as well. Finally, it should be noted that wind tunnel data, especially in the case of rather shallow boundary layers (Mulhearn, 1978), do not always agree well with full scale evidence.

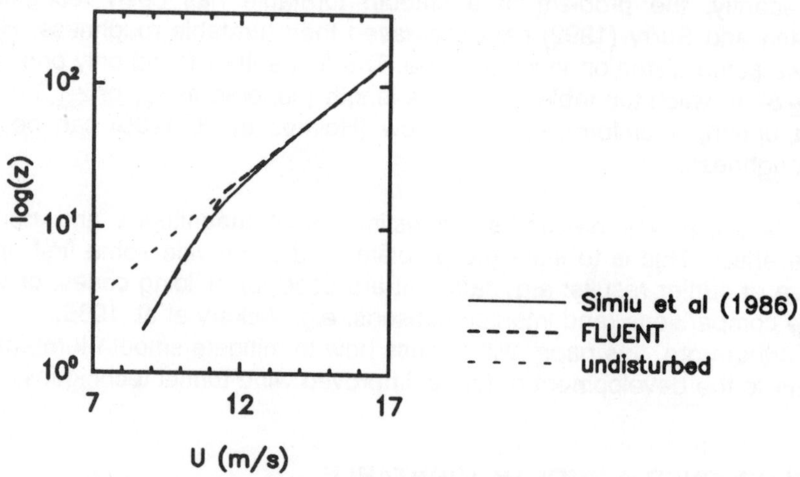

Figure 1 Wind speed 250 m downstream of decrease in surface roughness (representing conditions at centre of 1 m radius smooth turntable at scale 1:250), with z_o = 0.2 m and $z_{o,tt}$ = 0.002 m. Dashed line is prediction by CFD code FLUENT, solid line is analytical model (Simiu et al, 1986). Dotted line is extrapolation of undisturbed wind profile.

Finally, Maruyama's (1992) figures 5 and 6 show striking examples of wind profile and turbulence evolution (or degradation) within the 'turntable' IBL. However, he did not put it forward as an important issue.

3. NUMERICAL SIMULATION; TECHNIQUES AND VALIDATION

A numerical technique has been chosen for both practical (e.g. no wakes of upstream roughness elements) and financial reasons. The authors are well aware of the limitations of numerical techniques, especially in complex flows. They have looked for suitable experimental data - either for an independent estimate of smooth turntable effects or for validation of numerical estimates of smooth turntable effects - without success. Hence, the model is separately verified for internal boundary layer flow and obstacle flow, instead for the combination of the two. First, numerical simulation techniques will be discussed in more detail.

Numerical simulations have been carried out with a K-ε model (FLUENT), in which the model constants C_μ and σ_ε have modified values of 0.032 and 1.86 respectively. These modified constants are derived as follows.

The K-ε model computes the eddy viscosity ν_t from turbulent kinetic energy K and turbulent dissipation rate ε by:

$$\nu_t = C_\mu K^2 / \varepsilon \qquad (2)$$

In atmospheric surface layers and in wind tunnel boundary layers, the following relations apply:

$$\nu_t = U_* \kappa z \quad ; \quad \varepsilon = U_*^3/(\kappa z) \tag{3}$$

where z is the height, κ the Von Karman constant (0.4), and U* the friction velocity which also appears in the logarithmic wind profile. Substitution of the above expressions in eq. 2 yields:

$$K = C_\mu^{-0.5} U_*^2 \tag{4}$$

Typically $K/U^{*2} \approx 5.6$ in the lower 20% of uniform boundary layers, which yields a $C\mu = 0.032$ (Bottema et al, 1992). Richards et al (1993) impose the condition $\partial\varepsilon/\partial x = 0$ for equilibrium boundary layers as well. This yields the following equation:

$$C_{\varepsilon 2} - C_{\varepsilon 1} = \frac{\kappa}{C_\mu \sigma_\varepsilon} \tag{5}$$

where $C_{\varepsilon 1}$, $C_{\varepsilon 2}$ are K-ε model constants. Bottema et al (1992) choose to modify C_μ only, as they found that a modified $C_{\varepsilon 1}$ resulted in poor predictions of obstacle flow. However, later tests with modified σ_ε revealed no detrimental effects on model performance for obstacle flow. Hence, σ_ε is modified according to eq. 5 (as in Richards et al, 1993) which yields $\sigma_\varepsilon = 1.86$.

It is important to note that these model constant modifications are derived for the logarithmic surface (or constant flux) layer. They are not necessarily valid elsewhere. It should also be noted that only mean pressures can be computed, not peaks.

For the simulations of figs. 1, 3 and 4, and table 2 and 3 the dimensions of the computational domain were (length, width, height) 800 x 800 x 400 m. A rectangular, staggered grid is used; the number of grid points (x,z) is 34 x 34 x 19. Grid spacing is about 0.8 m near the ground and near walls, 14 m at the roughness change, and 50-100 at the edges of the computational domain. Grid expansion factors are less than 2.

Formulas for inflow conditions at the upsream side of the computational domain are given in (Bottema et al, 1992; Bottema, 1993); furthermore U(z=10m) = 10 m/s, and z_o = 0.2 m. Ground surface roughness equals approach flow z_o, except for 150 m < x < 650 m (the full scale 'turntable'), where the roughness ($z_{o,tt}$) is 0.2, 0.02 or 0.002 m. Building surface roughness is 0.002 m. All roughnesses are fixed by using logarithmic wall functions.

Various obstacles with a height H of 10 m were tested (normal approach flow). With the resulting H/z_o = 50, it was just possible to construct a computational grid without severe grid distortions or severe violation of the $20z_o$-requirement (below $20z_o$ wall functions are not valid as individual roughness elements determine the flow).

Figure 1 shows the validation for IBL flow. Numerical data (dashed line) show excellent agreement with the analytical model of eq. 1. For $z \approx h_{IBL}$, errors are up to 3% as FLUENT does not predict such a sharp kink in the wind profile (as expected). A separate check on turbulent kinetic energy could not be made by lack of suitable turbulence data for rough to smooth IBL's.

Bottema et al (1992) give validation results for a number of obstacles. Numerical data (mean wind speed U, turbulent energy K, pressure coefficients C_p) were within 10-20% of experiments. Table 1 gives pressures for an obstacle with W/H = 25, L/H = 0.5, $H/z_o = H/z_{o,tt}$ = 300. Grid properties are similar to those of the present simulations.

Later, Bottema (1993) found that computed pedestrian wind speeds around a high rise building were smaller than experimental data (Maruta, 1984) for normal flow, but that computed speeds were larger for oblique flow. Numerical diffusion would yield the reverse! A possible explanation is that the roughness elements on Maruta's turntable had a larger frontal area - and a larger roughness - for oblique flow. A surprising turntable roughness effect!

Table 1 Wind pressure coefficients C_p (referred to roof height wind speed) averaged over front and lee wall, for obstacle with W/H = 25, L/H = 0.5 and $H/z_o = H/z_{o,tt}$ = 300 (normal incidence). Blockage: 5%.

	C_p(front)	C_p(lee)	ΔC_p
experiment	0.42	-0.52	0.94
standard K-ε model (2-dim.)	0.53	-0.89	1.42
present modified K-ε model (2-dim.)	0.35	-0.78	1.13
present modified K-ε model (3-dim.)	0.38	-0.58	0.96

The present modified K-ε model performs much better than the standard K-ε model. For similar geometries (wide, thin obstacles) and similar grids, numerical C_p errors are expected to be about 10%, as in the present case.

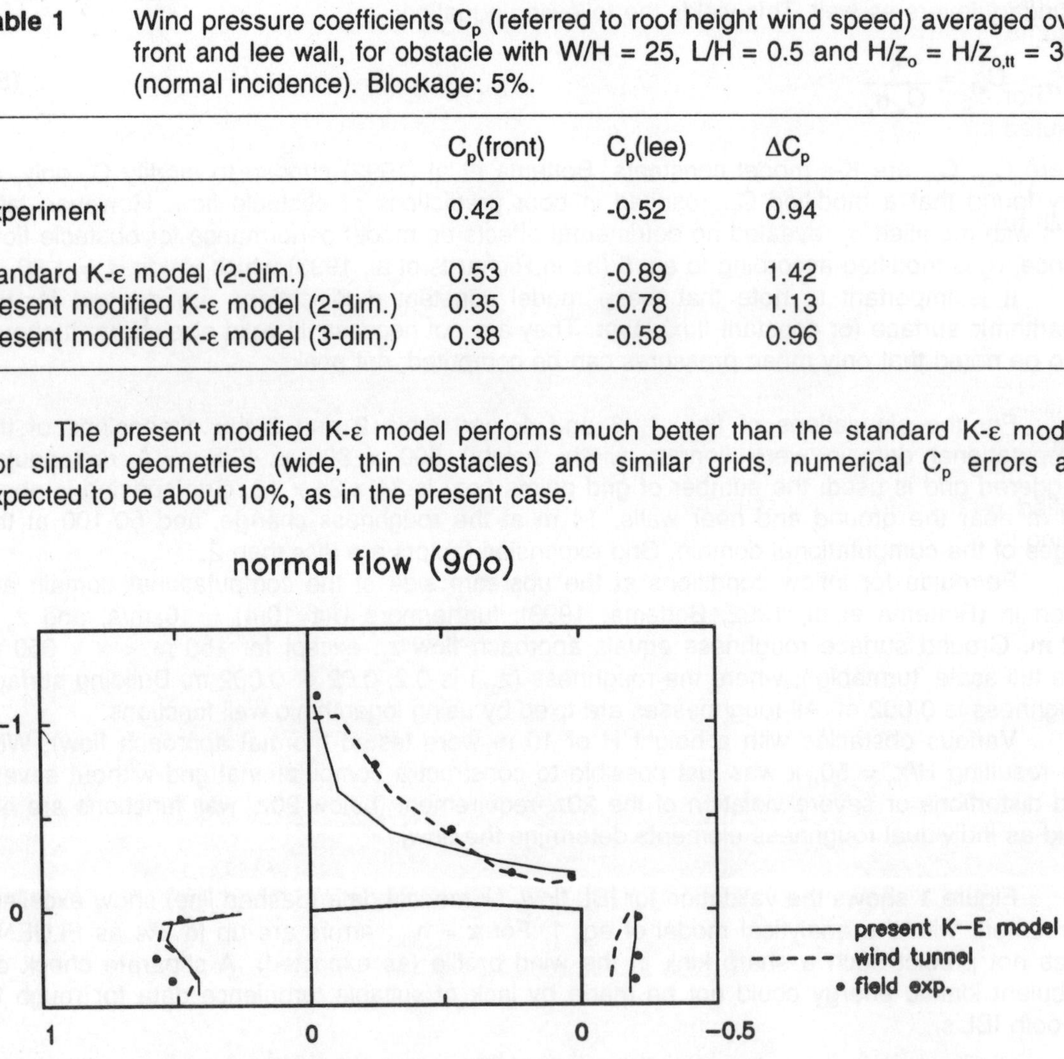

Figure 2 Pressure coefficients C_p at centre line of building (referred to approach flow wind speed at roof height) for Texas Tech building (L, W, H = 9, 14, 4 m; normal approach flow). Numerical $z_{o,tt} = z_o$ = 0.024 m; wind tunnel $z_{o,tt}$ (Surry, 1989) probably of same order as z_o. Solid line: numerical estimate; black dots: field experiment (Levitan et al, 1989); dashed line: wind tunnel exp.

Numerical results (present modified K-ε model) were also compared with field data for the Texas Tech University (TTU) building (Levitan et al, 1989). A 400 x 400 x 400 m domain has been used, with 32 x 18 x 20 grid points. Grid spacing near walls was about 0.3 m. The TTU building was approximated as a prism of 9 x 14 x 4 m. Approach flow roughness length z_o was 0.024, as was $z_{o,tt}$. Wind tunnel data (Surry, 1989) were compared with the same 'full scale' simulation, as earlier simulations showed that Reynolds effects are 1% or less.

Figure 2 compares numerical and experimental data for the TTU building (normal flow; 90°). For the frontal and leeward walls numerical data approximate full scale data even slightly better than wind tunnel results do. On the roof, the extremes are predicted well, but pressures fall off too rapidly. The reason for the smaller strong suction areas is not clear, but the same behaviour has been observed in PHOENICS simulations for Silsoe (Richards, 1992).

For oblique flow (60°) model performance is as good as for normal flow (90°; fig. 2). As measured pressure distributions are also similar, no separate graph is shown.

All in all, our model performs good enough to allow for at least some first estimates of smooth turntable effects on building pressures.

4. EFFECT OF SMOOTH TURNTABLE ON BUILDING PRESSURES

The effect of turntable roughness on building pressures was investigated for a number of buildings of 10 m height, and normal approach flow. Turntable roughness $z_{o,tt}$ was set at 0.2 m, 0.02 m, and 0.002 m, approach flow roughness z_o was either 0.2 m or 0.002 m.

For the smooth turntable case (z_o = 0.2 m and $z_{o,tt}$ = 0.002), figure 1 shows the modified wind profile at the building location, in the absence of the building. Note that the building is completely submerged in the internal boundary layer.

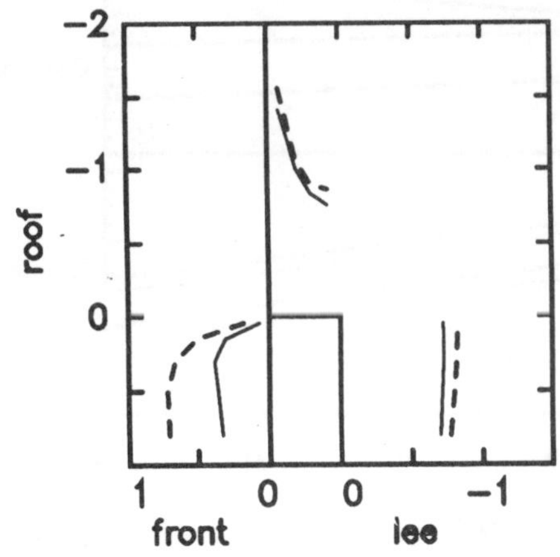

Figure 3 Pressure coefficients C_p (numerical estimate; referred to approach flow wind speed at roof height which is 10 m/s) for normal flow and 2-dim. building; H = 10 m, L = 5 m. Approach flow z_o: 0.2 m. Solid line: turntable $z_{o,tt}$: 0.2 m; dotted line $z_{o,tt}$ = 0.002 m. Fetch 250 m as in fig. 1 which gives wind profile at building location.

Figure 3 clearly shows the calculated effect of a smooth turntable on pressure coefficients. Pressure difference $\Delta C_p = C_{p,front} - C_{p,lee}$ is increased by 38%. This applies for a wide building.

Table 2 summarizes some results for other building geometries. For all these five buildings, the increase in ΔC_p is about 40%. For the 'thin' buildings, this can also be seen in figure 4. Trends as a function of W/H are only slightly affected. The C_p-values themselves however, are strongly affected for all W/H.

Table 2 Pressure coefficients on buildings (length L, width W) with H = 10 m for rough turntable ($z_{o,tt} = z_o = 0.2$ m; for smooth turntable $z_{o,tt} = 0.002$ m and fetch x = 250 m); blockage < 2.5%. All data are wall averages (fr=front; ro=roof), except for minimum roof pressure C_{pmin}. Approach flow wind speed at z=H is 10 m/s.

		rough turntable				smooth turntable			
W/H	L/H	$C_{p,fr}$	$C_{p,ro}$	C_{pmin}	$C_{p,lee}$	$C_{p,fr}$	$C_{p,ro}$	C_{pmin}	$C_{p,lee}$
0.5	0.5	0.45	-0.31	-0.70	-0.22	0.69	-0.37	-0.85	-0.26
2.0	0.5	0.45	-0.53	-1.01	-0.31	0.72	-0.61	-1.21	-0.35
∞	0.5	0.32	-1.00	-1.42	-0.72	0.63	-1.09	-1.58	-0.80
2.0	2.0	0.41	-0.34	-0.89	-0.23	0.66	-0.41	-1.14	-0.24
∞	2.0	0.29	-0.88	-1.50	-0.56	0.59	-0.99	-1.82	-0.61

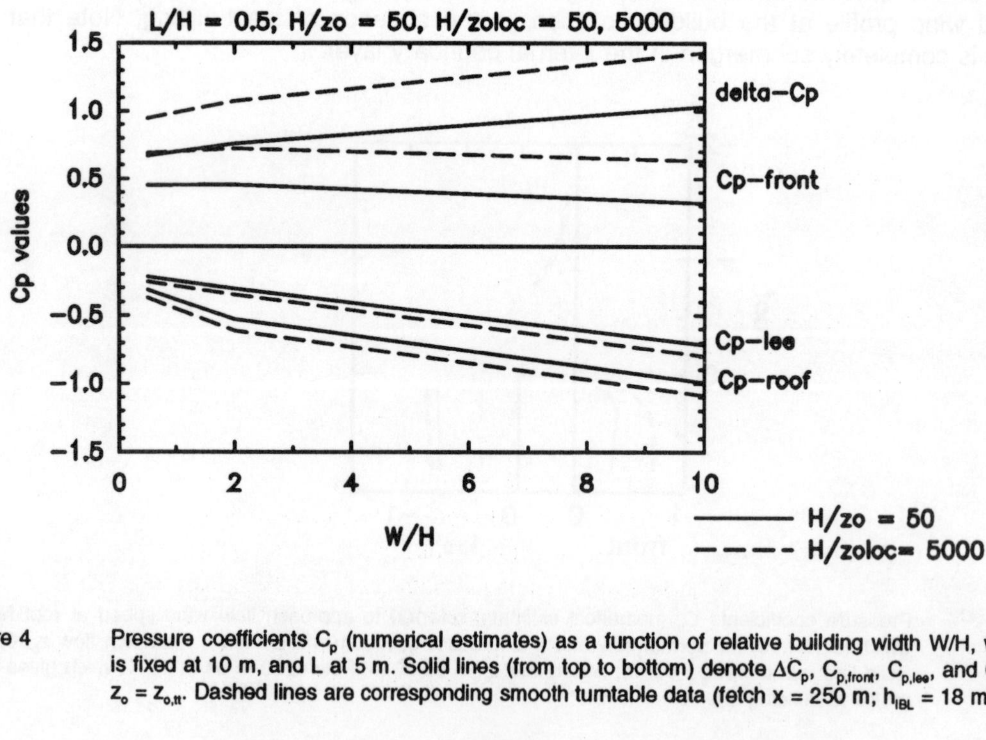

Figure 4 Pressure coefficients C_p (numerical estimates) as a function of relative building width W/H, where H is fixed at 10 m, and L at 5 m. Solid lines (from top to bottom) denote ΔC_p, $C_{p,front}$, $C_{p,lee}$, and $C_{p,roof}$ for $z_o = z_{o,tt}$. Dashed lines are corresponding smooth turntable data (fetch x = 250 m; h_{IBL} = 18 m).

Because of time constraints, the effect of h_{IBL}/H is not investigated yet. Significant effects are expected for $h_{IBL}/H < 1$ which is equivalent to $x/H < 15$. However, such fetches are not very common in low rise building testing as wakes of upstream roughness elements often are significant for these x/H-values.

A 40% error in building pressures is clearly unacceptable. Hence it is worthwhile to consider some strategies to mitigate smooth turntable effects.

Many researchers have tried to mitigate smooth turntable effects by increasing turntable roughness (carpet, smaller turntable roughness elements etc.). Table 3 shows that an increase of $z_{o,tt}/z_o$ from 0.01 to 0.1 results in only 25% reduction in error. For the buildings of table 3, effects of $z_{o,tt}$ on ΔC_p are expected to be below 10% only if $z_{o,tt}/z_o$ is larger than about 0.5. This is quite a strong demand.

Table 3 Effect of varying approach flow roughness z_o and turntable roughness $z_{o,tt}$ on wall averaged pressure coefficients C_p for two-dimensional building with H=10 m and L=5 m (blockage 2.5%). Undisturbed wind at z=H is 10 m/s (normal flow).

z_o	$z_{o,tt}$	C_p(roof)	C_p(front)	C_p(lee)	ΔC_p
0.2	0.2	-1.00	0.32	-0.72	1.04
0.2	0.02	-1.06	0.53	-0.78	1.31
0.2	0.002	-1.09	0.63	-0.80	1.43
0.002	0.002	-0.78	0.53	-0.58	1.11

Another approach to mitigate smooth turntable effects is to decrease z_o of the approach flow. In Surry's (1989) work on the TTU building, smooth turntable effects on ΔC_p are only about 13% (numerical estimates for normal and for oblique flow). This is mainly due to the fact that $H/z_o \approx 166$ for the TTU building, whereas $H/z_o = 50$ for the building of fig. 3. Furthermore, table 3 shows that the effect of decreasing z_o and $z_{o,tt}$ is smaller than that of decreasing $z_{o,tt}$ only. However, decreasing z_o has also an important disadvantage: errors in negative pressures will grow as small scale turbulence is (further) decreased.

The final option - probably the best - is to increase the model size as far as possible. This will solve some of the 'small scale turbulence problems' as well.

Two final points that should be emphasized are:
- the increase in ΔC_p is much stronger ($\approx$40%) than the increase of 16% in reference dynamic wind pressure (roof height; at building location in absence of building = over turntable centre)
- the effect of decreasing $z_{o,tt}$ only is stronger than the effect of decreasing z_o and $z_{o,tt}$

In a practical sense, this means that smooth turntable effects on C_p can not be compensated using a reference dynamic pressure over the turntable centre or by using pressure coefficients which are measured with lower approach flow z_o.

5. DISCUSSION

Table 3 shows that for a wide obstacle, smooth turntable C_p-values are consistently greater in magnitude than C_p values for a smooth approach flow where $z_o = z_{o,tt}$. This is also the case for obstacles of finite W/H. Part of this is caused by the roof height reference speed, which is increased by 7.6% over the smooth turntable. However, compared to the uniform smooth flow case ($z_o = z_{o,tt} = 0.002$ m), C_p-values for front and lee walls and roof in the smooth turntable case ($z_o = 0.2$ m; $z_{o,tt} = 0.002$ m) are increased by different rates. As the mean wind profiles up to 1.8 building heights are the same (see fig. 1), the only explanation left is the effect of turbulence. These turbulence effects are strongest (yield the largest increase in $|C_p|$) for suction areas on wide, thin obstacles (i.e. small L/H and large W/H).

As building pressures proved to be sensitive to turbulence, it is also important that our numerical model can simulate turbulence well. For flows around an obstacle with uniform surroundings, our model predicts the turbulent kinetic energy K within 20% (Bottema et al, 1992). Problems could occur in rough to smooth IBL's as σ_u/U^* and σ_v/U^* are increased (Tieleman, 1990b; Panofsky et al, 1984) over the fetch range of interest. This applies to K/U^*_2 as well, which - in the atmospheric surface layer - equals $C_\mu^{-0.5}$ (eq. 4). Hence, different values of the model 'constant' C_μ should be used for the approach flow and for the IBL! Clearly, this can be a source of errors. On the other hand, mean wind profiles in the IBL (fig. 1) behave as expected so that the errors are not expected to be dramatic.

Although model estimates for IBL wind profiles are good enough to make us fairly confident, it is still desirable to compare the present results with experiments in some way.

Surry et al (1986) compared wind tunnel data with field data for a mobile home. They noted that mean wind profiles were matched well, but that turbulence (σ_u) was too low by about 30%. As a result, their negative pressures ($|C_p|$) are about 40% too low. In table 3, the smooth turntable case ($z_o = 0.2$ m; $z_{o,tt} = 0.002$ m) differs from the uniform-smooth case ($z_o = z_{o,tt} = 0.002$ m) by an increased reference speed ($\approx +8\%$) and increased turbulence. It has already been noted (and it can be derived from data in table 3) that increased turbulence leads to larger suctions (provided that L/H is not too large). Hence, too low turbulence leads to smaller suctions. This is in qualitative agreement with Surry's results.

Wiren (1987) reports some experimental results on rectangular buildings (L, W, H = 90 x 350 x 60 mm) on a smooth turntable. H/z_o is not reported but should be of the order of 15 as roughness elements have the same size as the building model. Turntable radius is 1.4 m, and reference speed is taken at roof height at building location in absence of building.

Wiren (1987) has average C_p-values of 0.50 and -0.45 for the front and lee wall respectively; $\Delta C_p = 0.95$. If for the present numerical simulations reference speed is taken in the same way, C_p-values are 0.57/0.51 and -0.21/-0.52 ($\Delta C_p = 0.78/1.03$). The first number refers to a building with W/H = L/H = 2, and the second to a building with W/H = ∞ and L/H = 2 (note: dividing the above data by 0.86 returns values of table 2). Agreement is good for the frontal pressures, but the present lee pressures appear to be less negative (for W/H = 4, $C_p \approx -0.35$; Wiren has $C_p = -0.45$). The reason for this is not clear.

It is interesting to note that the Swedish building code data shown by Wiren prescribes a frontal C_p of 0.6, in accordance with smooth wind tunnel data. The Dutch wind loading code (NEN6702) has even a larger frontal C_p of 0.8 (both for wide and slender buildings). This is an indication that smooth turntable data are also used in building codes.

Finally, Kind (1986) found that roof peak pressures for oblique flow data changed little. Kind's x/H was between 20 and 60, corresponding with a h_{IBL}/H between about 1.2 and 3. As $h_{IBL}/H > 1$ (building submerged in IBL; hence there are only slow variations in reference speed and turbulence levels), and as numerical results indicate that sensitivity of roof pressures to $z_{o,tt}$ is relatively small, Kind's result is not surprising. For this reason, and because of time constraints, no additional validation was carried out.

The data discussed above suggest that the present numerical estimates of smooth turntable effects are at least reasonable. However, it is strongly recommended to set up an experimental programme for investigating, and further quantifying, smooth turntable effects. It should be noted that smooth turntables can also affect small scale turbulence, which has a strong influence on negative pressures (and peaks). Experiments are certainly the best way to investigate the effects of small scale turbulence.

6. CONCLUSIONS

6.1 Conclusions; numerical simulation

1. Tuning of the K-ε model constant C_μ to the approach flow value (Bottema et al, 1992) is essential, tuning of σ_ε (Richards et al, 1993) is desirable. With tuning, accuracy in prediction of pressure coefficients C_p is 10-20%, without only 40%. Accuracy for internal boundary layer (IBL) wind profiles is 5%.
2. Simulations with $H/z_o < 50$ (H: obstacle height; z_o: approach flow roughness) are not reliable: they require severe grid distortion or a first grid line too far in the layer where individual roughness elements dominate the flow (depth of layer: $20z_o$)
3. K-ε models can not exactly model rough to smooth roughness changes, and results should be interpreted with caution, despite excellent mean profile predictions.
4. Numerical modelers should try to avoid smooth 'turntable' effects as well.

6.2 Conclusions; smooth turntable effect

1. Smooth turntable effects (the effect of $z_{o,tt} < z_o$, where $z_{o,tt}$ is turntable roughness, and z_o approach flow roughness length) are especially relevant to wind tunnel tests for building ventilation and pedestrian comfort, but also to wind loading. Building or wind loading codes can be contaminated with smooth turntable data.
2. Smooth turntable effects are primarily dependent on the ratio of building height and approach flow roughness: H/z_o. If the fetch over height ratio $x/H > 20$, then ΔC_p ($C_{p,front}$-$C_{p,lee}$) increases by 40% for $H/z_o = 50$ and by 13% for $H/z_o \approx 170$. The fetch over height ratio x/H and the ratio of building height and internal boundary layer height h_{IBL}/H are rather important for $h_{IBL}/H < 1$ and $x/H < 15$. Their effects have still to be investigated.
3. The increase in ΔC_p by the use of a smooth turntable is larger than can be explained by the increase of the reference speed at roof height over the turntable. This reference speed is often used, but does only partly compensate for smooth turntable effects.
4. Pressure distributions on models on a smooth turntable are also different from pressure distributions for smooth - uniform - flow; even when corrected for reference speed effects as in (3). Hence, approach flow turbulence is important.
5. Trends as a function of building shape parameters (relative width and length) do not seem to be strongly affected.

6. Experimental data are needed for the following reasons:
- for further validation of the present work
- for estimates of smooth turntable effects outside the range of application of numerical models (i.e. for $H/z_o < 50$)
- to investigate smooth turntable effects on small scale turbulence which determines suction pressures and pressure peaks

6.3 Recommendations to mitigate smooth turntable effects

1. For H/z_o not too small, a carpet roughness can be used for both the approach flow and the turntable so that there are no smooth turntable effects at all.
2. The second best solution is to increase the obstacle size (and decrease h_{IBL}/H) as far as possible. This yields a better simulation of small scale turbulence as well
3. Many researchers increase turntable roughness. However, a $z_o/z_{o,tt}$ of 0.1 instead of 0.01 reduced smooth turntable effects only by a quarter. Moreover, turntable roughness elements can cause $z_{o,tt}$ to be dependent on approach flow angle!
4. A final solution which has several disadvantages, is to relax Jensen scaling by decreasing z_o of the approach flow.
5. Wake length of cubical upstream roughness elements is about 5H, see also Leene (1992). These wakes can be significant for short turntable fetches.

7. ACKNOWLEDGEMENT

Many thanks to the Group FAGO of Eindhoven University of Technology, who gave the first author the opportunity to do some further numerical work for this paper. Many thanks also to Prof. Wieringa (WAU-Meteorology), for giving the opportunity to spend some time in research for and writing of this article.

8. REFERENCES

Borges, A.R.J. and Saraiva, J.A.G. (1984), "Ventilation rates of two communicating low rise buildings as affected by terrain roughness", J. Wind Eng. Ind. Aerodyn. 15, pp 39-46

Bottema, M.; Wisse, J.A. and Leene, J.A. (1992), "Towards forecasting of wind comfort", J. Wind Eng. Ind. Aerodyn. 44, pp 2365-2376

Bottema, M. (1993), "Wind climate and urban geometry", PhD Thesis, Eindhoven Univ. of Technology, Eindhoven, NL, 212 p.

Holmes, J.D. and Carpenter, P. (1990), "The effect of Jensen number variations on the wind loads of a low rise building", J. Wind Eng. Ind. Aerodyn. 36, pp 1279-1288

Hussain, M. and Lee, B.E. (1980), "A wind tunnel study of the mean pressure forces acting on large groups of low-rise buildings", J. Wind Eng. Ind. Aerodyn. 6, pp 207-227

Kind, R.J. (1986), "Worst suctions near edges of flat rooftops on low-rise buildings", J. Wind Eng. Ind. Aerodyn. 25, 31-48

Levitan, M.L.; Holmes, J.D.; Mehta, K.C. and Vann, W.P. (1989), "Field measured pressures on the Texas Tech building", Proc. 8th Coll. on Ind. Aerodyn, Aachen, FRG, pp 13-24

Maruta, E. (1984), "The study of high wind regionsaround tall buildings" (in Japanese, translation available from BRE, Watford, UK), PhD Thesis, Nihon Univ., Tokyo, 200 p.

Maruyama, T. (1992), "Numerical simulation of boundary layer wind tunnel", J. Wind Eng. Ind. Aerodyn. 44, pp 2827-2838

Mulhearn, P.J. (1978), "A wind-tunnel boundary-layer study of the effects of a surface roughness change: rough to smooth", Bound.-Layer Meteor. 15, pp 3-30

Lawson, T.V. (1983), "The effect of fetch on wind profiles used in wind-loading calculations", J. Wind Eng. Ind. Aerodyn. 12, pp 363-364

Leene, J.A. (1992), "Building wake effects in complex situations", J. Wind Eng.Ind. Aerodyn. 44, pp 2277-2288

NNI (1990), "Loadings and deformations / TGB 1990 /", Netherlands Normalization Institute (NNI), NEN6702, Delft

Panofsky,H.A. and Dutton,J.A. (1984), "Atmospheric turbulence", John Wiley and Sons, New York,397p.

Plate, E.J. (ed.) (1982), "Engineering meteorology", Elsevier Sc. Publ. Comp. Amsterdam, NL, 740 p.

Reinhold, T.A. (ed.) (1982), "Wind tunnel modelling for civil engineering applications", Cambridge Univ. Press, New York, 688 p.

Richards, P.J. and Hoxey, R.H. (1992), "Computational and wind tunnel modelling of mean wind loads on the Silsoe structures building", J. Wind Eng. Ind. Aerodyn. 43, pp 1641-1652

Richards, P.J. and Hoxey, R.H. (1993), "Appropriate boundary conditions for computational wind engineering models using the K-ε model", J. Wind Eng. Ind. Aerodyn. 46-47, pp 145-153

Richardson, G.M. and Surry, D. (1992), "The Silsoe building: a comparison of pressure coefficients and spectra at model and full scale", J. Wind Eng. Ind. Aerodyn. 42, pp 1653-1664

Simiu,E. and Scanlan,R.H. (1986), "Wind effects on structures", John Wiley and Sons, New York, 590 p.

Surry, D. and Johnson, G.L (1986), "Comparisons between wind tunnel and full scale estimates of wind loads on a mobile home", J. Wind Eng. Ind. Aerodyn. 23, pp 165-180

Surry, D. (1989), "Pressure measurements on the Texas Tech building-II", Proc. 8th Coll. on Ind. Aerodyn., Aachen, FRG, pp 25-35

Tieleman,H.W.; Reinhold,T.A. and Marshall,R.D. (1978), "On the wind-tunnel simulation of the atmos pheric surface layer for the study of wind loads on low-rise buildings", J.Ind.Aerodyn. 3, pp 21-38

Tieleman, H.W. and Atkins, R.E. (1990a), "Effects on incident turbulence on pressure distribution on rectangular prisms", J. Wind Eng. Ind. Aerodyn. 36, pp 579-588

Tieleman, H.W. (1990b), Wind tunnel simulation of the turbulence in the surface layer, J. Wind Eng Ind. Aerodyn. 36, pp 1309-1318

Vickery P.J.; Surry, D. and Davenport, A.G. (1986), "Aylesbury and ACE, some interesting findings", J. Wind Eng. Ind. Aerodyn. 23, pp 1-17

Wiren, B.G. (1987), "Effects of surrounding buildings on wind pressure distributions and ventilation losses for single-family houses, part 2", The National Swedish Institute for Building Research, Gävle, Sweden, Bulletin R87:, March 1987, 33 pp (copies at Air infiltration and ventilation centre, Bracknell, UK)

9ICWE, 1995, New Delhi, India

A COMPARISON OF WIND RESPONSES THROUGH VARIOUS KINDS OF WIND TUNNEL TECHNIQUES ON A SUPER-TALL BUILDING

NAKAYAMA, M.[1], IDE, S.R.[2], SASAKI, Y.[1] and TANAKA, K.[1]

[1] *Research Engineer, Technical Research Institute, Fujita Corporation, 74 Ohdana-cho, Kohoku-ku, Yokohama, 223 Japan*
[2] *Graduate Student, Department of Civil Engineering, University of Ottawa, 770 King Edward Avenue, Ottawa, Ontario, K1N6N5 Canada.*

ABSTRACT

The wind response of a hypothetical super-tall building is examined through various wind tunnel techniques including a high-frequency force balance model (scale 1:833), a Single-Degree-Of-Freedom (SDOF) aero-elastic model (1:833) and a Multiple-Degree-Of-Freedom (MDOF) aero-elastic model (1:200). The building height is 500 m, it has a square cross section with bevelled corners, and it is tapered (ratio 2:1). In this paper, the differences of predicted wind response as determined via the above mentioned wind tunnel techniques are discussed. The results of this study can be summarized as follows. The wind responses obtained from the high- frequency force balance, SDOF and MDOF aero-elastic models generally agreed well. But around the resonance velocity with vortex shedding, the cross- wind response predicted from the high-frequency force balance measurement was considerably less than the aero-elastic wind response. Assuming this discrepancy is caused by the occurrence of negative aerodynamic damping, the amount of negative aerodynamic damping on this hypothetical building is inferred to be around -0.75 % as damping factor at the resonance velocity. If the original structural damping is large enough to suppress the amount of negative damping, the aero-elastic wind response can be predicted through a simple high-frequency force balance measurement. The displacement of the super-tall building can be expressed sufficiently by only the fundamental mode. However there might be some discrepancies at the upper part of the super-tall building by not considering the influence of higher modes. It was also verified through wind tunnel tests that corner-cuts and tapering of cross section can reduce wind load considerably.

1. INTRODUCTION

In recent years, building configurations have become taller and more slender. Engineers have to face a situation that the dominant design loads are wind loads. Even in Japan, the earthquake country, super-tall buildings, even higher than 500-meter high, have been tried to be designed.

The basic concepts in understanding the mechanism of the vibration of structures in turbulent wind was presented by Davenport [1] (1962), especially with respect to along-wind response of structures.

Various sophisticated techniques to predict wind response through wind tunnel tests have been developed. Typical kinds of aero-elastic models are reviewed by Isymov [2] (1982) including SDOF and MDOF aero-elastic models. The result of aero-elastic tests is realistic and more precise, but experimental cost is expensive and its measurement is time-consuming. It was recognized by Whitbred [3] (1975), Saunders & Melbourne [4] (1975) and other researchers that examining the general tendency of aero-elastic response from a view point of total fluctuating wind load characteristics acting on a building, such as power spectra, is more convenient especially in the cross-wind direction. A compact-type force balance with high natural frequency was utilized by Tschanz [5] (1982) and other researchers.

Another approach is to detect wind load through pressure measurements. Gumley [6] (1983) presented a general method for design and evaluation of transfer function for manifolded tubing system which can reduce the number of pressure transducers considerably. Kareem [7] (1982) estimated wind load acting on a prismatic column through statistical integration of pressure signal using relatively many pressure transducers. Fujii et al. [8] (1986) applied several hundreds pressure sensors electronically scanned for measurement of fluctuating wind load on a building model with square cross section.

It is generally recognized that wind load depends on the configuration of building especially on its cross section. Kwok et al. [9] (1988) pointed out that a small change of corner portion of cross section can reduce the resultant wind load significantly. It is helpful to reduce wind load by adopting some advantageous configurations of building within design conditions, as well as by installing a vibration control device on the objective building.

Considering that various kinds of wind tunnel techniques are available depending on the respective design stage, and super-tall buildings of more than 500 m high have been designed, it is meaningful to utilize almost all wind tunnel techniques to predict the response on a hypothetical super-tall building and to compare the differences of predicted results in a methodical fashion. An extensive wind tunnel study has been started on a hypothetical super-tall building, nominally 500 meter high, in collaboration with Fujita Corporation, University of Ottawa and AAL/NRCC (Applied Aerodynamics Laboratory, National Research Council, Canada). This study consists of the following two phases. The first phase corresponds to aero-elastic tests and a high-frequency force balance test in the collaborative work between Fujita Corporation and University of Ottawa. This also includes a preliminary test which was carried out to determine the configuration of the objective building. The second phase corresponds to pressure measurements on a stationary model and an oscillating model as a function of amplitude of building movement and reduced velocity in collaborative work between Fujita and AAL/NRCC. The scope of this study is presented in Table 1. In this paper, only the results of the first phase including some preliminary tests are presented. The results of Phase-2 are presented elsewhere [10], [11].

Table 1: SCOPE OF INVESTIGATION

Preliminary Test	Force balance models 1:833 scaled force balance test on proposed models
Phase-1 Fujita & Ottawa U.	1:833 scaled SDOF aero-elastic model 1:200 scaled MDOF aero-elastic model
Phase-2 Fujita & NRCC	1:833 scaled stationary pressure model 1:200 scaled oscillating pressure model

2. PRELIMINARY TEST FOR DESIGN OF THE OBJECTIVE SUPER-TALL BUILDING

At the stage to determine the configuration of a hypothetical super-tall building, several normal building types were proposed taking into account the possibility of a configuration that could be built someday in future. Therefore this research did not take into account city-scaled super-tall buildings, whose heights are more than 1000 m and which contain all functions for one whole city inside a building. During this selection phase, high-frequency force balance models (1:1000) were adopted to quickly obtain the general characteristics of the wind forces. In this preliminary test, two tactics were applied to a fundamental building with a square cross section: one is corner-cuts of cross section, and the other is tapering with height. The nine-type configurations of tested models are shown in Figure 1. The coordinate system is also shown in the same figure. Here after in this paper aerodynamic force coefficients denote the normalized over turning moment divided by reference moment:Mref. Mref is defined by $Q_H BH^2$, where Q_H is velocity pressure, B is representative width and H is the height of the building.

Wind tunnel tests were carried out in the same approaching flow described in detail at section 4.3. The effects of these tactics on reduction of mean and rms aerodynamic force coefficients are presented in Figure 2 and Figure 3. It can be seen that the tactics are very effective in reducing wind load, especially at the azimuth angle of around zero degree, where the largest wind loads are expected. As azimuth angle gets bigger, these effects get smaller. Figure 4 shows examples of power spectra of aerodynamic force at the angle of zero degree. It can be said that both tactics reduce the peak value of spectrum intensity at the vortex shedding frequency, and increase the Strouhal Number. Meanwhile, at frequencies in excess of those corresponding to vortex shedding, the intensity of the power of cross-wind force spectra is not changed. Because the higher frequency components of cross-wind forces are almost the same, these tactics are not as effective in improving habitability.

In general, it was confirmed that tapering and corner-cuts of cross section are very helpful to reduce design wind load around relatively higher reduced velocity ranges. It should be noticed that there may exist the optimal combination of these factors, such as tapering ratio and width of corner-cutoff cross section as pointed out by Shiraishi [12] (1986). In this study such an optimal solution was not the objective because it was just a hypothetical building.

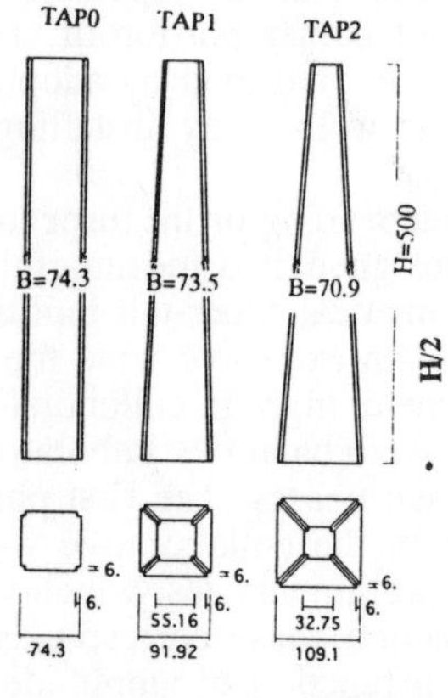

Tapering	Edge Configuration		
	SN	ST	SS
TAP0	TAP0SN	TAP0ST	TAP0SS
TAP1	TAP1SN	TAP1ST	TAP1SS
TAP2	TAP2SN	TAP2ST	TAP2SS

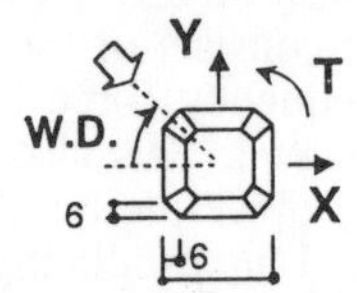

dimensions are in meter

Definition of Aero-dynamic Force Coefficient

$CM_{XX}=M_{XX}/M_{REF}\cdot 100$ (%)

$CM_{YY}=M_{YY}/M_{REF}\cdot 100$ (%)

$M_{REF}=Q_H\cdot B\cdot H$

where

CM_{XX}, CM_{YY} : Aero-dynamic Force Coefficient

M_{XX} : Overturning Moment around X-axis

M_{YY} : Overturning Moment around Y-axis

M_{REF} : Reference Moment ($M_{REF}=Q_H\cdot B\cdot H$)

Q_H : Velocity pressure at the building Top

B : Representative Width of the Building

H : Height of the Building

Fig.1 Nine-type Configurations of Tested Building in Force Balance Test

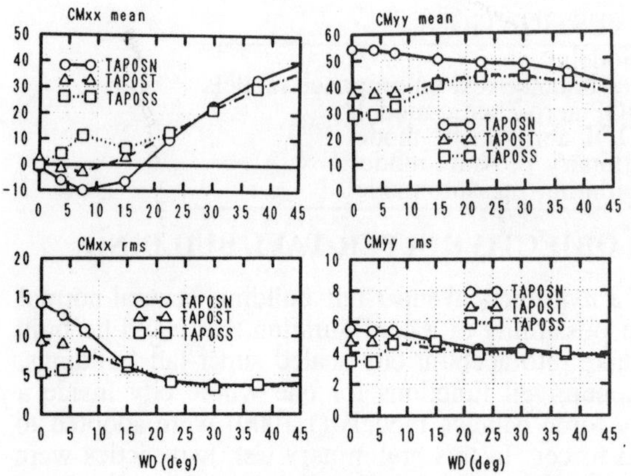

Fig.2 Effect of Edge Configuration on Aerodynamic Force Coefficients Acting on a Square Building

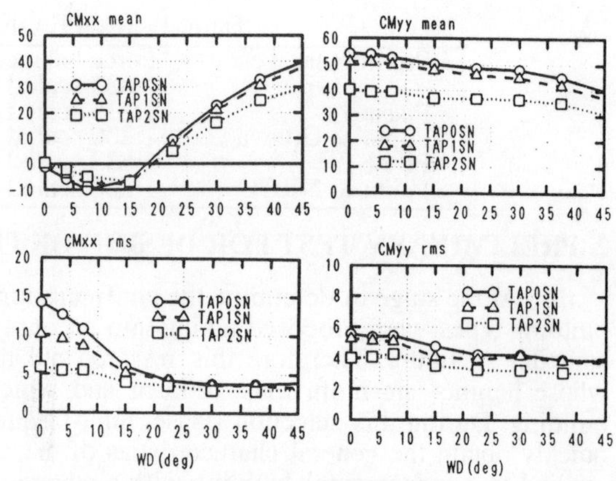

Fig.3 Effect of Tapering on Aerodynamic Force Coefficients acting on a Square Building

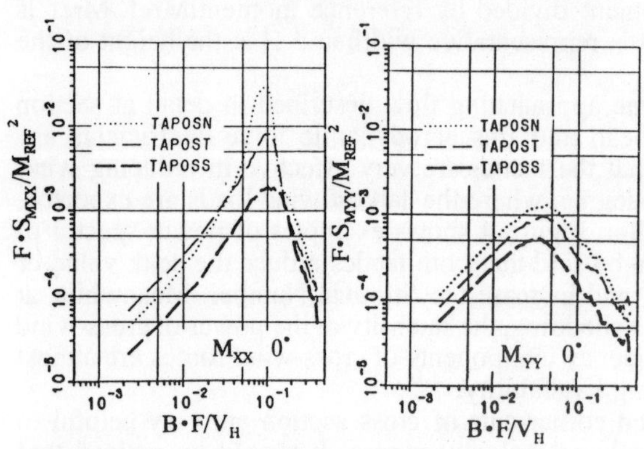

Fig.4 Effect of Edge Configration and Tapering on P.S.D. of Aerodynamic Force

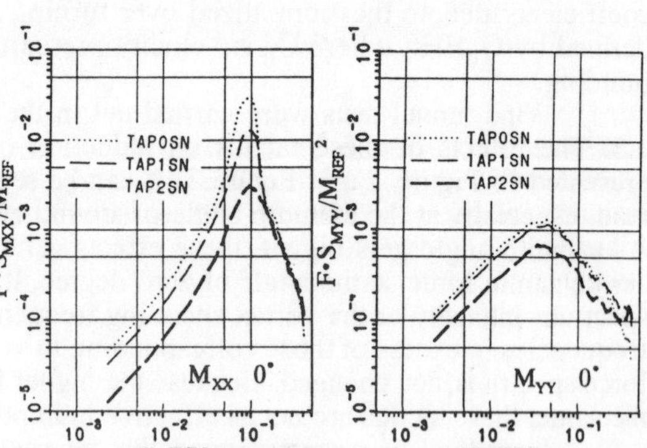

Fig.5 The Objective Hypothetical Super-tall Building

3. DESCRIPTION OF OBJECTIVE BUILDING

Through the preliminary test described in the previous section, a building shown in Figure 5 was proposed as the hypothetical objective super-tall building in this study. For the simplicity of the problem, fundamental vibration modes in X, Y and T (twisting) directions were assumed to be linearly proportional to the building height. The building was assumed to be homogeneous in mass, with average density of 150 Kg/m^3, and one percent structural damping for each fundamental vibration mode was set. The assumed fundamental dynamic characteristics are listed in Table 2.

This building was assumed to be located in a homogeneous and moderately built-up suburban area with the power-law exponent of 0.2. According to statistical data [13] on strong wind velocity in Tokyo, Japan, the yearly-maximum reoccurrence wind velocity (10 minute-averaged) for a return period of 100 year at the top of the building is about 65 m/sec.

Table 2: ASSUMED DYNAMIC CHARACTERISTICS

	X	Y	T
Fundamental frequency	0.111 Hz	0.100 Hz	0.156 Hz
Averaged Density	150 Kg/m^3	150 Kg/m^3	208 Kgf/m^3
Damping Factor	1.0 %	1.0 %	1.0%

Fundamental mode is linearly proportional to the height.

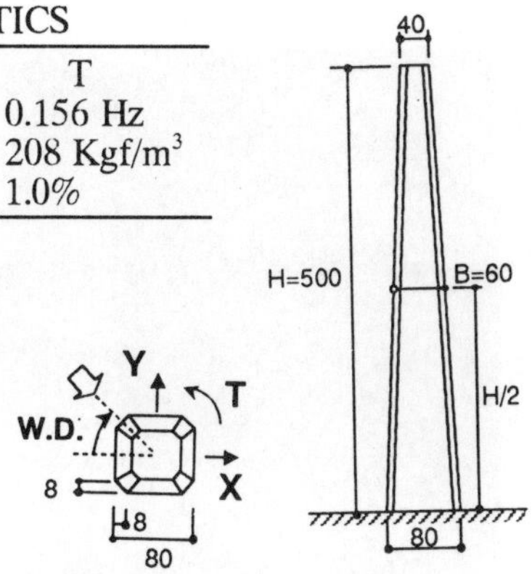

Fig.5 The Objective Hypothetical Super-tall Building

4. OUTLINE OF WIND TUNNEL TESTS ON THE OBJECTIVE MODEL

4.1. SDOF AERO-ELASTIC MODEL AND HIGH-FREQUENCY FORCE BALANCE MODEL

All the wind tunnel tests on 1:833 models were carried out in the boundary layer wind tunnel at Fujita Technical Institute in Japan. It has a cross section of 1.6 m x 1.2 m (Width x Height) and is 10 m long. The experimental setup for the aero- elastic test is schematically presented in Figure 6. The rigid model, which was designed to match the similarity of density ratio (density of air / average structural density), was made of balsa wood and was pivoted to be able to rotate around the X and Y axes. The aluminum rod connected to the rigid model was supported by two coil springs in X direction and four coil springs in Y direction. The model was also able to rotate around Z axis to simulate the twisting motion and the stiffness of twisting was provided by the four coil springs in Y direction. The reaction force at the respective coil spring was detected by a load cell at the end of each coil spring. The output signals of load cells were summed up to give rotation angle around X, Y and Z axes. It was possible to adjust the structural damping by changing the diameter of the damping disc and the length of the aluminum rod. However it was difficult to get the similar dampings in the sway and twisting modes. The measured dynamic characteristics are summarized in Table 3. The azimuth angle was set at zero to 45 degree with increment of 5 degree. Mean wind velocity at the building top was changed from 1.5 m/s to 13 m/s. Data acquisition was made at the sampling rate of 200 Hz and 8192 data per channel at each measuring point were collected. The model used in the SDOF aeroelastic test was the same one used in the high-frequency force balance test.

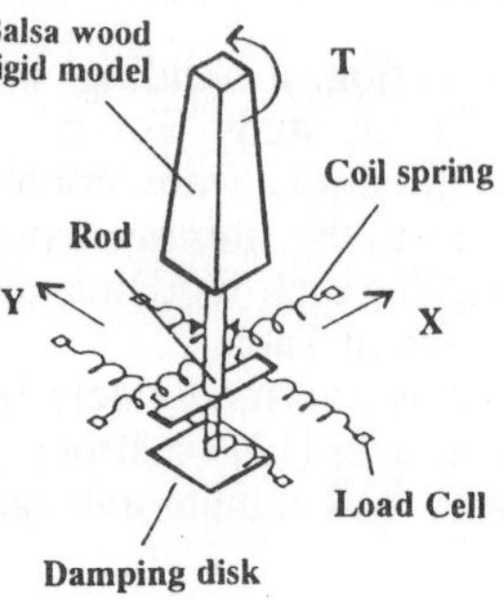

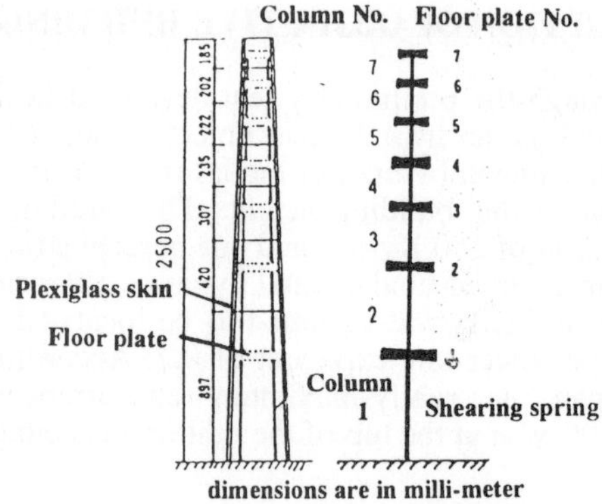

Fig.6 The SDOF Aero-elastic Model

Fig.7 The MDOF Aero-elastic Model

Photo 1 The MDOF Model Installed inside the Wind Tuunel

Photo 2 Closer Details of the MDOF Model

4.2. 1:200 MDOF AERO-ELASTIC MODEL

A schematic diagram of 1:200 MDOF model is shown in Figure 7. The model consisted of seven floor plates, four supporting columns for each floor and the surface plexiglass skins which defined the geometric configuration of the building. The floor plates were made of aluminum, whereas the columns were made of steel. The MDOF aero-elastic model was basically designed following the paper by Templin and Cooper [14] (1980). The MDOF aero-elastic model was designed to obtain the similar properties as for the SDOF aero-elastic model (e.g. density ratio and fundamental vibration modes,). Three sets of wire strain gauges were attached at the bottom of steel columns to detect the relative displacement (story draft) of each floor plate; twenty-one sets of strain gauge were used in all.

In a static calibration test, the stiffness of each floor was determined and each set of strain gauges was calibrated. Static loads were applied on the assembled model and displacement of floor plates were detected by fourteen Karman-type displacement meters. The natural frequencies and vibration modes were examined in the dynamic calibration test. The assembled model was excited by random-sinusoidal force using an electromagnetic shaker. The vibration modes were determined by rms displacement of floor plates in the resonance state. Damping factor of each vibration mode was determined from free vibration curves just after excitation force was stopped in the resonance state. The signal of free vibration was filtered through a band-pass filter.

Thin rubber films were attached to each gap of outer plexiglass skins in order to get the desired modal damping factor for each fundamental vibration mode. Consequently all structural damping factors became nearly one percent when the fundamental modal damping factor was adjusted to have one percent. The measured natural frequencies and vibration modes were almost identical with the ones calculated from the result of the static calibration test.

Wind tunnel tests were carried out in the 30 foot x 30 foot wind tunnel at NRCC. Wind azimuth angle was set at zero to 45 degree with increment of 5 degree and also at 90 degree. The mean wind velocity at the building top was changed from 3 m/s to 28 m/s (11 to 98 m/s in full scale). Data was acquired at the sampling rate of 500 Hz during 30 s giving 15000 data per channel; there were 21 channels in all. The measured dynamic characteristics are summarized in Table 4. The MDOF model is shown inside the NRCC wind tunnel (Photo 1) and with closer details in Photo 2.

Table 3: MEASURED DYNAMIC CHARACTERISTICS of the 1:833 SDOF Aero-elastic Model

	X	Y	T
Fundamental frequency	9.44 Hz	8.8 Hz	13.2 Hz
Damping Factor	X	Y	T
Medium Damping (MD)	1.03%	0.97%	2.42%
High Damping (HD)	1.75%	1.82%	2.88%
Very High Damping(VHD)	3.71%	3.38%	6.83%

Table 4: MEASURED DYNAMIC CHARACTERISTICS of 1:200 MDOF Aero-elastic Model

	Modal frequency (damping factor)		
	X	Y	T
1 st mode	6.25 Hz (1.27%)	5.70 Hz (0.97%)	11.3 Hz (1.39%)
2 nd mode	13.3 Hz (1.14%)	12.2 Hz (1.05%)	19.3 Hz (1.64%)
3 rd mode	22.1 Hz (0.90%)	20.1 Hz (1.21%)	29.9 Hz (1.57%)

4.3. APPROACHING FLOW

The building was assumed to be located in a homogeneous and moderately built-up suburban area with the power-law exponent of 0.2. The turbulence intensity around the building top was assumed to be about 4 to 5 percent. The turbulence characteristics near or above the atmospheric boundary layer height have not yet been precisely obtained. It was also desired that aerodynamic response included the buffeting effect by the turbulence in oncoming flow. The turbulent boundary layer was generated by four triangular spires and a roughness bar on the wind tunnel floor during both wind tunnel tests at Fujita and NRCC, as shown in Photo 1.

The velocity profiles for U component of the generated approaching flow are presented in Figure 8. It can be verified that both generated velocity profiles are substantially identical to the assumed one. The height of the generated boundary layer seems to be a little deeper than the ordinary value which would be expected for a flow with a power-law exponent of 0.2; ref. [13]. However, it was desired that the model building would be submerged in the boundary layer in order to include various components of wind force excitation.

Figure 9 shows the normalized velocity spectra at two thirds of model height in the small and big wind tunnels. Experimental velocity spectra agree well with the semi-empirical von Karman-type velocity spectrum, for which the equation is shown in the same figure. The spectrum length scale (L) in this equation, which is equal to the integral scale of auto-correlation function of fluctuating velocity, is estimated to be around 150 meters in full-scale at this height (333 m). This value of length scale seems to be a little smaller than the expected value, ref. [13], associated with this height (333 m) in full scale, but it is still larger than the representative model width (B) of 60 meter. It is confirmed from these figures that both approaching flows are substantially similar.

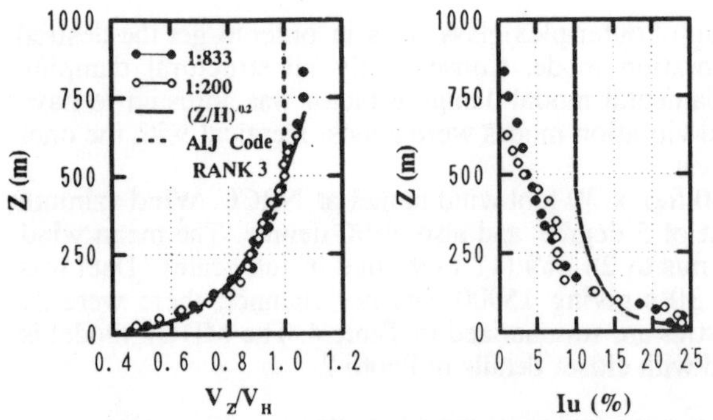

Fig.8 Vertical Distribution of Approaching Flow

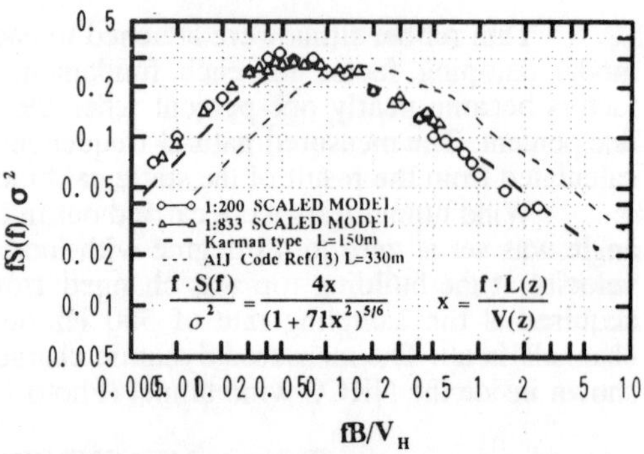

Fig.9 Power Spectral Density of Approaching Flow (U-comp, at 0.67H)

5. EXPERIMENTAL RESULTS AND DISCUSSION

5.1. GENERAL TENDENCY OF OVERALL WIND FORCE

Figure 10 presents the aerodynamic force coefficient obtained from the high- frequency force balance test on 1:833 model. The general tendency is similar to the results of the preliminary tests already shown in section 2. Figure 11 shows a plot of normalized power spectrum of along/cross-wind aerodynamic force at azimuth angle of 0,5 and 10 degree. The Strouhal number is about 0.13 at the azimuth angle of zero degree. The Strouhal number is defined as Bfp/V_H; where B is the representative width, that is, the width at one half of the building height; V_H is the wind velocity at the building top; and fp is peak frequency in the power spectra of cross-wind force.

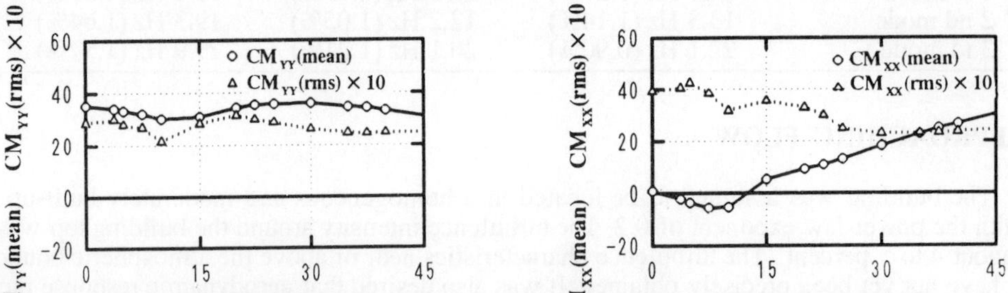

Fig.10 Aerodynamic Force Coefficients for the Objective Building by Force Balance Measurement (Definitions are shown in Fig.1.)

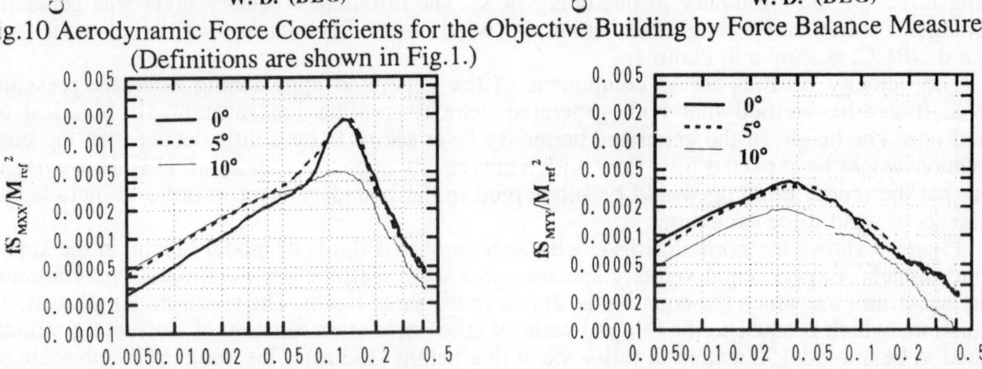

Fig.11 Normalized Power Spectral Density of Aerodynamic Force on the Objective Building (Definitions are shown in Fig.1.)

5.2. COMPARISON OF WIND RESPONSES

Here the response in X and Y direction is presented. The predicted response through high-frequency force balance test is computed through the method of ordinary spectral analysis in frequency domain. Figure 12 presents a global comparison of wind responses obtained from the high-frequency force balance, 1:833 SDOF model and 1:200 MDOF model. The mean, rms and maximum displacement at the building top are compared at the same building top velocity. The top displacement divided by the height of the building, is converted to rotation angle around X and Y axis. These results correspond to the responses at 1% structural damping and the building top velocity of 54. m/s in full scale.

The computed response from high-frequency force balance in this figure do not include any effects of aerodynamic damping. As a general tendency, these three kinds of results agree well as pointed out by Isyumov [2] (1982). The wind response of both aero-elastic models are almost identical to each other. But in the case of shallow azimuth angle (0 and 5 degree), the cross-wind responses (Y direction) of aero-elastic models are significantly greater than the computed response from high-frequency force balance test. The difference is almost a factor of two. Strong tendency towards vortex-induced oscillation was observed on aero-elastic models at these azimuth angles. The discrepancy at these shallow azimuth angles may be caused by the occurrence of negative aerodynamic damping in the case of cross-wind responses.

Figure 13 shows the relation between reduced velocity and displacement at building top for an azimuth angle of zero degree. In the case of along wind response (X-direction) from three kinds of tests correspond well. Moreover, if the aerodynamic damping effects are considered into force balance prediction, the agreement between the computed response from force balance test and aero- elastic responses gets better. The aerodynamic damping is estimated by the next equation:

$$h_a = \frac{\rho \int_0^H C_{FX}(z) U(z) b(z) \phi^2(z) dz}{2 \cdot m^* \cdot \omega_{X0}}$$

where

h_a : aerodynamic damping in the along-wind direction
ρ : air density
$U(z)$: local wind velocity at the height of z
$C_{FX}(z)$: local force coefficient at the height of z
$B(z)$: local width of the building
$\phi(z)$: the fundamental modal function (=z/H)
m^* : the fundamental modal mass
ω_{X0} : the fundamental modal circular frequency

At this stage, as the pressure distribution on the surface of the building was not known, the vertical along-wind-force distribution was assumed to be proportional to the local velocity pressure and local building width. The proportional constant (local force coefficient $C_{FX}(z)$) was derived to give the same overturning moment or shearing force in along-wind direction as the result of 1:833 force balance model. Subsequent values of local force coefficients were substantially identical (around 1.00), and gave almost the same aerodynamic damping. This assumption was confirmed by the results of pressure test in Phase-2 [10]. In the case of cross-wind predicted response, the discrepancy between the response determined from the force balance and both aero-elastic models was evidently shown around reduced velocity (U_Y) of eight, which coincides with the inverse of Strouhal number shown in the normalized power spectral density of cross-wind overturning moment. At the reduced velocity (U_Y) of around eight, the difference gets the largest and is about a factor of two. Assuming that this discrepancy is caused by the occurrence of negative aerodynamic damping, the total damping factor should be reduced to only a quarter of original or initial structural damping. As the original structural damping is one percent in this figure, the amount of negative aerodynamic damping factor could be evaluated as around -0.75 % at the reduced velocity of around eight. In the range of reduced velocity more than eight, it can be seen that the discrepancy between the computed response from force balance test and aero- elastic response get smaller. This implies that the amount of negative damping tends to decrease in this higher reduced velocity range, compared with those in the range of resonance velocity with vortex shedding.

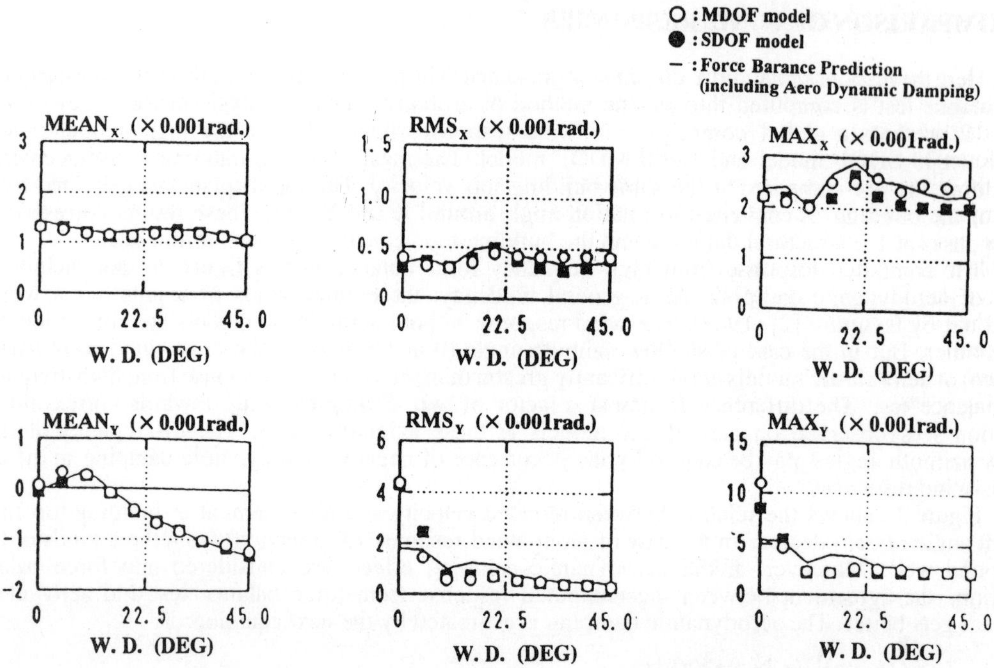

Fig.12 A Global comparison of Top Displacement through Force Balance, SDOF and MDOF Model Test
(Top displacement is divided by the height of building. Results by force balance do not include any aerodynamic damping effects, Damping Factor: 1%)

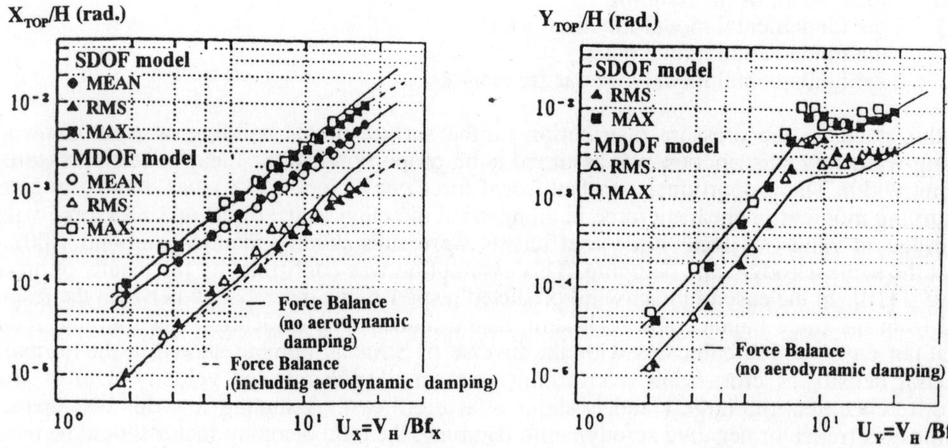

Fig.13 Comparison of Top Displacement through Force BAlance, SDOF and MDOF Model Test
(Top displacement is divided by the height of building. Azimuth angle: 0deg, Damping Factor: 1%)

5.3. EFFECTS OF STRUCTURAL DAMPING

It is possible to discuss the effects of structural damping on wind response as structural damping was changed in the case of SDOF model. Figure 14 shows the cross-wind responses of SDOF model when damping factor was changed from around 1% to 3%. In the same figure, the computed response from force balance test is also plotted without considering the effect of aerodynamic damping. As easily seen, the discrepancy between aero-elastic results and force balance prediction tends to get very small as damping factor increases. Judging from this figure, if structural damping was around more than 3% in this case, the force balance prediction got closer to aero-elastic response. This means that if it is possible to keep large original or initial structural damping at more than three percent, even if aerodynamic negative damping occurred, such effects could be suppressed by original structural damping. In such a case, the response of a super-tall building could be easily predicted by a simple high- frequency force balance measurement. As it is still difficult and uncertain to evaluate the precise amount of aerodynamic negative damping and of structural damping in advance, attempt to substantially increase structural damping is very significant to predict reliable aero-elastic wind response.

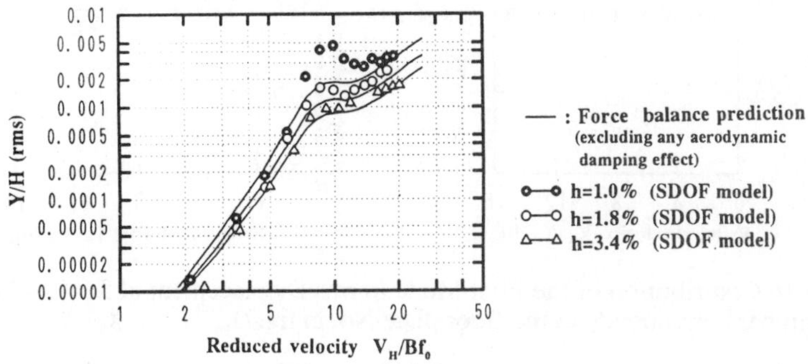

Fig.14 Effect of Damping Factor on Cross-wind Responses
(Top displacement is divided by the height of building. Results are those on SDOF model)

5.4. EFFECTS OF HIGHER MODES

The total displacement on MDOF model was defined as the displacement from ground and relative displacement (story drift) was the displacement between neighboring two levels. The relative displacement was proportional to stress of columns. As already shown in the previous section, since the displacement at the building top on SDOF and MDOF model were close to each other, the total displacement can be expressed simply by the fundamental or the first modal vibration.

Figure 15 presents the same fact from the vertical distribution of total displacement at each level on the MDOF model. The vertical distribution of mean and rms total displacement of floor plates coincide with the first mode obtained experimentally. But in case of relative displacement it should be certified whether this fact is also adoptable. The contribution of higher mode was derived from integration of power spectrum around respective natural frequency.

Figure 16 presents the ratio of the first modal rms total displacement to the one including all modes. The results of four floor levels; the bottom (the first), second, fifth and seventh (top) are plotted at the azimuth angle of zero degree. As adding remarks in Figure 16, it can be pointed out that vibration energy of the building concentrates in the first mode especially in the case of cross-wind response (Y direction) around at the reduced velocity of eight, which is close to resonance velocity with vortex shedding.

Figure 17 presents the ratio of the first modal rms relative displacement to the total rms displacement. The results of four columns; the bottom (the first), second, fifth and seventh (top) are plotted at the azimuth angle of zero degree. It is evident that in case of relative displacement, there is a possibility that the value evaluated by only the first mode could be considerably underestimated. Contribution of higher modes will be even intensified in case of considering the effects of higher modes on habitability, because it is known that human perception of vibration is directly connected to acceleration.

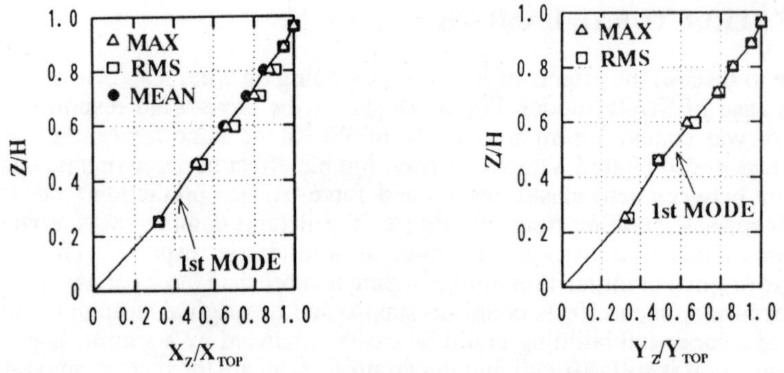

Fig.15 Vertical Distribution of Displacement on MDOF Model
(Displacement is normalized by top displacement. Azimuth angle: 0deg, V_H is 50m/s in Full scale)

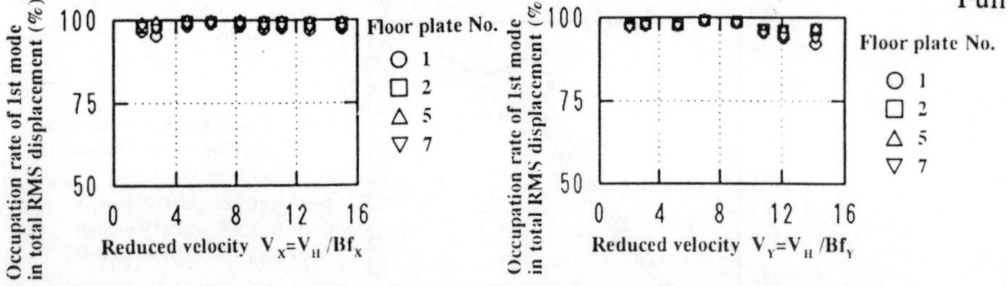

Fig.16 Contribution of the First Mode to rms Displacement at Several Levels
(Number corresponds to the floor plate No. in fig.7)

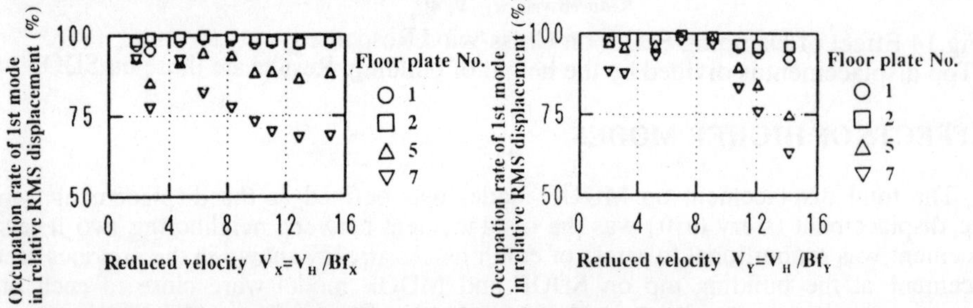

Fig.17 Contribution of the First Mode to rms Story Drift at Several Levels
(Number corresponds to the column No. in fig.7)

6. CONCLUSION

The wind response of a super-tall building whose height is supposed to be 500 meter, were examined through various wind tunnel techniques, including a high- frequency force balance model (1:833), a SDOF aero-elastic model (1:833) and a MDOF aero-elastic model (1:200). The objective hypothetical building has a tapered cross section and also corner-cuts of cross section. In this paper, the difference of predicted wind response through the above mentioned three kinds of wind tunnel techniques were discussed. The results of this study can be summarized as follows:

1) It was verified through wind tunnel tests that corner-cuts and tapering of cross section can reduce wind load considerably compared with the building with no treatments.

2) The wind response obtained from high-frequency force balance, SDOF and MDOF aero-elastic model generally agreed well with one another. However, around the resonance velocity with vortex shedding, predicted cross-wind response through high-frequency force balance measurement was considerably less than the actual aero-elastic response. Assuming this discrepancy was caused by the occurrence of negative aerodynamic damping, the amount of negative aerodynamic damping on this hypothetical building, was inferred to be around -0.75 % as damping factor at the resonance velocity.

3) If the original structural damping is big enough to suppress the amount of negative damping (in the case of this building, the value could be around four percent), the aero-elastic wind response can be predicted through a simple high frequency force balance measurement.
4) The total displacement of a super-tall building can be expressed sufficiently by considering only the fundamental mode but relative displacement (story drift) might be underestimated only by the fundamental mode, since it is affected by the contribution of higher modes. This tendency is most obvious at the upper part of a super-tall building.

ACKNOWLEDGMENTS

The authors would like to express their appreciation to Prof. H.Tanaka (University of Ottawa) for his active management in accomplishing phase-1 project. Also the authors would like to thank Mr. K.R. Cooper, Mr. M. Savage and Mr. A. Baskaran (working at NRCC); as well as to Mr. K. Kimura, Mr. J. Felbor and Mr. A. Y. Amicel (then graduate students at the University of Ottawa) for their great help in carrying out the wind tunnel test on the MDOF model. Thanks are extend to Mr. K. Yamamoto, Mr. Y. Tanagi at Fujita Technical Institute, and Mr. N. Hagiwara (staff at Fujita Technical Institute) for their help in the data acquisition and data processing in the wind tunnel experiment on 1:833 models.

REFERENCES

[1] Davenport, A. G. (1962), "The Wind Response of Slender, Line-like Structures to a Gust Wind", Proc. I.C.E. , Vol.23, No.ST3, June, 1962, 389-408.
[2] Isyumov, N. (1982), "The Aero-elastic Modeling of Tall Buildings", Proc. of Int. Workshop on Wind Tunnel Modeling Criteria and Techniques in Civil Engineering Applications, 373-407, 1982.
[3] Whitbred, R. E. (1975), "The Measurement of Non-steady Wind Forces on Small-scale Building Models", Proc. of the 4-th Conf. on Wind Effects on Buildings and Structures, Heathrow, 1975, 567-574.
[4] Saunders, J. W. and Melbourne, W. H. (1975), "Tall Rectangular Building Response to Cross-wind Excitation", Proc. of the 4-th Conf. on Wind Effects on Buildings and Structures, Heathrow, 1975, 269-379.
[5] Tschanz, T. (1982), "Measurement of Total Dynamic Loads Using Elastic Models with High Natural Frequencies", Proc. of Int. Workshop on Wind Tunnel Modeling Criteria and Techniques in Civil Engineering Applications, 296-312, April,1982.
[6] Gumley, S. J., "Tubing Systems for Averaging of Fluctuating Pressures", J. of Wind Engineering and Industrial Aerodynamics, 1983, 189-228.
[7] Kareem, A. (1982), "Measurement of Total Loads Acting Surface Pressures", Proc. of Int. Workshop on Wind Tunnel Modeling Criteria and Techniques in Civil Engineering Applications, 275-295, April, 1982.
[8] Fujii, K., Hibi, K. and Ueda, H. (1986) "A New Measuring System Using Electronically Scanned Pressure Sensors and Some Applications of the ESP System to a Square Building Shape", 9-th National Symposium on Wind Engineering (in Japanese), 1986, 313-318.
[9] Kwok, K. C. S., Wilhelm, P. A. and Wilkie, B. G. (1988), "Effects of Edge Configuration on Wind-induced Response of Tall Buildings", Eng. Struct., 1988, vol.10, April, 135-140.
[10] Fediw, A.A., Nakayama, M., Cooper, K.R., Sasaki, Y., Resende-Ide, S. and Zan, S.J. (1993), "Wind Tunnel Study of an Oscillating Tall Building", Proc. of 1st European and African Wind Engineering Conference, Guernsey, Sep.,1993.
[11] Cooper, K.R., Nakayama, M., Sasaki, Y., Fediew, A.A., Resende-Ide, S. and Zan, S.J., (1995), "Unsteady Aerodynamic Force Measurement on a Super-tall Building with a Tapered Cross-section", Proc. of 9-th ICWE, 1995, (to be published).
[12] Shiraishi, N., Matsumoto, M., Shirato, H., Ishizaki, H., Osada, M. and Matsui, T., "On Aerodynamic Stability Effects for Bluff Rectangular Cylinders by their Corners Cut", Proc. of 9-th National Symp. on Wind Engineering (in Japanese), 1986, 193-198.
[13] Architectural Institute of Japan, "Recommendations for Loads on Buildings", (in Japanese), 1993.
[14] Templin, J.T. and Cooper, K.R. ,"Design and Performance of a Multi-degree of freedom Aero-elastic Building Model", Proc. of the 4th Collq. on Industrial Aerodynamics, 123-135, June, 1980.

9ICWE, 1995, New Delhi, India

FLUCTUATING WIND PRESSURES ON BUILDINGS AND STRUCTURES OF CIRCULAR CROSS-SECTION AT HIGH REYNOLDS NUMBERS

UEMATSU, Y. and YAMADA, M.
Associate Professor *Professor*

Department of Architecture, Tohoku University, Sendai, Japan.

Abstract

Fluctuating wind pressures on buildings and structures of circular cross-section have been studied in a wind tunnel by using rough-walled models. Measurements were made of the RMS wind pressures on the cylinder surface and the Strouhal number in a turbulent boundary layer. The data gathered from a number of experiments concerned with this subject are also reviewed. The results indicate that the fluctuating wind pressures can adequately be predicted by a quasi-steady theory for actual structures in turbulent atmospheric winds. An empirical equation describing the behavior of the Strouhal number in relation to the aspect ratio and the relative surface-roughness of the cylinder is presented. Based on the results, a discussion is made of the design wind pressures, both positive and negative, for the wall cladding.

1. INTRODUCTION

The work described herein forms part of a study of the aerodynamic forces on buildings and structures of circular cross-section. The main purpose of this study is to construct models of the aerodynamic coefficients for full-scale application. Therefore, the transcritical regime is primarily considered in the present study; the terminology and the definition of the Reynolds-number regimes relevant to the flow past a circular cylinder are based on Nakamura and Tomonari (1982).

During the study, we have conducted a series of wind-tunnel experiments using rough-walled models to obtain a transcritical flow in the wind tunnel. The effects of aspect ratio and surface roughness on the aerodynamic coefficients were investigated. The results for the time-averaged aerodynamic forces were published in Uematsu and Yamada (1993). The present paper describes the results for the fluctuating wind pressures acting on the wall surface. Based on the results, we discuss the design wind pressures, both positive and negative, for the wall cladding. It should be mentioned that some basic results on this subject were published in Uematsu and Yamada (1994).

2. EXPERIMENTAL ARRANGEMENTS AND PROCEDURES

The experiments were conducted in an Eiffel-type wind tunnel with a working section 6.5 m long, 1.4 m wide and 1.0 m high. A turbulent boundary layer with a power law exponent of $\alpha = 0.25$ and a thickness of $z_G = 52$ cm was generated on the wind-tunnel floor using a set of turbulence generators installed at the entrance of the working section and a number of roughness blocks distributed on the floor. Fig. 1 shows the profiles of the mean velocity U, normalized by the free-stream velocity U_G, and the turbulence intensity I_u. A longitudinal velocity spectrum $S_u(f)$ of the boundary layer is shown in Fig. 2, where f is the frequency and σ_u the standard deviation of the longitudinal velocity fluctuation. The spectrum is generally consistent with the so-called Karman spectrum with a longitudinal length scale of $L_{ux} \approx 13$ cm.

The diameter D of the test cylinder was 6 cm and the aspect ratio H/D, with H being the cylinder height, was varied from unity to 5. Surface roughness was provided by gluing sand paper to the cylinder surface in order to obtain a transcritical flow within the limits of wind velocity range available in the wind tunnel. The standard roughness used is based on the equivalent sand roughness k_s. The relative roughness k_s/D of each cylinder was evaluated in the same way as Achenbach (1971). The value of k_s/D was in a range from 420 x 10^{-5} to 750 x 10^{-5}. Seven to fourteen pressure taps, depending on the H/D ratio, were provided along a generator of the cylinder and the pressures at any angular position θ were measured by rotating the cylinder about its axis. Each pressure tap was connected to a pressure transducer via 60 cm length of flexible vinyl tubing which had a small restrictor tube inserted in it to attenuate peaks in the frequency response. The amplitude response of the measurement system was within 5% of unity up to 100 Hz, and the phase response was close to linear over this range. The largest model caused a wind–tunnel blockage of about 1.3%. Since the blockage effect on the aerodynamic coefficients was thought to be insignificant, no correction was applied to the results.

The transition of the Reynolds–number regime relevant to the flow around a circular cylinder is promoted by the surface roughness of the cylinder as well as by the free–stream turbulence, and it can be identified from the base pressure coefficient C_{pb} versus Reynolds number curve (see Uematsu and Yamada (1993, 1994)). As mentioned above, we focus on the transcritical flow in the present study. Therefore, in the experiments, the variation of C_{pb} with the wind velocity was first measured at the mid–height ($z = H/2$) and the transition of the Reynolds–number regime was identified. Then, the pressure measurements were made at some transcritical Reynolds numbers. The Reynolds numbers, defined in terms of D and the wind velocity U_H at the model height ($z = H$), was in a range from 4.4 x 10^4 to 5.8 x 10^4. All the aerodynamic coefficients are defined in terms of the velocity pressure q_H (= $1/2\rho U_H^2$, with o being the density of the flow) of the approaching flow at $z = H$. The vortex shedding frequency f_v was determined from the surface–pressure spectra in the wake region.

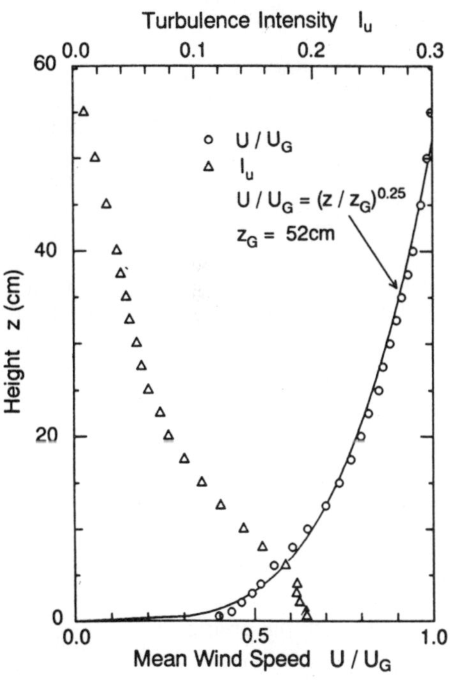

Figure 1. Profiles of mean velocity and turbulence intensity

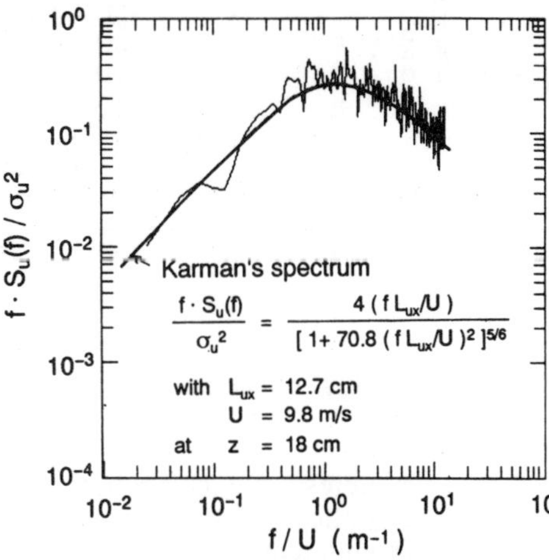

Figure 2. Power spectrum of wind velocity fluctuation

3. CHARACTERISTICS OF FLUCTUATING WIND PRESSURES

3.1 RMS pressure coefficient

Fig. 3 shows a sample result for the distribution of the RMS pressure coefficient C_p'. It can be seen that the value of C_p' on the front centerline is approximately proportional to the turbulence intensity of the approaching flow at the corresponding height. On the other hand, the value in the wake region is approximately constant regardless of height. The effect of surface roughness on the circumferential C_p'–distribution at the mid–height is illustrated in Fig. 4, in which θ represents the angle from the front stagnation point. It is found that the C_p' values in the front and wake regions are less sensitive to k_s/D. For relatively smooth cylinders, e.g., when $k_s/D = 420 \times 10^{-5}$, the distribution peaks at some angular position near θ_m which gives the minimum pressure coefficient C_{pm}; θ_m was about 70° in any case tested. However, with increasing k_s/D, this peak gradually subsides.

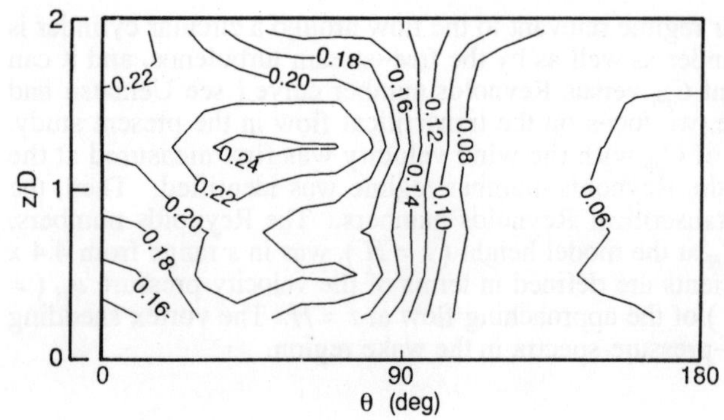

Figure 3. Distribution of RMS pressure coefficient C_p' ($H/D = 2$, $k_s/D = 560 \times 10^{-5}$)

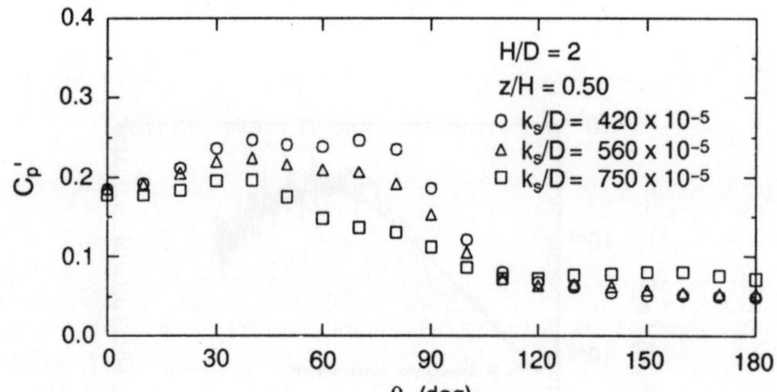

Figure 4. Influence of k_s/D on the circumferential C_p'–distribution at $z = H/2$ ($H/D = 2$)

It may be interesting to compare the experimental results with the predictions of a quasi–steady theory. The application of the two–dimensional quasi–steady theory yields the following expression for the circumferential C_p'–distribution:

$$C_p'(\theta) = \left\{ \left[2\frac{\sigma_u}{U} C_p(\theta) \right]^2 + \left[\frac{\sigma_v}{U} \frac{dC_p(\theta)}{d\theta} \right]^2 \right\}^{1/2} \qquad (1)$$

where σ_u and σ_v represent the RMS values of the longitudinal and the lateral turbulence, respectively. Fig. 5(a) shows a comparison between the experiment and the theory; z represents the height of pressure measurements. In the front and wake regions, the two results agree well with each other. However, in a region ranging from $\theta \approx 30°$ to $\theta \approx 100°$, the theoretical results are generally larger than the experimental ones. This difference is probably due to a relatively small ratio of the turbulence scale L_{ux} of the approaching flow to the cylinder diameter, i.e., $L_{ux}/D \approx 2$. Cook and Redfearn (1980) measured the pressures on a full-scale silo in an atmospheric wind; the L_{ux}/D ratio is expected to be of the order of 10 and greater. The results for the C_p'-distribution, quoted from Macdonald et al. (1988), are plotted on Fig. 5(b), together with the predicted result. Since the values of I_u ($= \sigma_u/U$) and I_v ($= \sigma_v/U$) are not presented in the publications, we obtained them by assuming that $C_p'(0°) = 2I_u \cdot C_p(0°)$ and $I_v = I_u/2$. In this case, the agreement between these two results appears to be satisfactory. Besides this measurement, several researchers published the results of wind–tunnel and full–scale measurements of the circumferential C_p– and C_p'–distributions in the transcritical regime. Using their results for the C_p–distribution, we computed the C_p'–distributions. Table 1 shows a comparison between the experiment and the theory for the maximum C_p' value, together with the experimental conditions. Melbourne et al. (1983) used a model of a tapered reinforced concrete stack 256 m high; the mean diameter of the stack is used to define the aspect ratio. Sabransky and Melbourne (1987) dealt with a grain storage silo with a conical roof. Ueda et al. (1991) made a full–scale measurement using a circular building with $H = 149.35$ m and $D = 38.2$ m. The Reynolds number is defined in terms of the wind velocity measured at a level of 10.5 m above the roof. Except for this value, all the values of Re and C_p' in the table are defined in terms of the wind velocity U_z or the velocity pressure q_z ($= 1/2\rho U_z^2$) at the height of pressure measurements. It is found that the agreement between the experiment and the theory is fairly good, except for Macdonald et al.'s measurements (1988, 1990) in which the L_{ux}/D ratio was relatively small (≈ 2) and the maximum C_p' value was observed at $\theta \approx 0°$, not on the side of the cylinder. From these results, it is reasonable to expect that the quasi–steady theory can yield a reliable prediction of the C_p'–distribution for actual structures where the L_{ux}/D ratio is fairly large.

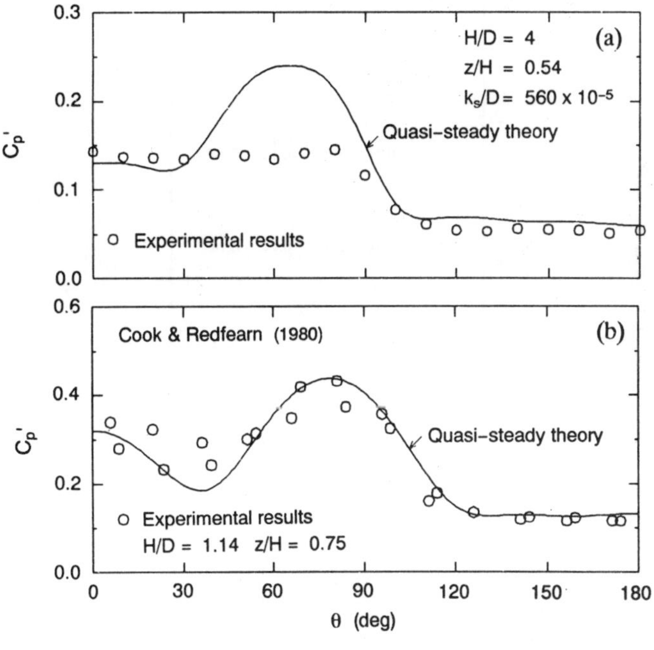

Figure 5. Comparison between experiment and quasi–steady theory for the circumferential C_p'–distribution. (a) Present result; (b) Cook and Redfearn's result

Table 1 Comparison between experiment and quasi-steady theory for the maximum C_p' value; z represents the height of pressure measurements.

Reference	Measurement	H/D	z/H	Re	I_u	Maximum C_p' value	
						Measurement	Theory
Melbourne et al. (1983)	Wind tunnel	15.5	0.83	1.1×10^6	0.078	0.26	0.29
Sakata et al. (1984)	Wind tunnel	1.82	0.55	1.2×10^5	0.090	0.29	0.30
Sabransky et al. (1987)	Wind tunnel	1.16	0.80	9.7×10^4	0.156	0.34	0.33
Macdonald et al. (1988)	Wind tunnel	1.00	0.81	$> 1 \times 10^5$	0.17	0.29	0.51
Macdonald et al. (1990)	Wind tunnel	1.00	0.81	1.3×10^5	0.19	0.38	0.60
Ueda et al. (1991)	Full scale	3.90	0.65	1.7×10^7	—	0.38	0.35

3.2 Strouhal number

Fig. 6 shows a typical pressure spectrum $S_p(f)$ measured in the wake region. The spectrum in the windward region was found to be similar in shape to that of the wind velocity fluctuation. On the other hand, a prominent peak was observed at a frequency f_v in the spectra measured on the side and leeward faces, except in the top region up to $z'/D \approx 1$, with z' being the distance from the top. In the case where $H/D = 1$, no predominant frequency was detected in the pressure fluctuations. The spectral peak is thought to be associated with the vortex shedding from the cylinder. The Strouhal number S_t ($= f_v \cdot D/U_H$) obtained from the pressure spectra are illustrated in Fig. 7, together with the locations of the pressure measurements, for a cylinder with $H/D = 4$ and $k_s/D = 750 \times 10^{-5}$. Within the limits of the present experiments, no axial variation of S_t was observed.

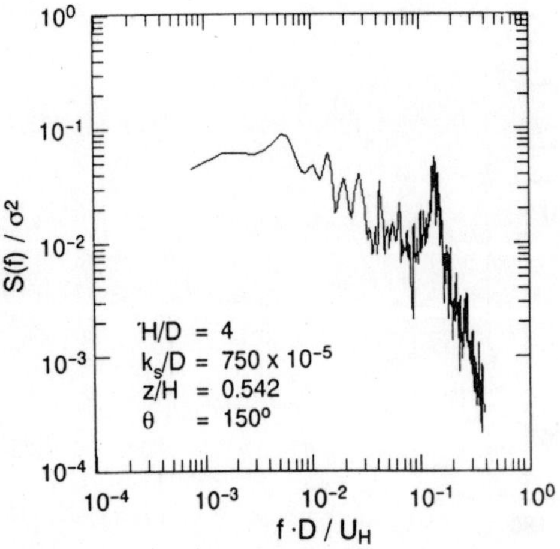

Figure 6. Pressure spectrum measured in the wake region

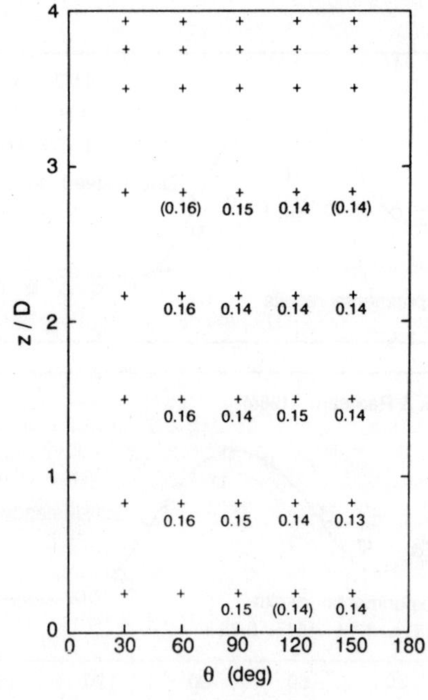

Figure 7. Locations of pressure measurements (+) and Strouhal numbers

Many researchers measured S_t in the transcritical regime under two-dimensional conditions. Fig. 8 shows the results compiled by Basu (1986). A general trend of increasing S_t with reducing k_s/D is apparent, though this trend is poorly defined for smoother cylinders with $k_s/D < 10^{-5}$. Based on these results, we obtained an empirical equation for S_t as a function of k_s/D using the least squares method; this equation is represented by the solid line in the figure.

In the cantilevered cylinder case, the Strouhal number is affected by H/D also. To examine its effect, the value of S_t is normalized by the corresponding two-dimensional value $S_{t\infty}$ for the same k_s/D ratio. Plotted on Fig. 9 are the $S_t/S_{t\infty}$ ratios for the subcritical and transcritical regimes against H/D. The results published in the previous papers are also included. In the construction of the figure, it was necessary to assume a value of $S_{t\infty}$; a value of 0.20 was used for the subcritical flow regardless of k_s/D, while the value presented by the empirical equation obtained above was used for the transcritical flow. It is interesting to note that the reduction in S_t used here can make the data for the both regimes collapse into a relatively narrow region. The solid line in the figure indicates a representative equation obtained by using the least squares method.

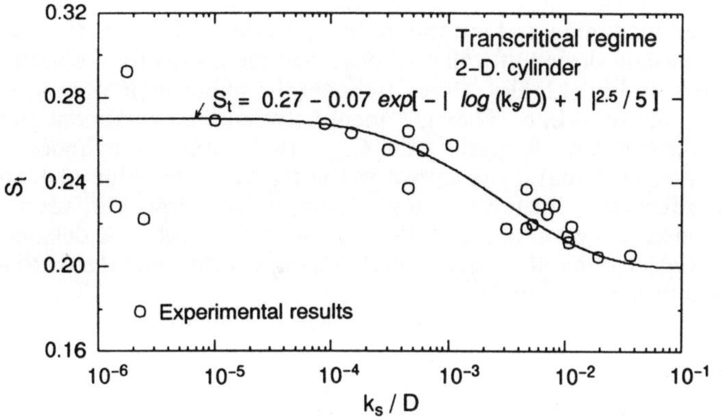

Figure 8. Variation of Strouhal number S_t with k_s/D for two-dimensional cylinders

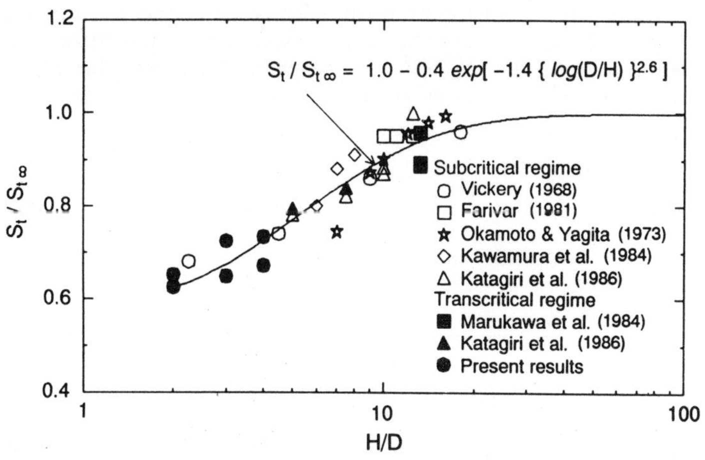

Figure 9 Variation of S_t with H/D

4. DESIGN WIND PRESSURES FOR WALL CLADDING

4.1 Formulas for the external pressure coefficient

The net wind load acting on the cladding is given by the difference between the external and the internal pressure. In this study, only the external design pressure is discussed. Since a circular cylinder is an axisymmetric body, it is necessary in designing the wall cladding to specify the maximum and the minimum peak pressure coefficient at any location. According to the results presented in Section 3.1, the C_p' value generally peaks at $\theta \approx 0°$ and at $\theta \approx \theta_m$ where the time-averaged pressure coefficient becomes the maximum and the minimum, respectively. This suggests that the positive peak pressure becomes the maximum at $\theta \approx 0°$, while the negative peak pressure becomes the minimum at $\theta \approx \theta_m$; the results of Sakata et al.'s experiments (1985) clearly indicates that this inference is valid.

In our previous papers, we provided quantitative descriptions for the time-averaged aerodynamic coefficients as functions of H/D and k_s/D. The relationship between k_s and the representative physical height k of surface roughness was also examined for some typical roughness types, such as sharp-edged vertical ribs and closely-packed spheres. Fig. 10 shows the procedure for obtaining the value of each aerodynamic coefficient when the H/D and k/D ratios and the roughness type are given. (Regarding the detailed discussion of the procedure, see Uematsu and Yamada (1993, 1994).) Among these aerodynamic coefficients, the maximum and the minimum pressure coefficient, C_{ps} and C_{pm}, are the most important for the wind resistant design of wall cladding. Sample results for the vertical distributions of C_{ps} and C_{pm} are shown in Fig. 11; the circles indicate the experimental results, while the predicted result is represented by the solid line. The agreement between the experiment and the prediction is fairly good for the C_{ps}-distribution. Regarding the C_{pm}-distribution, the cylinder is divided into two parts, i.e., the top and the lower region, as shown in the figure. The value in each region was assumed to be constant regardless of z for the purpose of simplicity. However, such a simple assumption generally results in conservative values, in particular, at lower levels. A detailed examination of the experimental results suggested that the value slightly changes with z and the distribution is approximately represented by the following equation:

$$C_{pm} = C_{pm0} \cdot \left(\frac{z}{H_r}\right)^{0.8 \cdot \alpha} \tag{2}$$

where C_{pm0} represents the minimum pressure coefficient predicted by the previous formulas for each region; and H_r the reference height defined as H for the top region and as $H-2D$ for the lower region. The result predicted by the above equation is represented by the dashed line in Fig. 11(b). It can be seen that the agreement with the experiment is better than the previous one.

The results in Section 3.1 indicate that the C_p'-distribution can adequately be predicted by the quasi-steady theory for actual structures in turbulent atmospheric winds. According to this theory, the C_p'-values at $\theta = 0°$ and $\theta = \theta_m$ are given by $2I_u \cdot C_{ps}$ and $2I_u \cdot |C_{pm}|$, respectively. Hence, the external design pressure, P_{ex}^+ and P_{ex}^-, for the wall cladding may be given by the following equations:

$$\text{Positive pressure: } P_{ex}^+ = q_H \cdot G_c \cdot C_{ps} \tag{3.a}$$

$$\text{Negative pressure: } P_{ex}^- = q_H \cdot G_c \cdot |C_{pm}| \tag{3.b}$$

where G_c represents a gust loading factor. The probability density function of the pressure fluctuations in the region before the flow separation line is approximately represented by the normal distribution in the same manner as the wind velocity fluctuations (see Sakata et al. (1985), for example). Hence, the gust loading factors for the positive and negative pressures are both given by

$$G_c = 1 + 2g \cdot I_u \tag{4}$$

where g is a peak factor which depends on the size and the natural frequency of the cladding as well as on the design wind speed; $g = 3.5$ is a representative value for ordinary cladding.

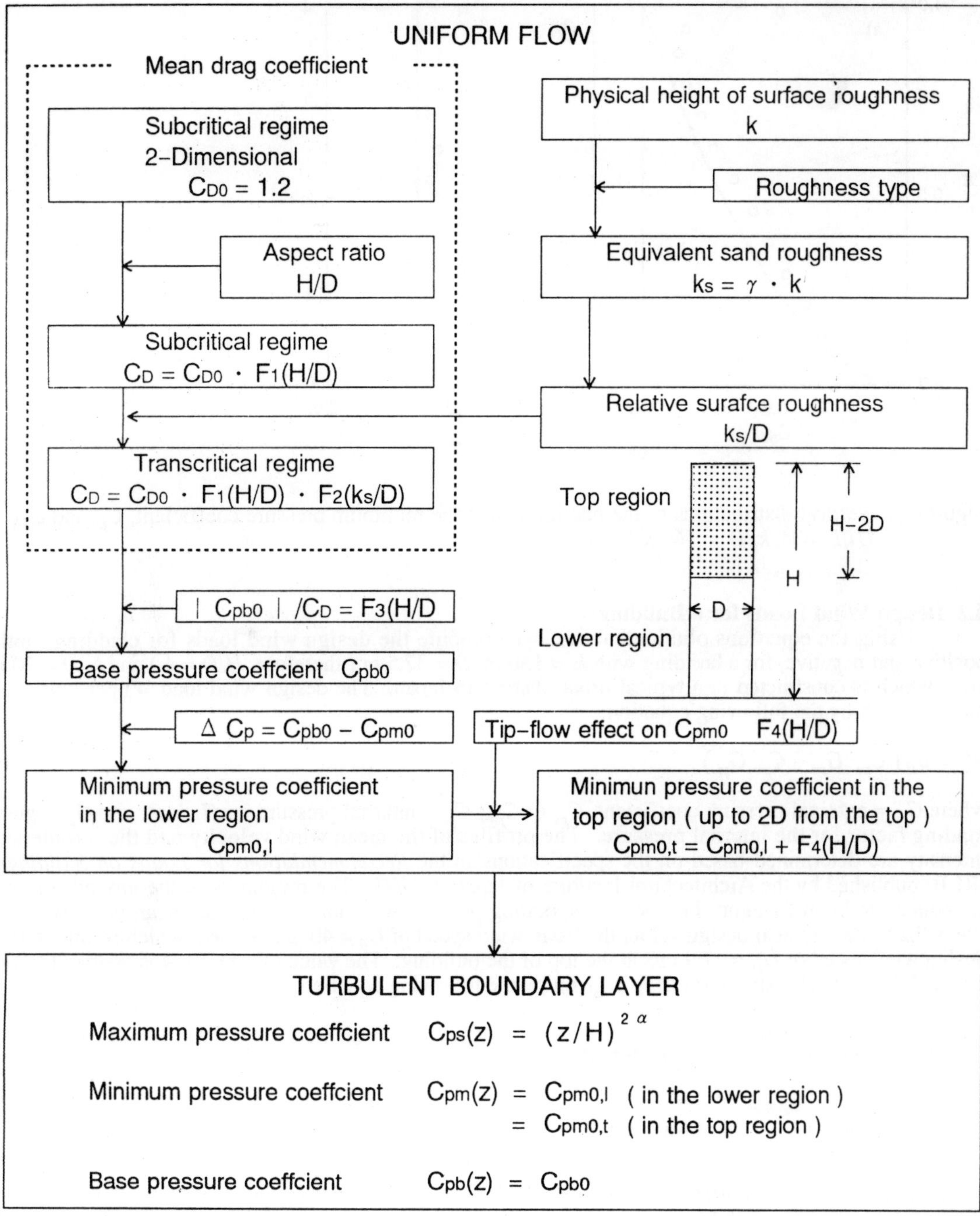

Figure 10. Procedure for obtaining the time-averaged aerodynamic coefficients for given *H/D* and *k/D* ratios

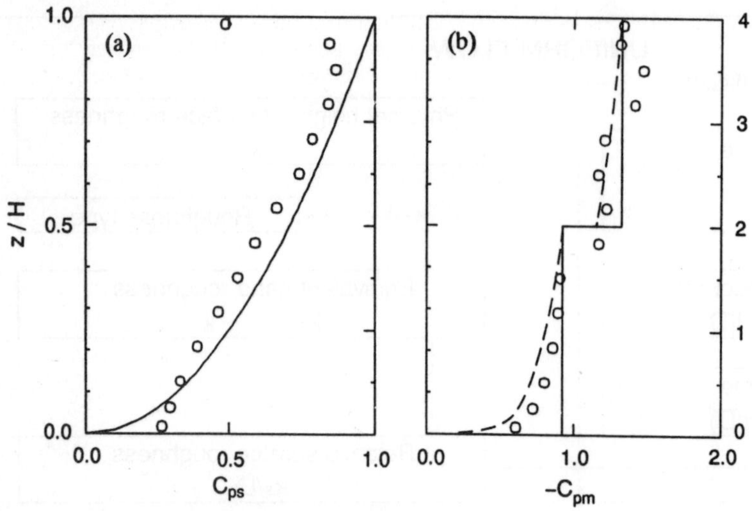

Figure 11. Vertical distributions of the maximum and the minimum pressure coefficient, C_{ps} and C_{pm} (H/D = 4, k_s/D = 560 x 10^{-5})

4.2 Design Wind Loads for a Building

Using the equations obtained above, we compute the design wind loads for cladding, both positive and negative, for a building with H = 150 m, D = 37.5 m (therefore, H/D = 4) and k_s/D = 70 x 10^{-5}, which is constructed in a typical urban district in Japan. The design wind load W_c per unit area may be given by the following equation:

$$W_c = q_H (C_{pe} \cdot G_{pe} - C_{pi} \cdot G_{pi}) \tag{5}$$

where C_{pe} = external pressure coefficient, C_{ps} or C_{pm}; C_{pi} = internal pressure coefficient; and G_{pi} = gust loading factor for the internal pressure. The profiles of the mean wind velocity and the turbulence intensity are determined based on the specifications in the *Recommendations for Loads on Buildings* (RLB) published by the Architectural Institute of Japan in 1993. The roughness of the ground surface is assumed to be in Category IV; the corresponding power law exponent α for the mean velocity profile is 0.27. As a typical design value, the basic wind speed of U_0 = 40 m/s is used, which results in the design wind speed of U_H = 47.9 m/s at the top of the building. The value of g in Eq. 4 is assumed to be 3.5, and the vertical distribution of I_u is given by the following equation:

$$\begin{aligned} I_u &= 0.1 \cdot \left(\frac{z}{z_G}\right)^{-\alpha-0.05} \quad (\text{for } z > z_b) \\ &= 0.1 \cdot \left(\frac{z_b}{z_G}\right)^{-\alpha-0.05} \quad (\text{for } z \le z_b) \end{aligned} \tag{6}$$

where z_G and z_b are the constants which depend on the roughness of the ground surface. For Category IV considered here, z_G = 550 m and z_b = 10 m. The internal pressure coefficient is assumed to be −0.4 for the positive external pressure and zero for the negative external pressure and the gust loading factor for the internal pressure is assumed to be 1.3, as specified in the RLB.

The computed result is shown in Fig. 12. The positive pressure increases with increasing z, approximately proportional to $z^{\dot{\alpha}}$. This is due to a fact that the time−averaged wind pressure is proportional to $z^{2\alpha}$, while the value of I_u varies inversely as $z^{\alpha+0.05}$. On the other hand, the negative pressure does not change significantly in the vertical direction in each region. Therefore, the magnitude of the negative pressure is rather large at lower levels, while both pressures become comparable to each other in a region near the top.

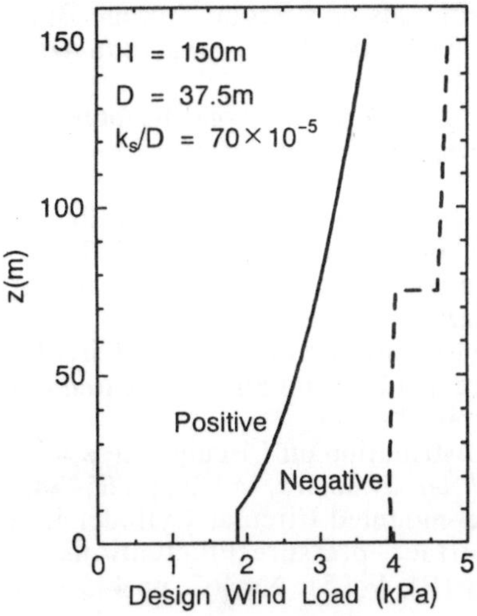

Figure 12. Vertical distribution of the design wind pressures for a building (H = 150m, D = 37.5m, k_s/D = 70 x 10^{-5}, urban district)

5. CONCLUDING REMARKS

The RMS pressure coefficient can adequately be predicted by a quasi-steady theory for actual structures in turbulent atmospheric winds where the turbulence scale would be much greater than the diameter of the cylinder. A quantitative description for the Strouhal number is presented as a function of the aspect ratio and the relative surface roughness, H/D and k_s/D. The design pressures, both positive and negative, for the wall cladding can be given by the product of the maximum or the minimum mean pressure and a gust loading factor. The maximum and minimum pressure coefficients are presented as functions of H/D, k_s/D and the power law exponent α of the mean wind velocity profile. The gust loading factor can be determined from the turbulence intensity of the approaching flow.

6. REFERENCES

Achenbach, E. (1971), "Influence of Surface Roughness on the Cross-flow around a Circular Cylinder", *Journal of Fluid Mechanics*, Vol.46, Part 2, pp.321–335.

Basu, R.I. (1986), "Aerodynamic Forces on Structures of Circular Cross-section. Part 2. The Influence of Turbulence and Three-dimensional Effects", *Journal of Wind Engineering and Industrial Aerodynamics*, Vol.24, pp.33–59.

Farivar, Dj. (1981), "Turbulent Uniform Flow around Cylinders of Finite Length", *AIAA Journal*, Vol.19, No.3, pp.275–281.

Katagiri, J., Marukawa, H., Nakamura, O., Tamura, Y. and Ohkuma, T. (1986), "The Characteristics of Aerodynamic Forces Acting on Cantilevered Circular Cylinders", *Proceedings of the 9th National Symposium on Wind Engineering*, Tokyo, December, 1986, pp.103–108.

Kawamura, T., Hiwada, M., Hibino, T., Mabuchi, I. and Kumada, M. (1984), "Flow around a Finite Circular Cylinder on a Flat Plate (Cylinder Height Greater than Turbulent Boundary Layer Thickness)", *Bulletin of the Japan Society of Mechanical Engineers*, Vol.27, No. 232, pp.2142–2151.

Macdonald, P.A., Kwok, K.C.S. and Holmes, J.D. (1988), "Wind Loads on Circular Storage Bins, Silos and Tanks: I. Point Pressure Measurements on Isolated Structures", *Journal of Wind Engineering and Industrial Aerodynamics*, Vol.31, pp.165–188.

Macdonald, P.A., Holmes, J.D. and Kwok, K.C.S. (1990), "Wind Loads on Circular Storage Bins, Silos and Tanks: II. Effect of Grouping", *Journal of Wind Engineering and Industrial Aerodynamics*,Vol.34, pp.77–95.

Marukawa, H., Tamura, Y., Sanada, S. and Nakamura, O. (1984), "Lift and Across-wind Response of a 200 m Concrete Chimney", *Journal of Wind Engineering*, JAWE, No.19, pp.37–52 (in Japanese).

Melbourne, W.H., Cheung, J.C.K. and Goddard, C.R. (1983), "Response to Wind Action of 265-m Mount Isa Stack", Journal of Structural Engineering, ASCE, Vol.109, No.11, pp.2561–2577.

Nakamura, Y. and Tomonari, Y. (1982), "The Effects of Surface Roughness on the Flow Past Circular Cylinders at High Reynolds Numbers", *Journal Fluid Mechanics*, Vol.123, pp.363–378.

Okamoto, T. and Yagita, M. (1973), "The Experimental Investigation on the Flow Past a Circular Cylinder of Finite Length Placed Normal to the Plane Surface in a Uniform Stream", *Bulletin of the Japan Society of Mechanical Engineers*, Vol.16, No. 95, pp.805–814.

Sabransky, I.J. and Melbourne, W.H. (1987), "Design Pressure Distribution on Circular Silos with Conical Roofs", *Journal of Wind Engineering and Industrial Aerodynamics*, Vol.26, pp.65–84.

Sakata, I., Kiya, M., Nagai, H. and Arie, M. (1985), "A Surface-mounted Circular Cylinder in a Turbulent Boundary Layer (Probability Density of Surface-pressure Fluctuations)", *Transactions of the Japan Society of Mechanical Engineers* (B), Vol.51, No.462, pp.445–451 (in Japanese).

Sakata, I., Kiya, M. and Arie, M. (1984), "A Surface-mounted Circular Cylinder in a Turbulent Boundary Layer (Flow Visualization and Surface-pressure properties)", *Transactions of the Japan Society of Mechanical Engineers* (B), Vol.50, No.457, pp.2050–2058 (in Japanese).

Ueda, H., Hibi, K. and Fujii, K. (1991), "Wind Pressure on Cylindrical Building", *Summaries of Technical Papers of Annual Meeting*, AIJ, Structure I, pp.151–152.

Uematsu, Y. and Yamada, M. (1993), "Effects of Aspect Ratio and Surface Roughness on the Time-averaged Aerodynamic Forces on Cantilevered Circular Cylinders at High Reynolds Numbers", *Proceedings of the 3rd Asia-Pacific Symposium on Wind Engineering*, Hong Kong, December 13–15, 1993, Vol.II, pp.667–672.

Uematsu, Y. and Yamada, M. (1994), "Aerodynamic Forces on Circular Cylinders of Finite Height", *Journal of Wind Engineering and Industrial Aerodynamics*, Vol.51, pp.249–265.

Vickery, B.J. (1968), "Load Fluctuations in Turbulent Flow", *Journal of the Engineering Mechanics Division*, ASCE, Vol.94, No. EM1, pp.31–46.

WIND TUNNEL STUDY ON DETERMINATION OF BASE MOMENTS OF A PARALLELLOPIPED BUILDING

ARUNACHALAM, S. and SELVI RAJAN, S.

Scientists, Structural Engineering Research Centre, Madras, India

Abstract

The performance of a scaled parallellopiped building model to dynamic action of wind was studied in a boundary layer wind tunnel. The model was tested for two different terrain conditions ,viz, open terrain (with $\alpha = 0.13$) and sub-urban terrain(with $\alpha = 0.23$). The dynamic base moments of the model were measured using a high frequency force balance for different wind directions and for the terrain conditions. The mean and the rms base moment coefficients were found to be considerably varying with wind direction. The results showed significant correlation between the three components of the base moments. The coherence values obtained in the present experiment were compared with the values reported in the literature.

1. INTRODUCTION

The increasing modern trend for design and construction of tall, slender, and lightweight structures such as high-rise buildings, cooling towers, transmission-line and communication towers has shown that such structures are highly susceptible to dynamic action of wind. Wind tunnel testing of structural models has been recognised internationally as an established procedure for investigating the wind-induced response of different types of structures. Conventionally, aeroelastic qmodels have been extensively used by researchers for experimental assessment of unsteady wind loads and response of structural models (Cermak, 1979, Whitebread, 1967, Isyumov, 1982). However, an aeroelastic model test has a disadvantage that any change in mass distribution and/or stiffness developed during design modifications of the model, requires repetition of the entire test in order to study the revised response characteristics, which is, in general time consuming. Hence as an alternative method, aerodynamic model test using high frequency force balances was developed subsequently and use of this type of model test has become more popular in boundary layer wind tunnels (Whitebread, 1967,Cermak et al, 1970, Tschanz, 1983, Boggs, 1967). The response of a structure can be conveniently calculated using spectral analysis, once the atmospheric aerodynamic wind loading is measured in a wind tunnel. The correlation of base moment components of wind loads on structures has been studied by several researchers and it has been reported that contrary to the above assumption the base moments are highly correlated (Reinhold and Kareem, 1986, and Thoroddsen et al, 1985). The authors have carriedout an experimental investigation on the performance of a scaled parallelogram building model and studied the correlation of the three components Mx , My and Mz using a high frequency force balance (HFFB) for two different terrain conditions in the Industrial Aerodynamics Wind Tunnel facility available at Colorado State University, Fort Collins, USA (SelviRajan & Arunachalam, 1993). The details of the experimental procedure for the simulation of atmospheric boundary layer and determination of base moments are described in this paper.

2 EXPERIMENTAL PROCEDURE

2.1. Simulation of the atmospheric layer (ABL)

In the present experiment, two types of different terrain conditions were simulated and the response of the building model was studied. The simulation of smooth terrain was accomplished using four numbers of Standen-spire type of vortex generators placed at the entrance of the test section followed by two frames of inclined horizontal vanes at a distance of 78.5 cm. Downwind of these horizontal vanes, two wooden fences of height 24 cm and 12 cm spaced at 195 cm apart were also provided. A saw-tooth roughness was also kept at a distance of 236 cm downwind of the lower fence. The test section floor was covered with steel chain which acted as floor roughness elements. The above simulation was also used for the simulating of the rougher terrain, with only difference of replacing the chain on the test section floor by wooden roughness cubes of 5 cm size and also by removing the saw-tooth fence.

2.2 Building Model

One of the objectives of the present experiment was to study the interdependence between the three components of the dynamic base moments Mx, My and Mz by measuring their values using a high frequency force balance. For the present study, a parallellopiped model was selected for the following reasons. The inclination of the faces of a building with respect to its central axis in plan, plays a significant role in the contribution of correlation between the overturning moments about the central axis. For example, in the case of a building with a rectangular shape in plan, the building faces are parallel either to x or y axis. Further the wind loads, acting on the building are normal to the faces. Hence it may be inferred that any correlation between Mx and My is primarily due to the organisation of the flow around the building (depending on aspect ratio, etc.). On the other hand for a building with a non-rectangular shape in plan, such as a parallelogram, some of the faces of the building are inclined to central axes in plan and because of the geometry of the building in plan (in addition to effects of aspect ratio, etc.) the inclined faces contribute additionally to the correlation of the overturning moments about these two axes. The selected model had dimensions of 7.62 cm x 15.24 cm in plan and a height of 40.6 cm . It was made of white Rohacell Styrofoam material, which is light in weight, having a density of 31 kg/m^3. The model was stiffened with a central aluminium tube, and was fitted to the central aluminium stem of the force balance. The model was rigidly connected to the force balance reaction plate as shown in Fig.1. The force balance which had been developed for studies on aerodynamic models at Engineering Research Centre at CSU was used in this experiment. For the present model, the ratio of the height to least horizontal dimension was slightly more than 5 and hence it was assumed that the aeroelastic effects could be conveniently neglected (Thoroddsen et al, 1988).

2.3 Measurement of data

A standard pitot static tube was used to measure the reference dynamic pressure, andthe fluctuating longitudinal components of the wind at different heights were measured using a hot-wire anemometer. The data from the hot-wire anemometer and the pitot static tube were sampled at a rate of 500 Hz and 30 Hz respectively for a period of 20 seconds. The boundary layer mean velocity and turbulence intensity profiles corresponding to the above terrains are shown in Figs. 2(a) and 2(b)

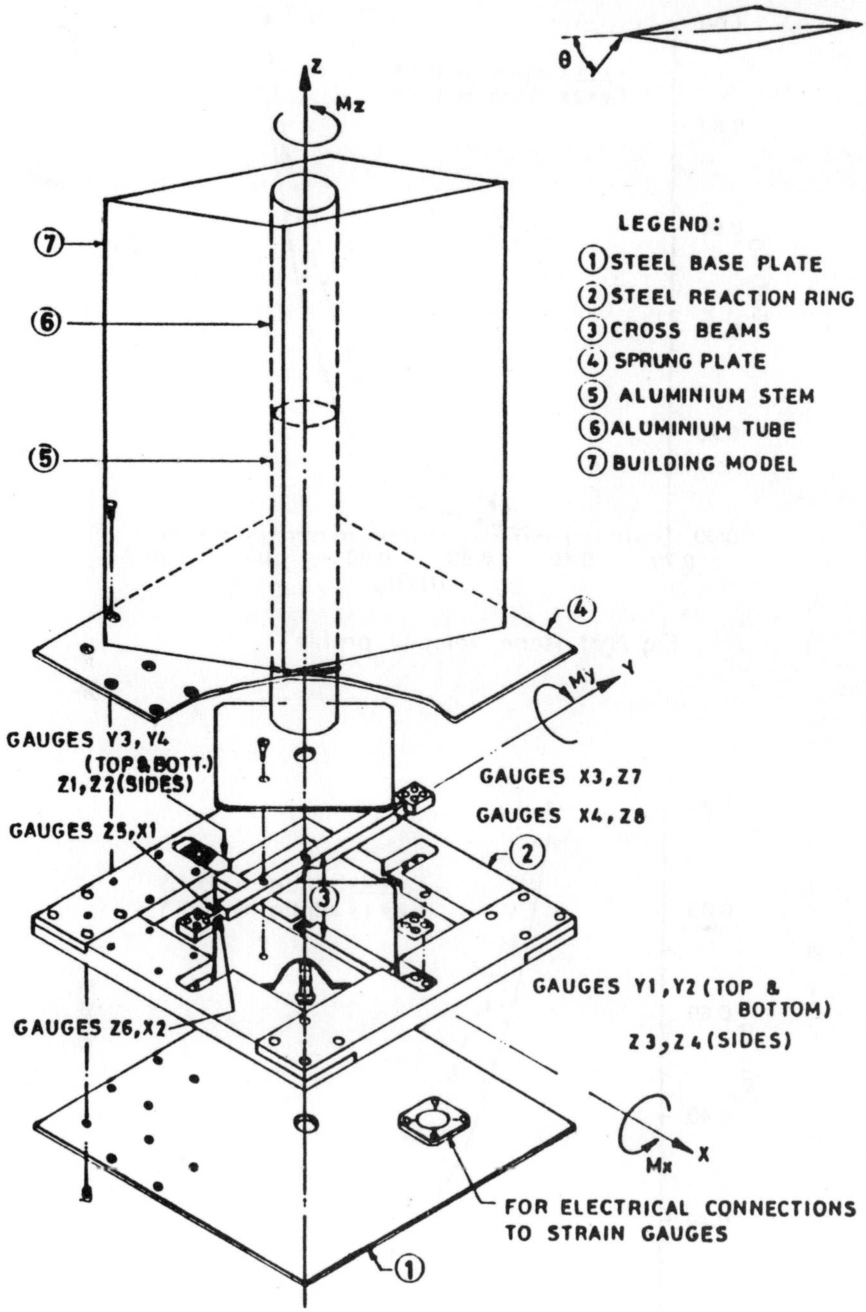

Fig.1 Force balance with the model

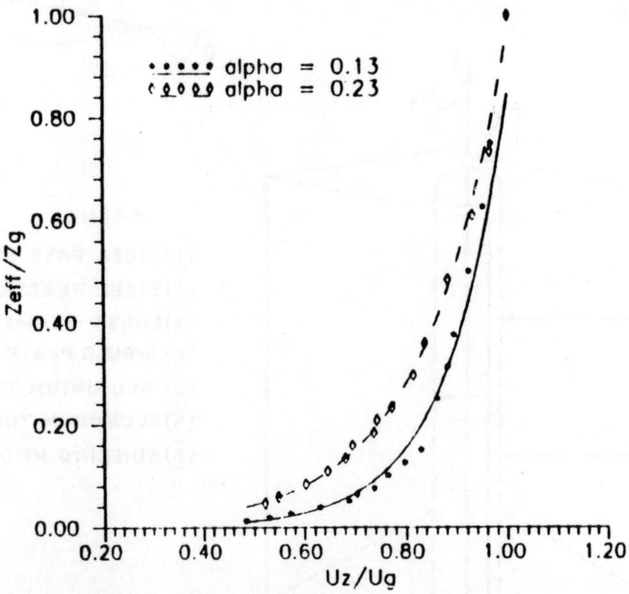

Fig.2(a) Mean velocity profile

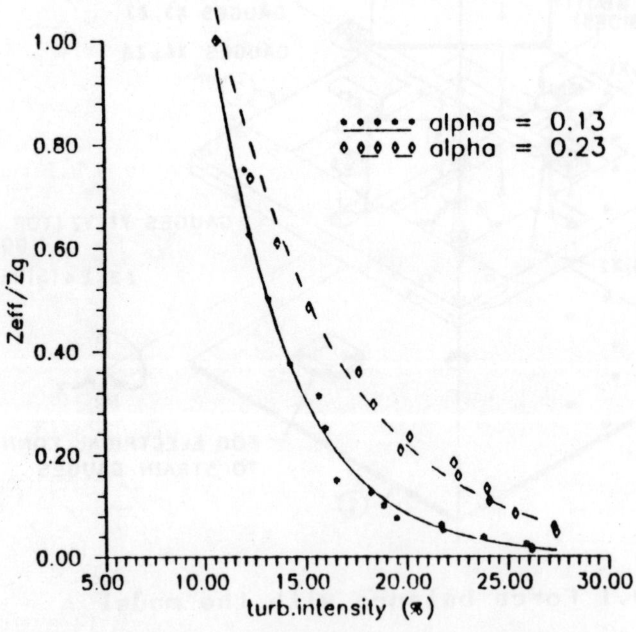

Fig.2(b) Turbulence intensity profile

respectively. Velocity time series at different heights were also taken for computation of time and length scales of turbulence. The model was kept at the centre of the turn-table and using the HFFB with flat frequency response up to 102 Hz, the mean and dynamic base moments were measured for the two overturning moments, Mx and My and the torsional moment, Mz. These data were measured for different wind directions in both the terrain conditions. The results were presented in the form of non-dimensional coefficients.

3. DISCUSSIONS

The model was tested for two different terrain conditions, namely, for open terrain with power law exponent, $\alpha = 0.13$ and suburban terrain with $\alpha = 0.23$. The results of the simulated mean velocity and turbulent characteristics of ABL were found to be in good agreement with the values recommended in the literature (Counihan, 1975, Simiu and Scanlan, 1986). The values of the turbulence intensities at Z = 30 m in full scale corresponding to the above smooth and rough terrains were found to be equal to 17.2% and 27.3% ,respectively. The length scale values for smooth terrain increased with height and reached a maximum value at about one third height of the boundary layer beyond which it was decreasing upto Z/d =0.5, and remaining constant for higher elevations. Similar observations were reported by Blessmann and Reira (1983) and Thoroddsen (1986). In the case of rougher terrain similar trend was obtained for the variation of (Lu(Z)/d) with height (Z/d); however, the position of maximum value of Lu(Z)/d) was shifted to Z/d = 0.45. The maximum values of Lu(Z)/d for smooth and rough terrains were found to be 0.58 and 0.55, respectively. Since the value of d was same in both cases, it was clear that in the lower regions of the ABL, Lu(Z) at a given height had a higher value in the smooth terrain than in the rough terrain. Similar trend was reported by Counihan (1975). The measured spectra at Z = 60 cm in both terrains were showing excellent agreement with Karman spectrum and the measured spectra at lower heights (say at Z=40 cm) showed good agreement with Simiu spectrum. Measured spectra at Z=40 cm and 60 cm for the two terrains are shown in Fig. 3(a) and 3(b), respectively. Since winds at larger heights contribute more to the base moments , Karman spectrum was selected in this case. By matching the reduced frequency of the Karman spectrum with that of the measured spectrum, the model scales were found to be equal to 1:350 and 1:500 for smooth and rough terrains, respectively.

Data for the mean and dynamic base moments, Mx, My and Mz were collected using the high frequency force balance for both the terrain conditions, for every 22.5 degree increment of the angle of incident wind direction. A sampling period of 50 sec was used for computing the mean and the rms values of base moments. The mean dynamic wind pressure of the incident wind was monitored using a pitot static tube at the height of the building at a distance of 150 cm upwind of the building. This reference wind velocity and the mean and rms values of Mx, My and Mz were measured on-line for each angle of incident wind direction.

3.1 Base moment coefficients

The actual base moment values were expressed in a non-dimensional form using the characteristic building dimensions and the reference dynamic wind pressure as follows:

$$C_{Mi} = M_i \Big/ \left(0.5 * \rho * \bar{U}^2 * D_i * H_i * L_i\right) \quad (1)$$

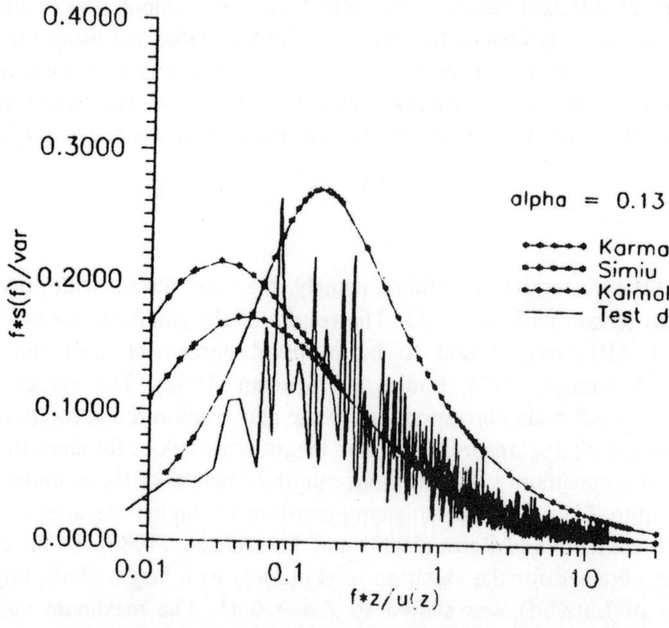

Fig.3(a) Velocity spectrum at z = 40cm.

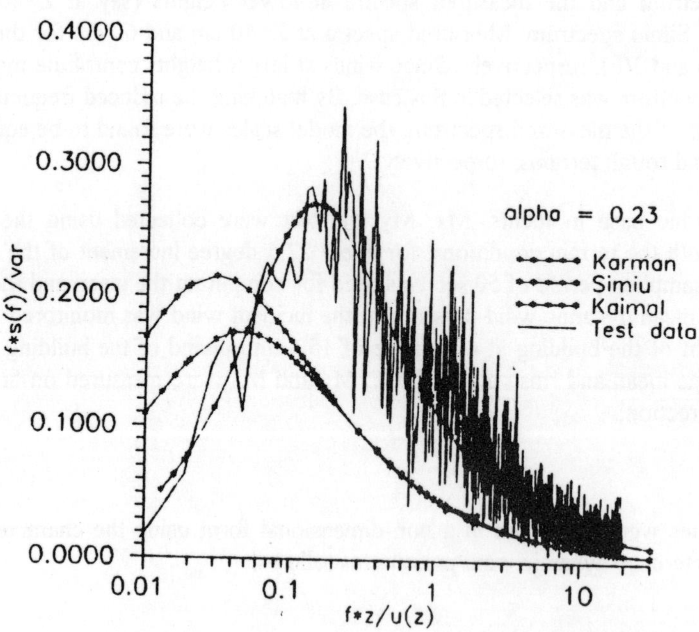

Fig.3(b) Velocity spectrum at z = 60cm.

where C_{Mi} represents Cmx or Cmy or Cmz as the case may be;
$\overline{U}$ is the mean velocity at the building height
ρ is the mass density of air.

A value of peak factor, g = 3.8 was assumed in deducing the maximum and minimum peak moment coefficients from the mean and rms moment coefficients (Isyumov, 1982).

3.2 Mean moment coefficients

In the present model system, X-axis was always the line joining the longer diagonal of the parallelogram and $\theta=0^o$ coincided with the X-axis. The variation of all the mean moment coefficients with different wind directions, for α=0.23 and α=0.13 are shown in Figs. 4(a) and 4 (b). It is clear that the values are dependent on wind direction. The Mx mean moment coefficients were symmetrical about 180° as expected. The scatter in the value of Mx moment coefficients was the largest and it varied from -0.554 to +0.586 for α=0.23 case and from -0.849 to +0.526 for α=0.13 case. These two positive and negative values occurred at $\theta=90^o$ and $\theta=270^o$, respectively. This is explained by the fact that the area of wind force Fy causing overturning moment Mx was the largest and hence resulted in higher magnitudes of mean Mx moment coefficients.
The variation of Mx mean moment coefficients with wind direction was similar in both the terrains. The maximum and minimum values of mean My coefficients were equal to +0.566 and -0.248 and they occurred at wind angles of $\theta=360^o$ and 135°. The corresponding values in smooth terrain were equal to +0.518 and -0.384 occurring at $\theta=45^o$ and $\theta=180^o$ respectively. As seen from Fig. 4(b), the maximum and minimum values of Mz mean moment coefficients were equal to +0.242 and -0.404 and they occurred at wind angles of $\theta=90^o$ and 22.5^o , respectively, for α=0.23 case. The present model being a parallelogram type,with inclined faces with respect to its centroidal axes, a larger amount of torsion could be expected particularly when the wind direction was parallel to one of its inclined faces. Since the long diagonal axis was kept as X-axis the shorter inclined face subtended an angle of 22.5^o with X-axis. Accordingly, the highest torsion occurred at $\theta=22.5^o$ and this agreed with experimental data. Similar observations were also reported by others (Thoroddsen et al, 1988).

3.3 RMS moment coefficients

The test data showed that the RMS moment coefficients were also highly dependent on wind direction and are as shown in Fig.5(a) and 5.(b). It was observed that for α=0.23 case, the maximum Mx and Mz RMS moment coefficients occurred at $\theta=22.5^o$ or when the flow was parallel to one of its slanting faces. The maximum My RMS moment coefficient occurred at $\theta=45^o$. In the case of smooth terrain, the maximum values of Mx and My were found to occur at $\theta=22.5^o$. Further the Mz RMS moment coefficients exhibited symmetry about $\theta=180^o$ in both terrains. Hence for the smooth terrain, Mz RMS value at $\theta=22.5^o$ was assumed equal to 0.0361 based on its value at $\theta=205^o$ and this value was very nearly equal to the maximum value occurring at $\theta=157.5^o$. Hence it is reasonable to state that RMS values of moment coefficients had maximum values when the wind flows parallel to one of the inclined faces.

3.4 Correlations of base moment components

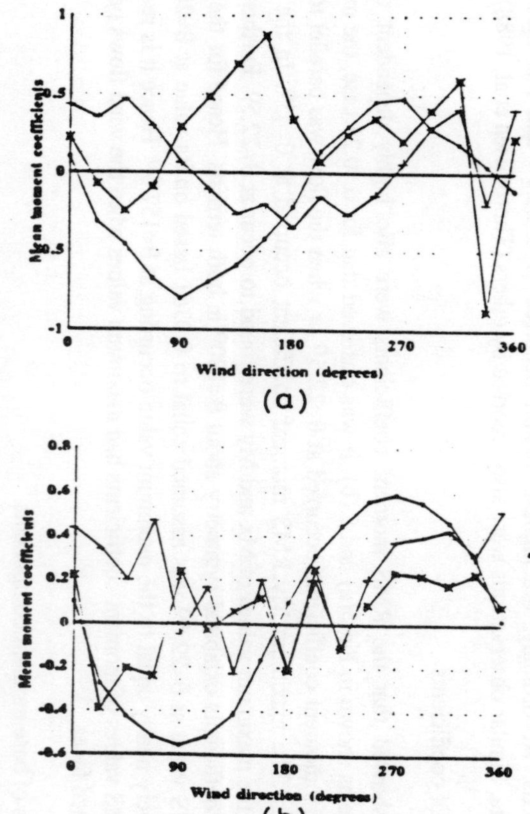

Fig. 4 Mean moment coefficients for two terrains

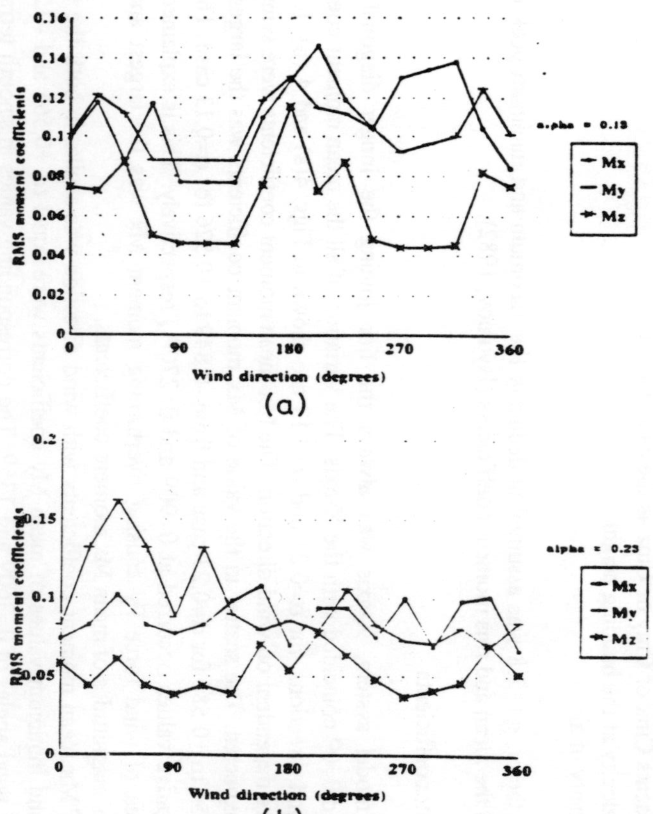

Fig. 5 RMS moment coefficients for two terrains

The components of the base moments can be expected to be correlated since they emanate from the same physical loading. For example, for a square building when the wind direction is diagonal to one of its central axis, the loads due to buffeting will be correlated in the X and Y direction. For the present parallelogram model, the variation of cross correlation between Mx, My and Mz components is shown in Fig.6 for the rough terrain. It was seen that the correlation was varying depending on wind direction. The maximum correlation was for Rxy equal to 0.683 corresponding to a wind angle of 112.5. The maximum values of Rxz and Ryz were found to be 0.5569 and 0.3922 occurring at wind directions of 90 and 0 degrees respectively. Similarly the maximum coherence curve indicated the dependence between the components in the frequency domain. For the present case the maximum coherence between Coh_{xy}, Coh_{xz} and Coh_{yz} components for different wind directions is shown in Fig. 7. The maximum value of Rxy was found to be equal to 0.9129 corresponding to a wind direction of 45°. The maximum value of Coh_{xz} and Coh_{yz} were respectively equal to 0.8723 and 0.9213 both occurring at $\theta=180^{o}$. In other words this implied that for a parallelogram model, when the wind is parallel to the longer diagonal, the maximum coherence occurred between the torsional component and the lateral components. The coherence values obtained in the present experiment can be compared with the results reported by Thoroddsen et al, (1988), wherein a parellelogram model of similar dimensions was tested. Considering the definition of axes used and referring to Figures 5(a) through 5(c) of the above reference, it may be said that the maximum coherence values obtained in the present study with $\theta = 22.5^{o}$ correspond to respective values at $\theta = 67.5^{o}$ of the above reference. Since the nearest value of wind direction included in Fig.5 of the above reference is 60°, as seen from the following table, there is a good agreement between the maximum coherence values in these two cases.

Table.1 Comparison of Maximum Value of Coherence

Coherence parameter	Based on present study($\theta=22.5^{o}$)	Based on Fig.5 of Thoroddsen et al, 1988 ($\theta = 60^{o}$)
Coh_{xy}	0.8614	0.8670
Coh_{xz}	0.8165	0.8400
Coh_{yz}	0.7476	0.7830

4. CONCLUSIONS

The performance of a scaled parallellopiped building model to the dynamic action of wind was studied in a boundary layer wind tunnel. The model was tested for two different terrain conditions, viz,open terrain (with $\alpha = 0.13$) and suburban terrain (with $\alpha = 0.23$). The results of the simulated mean velocity and turbulent characteristics of ABL were found to be in good agreement with the values recommended by Counihan and others. By matching the measured spectra with the Karman spectrum, the model scales were found to be equal to 1:350 and 1:500 for smooth and rough terrains respectively. The dynamic base moments of the model were measured using a high frequency force balance for different wind directions and for the above two terrain conditions. The mean and the rms base moment coefficients were found to be considerably varying with the wind direction. Since the present model was a parallelogram with inclined faces with respect to its centroidal axes, the

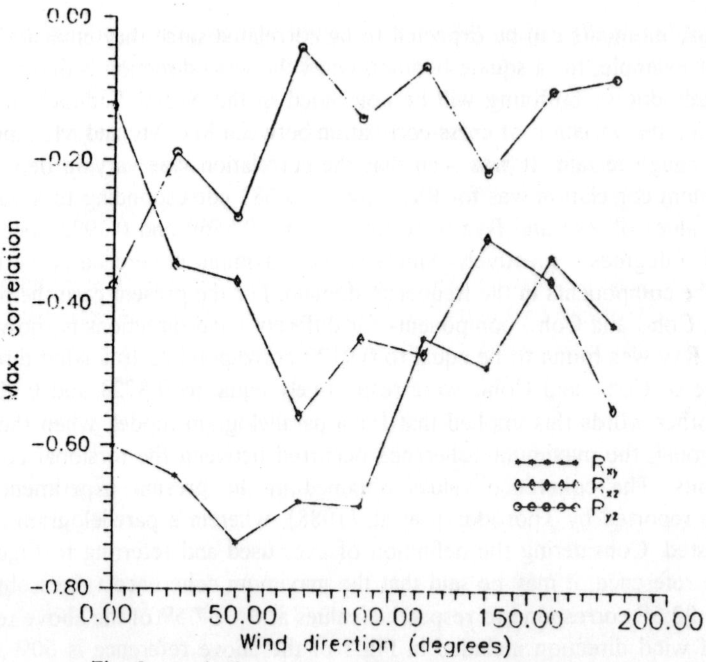

Fig.6 Variation of maximum correlation

Fig.7 Coherence between base moments

maximum torsion occurred at $\theta = 22.5^{o}$, when the wind direction was parallel to one of its slanting faces. The results clearly showed that the three components were correlated with each other. The maximum values for Rxy, Rxz and Ryz were found to be equal to 0.683 at $\theta = 112.5^{o}$,0.5569 at $\theta = 90^{o}$ and 0.3922 at $\theta = 0^{o}$, respectively. The coherence values obtained in the present experiment showed good agreement with the results reported in the literature (Thoroddsen et al, 1988).

5. ACKNOWLEDGEMENTS

This paper is published with the kind permission of the Director, SERC,Madras. The authors would like to thank Prof.J.E.Cermak, University Distinguished Professor, Colorado State University, USA., for his constant encouragement and guidance in carrying this experiment.

6. REFERENCES

Blessmann,J., and Reira,J.D., (1983)"Wind Excitation of Neighbouring Tall Buildings," Jl.of Wind Engg. and Industrial Aerodynamics, Vol.18, 1983.

Boggs, D.W., and Peterka, J.A., (1987) "The Application of Aerodynamic Model Tests to Wind Loading of Tall Buildings," Structures Congress, Orlando, Florida, 17-20, Aug., 1987.

Cermak, J.E., Sadeh, W.Z., and Hsi, G., (1970) "Fluctuating Moments on Tall Buildings Produced by Wind Loading", Wind Loads on Buildings and Structures, Ed. H.C.S. Thom, Building Science Series 30, NBS, 1970.

Cermak, J.E.,(1977) Wind Tunnel Testing of Structures , Journal of Engineering Mechanics Division, ASCE, Vol.103, No.EM6, Dec.1977.

Cermak, J.E., (1985) "Boundary Layer Wind Tunnel Testing", Section 4 in "Design Wind Loads for Buildings and Boundary Layer Wind Tunnel Testing," AAMA, Aluminium Curtain Wall Series, No.11, 1985.

Counihan, (1975) "Adiabatic Atmospheric Boundary Layer: A Review and Analysis of the Data for the Period 1880-1972" Atmospheric Environment, 5, 1975.

Isyumov, N., (1982) "The Aeroelastic Modeling of Tall Buildings", "Wind Tunnel Modeling for Civil Engineering Applications", Ed., T.A. Reinhold, Cambridge University Press, Cambridge, 1982.

Kareem, A., (1985) "Lateral-Torsional Motion of Tall Buildings to Wind Loads," Journal of the Structural Division, ASCE, Vol.111, No.11, November, 1985.

Reinhold, T.A., and Kareem, A., (1986) "Wind Loads and Building Response Predictions using Force Balance Techniques," Engg. Mechanics Division, Ed. Gary C. Hart and R.B. Nelson, ASCE, 1986.

Selvi Rajan, S., and Arunachalam, S.,(1993) "Experimental Determination of Base Moments on a

Parallellogram Building Model using a High Frequency Force Balance", UNDP Training Report TR-5/92, Structural Engineering Research Centre, Madras, India, March 1993.

Simiu, E., and Scanlan, R.H.,(1986) "Wind Effects on Structures," Second Edition, John Wiley & Sons, 1986.

Thoroddsen, S.T., Cermak, J.E., and Peterka, J.A., (1985)"Mean and Dynamic Wind Loading caused by an Upwind Structure," Proc. Fifth National Conference on Wind Engg., Texas Tech. University, Lubbock, Texas, Section 4A, 1985.

Thoroddsen,T.S., (1986) "Wind Loading on High-rise Buildings Affected by an Upwind Structure", M.S. Thesis Report, Colorado State University, USA, 1986.

Thoroddsen, S.T., Peterka, J.A., and Cermak, J.E., (1988) "Correlation of the Components of Wind Loading on Tall Buildings,"Journal of Wind Engg. and Industrial Aerodynamics, No.28, 1988.

Tschanz, T., (1983) "The Base Balance Moment Technique and Applications to Dynamic Wind Loading of Structures,"Ph.D. Thesis report, University of Western Ontario, London, Ontario, Canada, 1983.

Whitebread, R.F., (1967) "The Measurement of Non-steady Wind Forces on Small Scale Building Models," Proc. Fourth International Conference on "Wind Effects on Buildings and Structures", Proc. International Research Seminar held at the National Research Council, Ottawa, Canada, Vol.1., Sept., 1967.

9ICWE, 1995, New Delhi, India

WIND INDUCED RESPONSE OF TENSION LEG PLATFORM

AHMAD, S.[1], ISLAM, N.[2] and ALI, A.[1]
Asstt. Professor Senior Lecturer Research Scholar

[1] *Dept. of Applied Mechanics, Indian Institute of Technology, New Delhi (India)*
[2] *University Polytechnic, Jamia Millia Islamia, New Delhi, India*

Abstract

Dynamic response studies of Tension Leg Platform (TLP) are carried out in time domain to investigate the influence of non-linearities due to hydro-dynamic drag force, variable cable tension, variable submergence, long excursions and fluctuating wind together with the effect of coupling. The sea state is characterised by Pierson Moskowitz spectrum while the fluctuating wind has been estimated using Emil Simiu's wind spectrum for compliant offshore structures. Random wind and waves are modelled by Monte Carlo simulation. Probability Spectral Density Functions (PSDF) are plotted to highlight the statistical characteristics of the structure.

1. INTRODUCTION

As the oil and gas reserves are found to be in deeper water where fixed platforms can not be economically constructed; the concept of compliant platform has been introduced. Offshore compliant structures in essence, attract lesser environmental loading. Due to their compliant behaviour large excursions are introduced. Tension Leg Platform (TLP) is one of the compliant structures, well established in offshore oil industry. These structures have more buoyancy than their weights and hence are anchored to the sea bed by means of taut tethers. These mooring cables (tethers) are subjected with large tensions due to huge buoyancy and are also subjected to tension fluctuations due to variable submergence, wind and wave induced platform motions and ballasting of the structure. The instantaneous tension in the tethers act as restoring force. Owing to this high tension the vertical motions of TLP such as heave, pitch and roll are almost restrained while the horizontal motions such as surge, sway and yaw may considerably be large.

As shown in Fig. 1 TLP's substructure below Still Water Level (SWL) has a massive hull made up of huge hollow horizontal and vertical cylinders responsible for higher buoyancy. Super structure above the SWL has all the essential components similar to any offshore platform. The super structure attracts high wind induced loading while the sub-structure is simultaneously subjected to hydro-dynamic loading. The serviceability and serviveability of TLP under strong wind and wave loading need a thorough response investigation.

Wind induced loading on TLP has been recognised as quite significant. Studies on TLP [1,2,3,5] considered wind induced oscillatory compliant platform motion. However, all the non-linearities involved were not taken into account. Emil Simiu [4] presented the fluctuating wind modelling at low frequencies by introducing alternative model spectrum. Kareem [2] also proposed a new spectral description of the longitudnal wind velocity fluctuation over the ocean.

In the present study TLP's sensitivity to dynamic effect of wind has been investigated. The influence of various non-linearities due to hydro-dynamic drag force, variable cable tension, variable submergence, long excursions and fluctuating wind on TLP's response have been investigated. In order to overcome higher degrees of nonlinearities time domain analysis has been adopted employing New Mark's β integration scheme. To estimate the hydro-dynamic loading Pierson Moskowitz wave spectrum has been employed. The Emil Simu's wind spectrum [4] for ocean environment has been employed for estimation of fluctuating wind component. Random wind and waves are modelled by Monte Carlo Simulation technique. The present study aims at the investigation of TLP's response influenced by the effect of coupling among various degrees of freedom as well as by the inherent non-linearities under random wind and wave loading.

2. DEVELOPMENT OF MATHEMATICAL MODEL FOR WIND FORCES

The mathematical model is composed of a square deck supported on four cylindrical columns at each corner which are interconnected at their bottom by four cylindrical pontoons and anchored to the foundation at the sea bed by four steel tethers (Fig 1). Assuming TLP as a rigid body its motion is described by six degrees of freedom comprising of three translations (Surge, Sway and Heave) along X, Y, Z axes and three rotations (Roll, Pitch and Yaw) about the same axes. The super-structure of the TLP exposed to wind has been modelled as shown in Fig 2

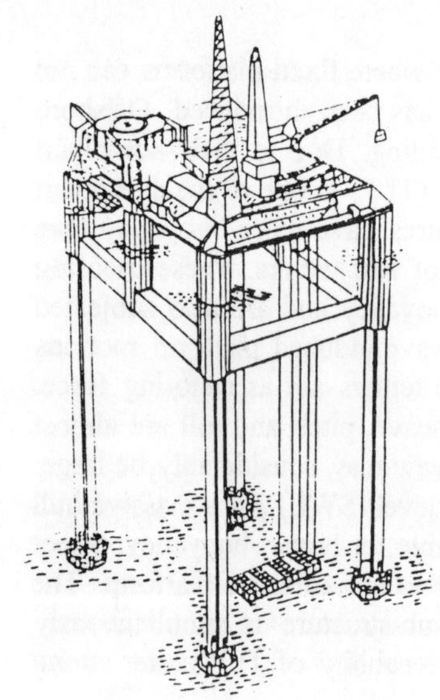

Fig 1–Tension leg platform (TLP)

All dimensions are in metre

A.C. = Aerodynamic Centre

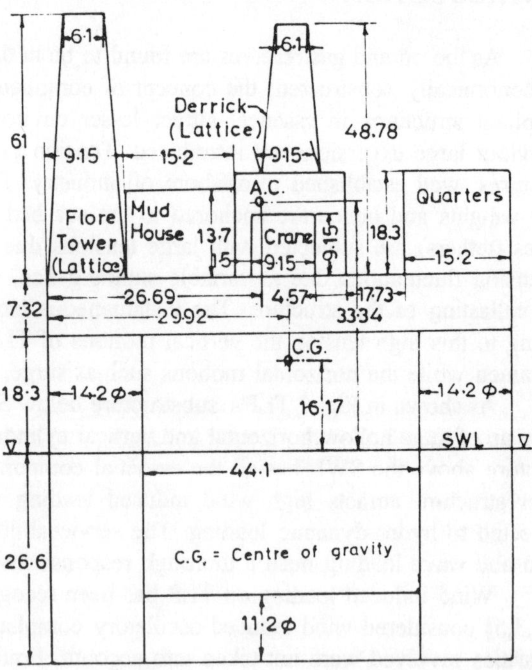

Fig-2 TLP Super Structure Model for Wind Load.

whose equivalent projected area for wind load is 2572 sq. metre. For methematical idealization its is assumed that the total wind force is to be concentrated at the aerodynemic centre (A.C.) while the total mass of the TLP is to be lumped at its centre of gravity (C.G.). As shown in Fig. 2 the C.G. is loacted at 16.17 metre above SWL while the AC is located at 33.34 metre above SWL and is 3.23 metre horizontally off the C.G. Owing to this eccentricity of AC and unsymmetry of the projected area yawing and pitching moments will also be introduced.

Apart from overall super structure the ancillary structures like lattice tower, mud houses, living quarters, derrick, halipad are considered for aero-dynamic coefficient estimation. The latticed components are replaced by equivalent projected areas. The aerodynamic forces result from the interaction of the wind with the super structure of platform which acts as a complex bluff configuration. Most of the wind-ward surfaces experience positive pressure, whereas the side and leeward surfaces experience negative pressure. The synthesis of pressure acting on the entire super structure results aerodynamic forces.

The development of the complete deterministic dynamic analysis model essentially involves the formulation of a non-linear coupled stiffness matrix. This matrix takes into account the instantaneous behaviour of the tethers and fluctuating buoyant forces. The tethers are therefore, mathematically idealized to behave like springs whose stiffness fluctuate with time.

2.1 DERIVATION OF STIFFNESS MATRIX

Based on the above mathematical model the stiffness matrix is formulated by considering the static equilibrium as

$$W + 4T_o - F_B = 0 \tag{1}$$

Surge Stiffness

For an infinitesimal displacement x the forces acting on the platform are shown in Fig. 3. Summation of forces in the horizontal and vertical direction yields (2) and (3)

$$K_{SS} = 4(T_0 + T_1) \text{ Sin } \tau_x \text{ and} \tag{2}$$

$$K_{HS} = 4\left(EA - T_0\right)\left(\frac{x^2}{2l^2}\right) \tag{3}$$

The summation of moments about y-axis yields

$$k_{PS} = - K_{SS}.h \tag{4}$$

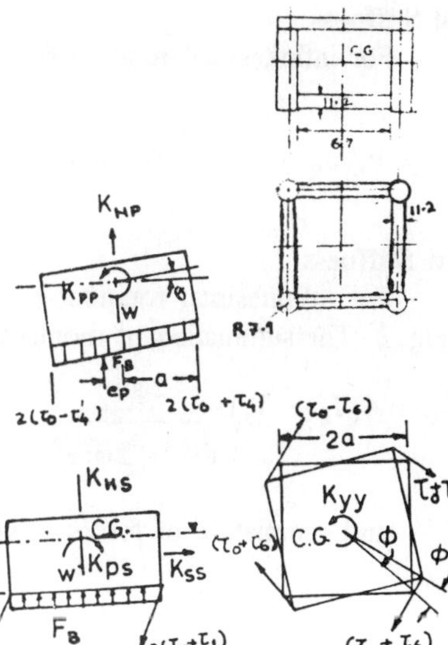

Fig 3-Stiffness Matrix Generation of TLP

where, K_{AB} means the force in the degree of freedom `A' due to an arbitrary displacement in the direction of degree of freedom `B' with all other degrees of freedom restrained.

Sway Stiffness

For an infinitesimal displacement 'y' following the above procedure, we get

$$K_{WW} = 4\left(T_0 + \frac{EA}{2l^2}.y^2\right)\frac{y}{l}; \qquad K_{HW} = 4(EA - T_0)\left(\frac{y^2}{2l^2}\right) \tag{5}$$

$$K_{RW} = -K_{WW}.h \tag{6}$$

Heave Stiffness

Considering the equilibrium of forces in Z direction, we obtain

$$K_{HH} = 4T_3 + \delta F_B \tag{7}$$

Pitch Stiffness

For an infinitesimal rotation 'θ' about Y-axis, the forces acting on the platform are shown in Fig. 3. the summation of moments about Y-axis yields

$$K_{PP} = F_B ep + 2a^2\frac{AE}{l}.\theta \quad \text{and} \quad K_{HP} = 0 \tag{8}$$

Roll Stiffness

For infinitesimal rotation θ' about X-axis following the above steps one obtains

$$K_{RR} = F_b.e_R + 2a^2\frac{AE}{l}\theta' \quad \text{and} \quad K_{HR} = 0 \tag{9}$$

Yaw Stiffness

For infinitesimal rotation 'φ' about Z-axis, the forces acting on the platform are shown in Fig. 3. The summation of moments about Z-axis yields

$$K_{YY} = 4(T_0 + T_6)\frac{2\varphi a^2}{\sqrt{l^2 + 2\varphi^2 a^2}} \tag{10}$$

and summation of forces in vertical direction gives

$$K_{HY} = 4T_0\left[\frac{1}{\sqrt{l^2 + 2\varphi^2 a^2}} - 1\right] + 4T_6\left[\frac{1}{\sqrt{l^2 + 2\varphi^2 a^2}}\right] \tag{11}$$

where, $T_6 = \frac{AE}{l^2}\varphi^2 a^2$

The resulting stiffness matrix is given by [K]

$$[K] = \begin{bmatrix} K_{SS} & 0 & 0 & 0 & 0 & 0 \\ 0 & K_{WW} & 0 & 0 & 0 & 0 \\ K_{HS} & K_{HW} & K_{HH} & K_{HP} & K_{HR} & K_{HY} \\ K_{PS} & 0 & 0 & K_{PP} & 0 & 0 \\ 0 & K_{RW} & 0 & 0 & K_{RR} & 0 \\ 0 & 0 & 0 & 0 & 0 & K_{YY} \end{bmatrix} \begin{matrix} \text{S (SURGE)} \\ \text{W (SWAY)} \\ \text{H (HEAVE)} \\ \text{P (PITCH)} \\ \text{R (ROLL)} \\ \text{Y (YAW)} \end{matrix} \qquad (12)$$

2.2 MASS MATRIX

The total mass of the structure [M] consists of structural mass and added mass. The added mass consist time invariant added mass upto mean water level and added mass due to variable submergence of the platform. The mass matrix is diagonally populated consisting of total mass w.r.t. X, Y and Z axes and moment of inertias about the same axes.

3. EQUATION OF MOTION

Dynamic equilibrium of inertia, damping, restoring and exciting forces are described as

$$[M]\{\ddot{x}\} + [C]\{\dot{x}\} + K\{x\} = \{F(t)\} \qquad (13)$$

where, $\{x\}$, $\{\dot{x}\}$ and $\{\ddot{x}\}$ are structural displacement, velocity and acceleration vectors and $\{F(t)\}$ is exciting force vector.

The damping matrix [C] is estimated as

$$[C] = \alpha\,[M] + \beta\,[K] \qquad (14)$$

where, α and β are constants whose appropriate values are cosidered.

The time varying non linear force in equation (16) consist of wind and wave induced forces. Hydro-dynamic force consisting of drag and inertia components has been estimated by Modified Morision's equation as under

$$F = 0.5\,\rho_w C_D D U_r |U_r| + \frac{\pi}{4} D^2 \rho_w [\,C_m(\dot{u}_w) - (C_m - 1)\ddot{x}\,] \qquad (15)$$

in which ρ_w is the mass density of water, C_D and C_m are hydrodynamic drag and intertia coeifficients, F is Hydro dynamic force per unit length of TLP's cylinderical structure, D is Dia of the cylinder, U_r is the relative velocity between cylinder and water and $\dot{u}_w$ is water particle acceleration.

Water particle velocity and acceleration due to random sea state are determined by Monte Carlo Simulation of Pierson Moskowitz Long Crested Sea Spectrum.

The total aero-dynamic load acting on the platform can be expressed as

$$F_u(t) = 0.5\ \rho_a C_a A_a \left[\overline{u(z)} + u_{eq}(t) - \dot{x}(t) \right]^2 \tag{16}$$

in which ρ_a denotes air density, C_a is aerodynamic drag coefficient, A_a is the projection of wind exposed area of the platform on a plane normal to the mean wind speed, $\overline{u(z)}$ is mean wind speed at an elevation z and u_{eq} is the fluctuating wind speed.

4. AERODYNAMIC LOADING

The wind induced loading is estimated due to mean and fluctuating wind components. The mean wind speed description is assumed to be governed by logarithmic law as under

$$\overline{u(z)} = \overline{u(z_{ref})} * l_n\left(\frac{z}{z_o}\right) \Big/ l_n\left(\frac{z_{ref}}{z_o}\right) \tag{17}$$

The roughness length (z_o) over the sea surface is usually provided by specifying the value of sea drag coefficient (C_D) given as

$$C_{D_{sea}} = k \Big/ l_n\left(\frac{10}{z_o}\right) \tag{18}$$

in which K = Von Karman constant, its value is taken as 0.4

The equation (18) gives a conservative estimate of the mean wind. It is due to the presence of water waves, the wind profiles are distorted at low wind velocities.

The fluctuating wind speed u'(y, z, t) is assumed to be governed as sigle point description and random variation of velocity fluctuation is assumed to be coherent all through the projected area. This pulsating velocity component is estimates by Fourier synthesis of Emil Simiu's wind spectrum (Fig. 4) for longitudinal velocity fluctuations whose detailed expressions are given in Appendix I of reference [4].

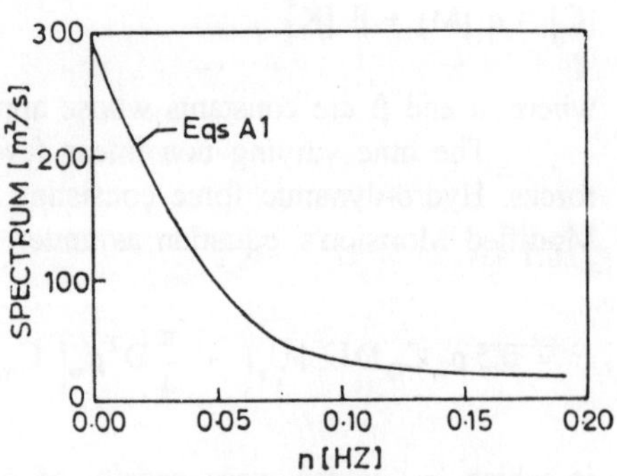

Fig 4- Emil Simiu's Wind Spectrum

To estimate the pulsating wind velocity component using Monte Carlo technique the total area of the Emil Simiu's spectrum is divided into finite number of strips. Each strip of width Δn_i and ordinate $S_u(z, n_i)$ represents the energy content of sinusoidal wave whose velocity (u_{eqi}) is given as

$$u_{eqi} = \sqrt{2 * s_u(z, n_i)\Delta n_i} \tag{19}$$

The spectral ordinate for the ith frequency (n_i) may be obtained using equation (A1) of Appendix I of reference [4].

The summation of velocities of k such strips approximates the instantaneous velocity. All these frequencies (n_i ; $i \equiv 1, \ldots, k$) are at random phases apart; expressed as follows

$$u'_{eq} = \sum_{x=1}^{k} u_{eq} \cos(2\pi n_i t + \varphi_i) \tag{20}$$

The phase angle Φ_i is generated by random sampling from an uniform distribution in the interval $0 < \Phi_i < 2\pi$. This distribution may be obtained by the generation of random numbers. Finally the time history of velocity fluctuations may be generated by equation (20)

5. SOLUTION TECHNIQUE

TLP is a highly non-linear system subjected to random wind and wave loadings. The solution of governing equation of motion, therefore, demands a time integration scheme taking care of all the the non-linearities involved. Presently New Mark's- β time integration technique is employed. The integration procedure iteratively analyses the entire structure at every time station in order to estimate the complex loading realistically. The time step Δt is judiciously chosen taking into account the natural time periods in all the six modes are given as follows

Surge and Sway	92.5 seconds
Heave	2.0 seconds
Pitch and Roll	2.2 seconds
Yaw	72 seconds

6. NUMERICAL STUDIES

For numerical studies the geometrical and mechanical characteristics of TLP together with aerodynamic and hydro-dynamic characteristics are given follows.

6.1 Geometrical and Mechanical Characteristics

Tether length	590.0 m
Tether diameter	0.40 m
Axial stiffness	5806 t/m
Weight of tether	535 t
Radius of Gyration along x,y Rx = Ry	29.15 m
Radius of Gyration along z axis = Rz	32.1m
Dia.of Corner Column	14.2 m
Dia.of horizontal cylinder	11.2 m

Hull width, Height, Length (m)	8.3, 11.2, 44.1
Centre to Centre spacing	58.3 m
C.G.	16.17 m above SWL
Mass of Platform	20950 t
Total Tether pre-tension	33400t

6.2 Hydrodynamic and Aerodynamic Characteristics

Drag Coefficient C_D	0.8
Inertia Coefficient C_m	1.8
Mass density of sea water	1030 kg/m^3
Random sea spectrum	Pierson Moskwitz
Zero up crossing period T_z	15 seconds
Significant water height H_2	15 metre
Coefficient of wind Drag $C_{D_{sea}}$	0.002
Mean wind speed $\overline{u(z)}$	45 m/sec
Mass density of air ρ_a	1.25 Kg / m^3
Total equivalent area for wind load	2572 m^2
Coordinate of aerodynamic centre w.r.t. C.G.	Z = 17.17 m
	X = -3.23 m
Length of Longitudinal turbulence (L_u)	180 m
Emil Simiu's sea site wind spectrum :-	
ß	6.0
f_m	0.07
f_s	0.2
Decay coefficient (c_y)	16.0
Von Karman constant = K	0.4
z_o	0.0001266 m
z	35 m
u(35)	45 m
u_*	1.76 m/s

7. RESULTS AND DISCUSSIONS

A typical TLP platform of size 67m x 67m x 64m for water depth of 600m has been considered. The mass of the platform is 20.95 x 10^6 Kg and total cable tension is 3.34 x 10^5 KN. Due to the stochastic nature of the response, results are presented by the statistical characteristics in the following tables (displacements are in m and rotations are in radian)

Table 1. Comparison of Uncoupled Response Under Random Wind and Long Crested Sea

Mode	Mean + Random Wind + Long Crested Sea (Case - 1 A) α = 0, VS = 0, Uncoupled				Long Created Sea (Case - 1.B) α = 0, VS = 0, Uncoupled			
	Max	Min	Mean	S.D.	Max	Min	Mean	S.D.
Surge	17.49	8.00	11.65	2.24	12.05	-10.36	0.0032	3.742
Sway	-	-	-	-	0	0	0	0
Heave	0.068	-0.0678	-0.00013	0.025	0.0277	-0.034	-0.0015	0.00981
Pitch	0.0107	-0.0106	-0.0001	0.0043	0.0024	-0.0026	0.00	0.00074
Roll	-0.0026	-0.0026	-0.0001	0.00034	-	-	-	-
Yaw	-0.2755	-0.3302	-0.03	0.157	-	-	-	-

Table 2. Comparision of Coupled Response Under Random Wind and Long Crested Sea

Mode	Mean + Random Wind + Long Crested Sea (Case - 2 A) α = 0, VS = 0, Coupled				Long Crested Sea (Case - 2.B) α = 0, VS = 0 ; Coupled			
	Max	Min	Mean	S.D.	Max	Min	Mean	S.D.
Surge	17.48	6.00	11.65	2.35	12.05	-10.26	0.026	3.845
Sway	-	-	-	-	-	-	-	-
Heave	0.022	-0.0355	-0.140	0.058	0.11418	-0.1666	0.0033	0.0029
Pitch	0.006	-0.007	-0.0005	0.002	0.0024	-0.00289	0.0	0.00078
Roll	0.0033	-0.0034	-0.0001	0.00014	-	-	-	-
Yaw	0.337	-0.410	-0.033	0.170	-	-	-	-

Table 3. Comparison of Coupled Response Under Random Wind & Long Crested Sea With Variable Submergence (VS)

Mode	Mean + Random Wind + Long Crested Sea + VS Case - 3 A) α = 0, VS = 1 , Coupled				Long Created Sea (Case - 3.B) α = 0, VS = 1 , Coupled			
	Max	Min	Mean	S.D.	Max	Min	Mean	S.D.
Surge	17.48	6.00	11.638	2.36	18.41	- 14.3	0.332	4.7
Sway	-	-	-	-	-	-	-	-
Heave	0.022	-0.0354	-0.149	0.058	0.228	-0.367	0.0039	0.054
Pitch	0.0055	-0.006	-0.00047	0.0023	0.0025	-0.0034	0.00	0.0008
Roll	0.0033	-0.0034	-0.000	0.0014	-	-	-	-
Yaw	0.3275	-0.406	-0.036	0.173	-	-	-	-

As in the realistic sea environment TLP behaviour in the present study is investigated for mean and random wind together with Long crested wave loading (Case A). In order to study the wind induced response, results in case A are compared with results due to random wave loading (Case - B) and described as follows

Effect of Coupling

Generation of stiffness matrix as described above takes into account the coupling of various degrees of freedom. The heave response is significantly influence by the coupling as shown in case 1A and 2A of table 1 and 2 respectively. The heave is -0.0678 metre in uncoupled case while it is -0.355 metre in coupled case, roughly of the order of 80%. The negative heave indicates the effect of large surge offset of TLP under the wind that causes large coupled submergence of the platform. Infact the mean wind pushes the structure off the zero position to a significant offset. This is the new position about which TLP oscillates due to fluctuating wind and wave. The standard deviation (SD) of coupled heave response is almost the double of that due to uncoupled one. This result is also shown in Fig. 7, 8 by comparison of Probability Spectral Density Function (PSDF) of heave response time histories for case 1A and case 2A. Energy content of heave response at different frequencies varies showing small fluctuations due to nonlinearity involved. The maximum vertical load is attracted at the frequency of 0.15 rad/sec (Fig. 7, 8). The heave is most significantly influenced as it is coupled with all the degrees of freedom (See stiffness matrix derivation).

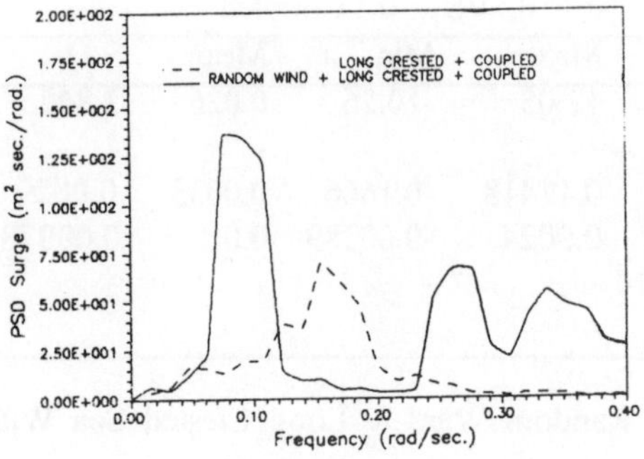

Fig–5 EFFECT OF WIND ON COUPLED SURGE RESPONSE

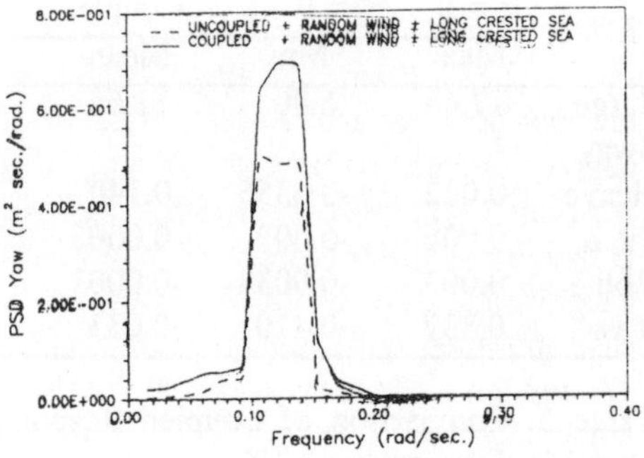

Fig6 EFFECT OF COUPLING ON WIND INDUCED YAW

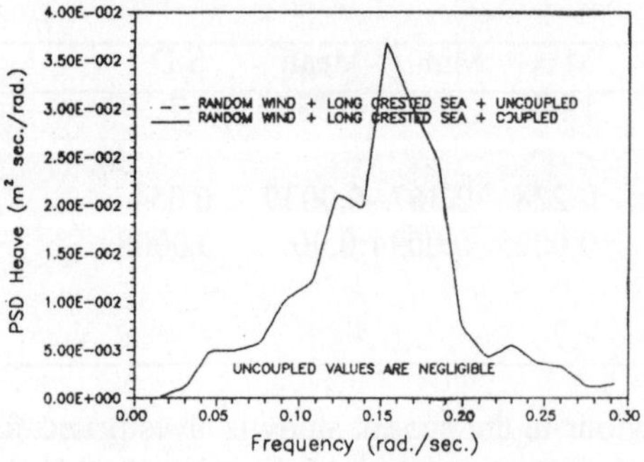

Fig–7 EFFECT OF COUPLING ON WIND INDUCED HEAVE

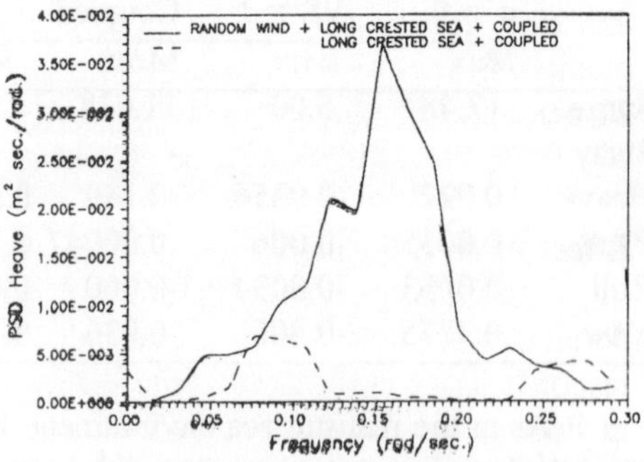

Fig8 EFFECT OF WIND ON HEAVE RESPONSE

Effect of Variable Submergence

Effect of VS is insignificant in the presence of wind loading except in the case of heave response (Fig. 10).

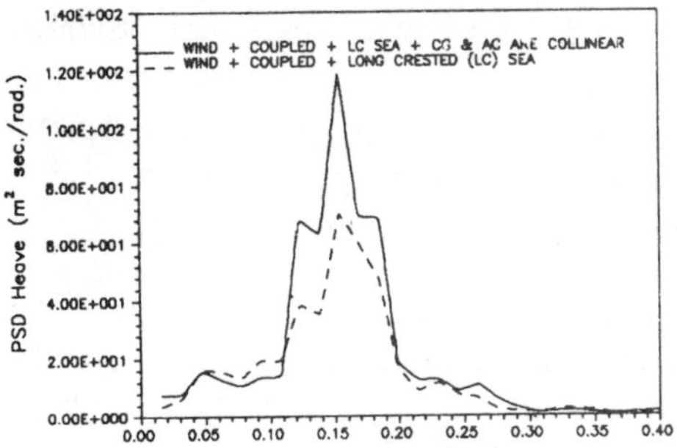

Fig9 EFFECT OF COUPLING ON WIND INDUCED SURGE (Aerodynamic Centre & C.G. are collinear)

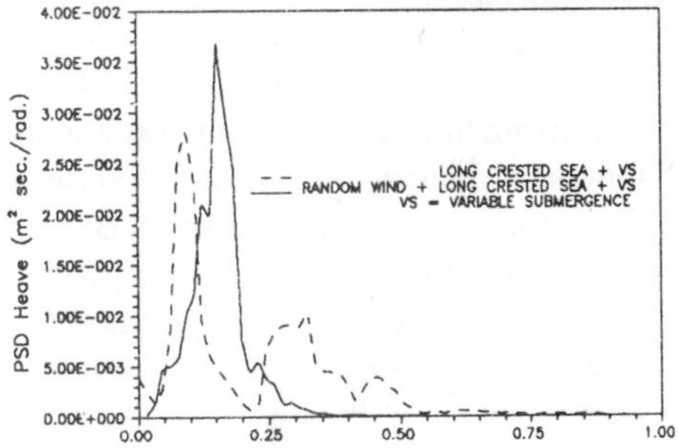

Fig-10 EFFECT OF WIND ON COUPLED HEAVE RESPONSE

Effect of Eccentricity of Aerodynamic Centre

The TLP super structure is idealised as shown in Fig. 2. The projected area is such that the aero dynamic centre (A.C.) is horizontally and vertically off the centre of gravity. All the six degrees of freedom refer to the C.G. Thus wind induced forces and moments are also refereed w.r.t. the C.G. These forces are at zero degree of incidence to the TLP structure. Sway response is therefore, insignificant as shown in Table 1, 2 and 3. Due to the location AC w.r.t the C.G. a significant Yaw motion is activated. The maximum value of yaw is 18.8 degrees in uncoupled case and goes upto 23 degrees for coupled degrees of freedom as given in Table 1 and 2 (Case 1A and 2A). Moments about X and Y axes (Roll and Pitch) are also activated but by a small amount. The pitching moments due to wind and waves are time varying and opposite in nature leading to reversal of stresses. That is why wind induced moments are important from fatigue point of view. The same tables (Case A and B) show that surge response is significantly attenuated (S.D. reduced from 3.74 to 2.24) when wind acts at the given aerodynamic centre together with long crested random sea. The same findings may be strengthened by studying the response under wind and wave loading when A.C. and C.G. both are lying on Z-axis (Fig. 9) in this case the entire wind force gets added up the wave induced surge force. No significant moment is activated. The total energy content is used to induce surge response that increases by 21 per cent.

Effect of Non-linearities

In the present study large mean wind velocity (45 m/sec.) with its fluctuating components dominates the major non-linearity due to VS. That otherwise has significant effect on heave response as given in table 2 and 3, Case 2B and 3B respectively. There is 50 percent increase in heave response due to the inclusion of VS effect. Furthermore, non-linearities like tether tension variation and drag cause small fluctuations in the response behaviour (Fig. 5, 6, 7, 8, 9 and 10). These small fluctuations may get enhanced if its frequency synchronizes with the frequency of some large force component. Super and Sub-harmonic responses may be activated in such cases. Fatigue life may seriously be effected due to the non-linear behaviour of TLP.

8. CONCLUSIONS

The results of study lead to the following conclusions

1. The magnitude of the wind force and the location of aerodynamic centre significantly influence the surge and yaw responses of TLP under long crested random sea.
2. The coupling of various degrees of freedom substantially enhances the heave response.
3. Wind induced pitch and roll responses are normally well within the permissible limits however, their cyclic nature is important from the fatigue point of view.
4. Emil Simiu's wave spectrum is found to be well suited for low frequency response assessment of TLP.

9. REFERENCES

1) John, A.M. and Steven, J.L. (1982), "Evaluation of Hutton TLP rsponse to Environmental Loads", Proceedings of Offshore Technology Conference, Huston, Texas, May 3-6, 1982.

2) Kareem, A. (1985),"Wind Induced Response Analysis of TLP",Journal of Structrual Engineering January- 1985, Vol-111, No.1.

3) Salveen et al. (1982),"Computations of Nonlinear Surge Motions of TLP", Proceedings of 14th Annual Offshore Technology Conference, Huston, Texas, May 3-6, 1982.

4) Simiu, E. and Stefan, D.L. (1984),"Turbulent Wind and TLP Surge" Journal of Structural Engineering, April-1984,Vol. 110,No.4.

5) Vickery, B.J. (1982)," Wind Loads on Compliant Offshore Structures", Procedings of Ocean Structural Dynamics Symposium, Oregon State University, Dept. of Civil Engg., Corvallis, Oreg, Sept 1982.

APPENDIX I of reference [4]

$$\frac{nS_u(z,n)}{u_*^2} = \begin{cases} a_1 f + b_1 f^2 + d_1 f^3 & 0 < f \le f_m) \\ c_2 + a_2 f + b_2 f^2 & f_m < f < f_s \\ 0.26\, f^{-2/3} & f \ge f_s \end{cases} \tag{A1}$$

Abbreviations used

A.C. Aerodynamic Centre
w.r.t. with respect to
VS variable submergence effect.

9ICWE, 1995, New Delhi, India

OVALLING VIBRATIONS OF CANTILEVERED CYLINDRICAL SHELLS: EXPERIMENTAL RESULTS AND A STABILITY CRITERIA

MAZOUZI, A. and LANEVILLE, A.

Département de génie mécanique, Université de Sherbrooke, Sherbrooke, Québec, Canada, J1K 2R1

Abstract

This article deals with an experimental investigation of the ovalling oscillations of clamped-free cylindrical shells exposed to a two-dimensional cross-flow. A new criteria including the Scruton number is proposed to predict the onset velocity of the instability.

1. INTRODUCTION

Data on the ovalling oscillations of cantilevered cylindrical shells in cross-flow are very few and, in addition, do not allow to develop a stability criteria based on damping, the shell modal damping value being unavailable in most of the data sets. Cylindrical shells are light structures susceptible to the phenomenon of ovalling, an instability resulting from the reduction of the shell modal damping by the aerodynamic damping generated by the flow. In this paper experimental data are presented in the case of two-dimensional cantilevered shells and an empirical criteria to predict the ovalling onset flow velocity is proposed. This criteria is based on a balance between the negative aerodynamic damping and the shell structural damping.

2. EXPERIMENTAL SET-UP

The experimental investigation was performed in the low-speed wind-tunnel laboratories of the Université de Sherbrooke. The testing of the different shells included two phases; (i) the experimental determination of the shell modal damping and frequency in still air using a mechanical excitation and (ii) in the wind tunnel, the measurement of the shell response to different air velocities as well as the associated vortex shedding frequency. The shell vibration modes (n,m) (circumferential,axial) were identified and recorded in both phases.

Five cylindrical shells with identical diameter (D=120mm) were manufactured from thin aluminiun foil with constant thickness (h=D/944) but different span (3D ≤ L ≤ 7.5D) to be exposed to the flow of a closed circuit wind tunnel in its 1,2m x 1,2m test section. Figure 1 shows a schematic view of the experimental set-up. The cylindrical shell was clamped at its base to a turntable (external tunnel balance) isolated from the tunnel floor and free at its top, the proximity of which was maintained within 0,02D from an adjustable false ceiling in order to reduce tri-dimensionnal effects. The shell clamped portion extended 10 cm over the wind tunnel floor, well away from the boundary layer.

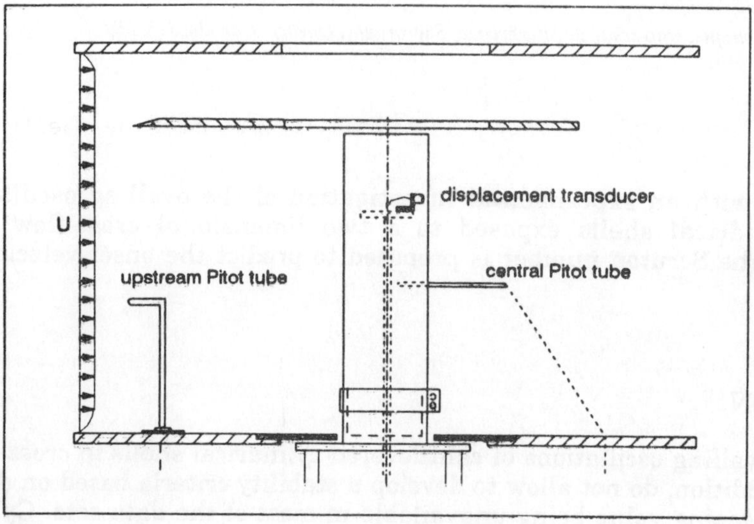

Figure 1. Schematic of the cylindrical shell mounted in the wind tunnel.

Two Pitot tubes were used, the first, located upstream of the shell, to measure the oncoming undisturbed air velocity, and the second, positioned outside the shell wake, vis-à-vis the shell central axis, to measure the air velocity and the vortex shedding frequency. Both Pitot tubes were connected to differential pressure transducers via flexible tubing, the length of which was chosen to preserve a sufficiently large dynamic response.

A non-contacting displacement transducer of the capacitive type was installed on a plateau inside the shell in order to define its response. The vibration mode (n,m) was identified by rotating and translating the plateau or by a direct visualization with a stroboscopic light through a window in the tunnel ceiling.

For each of the five shells, the modal damping was determined with the shell mounted in the wind-tunnel, under no wind loading. The shell remained with the same clamping conditions during the testing. A small shaker, linked to the shell at its base near the clamping area, was computer controlled to apply a constant energy level excitation, and to insure a high precision for the excitation frequency. The shell response was detected with the displacement transducer and its analog output, fed to the computer via the data acquisition system.

The ability of a light structure such as a cylindrical shell to damp flow induced vibrations can be estimated using the Scruton number (Sc_{nm});

$$Sc_{nm} = 2\pi \frac{\rho_s}{\rho} \frac{h}{D} \delta_{nm} \tag{1}$$

where δ_{nm} is the logarithmic decrement for the mode (n,m), n,m are the circumferencial and the axial wave numbers respectively, ρ_s, the shell density and ρ, the fluid density.

The shell air loading flow conditions can be characterized by the following criteria:

Reynolds number Re $\leq 1.5 \times 10^5$
Turbulence intensity < 0,5%
Solid blockage = 10%

For each shell, the testing approach consisted in increasing progressively the air velocity and then measuring the shell response as well as the shedding frequency. As the shell entered in a given vibration mode, the corresponding air velocity was defined as its threshold velocity, U_{Thr}. The first threshold velocity encountered by the shell was defined U_{cr}, the critical onset ovalling velocity.

3. EXPERIMENTAL RESULTS

In the tests, the shells developed vibrations in the axial mode (m=1) for different circumferential modes ($2 \leq n \leq 5$), switching from one circumferential mode to another as the wind velocity was varied. At the onset of ovalling, i.e. at the air speed ($U_{Thr}=U_{cr}$) for which any given mode appears, the circumferential modes are not coupled and the mode (3,1) is the most frequent. This behaviour has also been observed for shells made of different materials, with different geometries (L/D and h/D) and with different end conditions by Païdoussis et al.(1982a, 1988). This observation allows for the hypothesis of uncoupled circumferential modes at the onset of ovalling. For air speeds larger than U_{cr}, other circumferential modes can develop.

The shell modal frequencies do not differ by more than 2.5%, as shown in Table 1, whether exposed to the air flow or excited by the mechanical shaker. This slight difference can be attributed to a modification in the system rigidity (shell and enclosed air). The ratio, r, of the modal frequency to the vortex shedding frequency is not necessarily an integer as the shell enters in ovalling vibrations, i.e. at the threshold velocity ($U_{Thr} = U_{cr}$). For the subsequent modes occurring at a velocity $U > U_{Thr}$, the r value becomes closer to an integer as indicated in Table 1. The absence of lock-in and the poor matching between the shell modal frequency and any multiple of the vortex shedding frequency support again the conclusion that the subharmonic excitation by vortex shedding is a condition neither necessary nor sufficient for the onset of ovalling at U_{cr}.

Table 1. Experimental Data.

Shell no	Mode (n,m)	D [mm]	L [mm]	h [mm]	Mechanical experiments		Wind tunnel experiments			$\frac{U_{Thr}}{f_{nm} D}$	Sc
					δ_{nm}	f_{nm}[Hz]	f_{nm}[Hz]	U_{Thr}[m/s]	f_{nm}/f_v		
L1	(2,1)	120	910	0.127	0.098	47.0	47.0	7.8	4.00	1.38	1.43
•	(3,1)	"	"	"	0.040	64.5	64.5	5.0	8.06	0.64	0.58
L2	(2,1)	120	810	0.127	0.090	56.0	-	-	-	-	-
•	(3,1)	"	"	"	0.040	66.5	66.5	5.1	8.86	0.64	0.58
•	(4,1)	"	"	"	0.038	121.0	118.0	12.6	7.00	0.89	0.55
L3	(2,1)	120	610	0.127	0.080	96.0	-	-	-	-	-
•	(3,1)	"	"	"	0.048	76.6	76.5	6.6	7.75	0.72	0.70
•	(4,1)	"	"	"	0.026	122.5	120.5	9.5	8.88	0.66	0.38
L4	(3,1)	120	410	0.127	0.080	121.5	122.0	18.0	5.00	1.23	1.16
•	(4,1)	"	"	"	0.040	131.0	132.0	10.75	8.23	0.68	0.58
L5	(5,1)	"	360	0.127	*	*	195.0	18.0	*	*	*

Shell parameters: Aluminium foil; Young's modulus E=68,96 GPa; density ρ_s=2643 Kg/m^3; Poisson's ratio ν=0,3; fluid density ρ=1.204 Kg/m^3; (*) Denotes visualisation test only.

When two-dimensional conditions prevail, the aspect ratio L/D should influence only the shell modal dampings and frequencies, or the Scruton number. Since ovalling vibrations begin when the shell modal damping is dissipated by the energy from the flow field, δ_{nm} (or Sc_{nm}) is expected to correlate with U_{Thr} (or $U_{Thr}/f_{nm}D$). Figure 2, that includes all the data available for cantilevered two-dimensional cylindrical shells, confirms this hypothesis and reinforces the thesis proposed by Païdoussis and his collaborators (1982, 1988). It should be noted that the data for the onset of all ovaling modes are included. Three data points from Sharma & Jones (1970) as well as two data points from Suen (1980) have been added using some assumptions. A linear regression of the data gives the following equation:

$$\frac{U_{Thr}}{f_{nm}D} = 0.7\, Sc_{nm} + 0.3 \qquad (2)$$

The level of correlation of the data 0.9, is fairly high.

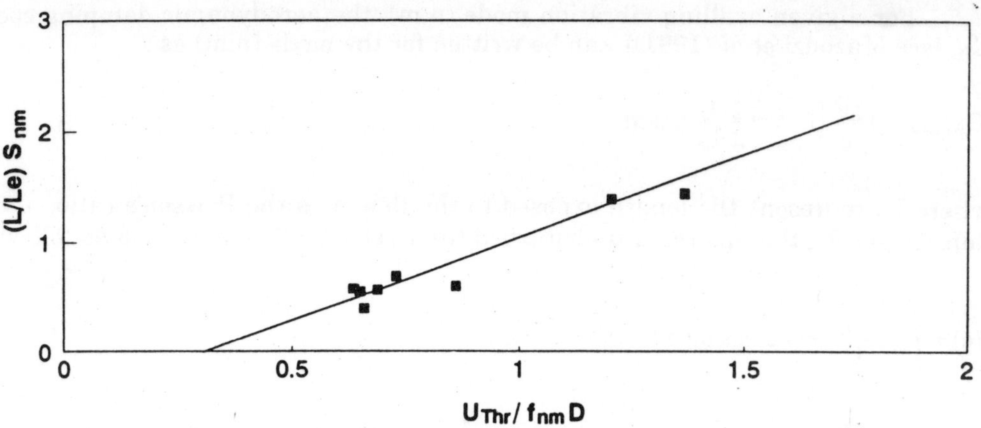

Figure 2. Variation of the non-dimensional flow velocity $U_{Thr}/f_{nm}D$ as a function of the Scruton number Sc_{nm}. The cicumferencial mode is identified in Table 1.

4. EMPIRICAL CRITERIA

The earliest criteria for the ovalling onset, set by the British Standard Institute (1966), is based on a matching between the Strouhal frequency and one of the shell modal frequencies. The resultant criteria corresponds essentially to a modified Strouhal number for a static circular cylinder. A second criteria, proposed by Païdoussis & Helleur (1979) and extended by Katsura (1985) links the threshold flow velocity for the instability onset, U_{Thr}, to the shell geometrical dimensions (h,D) and material properties (E,ν,ρ_s). Uematsu & Uchiyama (1985) offer a third criteria that ties U_{Thr} to the wall pressure fluctuations, the wall frequency oscillations and the phase between the two.

Since damping plays a critical role in a flutter type phenomenon as ovalling, the shell geometrical properties and natural frequencies are then believed to be only partial components of an instability criteria. An instability criteria based on the Scruton number (or modal damping) to predict the flow velocity at the onset of ovalling can be developed.

In the case of harmonic oscillations, the structural damping coefficient can be replaced by an equivalent viscoelastic damping coefficient (Ewins (1984)) for which the same amount of energy is dissipated by vibration cycle. It follows then :

$$(C_{eq})_{nm} = D\, h\, L\, \rho_s\, \delta_{nm}\, \omega_{nm} \tag{3}$$

where ω_{nm} represents the angular modal frequency of the shell.

For a given ovalling vibration mode (n,m), the aerodynamic damping coefficient, C_A, (see Mazouzi et *al* (1991)), can be written for the mode (n,m) as :

$$(C_A)_{nm} = \rho \frac{D^2}{h}\left(1 - \nu^2\right)\frac{2}{\pi}\frac{L}{L_e} U\, f(n) \tag{4}$$

where L_e represent the length exposed to the flow; ν is the Poisson's ratio, ρ, the fluid density and U, the upstream undisturbed fluid velocity; f(n) is defined as follows:

$$f(n) = \int_0^{\pi/2} \frac{V_\theta^o}{U} \sin n\theta \cos n\theta \, d\theta \tag{5}$$

The term V_θ^o represents the tangential steady flow velocity on the shell wall and can be described using the model proposed by Parkinson and Jandali (1969).

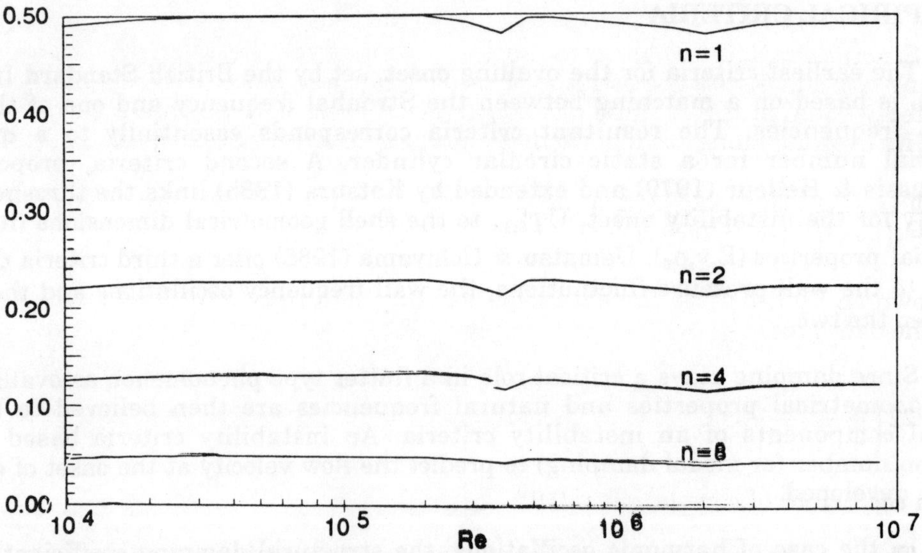

Figure 3. The aerodynamic function f(n) and its variation with Reynolds number and circumferential modes (n,1)

The function f(n) takes into account the combined influences of the Reynolds number and circumferential mode on the aerodynamic damping. With the observation that prior to ovalling onset ($U < U_{Thr}$) at any circuferential mode n, the cross-section of the shell remains circular (n=1), the aerodynamic damping coefficient becomes

$$(C_A)_{nm} = (C_A)_{1m} = \rho \frac{D^2}{h}(1 - \nu^2)\frac{2}{\pi}\frac{L}{L_e} U f(1) \tag{6}$$

and is independant of the mode n and a linear function of the flow velocity for a given shell. The function f(1) does not vary significantly for the range $10^4 \le Re \le 2 \cdot 10^5$ as shown in Figure 3.

The ratio of the damping coefficients is then:

$$\left(\frac{C_{eq}}{C_A}\right)_{nm} = \left(\frac{\pi}{2}\frac{h}{D}\frac{1}{(1-\nu^2)}\frac{1}{f(1)}\frac{L}{L_e}\right)\left(\frac{\rho_s}{\rho}\frac{h\ \delta_{nm}\ \omega_{nm}}{U}\right) \tag{7}$$

and at the onset of ovalling mode (n,m), the flow velocity corresponds to the threshold velocity ($U = U_{Thr}$) and the ratio of the damping coefficients becomes:

$$\left(\frac{C_{eq}}{C_A}\right)_{nm} = \left(\frac{\pi}{2}\frac{h}{D}\frac{1}{(1-\nu^2)}\frac{1}{f(1)}\frac{L}{L_e}\right)\left(\frac{S_{nm}}{(U_{Thr}/f_{nm}D)}\right) \tag{8}$$

Since equation (8) is essentially a comparison between the two modelled damping coefficients involved at the onset velocity, and since both real coefficients are admittedly equal at the ovalling onset for a given shell, with any end conditions and for any of its ovalling modes, it follows that equation (8) must be a constant, or

$$K_s = \left(\frac{h}{D}\frac{1}{(1-\nu^2)}\right)\left(\frac{L}{L_e}\frac{S_{nm}}{(U_{Thr}/f_{nm}D)}\right) \tag{9}$$

As shown on Figure 4, all the data available correlate well for different shells. A linear fit of the data gives

$$K_s = 1.22\left(\frac{h}{D}\frac{1}{(1-\nu^2)}\right) \tag{10}$$

with a coefficient of correlation of 0.96.

The resultant experimental boundary between stability and instability of the shell is then defined as:

$$\frac{L}{L_e} S_{nm} = 1.22\,(U_{Thr} / (f_{nm} D)) \tag{11}$$

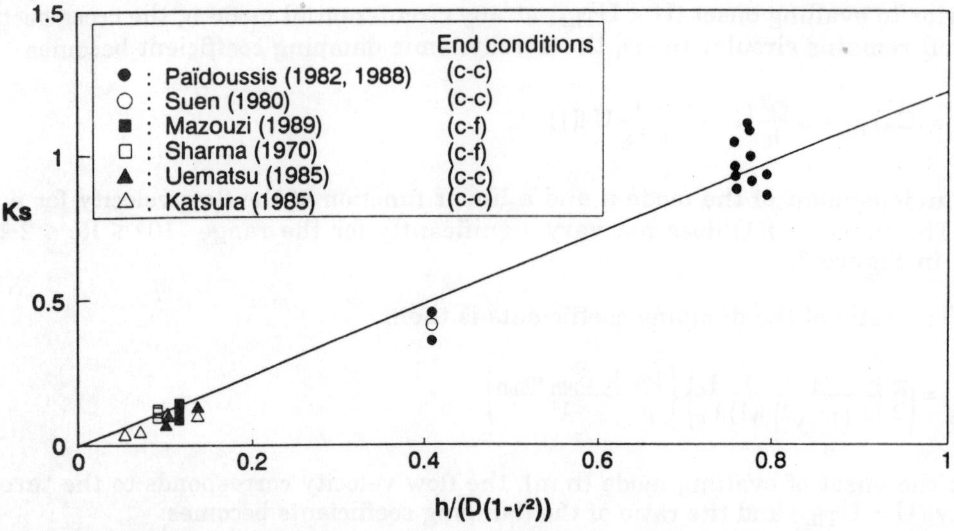

Figure 4: Comparison of the stability limit with the experimental data of clamped-free and clamped-clamped cylindrical shells.

For any shell, the region on the left of the boundary (Figure 5) is one of stability and is seen to include little experimental evidence of instability. The boundary is somewhat conservative. In lieu of equation (9), the following correlation

$$\frac{L}{L_e} S_{nm} = 1.45\,(U_{Thr} / (f_{nm} D)) - 0.43 \tag{12}$$

describes more accurately the experimental data with a correlation coefficient of 0.93 for all the shells (Figure 5). Once reorganised, equation (11) predicts the ovalling onset flow velocity according to:

$$U_{Thr} = f_{nm} D \left(0.7 \frac{L}{L_e} S_{nm} + 0.3 \right) \tag{13}$$

This last result indicates that a shell is unconditionally stable for any value of structural damping, even **null** damping if the flow velocity U is less than $0.3\, f_{nm}\, D$. In terms of vortex shedding frequency, the shell natural frequency f_{nm} should be at least 17 times larger than the vortex shedding frequency of the **stationary** shell in order to have an **unconditionally** stable shell.

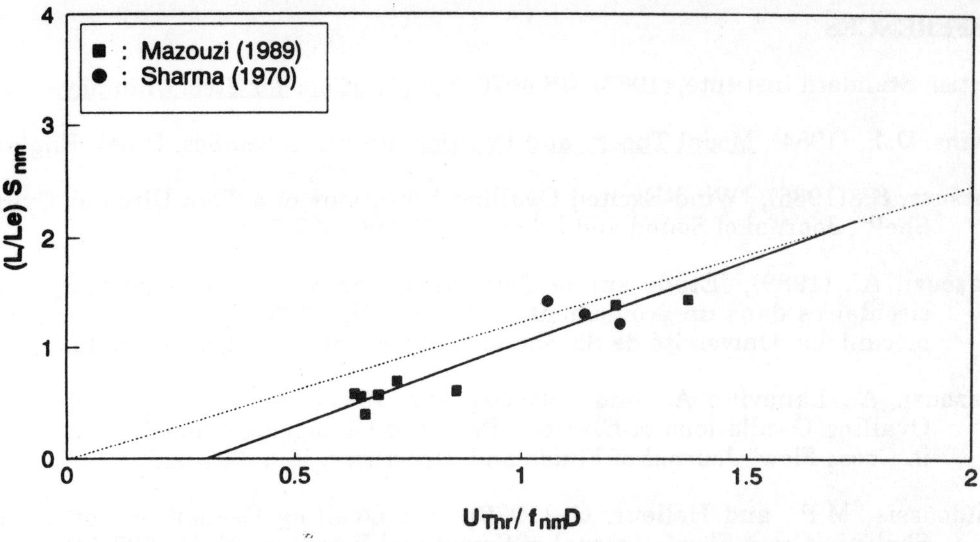

Figure 5. Variation of the non-dimensional flow velocity $U_{Thr}/f_{nm}D$ as a function of the Scruton number Sc_{nm}.

5. CONCLUSION

The results of an experimental investigation on the ovalling of five clamped-free cylindrical shells have indicated the followings:

- at the critical air velocity, a single circumferential mode is coupled to an axial mode. For subsequent threshold velocities, two circumferential modes can co-exist but usually this is not the case;

- at the critical air velocity, the shell modal frequency does not correspond to a multiple of the measured Strouhal frequency. For subsequent threshold velocities or ovalling modes, this correspondence is better;

- for shells exposed to a two dimensional flow, the Scruton number and the reduced ovalling onset velocity show a linear relationship with a high level of correlation.

- the prediction of the flow velocity for which the ovalling of the shell onsets is based on a balance between the shell structural damping and the negative aerodynamic damping. The non-dimensional onset flow velocity (U_{Thr} / f_{nm} D) is a linear function of the Scruton number S_{nm}.

- according to the data correlation, a cylindrical shell exposed to a cross-flow would be unconditionally stable for flow velocities less than 0.35 f_{nm} D.

REFERENCES

British Standard Institute, (1966), BS 4076, Specifications for Steel Chimneys.

Ewins, D.J., (1984), Modal Theory and Practice, Research Studies, Press, England.

Katsura, S., (1985), "Wind-Excited Ovalling Vibrations of a Thin Circular Cylindrical Shell", Journal of Sound and Vibration, V. 100, 527-550.

Mazouzi, A., (1989), "Étude sur le flottement transversal des coques cylindriques circulaires dans un écoulement uniforme", thèse Ph.D., Département de génie mécanique, Université de Sherbrooke, Sherbrooke (Québec), Canada.

Mazouzi, A., Laneville A., and Vittecoq, P., (1991) "An Analytical Model of the Ovalling Oscillations of Clamped-Free and Clamped-Clamped Cylindrical Shell in Cross-Flow" Journal of Fluids and Structures, V. 5, 605-626.

Païdoussis, M.P., and Helleur, C. (1979), "On Ovalling Oscillations of Cylindrical Shells in Cross-Flow", Journal of Sound and Vibration, V. 63, 527-542.

Païdoussis, M.P., and Wong, D.T. - M (1982a), "Flutter of Thin Cylindrical Shells in Cross-Flow", Journal of Fluid Mechanics, V. 115, 411-426.

Païdoussis, M.P., Price, S.J. and Ang., S.Y., (1988), "Ovalling Oscillations of Cylindrical Shells in Cross Flow : A Review and Some New Results", Journal of Fluids and Structures, V.2, 95-112.

Païdoussis, M.P., Price, S.J. and Suen H.C., (1982b), "An Analytical Model for Ovalling Oscillation of Clamped-Clamped Cylindrical Shells in Cross-Flow", Journal of Sound and Vibration, V. 83, 555-572.

Parkinson, G.V. and Jandali, T. (1969), "A Wake Source Model for Bluff Body Potential Flow". Journal of Fluid Mechanics 40, 577-596.

Sharma, C.B. and Johns, D.J., (1970), "Wind-Induced Oscillations of Circular Cylindrical Shells: An Experimental Investigation", Loughborough University of Technology Report TT7001.

Suen, H.-C., (1980), "Flutter of Thin Shells in Cross-Flow", M. Eng. Thesis, Dept. of Mechanical Engineering, McGill University, (Québec), Canada.

Uematsu, Y. and Uchiyama, K., (1985), "An Experimental Investigation of Wind-Induced Ovalling Oscillations of Thin Circular Cylindrical Shells", Journal of Wind Engineering and Industrial Aerodynamics, V. 18, 229-243.

9ICWE, 1995, New Delhi, India

DETERMINATION OF AERODYNAMIC COEFFICIENTS OF DIFFERENT RECTANGULAR PROFILES FOR WIND INDUCED EXCITATION

ZILLER, C.

Institute for Steel Construction, Department of Wind Engineering, RWTH Aachen, Germany

Abstract

The aerodynamic instabilities such as vortex-induced oscillation, galloping and flutter of different rectangular profiles should be investigated in order to assure the safety of buildings and structures containing those elements.

The present study provides aerodynamic coefficients for the calculation of vibrations caused by galloping oscillations either in the vertical mode or torsional mode in natural wind. The investigated section types are rectangular slim building components (l/d ≥ 10) and they correspond to forms common in practical use. By performing wind tunnel tests, stability parameters such as lift and moment coefficients are determined. Freely vibrating section models are used for the wind tunnel tests. The coefficients for various on-flow angles, for variations of the degree of turbulence and for different geometric dimensions of the profiles are measured on these. Within this study the nonlinear relation of the aerodynamic damping of a system and the wind velocity is demonstrated and discussed.

Comparisons between observed torsional flutter responses of existing structures and the results of the section model tests are made. A case of a crane damage and of a failure of a coal mining machine is reported. In both cases the failure was caused by torsional mode vibrations of the pendant bars (long and slender steel bars) in natural wind. Calculations with the determined aerodynamic coefficients for galloping especially in the torsional mode on slim building components such as the pendant bars are made and compared to the observations and to the results of the full scale measurements made on the damaged structures.

Notation

b	width of the profile
C	spring constant
d	depth of the profile
F(t)	lateral component of aerodyn. force
f_e	natural frequency
I	turbulence intensity
k	coefficient of viscous damping
l	length of the profile
M(t)	aerodynamic moment
m	mass
t	time
u_{oo}	wind velocity (undisturbed flow)
u_0	onset velocity
u_{red}	reduced wind velocity ($u_{red} = u_{oo}/f_e*d$)
y	lateral displacement

α on-flow angle
δ_a log. aerodynamic damping
δ_s log. structural damping
φ torsional displacement
ρ air density
σ_y^* stability parameter bending mode
σ_φ^* stability parameter torsional mode

1. INTRODUCTION

Slim structural designs are known to be sensitive to wind induced vibrations such as vortex shedding, galloping and torsional flutter. Such vibrations have repeatedly led to damage so that reconstruction has to be undertaken. Some buildings and their slim structural parts are beeing built more and more frequently with material saving and filigree methods. This increases the susceptibility to vibration and the probability of wind-excited vibrations occuring becomes larger. To predict the response of those structures to aerodynamic instability effects and to assure their safety it is necessary to perform wind tunnel tests.

The aim of this study is to determine aerodynamic coefficients which describe the response of slender rectangular structures in natural wind. The influence of the degree of turbulence, the geometric dimensions of the profiles and of the on-flow angle on those aerodynamic instability effects is determined and discussed. Galloping vibrations are dependent on turbulence to a considerably greater extent compared to the influence of turbulence effects for vortex induced vibrations. It is tried to show a tendency. But whether a profile is stabilized or destabilized in turbulent flow is highly dependent on the section shape and on the mode of vibration.

The wind tunnel tests are performed using freely vibrating section models. It is determined that the measured stability parameters for galloping vibrations partly deviate from the known statically determined coefficients. This shows that the aerodynamic force terms which are linked to the vibration movement of the system may not be neglected. The available static coefficients usually lead to conservative results. But there are exceptions in which these static coefficients identify a system as stable although it can become unstable. For this reason the stability coefficients should only be determined from aeroelastic model tests.

2. THEORETICAL BACKGROUND

Vertical Mode Oscillations:

Figure 1 shows an arbitrarily shaped two-dimensional bluff cylinder with a linear spring and viscous damping restrained to a translational degree of freedom in a smooth uniform flow with the velocity u_{∞}. The system is a subject to a vertical aerodynamic force F(t).

The equation of motion for the system can be written as:

$$m\,\ddot{y} + k_y\,\dot{y} + C_y\,y + F(t) = 0$$

$$F(t) = \frac{\rho}{2}\,u_{\infty}^2\,d\,c_y(\alpha) \qquad (1)$$

To include the experimental determined steady lift coefficient $c_y(\alpha)$ in an analytical way, the following polynomial form can be used:

$$c_y(\alpha) = \sum_{n=1}^{N} K_n \left(\frac{\dot{y}}{u_{\infty}}\right)^n \qquad (2)$$

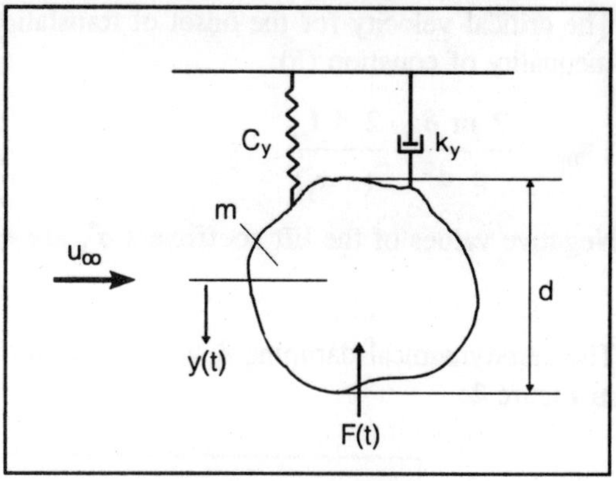

Figure 1 Aerodynamically excited motion of a bluff body in translational degree of freedom

In the instability point where the aerodynamic vibrations start off and the amplitudes of oscillations are small, it is possible to use the following linearized relation:

$$K_1 = \frac{dc_y}{d\alpha} = \sigma_y^* \qquad (3)$$

With $\omega_{ey} = 2\,\pi\,f_{ey} = \sqrt{\frac{C_y}{m}}$ and $\delta_{sy} = \frac{k_y}{2\,m\,f_{ey}}$ the equation of motion then becomes:

$$\ddot{y} + \left(\delta_{sy} + \frac{\rho\,u_{\infty}}{4\,m\,f_{ey}\,d}\,d^2\,\sigma_y^*\right) 2\,f_{ey}\,\dot{y} + \omega_{ey}^2\,y = 0 \qquad (4)$$

The model will be unstable in translational oscillations if the damping term in equation (4) is smaller or equal than zero:

$$\delta_{sy} + \frac{\rho\,u_{\infty}\,d^2}{4\,m\,f_{ey}\,d}\,\sigma_y^* \le 0 \qquad (5)$$

where the aerodynamic damping is:

$$\delta_{ay} = \frac{\rho\,d^2}{4\,m}\,\frac{u_{\infty}}{f_{ey}\,d}\,\sigma_y^* \qquad (6)$$

The critical velocity for the onset of translational galloping vibrations can be determined from the inequality of equation (5):

$$u_{0y} = \frac{2\ m\ \delta_{sy}}{\rho\ d^2}\ \frac{2\ d\ f_{ey}}{(-\ \sigma_y^*)} \tag{7}$$

Negative values of the lift coefficient σ^*_y are essential for instability.

The aerodynamical damping δ_a is a function of the on-flow velocity u_{oo} and is shown in principle in Figure 2:

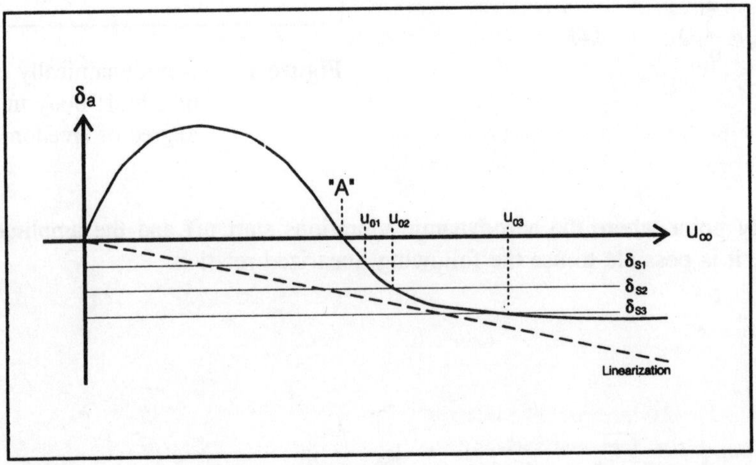

Figure 2 Aerodynamic damping δ_a as a function of the on-flow velocity u_{oo}

This relation has to be determined with experimental p·.cedures and it shows the nonlinear character of the onset of galloping oscillations. For different structural damping values δ_s of a vibrating system the curve or the x-axis have to be shifted as shown in Figure 2. This nonlinear relation is the reason for the following observation: an increase of the structural damping from δ_{s1} to δ_{s2} leads to an insignificant change of the onset velocity u_0 but for another increase of δ_s with the same step size (from δ_{s2} to δ_{s3}) the onset velocity is changing significantly.

In order to determine the instability parameter σ_y^* using equation (6) it is necessary to take this nonlinear relation of δ_a and u_{oo} into account. A very conservative method for the determination of this instability parameter (which is the method used to determine the instability parameters within this investigation) is to linearize the relation as shown with the dashed line in Figure 2. The linearized equation can be written as:

$$\delta_{ay} = (-\ n)\ u_{oo} \tag{8}$$

where (-n) is the negative slope of the line.

Equation (6) then becomes:

$$\sigma_y^* = \frac{4\ m\ f_{ey}}{\rho\ d}\ (-\ n) \qquad (9)$$

Below the point "A" in the diagram of Fig. 2 the linearization indicates instability while the real nonlinear curve shows that in this range the profile is stable. Thus, the linearization has to be handled with care.

Torsional Mode Oscillations:

The same analytical considerations done for the vertical mode oscillations have to be made for the torsional mode oscillations. The procedure is the same and should be therefore shown briefly.

The equation of motion for a system restrained to a torsional degree of freedom as shown in Figure 3 is:

$$\Theta\ \ddot{\varphi} + k_\varphi\ \dot{\varphi} + C_\varphi\ \varphi + M_\varphi(t) = 0$$

$$M_\varphi(t) = \frac{\rho}{2}\ u_{oo}^2\ d^2\ c_M(\alpha) \qquad (10)$$

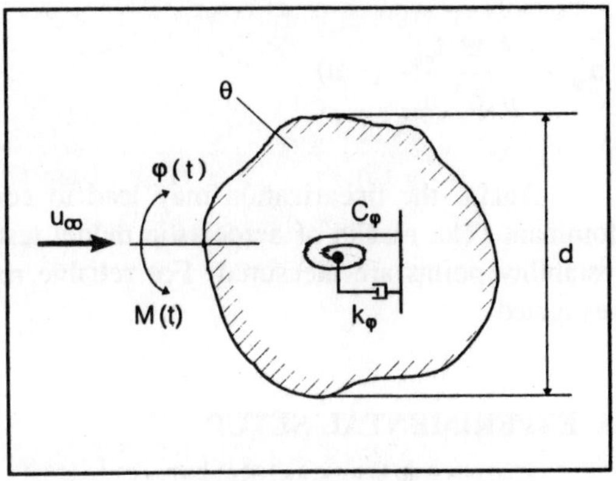

Figure 3 Aerodynamically excited motion of a bluff body in torsional degree of freedom

By following the procedure of linearization for the instability point for the vertical mode of oscillation (equation (3)) and with:

$$\omega_{e\varphi} = 2\ \pi\ f_{e\varphi} = \sqrt{\frac{C_\varphi}{\Theta}}$$

$$\delta_{s\varphi} = \frac{k_\varphi}{2\ \Theta\ f_{e\varphi}}$$

the equation of motion (equation (10)) becomes:

$$\ddot{\varphi} + \left(\delta_{s\varphi} + \frac{\rho}{2}\ \frac{d^3\ u_{oo}}{2\ \Theta\ f_{e\varphi}}\ \sigma_\varphi^*\right) 2\ f_{e\varphi}\ \dot{\varphi} + \omega_{e\varphi}^2\ \varphi = 0 \qquad (11)$$

The point of instability is reached when:

$$\delta_{s\varphi} + \frac{\rho\ d^3\ u_{oo}}{4\ \Theta\ f_{e\varphi}}\ \sigma_\varphi^* \leq 0 \qquad (12)$$

and the last term in this equation is the aerodynamical damping for the torsional mode of vibration, $\delta_{a\varphi}$:

$$\delta_{a\varphi} = \frac{\rho\ d^3\ u_{oo}}{4\ \Theta\ f_{e\varphi}}\ \sigma^*_\varphi \tag{13}$$

Torsional mode galloping starts off when:

$$u_{0\varphi} = \frac{2\ \Theta\ \delta_{s\varphi}}{\rho\ d^4}\ \frac{2\ f_{e\varphi\ d}}{(-\ \sigma^*_\varphi)} \tag{14}$$

The principal of the relation of the aerodynamic damping for the torsional mode $\delta_{a\varphi}$ and the on-flow velocity u_{oo} is the same as described for the translational mode of oscillation (see Figure 2). Figure 2 is valid for either vertical or torsional galloping modes. Therefore, the same linearizations have to be done to determine a conservative stability parameter σ^*_φ using equation (13):

$$\sigma^*_\varphi = \frac{4\ \Theta\ f_{e\varphi}}{\rho\ d^3\ u_{oo}}\ (-\ n) \tag{15}$$

Again: the linearization may lead to confusing results if the nonlinearity of the profile is dominant. The results of aeroelastic model tests depend on the damping of the model if only the instability points are measured. For reliable results the whole δ_a - curve versus u_{red} has to be investigated.

3. EXPERIMENTAL SETUP

<u>Experimental Setup:</u>

For the wind tunnel tests the small Eiffel type wind tunnel of the Department of Wind Engineering of the Technical University Aachen, Germany with a working section of 0,4 m width and 0,4 m height was used. Turbulence was generated by a grid installed in the upstream of the working area. The turbulence intensity was varied in four steps, from the lowest level (smooth flow) I_0 = 2,6% up to I_3 = 10,8%. The mean on-flow velocity of the wind with a maximum of 10 m/s was measured just above the model using a thermo-anemometer (Dantec Flow Master).

The experimental tests were carried out with section models between large endplates (section-model-technique). The models were elasitically supported as shown in Figure 4:

Coil springs are supporting the model as to satisfy dynamic similarity of oscillations in the vertical and torsional degree of freedom. An axis mounted in the center of gravity of the model is fixed to a rigid bar by screws allowing the adjustment of different on-flow angles, α. Varying the coil spring stiffness or its location on the rigid bar gives the possibility to modify the systems frequency in both, the vertical (f_{ey}) and the torsional ($f_{e\varphi}$) mode of oscillation. The structural damping of the system (δ_{sy} and $\delta_{s\varphi}$) was varied by modifying the mounting of the section model. The amplitudes of oscillation, the frequencies and the damping parameters of both vibration modes of the system were detected by strain gauges which are fixed on flat springs that are connetcted to the coil springs as shown in Figure 4.

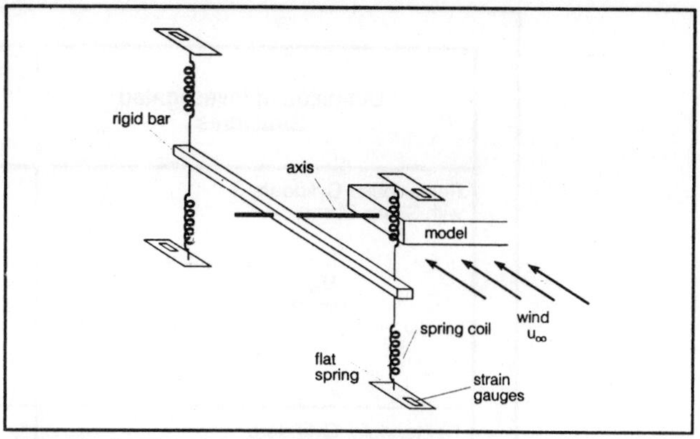

Figure 4 Model Support System

This dynamic experimental setup is usually prefered to measure the aerodynamic coefficients which describe the mechanisms of vortex shedding. In the case of galloping and torsional flutter the lift and moment coefficients are usually determined using static arrangements. But it is known that vibrations affect the aerodynamic coefficients and therefore they can not be neglected. To find more realistically describing instability parameters it is necessary to perform the experiments using aeroelastic models. Due to the effects of the aerodynamic force terms the lift and moment coefficients found within this investigation differ significantly from coefficients that are determined with static measurement techniques.

Models:

The wind tunnel models are rectangular cylinders with different b/d ratios. For the cylinders with b/d = 1 and b/d = 2 measurements with a variation of the edge curvature (r/d ratio) are performed. The geometric variation of the investigated structures are given in Table 1.

Within this study the geometric dimensions as given in Table 1, the on-flow angle α and the degree of turbulence I is varied.

Definition of Investigated Structures	Geometric Dimensions	
Rectangular Cylinders with Sharp Corners 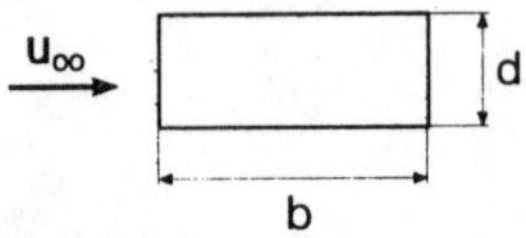	b/d	1 2 2,5 3 4 6 8
Rectangular Cylinders with Rounded Corners 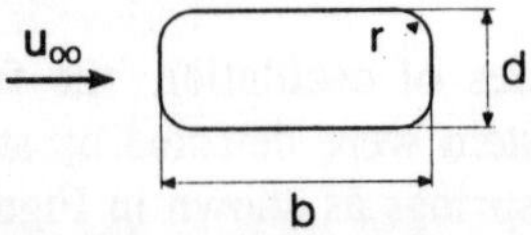	b/d	r/d
	1	0,05 0,10 0,15 0,25
	2	0,25

Table 1 Investigated Structures

4. RESULTS

Figure 5 illustrates the values of the linearized Stability parameter of bending mode σ_y^* for a rectangular cylinder with b/d = 2 versus the on-flow angle α for different turbulence intensities. Negative values of the stability parameter σ_y^* characterize the unstable regions of the profile. Hence, vertical galopping vibrations may occur for $0^\circ \leq \alpha \leq 10^\circ$ and $75^\circ \leq \alpha \leq 90^\circ$.

For small angles α, the stability parameter shows larger values in smooth flow then in turbulent flow. For the highest degree of turbulence I_3 = 10,8% the profile is even stable in regions where it will oscillate in smooth flow. In this case, the increase of the turbulence intensity has a stabilizing effect.

Figure 6 shows the values of the stability parameter of torsional mode σ_φ^* for the same model. Instable values can be found for high turbulence intensities and small angles of attack. In opposite to the vertical galloping oscillations torsional galloping may occur in turbulent flow, i.e. the turbulence has a destabilizing effect.

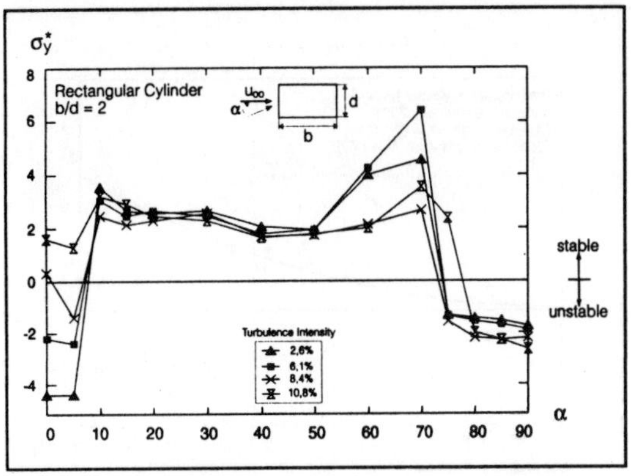

Figure 5 Stability Parameter σ^*_y of Vertical Mode Galloping of a Rectangular Cylinder (b/d=2) with sharp corners for different Turbulence Intensities

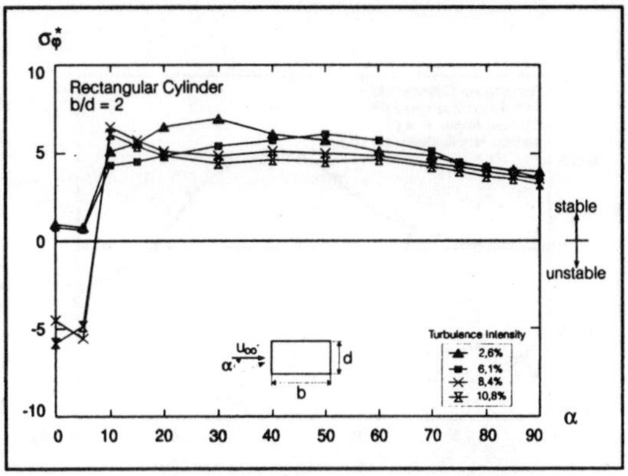

Figure 6 Stability Parameter σ^*_φ of Torsional Mode Galloping of a Rectangular Cylinder (b/d=2) with sharp corners for different Turbulence Intensities

In Figure 7 the moment coefficient for the profile with b/d = 8 is presented. Vertical galloping oscillations did not occur. The profile is stable for each on-flow-angle and for the investigated turbulence intensities. The profile shows torsional galloping instability for 0° ≤ α ≤ 15°. In this case the turbulence intensity has a stabilizing effect, as well as a destabilizing effect depending on the angle α.

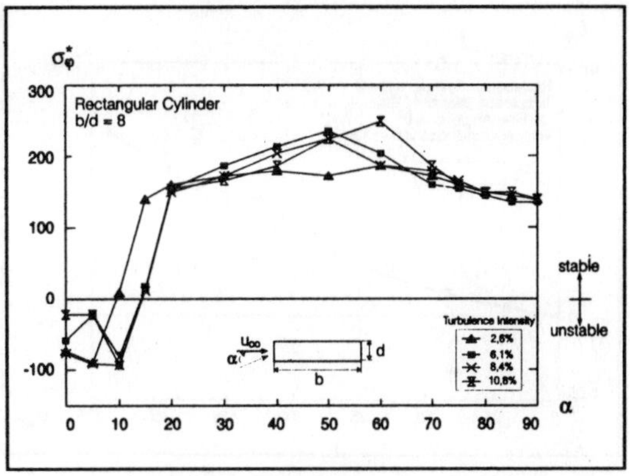

Figure 7 Stability Parameter σ_φ^* of Torsional Mode Galloping of a Rectangular Cylinder (b/d=8) with sharp corners for different Turbulence Intensities

Figures 8a, b and 9a, b are showing the aerodynamic damping characteristic for vertical and torsional oscillations in low turbulence flow (I_0 = 2,6%) for the profiles with b/d = 2 and b/d = 8 with an on-flow angle of α = 0°.

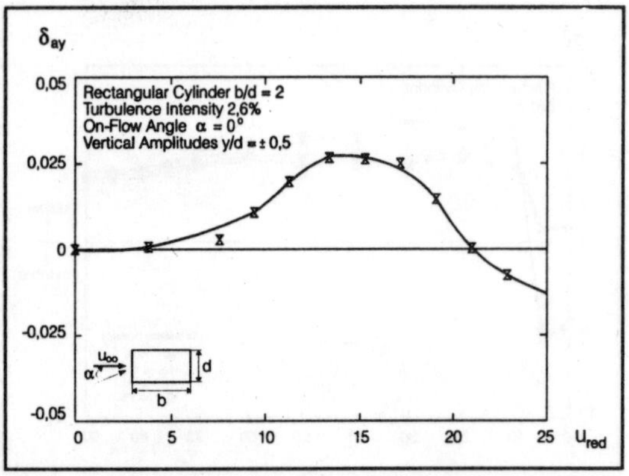

Figure 8a Aerodynamic Damping for the Vertical Mode versus Reduced Velocity for the Profile with b/d=2 and sharp corners

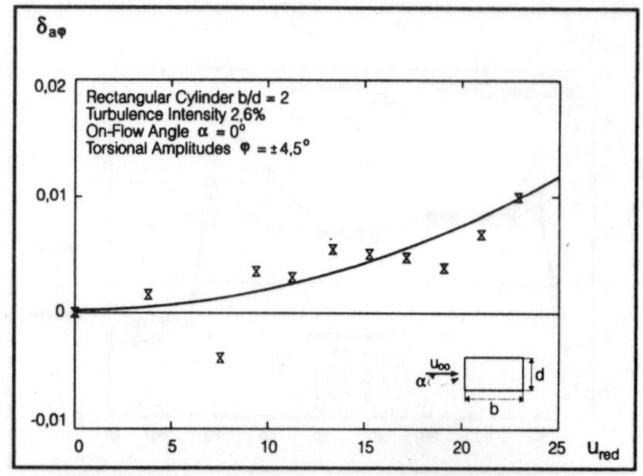

Figure 8b Aerodynamic Damping for the Torsional Mode versus Reduced Velocity for the Profile with b/d=2 and sharp corners

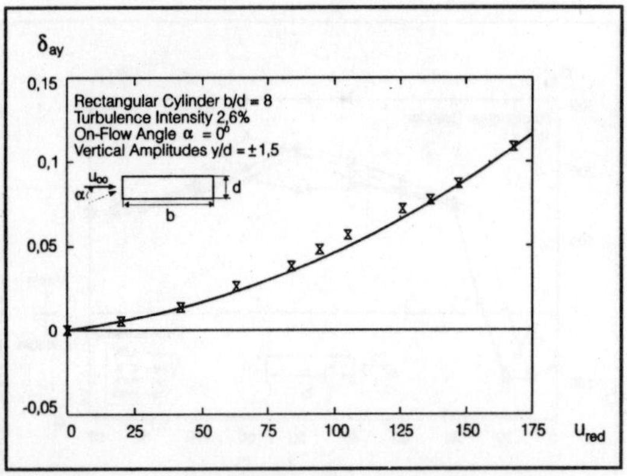

Figure 9a Aerodynamic Damping for the Vertical Mode versus Reduced Velocity for the Profile with b/d=8 and sharp corners

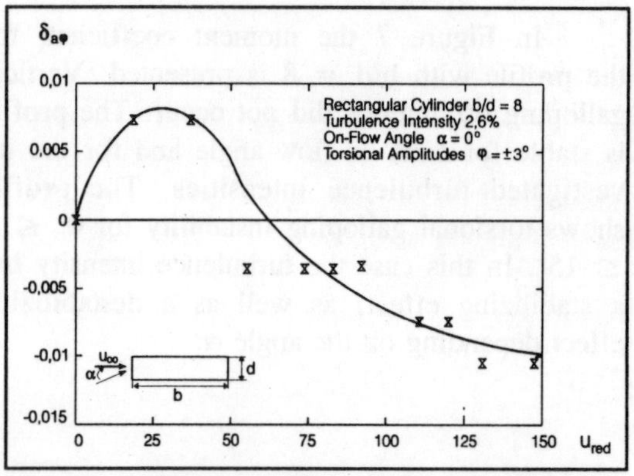

Figure 9b Aerodynamic Damping for the Torsional Mode versus Reduced Velocity for the Profile with b/d=8 and sharp corners

For Figure 8b and 9a the slope of the curves is positive, the aerodynamic damping δ_a does not change its sign, i.e. the profile is stable for this mode of oscillation which verifies the results described before.

In Figure 8a and 9b the aerodynamic damping values are becoming negative. That means there may be a point where the aerodynamic damping exceeds the structural damping of the system. The damping term in the equation of motion (vertical mode: equation (4), torsional mode equation (11)) will be zero or less then zero and thus, the structure is induced to vibrate.

From Figure 8a and 9b one can see the strong nonlinear characteristics of the aerodynamic damping with a stable range below a specific value of u_{red}.

Figure 10 and 11 are showing the influence of the geometric dimensions on the stability parameters σ_y^* and σ_φ^*.

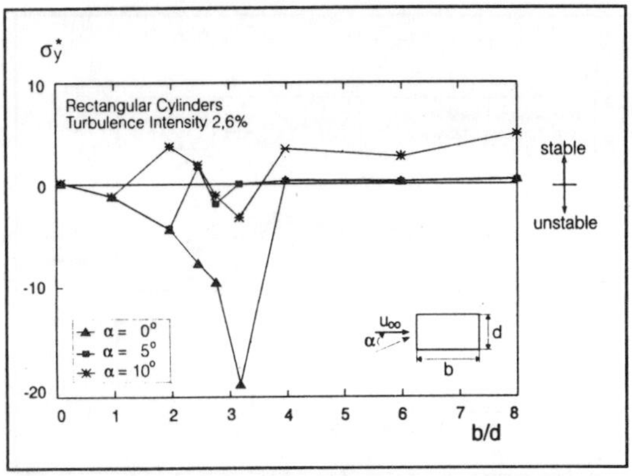

Figure 10 Stability Parameter σ_y^* of Vertical Mode Galloping of Rectangular Cylinders with sharp corners for different Angles α

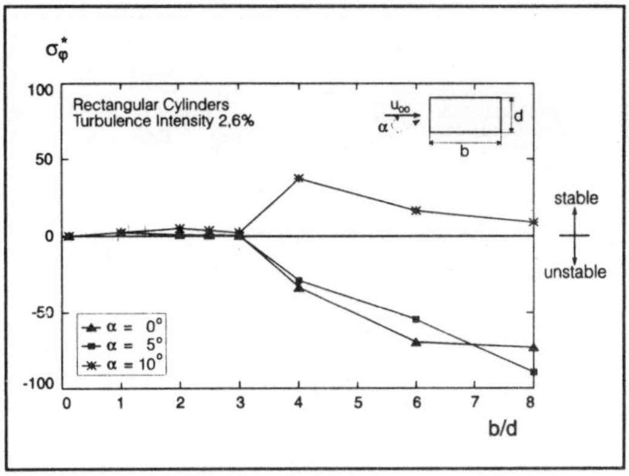

Figure 11 Stability Parameter σ_φ^* of Torsional Mode Galloping of Rectangular Cylinders with sharp corners for different Angles α

Rectangular structures with a b/d ratio smaller than $b/d \leq 4$ are sensitive to galloping vibrations in the vertical mode while structures with $b/d \geq 3$ are usually vibrating in the torsional mode. For the stability parameter σ_y^* (Figure 10) the maximum of instability can be found for b/d ratios near b/d = 3. Stabilization have been observed for ratios smaller and larger than b/d = 3. The stability parameter σ_φ^* (Figure 11) are increasing with an increasing b/d ratio. It is to assume that the point of maximum instability can be found for b/d ratios larger than b/d = 8.

The results of the wind tunnel tests are showing that the turbulence effects have a considerable influence on the stability parameter σ_y^* and σ_φ^* . For the vertical galloping vibrations the increase of the turbulence intensity has a stabilizing effect. The onset point of the galloping oscillations is shifted to higher wind velocities. Some structures even become stable for turbulent flow while they are instable in smooth flow.

For the stability parameter σ_φ^* the same observations can be made for large b/d ratios, but for small b/d ratios the values tend to decrease in turbulent flow (Figure 6). An increase in turbulence intensity has a destabilizing effect. The degree of turbulence drastically changes the magnitude and even the sign of σ_φ^*. These observations are discussed in detail by Matsumoto et al. The changes are mainly caused by the turbulent effect on the flow separation from the leading edge. A bluff body in turbulent flow tends to show similar aerodynamic properties to those of a more slender body. It can be stated that bluff bodies (small b/d ratios) are destabilized with increasing turbulence intensity while bodies with a large b/d ratio are stabilized for $\alpha \leq \pm 10°$.

Figure 12 and Figure 13 are showing the lift coefficient for rectangular cylinders with rounded corners for different turbulence intensities.

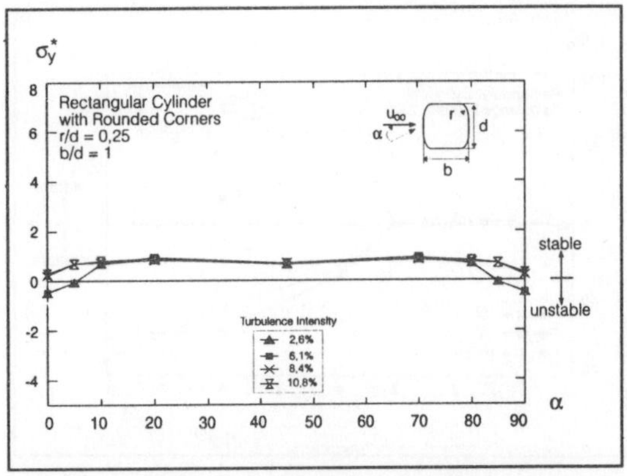

Figure 12 Stability Parameter σ_y* of Vertical Mode Galloping of a Rectangular Cylinder with b/d=1 and r/d=0,25 for different Turbulence Intensities

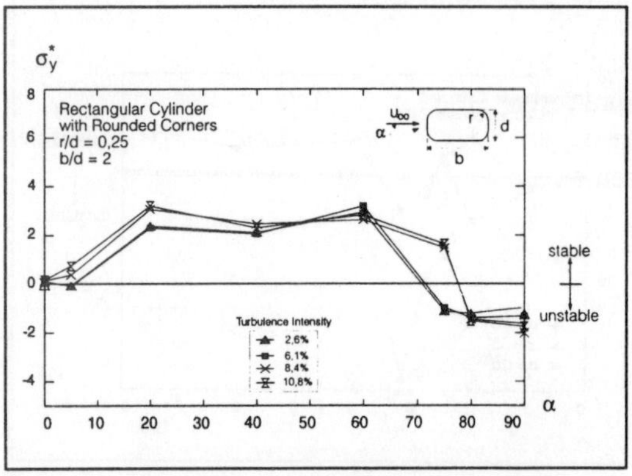

Figure 13 Stability Parameter σ_y* of Vertical Mode Galloping of a Rectangular Cylinder with b/d=2 and r/d=0,25 for different Turbulence Intensities

The results are showing that the turbulence effects for both profiles (b/d = 1 and b/d = 2) are relatively small compared to the results for the profiles with sharp corners.

For a structure with b/d = 1 and r/d = 0,25 negative lift coefficients are only observed for an on-flow angle of $\alpha = 0°$. The magnitude of these values is small and thus, the onset velocity is high.

By comparing Figure 13 to Figure 5 it becomes obvious that a rounded corner stabilizes a structure. The regions and the magnitude of instability is significantly decreased by the edge curvature. For small on-flow angles galloping vibrations are even prevented.

Torsional galloping vibrations have not been observed for the investigated structures with rounded corners.

Figure 14 shows the stabilizing effect of increasing r/d ratios for a profile with b/d = 1 and different on-flow angles. The curve ends with values for round cylinders (r/d = 0,5) which are known to be stable for galloping vibrations. Thus, it is obvious that rounding of the edges of rectangular cylinders is an effective method to inhibit galloping vibrations.

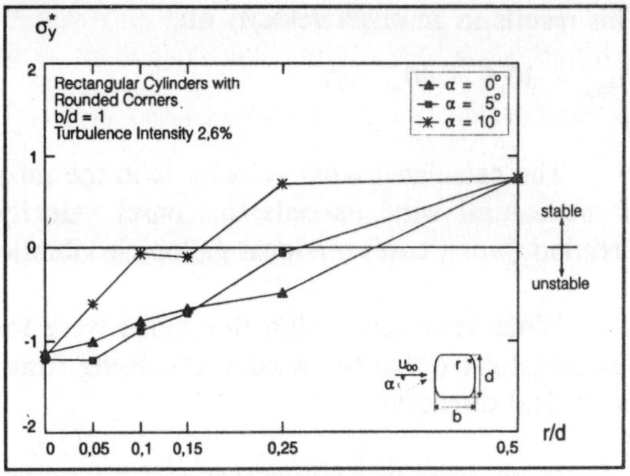

Figure 14 Stability Parameter σ_y^* of Vertical Mode Galloping of Rectangular Cylinders with b/d=1 for different On-Flow Angles α

5. FULL SCALE OBSERVATIONS

Two cases of damage due to wind induced vibrations in the torsional mode have been reported in the last time, the collapse of a crane and the failure of a coal mining machine. In both cases vibrations of the pendant bars, which are long and slender steel bars with a large b/d ratio, have been observed. Investigations in both cases have revealed that the pendant bars suffered fatigue failure following a period of wind induced torsional oscillations. For both cases of damage situations with high wind velocities have been reported.

The question that arises now is whether the observations made on the cases of damage and the results of the wind tunnel tests agree.

In the case of the crane damage the b/d ratio is b/d = 8 (b = 250 mm, d = 30 mm). As shown in Figure 7 and Figure 11 a structure with these geometric dimensions is sensitive to torsional galloping vibrations especially when the wind acts in the direction of the long side of the profile (small on-flow angles α).

With equation (14) and Figure 7 the onset velocity $u_{0\varphi}$ of the collapsed crane bar can be calculated. The structural parameters for this pendant bar are as follows:

$$\Theta = \frac{m}{12}(b^2 + d^2) = 0{,}311 \text{ kg m}^2 / \text{m}$$

$$f_{e\varphi} = 13{,}1 \text{ Hz}$$

$\delta_{s\varphi} = 0{,}006 \div 0{,}008$ (assumed damping value for undamped, prestressed steel)

Using the linear theory, the worst case stability parameter σ_φ^* can be taken from Figure 7:

$$\sigma_\varphi^* = -95$$

This results in an onset velocity of:

$u_{0\varphi}$ = 30,5 ÷ 40,7 m/s

The calculated wind velocity is in the range of probable wind velocities. When the velocity of the natural wind exceeds this onset velocity (even with gusts) and acts in the unfavorable direction (worst case) torsional galloping vibrations will occur.

Wind velocities within this range were measured at the site of the damage. In addition, it was investigated that the wind was coming from west and the pendant bar profile was attacked in the critical direction.

In an additional investigation of the crane damage the damping of the pendant bar installed on the reconstructed crane was measured. These measured damping values are therefore representative of those exhibited by the original pendants at the time of the failure. The measured damping values are in the range of $0{,}0022 \leq \delta_{s\varphi} \leq 0{,}0046$.

Following the conservative linearization, described in chapter 3. with Figure 2 and equation (8), this low damping value would result in an onset velocity of:

$u_{0\varphi}$ = 11,65 ÷ 24,36 m/s

But observations made for the damaged crane when the wind velocity was in the range of 10 ÷ 20 m/s showed no torsional flutter responses. This effect can be explained with the nonlinear damping character discussed in the theoretical backgrounds (chapter 3.). Figure 15 presents the explaination: using the linearized theory the result will be very conservative ($u_{0\varphi}$ = 10,8 m/s). But following the realistic character of the curve the onset velocity is larger ($u_{0\varphi}$ = 27,5 m/s).

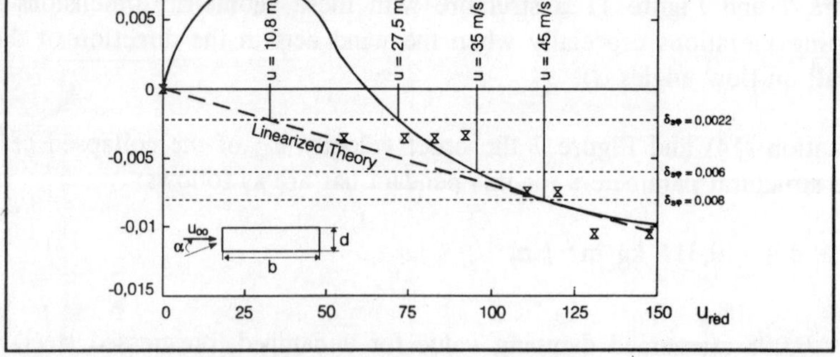

Figure 15 Aerodynamic Damping versus Reduced Wind Velocity for the Damaged Crane Bar

These considerations demonstrate the importance of the enclosure of the nonlinear response of structures in natural wind. For the design of structures or building components the linearized theory is sufficient and safe. But in cases of damage or even failure it may be necessary to eveluate the exact and realistic character of wind responses.

However, the stability parameters determined within this investigations and the observations and measurements of the collapsed crane bar show a very good agreement.

6. CONCLUSIONS

The aim of this research work was to investigate rectangular profiles with regard to their vibration behaviour in natural wind (specially galloping vibrations).

Wind tunnel tests were carried out with freely vibrating section models. The measured stability parameters σ_y^* (lift coefficient) and σ_φ^* (moment coefficient) differ in part considerably from the available statically determined coefficients. This is due to the aerodynamic force terms linked with the vibration movement, which can not be measured with static measurements. Wind tunnel tests for establishing these coefficients should, therefore, only be carried out with a dynamic experimental setup.

An important point of the investigation was the influence of the turbulence intensity. Galloping vibrations are dependent on turbulence. It is difficult to find a clear concept whether a profile is stabilized or destabilized by turbulence effects. However, it was tried to show a tendency.

Comparisons between observations of torsional flutter responses of a crane and the results of the section model tests are made. The agreement between the results of full scale measurements and calculations of the vibrational response with the stability parameters determined in wind tunnel tests are satisfying.

7. REFERENCES

König and Bremer, (1980), "Torsional Flutter of Tension Bars", Final Report No. 530-3/80, Institute for Light Construction, Dresden, Germany (in German).

Matsumo, M., Shirato, H., Hirai, S., (1992), "Torsional Flutter Mechanism of 2-D H-Shaped Cylinders and Effect of Flow Turbulence", Journal of Wind Engineering and Industrial Aerodynamics, Vol. 41-44, pp 687-698.

Modi, V.J., Slater, J.E., Vancouver, B.C., (1974), "Quasi-Steady Analysis of Torsional Aeroelastic Oscillators", Flow-Induced Structural Vibrations, Springer Verlag, pp 355-372.

Reinold, T.A., Brinch, M., Damsgaard, A., (1992), "Wind Tunnel Tests for the Great Belt Link", Aerodynamics of Large Bridges, A. Larsen (ed.), Balkema, Rotterdam. ISBN 90 5410 0427.

Ruscheweyh, H., (1982), "Dynamic Wind Effects on Structures", Vol. I/II, Bauverlag, Germany (in German).

Ruscheweyh, H., (1992), "Systematic Investigation of Wind Induced Vibrations of Slender Structures in order to establish Specifications for Constructions (e.g. DIN 18809 and Eurocode)", Final Report of the Research Project No. FE15.184.R89G, Institute for Steel Construction, Department of Wind Engineering, RWTH Aachen, Germany (in German).

Schnakenberg, C., (1993), "Systematic Investigation of Slender Structures in Respect to their Excitation by Natural Wind (Specially Vibrations in the Torsional Mode) with Wind Tunnel Tests", Institute for Steel Construction, Dep. of Wind Engineering, RWTH Aachen, Germany (in German)

Simui, E. and Scanlan, R.H., (1978), "Wind Effects on Structures", Wiley Intersience Publication, John Wiley & Sons, New York / Chichester / Brisbane / Toronto

THEME V
Field/Prototype Measurements

9ICWE, 1995, New Delhi, India

LOCAL SEVERE SUCTIONS ON THE SIDE OF A PRISM MODEL ON A FIELD

OKUDA, Y.[1], KATSURA, J.[2] and KAWAMURA, S.[1]
Research Associate *Professor* *Professor*

[1] *Department of Architecture and Building Engineering, Faculty of Engg., Osaka City University, Osaka, 558, Japan.*
[2] *Disaster Prevention Research Institute, Kyoto University, Uji, 611, Japan.*

Abstract

The wind pressures on a three-dimensional prism model on an outdoor field were measured with multi-simultaneous data acquisition system. From the results of the outdoor measurements, it was shown that the local severe suctions on the side of the prism model were mainly caused by two typical types of conical vortices. These local severe suctions were indicated in the similar wind pressure distributions to the local severe suctions caused by two types of conical vortices, that is, a standing conical vortex on the upper side and an inverted conical vortex on the lower side, which were shown by Okuda et al(1993) in wind tunnel experiments. The occurrence conditions of the conical vortices in the natural wind were not a little different from those of the conical vortices in the wind tunnel. The differences were expected to be mainly caused by non-stationary phenomena in natural wind.

1. INTRODUCTION

Wind pressure measurements have been conducted to show local severe suctions on the surface of a building model in wind tunnel experiments or on the surface of the full scale building. Two kinds of suctions can be found in typical cases; on the roof of low rise buildings and on the side of high rise buildings. Ackert(1965) showed that the former suctions were caused by a pair of conical vortices on the roof at the diagonal angle. The mechanisms of the latter suctions, however, have not been clarified yet owing to the complexities of the phenomena. Melbourne(1993) showed various examples of the local severe suctions measured in wind tunnel studies. He considered that the condition for the occurrence of the suctions were the freestream turbulence and an edge discontinuity, although the mechanisms causing the suctions were not clear. From the results of wind pressure measurements on a prism model or flow visualization around the prism model in a wind tunnel, Okuda et al(1993) deduced that local severe suctions on a side of the prism model were caused by two types of conical vortices. Outdoor measurements were planned to investigate the following questions. Are similar conical vortices formed on the similar prism model even in natural wind conditions? Do the conditions on the occurrence of the conical vortices in the natural wind accord with the conditions of them in the wind tunnel?

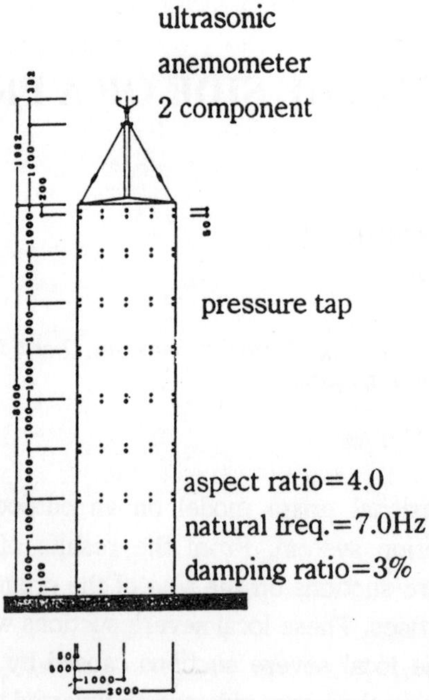

Fig.1 Prism model in natural wind. (mm)

run	date and time	chn	Δt	pts	files
01	Sep.04 93 00:44-02:46	48	0.05	256	583
02	Sep.04 93 02:47-04:46	48	0.05	256	561
03	Sep.04 93 04:48-06:45	48	0.05	256	552
04	Sep.04 93 06:46-08:48	48	0.05	256	579
05	Dec.27 93 18:33-19:49	48	0.05	512	179
06	Dec.31 93 13:48-15:56	48	0.05	512	300
07	Dec.31 93 15:56-17:26	48	0.05	512	209
08	Dec.31 93 17:29-19:37	48	0.05	512	300
09	Dec.31 93 20:06-21:44	48	0.05	512	229
10	Jan.02 94 14:23-16:31	48	0.05	512	300
11	Jan.17 94 15:48-17:55	48	0.05	512	300
12	Jan.17 94 17:56-20:03	48	0.05	512	300
13	Jan.17 94 20:04-22:12	48	0.05	512	300
14	Jan.17 94 22:12-00:20	48	0.05	512	300
15	Jan.26 94 08:45-10:52	48	0.05	512	300
16	Feb.01 94 10:13-12:21	48	0.05	512	300
17	Feb.01 94 12:21-14:29	48	0.05	512	300
18	Feb.01 94 16:15-17:06	48	0.05	512	300
19	Feb.02 94 22:38-00:46	48	0.05	512	300
20	Feb.03 94 00:49-02:56	48	0.05	512	300
21	Feb.03 94 03:05-05:12	48	0.05	512	300
22	Feb.03 94 05:32-07:39	48	0.05	512	300
23	Feb.03 94 09:02-11:09	48	0.05	512	300
24	Feb.03 94 11:10-13:18	48	0.05	512	300
25	Mar.10 94 15:22-17:29	48	0.05	512	300
26	Mar.10 94 17:30-17:50	48	0.05	512	47
27	Mar.24 94 14:07-16:15	48	0.05	512	300
28	Mar.24 94 16:16-16:24	48	0.05	512	20

※ chn : number of channels
Δt : sampling interval
pts : number of points per data file
files : number of continuous data files

Table 1 List of acquired data in natural wind.

Using 180 manometers, VTR and an image processing system(PIAS LA-500), Okuda et al(1993) measured simultaneously the wind pressures on the whole sides of the prism model and was able to catch the non-stationary phenomena of the local severe suctions on the side of the prism in the natural wind. The authors conducted the simultaneous measurements of the wind pressures with 45 differential pressure transducers in order to pursuit more transient phenomenon causing the local severe suctions on the side of the prism. Comparing the results of wind tunnel experiments and outdoor measurements, the authors considered the local severe suctions on the side of a three dimensional prism in natural wind.

2. EXPERIMENTAL PROCEDURES

2.1 Prism model and measuring instruments

A prism model, 8m high and 2m wide, was installed on a flat observation field of Shionomisaki Wind Effect Laboratory of the Disaster Prevention Research Institute, Kyoto University. Each side of the prism model was correctly aligned to the four points of the compass. The structure of the prism model was made of steel. The wall of it was made of reinforced lauan plywood, 21mm thick, and was waterproofed. The natural frequency of the prism was about 7Hz and the damping ratio of the model was about 3%. 45 pressure taps, each diameter of which was 16mm, were made on each side of the prism model (see Fig.1). Wind pressures on the pressure taps were measured through vinyl tubes connected to differential pressure transducers(Copal P3000S-501D). The calibre and the length of the vinyl tube were 4mm by 50cm. The reference wind direction and wind velocity were measured at the height of about 2m upper from the top of the prism with an ultrasonic anemometer(Kaijo DA-210). Inner fluctuating pressures in the model were measured with a pseudo-constant pressure reservoir developed by Katsura(1988) and was used in order to correct the wind pressures on the wall of the prism.

2.2 Data acquisition system

Generally, the data can never be measured again using the same conditions at the acquisition of the data in the natural wind. The lack of data sequences, therefore, must not occur on the outdoor measurements. In order to investigate the unsteady flow field around a body in natural wind, wind pressures' data on so many channels need to be measured simultaneously that wind pressure distributions can be drawn. Okuda et al(1994) made a multi-channel data acquisition system for outdoor measurement. This system was composed of a personal computer(NEC PC-9801DA), two A/D-converter boards(Canopus ADJW-98), a hard disk(Caravelle AV-570SDA) and a magnet optical disk(ICM MO-4120). Analogue data was converted to digital data on "Interrupt mode" using the handler soft for the A/D-converter board(Canopus CHS-ADT), which enable parallel like A/D-converting and data processing on a user program. This system, therefore, allowed the making of continuous data for a long time to be recorded onto the external memory media, the hard disk and so on. The sampling rate of the A/D-converting was proportional mainly to the speed of the writing out of the DMA(Direct Memory Access) buffer to the hard disk. The minimum sampling interval was extended to around 0.003sec in the case of 64 channels. The size of each data file saved on the hard disk can be simply changed to any size on the user program, but the size of the data file should be divided onto a smaller size for convenience in processing the data file on the stage of post data processing. Data files were copied from the hard disk onto the magnet optical disk. The analogue data of wind pressure on the prism, wind direction and wind velocity were A/D-converted and measured continuously and simultaneously with the condition that the sampling interval was 0.05sec and the cut-off frequency of the low-pass analogue filter was 5Hz.

3. RESULTS AND CONSIDERATIONS

3.1 Local severe suctions on side of prism

Locating 45 differential pressure transducers on the north or west side of the prism model, wind pressures were measured simultaneously on one side of the model in natural wind. The total number of the acquired data with this system was 28-run, and the total length of the data was longer than 60 hours, for seven months from September 1993 to March 1994, as shown on Table 1. Figs.2(a)(b) show typical examples of time series of wind pressure distributions on the west side of the prism in natural wind. The date was September 4, 1993, when typhoon 9313 moved up north over the Japan Sea. The main wind direction was south. The interval of each

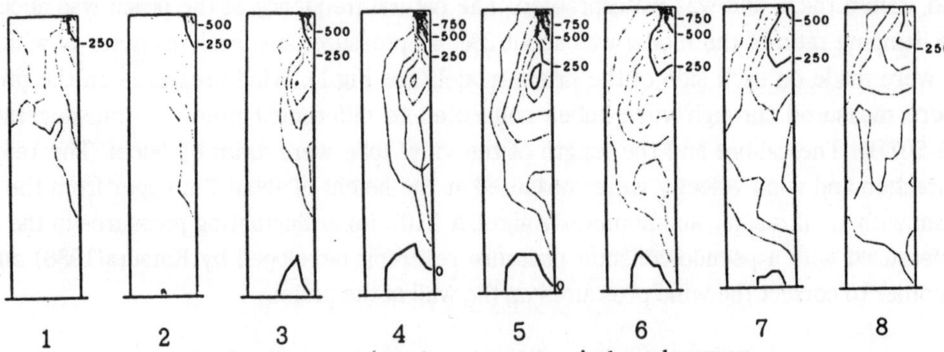

(a) Local severe suction at upper windward corner.

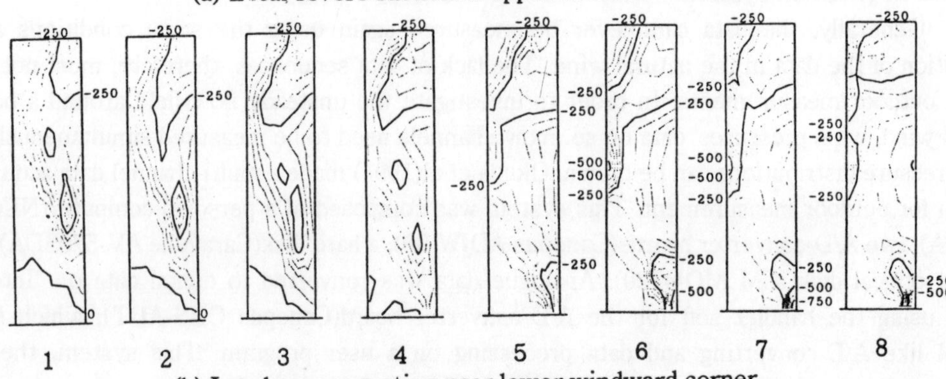

(b) Local severe suction near lower windward corner.

Fig.2 Time series of wind pressure distributions. (Pa)

frame was 0.1sec on Fig.2(a) or 0.05sec on Fig.2(b). The local severe suctions were formed at the upper windward corner called "corner A" as shown in Fig.2(a), or near the lower windward corner called "corner B" as shown in Fig.2(b). The wind pressure distributions of the former suction were rapidly formed to be inclined leeward and downward from corner A, and disappeared around 1sec later. The wind pressure distribution of the latter suction suddenly occurred to be inclined leeward and upward from corner B. The suction immediately reached near -1000Pa at corner B and disappeared rapidly about 0.2sec later. The phenomena of these suctions on the side of the prism model were usually non-stationary in natural wind.

3.2 Two types of conical vortices

Okuda et al(1993) showed from the results of wind tunnel experiments that the local severe suctions on the side of a three dimensional prism were caused by two types of conical vortices. One was a standing conical vortex on corner A of the prism at the glancing angle. The other was an inverted conical vortex on the middle or lower side of the prism in the case where

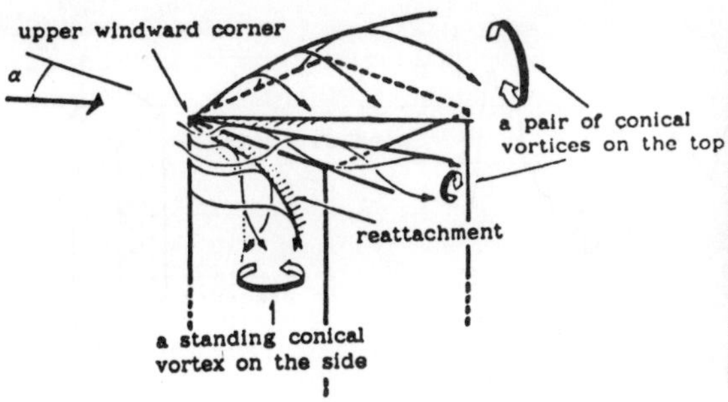

(a) Sketch.

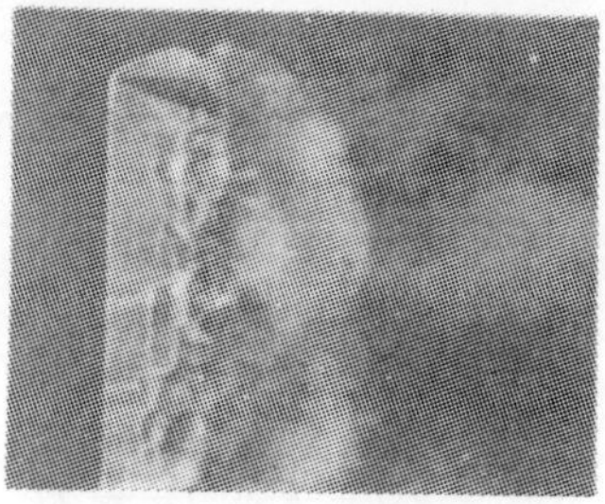

(b) Flow visualization.

Fig.3 Standing conical vortex in wind tunnel (Okuda et al(1993)).

the wind perpendicularly attacked the front of the prism. No vortex phenomena except these conical vortices were found to cause any local severe suction on the side of the prism model in the wind tunnel. Investigating the time series data of the wind pressures in detail, the authors found that almost all of the local severe suctions on the side of the prism model in natural wind indicated similar wind pressure distributions to Fig.2(a) or (b). It was expected that the local severe suctions were caused by these conical vortices.

3.2.1 Standing conical vortex

New knowledge on the standing conical vortex from the results of wind tunnel experiments is summarized as follows. As shown in Fig.3(a), the conical vortex was formed between the leading edge of the side and the region of reattachment leeward. The turning radius of this vortex became smaller near corner A, where a larger suction appeared, and the greatest suction almost certainly occurs at corner A. The wind pressure distribution at the vicinity of the corner A, therefore, was similar to Fig.2(a). When the wind direction was fixed against the prism model, the quantity of flow into this vortex from the leading edge of the side was evenly balanced with the quantity of flow out of this vortex, which became a standing vortex on the upper side. If the wind velocity did not change, the suction near corner A became fixed and the fluctuating suction was zero. The flow at the stagnation point on the front of the prism spread out radially on the surface forming laminar boundary layers. The laminar boundary layers were separated from the edge between the front and the side or between the front and the top. It was, however, expected that the boundary layers on the surface of the front were not able to separate over corner A, from the results that the white smoke of $TiCl_4$ (Tetra-chloride Titanium), as shown in the flow visualization photograph of Fig.3(b), was cut into the upper and lower parts by corner A. The lower part of the upper leeward region of the side was rolled inside to be formed into a standing conical vortex on the upper side. This standing conical vortex was, therefore, formed only by the separated shear layer from the front, but the other flow from the top or over corner A never acted on this vortex.

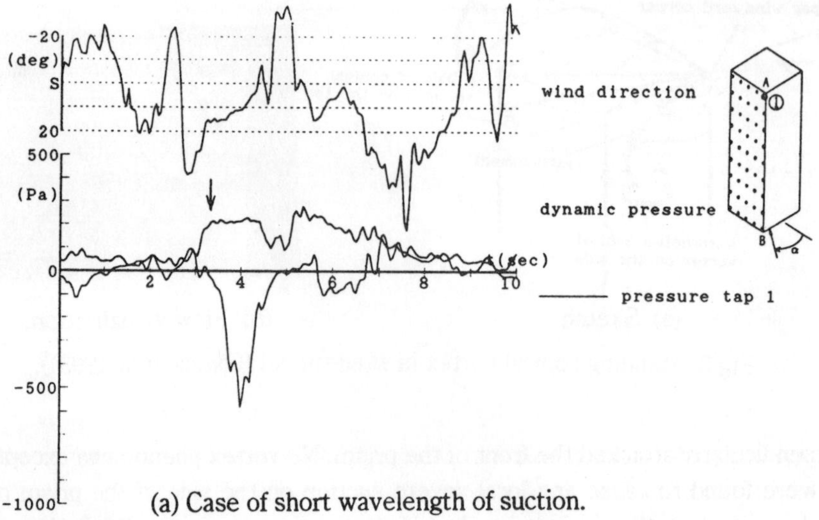

(a) Case of short wavelength of suction.

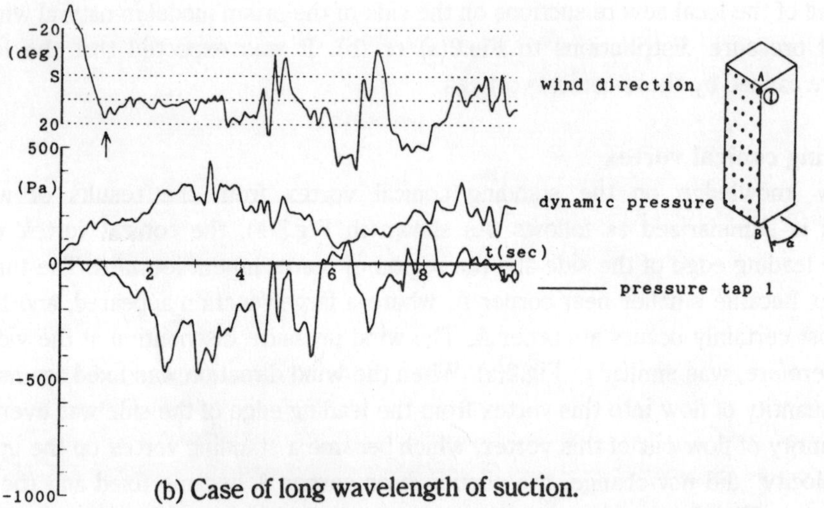

(b) Case of long wavelength of suction.

Fig.4 Time series of wind direction, dynamic pressure and wind pressure near upper windward corner. (Pa)

The angle α of attack when this conical vortex was formed, was between 13 and 15 degrees in wind tunnel experiments, but varied among 7 and 12 degrees in the outdoor measurements. At this time the difference between the value in the wind tunnel experiments and in the outdoor measurements can not be estimated simply owing to the precision or the problem of the location of the ultrasonic anemometer, but the values in wind tunnel experiments seemed to be a little smaller than those in the outdoor measurements.

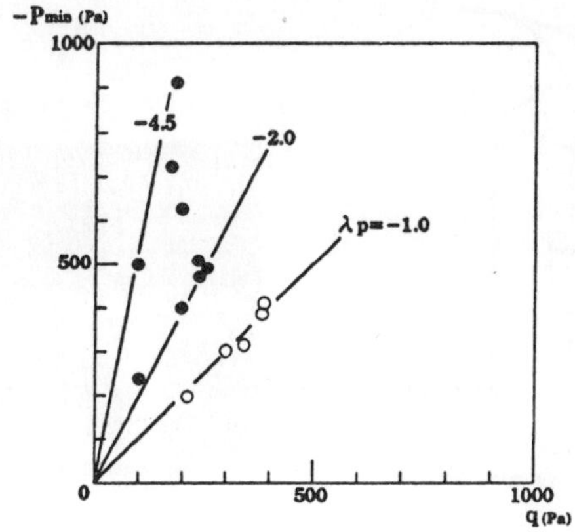

Fig.5 Relation between peak pressure -Pmin and dynamic pressure q at upper windward corner. (Pa)

The wavelength of the suction caused by this conical vortex was about 1.0 to 2.0sec in the case of the shortest wavelength of the suction. Fig.4(a) shows an example of the time series data of the wind direction, dynamic pressure and wind pressure at corner A in the case of the shortest wavelength of the suction. When the dynamic pressure was almost constant at the value of 200Pa (see an arrow in Fig.4(a)), the suction occurred more strongly than -500Pa at corner A. The wind direction was varied largely before and after the suction. It was about +15 degrees from a southern direction at the moment of the occurrence of the suction, and then moved to the minus direction. On the other hand, Fig.4(b) shows an example in the case of the longer wavelength of the suction. When the wind direction was fixed at the value of +12 degrees from a southern direction (see an arrow in Fig.4(b)), the wind pressure at corner A fluctuated nearly proportionally, or in other word almost quasi-steadily, to the dynamic pressure. Fig.5 shows the relationship between the peak pressure -Pmin and the dynamic pressure q when the local severe suction occurred at pressure tap No.1. The peak pressure ratio λp, which equals -Pmin/q, was -2.0 to -4.5 in the case that the wavelength of the suction was about 1.0 to 2.0sec, shown as closed signs in the figure. This value of λp was much stronger than the value of Cp, which was about -1.2 to -1.4 in wind tunnel experiments(see Okuda et al(1993)). As the wavelength of the suction was extended, the value of the peak pressure ratio λp became close to the value of Cp in the wind tunnel experiments. The peak pressure ratios λp were shown as open signs in the figure in the case that the wavelength of the suction was 3 to 10sec. When the wind direction varied little or was fixed at a glancing angle for a long time, a conical vortex was expected to be nearly constantly formed on the upper side of the prism even in the natural wind, which was similar to the standing conical vortex formed in the wind tunnel. On the other hand, in the case where the wind direction varied for a short time, this conical vortex was expected to occur on the upper side for a brief moment. If the wind direction was immediately changed, the vortex's point of contact remained

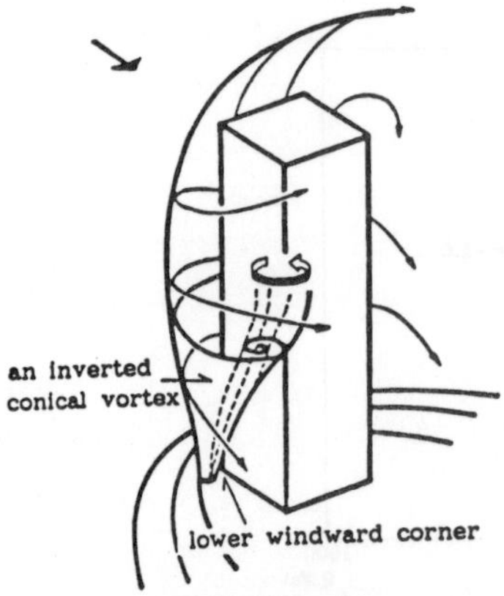

(a) Sketch.

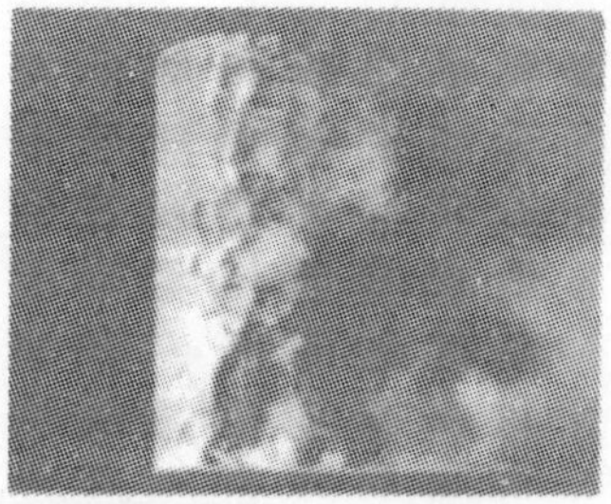

(b) Flow visualization.

Fig.6 Inverted conical vortex in wind tunnel(Okuda et el(1993)).

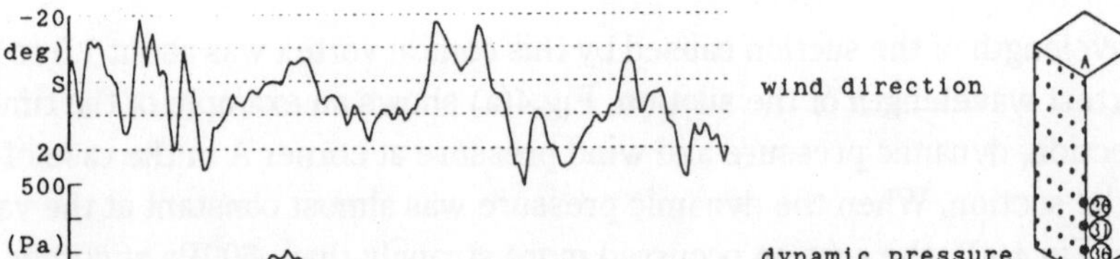

Fig.7 Time series of wind direction, dynamic pressure and wind pressure on windward edge. (Pa)

attached onto the corner A and the other part of the vortex flew leeward, which stretched the conical vortex into a more slender shape. Then the turning radius of the vortex became smaller and a larger suction occurred sharply in the vortex. The wavelength of the suction was reduced to the distance scale, the run of wind proposed by Katsura(1993), 60 to 100D (D: width of model) in the former case or 10 to 20D in the latter case.

3.2.2 Inverted conical vortex

Figs.6(a)(b) show the sketch and the flow visualization photograph of the inverted conical vortex in the turbulent boundary layer of wind tunnel experiments, where the separated shear

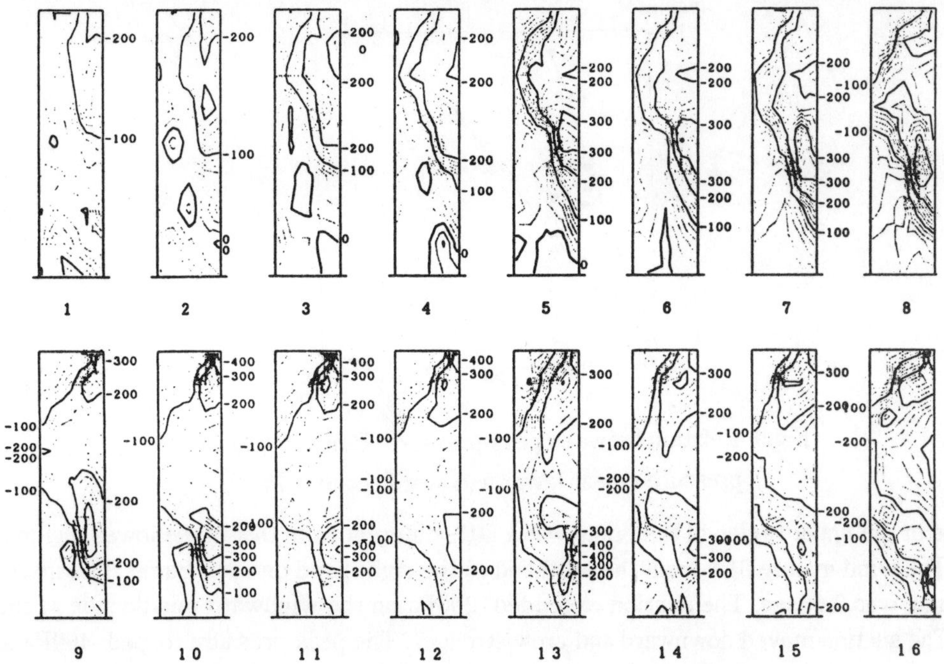

Fig.8 Time series of wind pressure distributions. (Pa)

layer from the windward edge on the side of the prism model was rolled inside over the lower side to form the inverted conical vortex, which was inclined leeward. This inverted conical vortex was nearly periodically formed on the lower side of the prism and was shed into the wake of the prism. Okuda et al(1990) showed that the period of this vortex was accorded with that of von Karman vortex in the wake of the prism in wind tunnel. Okuda et al(1992) added the results of wind tunnel experiments that the periodicity of this conical vortex became weaker as well as that of von Karman vortex, if the turbulence increased in the approaching flow against the prism.

The angle α of attack at occurrence of this conical vortex was 0 to 5 degrees in these outdoor measurements as well as in the wind tunnel experiments. Any local severe suction, however, did not occur on the lower side, unless the angle α of the attack was fixed smaller than 5 degrees for a certain interval in natural wind conditions. If the wind direction was almost fixed for more than 20D in the distance scale proposed by Katsura(1993), some local severe suctions occurred on the lower side of the prism, but the wavelength of each suction was very short, around 5D. It was much shorter than the wavelength of the suction caused by the standing conical vortex in the previous section. The wave form of the suction due to the inverted conical vortex indicated sharply on the spike shape as shown in Fig.7.

The strongest suction caused by the inverted conical vortex occurred not always at corner B, but some local severe suctions occurred leeward and upward from corner B. Fig.8 shows another example of the wind pressure distributions, showing that two types of suctions simultaneously occurred on the side of the prism in natural wind, which indicated that the wind

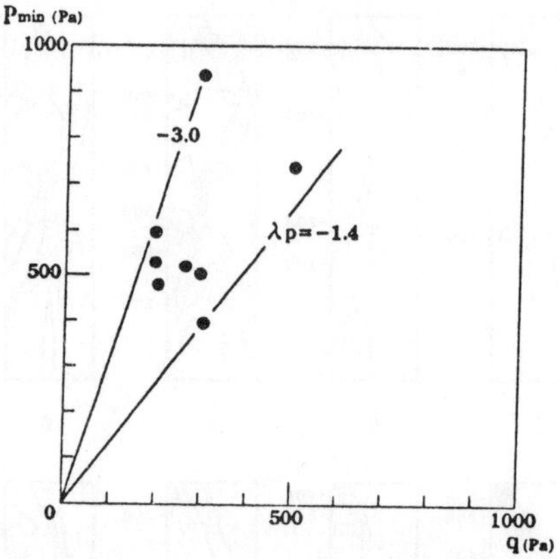

Fig.9 Relation between peak pressure -Pmin and dynamic pressure q near lower windward corner. (Pa)

direction of the upper region of the prism was a little different from that of the lower region of the prism. The wind mainly flew from the left hand to the right hand on each frame. The interval of each frame was 0.05sec. The suction exceeded -200Pa on the windward middle side at the 2nd frame. The suction moved downward and grew stronger. The peak pressure topped -400Pa on the lower windward edge at the 13th frame, but not at corner B. Okuda et al(1993) indicated from the results of the flow visualization experiments in the wind tunnel that the tip of the inverted conical vortex moved downward and leeward on the side but not always fixed at corner B. On the contrary in the 10th, 11th and 12th frames, the suctions was the strongest at the corner A and topped -400Pa, because the point of the standing conical vortex was anchored at corner A.

Fig.9 shows the relationship between the peak pressure -Pmin near the corner B and the dynamic pressure q. The peak pressure ratio λp varied from -1.4 to -3.0. The reason that the strong suction occurred on the lower side of the prism was expected to be caused by the vortex stretching at the release into the wake behind the prism as shown by Okuda et al(1990) in wind tunnel experiments.

4. CONCLUSIONS

The following conclusions were acquired.

1) It was confirmed that the local severe suctions occurred on the side of the prism, which was caused by two types of conical vortices, a standing conical vortex on the upper side and an inverted conical vortex on the lower side of the prism even in natural wind as well as in wind tunnel.

2) The standing conical vortex in natural wind conditions were formed when the angle α of attack varied from 7 to 12 degrees and the distance scale, instead of the time scale, was longer than 10D, when the wind direction was fixed at the angle. The peak pressures were indicated at the upper windward corner. In the case where the distance scale was about 15D, the peak

pressure ratio λp was -2.0 to -4.5 at the upper windward corner and much larger than the Cp in the wind tunnel experiments. When the distance scale became longer, the peak pressure ratio λp was close to the Cp in wind tunnel experiments.

3) The inverted conical vortex was formed in natural wind conditions when the angle α of attack was around 0 degrees and the distance scale was longer than 20D. The peak pressures did not always appear at the lower windward corner. The peak pressure ratio λp was varied between -1.4 and -3.0 on the lower side on the prism.

ACKNOWLEDGEMENTS

The authors are very grateful to technical staffs Mr. T. Ozaki and Mr. S. Kawauchi in Disaster Prevention Research Institute, Kyoto University on their technical supports. This research has been partly supported under the Grant of Science Research Expense by the Ministry of Education, Culture and Science.

REFERENCE

Ackert J.(1965),"Anwendungen der Aerodynamik im Bauwesen", Zeitschrift fur Flugwissenschaften, Vol.13, pp109-122 (in Germany).

Katsura J.(1988), "A Pseudo-Constant Pressure Reservoir for Indoor Pressure Measurements", Annuals of Disaster Prevention Research Institute, Kyoto University, No.31 B-1, pp393-398 (in Japanese).

Katsura J.(1993), "Gust Response of Flow Pattern Formed around a Model Building, Due to a Wind Pressure Measurement", Journal of Structural and Construction Engineering, Transactions of Architectural Institute of Japan, No. 451. September., 1993, pp65-78 (in Japanese).

Melbourne W.H.(1993), "Turbulence and the Leading Edge Phenomenon", Journal of Wind Engineering and Industrial Aerodynamics, Vol.49, pp45-63, Elsevier.

Okuda Y. and Taniike Y.(1990), "Vortex Structure around a 3-Dimensional Prism, Vortices over a Side Surface", Proceedings of the 11th National Symposium on Wind Engineering, Tokyo, Japan, December 6-7, 1990, pp125-130 (in Japanese).

Okuda Y. and Taniike Y.(1992), "Inverted conical Vortex over Side Face of a Three-Dimensional Prism", Proceedings of the 12th National Symposium on Wind Engineering, Tokyo, Japan, December 3-4, 1992, pp167-172 (in Japanese).

Okuda Y. and Taniike Y.(1993), "Conical Vortices over Side Face of a Three Dimensional Square Prism", Journal of Wind Engineering and Industrial Aerodynamics, Vol.50, pp163-172, Elsevier.

Okuda Y., Katsura J., Kawamura S. and Ozaki T.(1993), "Wind Pressure Distributions on a Three-Dimensional Prism Model in Natural Wind", Journal of the Visualization Society of Japan, Vol.13, Supplement. No.1, pp63-66 (in Japanese).

Okuda Y., Katsura J. and Kawamura S. (1994), "Measurement of Wind Pressures on a Three-Dimensional Prism in Natural Wind, Multi-Channel Data Acquisition System for Outdoor

Measurement", Summaries of Technical Papers of Annual Meeting Architectural Institute of Japan, B Structures 1, 1994, pp229-230 (in Japanese).

9ICWE, 1995, New Delhi, India

FULL-SCALE MEASUREMENTS OF WIND-INDUCED VIBRATION OF TRANSMISSION LINE SYSTEM IN MOUNTAINEOUS AREA

KATOH, T.[1], MOMOMURA, Y.[2], IZAWA, K.[3], OKAMURA, T.[4] and OHKUMA, T.[5]

[1] *Researcher, Izumi Sohken Eng. Co., Ltd., 4-1-10, Hacho-bori, Chuoh-ku, 104 Tokyo, Japan.*
[2] *Manager, Izumi Sohken Eng. Co., Ltd., 4-1-10, Hacho-bori, Chuoh-ku, 104 Tokyo, Japan.*
[3] *Manager, Tomoe Co., 3-4-5, Toyosu, Kotoh-ku, 135, Tokyo, Japan*
[4] *Senior Researcher, Tomoe Co., 3-4-5, Toyosu, Kotoh-ku, 135, Tokyo, Japan.*
[5] *Professor, Department of Architectural Engg., Kanagawa University, 3-27, Rokkakubashi, Kanagawa-ku, Yokohama, 221, Japan*

Abstract

This paper clarifies the characteristics of wind and wind-induced vibration of an electrical transmission tower installed on a mountainous area, based on full-scale data obtained at wind speed of less than 25m/s. The data was collected between 1991 and 1993. As a result, it can clearly be seen that difference of wind direction has a strong influence on wind characteristics. The vibration characteristics of the tower with conductors are strongly influenced by the behavior of the conductors. The aerodynamic damping of the conductors plays an important role in their total damping. The response values for acceleration, strain of the tower with conductors , and tensile forces of conductors increase in proportion to the power of the wind speed. In general, the value of each power index is less than 2.

1. INTRODUCTION

Recently in Japan, large-scale towers for electric power transmission have mostly been constructed in mountainous areas, due to geographical conditions. The local wind must be evaluated for the structural design of towers in mountainous areas. The Japanese design standards for transmissions (JEC-127,1979) give a recommendation regarding local wind in mountainous areas. However, a design procedure for local wind has yet to be clarified in concrete terms. Under this situation, the wind load is investigated empirically and analytically (e.g., Momomura et al.,1991). However, the validity of the results cannot be adequately confirmed, owing to the insufficiency of full-scale measurement data (e.g., Mitsuta et al., 1982) obtained in mountainous areas.

In this study, the wind-induced vibration of transmission steel towers and the wind characteristics in a mountainous area were clarified from the results obtained from full-scale measurements. The measurements were made only at a site where a tower stands on the top of a mountain with a height of 1243m above sea level.

2. OUTLINE OF FULL-SCALE MEASUREMENTS

2.1 Topographic Shape of Area Surrounding Measuring Point

The measuring point is located at lat. 35° 36'N and long. 138° 47' E, on the top of mountain with a height of 1243m above sea level. This area is situated in a prominent mountainous area of Japan having many peaks with heights of 1000m~ 3000m. With regard to the area within a radius of 10km from measuring point, there are a few mountains which are higher than the point on the SW and NE sides. Furthermore, there are many complicated valleys situated on the NW and SE sides. Figure 1 shows the topographic shape of the area within a radius of 1500m from the measuring point. It is situated on the ridge-line extending out to the east and west. The topographic feature of the northern area within a radius of 500m is a valley with a conical shape surrounded by ridge-lines each of which extends out toward the north in both easterly and westerly directions. Compared with the north side, the topographic shape of the south is a flat slope. The angle of inclination of the slope in both northerly and southerly directions is approximately 25° except for some small uneven shapes on the slope. The northern and southern areas are overgrown with trees standing approximately 10m and 20m high, respectively.

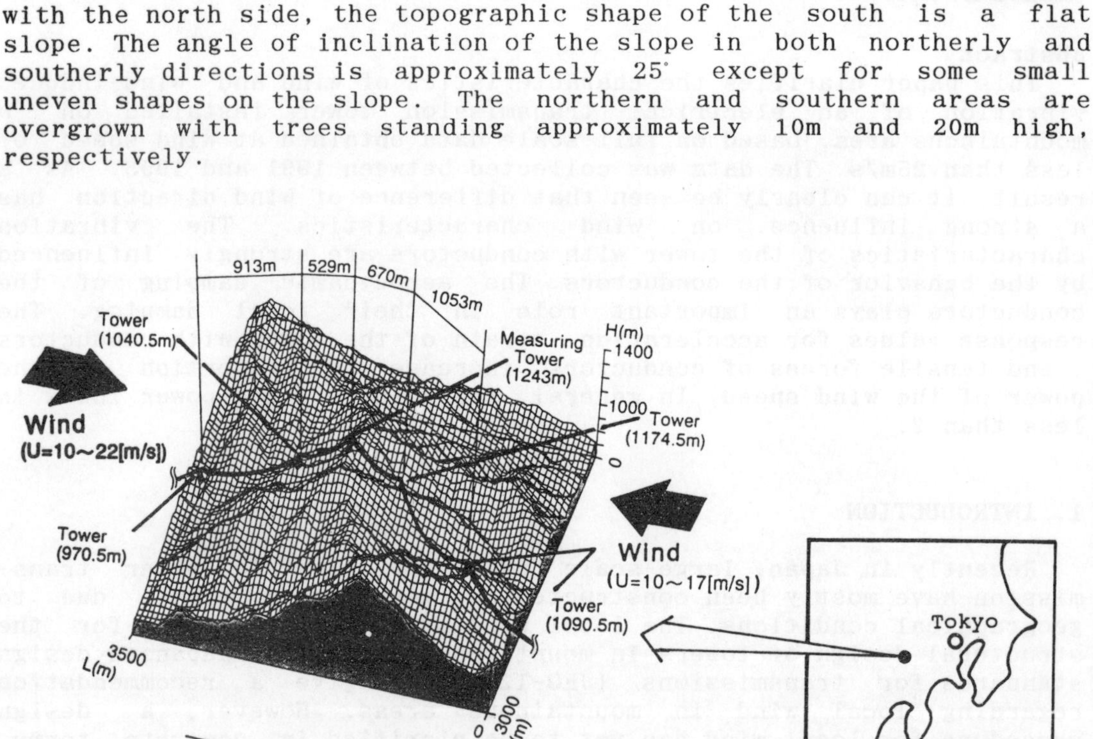

Figure 1. Topographic Shape of Area Surrounding Measuring Point.

2.2 Measuring Tower and Measuring Details

The measuring tower constructed on the top of the mountain is a tension type steel tube tower with a height of 96m. This measuring tower is connected to the adjacent northern and the southern towers using 8-bundled conductor lines at each arm of the towers through the insulators(Figure 1.). The span of the adjacent tower is 670m on the north side and 529m on the south side. The angle of inclination of the line chord is 6° in the case of the north side and 21 in the case of the south side. Horizontal component of the initial tensile force of the 8-bundled conductor is 29× 8kN in temperature condition of 15° C. The conductor is made of 7 steel wires at its center and 54 aluminum wires. Its diameter is 3.4cm and its weight is 22.7N/m.

Measurements were carried out for wind speed, wind direction, acceleration of the tower, strain of the tower and tensile forces of the conductors. The measurements were started by the auto trigger system based on the mean wind speed during a 10 minute period. Each run of the measurement continued for 1800sec. Sampling time of the data was 0.02sec. The average time for each run of analysis was 600sec. The frequency regions for the analysis are shown in Figure 2.

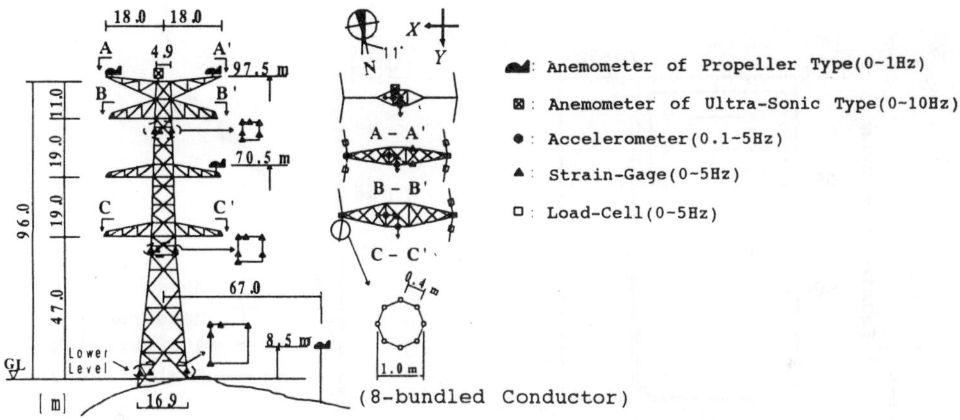

Figure 2. Measuring Tower and Items.

3. Structural Dynamic Characteristics of Measuring Tower

A measurement of the microtremor and a test of free vibration were done in order to understand the structural dynamic characteristics of the measuring tower. The microtremor measurements were carried out for the "tower without conductors" and for the " tower with conductors". During the microtremor measurement for the tower without conductors, the mean wind speed at the tower's top was under 2m/s or less. As for the tower with conductors, it was approximately 5m/s. In the test of free vibration for the tower without conductors, vibration was induced by artificial excitation. The critical damping ratio was estimated from the time series of free vibration. In addition, it was estimated by the random decrement technique(Tamura et al.,1993)(hereafter referred to as R.D.T) using the time series of microtremor induced response.

3.1 Tower without Conductors

Figure 3 shows the eigenvalues for the tower without conductors obtained from the microtremor measurement. There is only one peak at 1Hz in the Fourier spectrum of the tower's displacement at a height of 85m in both X and Y-direction. These vertical vibration mode shapes are similar to each other. The values almost correspond to the values computed by Ohkuma et al.(1993).

The critical damping ratio of the tower estimated from the free vibration test is approximately 0.1% in X-direction. In Y-direction, it is approximately 0.3%. The R.D.T results to approximately 0.1% in each direction.

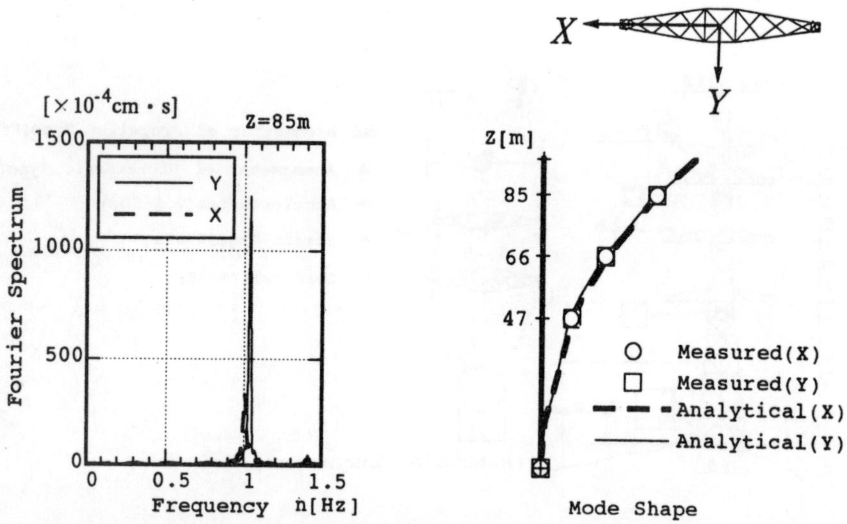

Figure 3. Eigenvalues for Tower without Conductors obtained from Microtremor Measurement.

3.2 Tower with Conductors

Figure 4 shows the eigenvalues for the tower with conductors gained from the microtremor measurement. The Fourier spectrum of the tower's displacement at the height of 85m has many peaks in the frequency range of 0.1 to 1.0Hz in both X and Y-directions. This is different from the structural characteristics of the tower without conductors. Referring to the eigenvalues computed by Ohkuma et al.(1993), those values are caused by the behavior of the conductors. The peak at 0.23Hz is largest among all of these peaks in both X and Y-directions. The vibration mode shape of the conductors at 0.23Hz shows three loops in plane. Each mode shape of the tower in the range of 1Hz or less almost corresponds to the fundamental mode of the tower without conductors. The numbers of natural frequency shown in Figure 4 are for convenience.

As a result of the R.D.T, the critical damping ratio at 0.23Hz is approximately 3.3% in X-direction, and 1.7% in Y-direction. These values are larger than those for the tower without conductors.

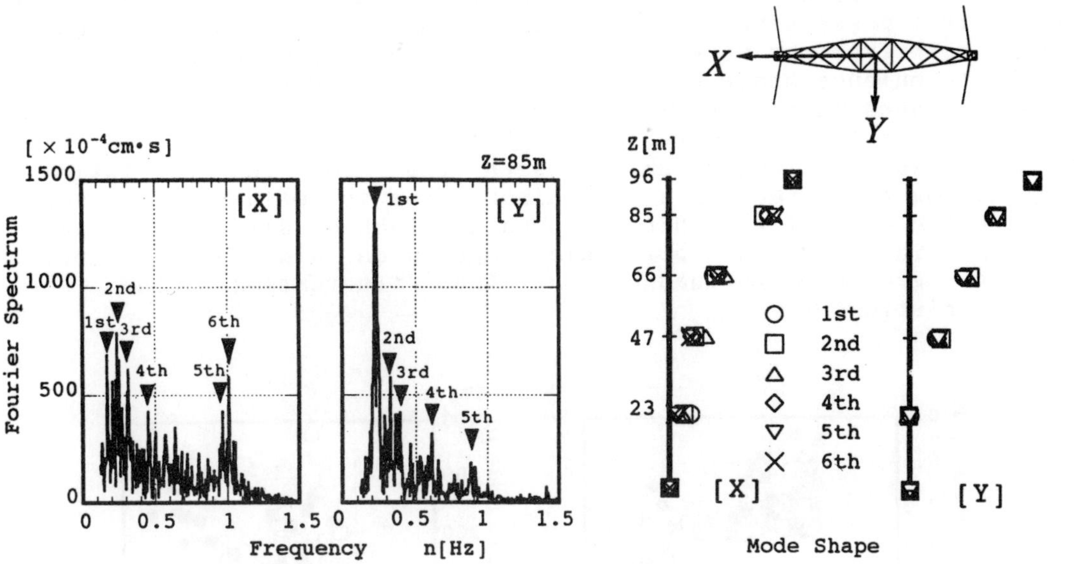

Figure 4. Eigenvalues for Tower with Conductors obtained from Microtremor Measurement.

4. WIND CHARACTERISTICS

The wind characteristics of the wind direction component(u-component) during a typhoon are compared with those during a non-typhoon. After this, the wind characteristics in each direction are compared to each other.

4.1 Wind Characteristics at Tower's Top

The characteristics of fluctuating wind speed u-component gained using a three-dimensional ultra-sonic anemometer at the tower's top (Z=97.5m) are described. The vertical angle of attack was approximately 20° regardless of wind speed and wind direction. Figure 5 shows the characteristics of the fluctuating wind speed u-component at the tower's top.

In comparison between the characteristics of the fluctuating wind speed during a typhoon and those during a non-typhoon in the case of the wind direction being S, they are not in general different from each other. The wind turbulence intensity decreases with an increase in the wind speed. The dispersion of the turbulence scale is large, and the value generally scatters over a range of 100 to 300m regardless of the wind speed. The power spectral shape of the fluctuating wind speed corresponds to a Karman's type.

Comparing the wind turbulence intensity of wind with the wind speed of 20m/s or more during Typhoon 9119, where the wind direction was SSW, with that during Typhoon 9313, with a wind direction of S, the value for Typhoon 9119 was almost twice the value for Typhoon 9313. The turbulence intensities measured at a height of 187m in an urban field possessing similar ground roughness in all directions during both Typhoons were not different from each other.

The turbulence scale was approximately 50m regardless of the wind speed in cases where the wind direction was NNW. The value is smaller than that for the wind direction S~ SSW.

As a result, it can clearly be seen that the characteristic of fluctuating wind speed in a mountainous area is influenced by even a small difference of wind direction rather than by different type of typhoons. Therefore, a topographic shape in an along-wind direction has a greater influence on the fluctuating wind speed characteristics.

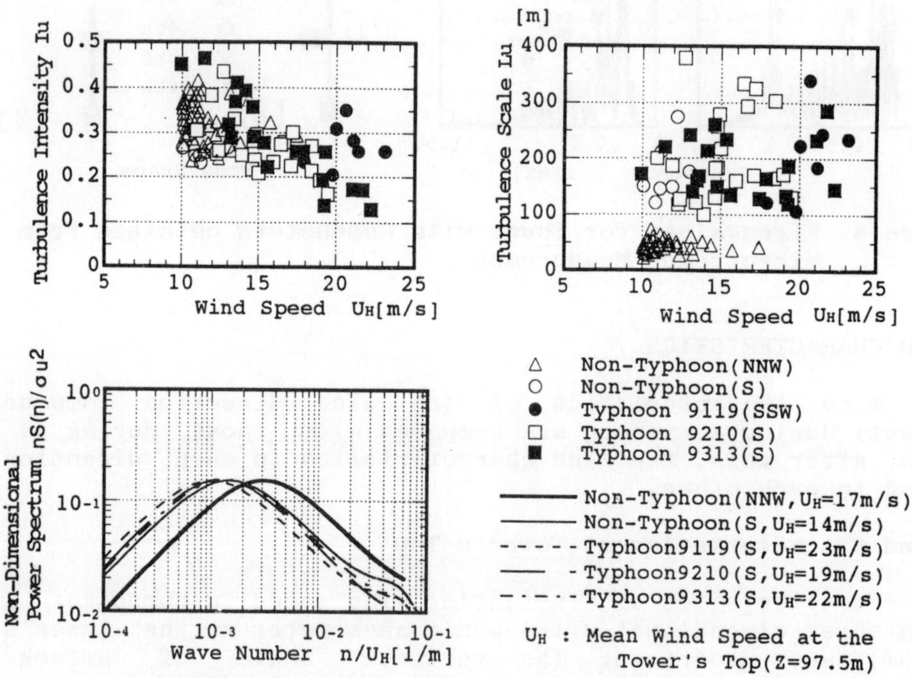

Figure 5. Fluctuating Wind u-component Characteristics at Tower's Top.

4.2 Spatial Characteristics

Vertical distribution of mean wind speed and turbulence intensity, and spatial correlation of fluctuating wind speed are described using data obtained at three levels of the tower by an anemometer of propeller type. The length constant of the anemometer is 3.8m. The characteristics of the fluctuating wind speed obtained by the propeller type anemometer is almost the same as those obtained by the ultra-sonic in a range of wave number(n/U) 0.01 or less.

Figure 6 shows the vertical distribution of mean wind speed and the turbulence intensity, which are normalized by the values at the top of the tower(Z=97.5m). The shapes of the vertical distribution of mean wind speed in any wind direction are almost similar to each other. The wind speed value at Z=70.5m is larger than that at Z=97.5m. The distribution shape of the turbulence intensity is different from that of wind speed. The value at Z=70.5m is the smaller than that at Z=97.5m. The vertical changing ratio of the turbulence intensity distribution in wind direction NNW is larger than that in S or SSW.

Figure 7 shows the co-coherence in vertical Cov(n*) of the fluctuating wind speed and the transverse co-coherence in horizontal Cot(n*). Reduced frequency n* in Figure 7 is defined by the projected length Lij and the averaged value Ua of the mean wind speed between the two points. When comparing the Cov(n*) in each case, tendencies in which the value of Cov(n*) decreases with an increase of n* are similar to each other. However, the value of Cov(n*) in wind direction NNW is smaller than the value in wind direction S in a low frequency range. Cot(n*) in NNW is smaller than Cot(n*) in S at n*=0. However,the value of Cot(n*) in NNW decreases gradually with an increase of n* in comparison with the values in S.

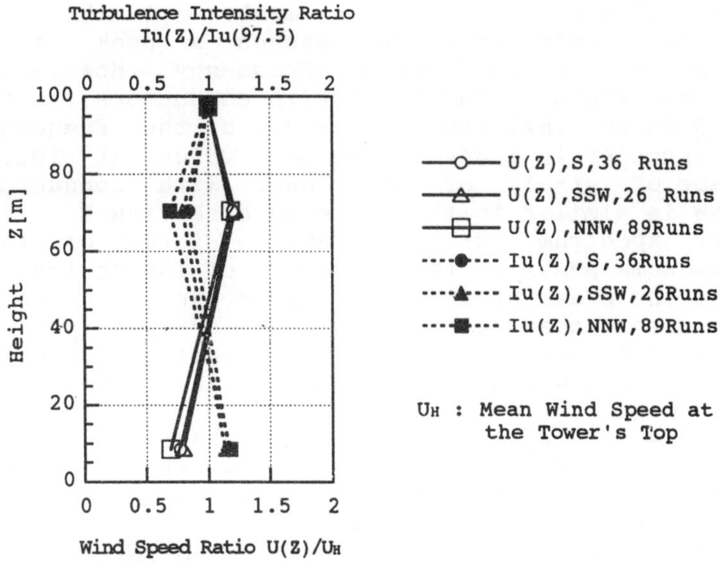

Figure 6. Vertical Distribution of Mean Wind Speed and Turbulence Intensity.

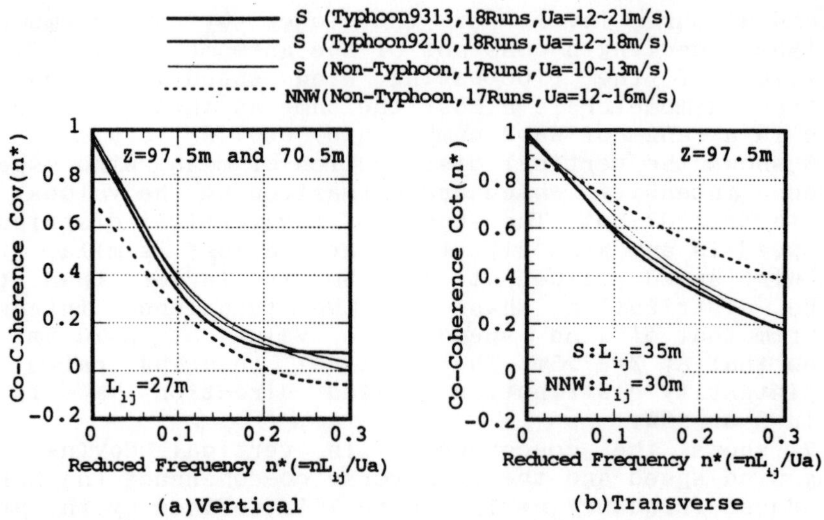

Figure 7. Co-Coherence of Fluctuating Wind Speed.

5. RESPONSE CHARACTERISTICS

5.1 Spectral Characteristics of Response

Figure 8 shows the power spectra for the response strain of the main member of the tower, the tensile force of the 8-bundled conductor and the response acceleration of the tower.

As for the strain of the main member of the tower without conductors, the power spectrum reaches a peak at about 1Hz corresponding to the fundamental frequency. However, the power spectrum of the strain of the tower with conductors reaches peaks at 0.23, 0.33, 0.68 and 1Hz. The peak values at the frequency of less than 1Hz are greater in general than the value at 1Hz. The power spectral shape of strain of the tower with conductors at wind direction NNW is similar to that at wind direction S.

The power spectrum for the tensile force of the 8-bundled conductor has some peaks at the frequency of 1Hz or less. These peaks correspond to the peaks of the main member strain on the whole.

The power spectrum of the response acceleration at Z=85m in Y-direction has more peaks than that in X-direction at the frequency of 0.7Hz or less. The power spectral shape of the response acceleration at Z=85m is similar to the shape at Z=47m.

Figure 9 shows the spanwise co-coherence of response acceleration of the tower with conductors in Y-direction at the two points of Z=85m and Z=47m. The value of the co-coherence is over +0.8 at the peak frequency of response acceleration.

Figure 10 shows the co-coherence of tensile force at points ① and ② of the 8-bundled conductor at the height of Z=85m. The value is relatively large at the peak frequency of the tensile force, and is smaller at a higher frequency.

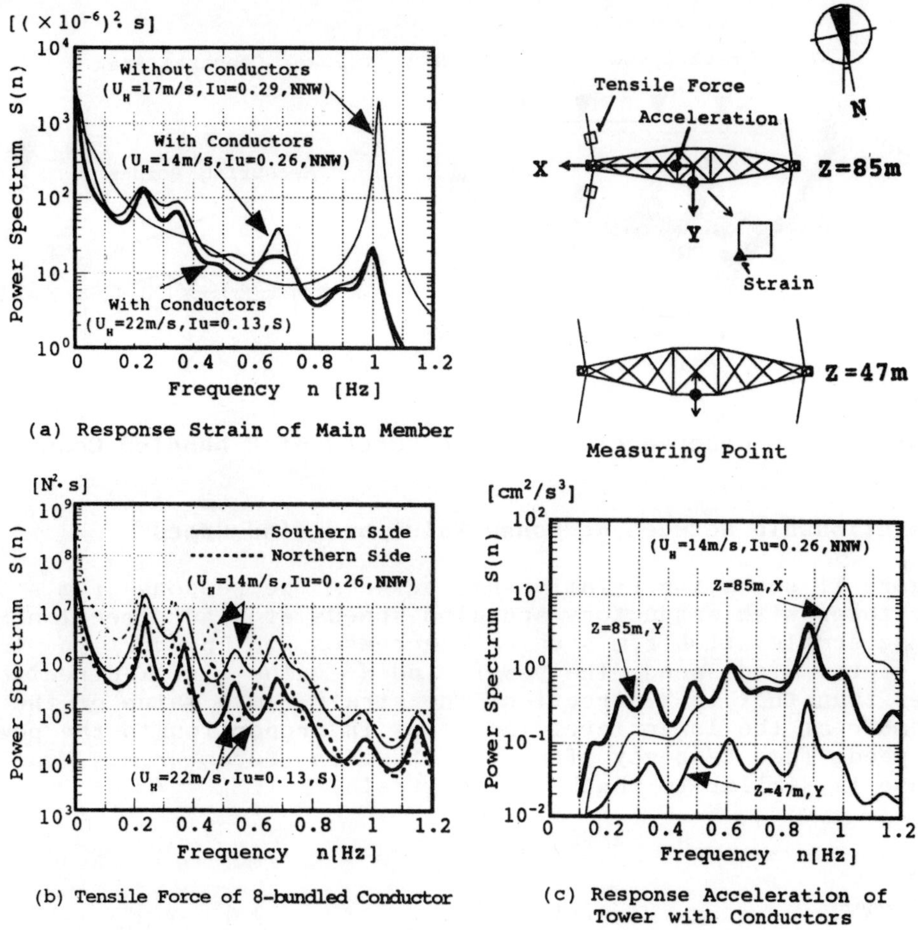

Figure 8. Power Spectrum of Wind-induced Vibration.

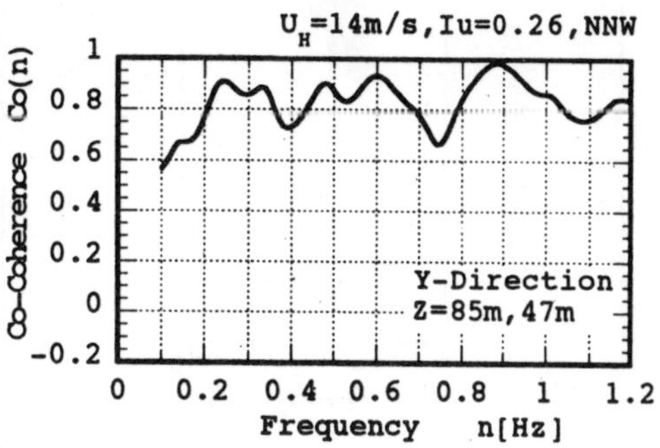

Figure 9. Co-Coherence of Response Acceleration of Tower.

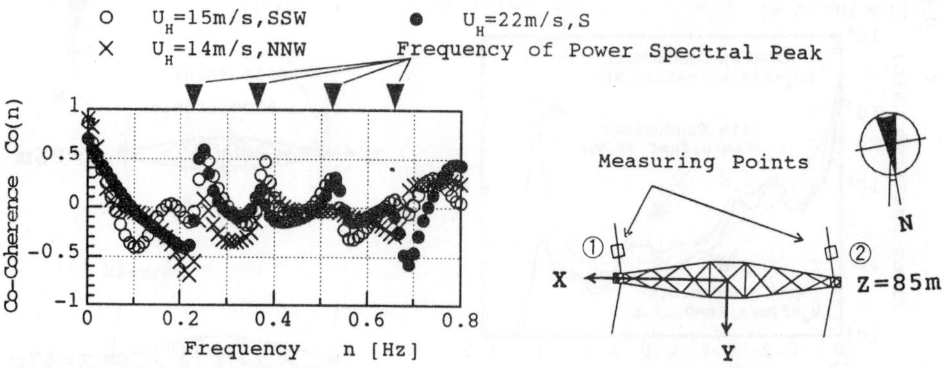

Figure 10. Co-Coherence of Tensile Forces of 8-bundled Conductor.

5.2 Relationship between Response Value and Wind Speed

Figure 11 shows the relationship between the response r.m.s values of the tower with conductors and wind speeds at wind direction S. The response acceleration r.m.s value increases in proportion to the power of the wind speed. The power index value in Y-direction is smaller than that in X-direction. The strain r.m.s value of the main leg member at the lower level increases in proportion to the power of wind speed. The tendency for the tensile force r.m.s value to increase is similar to that for the strain.

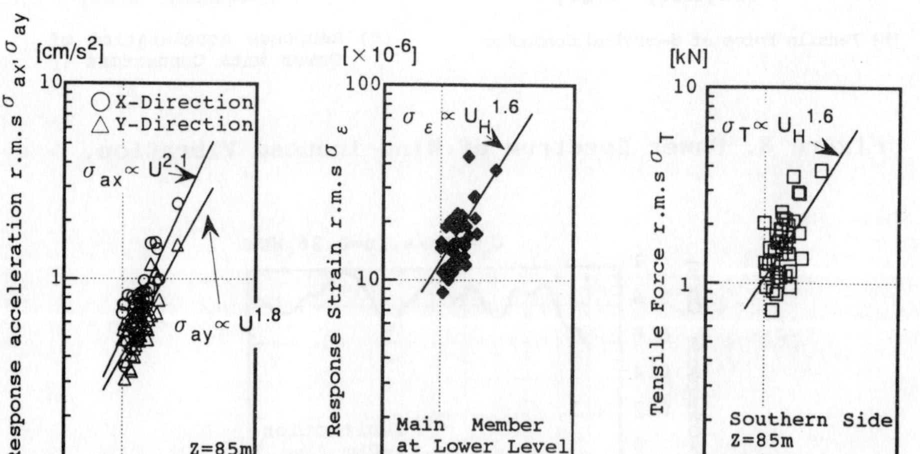

Figure 11. Relationship between Response r.m.s Value and Wind Speed.

5.3 Damping of conductor during wind-induced Vibration

As a result mentioned above, it can clearly be seen that the response of the tower with conductors is greatly influenced by the behavior of the conductors. Aerodynamic damping acting on the conductors plays an important role in the behavior of the conductors (Davenport.,1988). The critical damping ratio of the 8-bundled conductor during wind-induced vibration was estimated by using the R.D.T. The tensile force of the 8-bundled conductor at Z=85m(See Figure 10.) was used as data obtained when the wind speed was in a range of 10~ 22m/s, and the wind direction was S. This data was divided into six groups for the investigation at intervals of 2m/s for the wind speed. Critical damping ratio was estimated for predominant frequencies of the 8-bundled conductor which are 0.24, 0.37 and 0.54Hz.

Table 1 shows the number of times for the superposition of samples in the R.D.T. The value estimated by the superposition 200 times disperses within a range of approximately 20% of the value estimated by the superposition done 1500 times.

Figure 12 shows the relationship between the critical damping ratio and the wind speed. In consideration of the dispersion varying with the number of times for the superposition, the change of the critical damping ratio for wind speed may be negligible. With an increase in frequency, the critical damping ratio becomes smaller. Referring Iwama et al.(1982) study that the structural damping ratio of the conductors is 0.4%, it can be clarified that aerodynamic daming plays a large part in critical damping ratio of conductors.

Table 1. Number of Times for Superposition in R.D.T.

Natural Freq.	Mean Wind Speed at the Tower's Top					
	10~ 12m/s	12~ 14m/s	14~ 16m/s	16~ 18m/s	18~ 20m/s	20~ 22m/s
0.24Hz	1246	1534	987	394	571	201
0.37Hz	1863	2488	1576	699	864	372
0.54Hz	2665	3459	2323	1071	1380	538

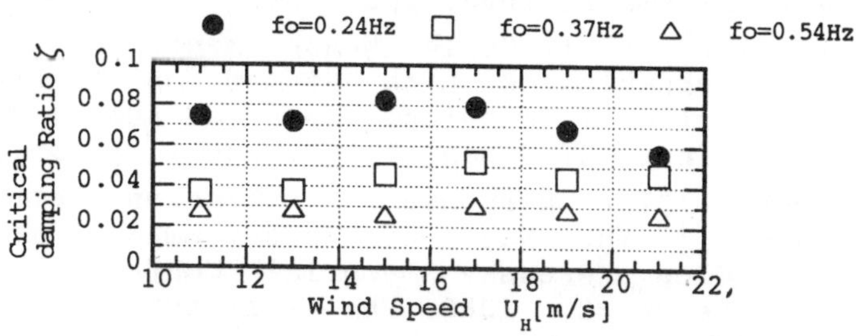

Figure 12. Damping of 8-bundled Conductor During Wind-Induced Vibration.

6. CONCLUSION

Main results of the full-scale measurements are as follows.

(1) Difference of wind direction has a strong influence on the wind characteristics.

(2) Vertical profiles of the mean wind speed and the turbulence intensity are influenced by topography

(3) Vibration mode shape of the tower with conductors in a frequency range of 1Hz or less is similar to the fundamental mode shape of the tower without conductors.

(4) Vibration characteristics of the tower with conductors are greatly influenced by the behavior of the conductors.

(5) The aerodynamic damping of the conductors plays an important role in the total damping of the conductors.

Valuable data which can be utilized for empirical and analytical studies in the future were obtained from the full-scale measurements, although the measurements were made only at a single point situated in a mountainous area.

ACKNOWLEDGEMENTS

The authors would like to acknowledge the financial support given for the full-scale measurements by Tokyo Electric Power Co.,Inc.

REFERENCES

The Japanese Electrotechnical Committee. (1979),"Design Standards on Structures for Transmissions", Published by Denkishoin.

Momomura,Y., Marukawa,H., Ohkuma,T.(1991), "Wind-Induced Vibration of Transmission Line System", Proc. of the 8th International Conference on Wind Eng., Canada, July,1991, Part3, pp2035~ 2046.

Mitsuta,Y. et al.(1982), "Spatial Distribution of Wind Turbulence over Complex Terrain", Proc. of 7th National Symposium on Wind Eng., Tokyo, Japan, December, 1982, pp45~ 50.

Tamura,Y. et al.(1993), "Evaluation of Damping Ratio of Randomly Excited Buildings using the Random Decrement Technique", J. of Structural and Construction Eng., Architectural Institute of Japan, No.454, December, 1993, pp29~ 38.

Ohkuma,T. et al.(1993),"Full-Scale Measurements of Wind-Induced Vibration of Transmission Line System on Mountainous Area (part.1)", Summaries of Technical papers of Annual Meeting 1993 , Architectural Inst. of Japan, Structures 1, B, pp1149~ 1150.

Davenport,A.G.(1988), "The Response of Tension Structures to Turbulent Wind : The Role of Aerodynamic Damping", 1st International Oleg Kerensky Memorial Conference on Tension Structures, June 20-22, 1988, The Institution of Structural Engineers, London, England.

Iwama,T. et al.(1982), J. of Steel Tower,No.53,1982, pp277~ 299. (in Japan).

FULL-SCALE FIELD EXPERIMENTS ON A 101 m HIGH LATTICE TOWER

SHANMUGASUNDARAM, J., GOMATHINAYAGAM, S., HARIKRISHNA, P., ARUMUGAM, M. and VENKATESWARLU, B.

Scientists, Structural Engineering Research Centre, Madras, India

Abstract

The characteristics of wind and the response of a 101 m high lattice tower are studied by instrumenting the tower with anemometers and accelerometers at various levels. The experimental structure is located in a sub-urban area in Madras very near to the sea coast. The predominent wind in this particular site is in the South-East direction. The mean values of power law coefficient, surface roughness length, shear friction velocity, surface drag coefficient, turbulence intensity and length scales are calculated and compared with values available in the literature for similar type of terrain.Further the turbulence power spectrum at different heights are evaluated and compared with spectra of Davenport, Harris, and Shiotani.The measured fundamental frequency of the tower is calculated and compared with the theoretical (3-D FEM) values.

1. INTRODUCTION

The characteristics of wind and structural response to wind action can be best studied by conducting model tests in atmospheric boundary layer wind tunnels or by conducting full-scale field experiments. Sometimes it is very difficult to reproduce the exact field conditions such as terrain and turbulence characteristics in model studies. So full scale field experiments are resorted to. Since they are expensive and time consuming, limited number of field experiments have been carried out in India. Structural Engineering Research Centre (SERC), Madras, has taken a lead in conducting full-scale field experiments on wind characteristics on a 101 m high lattice tower (Fig. 1), which is situated in a semi-urban area near the coast,[Shanmugasundaram, et al (1993)]., in Madras. The main objective of the experiment is to measure the wind speed and direction at different levels in order to determine the wind and terrain characteristics which form the primary data for estimating wind load on structures. The acceleration of the tower corresponding to these wind speeds was also measured to obtain the structural response. Details of this experiment are briefly presented in this paper. This paper also includes the comparison of the test results with other results published in the literature.

2. TEST STRUCTURE

The test structure for carrying out the full-scale field experiment is a free standing square lattice tower situated on the sea coast in Madras. It is made of mild steel angle sections with bolted

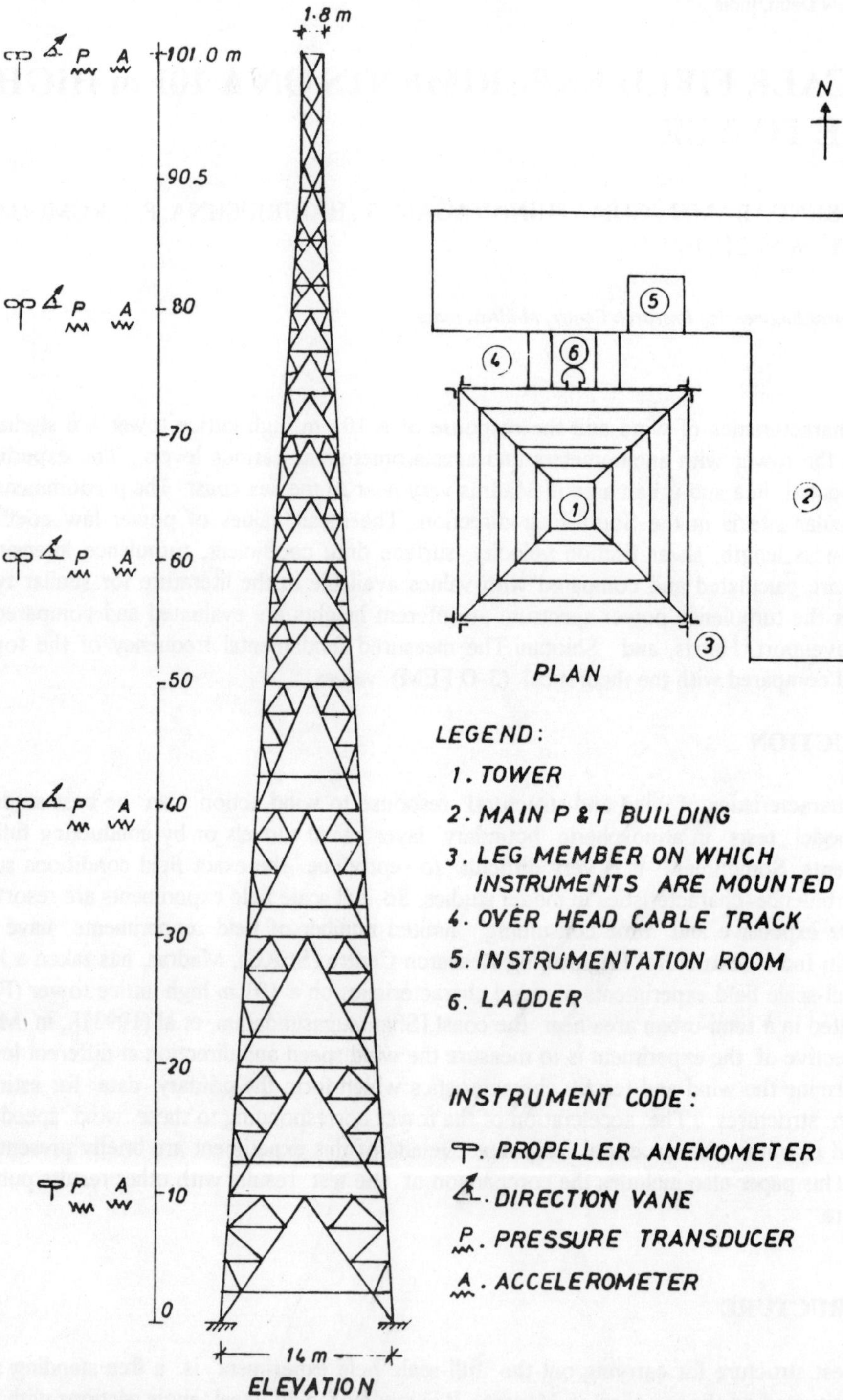

Figure 1. Schematic plan and side elevation of structure

connections. The height of the tower is 101 m above the ground level, and the widths at top and base are 1.8 m and 14.0 m, respectively. The structure is mounted with 11 dish antennas of varying diameters at different heights starting from 60 m from ground for communication purpose. The east and north sides of the test structure are shadowed by a building, up to a height of 30 m. The neighbouring buildings and structures surrounding the tower are at a height varying from 10 m to 30 m. Since the structure is a microwave tower used for telecommunications, no drilling of holes on the tower members was permitted.. Hence a special frictional clamping system has been designed and fabricated at SERC for mounting the sensors. The details of the clamping arrangement for the instrumentation are shown in Fig.2.

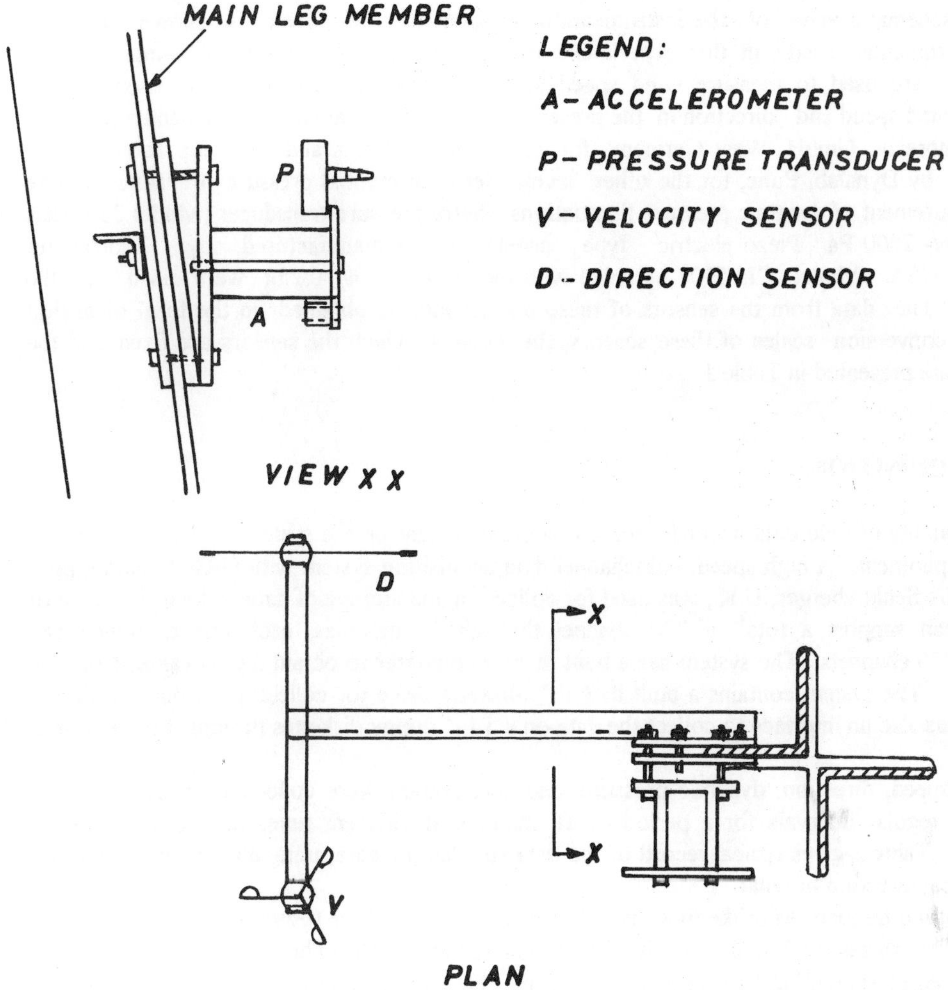

Figure 2. Details of clamping arrangement for the instrumentation

3. INSTRUMENTATION

Different instruments were fixed to the south east leg of the tower at 5 different levels (10, 40, 60, 80 and 100 m) for the collection of the following data.

i) Wind velocity - 5 levels(10 m,40 m,60 m,80 m, 100 m)
ii) Wind direction - 4 levels(40 m, 60 m, 80 m, 100 m)
iii) Dynamic pressure - 5 levels(10 m, 40 m, 60 m, 80 m, 100
iv) Acceleration of the tower - 5 levels(10 m,40 m,60 m,80 m,100 m)

The schematic view of the instrumentation system of the tower is shown in Fig.1 Various instruments used in this experiment are briefly explained in the following sections. Anemometers are used to measure wind speed and direction.The anemometers used for the collection of wind speed and direction in the present study are 3 cup anemometers manufactured by M/s.Wilh Lambrecht, GmbH, West Germany, for 10 and 40 m levels and 3 cup anemometers manufactured by Dynalab, Pune, for the other levels. Setra differential pressure transducers were used for measurement of dynamic pressure fluctuations. Setra pressure transducer (Model 239). has a range of +/- 2000 Pa. Piezo electric type accelerometers manufactured by Vibrometer Corporation, USA (Model CI 510 M 101) with the range of +/- 0.25g were used in the present study. The data from the sensors of these instruments is obtained in the form of analog signals. The conversion scales of these sensors, the levels at which the sensors are fixed and the other details are presented in Table 1.

4. DATA ACQUISITION

The quality of field data depends to a considerable extent on the system of data acquisitions used in the experiment. A high speed multichannel data acquisition system with ORION data logger supplied by M/s Schlumberger, U.K., was used for collection and storage of data. The main frame of data logger can support a total of 200 channel through 10 modules, each one of which can accommodate 20 channels. The system has a built in A/D converter to obtain digital values from the analog signals. The system contains a built in 3 1/2" diskette drive for collection of data on floppy diskettes. It has also an interface to collect the data on a 5 1/4" floppy diskettes through a tracker unit.

Wind speed, direction, dynamic pressure, and acceleration were collected through the data acquisition at regular intervals for a period of 18 minutes at different times during day time on different days. Table 2 gives typical record of the data such as the date, time, maximum wind speed and the principal direction of wind.

The natural frequencies of the first three modes of vibration of the tower were calculated to be below 5 Hz. This makes the Nyquist frequency to be greater than 5 Hz. The corresponding sampling interval Δt expected should be the inverse of the double the Nyquist frequency i.e.$\Delta t = 1/2.N_f = 0.1$ second. Thus to get the required sampling rate of 10 Hz , the data were collected in a compact binary coded format on the 3 1/2" diskettes. These data were then decoded and converted to the standard engineering units for further analysis at the Laboratory.

Table 1.Conversion Scales for Different Sensors

Measurement	Instrument	Measuring Range	Output (V)	Conversion Scale	Levels (m)	Sampling Rate (Hz)
Wind Speed	Dyna Lab (3-cup) Anemometer	0 - 240 kmph 0 - 66.67 m/s	0 - 3.0 V	1Volt=80 kmph =22.22 m/s	60,80,100	10
	Lambrecht Anemometer	0 - 60 m/s 0 - 216 kmph	0 - 10.0 V	1 Volt=6 m/s =21.6 kmph	10,40	10
Wind Direction	Dyna Lab	0 - 360	0 - 2.5 V	1 Volt=144	60,80,100	10
	Lambrecht	0 - 360	0 - 8.0 V	1 Volt=45	10,40	10
Pressure (differential)	Setra Pressure Transducer	0 -2000 Pa	0 -2.5 V	1 Volt=800 Pa	10,40,60 80,100	10
Acceleration	Vibrometer Accelerometer	0 0.25 g 0 - 2.45 m/s	0 -12.0 V	1Volt=0.125 =1.226 m/s	10,40,60 80,100	10

Table 2. Typical Wind Data on Different Dates

Date	Time	Duration (min.)	Principal Direction	Maximum Velocity (m/sec)
3.5.93	15.50	18	SE, SOUTH	68.0
12.5.93	12.35	18	SE, SOUTH	53.0
26.5.93	14.50	18	SE, SOUTH	37.0
27.5.93	15.20	18	SW, WEST	17.5
10.6.93	12.30	18	SE	15.0
28.6.93	15.35	18	SE	22.5
7.7.93	16.05	18	SE	18.5
12.7.93	11.30	18	SW, WEST	34.0
21.7.93	15.40	18	SE, SOUTH	27.0
27.7.93	15.15	18	NW	21.0
30.7.93	12.55	18	NW, WEST	29.0
2.9.93	15.40	18	NE, EAST	16.0
16.9.93	15.10	18	SOUTH, SW	27.5

(SE - South East; NW - North West;
NE - North East; SW - South West)

5. DATA ANALYSIS

As seen in Table 2, the wind at the test site is predominant in the south-east direction. Hence these data are presented, and analysed for discussion in this paper. The analysis of these data includes determination of the wind characteristics such as the mean wind velocity and its variation with height, power law coefficient, turbulence length scales and the spectrum, and the terrian characteristics such as the aerodynamic roughness height, and the zero-plane displacement.

6. MEAN VELOCITY AND ITS PROFILE

The mean wind speed (averaged over 15 minutes of duration) of the records in the south-east direction varied from 5.43 m/s to 11.4 m/s at 100 m level, while the mean wind direction varied from 120° to 157°. A typical set of wind velocity traces at 60m, 80m, and 100m levels are given in Fig. 3. The corresponding mean velocity profile is presented in Fig. 4. The values of power law coefficient (α), the surface roughness length (z_o), the shear friction velocity (u_*), the surface drag coefficient (K), and turbulence intensity (I_z) along the height(Fig. 5.) for the present test terrain are given in Table 3.

7. TURBULENCE LENGTH SCALE

In the present study, the turbulence length scale values are calculated based on all the three methods suggested by Flay et al, (1988). The values of length scales at 60 m height for different runs of the south-east direction are given in Table 3. A large scatter of these values was observed in different runs. However, it was observed that the values of the length scale increases with height (Fig. 6).

8. TURBULENCE POWER SPECTRUM

In order to evaluate the realistic wind loads on structures which are dynamically sensitive, the power spectral density of the wind velocity is required. Various authors have come out with different empirical expressions for the turbulence spectrum based on field experiments. In the present study, attempts are made to evaluate the turbulence spectrum of wind in a coastal town based on field measurements.Fig.7 shows the turbulence spectrum for the south-east direction.

9. TERRAIN CHARACTERISTICS

It may be noted that the value of zero plane displacement is obtained as 7.0 m for the south-east direction. This zero plane displacement value has been obtained based on the best fit curve for calculating the surface roughness height, z_o. Also on inspection of the test site, the above values of the zero plane displacement taken for the analysis are found to be reasonable.

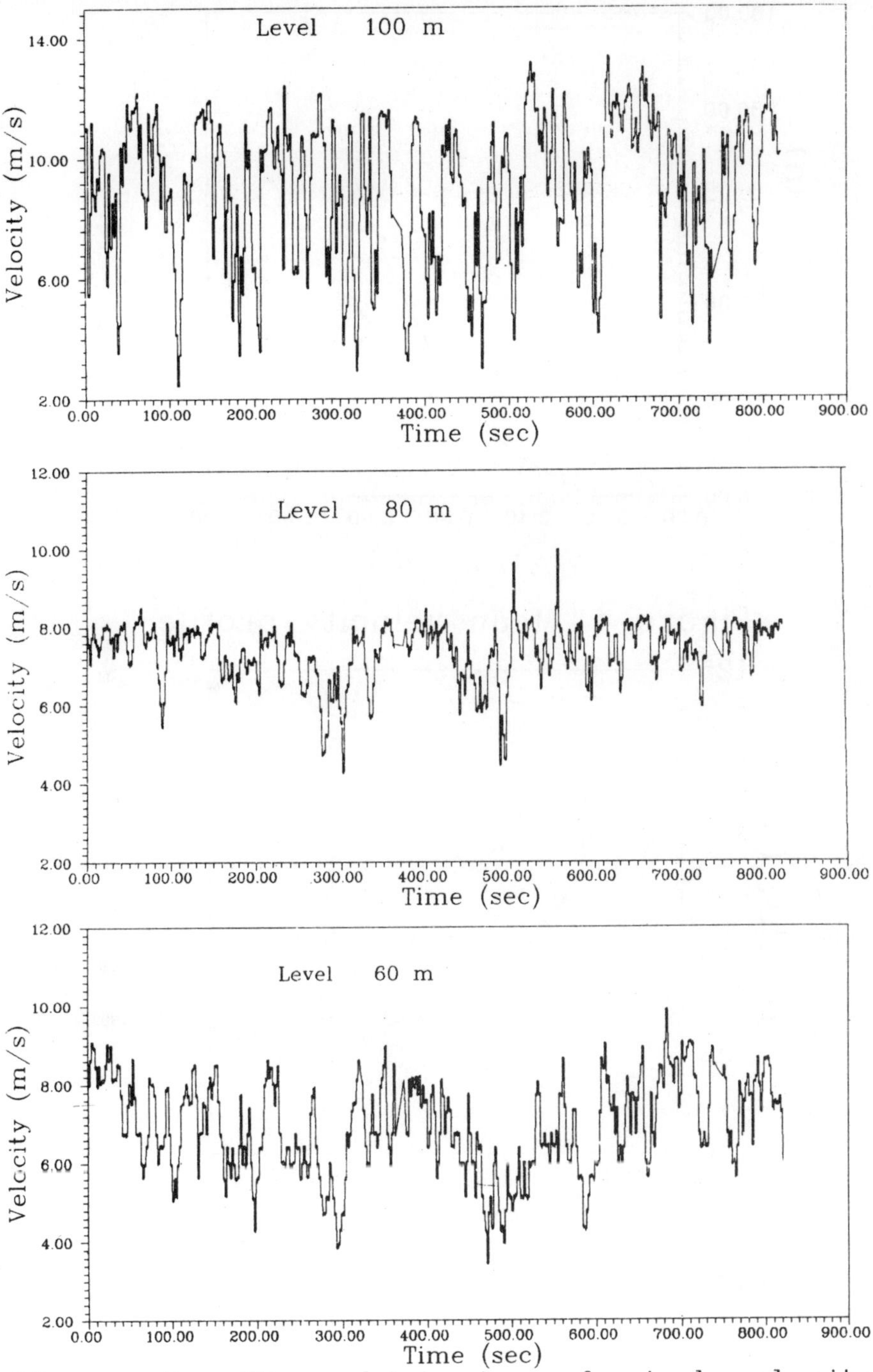

Figure 3. Time histories of wind velocity

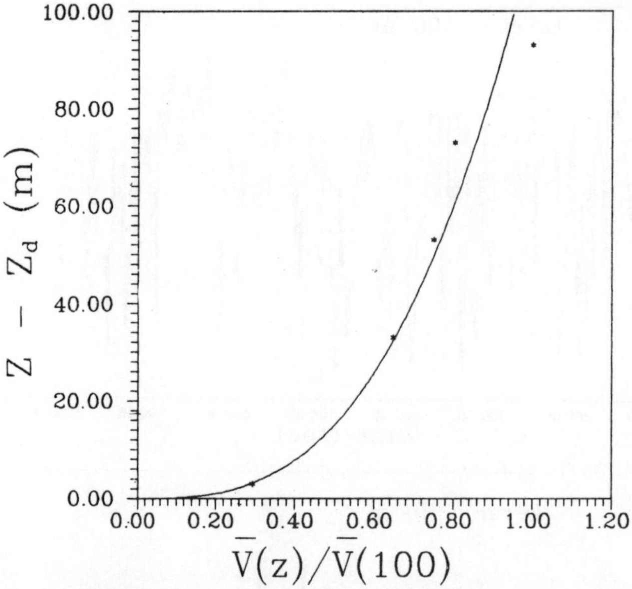

Figure 4. Mean velocity profile

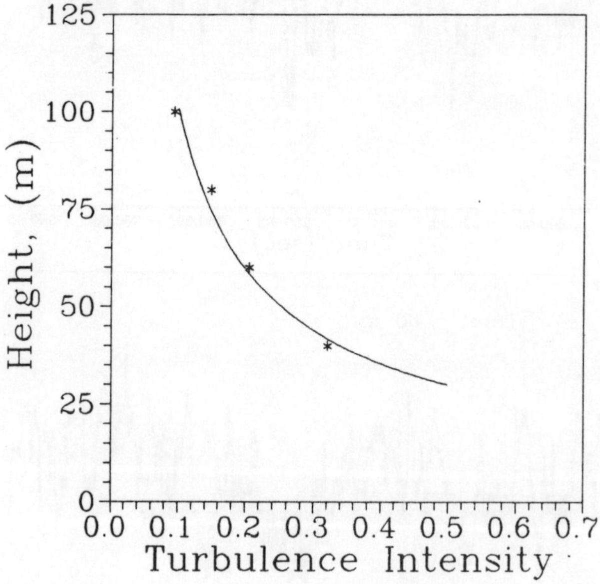

Figure 5 Turbulence intensity profile

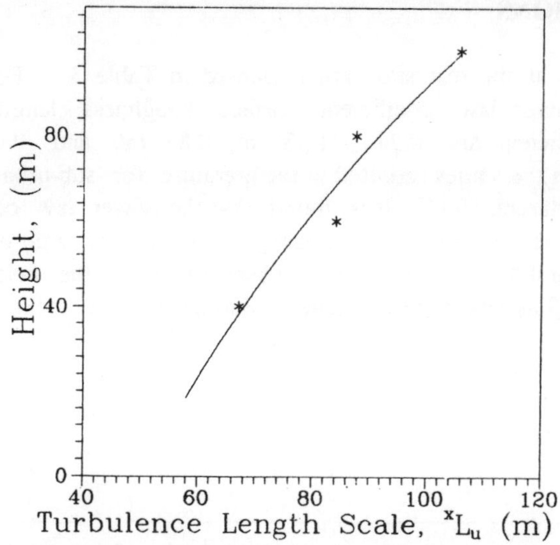

Figure 6.Turbulence length scales with height

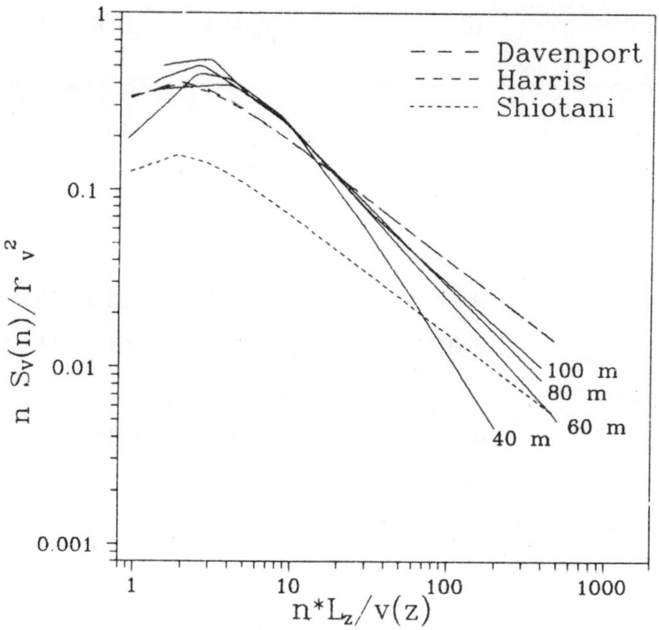

Figure 7 Comparison of turbulence spectra

10. RESULTS AND DISCUSSIONS

The wind characteristics at the test site are presented in Table 3. For the south-east direction, the mean values of power law coefficient, surface roughness length, shear friction velocity and surface drag coefficient are 0.315, 1.38 m, 0.83 m/s and 0.045, respectively. These values are compared with the values reported in the literature for sub-urban terrain and are given in Table.4 [Shanmugasundaram, 1993]. It is found that the power law coefficient for the experimental terrain is 0.32 which is closer to Category 3 (sub-urban) value of 0.28, than the terrain Category 4 (urban) value 0.40. Further it is observed, that the field measured value agrees well with the value of 0.33 as reported by Duchene-Marullaz,(1975) in his experiments on a sub-urban area.

Table 4. Terrain Characteristics

	Power Law Coefficient α	Roughness Length Z_0 (m)	Surface Drag Coefficient K
Experimental Values:	0.32	1.38	45.4
Duchene-Marullaz:	0.33	1.35	39.9
E.Simiu:			
Suburban terrain	0.28	0.8 - 1.20	25.1 - 35.6
Urban terrain	0.40	2.0 - 3.00	61.8 - 110.4
Davenport:			
Suburban terrain	0.28	0.4	15.0
Urban terrain	0.40	1.6	50.0

The mean values of turbulence intensity for the south-east direction are 0.85, 0.23, 0.16, 0.12 and 0.19 for 10 m, 40 m, 60 m, 80 m and 100 m above ground, respectively. As seen in Fig. 5, the turbulence intensity decreases as the height increases, except at 100 m level. This may be due to the presence of three dish antennas of 3.66 m diameter at 100 m level which makes the wind more turbulent locally in this region than expected. Further the experimental values of turbulence intensity at 40 m and 60 m level, 0.23 and 0.16, agree well with the values of 0.24 and 0.20 as reported by [Duchene Marullaz (1975)] in his study. The average values of turbulence length scales obtained in this experiment are very much less than the values reported by Duchene Marullaz(1975) and Shiotani(1967).

Table 5. Turbulence Length Scales

Author	Turbulence length scale $^{x}L_{u}$ (m)	
	At 40 m	At 60 m
Duchene Marullaz	145	190
Shiotani	155	-
Present experiment	42	51

As seen in Table 5, the average values of the turbulence length scales obtained in this experiment are much less than the values reported by Duchene-Marullaz (1975) and Shiotani (1967). This difference may be due to low mean velocities of 3.9 to 6.0 m/sat 40 m level and 4.2 m/s to 7.1 m/s at 60 m level. These velocities are very much less compared to the mean velocities of about 15 m/s at 40 m 18 m/s at 60 m heights as reported by Duchene-Marullaz.

The turbulent wind spectrum at different heights are plotted with spectra of Davenport, Harris and Shiotani (Fig.7). It is clearly seen from Fig.7 that the measured spectrum agrees reasonably well with the spectra proposed by Davenport and Harris. The spectral values of this experiment are found to be more in the low frequency region and less in the high frequency region compared to the values proposed by Davenport and Harris.

The measured dominant frequency of the tower from accelerograms was 1.13 Hz which compares well with the theoretical prediction of 1.127 Hz using 3-D FEM analysis.

11. CONCLUSION

The experimental investigation described in this paper shows the viability of measuring wind and response characteristics of existing wind-sensitive tower without affecting its operational function. Within the limitations of the data collected in this experiment, the following observations are made. The characteristics of wind in the South-east direction are as below:

During the period of this investigation, the mean wind velocity ranges from 5.43 m/s to 11.4 m/s at 100 m height above the ground level at the test site. The mean wind speed increases with height following the power law with a coefficient of 0.315. The turbulence length scales are much lower because of the low mean velocities at the test site. Based on the results of the terrain characteristics, the site of this investigation may be classified as suburban terrain.

The fundamental frequency of the tower measured in this experiment is very close to the computed value.

12. ACKNOWLEDGEMENTS

The authors sincerely thank the Director, SERC, for permission to present the paper. The authors also record their acknowledgements to the Chief General Manager (Technical), P&T, Madras and the Divisional Engineer(Microwave) and staff for their cooperation to instrument the microwave tower.

13. REFERENCES

Duchene-Marullaz, P. (1975), "Full-Scale Measurements of Atmospheric Turbulence in a Suburban Area", Proceedings of the Fourth International Conference on Wind Engineering, 1975.

E.S.D.U.,(1974) "Characteristics of Atmospheric Turbulence Near the Ground", Engineering Sciences Data Unit, Nos.74030 and 74031, 1974.

Flay, R.G.J.,(1988) and Stevenson, D.C., "Integral Length Scales in Strong Winds Below 20 m", Journal of Wind Engineering and Industrial Aerodynamics, 1988.

Hino, M.(1967), "Spectrum of Gusty Wind", Proceedings of the Third International Conference on Wind Effects on Buildings and Structures, Tokyo, Japan, 1971.

IS:875 (Part-3)-1987 (1989), "Indian Standard Code of Practice for Design Loads (Other than Earthquake) for Buildings and Structures, Part 3, Wind Loads", Bureau of Indian Standards, New Delhi, 1989.

Shanmugasundaram, J. (1993), "Field Experiments on Wind characteristics and Structural Response to Wind Force," Workshop on Design and Construction of Cyclonic and Other Extreme Wind Disaster Resistant Structures, SERC, Madras, February, 1993.

Shanmugasundaram, J., Gomathinayagam, S., Harikrishna .P., Arumugam, M. and Venkateswarlu, B.(1994), Final Report on "Full Scale Field Experiments on 101 m High Microwave Tower", March, 1994.

Shiotani, M.and Arai, H.(1967) , "Lateral Structures of Gusts in High Winds", Proceedings of International Seminar on Wind Effects on Buildings and Structures, Ottawa, Canada, 1967.

Venkateswarlu, B. (1992), "Wind Characteristics for Application in Structural Design", Course on Design of Wind Sensitive Structures, SERC, Madras, January, 1992.

Venkateswarlu, B., Harikrishna, P., Selvi Rajan, S., and Satish Ram Kumar, M.(1993), "Dynamic Response of a Lattice Tower Under Wind Loading", accepted for publication in the Journal of computers and structures, USA

WIND LOADS ON FREE-STANDING WALLS: A FULL-SCALE STUDY

ROBERTSON, A.P.[1], HOXEY, R.P.[1], SHORT, J.L.[1], FERGUSON, W.A.[2] and OSMOND, S.[2]

[1] *Silsoe Research Institute, Wrest Park, Silsoe, Bedford, MK 45 4HS, U.K.*
[2] *Building Research Establishment, Garston, Watford, WD2 7JR, UK*

Abstract

Recent developments in the wind loading codes for the UK, Australia and Europe have introduced new, more onerous, pressure coefficient data for the design of free-standing walls. These data, which are derived from wind tunnel studies, are reviewed briefly. In 1993, a 3-year research programme was initiated to undertake an independent, full-scale study of the wind pressures on free-standing walls in order to determine reliable design data. The full-scale, variable-geometry experimental facility which includes automatic, rapid, data-logging instrumentation is described. Results obtained in the first year are presented which reveal the significance of wall length, wind direction and the presence of return corners, and priorities for the future are outlined.

1. INTRODUCTION

Strong interest has arisen recently in determining reliable pressure coefficient data for use in the design of free-standing (or boundary) walls. Concern was aroused in the UK, principally within the masonry industry, following the release in September 1991 of the Draft-for-public-comment BS 6399: Part 2: Code of practice for wind loading, British Standards (1991), because the draft contained new pressure coefficients derived from wind-tunnel tests which were much higher than the value of 1.2 used hitherto in design practice. The same new pressure coefficients are contained in the Draft Eurocode 1: Part 2.3: Wind loads, British Standards (1993). A deficiency exists in the current UK wind loading code CP3: Chapter V: Part 2, British Standards (1972), in that it provides no guidance on appropriate pressure or force coefficients for use in the design of free-standing walls. Its guide document, the 'Wind loading handbook', Newberry and Eaton (1974), however, does provide a set of force coefficients for walls as a function of length/height ratio and end condition (i.e. end free or end obstructed), but these are based on old wind-tunnel tests.

The Australian Standard for wind loads, AS 1170.2, when it was revised in 1989, Standards Australia (1989), also introduced new pressure coefficients for free-standing walls which for oblique winds again gave values well in excess of 1.2. These data were derived predominantly from the same wind-tunnel results as were the Draft BS 6399: Part 2 and Eurocode 1: Part 2.3 data. The Australian Standard data were revised again, however, in an Amendment in January 1993 which gave reduced pressure coefficients for shorter length walls (length < 5 heights). A further source of recent data is the Engineering Sciences Data Unit, ESDU 89050, which is again based principally on the same wind-tunnel results as were the data in the draft UK and European codes, but its more detailed data provide some apparently conflicting design values compared with those of the draft codes. Owing to the uncertainty over the reliability of the various data sets and the fact that many boundary walls are blown over each year, often causing injuries and even deaths (Cook, 1985), an independent full-scale test programme was initiated in June 1993 by the Building Research Establishment and Silsoe Research Institute to critically appraise the wind-tunnel and code data, and establish reliable design data. The study would investigate the effect of wind direction, wall length and the presence of return corner walls. If weather conditions and resources permitted, the effect of shelter and of an opening in the wall would also be investigated.

After reviewing briefly the data mentioned above, the specially-designed full-scale test facility is described, and selected results obtained during the first year of measurements are presented. These help to formulate priorities for the second phase of measurements for 94/95 and these are outlined.

2. REVIEW OF CURRENT PRESSURE COEFFICIENT DESIGN DATA

2.1 Current UK practice

This is described in the design guide by Korff (1984) which recommends simply the use of a net pressure (or force or drag) coefficient of $C_p = 1.2$ for the design of all free-standing walls.

2.2 UK Wind loading handbook

In the absence of specific data in the current UK Wind loading code CP3: Chapter V: Part 2 (1972), its guide document, Newberry and Eaton (1974), gives net pressure coefficients for walls with different edge conditions. The data for a free-standing wall with free ends are as shown in Table 1.

Table 1. Net pressure coefficients, C_p, for a free-standing wall with free ends given in the Wind loading handbook

L/h	2	4	10	20	40	100	∞
C_p	1.20	1.24	1.32	1.38	1.62	1.78	2.00

2.3 Draft BS 6399: Part 2: Code of practice for wind loading

This draft code contains simplified and detailed net pressure coefficients for walls with and without return corners (perpendicular runs of wall at the wall ends). The detailed data give coefficients in relation to wind direction. The simplified data give worst-case values irrespective of wind direction and are shown in Table 2. The high values near the free end are commented to occur when the wind approaches at approximately 45° on to the free end. The reduction in loads near the end of the wall due to return corners is apparent. The data are derived principally from wind-tunnel research conducted at Oxford University, Letchford (1985, 1986). The same data are contained in Draft Eurocode 1: Part 2.3: Wind loads (1993).

Table 2. Simplified net pressure coefficients, C_p, for a free-standing wall with and without return corners given in Draft BS 6399: Part 2

	Distance from end	0 - 0.3h	0.3h - 2h	2h - 4h	> 4h
C_p	Without return corners	3.4	2.1	1.7	1.2
	With return corners	2.1	1.8	1.4	1.2

2.4 AS 1170.2: Australian Standard for wind loads

This wind loading code was fully revised in 1989 and introduced new net pressure coefficient data for walls for winds at 0° (normal), 45° and 90° to the wall. These data, shown in Table 3, were derived partly from the same wind-tunnel research, Letchford (1985, 1986), as were the data contained in Draft BS 6399: Part 2, and partly from wind-tunnel tests at CSIRO, Australia by Holmes (1986 I, 1986 II). Appendix E of the standard comments that when an adjacent wall with a length greater than 2h runs at right angles to a free end, forming a corner, reduced wind loads may be used for the 45° and 90° wind directions. It is suggested that the values given for 2h - 4h be extended up to the corner, ie they should apply over the distance 0 - 4h from the windward corner.

Table 3. Net pressure coefficients, C_p, for a free-standing wall given in Australian Standard for wind loads

Distance from windward end		0 - 2h	2h - 4h	> 4h
C_p	Wind direction = 0°	1.2	1.2	1.2
	Wind direction = 45°	2.4	1.2	0.6
	Wind direction = 90°	±1.0	±0.5	±0.25

In 1993, however, an amendment to the Standard reduced the coefficients for shorter length walls ($0.5 \leq L/h < 5$) for the 45° wind direction, giving instead $C_p = 1.7$ over the half-length 0 - L/2 from the free-end, and $C_p = 1.1$ over the downwind half-length L/2 - L.

2.5 Engineering Sciences Data Unit (ESDU) 89050

This publication gives extensive data for free-standing walls with and without return corners. The data are again based partly on the wind-tunnel research by Letchford (1985, 1986), partly on wind-tunnel work at CSIRO in Australia, Holmes (1985, 1986 II), and partly on earlier studies by other workers. It also gives information on the height of the stagnation streamline and on the height to the centre of the net pressure acting on the wall or element thereof.

For an infinitely long wall without return corners under a normal wind ($\phi = 0°$), ESDU gives a mean net pressure coefficient of 1.15. For a non-normal wind direction ϕ, this coefficient value reduces by the factor $(\cos\phi)^{3/2}$.

For a panel of a semi-infinite wall under a 45° wind, the loading on the panel is given as a function of its length from the windward end, as summarised in Table 4.

Table 4. Net pressure coefficients, C_p, for a panel of a semi-infinite free-standing wall given in ESDU 89050

Length of panel from windward end		0 - 0.3h	0 - 0.5h	0 - 2h	0 - 4h	0 - 6h
Without return corners	C_p for $\phi = 45°$	2.60	2.67	2.05	1.55	1.35
	C_p max for all ϕ	2.70	2.78	2.23	1.77	1.62
With return corners	C_p max for $\phi = 45°$	1.92	1.92	1.87	1.55	1.36

With no return corners, the maximum load on the panel is indicated by ESDU 89050 to arise for a wind direction of between approximately 20° and 37° from the normal to the wall, depending on the panel length. Thus the maximum C_p values for all ϕ given in Table 4 do not arise for a common wind direction. Again, the reduction in loads near the end of the wall due to the return corner is apparent.

For walls of finite length without return corners, ESDU 89050 gives 'whole wall' mean net pressure coefficients for a normal wind direction as summarised in Table 5. These 'whole wall' coefficients apply to the entire finite length of the wall.

Table 5. Net pressure coefficients, C_p, for free-standing walls of finite length and for a normal wind given in ESDU 89050

L/h	0.5	1.0	2.0	4.0	6.0	> 10
C_p ($\phi = 0°$)	1.2	1.18	1.13	1.07	1.06	1.15

The effect of different wind directions on the 'whole wall' pressure coefficients is indicated to increase loads by a maximum of only some 8% (for L/h = 3 and $\phi = 40°$) but to reduce loads by up to some 60% (for L/h = ∞ and $\phi \geq 60°$). Further data give the increased loadings acting on sub-panels of varying length from the end of a wall of finite length. These increased sub-panel loadings are given for the wind direction $\phi = 45°$.

Comparing the data in Tables 4 and 5 shows that substantially greater loads are given to act on a panel of a given length at the end of a semi-infinite wall than are given to act on a wall of total length equal to that of the panel.

2.6 Comparisons between the data sets

The Wind loading handbook gives a contrary indication to the more recent sources listed here in that it gives net loadings per unit length which increase (rather than decrease) with increasing length. The C_p value of 1.2 advocated by the Design of free-standing walls, Korff (1984), is equal to the lowest value given in the Wind loading handbook, Draft BS 6399: Part 2 and the Australian Standard, and is very similar to the minimum value given in ESDU 89050 for a wall of finite length (≈ 1.15).

By averaging the coefficient values over the different zones for which data are provided, it is possible to compare the worst-case net loadings given by Draft BS 6399: Part 2, the Australian Standard and ESDU 89050, as shown in Table 6.

Table 6. Comparison of worst-case net pressure coefficients, C_p, from three sources for walls without return corners

L/h	2.0	4.0	6.0	10
Draft BS 6399: Part 2	2.30	2.00	1.73	1.20
Australian Standard	2.40	1.80	1.40	1.20
ESDU 89050 Finite wall	1.13	1.14	1.10	1.15
ESDU 89050 End panel (semi-infinite wall)	2.23	1.77	1.62	≈1.5

The principal observation here is the lower overall net loadings indicated by ESDU 89050 for walls of finite length. The closer agreement between the ESDU 89050 data for an end panel of a semi-infinite wall and the data in the Draft BS 6399: Part 2 and the Australian Standard suggests that the two sets of code data relate to very long walls. When applied to much shorter walls, these data exceed the ESDU 89050 finite-length wall data by a factor of 2 or more. This appears to be anomalous, particularly as all three data sets draw from a common source of wind-tunnel results. The 1993 amendment to the Australian Standard which reduced coefficients for shorter walls partly redresses this contrast.

3. THE FULL-SCALE TEST FACILITY

3.1 The free-standing wall structure

The experimental wall (Fig. 1) has a height of 2 m and is 215 mm thick (representing typical free-standing masonry walls). It was erected on a specially prepared concrete foundation but, for experimental convenience, was constructed not of bricks but of 2 m high steel posts at 2 m centres and square, hollow, timber panels which located snugly between pairs of posts. The panels were readily removable, enabling the length of the wall to be changed from 2 m (one panel) to 18 m (nine panels) in 2 m increments (Fig. 2). At either end of the 18 m long wall, a corner could be formed as required by introducing one or two wall panels to form a perpendicular run of wall of 2 m or 4 m in length. The wall was constructed on the flat and exposed site at Silsoe, near to the Silsoe Structures Building which has been the focus of several years of full-scale wind loading research, including classification of the boundary layer wind structure for the site, Hoxey (1992).

Figure 1. General view of the experimental wall and reference-mast instrumentation

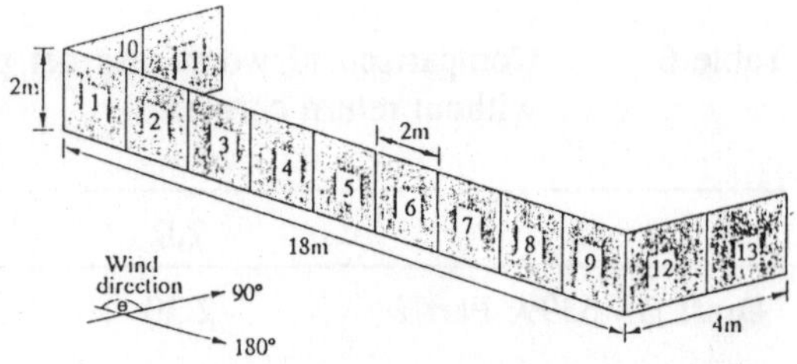

Figure 2. Schematic of the experimental wall

3.2 Instrumentation

One of the thirteen panels contained 15 tappings on each face located at corresponding positions in 3 rows of 5 columns. All the pressure transducers, switching solenoids and power supplies were mounted inside the panel. A sonic anemometer, static probe and directional pitot tube were mounted at 2 m height on a reference mast positioned some 15 m off one end of the wall.

The instrumented panel was mounted on 2 roller-bearing blocks, one positioned at each end of the base of the panel such that it could rotate freely about an axis just above the ground. The panel was restrained on its top edge by a precision 1 kN load-cell mounted on a stiff steel beam which spanned between adjacent posts. The beam and load-cell were recessed in a rebate running along the top edge of the panel which was covered by a capping plate such that this instrumentation was also neatly concealed within the instrumented wall panel. The load-cell monitored the force required to counter the overturning moment generated on the panel by the wind. Two adjusting screws enabled the load-cell to be zeroed in-situ whenever required.

A data-logger and PC were housed in a small box positioned behind the wall and 6 m to one side of the instrumented panel. When the wind speed and direction complied with prescribed settings, the data-logger automatically recorded at 5 Hz the outputs from the 30 pressure transducers, the load-cell and the sonic anemometer for 1 hour. The pressure transducers were automatically zeroed and calibrated at the start and end of each hour using the pitot and static probe pressures.

4. RESULTS

4.1 General

The majority of the 500 hours of measurements made in the first year have been with the instrumented panel positioned at one end of the wall (position 1, Fig. 2) and with different arrangements of panels in the other positions.

4.2 Effect of wind direction θ

Figure 3 shows a plot of mean pressure coefficient, C_p, against wind direction, θ, for the centre tap of the instrumented panel when in position 1 and when nine panels were installed in the wall (i.e.

18 m long wall) but no returns were present. Some 800 data points are plotted where each corresponds to a 10-minute record. Transpositioning of front and back pressures has been applied in order to increase the number of data points in the plot. The maximum coefficient on the windward face and the minimum coefficient on the leeward face occur when the wind approaches the windward end at 30° - 45° from the normal to the wall. Pressure coefficient contour plots are shown in Fig. 4 for the 130° (i.e. 40° from the normal) wind direction which produces the maximum load on the panel. The overall net coefficient for the end panel in this case is 2.52, with the centre of pressure at 0.51h. However, loads are strongly wind direction dependent, as is illustrated in Fig. 5 where the mean moment coefficient, C_m, is plotted. The mean moment coefficient is given by $C_m = M/qAh$ where M is the overturning moment on the panel (obtained from the force measurements), q is the mean wind dynamic pressure, A is the facial area of the panel, and h is the wall height. The overall loading on the panel is sensitive to wind direction with the peak load again being shown to occur for a wind approaching the end at 40° from the normal to the wall. The C_m computed from the pressures recorded at all 30 taps are very similar to the measured C_m plotted in Fig. 5. If the net pressure is uniformly distributed vertically, then the net, or force, coefficient is twice the moment coefficient, i.e. $C_p = 2C_m$.

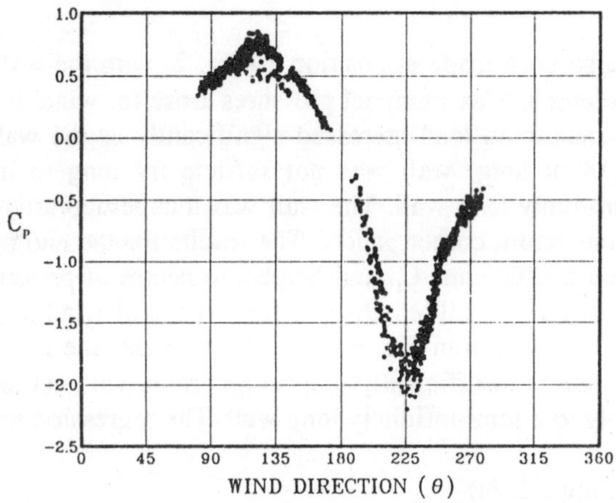

Figure 3. Mean pressure coefficients at the centre of panel position 1 for the 9h long wall

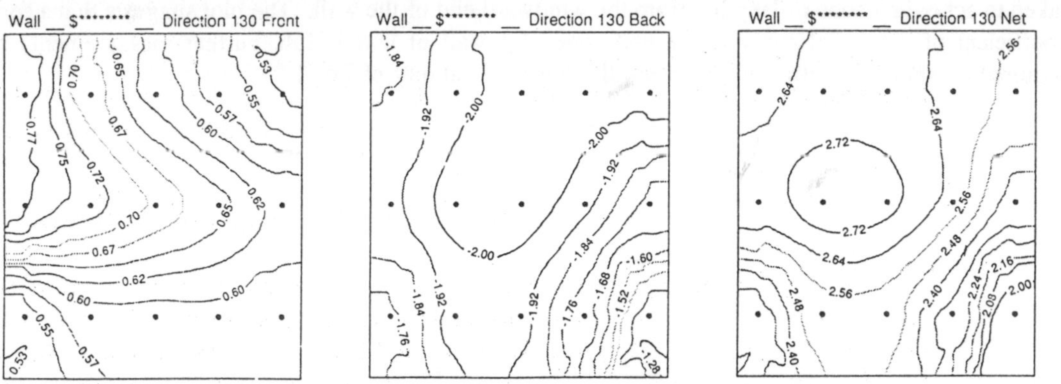

Figure 4. Front-face, rear-face and net pressure coefficient contour maps for panel position 1 of the 9h long wall and for a 130° wind direction

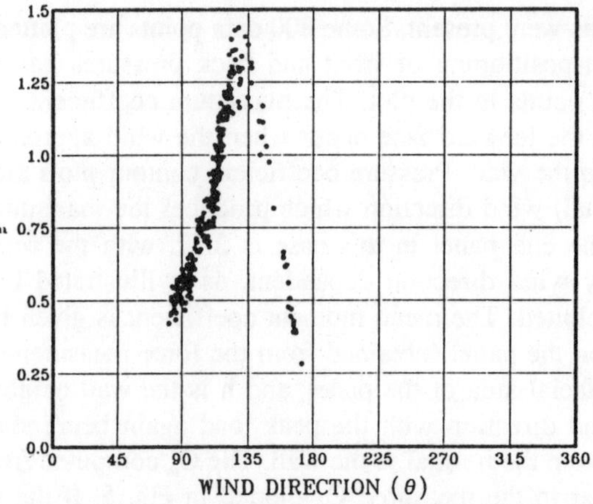

Figure 5. Measured mean moment coefficients for panel position 1 of the 9h long wall

4.3 Effect of length L

Measurements have also been made at position 1 (Fig. 2) with the wall shortened to 5 panels (10 m) and 3 panels (6 m) in length. Maximum net pressures arose for wind directions of 130° - 140°. The results showed that the maximum load increased significantly as the wall length increased and revealed that the 9 panel (18 m long) wall was not sufficiently long to indicate the asymptotic maximum load for a semi-infinitely long wall. The wall was thus temporarily extended to 13 panels (26 m) in length using the four return corner panels. The results for the end panel loads are listed in Table 7 which gives the force coefficients, C_f, and heights to centre of pressure, z_{cp}, as a proportion of the wall height, h, for the different wall lengths and for a range of wind directions. The maximum C_f values are plotted against wall length in Fig. 6. This plot provides the rather surprising indication that even a wall of 13h in length is insufficiently long to ascertain, with certainty, the maximum end panel loading that would apply to a semi-infinitely long wall. The regression line is well described by

$$C_f = 2 \log_{10}(2L/h) \tag{1}$$

where L and h are the wall length and height respectively, and C_f is the maximum force coefficient taken to act over a zone of length h from the windward end of the wall. The plot suggests that a force coefficient of 3.4 could apply to the end zone of a wall of length 25h. Further measurements are required on shorter length walls to check the range of validity of Eq. 1.

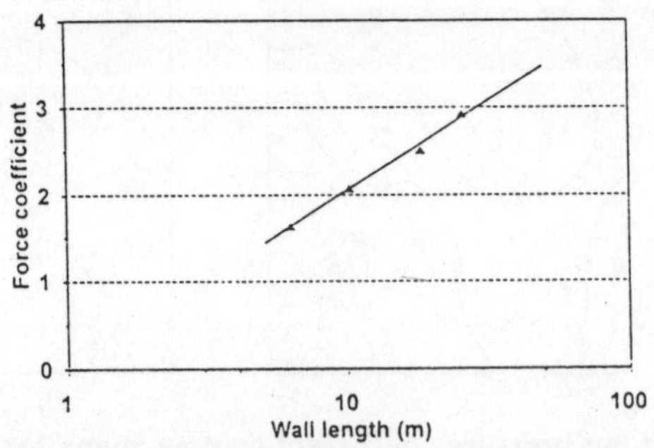

Figure 6. Maximum C_f (end panel) against wall length

Table 7. **Force coefficients, C_f, and heights to centre of pressure, z_{cp}, for panel position 1 of the 3h, 5h and 9h long wall**

Wind direction	Wall length/height (L/h)							
	3		5		9		13	
θ	C_f	z_{cp}/h	C_f	z_{cp}/h	C_f	z_{cp}/h	C_f	z_{cp}/h
90°	0.98	0.53	-	-	-	-	-	-
100°	1.03	0.52	-	-	-	-	-	-
110°	1.14	0.51	-	-	1.52	0.51	1.77	0.51
120°	1.35	0.51	1.65	0.51	2.22	0.51	2.73	0.52
130°	1.57	0.51	2.08	0.51	2.52	0.51	2.93	0.53
140°	1.64	0.51	2.00	0.52	2.29	0.52	2.36	0.53
150°	1.45	0.51	1.49	0.52	1.73	0.52	1.60	0.52
160°	-	-	0.87	0.51	1.10	0.50	1.03	0.51
170°	-	-	0.37	0.50	-	-	-	-

4.4 Effect of return corner

The effect of a return corner has been investigated for the case of the 3-panel wall with the instrumented panel in position 1. Two panels were added at positions 10 and 11 (Fig. 2) to give a perpendicular run of wall of length 2h adjacent to position 1. The results for the with and without return corner cases are given in Table 8. Measurements have not yet been made for winds blowing into the corner opening (i.e. for wind directions of 270° - 360°, Fig. 2).

Table 8. Force coefficients, C_f, and heights to centre of pressure, z_{cp}, for the end panel of the 3h long wall with and without adjacent return corner of length 2h

Wind direction	With return corner		Without return corner	
θ	C_f	z_{cp}/h	C_f	z_{cp}/h
90°	0.85	0.53	0.97	0.53
100°	1.02	0.53	1.03	0.52
110°	1.16	0.52	1.14	0.51
120°	1.20	0.51	1.35	0.51
130°	1.14	0.50	1.57	0.51
140°	0.97	0.50	1.64	0.51
150°	0.70	0.50	1.45	0.51
160°	0.35	0.49	-	-

Thus the return corner reduces the maximum loading by some 27% for this case. Further measurements are needed to investigate the magnitude of load reduction for longer walls when the end panel loads are considerably greater.

4.5 Effect of position along wall

Limited measurements have been made so far with the instrumented panel in positions other than position 1. Those that have been made show the maximum load for an oblique wind to reduce. More detailed results and comparisons will be presented once further measurements have been completed.

4.6 Height of centre of pressure z_{cp}

In all cases so far tested for panel position 1, z_{cp} was close to the mid-height of the wall. Typically, $0.49 \leq z_{cp}/h \leq 0.53$ was found which is very much in accordance with the findings of recent wind-tunnel studies.

5. FURTHER DISCUSSION ON OTHER RELATED WORK

5.1 Effect of wall height h

The effect of walls of different height submerged in different regions of the turbulent boundary layer has been considered by Holmes (1986 I) and Letchford and Holmes (1994). Quoting the experimental wind-tunnel work of Ranga Raju et al (1976) and Good and Joubert (1968), it is concluded that provided the force or drag coefficient is based on the mean wind velocity at the wall height, h, the force coefficient is, for practical purposes, insensitive to the Jensen Number, h/z_o, over the range 40 - 1000 (where z_o is the surface roughness height parameter). This is consistent with the design data reviewed in Section 2 where, in all cases, the data given are independent of wall height.

5.2 Effect of shelter

The influence of a parallel wall or building on the loads on a free-standing wall is rather complex since the shelter effect depends on both the spacing and the relative heights of the two structures, and also on the permeability of the sheltering structure. Some wind-tunnel investigations have been conducted, Letchford and Holmes (1994), and also some two-dimensional computational analyses, Robertson et al (1993), which indicate that for two walls of the same height, the loadings on the upwind wall are increased by some 12 - 15% for a spacing of 2h, and by 27% for a spacing of 0.5h compared with those on the isolated wall, due to increased suctions on the lee side. These findings may explain the high force coefficients of 1.38 and 1.78 measured by Robertson et al (1993) on two 5 m high prison walls under a normal wind since buildings of 5 m or more in height were present on the lee side of both walls. Shelter factors are given in Draft BS 6399: Part 2 and Draft Eurocode 1: Part 2.3 for a full range of solidities of the sheltering fence and for a range of spacings, but these data convey no indication of the increased loadings discussed above for the upwind wall when the spacing is less than 2h. The draft code data are stated to be applicable when the upwind wall or fence is equal in height or taller than the wall in question. Little is known about the shelter provided with walls of different height.

5.3 Effect of thickness/height ratio d/h

Whilst free-standing walls are likely to have a d/h of around 0.1, a parametric study of this has been presented by Robertson et al (1993). This two-dimensional computational investigation considered d/h ratios of 0.05, 0.1, 1.0, 2.0 and 4.0 so as to embrace geometries from thin walls to typical building proportions. The results showed the windward face pressures to be relatively insensitive to d/h and the leeward suctions to steadily reduce with increasing d/h, resulting in net pressures reducing by 35% over the range of d/h investigated.

5.4 Effect of porosity

Cook (1990) states that with significant porosity (solidity < 0.8), the local high load regions (such as at the end of a wall under an oblique wind) disappear. As a result, Draft BS 6399: Part 2 and Draft Eurocode 1: Part 2.3 stipulate a net pressure coefficient value of 1.2 for a free-standing wall with a solidity of 0.8, irrespective of position along the wall and wind direction. For lower solidities it is stated that the wall should be treated as a lattice frame. Richardson (1987), conducted full-scale measurements on an artificial windbreak with a solidity of 0.5 and proposed a force coefficient of 0.97.

5.5 Effect of a gap in a wall

No information is available on the effect of a gap in a long wall - neither on how big a gap needs to be to create two aerodynamically distinct walls, nor on the severity of loads at the ends of the wall adjacent to the gap (where funnelling of the wind may exacerbate loadings).

5.6 Effect of capping profile

No information is available on the effect of the cross-sectional profile of the top edge of a wall. This may influence the flow separation and, consequently, the wind loading.

6. RESEARCH PRIORITIES

The priorities for research on the full-scale experimental wall at Silsoe are:

- monitoring the loads at positions remote from the end, up to the mid panel position, for walls of different length;
- investigating the effect of a corner return on a long wall;
- monitoring a short (L=h) wall configuration;
- investigating the effect of a gap in a long wall.

7. CONCLUSIONS

A specially designed full-scale experimental facility for researching the wind loads on free-standing walls has been described. The versatile modular construction enables studies to be made of the effect of wall length, the presence or absence of return corners, and the introduction of a gap in a wall. Reliable and automatically controlled instrumentation mean that high quality data are acquired rapidly (500 hours in 6 months). The facility was designed and constructed in 1993 and results have been presented from the first year of a 3 year programme of measurements which illustrate the severity of loads which can arise at the end of a wall, and the dependence loads have on wind direction, wall length and the presence of a return corner. These results and those to be obtained from the research priorities being formulated for the remaining 2 years of measurements will enable fully reliable design data to be derived.

8. REFERENCES

Australian Standards (1989), "SAA Loading Code, Part 2: Wind Loads", AS 1170.2, Standards Association of Australia, North Sydney, NSW.

British Standards (1972), "CP3, Code of Basic Data for the Design of Buildings, Chapter V, Loading, Part 2, Wind Loads", British Standards Institution, London.

British Standards (1993), "Draft Eurocode 1: Basis of Design and Actions on Structures, Part 2.3: Wind Actions," CEN/TC250/SCI/1993/N118B, British Standards Institution, London.

Cook, N.J. (1985), "The Designer's Guide to Wind Loading of Building Structures, Part 1", BRE/Butterworths, London.

Cook, N.J. (1990), "The Designer's Guide to Wind Loading of Building Structures, Part 2, Static Structures", BRE/Butterworths, London.

Good, M.C. and Joubert, P.N. (1968), "The form drag of two-dimensional bluff plates immersed in turbulent boundary layers", Journal of Fluid Mechanics, Vol. 31, pp 547-582.

Holmes, J.D. (1985), "Wind Loading on Free-Standing Walls", Internal Report 85/13, CSIRO Division of Building Research, Melbourne.

Holmes, J.D. (1986 I), "Pressure and Drag on Surface-Mounted Rectangular Plates and Walls", Proceedings of the 9th Australasian Fluid Mechanics Conference, Auckland, 8-12 December 1986.

Holmes, J.D. (1986 II), "Wind Tunnel Tests on Free-Standing Walls at CSIRO", Internal Report 86/47, CSIRO Division of Building Research, Melbourne.

Hoxey, R.P. and Richards, P.J. (1972), "Structure of the Atmospheric Boundary Layer below 25 m and Implications to Wind Loading on Low-Rise Buildings", Journal of Wind Engineering and Industrial Aerodynamics, Vol. 41-44, pp 317-327, Elsevier.

Korff, J.O.A. (1984), "Design of Free-Standing Walls", The Brick Development Association, Windsor.

Letchford, C.W. (1985), "Wind Loads on Free-Standing Walls", Report OUEL 1599/85, Department of Engineering Science, University of Oxford, Oxford.

Letchford, C.W. (1986), "Further Studies of Wind Loads on Walls", Report OUEL 1620/86, Department of Engineering Science, University of Oxford, Oxford.

Letchford, C.W. and Holmes, J.D. (1994), "Wind Loads on Free-Standing Walls in Turbulent Boundary Layers", Journal of Wind Engineering and Industrial Aerodynamics, Vol. 51, pp 1-27, Elsevier.

Newberry, C.W. and Eaton, K.J. (1974), "Wind Loading Handbook", Department of the Environment/BRE/HMSO, London.

Ranga Raju, K.G.; Loeser, J. and Plate, E.J. (1976), "Velocity Profiles and Fence Drag for a Turbulent Boundary Layer along Smooth and Rough Plates", Journal of Fluid Mechanics, Vol. 76, Part 2, pp 383-399.

Richardson, G.M. (1987), "A Permeable Windbreak: Its Micro-Environment and its Effect on Structural Loads", Journal of Agricultural Engineering Research, Vol. 38, pp 65-76.

Robertson, A.P.; Hoxey, R.P. and Richards, P.J. (1993), "Design Code, Full-Scale and Numerical Data for Wind Loads on Free-Standing Walls", Proceedings of the 1st IAWE European and African Regional Wind Engineering Conference, Guernsey, September 1993, pp 185-197.

9ICWE, 1995, New Delhi, India

FULL-SCALE MEASUREMENTS ON A 3.5 M DUMMY MIRROR FOR THE NEW VERY-LARGE TELESCOPE (VLT) IN LA SILLA, CHILE

HORTMANNS, M.[1] and NOETHE, L.[2]

[1] *Institute of Steel Construction, Dept. Wind Engineering, T.U. Aachen, D-52074 Aachen, FRG*
[2] *European Southern Observatory, ESO, D-85748 Garching/Munich, FRG.*

Abstract

In December 1990 the council of the European Southern observatory decided that the world's largest optical telescope, the *16-meter* equivalent Very Large Telescope, shall be placed on Cerro Paranal, an isolated mountain top at *2664 m* altitude in the central part of Chile's Atacama desert. The VLT consists of four *8 m* diameter telescopes with mirrors with a constant thickness of only *175 mm*. Therefore the primary mirrors are very flexible and the wind pressure has a direct impact on the shape of the mirror, possibly generating deformations. During 1991 and 1992 wind pressure measurements with a *3.5 m* dummy mirror were performed at La Silla in the NTT-building and in an inflatable dome. The results are in good agreement with results of wind tunnel tests.

1. INTRODUCTION

The latest generation of reflective telescopes will be equipped with mirrors which have diameters in the order of eight meters. In some telescopes these mirrors will be continuous menisci with a constant thickness of the order of only *200 mm*. Two other features which characterize the new generation of telescopes are their capability to change and optimize on line their optics which is a technique called active optics, and the tendency to more open buildings. The latter allows for a free flow of air to reduce temperature inhomogeneities which severely affect the image quality. On the other hand the air flow may generate pressure variations on the front surface of the extremely flexible mirrors, leading to unacceptable deformations. The r.m.s. values of these deformations due to wind pressure should not exceed *100 nm*. The design of the telescope building has to find a compromise between the effects of the air flow on the mirror and the capability to remove temperature inhomogeneities. For the study of the effects of wind pressure on the mirror only some small scale wind tunnel tests have been done. We have therefore measured the pressure fields on a large *3.5 m* dummy mirror located at the entrance slit of the New Technology Telescope (NTT) of the European Southern Observatory in Chile. While the effect of constant or quasi constant pressures can be corrected by a closed loop of the active optics control system, the more higher frequencies of the pressure distribution above the basic cycle (30 sec) of the active system may become important.

2. FULL SCALE EXPERIMENT AND INSTRUMENTATION

2.1 PURPOSE OF MEASUREMENTS

For the wind pressure problem some data are available from wind tunnel tests but no systematic measurements have been done to measure the direct influence of the wind on the primary mirror. Therefore the goal of the measurements was, to determine the influence of various parameters (wind screen elevation, air flow through the dome, wind speed) on the image quality, and to measure the direct influence of the wind on the primary mirror and the telescope structure. The central question is under which conditions the VLT can or has to be operated in the open air. There are two competing effects. On the one hand it is the deterioration of the image quality because of the elastic deformations of the primary mirror and the telescope structure owing to wind pressure variations, and on the other hand it is the improvement of the image quality by allowing a free flow of air through the dome, thereby removing thermal inhomogeneties.

The main focus of this study is the investigation of the wind induced pressure field and its effects on the primary mirrors of the VLT.

2.2 EXPERIMENTAL SETUP

A first approximation has shown that a wind induced pressure of the order of $\Delta p = 1\ N/m^2$ on the surface may change the shape of the mirror significantly. This corresponse to a wind speed in the range of *1 m/s*. Therefore a high accuracy of the pressure data is necessary. In order to include the effect of Reynoldsnumber and the natural wind on site the experiments have been done in full-scale with a large model. A wooden mirror with 3.5m diameter, which is equivalent to the mirror of the NTT (New Technology Telescope) in La Silla, was used, because full-scale measurements at the VLT are not yet possible. The dome of the NTT is similar to the planned dome of the VLT. Therefore the wooden mirror was located in the NTT close to the telescope tube. This test arrangement is equivalent to a model test for the VLT with a scale of 1:2,3. A sketch with the position of the dummy in the NTT building is shown in Figure 1.

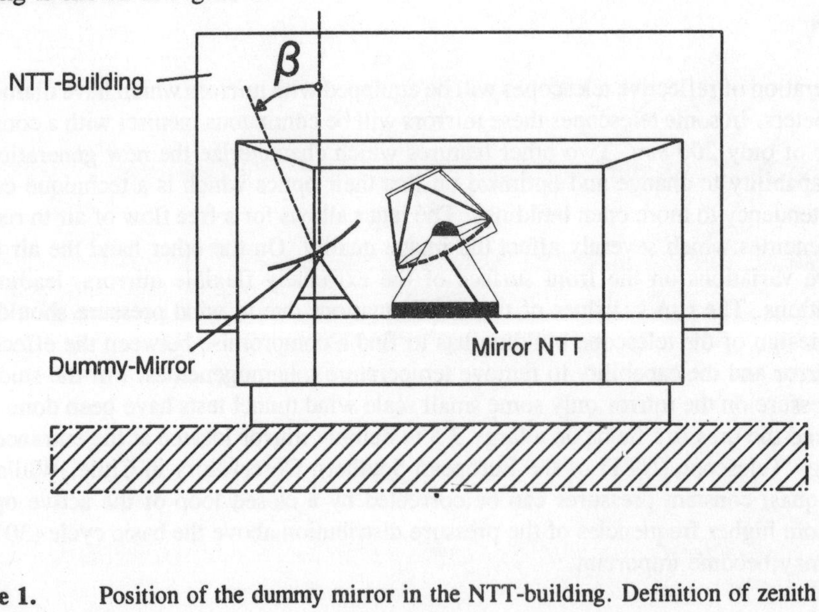

Figure 1. Position of the dummy mirror in the NTT-building. Definition of zenith angle β (inclination).

The $3,5m$ NTT dummy mirror contains 13 pressure sensors. The sensors were located on circles with radii of *0, 750 and 1500 mm*. Figure 2 shows the distribution of the sensors on the surface of the dummy. The sensors are piezo-resistive sensors which are able to measure static and dynamic pressurs. The eigenfrequency of the sensors are higher than *20 kHz*. For the frequency range, which have to be taken into account (*0 - 100 Hz*), the characteristic of the sensors is linear. Their accuracy is mainly influenced by the quality of the amplifier. During the tests an accuracy of the r.m.s. of the pressure variation of approximately $1.5\ N/m^2$ could be attained. This accuracy is sufficient, because measurements will be made at high wind speeds. The sensors have been calibrated in La Silla after their installation into the body of the mirror. For this calibration a high precicion pressure transducer (Müller-Instruments) with an accuracy of $0,1\ N/m^2$ has been used.

At the NTT a sonic anemometer was installed at the top of the building. It measured the horizontal component v_h, the vertical component v_v and the direction of the horizontal component. The characteristics of the sonic anemometer allowed measurements up to wind speeds of 60 m/sec. The windspeed accuracy are *1%* and the sampling rate are *168 Hz* which was higher than the sampling rate of the measurements.

Another anemometer was installed in front of the dummy. This anemometer has not been used to measure the reference windspeed. It was used to get informations about the windspeed close to the surface of the dummy.

The minimum windspeed at the top of the NTT-building for a sufficient signal to noise ratio was $v_h \approx 10\ m/sec$.

The inclination of the dummy could be set from $\beta = 0° \text{ to } 90°$. For $\beta = 90°$ the dummy is pointing to the horizont.

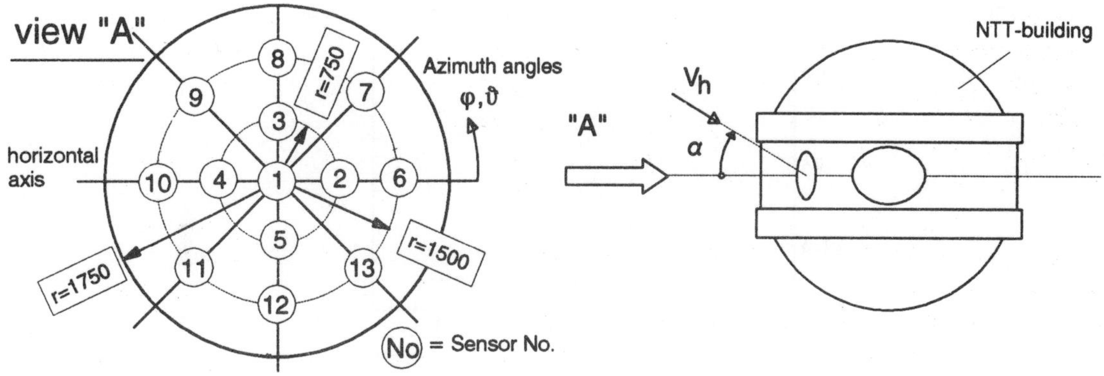

Figure 2. Distribution of the sensors on the surface of the dummy and definition of the azimuth angles α, φ und ϑ

The following definitions for angles are used:

Azimuth angle α — Angle between the plane in which the optical axis moves inclination and the average direction of the horizontal wind speed v_h

Zenith angle β	Angle between the axis of the dummy and the zenith (see Figure 1)
Azimuth angle φ	Angle between the axis of the sensors 1, 2, 6 and the axes of the other sensors (s. Figure 2)
Azimuth angle ϑ	Angle between the axis of the sensors 1, 2, 6 and the orientation axis of the non rotationally symmetric modes 1, 2 and 3 (s. Figure 2 and equation 1)

All pressures and wind data were measured in parallel at *50 Hz* over *82 sec*. The measurements have been carried out with a computer.

2.3 SENSITIVITY OF THE MEASUREMENTS

The average of the r.m.s. of the pressure variations of noise at individual sensors is approximately $1.5\ N/m^2$. This value is always subtracted quadratically from the other r.m.s. values.

Figure 3 shows the power spectrum, νS, of the noise measured at a single sensor in a closed environment. Above frequencies ν of $0.25\ Hz$ ($\triangleq \log(\nu) = -0,6$) the slope is equal to one indicating statistical white noise.

For $\nu > 10\ Hz$ we have $S \approx 0,1\ Pa^2/Hz$ for the individual sensor. The deformations calculated from the measurements of the noise alone would be of the order of *125 nm r.m.s.*.

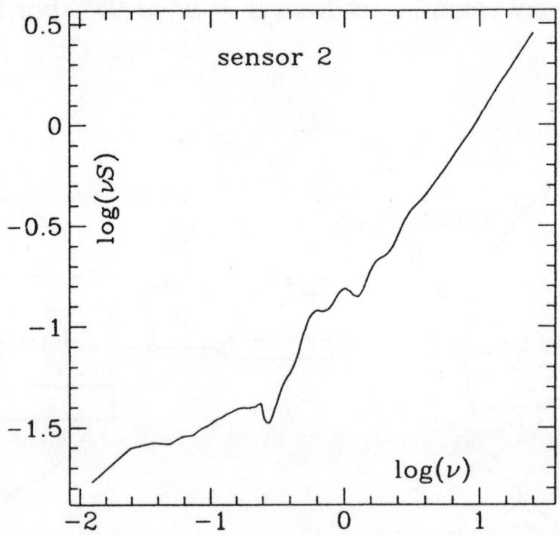

Figure 3. Power spectrum of sensor 2 in a closed environment

3. REFERENCE WIND SPEED AT THE TOP OF THE BUILDING

Figure 4 shows the r.m.s. σ_v of the wind speed (σ_v = average value of r.m.s of the three windspeed components) at the top of the building as a function of the average wind speed $\overline{v_h}$ ($\overline{v_h}$ = wind speed component in horizontal direction). The turbulence ratio, i.e. the ratio of σ_v to $\overline{v_h}$, is in most cases between 0.04 and 0.1. There are two measurements with turbulence ratios of the order of

0.3. The orientation of the building with respect to the wind may have an influence on this ratio. The two values with the large ratios were obtained for $\alpha < -20$, where α is the angle between the wind direction and the axis of the telescope building.

A typical power spectrum of the wind is shown in figure 4. For high frequencies it drops with a slope of approximately -2/3 which is in very good agreement with the *van Karman* power spectrum S expected for unobscured wind flow.

For frequencies above approximately 4 Hz the data are dominated by noise.

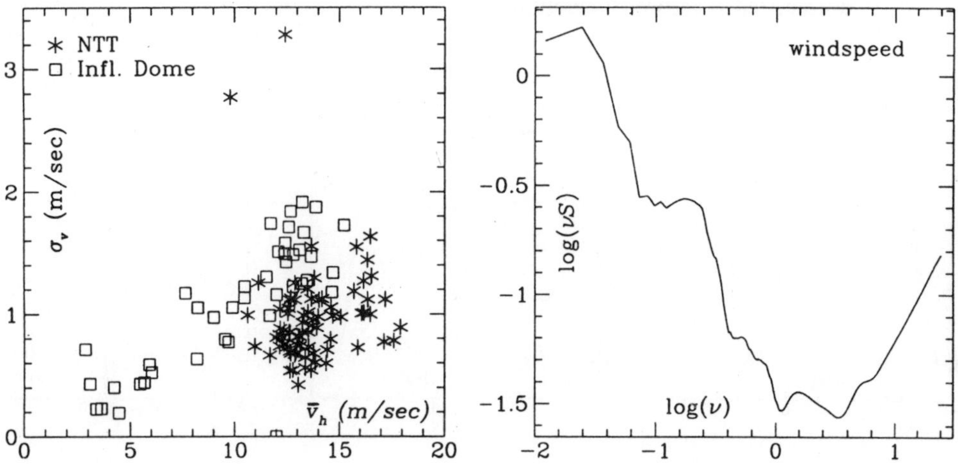

Figure 4. R.m.s. of wind speed variations versus average wind speed (left) and typical power spectrum of the free air flow

4. PRESSURES AT THE LOCATIONS OF THE 13 SENSORS

The table 1 shows the normalized average pressure coefficients $\overline{c_p}$ ($\overline{c_p}$ = average of the measured pressure data divided by $\frac{\rho}{2} \cdot \overline{v_h}^2$) and the averages $\overline{\sigma_p}$ of the r.m.s. values of the normalised pressure fluctuations at the 13 sensors for $\alpha \approx 0$, that is the mirror pointing into the wind, and various angles β between the axis of the mirror and the zenith.

Whereas the average pressure $\overline{c_p}$ depends strongly on the zenith angle β of the dummy, the pressure variations $\overline{\sigma_{c_p}}$, which is the variation coefficient, are virtually independent of β.

β	0°	15°	30°	45°	60°	75°	90°
$\overline{c_p}$	0.139	0.136	0.266	0.672	0.869	1.089	1.068
$\overline{\sigma_{c_p}}$	0.152	0.191	0.157	0.161	0.143	0.246	0.188

Table 1. Average pressure coefficients $\overline{c_p}$ and the averages $\overline{\sigma_{c_p}}$ of the r.m.s. values of the normalized pressure fluctuations

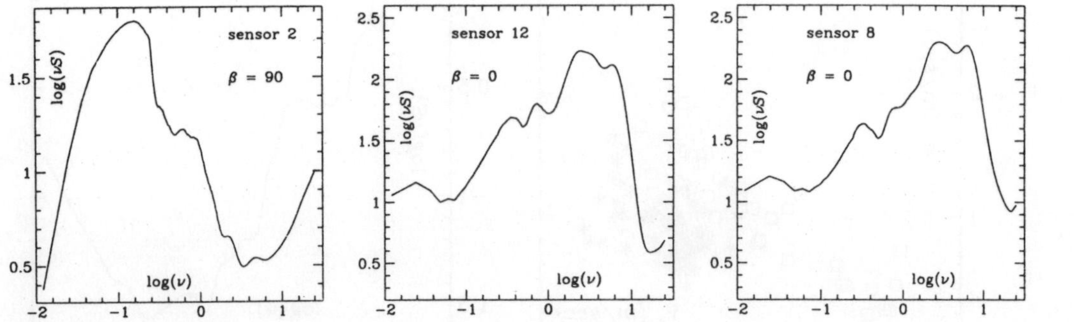

Figure 5. Power spectra of the pressures at the sensor 2 for $\beta = 90°$ and at the sensors 12 and 8 for $\beta = 0°$ and $\alpha = 0°$

Figure 5 shows the power spectra of the sensors 2 for $\beta = 90°$ and of the sensors 12 and 8 for $\beta = 0°$. The power spectra of the pressures at the locations of the 13 sensors depend strongly on β. For $\beta = 90°$ the highest energy density is at around *0.2 Hz*. For smaller β the turbulence is shifted to much higher frequencies.

A comparison of the two spectra on the right shows that they do not vary across the mirror if the mirror is pointing to the zenith. For an arbitrary choosen measurement the r.m.s. values for the pressure variations at the sensors 8 and 12 are for example $\sigma_{p,8} = 21.0 \frac{N}{m^2}$ and $\sigma_{p,12} = 19.3 \frac{N}{m^2}$.

A comparison with Figure 3 shows that the spectrum in Figure 5 on the left is probably noise limited for $\nu > 4$ *Hz* ($\triangleq \log(\nu) = 0{,}6$).

5. COMPUTATION OF THE DEFORMATIONS DUE TO WIND PRESSURES

The computation of the mirror deformations uses the fact that in the special case, where the functional dependence of the pressure field is identical to the one of a certain elastic or vibration mode of the mirror, the deformation is strictly proportional to this pressure field. The factor of proportionality depends on the stiffness of the mode which is proportional to the square of the eigenfrequency of this mode. The full deformation can therefore be computed by expanding the pressure field in an infinite series of orthogonal functions, namely the eigenmodes of the mirror, multiplying the expansion coefficients (modal coefficients) by the factors of proportionality to convert them to deformations and, finally, adding up these deformations.

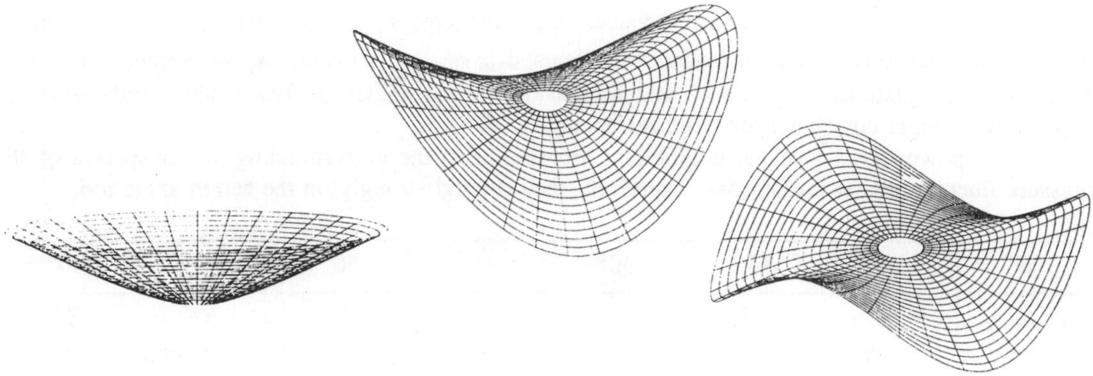

Figure 6. Scetch of the eigenmodes of rotational symmetries zero, two and three

Since the eigenfrequencies increase rapidly with the order of the modes, in practice only the lowest modes contribute significantly to the full deformations. Therefore, a knowledge of the content of a few lowest modes in the pressure distribution will be sufficient for the computation of the deformations.

Since the functional description of the eigenmodes is complicated, one can use the fact that the lowest eigenmodes are functionally very similar to simple polynomials. For each measurement the following polynomial was therefore fitted to the 13 data.

$$f_i = c_{0,0} + c_{0,2} r_i^2 + c_1\, r_i \cos(\varphi_i - \vartheta_1) + c_2\, r_i^2 \cos(2\varphi_i - \vartheta_2) + c_3\, r_i^3 \cos(3\varphi_i - \vartheta_3) \qquad (1)$$

where f_i describes the normalized pressure field for arbitrary locations (r_i, φ_i). The first and the third term describe the full body motions piston and tilt. The other three terms are similar to the lowest elastic modes of the rotational symmetries zero, two and three. The parameters $c_{0,0}$, $c_{0,2}$, c_1, c_2 and c_3 describe the moduli of the five terms and the parameters ϑ_1, ϑ_2 and ϑ_3 the orientations of the non rotationally symmetric terms. Alltogether this gives eight modal coefficients.

If the modal coefficients are treated as variables one can write down, for the 13 locations, 13 equations of the kind (1), where the left hand sides are given by the 13 measured pressures f_i. This set of equations is overdetermined. The modal coefficients are then calculated by a least square fit method, where the modal coefficients are choosen such that the sum of the squares of the differences between the measured values f_i and the calculated values, given by the right hand sides of the 13 equations, is minimized. This procedure is similar to a nonlinear regression analysis where the sum of the squares of the residuals has to become a minimum.

6. MODAL COEFFICIENTS

There is no clear correlation between the r.m.s. values of the variations of the modal coefficients and the zenith angle β. For all β they are of the same order of magnitude, between 0.05 and 0.2.

The following table gives absolut values of the averages A_c, called moduli, of the correlations between the variations of the five modal coefficients $c_{0,0}$, $c_{0,2}$, c_1, c_2 and c_3 (A_c = average value from 10 possible correlations) for $\alpha \approx 0$ and for various zenith angles β. The modal coefficients are apparently stronger correlated for large zenith angles β.

The power spectra of the coefficients are similar to the corresponding power spectra of the pressure fluctuations at the sensors. Therefore, they depend strongly on the zenith angle too.

β	0°	15°	30°	45°	60°	75°	90°
A_c *in %*	14	12	23	32	28	37	33

Table 2. Averages A_c of the moduli of the correlations between the variations of the five modal coefficients

7. MIRROR DEFORMATIONS

The table 3 shows, for $\alpha \approx 0$, the dependence of the r.m.s. of deformations of the lowest elastic modes of rotational symmetries zero, two and three and of the total deformations (in *nm*) on the zenith angle of the dummy. The data are normalised to an external wind speed of *10 m/sec*.

The deformations are only significantly smaller for $\beta = 0$ than for all other β. Otherwise, the independence of the deformations on the zenith angle reflects the fact that the pressure variations do not strongly depend on the zenith angle.

The bulk of the deformation is contained in the first elastic mode of symmetry two.

β	0°	15°	30°	45°	60°	75°	90°
$\sigma_{d,0}$	81	165	127	99	126	215	184
$\sigma_{d,2}$	318	738	802	672	594	825	595
$\sigma_{d,3}$	50	123	136	143	86	95	72
$\sigma_{d,total}$	327	766	824	685	613	858	626

Table 3. Dependence of the r.m.s. of deformations (in *nm*) of the lowest elastic modes of rotational symmetries zero, two and three and of the total deformations on the zenith angle of the dummy

Figure 7 shows for all measurements the correlation between the average $\overline{\sigma_p}$ of the r.m.s. values of the pressure variations at the sensors and the r.m.s. σ_d of the total deformations. The figure suggests a linear relationship between the two parameters. Therefore it is possible to predict deformations from pressure measurements with a single sensor.

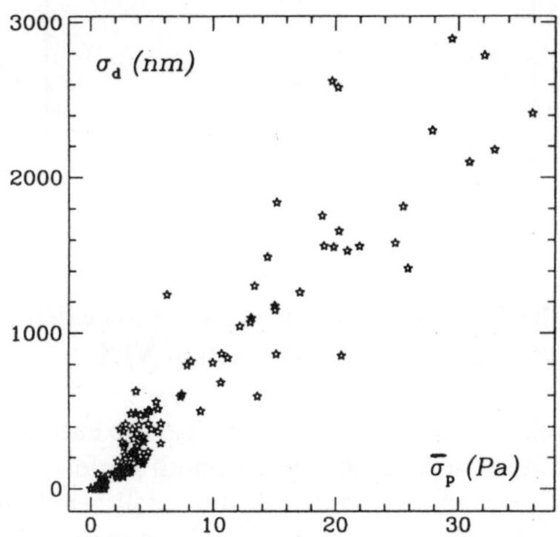

Figure 7. R.m.s. of the total deformations as a function of the average r.m.s. of the pressure fluctuations

8. COMPARISON WITH WIND TUNNEL TESTS FOR THE VLT-BUILDING

In one wind tunnel experiment [1] 152 pressure sensors were grouped in eight sectors on two rings on a 1/80 model of the mirror. By combining the measurements of the sectors in a appropriate way, one could measure at high frequencies the pressure fluctuations with distributions similar to the first modes of the rotational symmetries zero and two. The normalized pressure coefficients of these modes are roughly by a factor two smaller than those measured with the dummy in the NTT. Considering the huge difference in scale, this is an excellent agreement. In another wind tunnel measurement [2] with a 1/64 model of the foreseen VLT building, which included the structure and the mirror [2], four pressure sensors were placed on the mirror which measured the pressure differences between the front and the back surface of the mirror. The results were reliable up to full scale frequencies of *4 Hz*.

One interesting parameter is the frequency ν_{max}, where the energy density has its maximum. Figure 8 shows on the left the dependence of ν_{max} on the zenith angle β and the azimuth angle α. Apparently, when the mirror is not in the wake of the walls of the building, that is for $\alpha \leq 10°$, the energies are shifted to higher frequencies for small zenith angles.

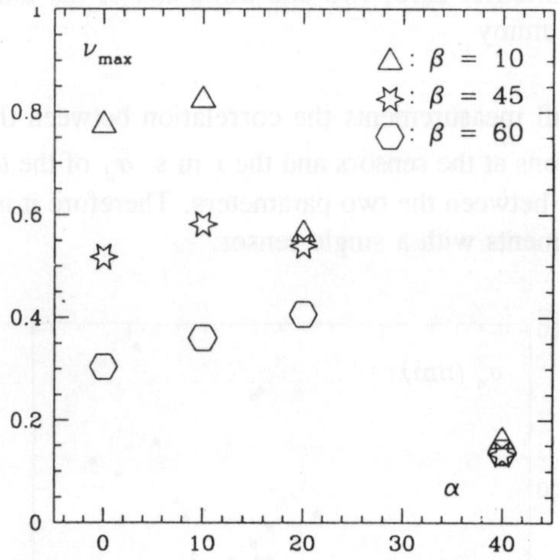

Figure 8. Dependence of the frequency ν of the highest energy density (left) on the zenith angle β and the azimuth angle α, wind tunnel test at VLT

If the mirror is in the wake of the walls of the building, the energy maxima are at low frequencies of $\nu \approx 0.15$ *Hz* and there is no dependence on the zenith angle.

The tendency to shift ν_{max} to higher frequencies for small zenith angles is the same as in the measurements with the dummy. The difference is that for the dummy the ratio between ν_{max} for small and ν_{max} for large zenith angles is of the order of 20, whereas it is only of the order of 2.5 in the wind tunnel experiment. The explanation for this discrepancy is possibly that the dummy was located virtually in the free air whereas the mirror in the wind tunnel experiment was placed at the centre of the building.

Similar in both experiments is the sharp cutoff for frequencies larger than ν_{max} if the azimuth angle α is small. For large alpha the slopes for $\nu > \nu_{max}$ are smaller and the spectra are similar to the van Karman type.

9. CONCLUSIONS

The following conclusions can be drawn from the measurements :

- The pressure patterns on the dummy are highly irregular. Therefore, to a good approximation, the amount of the most important modes contained in the pressure profile can be guessed from measurements of the r.m.s. at a single location.

To limit the r.m.s. of the deformations to *100 nm*, the pressure fluctuations at a single location on the dummy must be limited to approximately $\sigma_p \leq 1.0\ Pa$.

- The pressure variation can be replaced by wind speed measurements. The limit for the r.m.s. of the wind speed variations to fulfill the specification is approximately *0.5 m/sec*. For an open air operation, that is no shielding at all, the maximum tolerable wind speed is *2.5 m/sec*.

- Active optics can only partially correct wind buffeting, even under the favourable assumption, that the corrections can be done instantaneously based on a current deformation (that is a signal based on a measurement with negligible integration time). For small zenith and azimuth angles, no use of additional shielding by, for example, a wind screen, and correction rates of *3 Hz*, the reduction of the deformations is, according to the NTT measurements, only of the order of 15% and, according to the wind tunnel measurements, of the order of 40%.

- There is excellent agreement between the nearly full scale measurement with a 3.5 m dummy and wind tunnel experiments with scales of the order of 1/80. The discrepancies are at most of the order of a factor of two.

10. REFERENCES

[1] Hertig, J.-A. and Alexandrou, C. (1988), "Wind effects on the VLT - Wind Tunnel tests", Report: Ecole Polytechnique Federale de Lausanne, Lausanne, Switzerland

[2] Noethe, L. et. al. (1992), "Pressure measurements with a 3.5 m dummy mirror", internal report (11.12.1992), European Southern Observatory, ESO, Garching/Munich, FRG

[3] Ruscheweyh, H. and Öry, H. (1985), "Wind tunnel study for the ESO New Technology Telescope (NTT)", Report R 100/0185, T. U. Aachen, FRG

[4] "Wind studies for the VLT enclosure Paranal Chile", Report 1251-92051-02, European Southern Observatory, ESO, Garching/Munich, FRG

9ICWE, 1995, New Delhi, India

FULL-SCALE MEASUREMENT ON WIND-INDUCED NOISE IN HIGHRISE APARTMENTS

SUDA, K. and YOSHIOKA, K.
Research Head *Researcher*

Engineering Research Center, Sato Kogyo Co., Ltd., Atsugi, Kanagawa, Japan

Abstract

Since we can hardly find the precedent research on wind–induced noise in buildings, authors conducted full–scale measurements on wind–induced noise in order to make its physical characteristics clear and to grasp its affection on residential amenity. Based on the results from measurements at four highrise apartments, we discussed the relation betweeen wind–induced noise level in rooms and mean windspeed, the difference in wind–induced noise among rooms or buildings, and so on. Moreover, we discussed the quantitative indices to evaluate its affection on residential amenity.

1. INTRODUCTION

As an environmental factor of our residential amenity, wind–induced noise, that is to say, flow–induced noise around and in a building has not been so often researched. Its affection might be noticed in some situations, but its physical characteristics and degree of affection has not been reported [Berhault(1977)].

Authors conducted a series of full–scale measurements on wind–induced noise in highrise apartments in order to make its affection on residential amenity to be clear.

As to mention about the noise from air flow caused by wind, it might be assumed that generation mechanism and location of sound source can be known easily on site, but authors found that wind–induced noise in common is not such a discrete one, but composition of various flow–induced noise. In some cases, it is true that source location can be easily detected, as when wind–whistle at a letter–box is the dominating sound source. But measured wind–induced noise reported in this paper does not include such cases as one sound source is dominating or some unique phenomenon can be assumed, because we intend to reveal the characteristics of wind–induced noise detected commonly in usual highrise apartments.

2. MEASUREMENT PROCEDURE

Measurements were conducted at four highrise apartments called A, B, C and D, which has 10, 13, 11 and 14 stories respectively. Noise in the rooms and reference windspeed at the top of each apartment were measured. Measuring apartments are summarized in Table 1 and shown in Fig.1. They are almost similar in architectural specifications but are different in plan and elevation.

The measuring rooms are shown in Fig.1. Noise was measured at the center of each room and evaluated by equivalent noise level Leq averaged for 10 minutes.

$$Leq = 10 \log[\int_0^T\{ p^2(t) / p_0^2 \}dt / T]$$

$$= 10 \log[\int_0^T\{ E(t) / E_0 \}dt / T] \quad (1)$$

where Leq : equivalent noise level(equivalent contineous sound pressure level) (dB)
E : sound energy density(J/m^3) = $p^2/\rho c^2$
p : sound pressure(Pa)
ρ : air density(kg/m^3)
c : speed of sound(m/s)
p_0 : 2×10^{5} (Pa)
T : average time

Windspeed and wind direction were measured at a point over the penthouse roof level in each building except the apartment A. And mean windspeed and maximum instantaneous windspeed were evaluated for 10 minutes. Wind direction was obtained as the most frequent direction for 10 minutes. In the case of the apartment A, we used wind data measured over the neighbouring fire station and converted into the values at the top of the apartment A by means of power law.

The measurements were conducted during the short period after the apartments were finished and before they were in use. Since no residents were in the apartments during the measurements, any kind of human–induced noise indoors was not generated. Therefore, only the outdoor noise from road traffic noise etc. was measured with wind–induced noise as detected in the room.

3. RESULTS AND DISCUSSION

3.1 Detection of wind induced noise

Fig.2 shows an example of time variations both of noise level measured in a room at the apartment C and of windspeed at the top of the same apartment. From this figure, it can be noticed that Leq varied correspondingly with windspeed. It seems that noise level in this room was affected by wind–induced noise. The same data are plotted with Leq against the logarithm of mean windspeed Vmean in Fig.3. When windspeed was low, noise level in the room was affected by background noise (traffic noise, etc.) and dispersed widely. As mean windspeed exceeded 5m/s, wind–induced noise began to be effective on noise level in the room. It is observed that the minimum of dispersed noise level seems to increase with mean windspeed

Table 1 Details of four measuring apartments

Apartment	Stories	Height (m)	Surroundings	Anemometer height (m)	Stories of measuring rooms
A	10	25	Suburban	27	9th
B	13	39	Rural	46	13th
C	11	32	Suburban	38	7, 9 & 10th
D	14	29	Seaside	40	10 & 14th

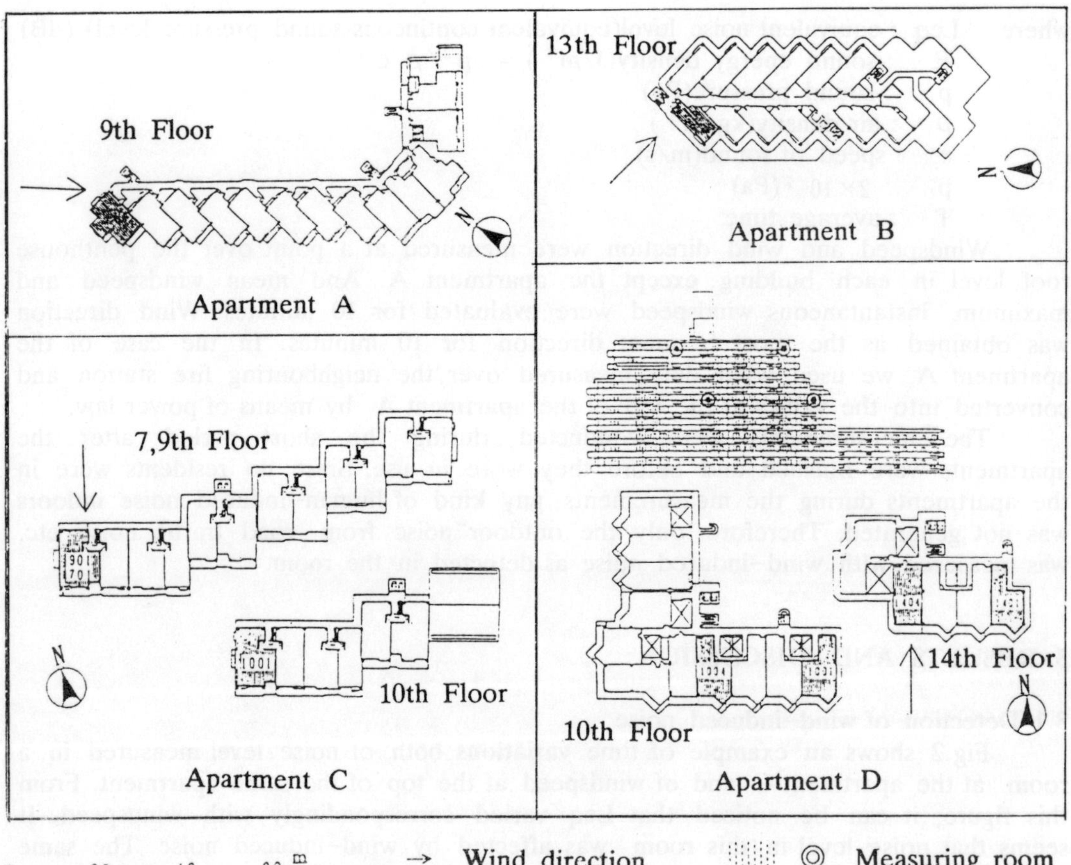

Fig.1 Plan and elevation of four measuring apartments

and it is supposed to be an addition of wind–induced noise in there. And when mean windspeed reached 8–9m/s, noise level began to exceed the upper level of background noise and the width of dispersion range in noise level became smaller, so it can be concluded that noise level in the room was dominated by wind–induced noise.

The relation of octave–band Leq with mean windspeed Vmean at the apartment D, as an example, is shown in Fig.4. It can be recognized that octave–band Leq also correlated with mean windspeed.

From the results mentioned above, it is confirmed that wind–induced noise can be detected easily in highrise apartments and are not so unusual.

3.2 Relation between wind–induced noise and windspeed

From Fig.4, it can be pointed out that each octave–band Leq of wind–induced noise increases almost linearly with the logarithm of mean windspeed. This relation can be expressed as Eq.2:

$$Leq = k + \alpha \ln(Vmean) \qquad (2)$$

Eq.2 means that power law is applicable to the relation between each octave–band sound energy density E and mean windspeed Vmean which are averaged for 10 minutes:

$$E = c\ Vmean^{\alpha} \qquad (3)$$

It was made sure from Fig.3 that the same relation was also applicable for all–pass Leq (non–weighted Leq and A–weightd Leq), which includes noise in wide frequency range. The constant k represents the strength for generation of flow–induced noise and the characteristics in the propagation process of noise. The power–law index α is closely related to the mechanism of flow–induced noise and must be an integer when generation mechanism is strictly related to the theoretical model [Lighthill(1952)]. But those indices resulted from measurements were not any integers. The reason for this phenomenon is supposed by the authors that wind–induced noise detected in the rooms must be the integration of flow–induced noises generated by various kinds of mechanism. For example, measured power–law indices α in the rooms at the apartment D are nearly 5.5 for non–weighted Leq and 6–7 for A–weighted Leq. Even for each octave–band Leq, they are not integers and they range from 4 to 6.

3.3 Difference of wind–induced noise among rooms

Figs.5 and 6 examplify the comparison of wind–induced noise level among rooms in the building. In these figures, wind–induced noise is represented by Leq at Vmean=20m/s, which has been calculated from the regression equation Eq.2 of each octave–band Leq.

In Fig.5, all three rooms are located at the corner of the apartment C. It seems that the wind–induced noise at three rooms have the similar characteristics in frequency domain and much the same in noise level.

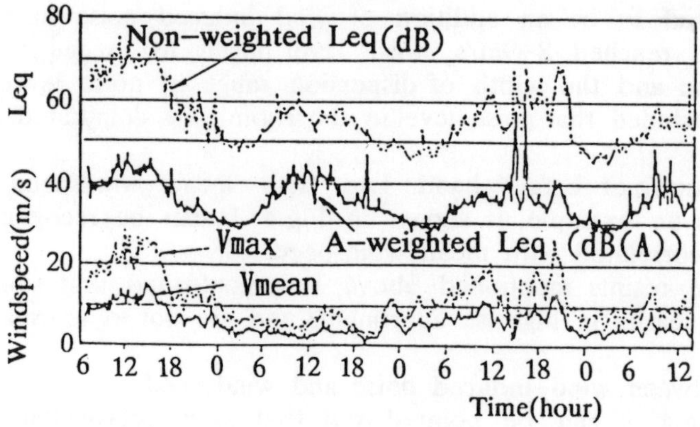

Fig.2 Example of time variation of noise level in room and windspeed(Apartment C)

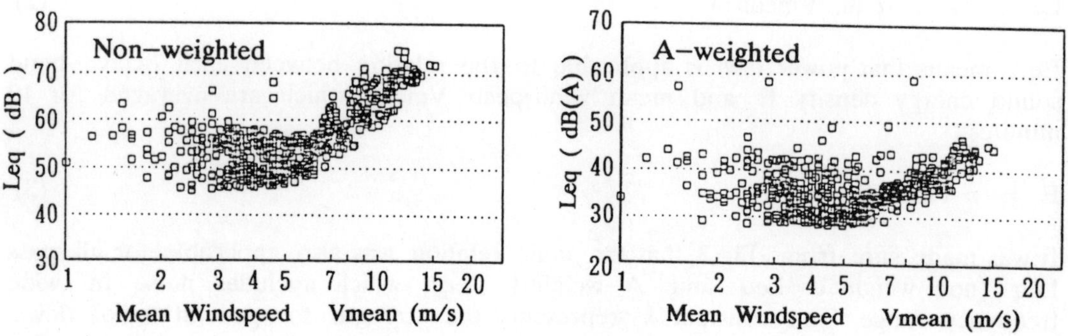

Fig.3 Relation between noise level in room and mean windspeed (Apartment C)

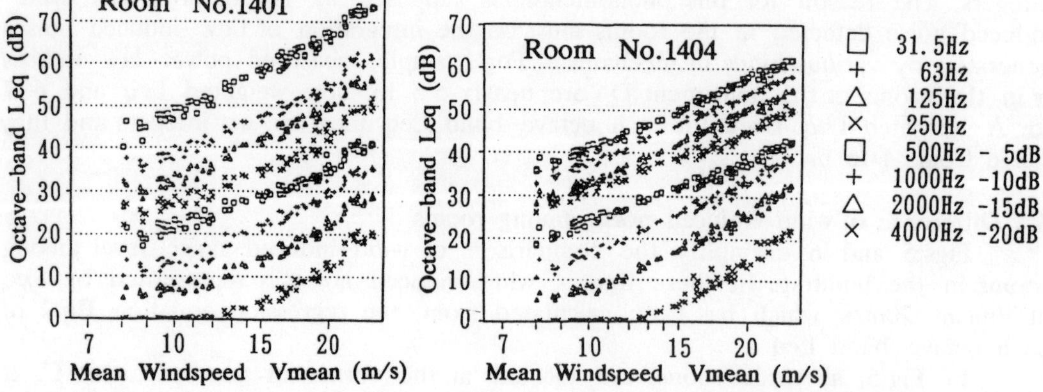

Fig.4 Relation between octave-band Leq in room and mean windspeed (Apartment D)

On the other hand, rooms at the apartment D shown in Fig.6 are located in different positions. Rooms no.1001 and no.1401 are located at the corner of the building and rooms no.1404 and no.1401 are on the top floor. In the two corner rooms wind–induced noise had similar characteristics in frequency domain and level difference was very little, like the apartment C. But some differences in the characteristics were recognized between the corner rooms and the others, especially at low frequency bands.

Regardless of room or building, wind–induced noise has a similar characteristics in frequency domain. Those spectra have a negative slope in the low frequency bands up to 500Hz and a gently broad peak which ranges from 500Hz to 1000Hz bands.

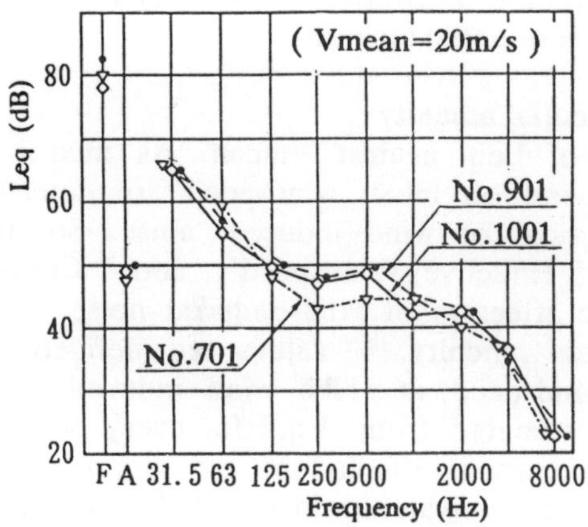

Fig.5 Difference of wind–induced noise among rooms in apartment C

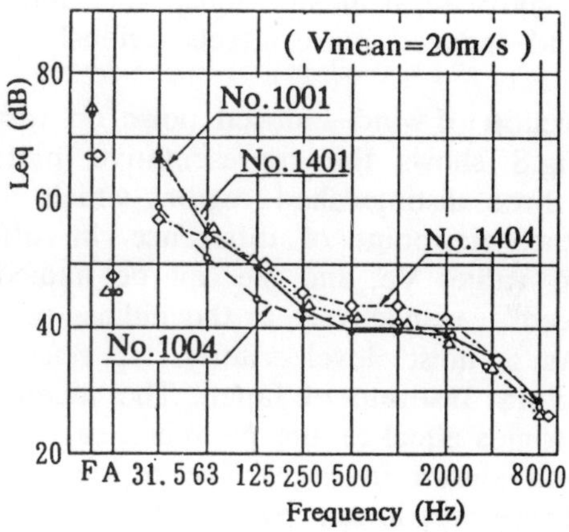

Fig.6 Difference of wind–induced noise among rooms in apartment D

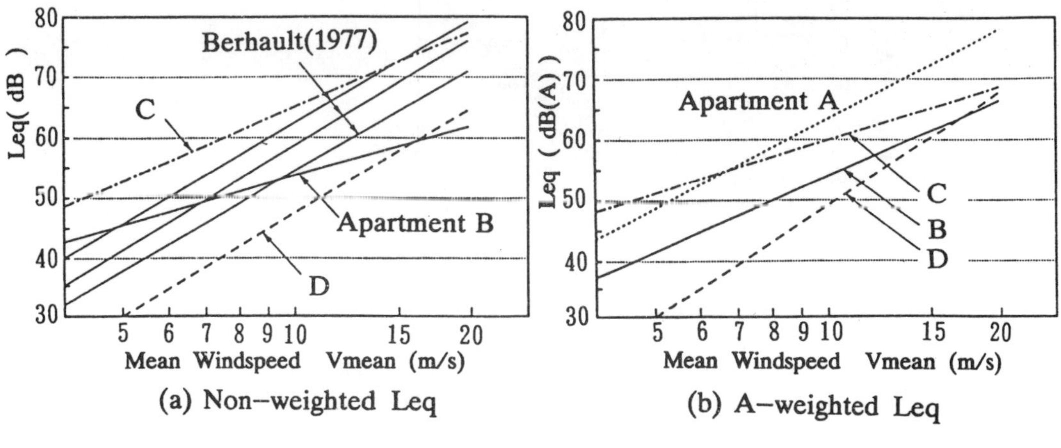

Fig.7 Difference of wind–induced noise among apartments

3.4 Difference of wind–induced noise among buildings

To compare wind–induced noise level among buildings, the regression lines at four apartments are shown in Fig.7, which represent Eq.2 for non–weighted and A–weighted Leq. As wind direction toward each building was not the same and the buildings have slightly different architectural datails and performances, the figure shows only the outline of the difference in wind–induced noise among buildings. In Fig.7, highest wind–induced noise level among the measuring rooms in each building is shown.

From this figure, it is recognized that wind–induced noise at four buildings differed considerably in the level and slightly in characteristics. The difference in level was beyond 10dB.

And then, we notice that in some apartments wind–induced noise level reached 40dB(A) at mean windspeeed under 10m/s. Therefore, it seems that in some cases wind–induced noise affects seriously on residential amenity.

3.5 Affection of wind–induced noise on residential amenity

Fig.8 shows the representative plots of Leq against Vmean. As mensiond earlier, three distinguished regions which depend on mean windspeed are detected from the viewpoint of difference in affection of wind–induced noise. So the bordered values Ve, and Vc can be named " Effective Windspeed" , and " Control Windspeed" respectively as the indices of the affection of wind–induced noise.

As a noise level criteria for residential amenity, N–values are defined by Architectural Institute of Japan. The mean windspeed, at which wind–induced noise level becomes equal to the N–value, can be estimated from Eq.2 for every octave–bands and shown in Fig.9. The residential amenity is affected by the minimum of them. Therefore, it can be defined as N–value equivalent windspeed V_N **. It is useful when we evaluate the affection of wind–induced noise on residential amenity and also when we detect most effective frequency region. It also can be said that the noise in this frequency region should be improved most immediately.

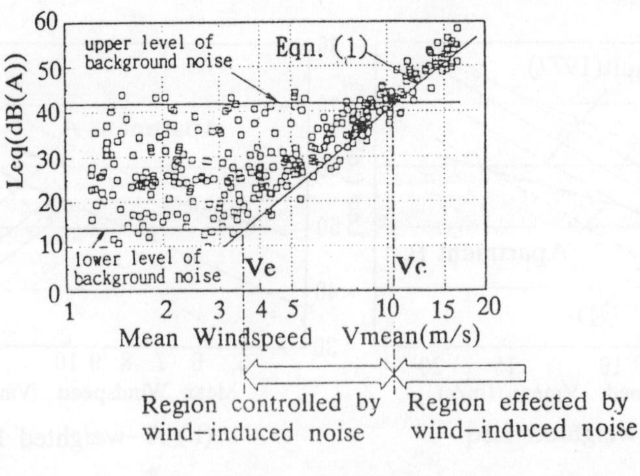

Fig.8 Representative plot of noise level against mean windspeed

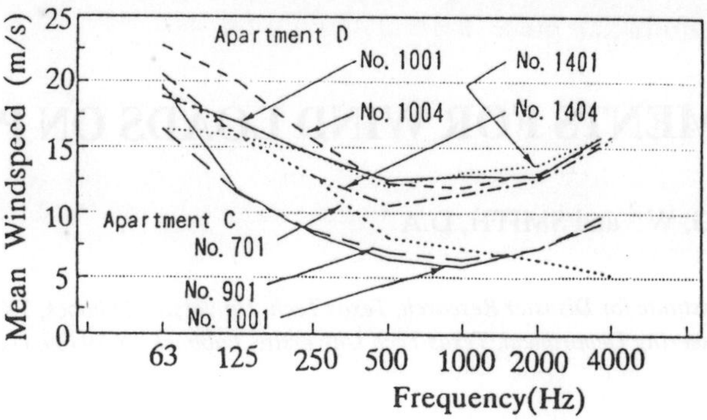

Fig.9 N–value equivalent windspeed $V_{N\ 30}$ at each frequency region

From Fig.9, it is found that such region was not the low frequency bands up to 250Hz, but the higher frequency bands ranged 500Hz to 1000Hz.

Practically speaking, it is proper that Ve, Vc or V_{N-**} should be evaluated for each wind direction. To do so, it is necessary that data on wind–induced noise should be gathered for every wind direction.

4. CONCLUSIONS

The sound energy of the wind–induced noise detected in highrise apartments increases in proportion to the power of mean windspeed. Therefore, wind–induced noise in highrise apartments exceeds the background noise easily and affects residential amenity in them considerably.

For the purpose of evaluating its affection on amenity, we introduced " Effective Windspeed" , " Control Windspeed" and N–value equivalent windspeed.

From the results of our measurements, some important characteristics are revealed on wind–induced noise detected in buildings. But it is necessary that further investigation should be continued, since the basic data on this subject might not be enough.

5. REFERENCES

Berhault,J.P.A.(1977), " Wind noise in buildings" , Wind Engineering 1(1)

Lighthill,M.J.(1952), " On sound generated aerodynamically, 1. General Theory" , Proc.R.Soc., London, A221

9ICWE, 1995, New Delhi, India

FIELD EXPERIMENTS FOR WIND LOADS ON PAVERS

McDONALD, J.R.[1], WANG, W.[2] and SMITH, D.A.[2]

[1] *Professor and Director of the Institute for Disaster Research, Texas Tech. University, Lubbock, TX 79409, U.S.A.*
[2] *Research Associate, Civil Engineering Department, Texas Tech University, Lubbock TX 79409, U.S.A.*

Abstract

The results of full-scale wind load measurements on roof pavers are presented in this paper as a benchmark for future wind tunnel and numerical studies. The current state-of-knowledge of roof paver behavior under wind load is also reviewed.

1. INTRODUCTION

Roof pavers are often preferred over roof gravel as ballast for single-ply loose-laid membrane roofing systems. The paver weight and/or interlock resistance must exceed wind uplift pressure if paver blow off is to be avoided. Although blow off has not been a serious problem in the past, at least one example was observed in Hurricane Andrew in South Florida.

A survey of previous studies on performance of roof pavers under wind pressure is carried out and presented in the next section. It is found that almost all of our knowledge of wind pressures on pavers is based on wind tunnel studies. Wind tunnel studies, however, have limitations because of difficulty in modeling Reynolds number, paver interlock resistance and flow resistance under the pavers. Full scale measurements are needed as benchmarks for calibrating wind tunnel and numerical results. To this end, a study was undertaken to measure wind pressures on full-scale pavers manufactured by Westile Corporation, Denver, Colorado, USA. Tests were conducted on the rotatable building at Texas Tech University's Wind Engineering Research Field Laboratory (WERFL) in Lubbock, Texas. Results presented in this paper include pressure coefficients (C_p) on top and bottom of three pavers, the net pressure coefficient (ΔC_p) on these pavers, summary statistics (minimum, mean and maximum) of C_p and ΔC_p as functions of angle of attack, mean and standard deviation of the extreme negative peaks of C_p and ΔC_p, time histories of C_p and ΔC_p and spectral analysis of C_P for a representative record.

2. LITERATURE REVIEW

Wind loading on a paver depends on the pressure distribution on both top and bottom of the paver. If there is sufficient permeability, the response of pressure underneath a paver to a pressure change above the paver is almost instantaneous. As a result, the net wind load on a paver may be significantly lower than that on an impermeable flat roof surface [Kind et al, 1979; Kramer et al, 1979; Bienkiewicz et al, 1992a]. The pressure distribution on top of pavers depends on many factors, including angle of attack of wind on the building, wind speed, turbulence intensity, building geometry and parapet height [Kind, 1986; Stathopoulos et al, 1987; Bienkiewicz et al, 1992b; Mehta et al, 1992; Ginger et al, 1993].

The pressure distribution underneath pavers is governed primarily by the resistance to flow between the paver (joints) and underneath the paver. If the joint resistance is much lower than the underneath resistance, as is the case for most loose-laid pavers, the pressure field will be similar to that on the top surface of the paver [Kind et al, 1979, 1982; Bienkiewicz et al, 1992a]. In this case, the pressure underneath a paver is determined mainly by the pressure on the perimeter of the top surface, and has approximately a linear distribution. Around the perimeter of the paver, the pressure on the top is nearly equal to the pressure at the bottom. The net pressure on the paver is thus the nonlinear part of the pressure on the top. The more nonlinear the top pressure is, the larger is the net wind load on the paver (Fig.1) [Kind et al, 1979]. When the joint resistance is much larger than the resistance underneath, the pressure distribution underneath the paver is nearly independent of the pressure gradient on the top [Amano et al, 1988; Bienkiewicz et al, 1992a]. There seems to be an optimal ratio of joint resistance to resistance underneath that minimizes the net wind uplift pressure on a paver [Amano et al, 1988]. Parapet height strongly affects the pressure fields on both top and bottom of pavers.

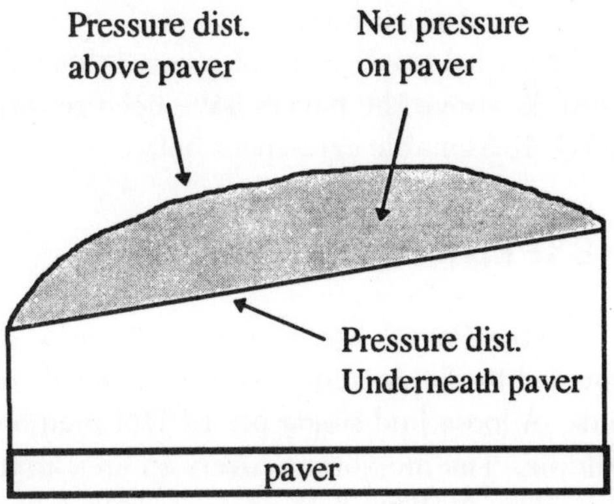

Fig.1 Net Wind Pressure on a Paver

When the net wind load on a paver is sufficiently large, it will be lifted into the flow and blown downstream. Almost all previous wind tunnel studies of paver failures were based on a quartering wind (angle of attack = 45°). For this wind direction, the first paver dislocation occurs along the roof edge about 1/6 of the building length downstream from the corner [Kind et al, 1979; Bienkiewicz et al, 1988]. The failure wind speed, which is usually defined as the mean wind speed at roof top when the first paver dislocation is observed, depends on many factors such as parapet height, turbulence intensity, paver size, paver orientation and the presence or absence of an interlock mechanism. Kind et al (1979) attributed 60% of the effective uplift force on a paver to the time-averaged component of the differential pressure and the remaining 40% to the fluctuating component. Increase in turbulence intensity causes a decrease in failure wind speed. Low parapets (< 15 cm) seem to reduce slightly the failure wind speed, but moderately high parapets (> 25cm) can increase the failure wind speed significantly. Adequate interlock between pavers and restriction on paver movement along roof edges can also increase failure wind speeds significantly. When an interlock mechanism is present, failure occurs with a sudden dislocation of a number of pavers at the same time [Kind et al, 1979, 1982; Bienkiewicz et al, 1988]. Kind et al [1988] correlated failure wind speed with building shape, parapet height, paver weight per unit area and interlocking resistance based on wind tunnel data. Gerhardt et al [1989] derived a formula for failure wind speed in terms of external pressure, relative paver size and paver weight per unit area.

Most wind tunnel experiments on roof pavers have been carried out in boundary layer wind tunnels, where model roof pavers are sized according to dynamic similarity requirements. Some error may be introduced owing to the difficulty of accurately modeling the interlock resistance, membrane elasticity and the details of paver configuration which govern the flow resistance underneath and between the pavers. Based on the findings of Kind [1977] and Meroney [1986] that paver movement is dominated by wind characteristics at roof top level, Pardo [1989] proposed a "Micro-zone" approach in which only a narrow layer of the boundary layer wind is blown across full-scale roofing systems in the wind tunnel. Amano et al [1988] studied a roofing system with relatively thick air layers underneath pavers. The prototype roofing system was tested in a wind tunnel designed for automobiles in uniform flow. While the pressure distribution patterns in the two studies look reasonable, no direct comparison with previous results from boundary layer wind tunnel tests was made by the authors. The merits and disadvantages of these testing methods need to be studied further.

Numerical predictions of pressure distribution underneath pavers from a known pressure distribution above the pavers have been reported by Amano et al [1988], Sun et al [1992] and Kind [1994]. Reasonable agreement between wind tunnel and computational results have been obtained.

3. EXPERIMENTAL SETUP

Westile Pavers are 41 cm × 33 cm × 3 cm thick and weigh 6.80 kg. A passive interlock is provided by sloping edges on the long side of the paver. Cavities underneath the pavers create air voids. A loose-laid single-ply EPDM membrane is installed on one corner of the WERFL field site building. The membrane covers an area approximately 5.5 m × 4.1 m. The edges of the membrane are sealed. A 10 cm high parapet extends along the two intersecting walls of the building. A grid of 12 rows of 13 pavers are placed over the single-ply membrane in a rectangular array. A total of six pressure taps are installed on three pavers, two for each paver. One tap measures the pressure on top of the paver, and the other measures the pressure underneath the paver at the same location. Figure 2 shows the paver installation and the shape of a paver. The three circles in Fig.2 show the location of the three pressure taps on top of the pavers. The locations of the pressure taps with respect to the building can be seen in Fig.3 along with a definition of angle of attack of the wind. The coordinates of the pressure taps are listed in Table 1. Pressure taps G21 through G24 are connected to Omega transducers; V10 and V17 are connected to Validyne transducers.

Terrain surrounding the WERFL field site building is flat and open. For more detailed descriptions of the terrain and data acquisition system, see Levitan et al [1992] and Mehta et al [1992], respectively.

Table 1: Locations of Pressure Taps*

	G21 / G22	G24 / G23	V10 / V17
location	top / bottom	top / bottom	top / bottom
x (m)	0.36	1.44	0.24
y (m)	0.34	0.34	1.25

* Distance measured from roof corner (Fig.3)

Fig.2 Westile Roofing Paver and Pressure Tap Installation

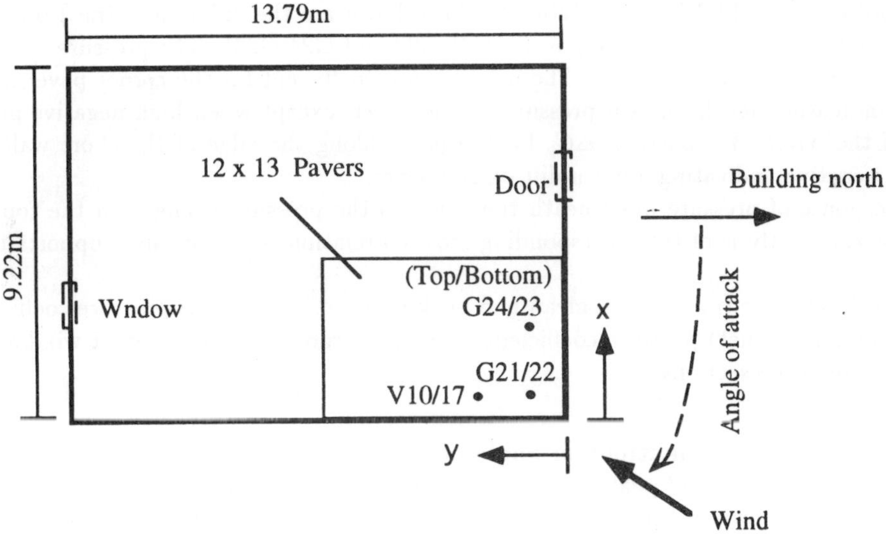

Fig. 3 Experimental Layout (Not to Scale)

4. RESULTS AND ANALYSIS

Data for this study is designated Mode 28, in which pressure and wind speed are sampled at a frequency of 30 Hz and 10 Hz, respectively. A total of 33 15-minute duration records that passed tests for stationarity were collected between March 17 and April 25, 1994. Table 2 lists record number, mean angle of attack, mean wind speed, turbulence intensity and roughness length for each record. Information regarding wind speed and direction is obtained with a UVW anemometer on the meteorological tower 3.96 meters above the ground. Measured pressures are converted to pressure coefficients by dividing by velocity pressure at roof height. Minimum, maximum, mean and standard deviation (rms) of the pressure coefficients are calculated. Spectral functions, mean and rms of the extreme distributions and cross-correlation functions of pressure coefficients are also computed for pressures on top and bottom of the three instrumented pavers. Representative results for taps G21, G22, G23 and G24 are presented below. Pressure coefficients from taps V10 and V17, which exhibit similar characteristics to taps G24 and G23, respectively, cross-correlation functions, and rms values are not presented here due to space limitation.

Note that negative pressure points away from the paver surface, and positive pressure points toward the surface. Since differential pressure coefficient ΔC_p is obtained by subtracting C_p on the bottom from C_p on the top, a positive ΔC_p indicates a net downward pressure and a negative ΔC_p indicates a net upward (uplift) pressure on a paver.

4.1 Time histories of C_p and ΔC_p

Figure 4 shows the pressure coefficient time histories of C_p for taps G21, G22, G24 and G23 from record number M28N250 (see Table 2). Also shown are the differential (net) pressures ΔC_p between the top and bottom of the pavers (G21-G22 and G24-G23). The pressure coefficients C_p on top of both pavers are negative. The net pressure coefficient for the corner paver is generally positive, indicating net downward pressure on the paver, except when high negative peaks occur on top of the paver. In contrast, ΔC_p for the paver along the edge of the short wall (Fig.3) is generally negative, indicating a net uplift on the paver.

The response of pressure underneath the paver to the pressure changes on the top is almost instantaneous. Analysis of the corresponding cross-correlation functions also supports this observation.

The peak negative (minimum), mean and peak positive (maximum) pressure coefficients (C_p) and the differential (net) pressure coefficients (ΔC_p) are compared for different wind angle of attack in the following sections.

4.2 Statistics of C_p as functions of angle of attack

Peak positive, mean, and peak negative values of pressure coefficients (C_p) for taps G21, G22, G24 and G23 are shown in Figs.5 (a, b, d and e), respectively. The following observations can be made: (1) The largest peak negative pressure coefficients occur on top of the pavers (taps G21 and G24). Values approach -12 for wind angle of attack between 30° and 40°. (2) Peak negative pressures underneath the pavers are about 50% to 80% of the peak negative values on top of the pavers for the worst wind directions ($30 \leq \theta \leq 60$). This indicates that pavers reduce the maximum suction on the roof membrane. (3) There is less scatter in the peak positive values of C_p than in the peak negative values. (4) Peak positive values of C_p are approximately zero and independent of angle of attack, θ, for $0 \leq \theta \leq 90$. (5) The mean C_p for the corner paver is also independent of angle of attack for $0 \leq \theta \leq 90$.

The relatively large peak negative pressures for an angle of attack near 30° is similar to that found by Mehta, et al [1992] on a bare roof. The high negative peak for tap G24 at $\theta = 60^o$ may be an outlier. Note that the mean pressure for this record is not the minimum compared with means of other records.

Table 2: Mode 28 Records for Westile Paver Study

	record number	mean angle of attack (^{o})	mean wind speed (m/s)	Turbulent intensity	Roughness length	Comments
1	M28N164	33.3	8.63	0.196	0.0135	
2	M28N165	33.6	8.85	0.178	0.0177	
3	M28N166	34.2	8.14	0.210	0.0170	
4	M28N167	40.7	7.96	0.253	0.0142	
5	M28N171	74.7	9.03	0.172	0.0167	no G24
6	M28N192	11.4	8.81	0.211	0.0556	
7	M28N211	13.7	10.2	0.214	0.0166	
8	M28N217	31.1	10.4	0.229	0.0199	
9	M28N219	28.1	8.85	0.204	0.0187	
10	M28N220	31.9	8.14	0.203	0.0324	
11	M28N241	33.8	8.27	0.234	0.0098	
12	M28N242	36.5	7.82	0.267	0.0084	no G23&G24
13	M28N243	32.9	8.63	0.243	0.0076	
14	M28N244	35.5	8.31	0.214	0.0156	
15	M28N245	32.5	8.67	0.232	0.0083	
16	M28N246	29.2	7.96	0.227	0.0086	
17	M28N247	35.9	8.76	0.185	0.0088	
18	M28N249	32.9	8.58	0.179	0.0114	
19	M28N250	45.3	8.67	0.186	0.0092	
20	M28N251	51.3	8.54	0.184	0.0166	no G24
21	M28N252	46.7	8.14	0.205	0.0128	no G24
22	M28N254	58.6	7.96	0.210	0.0180	no G24
23	M28N266	76.6	8.85	0.213	0.0151	
24	M28N267	62.8	9.03	0.244	0.0146	
25	M28N268	58.3	9.74	0.245	0.0258	
26	M28N270	79.2	10.5	0.196	0.0108	
27	M28N271	78.6	9.83	0.200	0.0233	
28	M28N368	74.6	13.0	0.192	0.0223	no V10&V17
29	M28N377	73.7	11.3	0.196	0.0107	no V10&V17
30	M28N378	75.2	11.3	0.216	0.0109	no V10&V17
31	M28N379	59.1	9.74	0.235	0.0146	no V10&V17
32	M28N380	63.6	10.5	0.208	0.0123	no V10&V17
33	M28N381	58.5	9.25	0.212	0.0116	no V10&V17

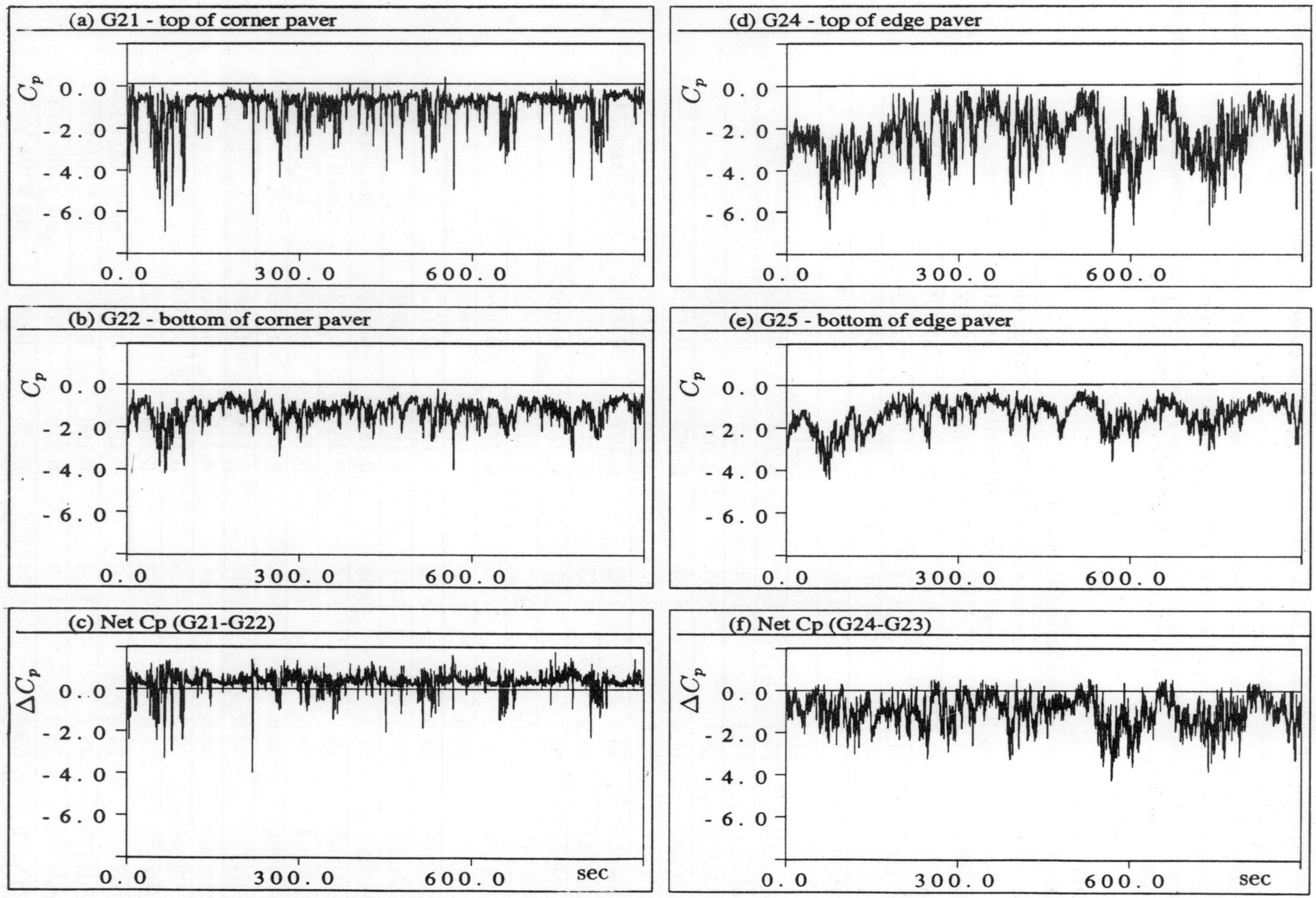

Fig.4 Time Histories of C_p and ΔC_p for the Corner and Edge Pavers
(Record 250, Mean Angle of Attack = 45.3^o)

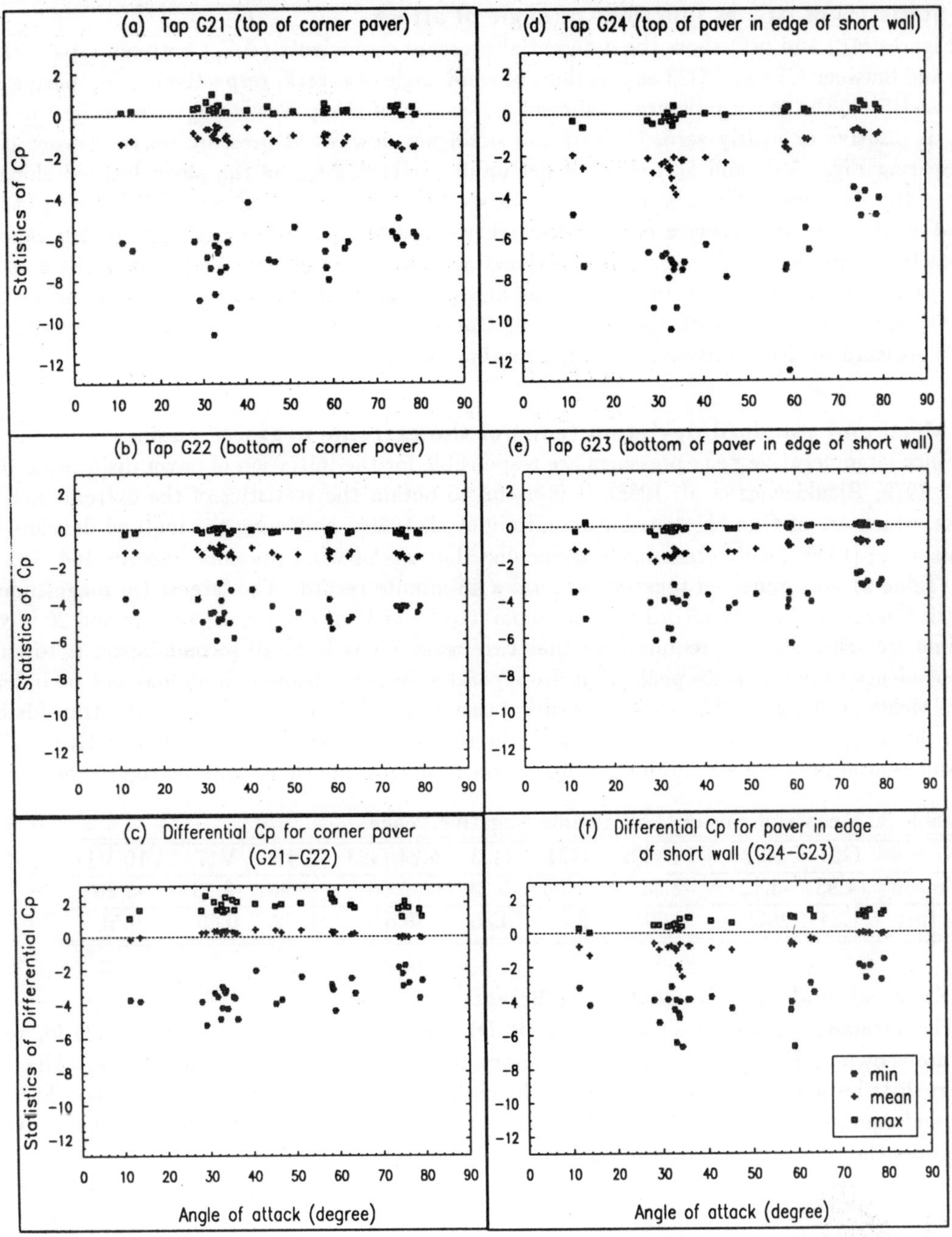

Fig.5 C_p and ΔC_p as Functions of Angle of Attack for the Corner Paver and the Paver in the Edge of the Short Wall (Fig.3)

4.3 Statistics of ΔC_p as functions of angle of attack

Figures 5(c) and 5(f) show the differential pressure coefficients (ΔC_p) between taps G21 and G22 and between G24 and G23 as functions of wind angle of attack, respectively. Comparing Figs. 5(a) and 5(c), the mean pressure coefficient C_p on top of the paver is negative, while the mean ΔC_p is positive or nearly zero, indicating a small net downward pressure on the corner paver. Comparing Figs. 5(d) and 5(f) shows a net uplift pressure ΔC_p on the paver located along the edge of the short wall. This observation is consistent with studies of paver blow off in wind tunnel [Kind et al, 1979; Bienkiewicz et al, 1988], where a paver along the wall edge, rather than the one at the corner blows off first. Figure 5 shows also that the net mean and peak uplift pressures on a paver are much smaller than the mean and peak uplift pressures on top of the paver. This observation is consistent with the concept that permeable roofing systems tend to have less net uplift pressure than the pressures on impermeable roofing systems.

4.4 Mean and standard deviation (rms) of the extreme negative peaks

Since large negative peak pressures are responsible for the initiation of paver dislocation [Kind et al, 1979; Bienkiewicz et al, 1988], it is useful to obtain the statistics of the extreme negative peaks, which are of Gumbel distribution [Simiu et al, 1986]. Following the method developed by Mehta et al [1995] for determining extreme distributions of wind pressure, records 164, 165 and 166 (Table 2) are connected together to form a 45-minute record. The largest (in magnitude) 20 negative peaks in the long record that are separated by at least one mean-crossing and 20 seconds in time are selected. The requirement that two peaks be at least 20 seconds apart is to ensure independence of peaks, since peaks that are separated by one mean-crossing may not be independent [Mehta et al, 1995]. The mean and rms of the 20 selected peaks in C_p and ΔC_p time histories from the 6 taps are listed in Table 3. Both the mean and rms of the extreme negative peaks of the differential pressure are more than 20% lower in magnitude than those on top of the paver.

Table 3: Mean and rms of the Extreme Negative Peaks

	G21	G22	G21-G22	G24	G23	G24-G23	V10	V17	V10-V17
mean	-4.93	-3.12	-2.54	-7.03	-3.24	-5.06	-6.96	-3.23	-5.20
rms	1.24	0.55	0.69	0.77	0.36	0.61	0.72	0.40	0.51

4.5 Spectral analysis of pressure coefficients

The normalized power spectral density (PSD) functions for taps G21 through G24 for record 250 are shown in Fig. 6. Ninety percent of the total energy is concentrated below 0.1Hz. The energy distribution of pressure underneath the paver is quite similar to that on the top. The pressure underneath the paver has a slightly greater percentage of energy concentrated below 0.02Hz than does the pressure on top of the paver.

5. DISCUSSION

It is difficult to compare the full-scale tests with available wind tunnel tests because of the difference in building and paver geometry and wind climate. In general, both wind tunnel and this full-scale study show that the largest net uplift on a paver does not occur at the roof corner but a short distance from the corner. The exact location of the maximum uplift depends on building geometry and parapet height. Wind tunnel measurements of the mean negative pressure on top and bottom of a paver are comparable to that measured in this full-scale study. However, the wind tunnel results for peak negative pressures on both top and bottom of a paver are much less than that measured in the field. Similar phenomenon has also been observed for pressure coefficients on a bare flat roof [Bienkiewicz et al, 1992b].

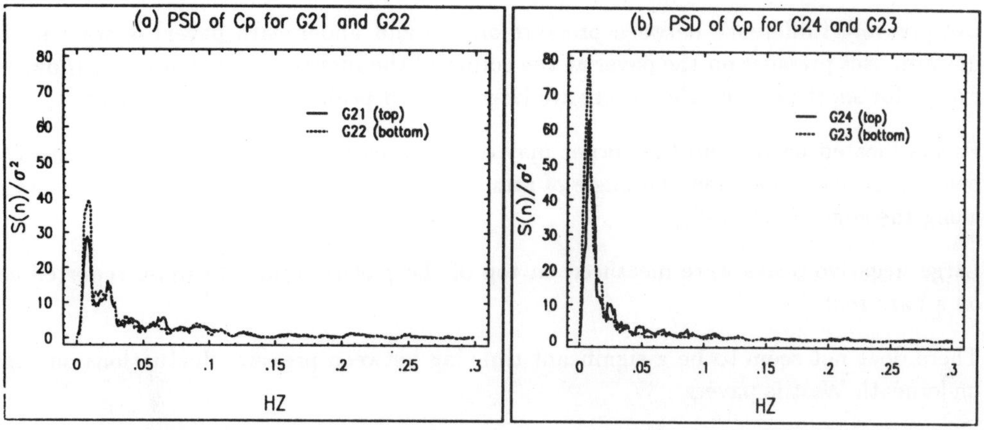

Fig.6 Normalized Power Spectral Density Functions of C_p from Pressure Taps G21 – G24 for Record 250 (mean angle of attack = 45.3^o)

Almost all the wind tunnel studies for paver loads consider a quartering wind. Recent research has, however, indicated that the largest negative peak in the separation region occurs at a wind angle of attack of about 30 degrees [Mehta et al, 1992; Ginger et al, 1993; Tieleman, 1993], which is confirmed by the full-scale measurements reported here.

From this and previous studies of wind loading on pavers, it seems clear that the net pressure on a loose-laid paver is significantly reduced due to the existence of a pressure field underneath the paver which instantly responds to the pressure change on top of the paver. The characteristics of the underelement pressure distribution depend primarily on the flow resistance between and underneath pavers and on the ratio of these two resistance. It is therefore important in designing pavers to leave proper gaps between and underneath pavers to allow pressure equalization between top and bottom of pavers. The optimum ratio of the joint to underelement flow resistance such that the net uplift wind pressure on pavers is minimized needs additional study. Holes in a paver theoretically should reduce the pressure difference between top and bottom, and thus should reduce the net wind uplift. However, too much permeability in a paver may cause freeze-thaw damage to the paver, reduce efficiency of insulation, and increase wind pressure on roof membranes. Proper paver orientation, interlock mechanism, edge attachment and use of parapets can significantly increase the wind uplift resistance of pavers. Heavier pavers may be used along edges of the roof, since this is where failure is most likely to occur.

6. CONCLUSIONS

A thorough review of the literature revealed that full-scale tests of wind loading on pavers for ballasted single-ply membrane roofing system are needed. Wind pressure measurements were made above and below full-scale Westile Corporation pavers at the Texas Tech University's Wind Engineering Research Field Laboratory (WERFL) building. The following general conclusions were reached from this study:

1. Net pressure (difference between pressure on top and underneath paver) is not necessarily upward. Net pressure on the paver at the corner of the roof is generally positive (downward) except for short periods when there are large upward peaks on top of the paver.

2. Pavers located away from the corner may or may not have mean net uplift pressures, depending on paver location and angle of attack of the wind. Largest net uplift appears to be along the edge of the wall.

3. Large negative peaks were measured on top of the pavers similar to those reported earlier on a bare roof.

4. There does not seem to be a significant time lag between pressure fluctuations on top and underneath Westile pavers.

5. The mean and rms of the extreme negative peaks of net pressure is more than 20% lower in magnitude than the mean and rms of the extreme negative peaks on the top of the paver.

7. ACKNOWLEDGMENT

We would like to thank the Westile Corp. Denver, CO, USA, for providing the Westile pavers and the Lydick-Hooks Roofing Co. Lubbock, TX, USA, for installing the EPDM materials.

8. REFERENCES

Amano, T.; Fujii, K. and Tazaki, A. (1988), "Wind loads on permeable roof-blocks in roof insulation systems." *J Wind Eng. & Industrial Aerodynamics*, v 29, p 39-48.

Bienkiewicz, B. and Meroney, R. N. (1988), "Wind effects on roof ballast pavers." *J Struct. Eng., ASCE*, v 114 n 6, p 1250-1267.

Bienkiewicz, B. and Sun, Y. (1992a). "Wind-tunnel study of wind loading on loose-laid roofing systems." *J Wind Eng. & Industrial Aerodynamics*, v 41-44, p 1817-1828.

Bienkiewicz, B. and Sun, Y. (1992b). "Local wind loading on the roof of a low-rise building." *J Wind Eng. & Industrial Aerodynamics*, v 45, p 11-24.

Gerhardt, H. J.; Kramer, C. and Bofah, K. K. (1989), "Wind loading on loosely laid pavers and insulation boards for flat roofs." *Proc. 6th US National Conf. on Wind Eng.*, v I, page B2-21 to B2-30, March 8-10, U. of Houston, Houston, Texas, USA

Ginger J. D. and Letchford C. W. (1993), "Characteristics of large pressures in regions of flow separation." *J Wind Eng. & Industrial Aerodynamics*, v 49, p 301-310.

Kind, R. J. (1977), "Further wind tunnel tests on building models to measure wind speeds at which gravel is blown off rooftops." *LTR-LA-189*, National Research Council of Canada, National Aeronautical Establishment, Ottawa, Canada, 32pp.

Kind, R. J. (1986), "Worst suction near edges of flat rooftops on low-rise buildings." *J Wind Eng. & Industrial Aerodynamics*, v 25, p 31-47.

Kind, R. J. (1994), "Predicting pressure distribution underneath loose laid roof cladding systems." *J Wind Eng. & Industrial Aerodynamics*, v 51, p 371-379.

Kind, R. J. and Wardlaw, R. L. (1979), "Model studies of the wind resistance of two loose-laid roof-insulation systems." *Laboratory technical report LTR-LA-234*, National Research Council, Canada.

Kind, R. J. and Wardlaw, R. L. (1982), "Failure mechanisms of loose-laid roof-insulation system." *J Wind Eng. & Industrial Aerodynamics*, v 9, p 325-341.

Kind, R. J; Savage, M. G. and Wardlaw, R. L., (1988), "Prediction of wind-induced failure of loose laid roof cladding systems." *J Wind Eng. & Industrial Aerodynamics*, v 29, p 29-37.

Kramer, C.; Gerhardt, H. J. and Kuster, H.-W. (1979), "On the wind-loading mechanism of roofing elements." *J Wind Eng. Industrial Aerodynamics*, v 4, p 415-427.

Levitan, M. L. and Mehta, K. C. (1992). "Texas Tech Field Experiments for Wind Loads." *J Wind Eng & Industrial Aerodynamics*, v 41-44, p 1565-1588.

Mehta, K. C.; Levitan, M. L.; Iverson, R. E. and McDonald, J. R. (1992). "Roof Corner Pressures Measured in the Field on a Low Building." *J Wind Eng. & Industrial Aerodynamics*, v 41-44, p 181-192.

Mehta, K. C.; Yeatts, B. B., Smith, D. A. and Attili, S. (1995), "Probability distribution of field measured peak pressures." To be presented in the 9th Int. Conf. on Wind Eng., January 9-13, New Delhi, India.

Meroney, R. N. (1986), "Wind-tunnel modeling of flow over bluff bodies." *CE Memorandum 86-87RNM-48*, Colorado State University, Fort Collins, CO., USA

Pardo, J. (1989), "Wind performance limits of roof ballast pavers." *Proc. 6th US National Conf. on Wind Eng.*, v I, page A5-1 to A5-10, March 8-10, U. of Houston, Houston, Texas, USA.

Simiu, E. and Scanlan, R. H. (1986), *Wind Effects on Structures* (2nd Edition). John Wiley and Sons, Inc., New York.

Stathopoulos T. and Baskaran, A. (1987), "Wind pressures on flat roofs with parapets." *J Struct. Eng., ASCE*, v 113, n 11, p 2166-2180.

Sun, Y and Bienkiewicz, B. (1992). "Numerical Simulation of Pressure Distribution Underneath Roofing Paver System." *J of Wind Eng., Japan Association for Wind Eng.*, No. 52, p 400-405. (1st Int Sym on Computational Wind Eng.).

Tieleman, H. W. (1993), "Pressures on surface-mounted prisms: the effects of incident turbulence." *J Wind Eng. & Industrial Aerodynamics*, v 49, p 289-300.

9ICWE, 1995, New Delhi, India

FIELD EXPERIMENTS FOR WIND EFFECTS ON LOW BUILDINGS

YEATTS, B.B.[1], MEHTA, K.C.[2], and SMITH, D.A.[3],

[1] *Manager, Wind Engineering Research Field Laboratory*
[2] *P.W. Horn Professor of Civil Engineering*
[3] *Research Associate, Department of Civil Engineering*
Texas Tech University, Box 41023 Lubbock, Texas, 79409 U.S.A.

Abstract

The Wind Engineering Research Field Laboratory (WERFL), established in 1987 at Texas Tech University, has permitted measurements of wind speeds and associated wind effects on a one-story test building. This paper describes a variety of experiments and measurements conducted at the WERFL. Data obtained in the field is analyzed and published in open literature. An outgrowth of some of the field measured data is development of analytical models. An overview of these items and other research possibilities at the WERFL are presented here.

1. INTRODUCTION

Damage documentation following wind related disasters such as Hurricanes Andrew and Iniki has demonstrated the need for a greater understanding of wind loads on structures. Most buildings are designed to resist wind loads which have been determined from wind tunnel tests. Full-scale research of wind-induced loads on structures helps to validate or modify values based on wind tunnel tests.

The Wind Engineering Research Field Laboratory (WERFL) at Texas Tech University (TTU) is providing an excellent opportunity to investigate wind loads on buildings at the full-scale level. The WERFL consists of a 9.1 x 13.7 x 4.0 m fully rotatable metal test building and a 49 m meteorological tower as shown in Figure 1. External and internal pressures, wind characteristics, and meteorological conditions data are collected at the WERFL. This paper presents an overview of a variety of experiments conducted and future studies planned at the WERFL. A brief description of the test facility and data collection and validation processes are also presented.

2. FIELD FACILITY

2.1 Surrounding terrain and wind climate

The WERFL is located on the TTU campus in Lubbock, Texas. The terrain surrounding the field site is flat and open. The immediate surrounding land is cultivated, while residential areas and open fields are found at distances of approximately 1000 m. The field site experiences strong winds frequently throughout the year. The passage of fronts in late autumn and spring brings sustained winds of 10-15 m/s. Thunderstorms are common in late spring and summer, and often create winds of even greater magnitude. Details of the terrain parameters and wind climate have been published (Levitan and Mehta, 1992).

Figure 1. Test building and meteorological tower.

2.2 Building and tower instrumentation

The test building is a prefabricated metal building which is anchored to a rigid frame undercarriage instead of a foundation. Hydraulic jacks at each corner are used to raise the building on wheels which ride on a circular steel rail. This combination allows the building to be rotated a full 360°, and therefore provides control over the wind angle of attack on the building. Over 100 pressure taps are installed on external surfaces of the test building. Almost 50 differential pressure transducers are connected to external surface taps. The transducers are attached to different tap location for various experiments (i.e., roof corner pressure or wall pressure studies). One pressure transducer is used to record the internal pressure for the test building. All transducer pressures are referenced to atmospheric pressure.

Wind speed, wind direction, and meteorological instrumentation are mounted at various heights on the tower, which is located 46 m west of the test building. UVW anemometers are mounted at the 2.5, 4, 10, 21, and 49 m levels, while a three-cup anemometer is mounted at the 1 m level. Temperature, relative humidity, and barometric pressure sensors are located at the 4 m level on the tower.

Data from all instrumentation is acquired by a 33 MHz 80486-based personal computer. The computer is housed in a fixed concrete block building located inside the rotatable test building as shown in Figure 2. LabTech Notebook software runs in a custom written shell to control the data acquisition system. All data is stored on 600 MB erasable optical disks. Details of the building, instrumentation, and data acquisition system have been published (Levitan and Mehta, 1992).

Figure 2. Test building under construction showing data acquisition room

3. DATA COLLECTION AND VALIDATION

3.1 Data collection modes

Several different data acquisition modes have been used to date. The main differences bewteen modes are sampling rate and duration of records and number of instruments. Table 1 summarizes the primary acquisition modes used over the past six years. Three modes were used for particular experiements as follows: M01 for transverse section pressures, M04 for roof corner pressures, and M29 (mostly low wind speeds) for atmospheric stability measurement. Acquisition modes M15, M22, and M28 have been used for various experiments (i.e., pressure transducers were moved around the building during each mode).

3.2 Data validation procedure

To attest the quality and validity of data recorded at the WERFL, two types of quality assurance programs are employed. The first involves daily field site checks and regular calibration of all instrumentation. An individual is assigned to perform a visual check of the field site each day of the week. The daily checklist includes a scan of all instrument channels for reasonableness, a check of the computerized data acquisition system, and a visual inspection of all instrumentation. Anything out of the ordinary is noted in a daily log.

Table 1. Data Acquisition Modes Used at WERFL

				Number and Sampling Rate of Instrumentation					
				Press. Transducers		Wind Instruments[1]		Miscellaneous[2]	
Mode	Dates	No. of Records Collected	Record Length (min)	Number	Sample Rate (Hz)	Number	Sample Rate (Hz)	Number	Sample Rate (Hz)
M01	1/89-6/89	246	15	11	10	6	10	3	1
M04	11/89-5/90	287	15	11	10	6	10	3	1
M15	4/91-9/92	850	15	47	40	9	10	7	1
M22	2/92-9/92	34	60	47	30	15	10	7	1
M28	1/94-pres.	580+	15	47	30	16	10	8	1
M29	9/93-5/94	1000	15	0	N/A	16	10	4	1

1 Wind instrumentation includes 3-cup and UVW anemometers and direction vanes.
2 Miscellaneous instrumentation includes sensors for temperature, barometric pressure, relative humidity, door and window openings, and purlin dislpacements.

The second type of quality assurance involves a three-stage validation program of data collected at the WERFL. Once data has been collected and stored on optical disks, it is processed by custom written software (i.e., converted to engineering units and corrections applied to various instruments). In the first stage, personnel check through summary statistics and time series plots for all instruments used in each data record. Particular care is given in coordinating checklist and calibration logs with summary statistics to note any problems that might have occurred with a particular instrument in a data record. The second stage of data validation is accomplished by an experienced group of wind personnel (different person for first and second checks), again checking through summary statistics and time series plots for each data record. Finally, summary statistics for each validated instrument are entered into a database. For each tap, the pressure coefficient versus angle of attack is plotted. Any outlying points are rechecked.

3.3 Mode M15 databases

All mode M15 data records have been validated and entered into a mode M15 database. The mode M15 pressure database contains validated summary statistics (including mean, rms, maximum, and minimum pressure coefficients) for all pressure taps which were monitored during data runs. Fifty-two external pressure taps and one internal pressure location were monitored for some or all of M15 data records. Most of the external pressure taps used were located on the roof with the heaviest concentration found near one roof corner. Wall pressures near the center of each wall and the pressure distribution along the centerline (perpendicular to the long axis) of the building were also recorded during many mode M15 data runs. Figure 3 depicts the variation of pressure coefficients (mean, rms, maximum, and minimum) recorded for tap 50501 during mode M15. Summary statistics for wind speed and direction at the 4 m height are included in the pressure database as well as summary statistics for meteorological conditions (temperature, barometric pressure, relative humidity, and air density).

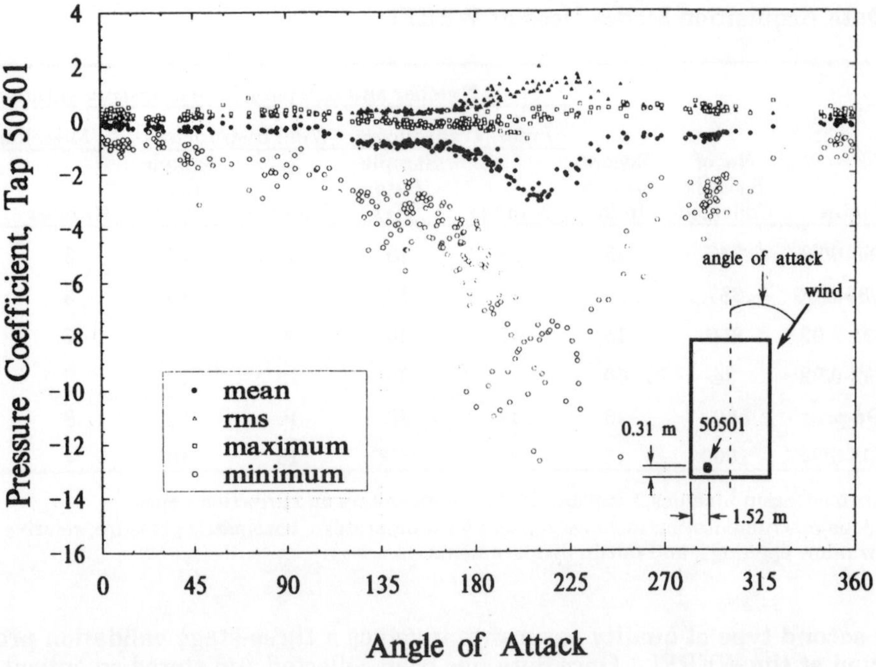

Figure 3. Mean, rms, maximum, and minimum pressure coefficients for tap 50501 during M15 data records

The mode M15 wind database contains summary statistics for all wind speed and direction instrumentation. The meteorological conditions are also stored in the wind database. Calculated log-law and power-law parameters are stored in both M15 databases. Data acquired in mode M28 is entered into databases similar to the M15 databases as it is validated.

4. OVERVIEW OF WERFL EXPERIMENTS

4.1 Pressure distribution along transverse section

One of the first experiments conducted at the WERFL was to measure and analyze surface pressure data in an area of the test building where pressures were predictable. Wind-induced pressures were measured at eleven points along a transverse section in the middle of the long dimension of the building. Pressures were measured at three locations on the windward wall, three locations on the leeward wall, and five locations on the roof. Data was analyzed for nineteen 15-minute duration records. An ensemble average of eleven records with an angle of attack of 90° (± 10°) and eight records with an angle of attack of 60° (± 10°) were obtained. These ensembles also provided a range. The data were analyzed to provide mean, root-mean-square (rms), and peak values of the pressure coefficients. The results were published in a paper (Levitan et al, 1991). This experiment and its results established the validity of the data measurement system used at the WERFL.

4.2 Wind pressures on a roof corner

A grid of nine pressure taps in an area of approximately 2.7 x 2.7 m in a roof corner measured wind-induced pressures. Data was acquired for a variety of wind angles of attack. These data were analyzed to obtain mean, rms, and peak pressure coefficients for each of the records. In addition, instantaneous averages of the nine pressure taps were obtained to assess the correlation of the pressures and area averaging effect. Results of this experiment were published in a paper (Mehta et at, 1992).

4.3 Internal pressures

The WERFL test building has two possible openings (other than building permeability): a window which provides a 2% wall opening when open and a door which provides a 5% opening. Experiments were conducted to determine the permeability of the test building, internal pressure for static opening conditions, and internal pressure for sudden opening conditions. Results of these experiments are published (Yeatts, 1992; Yeatts and Mehta, 1993a and 1993b).

4.4 Wind pressures above and below roof pavers

An area of 3.7 x 5.5 m in one roof corner of the test building is covered with a single-ply roof membrane. Commercially available concrete roof pavers are placed on top of the roof membrane as shown in Figure 4. The pressure is recorded at four pressure taps on the top surface of the pavers and four pressure taps on the surface of the roof membrane (below the roof pavers). The differential pressures on the roof pavers will provide the uplift forces acting on the pavers. Data for this experiment is under analysis at the time of writing this paper.

Figure 4 Concrete roof pavers installed in one corner of the test building.

4.5 Area averaging effects on pressures

The test building provides a unique opportunity to assess area averaged wind pressures acting on a structural member. A purlin (roof member) was instrumented to record its deflections during high winds. Six pressure taps were distributed in the wind effect area of this purlin to measure pressures at the six points on the roof surface. Simultaneous measurements of surface pressures at pressure taps, internal pressure, and deflections of the roof purlin provided area averaging effects. A total of twenty-two validated Mode 15 records were analyzed. Results of this experiment, which is sponsored by the Metal Building Manufacturer's Association and the American Iron and Steel Institute, are reported by Smith, et al (1994).

4.6 Static pressures around the test building

A basic study was pursued to assess the effect of the test building on static pressure in the area surrounding the building. The static pressure has implications in the accuracy of measurements of the external and internal pressures at the WERFL because differential pressure transducers are used and the reference pressure used for each transducer is the atmospheric pressure measured in an underground box located 23 m from the building. Static pressure data was obtained at the heights of 0.5, 4, and 6 m above ground surrounding the test building. Results show static pressures are affected more downstream of the building as compared to upstream of the building. The effect of the static pressure field created by the WERFL test building on the pressure coefficients measured is less than $\pm 0.10\ C_p$ at a distance of five times the height of the building. This magnitude of variation has a negligible effect on peak pressures measured. Details of the experiment and results are contained in a doctoral dissertation (Levitan, 1993).

4.7 Integral scales of wind turbulence

The longitudinal, lateral, and vertical integral scales of the wind turbulence at the WERFL have been studied. Three-cup anemometers were placed at a height of 4 m on five moveable poles as shown in Figure 5. The poles were placed in a line perpendicular to the mean wind direction with the spacings between the poles arranged to produce the maximum number of separation distances between the poles. Longitudinal and lateral integral scales were calculated from these moveable poles, while the vertical integral scale was calculated from the UVW anemometers located on the tower. Approximately 30 15-minute duration records were collected and analyzed. Results indicate a large range in the integral scales. The averages of the longitudinal, lateral, and vertical integral scales were found to be approximately 125, 23, and 15 m, respectively. Details of the experiment and results can be found in a master's report (Liu, 1994).

4.8 Flow visualization over the building

An experiment was conducted to determine the characteristics of the separation bubble created by wind flowing over the roof of the test building. A metal frame was used to create a 2 m tall by 4.3 m long vertical grid with yarn tufts spaced at 0.3 m intervals in each direction. The positions of the tufts, as controlled by the spatial wind variation, were recorded on videotape. Wind data from the tower and wind speeds near the roof were recorded simultaneously with the videotaping of tufts. A video editing machine allowed a frame-by-frame analysis of the videotape. By plotting the position of the tufts for each video frame, the dimensions and movement of the separation bubble can be determined. A master's thesis (Wagaman, 1994) contains the details and results of this experiment.

Figure 5. Moveable pole with 3-cup anemometer on top

Figure 6. Corner configuration of metal edge flashing on the test building

4.9 Wind-induced loads on metal edge flashings

Another current experiment involves the measurement of wind loads on metal edge flashings. Several configurations of edge flashings have been installed on the test building. Each configuration is instrumented with pressure taps both on the flashing and beneath the flashing. A view of a corner configuration for one type of edge flashing is shown in Figure 6 as installed on the test building. The pressure data collected will be used for input to a finite element analysis of the edge flashing. This will allow a detailed study of the failure mechanism and loadings to various parts of the metal edge flashing. Data collection and analysis are continuing on this experiment at this time. The metal edge flashing research is the first of its kind and gives an indication of the versatility of the WERFL test facility.

4.10 Pressure equalized rainscreen wall

One of the most recent experiments is the field testing of a Pressure Equalized Rainscreen (PER) facade system. A 85 x 85 x 7.6 cm deep PER facade element has been installed in the existing window opening of the WERFL test building as shown in Figure 7. Pressures are measured on the external face, facade cavity, and building interior for a range of venting areas. The net pressure on the rainscreen and the air barrier are determined from the collected data. Preliminary results indicate that the cavity pressure equalizes the pressure on the external rainscreen surface and the net pressure across the rainscreen is small. Data analysis for this experiment is continuing at the time of this writing.

Figure 7. Pressure Equalized Rainscreen facade element installed in existing window of test building

4.11 Development of analytical models

The data obtained at the WERFL over the past five years provides opportunities to develop analytical models for wind speed and wind-induced pressures. Some of the analytical efforts are presented below.

Stochastic modeling of wind speed: The Autoregressive Moving Average (ARMA) technique is used to obtain a stochastic model of wind speed measured at the WERFL. Three parameters are sufficient to model fluctuating wind speed. It is found that the three parameters are interrelated and that they are correlated with surface roughness z_0. The development of the model and results are published (Smith and Mehta, 1993a, 1993b, and 1995; Smith, 1993).

Frequency content of pressure fluctuations: Spectra are obtained for wind speed and associated pressures on the windward wall, leeward wall, and roof. It is found that the frequency content of wind speed and windward wall pressure are almost the same. However, the frequency contents of pressure fluctuations on the roof edge where the flow is separated and the roof corner where horizontal vortices exists are much higher than at other locations. Preliminary results are published in a paper (Thomas et al., 1993).

Probability distribution of peak pressures: Peak pressures, particularly in the separated flow regions on the roof, occur randomly. A Fisher-Tippett Type I distribution is reasonably applicable to independent peaks in a record. It is found that the location and scale parameters are not stable for 15-minute records. However, the parameters become stable for a 45-minute record. Results of this effort are assembled in a master's thesis (Attili, 1993).

5. FUTURE STUDIES AT WERFL

The possibilities of conducting studies at the WERFL are many. Some of the significant studies that are planned in the future are aerodynamic mitigation of high pressures, the effect of a nearby building, the effects of parapets and air conditioning units on the roof, reliability of low-rise buildings under wind loading, stochastic characterization of pressure fluctuations, the identification of admittance functions for wind pressures, the measurement of wind loads on column bases, and others. Field studies are time consuming, as well as expensive; therefore, the experiments need to be well planned and focused.

6. SUMMARY

Experiments conducted at the WERFL have provided good validated data of real life situations. A variety of specific experiments have been conducted and new ones are planned. The data obtained in the field are useful in helping to upgrade wind tunnel technology and improving wind load standards. The experiments and results obtained at WERFL cannot be all inclusive because it is not possible to control all variables in the field. However, the data collected in the field are necessary to provide accurate baseline data and aid in the understanding of wind effects in real life situations.

7. ACKNOWLEDGMENTS

Funding sources for the WERFL include the National Science Foundation (Grants BCS-8821163 and CES-8611601), the Metal Building Manufacturers Association, and the State of Texas. Many individuals have contributed to the design and operation of WERFL, and have conducted experiments. In addition to the authors, individuals who have conducted experiments are John Holmes of CSIRO, Australia; Henry Tieleman of VPI/SU; Chris Letchford of the University of Queensland, Australia; Hisashi Okada of the Ministry of Construction, Japan; and Marc Levitan of Louisiana State University. Their contributions and the continuing encouragement of Eleonora Sabadell at the NSF are appreciated.

8. REFERENCES

Attili, S. (1993), "Probability Distribution of Peak Pressures Measured in the Field," master's thesis submitted to the Department of Civil Engineering, Texas Tech University, Lubbock, Texas.

Levitan, M.L., Mehta, K.C., Vann, W.P., and Holmes, J.D. (1991), "Field Measurements of Pressures on the Texas Tech Building," *Journal of Wind Engineering and Industrial Aerodynamics*, Vol. 38, pp 227-234.

Levitan , M.L. and Mehta, K.C. (1992), "Texas Tech Field Experiments for Wind Loads Parts I and II," *Journal of Wind Engineering and Industrial Aerodynamics*, Vol. 41-44, pp 1565-1588.

Levitan, M.L. (1993), "Analysis of Reference Pressure Systems Used in Field Measurements of Wind Loads," doctoral dissertation submitted to the Department of Civil Engineering, Texas Tech University, Lubbock, Texas.

Liu, L. (1994), "Integral Scales of Wind at Wind Engineering Research Field Site," master's report submitted to the Department of Civil Engineering, Texas Tech University, Lubbock, Texas.

Mehta, K.C., Levitan, M.L., Iverson, R.E., and McDonald, J.R. (1992), "Roof Corner Pressures Measured in the Field on a Low Building," *Journal of Wind Engineering and Industrial Aerodynamics*, Vol. 41-44, pp 181-192.

Smith, D.A. (1993), "Stochastic Analysis of Wind Data," doctoral dissertation submitted to the Department of Civil Engineering, Texas Tech University, Lubbock, Texas.

Smith, D.A. and Mehta, K.C. (1993a), "Investigation of Stationary and Nonstationary Wind Data Using Classical Box-Jenkins Models," *Journal of Wind Engineering and Industrial Aerodynamics*, Vol. 49, pp 319-328.

Smith, D.A. and Mehta, K.C. (1993b), "Stochastic Models for Stationary and Nonstationary Data," Proceedings, Seventh U.S. National Conference of Wind Engineering (Los Angeles, California, June 27-30), Vol. II, pp 699-708.

Smith, D.A., Mehta, K.C., Yeatts, B.B., and Bhavaraju, S.V. (1994), "Area-Averaged and Internal Pressure Coefficients Measured in the Field," accepted for publication in the *Journal of Wind Engineering and Industrial Aerodynamics*.

Smith, D.A., and Mehta, K.C. (1995), "A General Autoregressive Model for Wind Data," to be presented at the Ninth International Conference on Wind Engineering, Dehli, India, January.

Thomas, G., Cochran, L., Cermak, J.E., and Mehta, K.C. (1993), "Comparison of Field and Wind-Tunnel Measured Spectra," Proceedings, Seventh U.S. National Conference of Wind Engineering (Los Angeles, California, June 27-30), Vol. II, pp 793-802.

Wagaman, A.S. (1994), "Full-Scale Flow Visualization Over a Low-Rise Building," master's thesis submitted to the Department of Civil Engineering, Texas Tech University, Lubbock, Texas.

Yeatts, B.B. (1992), "Internal Pressure for Buildings," master's thesis submitted to the Department of Civil Engineering, Texas Tech University, Lubbock, Texas.

Yeatts, B.B. and Mehta, K.C. (1993a), "Field Experiments for Building Aerodynamics," *Journal of Wind Engineering and Industrial Aerodynamics*, Vol. 50, pp 213-224.

Yeatts, B.B. and Mehta, K.C. (1993b), "Field Study of Internal Pressures," Proceedings, Seventh U.S. National Conference on Wind Engineering (Los Angeles, California, June 27-30), Vol. II, pp 889-897.

9ICWE, 1995, New Delhi, India

INTERNAL PRESSURES IN A LOW-RISE FULL SCALE BUILDING

GINGER, J.D., *Research Associate*
YEATTS, B.B. *Field Laboratory Manager* and
MEHTA, K.C. *Director*

Wind Engineering Research Center, Texas Tech University, Lubbock Texas, 79409, USA

Abstract

Theoretical analysis, numerical simulation and full scale measurements have been used to investigate internal pressures at the Texas Tech, WERFL low-rise test building. The mean and fluctuating internal pressure coefficients are small in the nominally sealed building, and increase with increasing windward/leeward open area ratio. The effects of building flexibility on the internal pressure fluctuations are accounted for in the study. Internal pressure fluctuations above a characteristic frequency are attenuated in the nominally sealed building. The internal pressure energy is increased close to the Helmholtz frequency in the building with a single dominant opening.

1. INTRODUCTION

Boundary layer wind tunnel and full scale wind engineering studies have mostly concentrated on the measurement and analysis of external pressures acting on the walls and roof of buildings. Studies have shown that, in many cases, low-rise building roof and wall failures in strong winds, have been the result of a combination of large internal pressure and external pressure acting in the same direction. Wind tunnel and theoretical studies have been carried out by Holmes (1979), Stathopoulos et al (1979), Kramer et al (1979), Liu et al (1981), Vickery (1986) and Harris (1990) on the mean and fluctuating internal pressures in nominally sealed buildings and buildings with openings in the envelope. Results from these tests form the basis of many of the latest wind load standards (i.e. AS-1170.2 (1989)).

The internal pressure is mainly controlled by the external pressure field around the building, and position and size of all openings connecting the interior of the building to the outside. Adjustable openings (vents) are used in tropical climates to control flow and ventilate internal spaces. The porosity (ε) of a building envelope is defined as the ratio of effective leakage area to the surface area of the building. Typical porosity in most nominally sealed, engineered buildings ranges from 10^{-4} to 10^{-3}. The porosity of the building controls the generation of internal pressure and ventilation in such buildings. With typical porosities being so small, a large opening in the building envelope would make a large difference to the internal pressure, and the flow through the building. Failure of (or open) doors or windows may generate significant interior pressures in strong wind conditions, and is an important consideration for ultimate design conditions.

The external and internal pressures and their interaction on a low-rise, full scale building are studied in detail at the Wind Engineering Research Field Laboratory (WERFL) at Texas Tech University. Womble et al (1995) have compared some of this full scale data with that obtained from a wind tunnel model of the WERFL building. A full scale WERFL study, on the internal pressures in the nominally sealed building and the building with dominant openings is presented in this paper. The internal pressure data are compared with theoretical analysis and numerical simulations.

2. EXPERIMENTAL SETUP AND PRESSURE MEASUREMENT

The WERFL has been described in detail by Levitan et al (1992a, 1992b). The WERFL consists of a 9.1 (B) × 13.7 (L) × 4.0 (H) m (30 × 45 × 13 ft) rotatable prefabricated metal building shown in Fig. 1. The test building is equipped with instruments to record both internal and external pressures during strong winds. Wind speed and direction are measured at different heights on a meteorological tower located 46 m (150 ft) West of the building. The approach terrain is typically open, the topography flat and the turbulence intensity $\sigma_U/\bar{U}$, at roof height (i.e. 4.0 m) is ~ 0.20. Located in the center of the building is a 3.1 × 3.1 × 2.4 m enclosure with an opening on the top, that houses the data acquisition and other electronic equipment. The nominal internal volume of the test building, V_I, is estimated at 470 m^3.

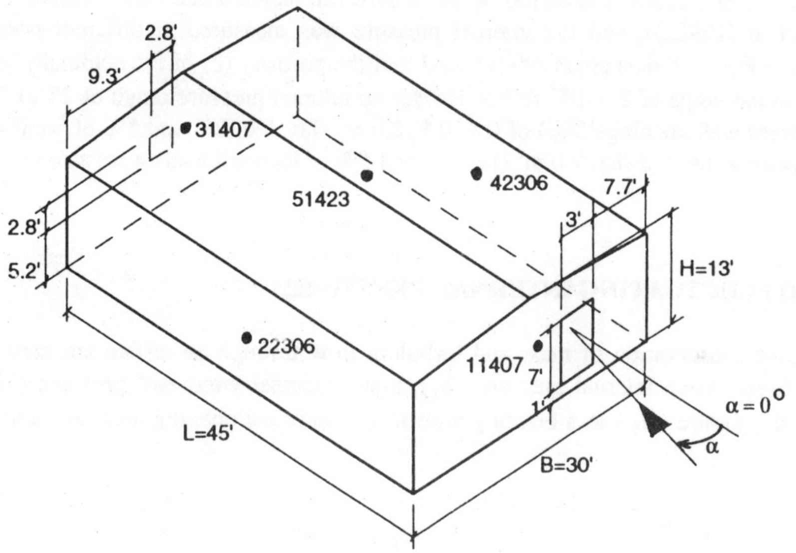

Figure 1. Windward and leeward wall openings and pressure tap locations •, on the full scale low-rise WERFL test building (all dimensions in ft).

Differential pressure transducers mounted adjacent to the taps are used to measure external pressures (p_E) and the internal pressure (p_I). The pressure transducers are connected to the taps through flexible tubes, via electrical solenoid valves and have a good response to 20 Hz. The pressure signals are low-pass filtered at 8 Hz, and sampled at 40 Hz (changed to 30 Hz for the latter tests) for a single run of 15 mins duration. The velocities are sampled at 10 Hz.

The mean, standard deviation, maximum and minimum pressure coefficients are defined as,

$$C_{\overline{p}} = \frac{\overline{p} - p_0}{(1/2)\rho \overline{U}^2}, \quad C_{\sigma_p} = \frac{\sigma_p}{(1/2)\rho \overline{U}^2}, \quad C_{\hat{p}} = \frac{\hat{p} - p_0}{(1/2)\rho \overline{U}^2}, \quad C_{\breve{p}} = \frac{\breve{p} - p_0}{(1/2)\rho \overline{U}^2}$$

where,

$\overline{p}, \sigma_p, \hat{p}, \breve{p}$ Mean, standard deviation, maximum and minimum pressure in a 15 min run
p_0 Reference pressure
ρ Density of air
$\overline{U}$ Mean wind speed at roof height (i.e. 4.0 m), over a 15 min run

The reference pressure (p_0) was obtained from a 1.4 × 1.1 × 1.0 m underground box containing a 10 mm diameter hole on the smooth ground surface. The box is located 23 m (75 ft) West of the building. The results are obtained when $\overline{U}$ exceeds 8 m/s, and are from averaging up to six runs.

Data obtained for wind orientations (α) of 0° ± 5°, and 180° ± 5° (i.e. wind flow perpendicular to the 30 ft sides of building) are presented in this paper. External pressures were measured at tap locations 11407, 31407, 22306 and 42306 on the windward, leeward and side walls, and 51423 on the roof (Levitan et al (1992a)), and the internal pressure was measured at different points within the building shown in Fig. 1. Yeatts et al (1993) found that the porosity (ε) of the nominally sealed WERFL test building is in the range of 2×10^{-4} to 3×10^{-4} for an internal pressure range of 25 to 100 Pa (0.5 to 2.0 psf). Windward wall openings (A_W) of 0.4, 0.8, 2.0 m^2 (i.e. 1%, 2% and 5% of windward wall) and leeward wall openings (A_L) of 0.8, 2.0 m^2 (i.e. 2% and 5% of leeward wall) were also tested for a range of (A_W/A_L) ratios.

3. MEAN AND FLUCTUATING INTERNAL PRESSURES

The laws of conservation of mass and turbulent flow through an orifice are used to obtain the relationship between mean internal pressure ($\overline{p}_I$), mean external windward pressure ($\overline{p}_W$) and mean external leeward pressure ($\overline{p}_L$) in a building with total windward opening area, A_W and total leeward opening area, A_L.

$$C_{\overline{p}_I} = \frac{C_{\overline{p}_W}}{1+\left(\frac{A_L}{A_W}\right)^2} + \frac{C_{\overline{p}_L}}{1+\left(\frac{A_W}{A_L}\right)^2} \tag{1}$$

Fluctuating wind velocities and the flow around the building generate a fluctuating external pressure field around the building. The internal pressure will respond in some manner to these external pressure fluctuations. The primary factors that influence the internal pressures generated by wind action are the external pressure field around the building, and position and size of all openings connecting the interior to the exterior. The response of internal pressure to changes in external pressure, is also dependent on the internal volume of the building and stiffness of the walls and roof of the building. Ginger et al (1994) found that the internal pressures measured at different points within the WERFL test building were spatially well correlated.

3.1 Effects of building flexibility

Vickery (1986) showed that there can be strong interactions between the building dynamics and the internal pressure in flexible, conventional structures. Ginger et al (1994) found that the natural frequency of vibration of the WERFL test building envelope is greater than the energy containing range of frequencies in the approach flow velocity fluctuations and the internal pressure fluctuations, in the nominally sealed building and in the building with dominant openings under consideration here. Hence the dynamic interactions between the pressure fluctuations and the building envelope are negligible, and the WERFL test building deflections are considered to follow the wind loads in a quasi-static manner.

When a building is flexible, the internal volume, V_I, will expand and contract with the changes in internal pressure. The internal pressure response is slowed as this additional change in volume must also pass through the openings. Vickery (1986), studied the effects of building flexibility in terms of the bulk modulus of air K_A, and the bulk modulus of the building K_B. The bulk modulus of the building is defined as the change in internal pressure Δp_I, required for a given change in internal volume ΔV_I;

$$1/K_B = \frac{(\Delta V_I/V_I)}{\Delta p_I} \tag{2}$$

Vickery (1986) showed that the internal pressure dynamics can be determined by replacing the internal volume of the building, V_I, by an effective internal volume of,

$$V_{Ie} = V_I\left(1+\left(K_A/K_B\right)\right) \tag{3}$$

The WERFL test building flexibility and interaction with the internal pressure were studied in detail by Ginger et al (1994). The walls and roof of the WERFL test building are constructed from flexible cladding material and the ratio of bulk modulus of the contained air to the bulk modulus of the building, K_A/K_B, was estimated at 1.5. Therefore the effective interior volume of the WERFL test building for use in the dynamic pressure analysis V_{Ie} is equal to $V_I \times \left(1+\left(K_A/K_B\right)\right) = 470 \times (1 + 1.5) = 1175\ m^3$.

3.2 Building containing a single dominant opening

If a building has an opening of area greater than about twice the background leakage area, then the external pressure at the opening has a significant influence on the internal pressure. In this case the satisfactory approach is to examine the equations of motion of air in a sealed building with a single dominant opening. However, Vickery et al (1992) found that the magnitude of internal pressure fluctuations are progressively reduced as the background leakage area is increased beyond 10% of the dominant opening.

Holmes (1979), used the principle of the Helmholtz resonator to study the motion of air in a building with a single dominant opening. An adiabatic process with a coefficient n, and an opening discharge coefficient k, is used to describe the time dependent motion of air in terms of internal pressure coefficient C_{p_I} and external pressure coefficient on the face containing the opening, C_{p_E} by Eq. 4. Eq. 4 is a second order differential equation of a single degree of freedom dynamic system with non-linear (square law) damping.

$$\frac{\rho l_e V_{Ie}}{n p_0}\ddot{C}_{p_I} + \frac{\rho V_{Ie}^2 q}{2k^2 n^2 A p_0^2} C_{p_I}\left|\dot{C}_{p_I}\right| + A C_{p_I} = A C_{p_E} \tag{4}$$

Here $q = (1/2)\,\rho \overline{U}^2$ is the dynamic pressure at roof height.

The first term in Eq. 4, represents the inertia of the slug of air moving in and out of the opening of area A. The mass of air is $\rho A l_e$, where the effective length of the slug l_e is equal to $\sqrt{\pi A/4}$. The second damping term represents the energy lost in the process and the third stiffness term represents the resistance to deflection of the air in the room. The term on the right hand side of Eq. 4 is the force generated by the change in external pressure, which forces the air in and out of the opening.

Holmes (1979) and Liu et al (1986) carried out wind tunnel model studies for a range of single dominant windward and leeward opening areas and showed that internal pressure resonance occurs close to the undamped Helmholtz frequency, $f_0 = \frac{1}{2\pi}\sqrt{\frac{nAp_0}{\rho l_e V_{Ie}}}$. They also showed that there was an increasing tendency for resonance to occur as the opening area increases relative to the internal volume. Eq. 4 shows that damping is reduced as the ratio of opening area to internal volume is increased. It should however be noted that the porous envelope of most full scale buildings helps attenuate the Helmholtz resonance.

3.3 Nominally sealed building

Openings in nominally sealed conventional building envelopes tend to be small and uniformly distributed, and the inertia term is negligible compared with the damping term in Eq. 4. Vickery (1986) and Harris (1990) studied the highly damped case with both windward and leeward openings, and showed that the response of internal pressure to changes in external pressure can be described by a characteristic response time T_C;

$$T_C = \frac{\rho V_{Ie} \overline{U} A_W A_L (\Delta C_{\overline{p}})^{1/2}}{n k p_0 (A_W^2 + A_L^2)^{3/2}} \tag{5a}$$

and a characteristic frequency f_C;

$$f_C = \frac{1}{2\pi}\left(\frac{1}{T_C}\right) = \frac{1}{2\pi}\left(\frac{n k p_0 (A_W^2 + A_L^2)^{3/2}}{\rho V_{Ie} \overline{U} A_W A_L (\Delta C_{\overline{p}})^{1/2}}\right) \tag{5b}$$

Here area A_W includes all openings on the surfaces having higher pressures than the interior causing an inflow of air while area A_L includes all openings on the surfaces having lower pressures than the interior causing an outflow of air and $\Delta C_{\overline{p}} = C_{\overline{p}_W} - C_{\overline{p}_L}$ is the net mean pressure difference between these "windward" and "leeward" surfaces. The interpretation of Eq. 5a and Eq. 5b by Vickery (1986) is that external pressure fluctuations above the frequency f_C, (time scales smaller than T_C) are attenuated and not passed effectively into the building. Conversely fluctuations at frequencies below f_C are transmitted through the skin but are attenuated if the bulk modulus of the building K_B is large compared to K_A.

4. RESULTS

Most wind load standards (i.e. AS-1170.2 (1989)) use the quasi-steady approach for designing low-rise buildings, where pressure coefficients are multiplied by the square of the design 3 sec gust wind speed to obtain peak external and internal pressures. The net wind loads on structural components and cladding are then determined by combining these peak external and internal pressures acting in the same direction. Ginger et al (1994) obtained net pressures on various parts of the nominally sealed WERFL test building, and the building with dominant openings. They found that the correlation between the external and internal pressure was low for most parts of the nominally sealed building For a building with a dominant windward opening, the highly fluctuating internal pressure was well correlated with the large suction pressures close to the windward roof edge, causing large net outward pressures. The magnitudes and characteristics of internal pressures and the external pressures on parts of the roof and wall in the nominally sealed building and the building with dominant openings are reported in this paper.

4.1 Nominally sealed building

The mean, standard deviation, maximum and minimum pressure coefficients obtained on the windward wall, leeward wall, side wall, roof and interior of the nominally sealed WERFL test building, for an approach wind flow normal to the 30 ft side of the building are given in Table 1.

Table 1. Mean, standard deviation, maximum and minimum pressure coefficients on the windward wall, leeward wall, side wall, roof and interior of the nominally sealed WERFL test building.

	Pressure coefficients			
Tap Location	$C_{\overline{p}}$	C_{σ_p}	$C_{\hat{p}}$	$C_{\breve{p}}$
Windward Wall	0.70	0.39	2.76	-0.22
Leeward Wall	-0.34	0.12	0.00	-0.85
Side Wall	-0.29	0.24	0.52	-1.95
Roof	-0.27	0.18	0.66	-1.72
Internal	-0.14	0.06	0.09	-0.38

For an approach wind flow normal to the 30 ft side of the building the windward wall experiences mean positive pressure and the other walls and roof experience mean suction pressures. The mean internal pressure coefficient was -0.14. The pressure fluctuations on the attached flow regions on the side walls and roof and the wake region on the leeward wall are progressively lower than the windward wall pressure. The internal pressure fluctuations are significantly lower than the external pressure fluctuations. The peak positive pressure coefficient on the windward wall was 2.76, and the peak negative pressure coefficients on the side wall, roof, leeward wall and building interior were -1.95, -1.72, -0.85 and -0.38 respectively.

A smoke flow study carried out in the nominally sealed WERFL test building indicated that leakage took place around the door and window clearances, and at the joints and connections on the walls and roof. Assuming the porosity is uniform over the whole WERFL test building envelope, for an approach wind flow normal to the 30 ft side of the building, the ratio of leakage areas on the windward wall, leeward wall, two side walls and roof is 1 : 1 : 3 : 3.5. Using mean pressure coefficients of +0.70 on the windward wall, -0.34 on the leeward wall, -0.29 on the side walls and -0.27 on the roof, and the "windward" to "leeward" leakage ratio, A_W/A_L of 1/7.5, $C_{\bar{p}_W} = 0.70$, $C_{\bar{p}_L} = -0.29$ and $\Delta C_{\bar{p}} = 1.0$.

For the WERFL test building of height H = 4.0 m and breadth B = 9.1 m, normal to the wind, the characteristic frequency f_C is determined by reducing Eq. 5b to Eq. 6 and applying the following typical values existing under experimental conditions, $V_{Ie} = 1175$ m^3, $\Delta C_{\bar{p}} = 1.0$, k = 0.65, mean wind speed at roof height $\bar{U} = 10$ m/s, $p_0 = 9 \times 10^4$ Pa, $\rho = 1.1$ kg/m^3, n = 1.4,

$$f_C = \frac{1}{2\pi}\left(\frac{1}{T_C}\right) = \frac{1}{2\pi}\left(\frac{\varepsilon H B n k p_0 r^{-1/2}(r + r^{-1})^{3/2}}{\rho V_{Ie} \bar{U} (\Delta C_{\bar{p}})^{1/2}}\right) \tag{6}$$

where, r = A_W/A_L, is the ratio of "windward" to "leeward" leakage areas. According to Eq. 6, the characteristic response frequency f_C, for the nominally sealed WERFL test building with uniform porosity ε, of 2.5×10^{-4} and r = 1/7.5 is 0.53 Hz.

The non-dimensional pressure spectra $fS_p(f)/\left(\frac{1}{2}\rho\bar{U}^2\right)^2$ for windward wall, leeward wall, side wall, roof and interior of the nominally sealed building are presented in Fig. 2. Ginger et al (1994), showed that the energy distribution in the windward wall pressure spectrum closely follows the approach 4.0 m wind velocity spectrum. Fig. 2 shows that the side wall, roof and leeward wall pressure spectra contain progressively less energy compared with the windward wall pressure, and the reduction is mainly over the lower frequency range (f < 0.5 Hz.). The internal pressure fluctuations contain much less energy compared with the external pressure fluctuations, and are significantly attenuated above 0.4 Hz with a sharp drop off. This agrees closely with the characteristic frequency f_C = 0.53 Hz, obtained from the theoretical analysis (i.e. Eq. 6).

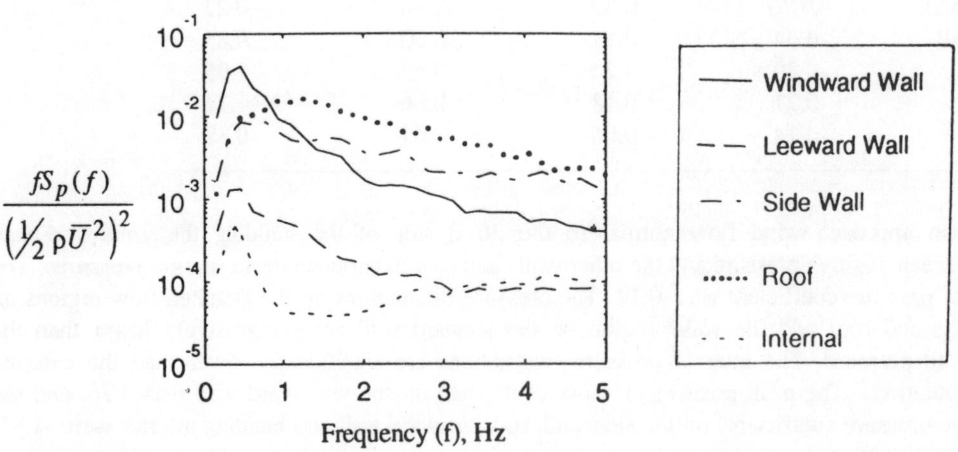

Figure 2. Windward wall, leeward wall, side wall, roof and internal pressure spectra on the nominally sealed WERFL building.

4.2 Building containing windward and/or leeward opening(s)

The mean, standard deviation, maximum and minimum pressure coefficients obtained on the windward wall, leeward wall and building interior, for a range of WERFL test building, windward and leeward wall opening A_W/A_L ratios (background leakage is neglected) are given in Table 2. For a single wall opening, the internal pressures closely follow the pressure on the wall containing the opening. In a building with windward and leeward wall openings, the mean and fluctuating internal pressures are influenced by the pressures on both the windward and leeward walls. Fig. 3 shows the variation of the mean, standard deviation, maximum and minimum internal pressure coefficients with A_W/A_L ratio (background leakage is neglected) and the mean internal pressure curve from Eq. 1 for $C_{\bar{p}_W} = 0.65$ and $C_{\bar{p}_L} = -0.30$. The experimental mean internal pressure data agree with the theoretical analysis of flow through an orifice. Wind tunnel results by Holmes (1979) also agreed with this analysis.

Table 2. Mean, standard deviation, maximum and minimum windward wall, leeward wall, and internal pressure coefficients for the WERFL test building with windward and/or leeward opening(s) (background leakage is neglected).

Wall openings		Pressure coefficients											
A_W	A_L	Windward Wall				Leeward Wall				Internal			
(m^2)	(m^2)	$C_{\bar{p}}$	C_{σ_p}	$C_{\hat{p}}$	$C_{\breve{p}}$	$C_{\bar{p}}$	C_{σ_p}	$C_{\hat{p}}$	$C_{\breve{p}}$	$C_{\bar{p}}$	C_{σ_p}	$C_{\hat{p}}$	$C_{\breve{p}}$
0.4	0.0	0.63	0.32	2.35	-0.15	-0.32	0.09	-0.05	-0.71	0.60	0.30	2.25	-0.18
0.8	0.0	0.62	0.42	2.76	-0.43	-0.43	0.13	-0.09	-0.95	0.61	0.42	2.74	-0.38
2.0	0.0	0.60	0.31	2.28	-0.16	-0.25	0.10	0.08	-0.61	0.60	0.31	2.20	-0.17
2.0	0.8	0.57	0.27	1.87	-0.11	-0.24	0.10	0.10	-0.46	0.46	0.24	1.90	-0.14
0.4	2.0	0.66	0.37	2.41	-0.28	-0.32	0.18	0.21	-0.90	-0.22	0.10	0.15	-0.64
0.8	2.0	0.65	0.36	2.29	-0.22	-0.28	0.16	0.16	-0.76	-0.13	0.10	0.36	-0.53
0.0	2.0	0.59	0.30	2.14	-0.20	-0.35	0.11	0.02	-0.78	-0.35	0.10	-0.02	-0.82

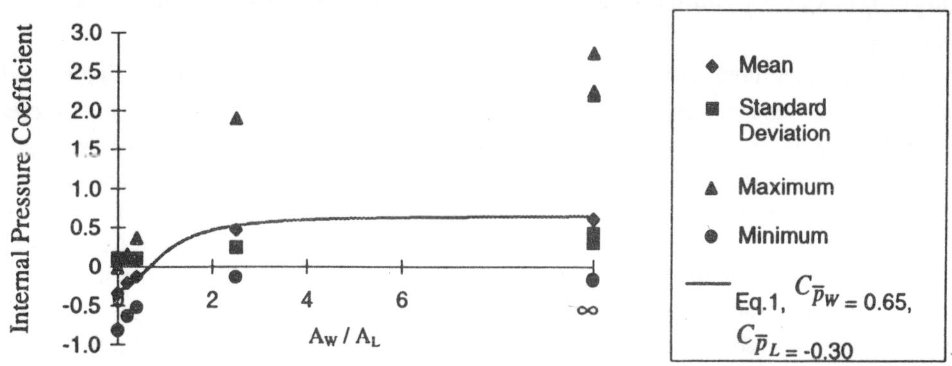

Figure 3. Mean, standard deviation, maximum and minimum internal pressure coefficients vs. A_W/A_L ratio for WERFL test building (background leakage is neglected).

4.3. Building containing a single dominant opening

The WERFL test building internal pressures were simulated using a finite difference numerical technique by applying the external pressures measured on the wall containing the single dominant opening (p_E) to the right hand side of Eq. 4. The density of air ρ, the barometric pressure p_0, and the mean velocity at roof height $\overline{U}$ are obtained from the WERFL field measurements. The effective internal volume V_{Ie} is 470 × 2.5 = 1175 m^3. The simulated internal pressures obtained for an adiabatic process (i.e. n = 1.4), and opening discharge coefficients k values between 0.10 and 0.70 were studied. The measured and simulated (for k = 0.65 and k = 0.15) internal pressure-time histories obtained for a portion of a single 15 min run with a 5% single windward wall opening (i.e. A_W = 2.0 m^2) are given in Fig. 4. The variation of internal pressures with time shows that the numerical scheme using k = 0.15, simulates the measured internal pressures better than the simulation using k = 0.65.

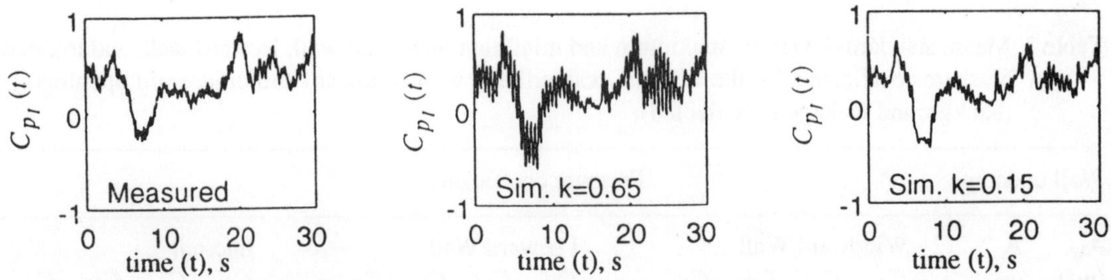

Figure 4. Portion of measured and simulated (using k = 0.65 and k = 0.15) internal pressure vs. time, in the building with 2.0 m^2, (5%) single windward wall opening.

The windward wall pressure spectrum and the measured and simulated (k = 0.65 and k = 0.15, 0.35) internal pressure spectra, $fS_p(f)/\left(\frac{1}{2}\rho\overline{U}^2\right)^2$ for the building with 1%, 2% and 5% single windward wall opening (i.e. A_W = 0.4 m^2, 0.8 m^2 and 2.0 m^2) are shown in Fig. 5 to Fig. 7. The leeward wall pressure spectrum and the measured and simulated (k = 0.65 and k = 0.10) internal pressure spectra, $fS_p(f)/\left(\frac{1}{2}\rho\overline{U}^2\right)^2$ for the building with 5% single leeward wall opening (i.e. A_L, = 2.0 m^2) are shown in Fig. 8.

In Fig. 5, the measured internal pressure spectrum, for the building with 1% single windward wall opening (i.e. A_W = 0.4 m^2, $A_W{}^{3/2}/V_{Ie}$ = 0.0002), shows an increase of internal pressure energy close to the Helmholtz frequency, f_0 of 1.34 Hz compared with the windward wall pressure fluctuations even though a distinct peak is not indicated. In this case the total background leakage area exceeds 10% of the windward opening area, the damping is increased and the resonance is attenuated, as described by Vickery et al (1992). The simulated internal pressure spectra indicate progressively higher peaks close to the Helmholtz frequency 1.34 Hz, for k = 0.35 and k = 0.65 respectively, with k = 0.35 closely agreeing with the measured spectrum.

In Fig. 6, for a 2% single windward wall opening (i.e. A_W = 0.8 m^2, $A_W{}^{3/2}/V_{Ie}$ = 0.0006), the simulated spectrum for k = 0.35 agrees with the measured internal pressure spectrum with a peak close to the Helmholtz frequency, f_0 of 1.58 Hz. Similarly, in Fig. 7, the measured internal pressure spectrum shows a peak close to 1.8 Hz, for the building with a 5% single windward wall opening (i.e. A_W = 2.0

m^2, $A_W^{3/2}/V_{Ie} = 0.0024$). The simulated internal pressure spectra indicate progressively higher peaks close to the Helmholtz frequency, f_0 2.00 Hz, for k = 0.15 and k = 0.65 respectively. The use of k = 0.15 to simulate the internal pressure provides better agreement with the measured internal pressure spectrum compared with using k = 0.65. As the opening area is increased from 1% to 5% the peak in the pressure spectrum is increased, and there is an increasing tendency for resonance to occur.

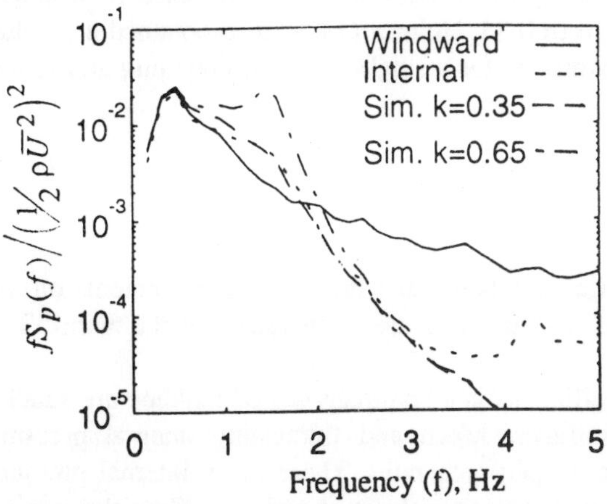

Figure 5. Windward wall, and measured and simulated (k = 0.35 and k =0.65) internal pressure spectra on the building with a single 1% windward wall opening.

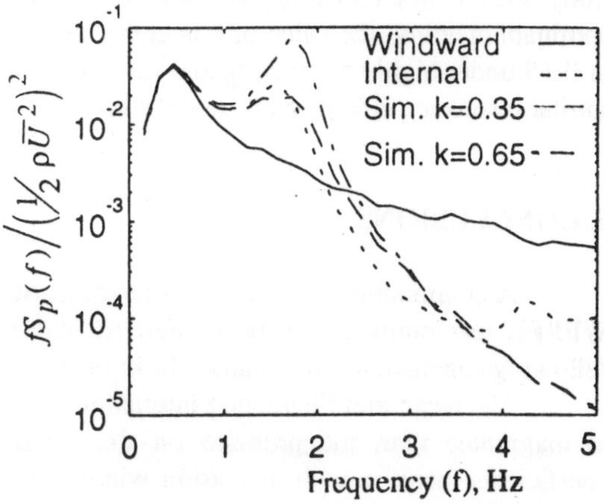

Figure 6. Windward wall, and measured and simulated (k = 0.35 and k =0.65) internal pressure spectra on the building with a single 2% windward wall opening.

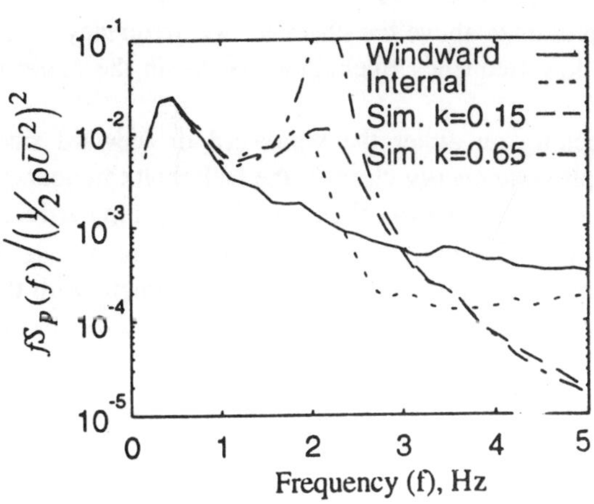

Figure 7. Windward wall, and measured and simulated (k = 0.15 and k =0.65) internal pressure spectra on the building with a single 5% windward wall opening.

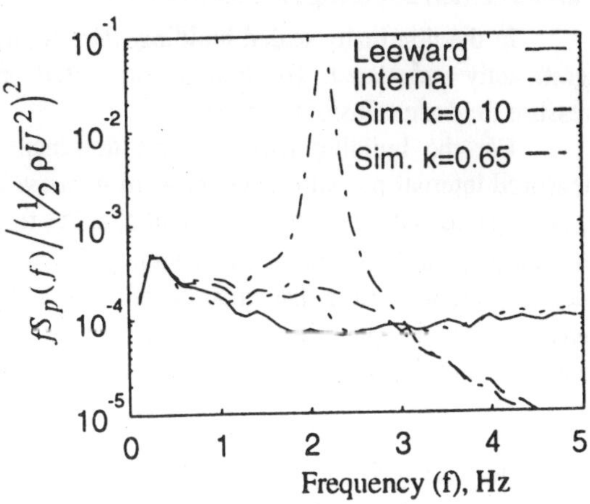

Figure 8. Leeward wall, and measured and simulated (k = 0.10 and k =0.65) internal pressure spectra on the building with a single 5% leeward wall opening.

In Fig. 8, the measured internal pressure spectrum shows a peak close 1.75 Hz, for the building with a 5% single leeward wall opening (i.e. $A_L^{3/2}/V_{Ie} = 0.0024$). The simulated internal pressure spectra for k = 0.10 and k = 0.65 indicate progressively higher peaks respectively close to the Helmholtz frequency, f_0 of 2.00 Hz. The use of k = 0.10 to simulate the internal pressure shows better agreement with the measured internal pressure spectrum compared with k = 0.65.

Under steady flow conditions the orifice discharge coefficient k is about 0.65. However, this study shows that under highly fluctuating and reversed flow conditions, as in the case of a single dominant opening, the value of k is in the range of 0.10 to 0.35. Holmes (1979) also obtained a k value of 0.15 under highly fluctuating and reversed flow conditions for a single windward opening and found similar resonance effects in his model study.

5. CONCLUSIONS

A combination of theoretical analysis, numerical simulation and full scale measurements on the WERFL test building has been used to investigate internal pressures generated by the wind. The following conclusions are obtained from the study;

The mean and fluctuating internal pressure coefficients in a nominally sealed building are smaller in magnitude than the pressure on the external surfaces. Mean and fluctuating internal pressure coefficients increase with increasing windward/leeward open area ratio. These mean internal pressure coefficients agree with the values obtained from the theoretical analysis of turbulent flow through an orifice.

The internal pressure fluctuations are influenced by the interaction of internal air volume with the building. The effect of building flexibility in slowing the internal pressure response is accounted for by increasing the nominal internal volume (V_I) by a factor of the ratio of bulk modulus of air (K_A) to the bulk modulus of the building (K_B). The experimental results are in good agreement with the theoretical analysis carried out using an effective interior volume of V_{Ie} equal to $V_I \times \left(1+\left(K_A/K_B\right)\right)$.

In the nominally sealed building, the internal pressures above the characteristic frequency f_C are significantly attenuated. The implication is that only low frequency fluctuations (< f_C) in the external pressure field are transmitted inside.

For the building with a single dominant opening on either the windward or leeward side, measured internal pressure data show an increase of pressure energy close to the Helmholtz frequency, f_0, compared with the pressure fluctuations at the opening. Increasing the area of the opening reduces the damping and increases the tendency for resonance. The time history and spectra of simulated internal pressure with a discharge coefficient, k, between 0.10 and 0.35 is in good agreement with the measured internal pressures. Holmes (1979) arrived at similar conclusions from his wind tunnel model study of a building with a single dominant windward opening.

6. ACKNOWLEDGMENTS

These studies were carried out as part of the program funded by the U.S. National Science Foundation (Grants BCS-8821163 and CES-8611601) and the State of Texas Advanced Research Program.

7. REFERENCES

AS-1170.2 (1989), "Australian Standard SAA Loading Code Part 2", Wind Loads, Standards Australia.

Ginger, J. D., Letchford, C. W. and Yeatts, B. B. (1994), "Internal Pressure Dynamics and Net Pressure on Low-rise Full Scale Buildings", (to be submitted to Jour. of Wind Engineering and Industrial Aerodynamics).

Harris, R. I. (1990), "The Propagation of Internal Pressures in Buildings", Jour. of Wind Engineering and Industrial Aerodynamics, Vol. 34, 169-184.

Holmes, J. D. (1979), "Mean and Fluctuating Internal Pressures Induced by Wind", Proc. 5th Int. Conf. on Wind Engg., Colorado USA, Vol. 1, 435-450.

Kramer, C., Gerhardt, H. J. and Scherer, S. (1979), "Wind Pressure on Block Type Buildings", Jour. of Industrial Aerodynamics, Vol. 4, 229-242.

Levitan, M. L. and Mehta, K. C. (1992a), "Texas Tech Field Experiments for Wind Loads Part I: Building and Pressure Measurement System", Jour. of Wind Engineering and Industrial Aerodynamics, Vol. 43, 1565-1576.

Levitan, M. L. and Mehta, K. C. (1992b), "Texas Tech Field Experiments for Wind Loads Part II: Meteorological Instrumentation and Terrain Parameters", Jour. of Wind Engineering and Industrial Aerodynamics, Vol. 43, 1577-1588.

Liu, H. and Rhee, K. H. (1986), "Helmholtz Oscillation in Building Models", Jour. of Wind Engineering and Industrial Aerodynamics, Vol. 24, 95-115.

Liu, H. and Saathoff, P. J. (1981), "Building Internal Pressure: Sudden Change", Jour. of The Engineering Mechanics Division, Vol. 107, No. EM2, 309-321.

Stathopoulos, T., Surry, D. and Davenport, A. G. (1979), "Internal Pressure Characteristics of Low-rise Buildings due to Wind Action", Proc. 5th Int. Conf. on Wind Engg., Colorado USA, Vol. 1, 451-463.

Vickery, B. J. (1986), "Gust Factors for Internal Pressures in Low rise Buildings", Jour. of Wind Engineering and Industrial Aerodynamics, Vol. 23, 259-271.

Vickery, B. J. and Bloxham, C. (1992), "Internal Pressure Dynamics with a Dominant Opening", Jour. of Wind Engineering and Industrial Aerodynamics, Vol. 41, 193-204.

Womble, J. A., Yeatts, B. B., Cermak, J. E. and Mehta, K. C. (1995), "Internal Wind Pressures in a Full and Small Scale Building", submitted to 9th International Conference on Wind Engineering, New Delhi, India, 9-13 January 1995.

Yeatts, B. B. and Mehta, K. C. (1993), "Field Experiments for Building Aerodynamics", Jour. of Wind Engineering and Industrial Aerodynamics, Vol. 50, 213-224.

SIMULTANEOUS FLOW VISUALIZATION AND PRESSURE MEASUREMENTS ON THE TEXAS TECH BUILDING

LETCHFORD, C.W.

Department of Civil Engineering, The University of Queensland, Brisbane, Australia 4072.

Abstract

Flow visualization has been synchronised with pressure measurements on the Texas Tech Building in an effort to determine the mechanism causing peak pressures in separated flow regions. For the **first time** at full scale, the flow patterns leading to the collapse of the separation bubble and subsequent generation of peak pressures have been observed. In general these observations confirm those made at model scale. For quartering winds a mechanism has been observed for peak pressure generation under delta wing vortices.

1. INTRODUCTION

Wind load information for design is ultimately derived from wind tunnel tests and it is important that every opportunity is taken to verify such tests by comparison with available full scale data. The Texas Tech University Wind Engineering Research Field Laboratory (WERFL) is currently providing extensive full scale data for such purposes and many wind tunnel laboratories are conducting comparative studies.

Whereas there has been generally excellent agreement for the vast majority of comparisons between model and full scale mean and fluctuating pressures on the WERFL, important exceptions remain. Cochran and Cermak (1992) report significant discrepancies between model and full scale fluctuating pressures near roof corners. Lin et.al. (1993) report similar differences at the same locations on the WERFL. Better agreement was claimed by Tieleman et.al.(1993), but only after extensive manipulation of the incident flow simulation. Similar discrepancies between model and full scale fluctuating pressures at roof corners on another building were reported by Milford et.al.(1992). The reason these exceptions are important is because the pressures in these regions are typically the largest in magnitude on the building and roofing failures are known to be initiated in these areas.

These discrepancies can be attributed to deficiences in (a) the incident flow simulation, (particularly turbulence characteristics) (b) instrumentation (tapping size, location and tubing frequency response) and linking each, (c) scaling problems associated with mismatch of the Reynolds Number of the flow.

Tieleman et.al.(1993) argue that deficiences in flow simulation, particularly lateral turbulence and fine scale turbulence are responsible. They found better agreement when these characteristics were better simulated at the expense of the more traditional mean velocity and longitudinal turbulence intensity and scale. The fine scale turbulence effect could be argued as a Reynolds Number effect because at the reduced geometric scale in the wind tunnel this turbulence is lost in viscous dissipation.

Instrumentation frequency response was considered by Letchford et.al.(1992) but it was demonstrated that this would not account for the significant differences being observed. Lin et.al.(1993) show data obtained from a Super Pressure Tap which correctly scaled the model pressure tap at full scale. The evidence from this indicates that tap size, although having some effect, is not causing the significant differences either.

Investigations of the distribution and correlation of such pressures at full scale were undertaken by Letchford et.al.(1993a & 1993b). It was *suggested* there that the same flow mechanism observed by Saathoff and Melbourne(1989) at model scale was responsible for peak pressure generation at full scale, namely; the intermittent break up of the separation bubble and the subsequent rolling up of the shear layer into a discrete vortex. However, other experimental studies by Kiya and Sasaki (1985) and Cherry et.al.(1984) indicated that the pressure fluctuations in the separation bubble were primarily due to freestream vortices convected *through* the shear layer. Clearly shear layer instability is an important parameter, strongly influenced by incident turbulence and one of the aims of the present study was to try and obtain a better understanding of this phenomenon and subsequent peak pressure generation.

However the major discrepancy between model and full scale fluctuating pressures has been found to occur near roof corners for oblique or quartering winds. For these winds, delta wing vortices form along the leading edges of the roof and to date no flow mechanism for generating these peak pressures has been postulated. The challenge for this study was to identify this flow mechanism.

2. EXPERIMENTAL METHODOLOGY

The Texas Tech University, Wind Engineering Research Field Laboratory (WERFL) has been fully described by Levitan and Mehta (1992a, 1992b). The building comprises a 9.1m wide by 13.7m long by 4.0m high rotatable metal framed structure with a smooth cladding. The building is equipped with pressure measurement instrumentation while approach wind flow characteristics are measured at various heights on a 49m meteorological tower located 46m to the west of the building.

For the present study a video camera was employed to record the passage of smoke and the movement of tufts on a grid set up on the roof of the WERFL. The video was mounted and operated from the roof in such a position as to obtain the best field of view of the grid. The smoke was generated by burning diesel fuel in an agricultural fogger placed towards the rear of the roof. The smoke was then piped to a point within the separation bubble where a baffle was employed to ameliorate the influence of the injected flow on the separation bubble. For the corner vortex flows the fogger was placed on the ground in front of the building with a pipe discharging the smoke just below roof level.

The grid of tufts consisted of black yarn fastened to nylon line strung on a metal frame approximately 4.6m long by 1.4m high. The yarns were positioned at 100mm vertical intervals to 600m then 150mm intervals to 900mm, with the top row 1200mm above the roof. Horizontally the nylon lines were spaced at 150mm intervals from the leading edge of the roof. For this study the grid was placed normal to, and centrally along the long wall.

Simultaneous pressure measurements were obtained in the MODE 98 data acquisition configuration which records pressures at 30Hz for 15 minutes. Pressures are reduced to coefficient form by dividing by the mean roof-top dynamic pressure recorded on the meterological tower. The exact locations of the pressure taps employed in this study are found in Levitan and Mehta (1992a) while Figure 1 shows the pressure tappings schematically and defines the angle of attack.

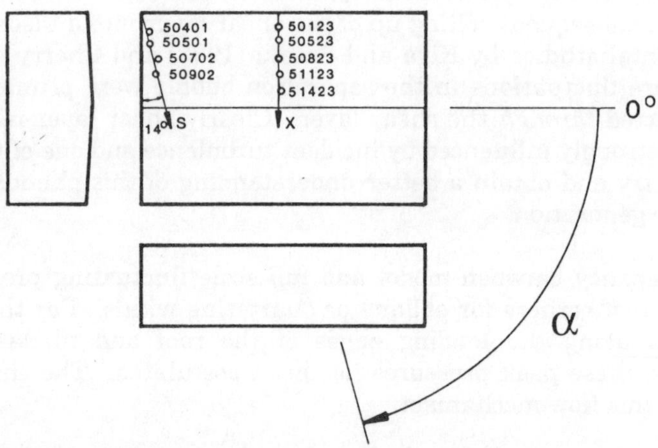

Figure 1. Tapping distribution and angle of attack definition.

The pressure and video records were synchronised by videoing the covering and uncovering of selected pressure taps during a run. These frames in the video and the corresponding points in the pressure record could be readily identified to allow synchronisation. It was found that the video camera, nominally recording at 30Hz, actually lost about 1 frame in 900 compared to the pressure record. This was corrected for in the analysis allowing a 'snap-shot' of the flow structure for each pressure record.

Some 3 hours of video recordings were obtained during February 1993 with weather conditions being mostly sunny and temperatures around 13°C. Measurements were obtained for wind normal and approximately 45° to the long wall, thus covering the separation bubble and delta wing vortex flows. Subsequent to this study a more detailed investigation, including wind speed measurements over the building, was undertaken by Wagaman (1993).

Of the 14 pressure data runs taken, half were stationary in wind speed and direction as defined by Levitan and Mehta (1992b). Only 4 of these runs proved to have all relevant information recorded and a summary of these is found in Table 1. It is worth noting that for the purposes of this study, stationarity was not of primary concern, but some culling of the enormous amount of data collected was required.

Table 1. Data Summary

Run	α deg	V_h m/s	G	I_u
M98N363	225	7.2	1.75	.21
M98N369	268	5.6	1.57	.19
M98N370	229	4.1	1.71	.23
M98N379	228	7.0	1.58	.20

Where α is the mean angle of attack, V_h the 15 minute mean velocity, G the gust factor and I_u the turbulence intensity, all at 4m (roof-top) height.

3. RESULTS AND DISCUSSION

Data analysis was undertaken by first viewing the pressure records for 'interesting events' - typically large magnitude fluctuations. Selected time histories were then obtained for adjacent pressure taps and a spatial and temporal picture of the large magnitude pressure fluctuation developed. After synchronisation, the video record was examined frame-by-frame in the vicinity of a particular event using a Panasonic AG650 video editing facility. Unfortunately it was not possible to obtain, with sufficient resolution, hard copies of each frame for flow analysis, and a painstaking manual recording of each tuft position was undertaken, typically at half second intervals. Edited video highlights will be presented in the oral presentation of this paper at the 9th International Conference on Wind Engineering.

3.1 Separation bubble study

Only one run (M98N369) with the mean wind direction approximately normal to the long wall was complete and the pressure coefficient distribution across the roof centreline for the full run is presented in Figure 2. It is apparent that mean pressure recovery occurs at approximately $x/h = 1.0$ indicating the mean length of the separation bubble. These results are consistent with the earlier work carried out by Letchford and Mehta (1993b).

Figure 3 shows successive cross correlation functions of roof pressures with the leading edge pressure tap 50123. As expected the maximum cross correlation occurs at increasing time delays, with those tappings outside the mean separation bubble being uncorrelated with the leading edge pressure fluctuations. A mean convection velocity of pressure fluctuations is estimated to be $\sim 0.5V_h$, somewhat higher than the $0.37V_h$ reported at model scale by Ginger(1992).

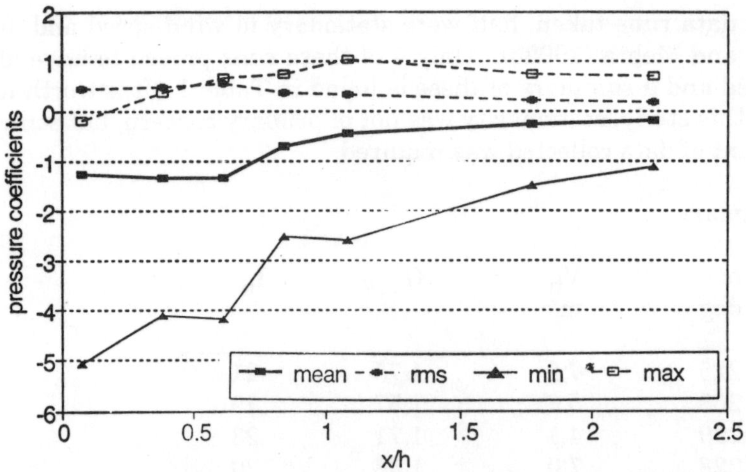

Figure 2. Pressure coefficient distribution across roof centreline for run M98N369.

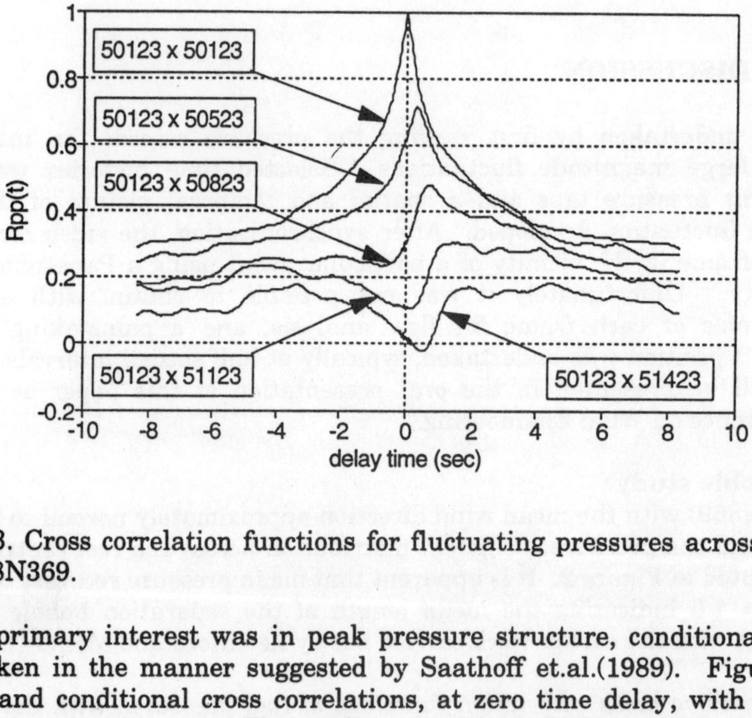

Figure 3. Cross correlation functions for fluctuating pressures across roof centreline for run M98N369.

As the primary interest was in peak pressure structure, conditional cross correlations were taken in the manner suggested by Saathoff et.al.(1989). Figure 4 compares the normal and conditional cross correlations, at zero time delay, with each tap being the conditioning signal in turn. The condition is triggered whenever the pressure there falls 3 standard deviations below the mean. It is evident that peak pressures are better correlated than time averaged fluctuations for upstream pressures and the opposite occuring for downstream pressures where the peaks actually become quite negatively correlated.

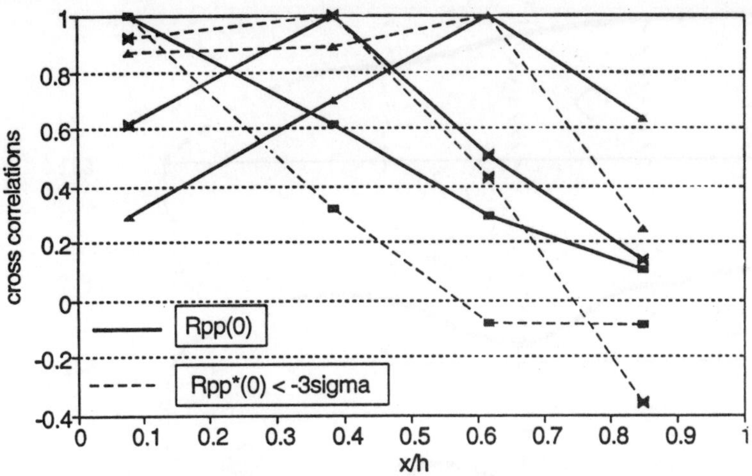

Figure 4. Normal and conditional cross correlations for run M98N369

To study the flow structure during peak pressure occurrences, four events were selected for analysis. Typical pressure time histories around one of these events is shown in Figure 5. The segment is 6 seconds in length and shows the simultaneous pressures at 4 tappings offset to facilitate spatial resolution. The 4 pressure taps are located 0.3, 1.5, 2.4 and 3.3m back from the leading edge. The peak suction in this event is typical of those in separation bubbles, ie., being intermittent and of short duration, typically less than half a second at the mid range value.

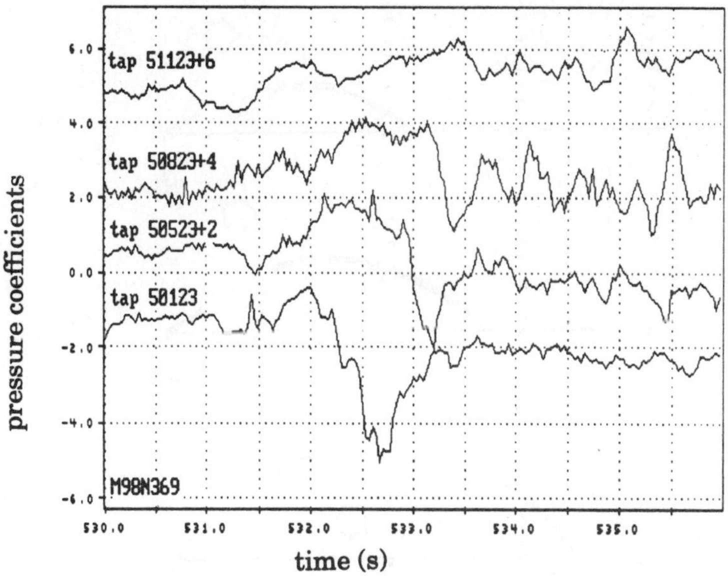

Figure 5. Pressure time histories under a separation bubble.

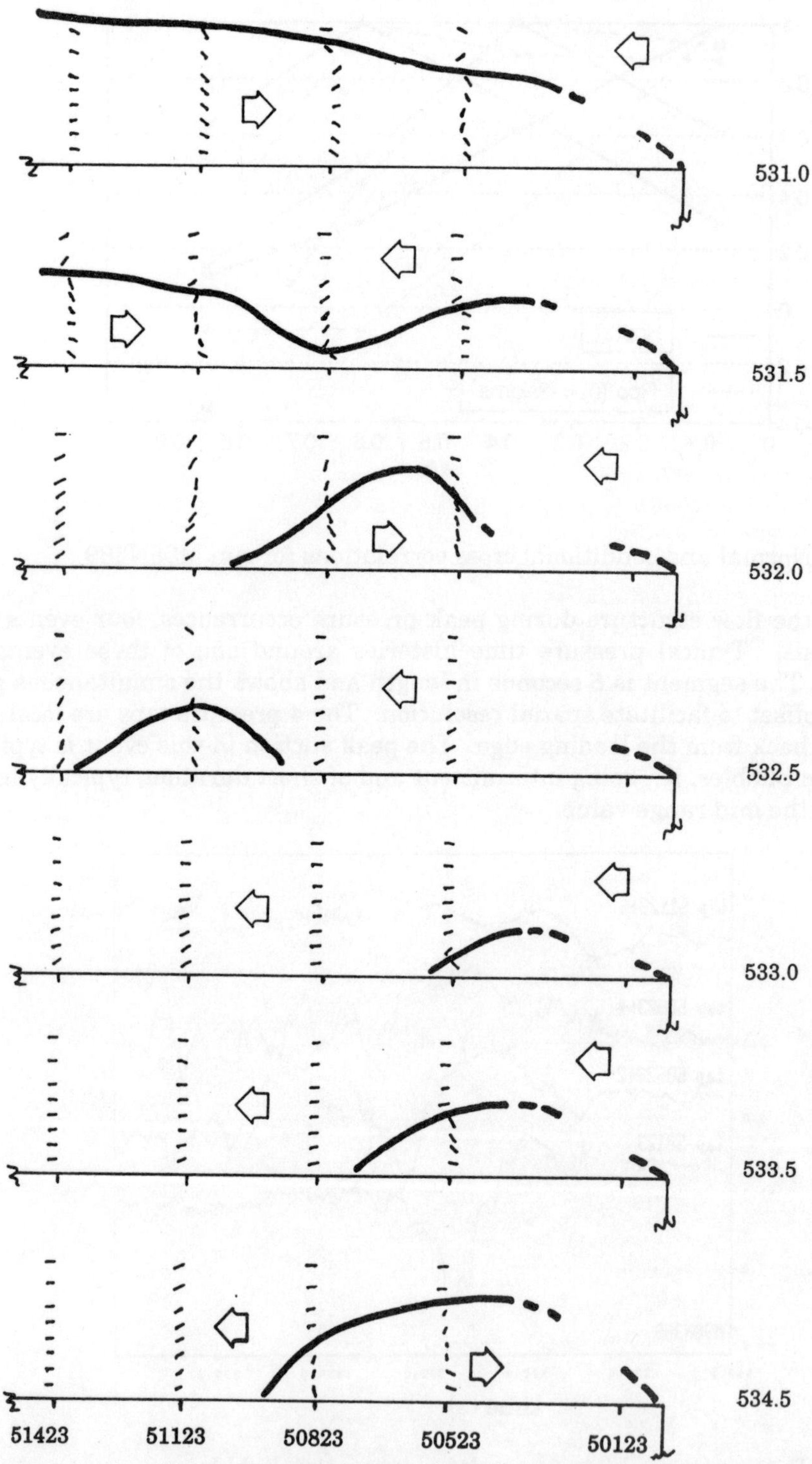

Figure 6. Flow structure for pressures in Figure 5.

It is interesting to note that the peak suction generally propogates spatially through the pressure record, which is consistent with the correlation study above and suggests a rather simple flow phenomenon causing the event. The synchronised video analysis of this event is presented as tuft sketches at half second intervals in Figure 6. The collapse and regrowth of the separation bubble is clearly in evidence. It is likely that this instability is caused by incident turbulence forcing the separated shear layer to prematurely reattach although no direct measurements of local freestream conditions were undertaken in the present study. The peak suctions are then associated with the passage of a discrete vortex formed from the severed shear layer and convected by the free stream. Pressure recovery occurs at those taps where the flow momentarily stagnates upstream and downstream of the vortex, but is most noticeable on the downstream side where the vortex directs flow onto the roof. As the vortex is convected downstream a similar pattern occurs at successive taps until the vortex dissipates and the separation bubble is reinstated. This flow structure is much clearer on the video. In essence these observations support both Saathoff et.al.(1989) and Kiya et.al.(1985) and Cherry et.al.(1984) in that an initial premature reattachment of the shear layer generates a peak suction near the leading edge and then this new flow structure is convected along the surface. The question remains what causes the shear layer instability? Quantitative flow field measurements would be required to explore this conundrum and Wagaman(1993) has made a start on this.

3.2 Delta wing vortex study

The three runs under quartering winds (α = 225°, 228° & 229°) all exhibit similar characteristics and the records of only one, M98N379, will be examined in detail here. Figure 7 shows the pressure distributions along the line defined as the ray s in Figure 1. This ray subtends 14° with the side wall and was found to be the location of maximum suctions for quartering winds in the extensive model study by Lin et.al.(1993). It is interesting to observe that the rms and minimum pressure coefficients under the delta wing vortex are approximately twice the magnitude of those under the separation bubble.

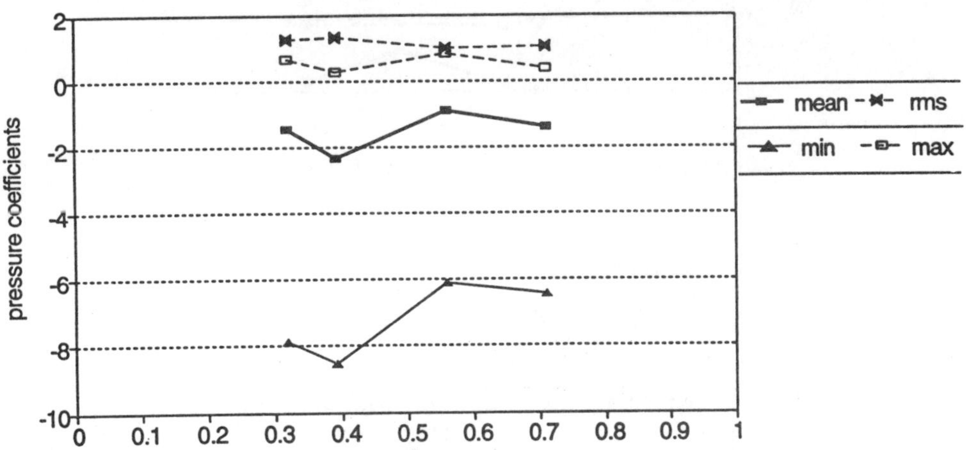

Figure 7. Pressure coefficient distribution beneath delta wing vortex for run M98N379.

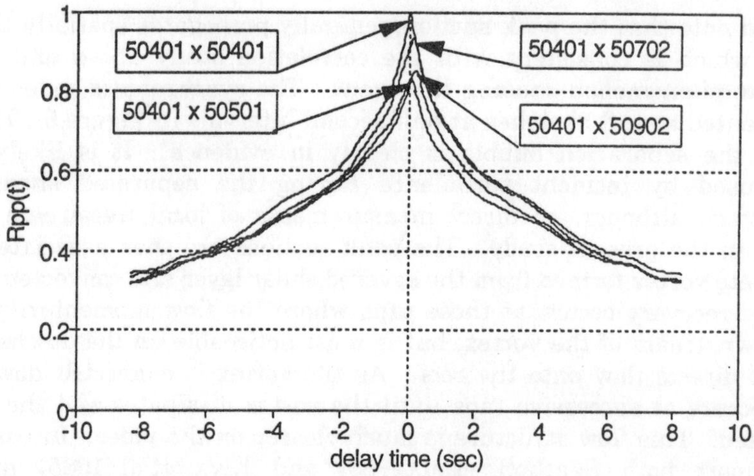

Figure 8. Cross correlation functions for fluctuating pressures beneath delta wing vortex for run M98N379.

Figure 8 shows successive cross correlation functions of roof pressures with the leading pressure tap 50401. In contrast to the separation bubble, here the maximum cross correlation occurs with negligible time delay and remains high (>0.8) over these four taps.

Once again the emphasis here is on peak pressures and so conditional cross correlations at zero time delay have been formed. The conditioning tap was 50401 and the trigger threshold was 3 standard deviations below the mean. Figure 9 shows both normal and conditional cross correlations and it is clear that peak pressures are better correlated than the time averaged fluctuations. Although not shown for clarity, there is very little difference between upstream and downstream tappings taking the condition. This contrasts with the separation bubble results presented in Figure 4.

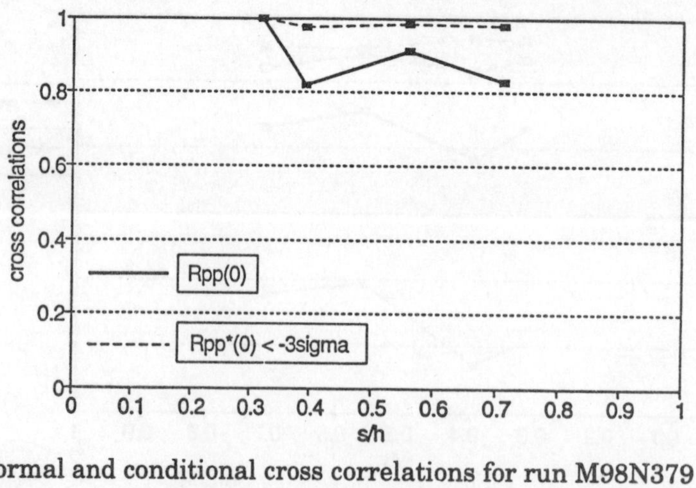

Figure 9. Normal and conditional cross correlations for run M98N379

In the three runs examined here there were many interesting flow events and a set of typical pressure time histories around one of these events in run M98N370 is shown in Figure 10. The segment is 12 seconds in length and shows the simultaneous pressures at 4 tappings offset to facilitate spatial resolution. The 4 pressure taps are located approximately 1.3, 1.6, 2.2 and 2.8m back along the ray s from the leading roof corner. The peak suction in this event is typical of those beneath delta wing vortices, ie., stable and relatively long, typically several seconds, in contrast to the peak suctions within separation bubbles.

The 3-dimensional nature of the delta wing vortex makes it much harder to present the flow structure causing peak pressures on paper. Video highlights will be the presented with the oral presentation of this paper. However, it is possible to say that peak pressures are associated with a tightening (or stretching?) of the vortex probably driven by incident turbulence. Quantitative velocity measurements just upstream of the roof would be required to verify this hypothesis, however Letchford et.al.(1993a) did find high cross correlations (>0.8) between upstream velocity and roof pressures beneath delta wing vortices.

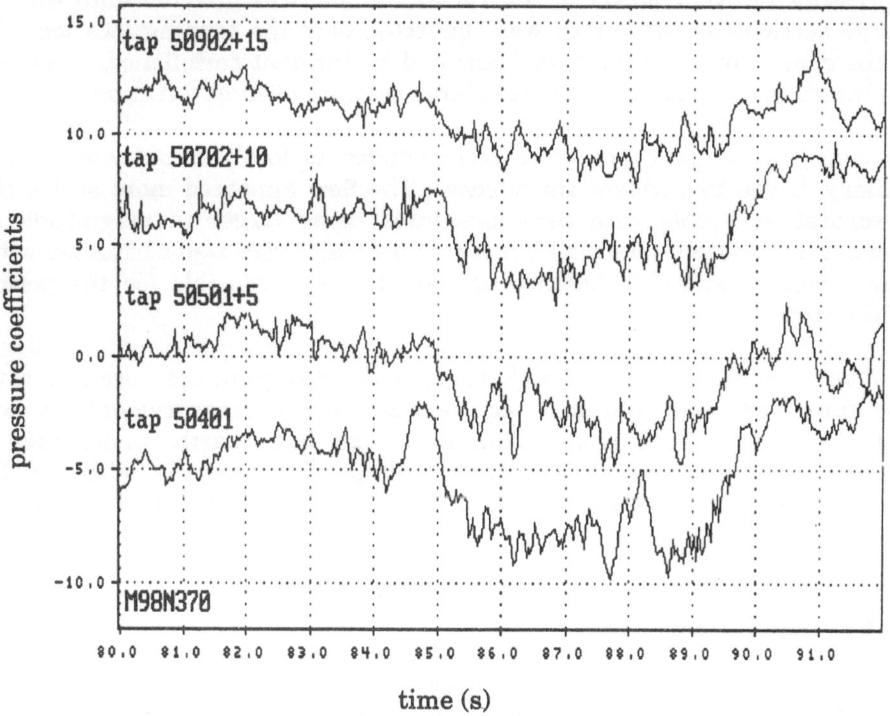

Figure 10. Pressure time histories beneath a delta wing vortex.

The sharp fall off in suction at the end of the peak pressure event illustrated in Figure 10 was also a reasonably common feature and generally corresponded to a complete sweeping away of the vortex. Incident turbulence, more particularly lateral turbulence, is probably the significant cause of this as Letchford et.al.(1993b) suggested. Vortex breakdown, the process that causes a progressive collapse of the vortex from the downstream section first, as occurs on delta wing aircraft at high angles of attack, did not appear to be the case here.

4. SUMMARY

Flow visualization and pressure measurements have been synchronised on the Texas Tech University WERFL in an effort to obtain a qualitative description of the flow structures generating peak roof pressures. Flow mechanisms for both normal and quartering winds, generating nominally 2D separation bubbles and 3D delta wing vortices respectively, have been studied.

For the separation bubble, it was observed that peak pressures corresponded with premature reattachment of the leading edge shear layer. The shear layer appeared to develop into a discrete vortex which was convected downstream leading to a pressure sequence of, pressure recovery ahead of the vortex as flow was directed onto the roof, followed by large suctions as the vortex was convected over the tapping location. The instability in the shear layer was probably initiated by incident turbulence, however it appeared to be intermittent leading to localized large suctions of short duration.

For the delta wing vortex, peak pressures corresponded to local strengthening of the vortex, most likely driven by incident turbulence. The flow appeared more stable than that for the separation bubble with peak pressures being larger in magnitude and significantly longer in duration. The peak pressures were also very well correlated within the delta wing vortex. Lateral turbulence appears to be responsible for the periodic destruction of the vortex.

The question of what initiates the *flow mechanism* involved in peak pressure generation remains to be fully addressed and will require quantitative measurements of wind velocities in the vicinity of the roof. This work has been partly undertaken by Wagaman(1993), however he has not analysed any pressure records with his flow observations and there remains a substantial data base to be analysed to resolve this issue.

5. ACKNOWLEDGEMENTS

The assistance of an Australian Government DITAC Bilateral Science and Technology Fellowship together with support from Professor K.C. Mehta and the Wind Engineering Research Centre at Texas Tech University are gratefully acknowledged. The assistance of Mr Scott Wagaman in data collection is also acknowledged.

6. REFERENCES

Cherry, N.J., Hillier, H. & Latour, M.E.M.P (1984), Unsteady measurements in a separated and reattaching flow, JFM 144, 13-46.

Cochran, L.S. & Cermak, J.E. (1992), Full- and model scale cladding pressures on the Texas Tech University experimental building, JWEIA 43, 1589-1600.

Ginger, J.D. (1992), Characteristics of large fluctuating pressures in regions of flow separation on roofs, PhD Thesis, The University of Queensland.

Kiya, M. & Sasaki, K. (1985), Structure of large-scale vortices and unsteady reverse flow in the reattaching zone of a turbulent separation bubble, JFM 154, 463-491.

Letchford, C.W., Sandri, P., Levitan, M.L. & Mehta, K.C. (1992), Frequency response requirements for fluctuating wind pressure measurements, JWEIA 40, 263-276.

Letchford, C.W., Iveson, R.E. & McDonald, J.R. (1993a), The application of the Quasi -steady Theory to full scale measurements on the Texas Tech Building, JWEIA 48, 111-132.

Letchford, C.W. & Mehta, K.C. (1993b), The distribution and correlation of fluctuating pressures on the Texas Tech Building, JWEIA 50, 225-234.

Lin, J.X., Surry, D. & Tieleman, H.W. (1993), The distribution of pressure near roof corners of flat roof low buildings, 1st IAWE European & African Regional Conference, Guernsey, UK, 20-24 September 1993.

Milford, R.V., Goliger, A.M. & Waldeck, J.L. (1992) Jan Smuts Experiment: Comparison of full scale and wind tunnel results, JWEIA 43 1589-1600.

Saathoff, P.J. and Melbourne, W.H. (1989), The generation of peak pressures in separating and reattaching flows, JWEIA 32, 121-134.

Tieleman, H.W., Surry, D. & Mehta, K.C. (1993) Full/model scale comparison of surface pressures on the Texas Tech Experimental Building, 1st IAWE European & African Regional Conference, Guernsey, UK, 20-24 September 1993.

Wagaman, S. (1993), Full-scale flow visualization over a low rise building, Master of Science in Civil Engineering Thesis, Texas Tech University.

AUTHOR INDEX